Periglacial Geomorphology

Periglacial Geomorphology

Colin K. Ballantyne

*Emeritus Professor in Physical Geography,
University of St Andrews, Scotland, UK*

WILEY Blackwell

This edition first published 2018

The right of Colin K. Ballantyne to be identified as the author of this work has been asserted in accordance with law.

Registered Offices
John Wiley & Sons, Inc., 111 River Street, Hoboken, NJ 07030, USA
John Wiley & Sons Ltd, The Atrium, Southern Gate, Chichester, West Sussex, PO19 8SQ, UK

Editorial Office
9600 Garsington Road, Oxford, OX4 2DQ, UK

For details of our global editorial offices, customer services, and more information about Wiley products visit us at www.wiley.com.

Wiley also publishes its books in a variety of electronic formats and by print-on-demand. Some content that appears in standard print versions of this book may not be available in other formats.

Library of Congress Cataloging-in-Publication Data has been applied for

ISBN: 9781405100069

Cover Design: Wiley
Cover Image: © Incredible Arctic/Shutterstock

Set in 10/12pt Warnock by SPi Global, Pondicherry, India

10 9 8 7 6 5 4 3 2 1

For Rebecca

Till a' the seas gang dry, my dear,
And the rocks melt wi' the sun:
And I will love thee still, my dear,
While the sands o' life shall run.

Robert Burns (1759–1796)

Contents

Preface

Periglacial geomorphology is the study of the landscapes, landforms, sediments and soil structures that have developed in cold nonglacial environments. It differs from other branches of geomorphology in that landscape development in periglacial environments is dominated or significantly influenced by the presence of perennially frozen ground (permafrost) and/or cyclic ground freezing and thawing over timescales ranging from diurnal to millennial. The periglacial realm encompasses at least 25% of the present surface area of the Earth, including 13–18% presently underlain by permafrost, and during the cold stages of the past 2.6 million years it extended over an additional 20%, encompassing vast swathes of mid-latitude regions that lay outside the maximum reach of successive Pleistocene ice sheets. The present periglacial domain includes high-latitude regions in both hemispheres, together with the higher parts of mountains and plateaus in middle and low latitudes.

Appreciation of the nature of periglacial processes and their geomorphological consequences underpins our understanding of the evolution not only of landscapes in present-day cold environments, but also of temperate mid-latitudes affected by periglacial activity in the recent geological past. Recognition of periglacial landforms and structures in former periglacial environments can also inform reconstruction of the nature and severity of past climate through comparison with the climatic controls on their present-day counterparts. Conversely, appreciation of the way in which mid-latitude periglacial environments adapted to rapid warming at the termination of the last glacial stage provides insights into the possible evolution of periglacial environments in response to future climate change. Predictions of future atmospheric warming in high latitudes and high mountain areas imply that the next few decades will witness radical changes in ground thermal regime, potentially leading to widespread degradation of permafrost and a cascade of landscape changes that have both local and global consequences. The study of periglacial processes and landforms is therefore not simply a component of our understanding of the landscapes of the Earth; we now understand the fragility of these vast silent areas of tundra, polar desert, mountains and boreal forest, and how climatic perturbation of periglacial landscapes may have consequences that affect the entire planet.

This book presents a comprehensive introduction to the processes that operate in present periglacial environments and how they contribute to our understanding of former periglacial environments. It includes accounts not only of the nature of frozen ground and frost-action processes, but also of the operation of slope, fluvial, aeolian and coastal processes in cold environments. It is organized in six parts. Part I (Chapters 1–2) introduces the historical and scientific context of periglacial geomorphology and the nature of periglacial environments. Part II (Chapters 3–5) addresses the physics of ground freezing and thawing, the characteristics of permafrost and the nature and origin of underground ice. Part III (Chapters 6–9) considers the characteristics, formation and significance of landforms, sediments and structures associated with permafrost, permafrost degradation and seasonal freezing and thawing of the ground. Part IV (Chapters 10–12) discusses rock weathering in periglacial environments, the periglacial processes operating on hillslopes and the distinctive landforms produced by rock breakdown and slope processes in cold environments. Part V (Chapters 13–15) is devoted to the operation of fluvial, aeolian and coastal processes in present and past periglacial environments. Finally, Part VI considers the employment of relict periglacial features to reconstruct past cold environments in mid-latitude regions (Chapter 16) and evaluates the response of periglacial environments to recent and predicted climate change (Chapter 17).

Most of the central chapters (6–15) follow a common format: a brief introduction defining terminology is followed by an account of relevant processes, a description of resulting landforms, deposits and/or sediment structures, and finally an outline of equivalent relict (Pleistocene) phenomena and their significance. This format allows consideration of relict periglacial features in association with their active counterparts.

The book is designed to be used by both undergraduate and graduate students studying geomorphology or

Quaternary science in the context of geography and geology degree programmes. It will be of use to all scientists whose research involves understanding of cold environments, whether from a geographical, geological, ecological, climatological, pedological, hydrological or engineering perspective. Its aim is to stimulate interest in and understanding of some of the most fascinating landforms and landscapes on Earth, the processes responsible for their formation, the part these processes played in mid-latitudes in the recent geological past and how these processes might be affected on a future warmer planet.

Colin K. Ballantyne
Blebo Craigs, Scotland

Acknowledgements

In preparing this book, I owe an enormous debt of gratitude to my longstanding friend and colleague Graeme Sandeman, cartographer in the School of Geography and Sustainable Development at the University of St Andrews. Graeme drafted or re-drafted all of the maps and diagrams, and the high quality of the figures herein is testimony to his skill, sound advice, professionalism, understanding and (not least) patience. I also owe a huge measure of thanks to Professor Julian Murton of the University of Sussex. Julian is the finest permafrost scientist in the UK, and enjoys an international reputation for outstanding research across a wide range of periglacial topics. Despite his busy schedule, he took time to read and comment on several chapters, and responded generously to queries and requests for illustrations and comments. Most of all, I have benefitted from his constant encouragement.

As publisher, John Wiley and Sons have showed remarkable patience as successive submission deadlines approached and passed. Kelvin Matthews provided sound advice and encouragement during the early stages of writing, and over the past three years Ramya Raghavan and Ashmita Thomas Rajapratapan dealt patiently with my numerous requests and queries. As production editor, Ramprasad Jayakumar professionally expedited the final stages of copyediting, proof preparation and revision.

This book took seven years to complete, as it was in competition with the demands of teaching and multiple other research projects. During this period, a small band of friends and colleagues helped in a variety of ways. The book benefitted from critical reading of particular chapters by Steve Gurney, Jim Hansom, Ole Humlum, Toni Lewkowicz and Brian Luckman, and from the response of numerous colleagues to requests for photographs or information. I particularly acknowledge the enthusiastic response of members of PYRN (the Permafrost Young Researchers Network) to a request for images of particular periglacial phenomena; all photograph contributions are acknowledged personally in the captions. My interest in commencing work on a periglacial text was stimulated by an invitation from Ole Humlum and Hanne Christiansen to teach on the Permafrost and Periglacial Environments course at UNIS on Svalbard in 2000. Over subsequent years, I have had the privilege of teaching (and learning from) some outstanding students at UNIS, an institution that offers a wonderful opportunity for study and research in the Arctic. Ole and Hanne have been generous hosts, and the long snowmobile excursions to remote areas have been highlights of my periglacial experiences.

I owe much to my teachers. Rob Price stimulated my interest in glacial and periglacial environments at the University of Glasgow, and Brian McCann at McMaster University provided me with the opportunity to carry out fieldwork on Ellesmere Island in arctic Canada, an unforgettable experience that hugely influenced my subsequent research career. As my PhD supervisor at the University of Edinburgh, Brian Sissons inculcated critical judgment and insistence on research integrity. The experience of working with Charlie Harris in co-writing *The Periglaciation of Great Britain* (1994) vastly broadened my periglacial horizons. I have subsequently had the privilege of research collaboration with numerous outstanding scientists working in diverse fields, and in particular I wish to acknowledge the debt I owe to John Matthews, Danny McCarroll and John Stone for numerous successful research collaborations, and for memorable days in the mountains of Scotland and Norway. My Geography colleagues at the University of St Andrews have provided support and collegiality for over 35 years. A special place in my thanks is reserved for Doug Benn, a staunch friend and wonderful colleague for over three decades, and now one of the most brilliant glaciologists in the world. *Slàinte*, Doug.

As the book grew, so did my children, Hamish and Kate, who have become accustomed to having a father who spends many hours working in a cabin at the bottom of the garden. I am deeply grateful to both for tolerating this eccentric behaviour and reminding me daily that the important things in life lie outside the cabin walls. My wife Rebecca has been a constant source of support in every possible way; without her love, encouragement, patience and sacrifice, this book would never have reached completion.

Colin K. Ballantyne
Blebo Craigs, Scotland

1

Introduction

1.1 The Periglacial Concept: Definitions and Scope

The term *periglacial* is used to describe the climatic conditions, processes, landforms, landscapes, sediments and soil structures associated with cold, nonglacial environments. *Periglacial geomorphology* is the study of the landforms developed under periglacial conditions, the processes responsible for their formation, modification and decay, and associated sediments and sedimentary structures. *Periglacial environments* are those in which cold-climate nonglacial processes have resulted in the development of distinctive landforms and deposits, usually related in some way to freezing of the ground. The term *periglaciation* describes the collective effects of periglacial processes in modifying the landscape, much as 'glaciation' describes the geomorphological effects of glacier ice.

The term 'periglacial' (literally 'bordering glaciers') is an etymological oddity. It was coined by Łozinski (1909, 1912) to designate a climatic zone of rock weathering by frost that occurs immediately outside the limits of present and former ice sheets, and to describe frost-weathered rubble ('periglacial facies') characteristic of this zone (French, 2000). Present usage of the term, however, contains no implication of present or former proximity to glacier ice, and some of the most extensive periglacial environments on Earth – in Canada, Alaska and northern Eurasia – are hundreds of kilometres distant from the nearest glacier. It has become, effectively, an elegant synonym for 'cold, nonglacial', whether applied geographically, climatically or geomorphologically.

Periglacial geomorphology is concerned primarily with developing our understanding of the physical and chemical processes that operate at the surface and near-surface of the Earth, and the nature, composition, evolution and distribution of landforms, sediments and sedimentary structures produced by such processes. It differs from other branches of geomorphology not only in that it focuses on cold, unglacierized environments, but also because landform development in such areas is dominated by freezing of the ground. This is manifest over wide areas in the presence of *permafrost* (ground in which the temperature remains below 0 °C for two years or more) and the associated existence of subsurface *ground ice*, together with a wide range of landforms and soil structures that reflect cyclic freezing and thawing of the uppermost layers of the ground over timescales ranging from diurnal to annual. However, a wide range of *azonal processes* that are common to most or all climates also affect periglacial environments: erosion and deposition by rivers, wind action, coastal processes and slope failure also shape the periglacial landscape, though often in particular ways that are conditioned by frozen ground, prolonged subzero winter temperatures or snow-cover (Berthling and Etzelmüller, 2011; Vandenberghe, 2011; Figure 1.1). The mechanisms by which these azonal processes occur are common to all environments, but the conditions under which they operate and their geomorphological effects are subtly to substantially different in cold climates.

In addition to its central focus on present-day processes and landforms, periglacial geomorphology is an integral component of *Quaternary science*, the study of how the environments and landscapes of the Earth have changed during the *Quaternary period*, an era of radical climatic shifts that encompasses the past 2.58 million years. Within the past million years, periods of pronounced global cooling triggered the growth of ice sheets that covered up to a third of the present land surface. During these *glacial stages*, periglacial conditions affected extensive tracts of mid-latitude landscapes ahead of the advancing ice, beyond the limits of ice-sheet advance, and during periods of ice-sheet retreat. As a result, periglacial landforms, sediment accumulations and soil structures developed in mid-latitude areas that now experience a temperate climate, and in favourable circumstances these are preserved as *relict periglacial features* (Figure 1.2). The distribution of relict periglacial landforms, deposits and sediment structures therefore provides evidence of former cold climatic conditions, often in the form of features indicative of former permafrost. Moreover, as some present-day periglacial landforms and soil structures presently occupy a fairly

Periglacial Geomorphology, First Edition. Colin K. Ballantyne.
© 2018 John Wiley & Sons Ltd. Published 2018 by John Wiley & Sons Ltd.

Figure 1.1 Periglacial landscape of the North Slope of Alaska (69°N), looking towards the Brooks Range. Though this area is underlain by permafrost and subject to severe winter freezing, the broad outlines of the landscape differ little from those in other environments. *Source:* Courtesy of Matthias Siewert.

well-defined climatic niche, identification of their relict counterparts can provide information on the nature of the climate at the time they were formed.

A further tenet of periglacial geomorphology is a concern for reconstructing long-term *landscape evolution*. This is a challenging area of research, particularly as many periglacial landscapes have been covered (and sometimes radically altered) by glacier ice for much of the past million years. However, periglacial landscapes that remained unglacierized throughout the Quaternary exist in a few locations, such as northern Yukon Territory in Canada (Figure 1.3) and the Dry Valleys of Victoria Land in Antarctica, and these provide tantalizing glimpses of how landscapes have evolved under prolonged periglacial conditions (e.g. French and Harry, 1992). The introduction of new techniques for establishing long-term rates of rock breakdown (e.g. Small *et al.*, 1999), coupled with numerical modelling of slope evolution (e.g. Anderson, 2002; Anderson *et al.*, 2013), promises to revolutionize our understanding of the rates at which periglacial landscapes evolve, and the forms they adopt as they do so.

Many periglacial landscapes, moreover, are highly sensitive to disturbance. This is particularly true of terrain underlain by ice-rich permafrost, which is prone to subsidence or slope failure if the ground temperature regime is altered (Figure 1.4). *Applied periglacial geomorphology* is that branch of the subject devoted to identification of sensitive terrain, the effects of human activity on such terrain and the geotechnical approaches to minimizing terrain disturbance. A major area of current concern is the effects of projected climate warming on permafrost environments, which has led to urgent research devoted to monitoring and modelling the thermal, geomorphological and geotechnical response of permafrost to recent and projected climate change (e.g. Nelson *et al.*, 2008; Harris *et al.*, 2009; Romanovsky *et al.*, 2010a; Callaghan *et al.*, 2011; Slater and Lawrence, 2013). A particular area of concern is that thaw of permafrost underlying arctic tundra environments and subarctic boreal forests will release greenhouse gases (carbon dioxide and methane) into the atmosphere, thus potentially accelerating global warming (Schuur *et al.*, 2015).

Finally, it is notable that periglacial phenomena are not confined to planet Earth. High-resolution imaging data obtained for Mars show landforms strikingly similar to some in terrestrial permafrost environments, suggesting that in the relatively recent geological past the Martian

Figure 1.2 Wedge-shaped structure in sand and gravel deposits near Lincoln, eastern England. This structure represents infill of the void left by thaw of an ice wedge that formed in permafrost during the last glacial period. *Source:* Courtesy of Julian Murton.

Figure 1.3 Landscape of northern Yukon, Canada, an area that escaped glaciation during the Pleistocene epoch and has evolved under periglacial conditions since the beginning of the Quaternary period. *Source:* Courtesy of Matthias Siewert.

Figure 1.4 Subsidence of buildings due to thaw of underlying permafrost in (a) Alaska and (b) Yakutsk. *Source:* Courtesy of (a) Matthias Siewert and (b) Robert Way.

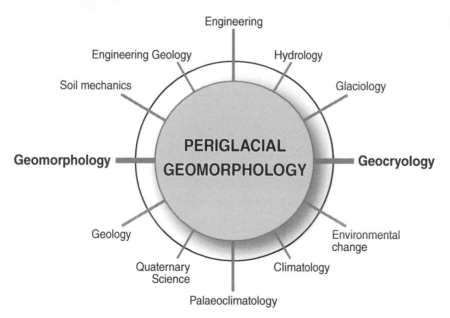

Figure 1.5 Relationship between periglacial geomorphology and cognate sciences.

surface was not only underlain by ice-rich permafrost, but may also have experienced surface or nearsurface freeze–thaw cycles that imply the existence, albeit transient, of liquid water (Balme *et al.*, 2013).

As this brief survey indicates, periglacial geomorphology is strongly integrated with several scientific disciplines. At the process-landform core of the subject lies the interaction of geomorphology and *geocryology*, the science of frozen ground, but there are also strong interactions with Quaternary science, climatology, hydrology and engineering geology, and weaker but important links with a range of other disciplines (Figure 1.5).

1.2 The Periglacial Realm

There are many different views regarding the geographical dimensions of what climatic geomorphologists have termed 'the periglacial zone' (Thorn, 1992; Berthling and Etzelmüller, 2011). This has variously been interpreted as constrained by the distribution of permafrost and/or deep seasonal ground freezing, or by climatic parameters such as the 0 °C mean annual air temperature isotherm. Such criteria are unrealistically restrictive from a geomorphological perspective. In this book, the present periglacial realm is considered to encompass all unglacierized parts of the Earth's land surface where frozen ground or freezing and thawing of the ground significantly influences landform development, so that the present operation of periglacial processes and the distribution of resultant landforms effectively defines the extent of the periglacial domain. Similarly, the extent of former periglacial environments that existed in temperate mid-latitudes during Quaternary cold stages can be defined by the distribution of relict periglacial landforms, deposits and soil structures.

Using similar criteria, French (2007) estimated that the present periglacial realm occupies at least 25% of the Earth's land surface, and that an additional 20–25% experienced periglacial conditions during Quaternary cold periods. Moreover, because air temperatures decline with increasing altitude as well as increasing latitude, active periglacial landforms can be found at high altitudes (>5000 m) even on subtropical mountains, where nocturnal clear-sky radiative cooling results in disturbance of the ground by shallow frost action. Subdividing the periglacial realm into distinctive periglacial environments is difficult, however, because individual categories merge into one another and their characteristics are influenced locally by relief and vegetation cover.

Several broad zones can nevertheless be distinguished on the basis of characteristic terrain types and landform assemblages. *Polar deserts* are areas of extreme cold and aridity, very limited vegetation cover and continuous permafrost at shallow depth, conditions typical of the northernmost circumpolar lands and some unglacierized parts of Antarctica. *Arctic tundra landscapes* are characterized by partial or complete tundra vegetation cover, and are usually underlain by continuous permafrost at shallow to moderate depths, typified by the landscapes of northern Alaska, northern Siberia, northernmost mainland Canada and much of the Canadian Arctic Archipelago. *Arctic and subarctic boreal forest landscapes*, often interrupted by wetlands, are underlain by continuous or discontinuous permafrost and extend in a huge belt across Eurasia and North America south of the tundra zone. *Maritime periglacial landscapes*, exemplified by Iceland, high plateaus in Scotland and subantarctic islands, are characterized by high winds, high precipitation, variable vegetation cover and patchy (or no) permafrost. Distinctive *alpine periglacial environments* occur on mid-latitude mountains above the treeline, where permafrost-free ground merges upwards into areas underlain by permafrost; they are typified by the European Alps, the Carpathians, the Canadian Rocky Mountains and the Southern Alps of New Zealand. A *mid-latitude high-plateau periglacial landscape* is represented by the Tibetan (Qinghai-Tibet) Plateau (28–40° N), much of which lies over 4000 m above sea level, incorporates several major mountain ranges and is extensively underlain by continuous permafrost in the interior and by discontinuous permafrost farther south. Finally, *subtropical high mountain periglacial landscapes*, notably found in the Andes, support a range of periglacial features related to diurnal freezing and thawing of ground above about 5000 m. Like all attempts to subdivide the Earth's surface, this broad categorization involves extensive generalization: even at a particular latitude, there may be significant differences in temperature and precipitation regime, often reflecting climatic contrasts between coastal environments and continental interiors, and some areas defy ready classification. The Svalbard archipelago (76–81° N), for example, is at similar latitudes to Ellesmere Island in the Canadian Arctic, but as a result of its maritime location experiences much less severe winters. Equally, the Jotunheimen Massif in Norway is difficult to classify: an area of alpine relief at subarctic latitude, but facing relatively mild, moist westerly airstreams from the North Atlantic. The climatic and terrain characteristics of the periglacial realm are considered further in Chapter 2.

1.3 The Development of Periglacial Geomorphology

Though the roots of periglacial geomorphology lie in the nineteenth century, the subject did not flourish as an independent discipline until midway through the twentieth. The development of periglacial research up to 1965 has been charted by French (2003), who described the various early accounts of permafrost and ground ice recorded by arctic explorers, gold miners and mining engineers, and Shiklomanov (2005), who summarized the early development of geocryology in Russia. It is arguable, however, that the origins of periglacial geomorphology as a scientific subdiscipline lie in the work of late 19th century geologists seeking to explain the origin of Pleistocene deposits and landforms in areas outside the limits of the Quaternary ice sheets (Ballantyne and Harris, 1994). Interest in frost action as a geomorphological agent appears to have been ignited at the beginning of the 20th century by several developments. One was the appearance of an influential account by Andersson (1906) of the rubble deposits of the Falkland Islands and of Łozinski's (1909, 1912) papers on the periglacial zone and 'periglacial facies'. Another was a scientific expedition to Svalbard following the International Geological Congress of 1910, which produced remarkably prescient work on various aspects of frost action (Meinardus, 1912; Högbom, 1914). A third was the appearance of reports by geologists working in Alaska and Yukon of a wide range of hitherto undocumented periglacial phenomena, such as frost polygons, ice wedges, rock glaciers and cryoplanation terraces (e.g. Capps, 1910; Cairns, 1912; Leffingwell, 1915; Eakin, 1916). Russian research of the era tended to focus on permafrost and ground ice phenomena (Shiklomanov, 2005).

Between 1920 and 1950, research in arctic and alpine environments produced several further seminal accounts of permafrost and periglacial processes and landforms (e.g. Shostakovitch, 1927; Elton, 1928; Poser, 1933; Sørensen, 1935; Paterson, 1940), and in both Europe and North America there was widening recognition of the importance of Pleistocene periglacial conditions in the development of mid-latitude landscapes, landforms and deposits. Particularly important developments during this era were the introduction of laboratory experimentation as a means of understanding freezing effects in soils (Taber, 1929, 1930; Beskow, 1935), the appearance of a monograph by Troll (1944) on alpine periglacial landscapes, and the publication of the first English-language treatise on permafrost (Muller, 1947), which drew substantially on earlier Russian research.

French (2003) identified three strands in the rapid evolution of periglacial geomorphology in the years after the Second World War. In Europe, energetic leadership was provided in France by André Cailleux and Jean Tricart, in Germany by Han Poser and Julius Büdel, and in Poland by Jan Dylik and Alfred Jahn. Much of these authors' research focused on Pleistocene periglacial features or, in the case of Tricart and Büdel, on climatic geomorphology, but collectively they forged periglacial research as a distinct entity, as evidenced by the publication of the first textbooks on the topic (Cailleux and Taylor, 1954; Tricart, 1963; Jahn, 1975), syntheses of Pleistocene periglaciation in Europe (e.g. Büdel, 1953; Poser, 1953–54), and the founding in 1954 of *Biuletyn Peryglacjalny*, the first journal devoted to periglacial research. In Great Britain, advances in periglacial research were sluggish by comparison, though research by Williams (1965, 1969) on the distribution of permafrost in England during the last glacial period bears comparison with similar work in continental Europe. A British textbook entitled *Glacial and Periglacial Geomorphology* (Embleton and King, 1968) proved excellent on the first topic but disappointingly pedestrian on the second.

North American research in the decades following 1945 looked north to the permafrost zone, where expansion of roads and settlements required a fuller understanding of permafrost terrain: permafrost distribution and thermal regime, the consequences of permafrost disturbance and the origins of permafrost phenomena such as thermal contraction cracking and ground ice formation (e.g. Muller, 1947; Hopkins, 1949; Lachenbruch, 1962). J.B. Bird's book *The Physiography of Arctic Canada* (1967) placed the periglacial geomorphology of the Canadian north in a broader landscape perspective, and R.J.E. Brown's monograph *Permafrost in Canada* (1970) summarized the findings of two decades of research on the topic. During the same period, permafrost scientists in the former Soviet Union made enormous advances in geocryological research, but much of this was inaccessible to researchers outside the Soviet Union until improved international relations allowed for the publication of key reports in English translations (e.g. Baranov, 1964; Kachurin, 1964; Shumskii, 1964; Tsytovitch, 1964; Kudryavstsev, 1965).

The third strand in postwar periglacial research was the beginning of detailed field and laboratory investigations of periglacial processes. An outstanding early proponent of such research was the Canadian geomorphologist Ross Mackay, who developed a masterful synthesis of physical theory and field testing to investigate a wide range of frozen-ground phenomena, most notably ice-cored hills (*pingos*) and thermal contraction cracking of permafrost; much of his work remains unsurpassed, and is described in later chapters. At roughly the same time, A.L. Washburn (1967, 1969) was carrying out long-term field investigation in NE Greenland of slope processes and frost action that forms the foundation of much current understanding of these topics (Hallet, 2008), and various geomorphologists were carrying out groundbreaking research on the processes operating in alpine periglacial environments (e.g. Rapp, 1960; Caine, 1974). Such work represents the beginning of the present era of periglacial research, with its central focus on process–landform relationships. Washburn's *Periglacial Processes and Environments* (1973) and its successor *Geocryology: A Survey of Periglacial Processes and Environments* (1979) are magnificent syntheses of the subject built on an encyclopaedic knowledge of the literature up to that time. *Geocryology* cites over 2100 references, and might justifiably be considered the Old Testament of periglacial research.

During the past three decades, there has been a shift in the focus of periglacial geomorphology, away from field mapping, classification and spatial differentiation of periglacial phenomena and towards a critical and technologically sophisticated re-evaluation of periglacial processes (André, 2009). Process-oriented field research has been transformed by the advent of instrumentation for simultaneous monitoring of such variables as snowcover, surface and subsurface ground temperature, soil moisture content, pore-water pressure and associated volume change and shear strain in rock and soil (e.g. Matsuoka, 2006; Harris *et al.*, 2007). Allied to these advances has been the development of geophysical techniques for the detection of permafrost and ground ice, and the delimitation of subsurface structures within periglacial landforms and deposits (e.g. Hauck and Kneisel, 2008; Kneisel *et al.*, 2008). Upscaling of data on landforms, processes and terrain characteristics to landscape scale has been accomplished using remote-sensing techniques and digital terrain modelling (e.g. Etzelmüller *et al.*, 2001; Hjort *et al.*, 2007), and major advances have been made in numerical modelling of ground thermal regime and certain geomorphological processes (e.g. Riseborough *et al.*, 2008). Laboratory simulation experiments, conducted both at full scale (e.g. Murton *et al.*, 2006; Harris *et al.*, 2008a) and at reduced scale in a geotechnical centrifuge (e.g. Harris *et al.*, 2005, 2008c), have transformed our understanding of the processes involved in thaw of ground ice, weathering of intact bedrock and particularly downslope movement of soils under cyclic freezing and thawing. New dating techniques, coupled with numerical modelling (e.g. Anderson, 2002; Anderson *et al.*, 2013), are transforming our understanding of how periglacial landscapes have evolved over long timescales. Overshadowing all these developments is an awareness that predictions of rapid warming in present-day arctic and alpine environments may bring about geomorphological changes at a rate and scale not seen on Earth since the end of the last glacial stage (e.g. Harris *et al.*, 2009; Callaghan *et al.*, 2011).

These developments have forced a reassessment of many long-established views concerning, for example, the efficacy of frost action in rock breakdown (e.g. Hall *et al.*, 2002), the erosional potential of late-lying or perennial snowbeds (e.g. Thorn and Hall, 2002) and the role of viscous flow in downslope movement of soil during thaw (e.g. Harris *et al*, 2003). Many of these changes in the understanding of periglacial processes can be traced through a comparison of the three editions of Hugh French's book *The Periglacial Environment* (1976, 1996, 2007), the papers published in *Permafrost and Periglacial Processes* (a journal founded in 1990 and now the main vehicle for publication of periglacial research), the proceedings of successive International Permafrost Conferences, and the mainstream geomorphological and Quaternary science journals.

1.4 Periglacial Geomorphology: The Quaternary Context

Though there are occasional references to periglacial sediments or structures in ancient sedimentary rocks (e.g. Deynoux, 1982; Williams, 1986), almost all periglacial research is devoted to present-day periglacial processes and landforms, or to relict periglacial features that formed under cold conditions during the Quaternary period, colloquially known as 'the Ice Age', which spans the last 2.6 million years or so. The Quaternary period is subdivided into two geological epochs, the *Pleistocene* (~2.6 Ma to 11.7 ka) and the *Holocene* (11.7 ka to the present). Note that 'years' are here expressed by *a* (Latin *annus*), thousands of years as *ka* and millions of years as *Ma*; these abbreviations may refer to a span of time, but are usually employed here in the sense of 'years before present'. The Quaternary period is further subdivided into the *Early Quaternary* (= Early Pleistocene, ~2.6 to ~0.9 Ma), the *Middle Quaternary* (= Middle Pleistocene, ~0.9 Ma to ~130 ka), and the *Late Quaternary* (= Late Pleistocene plus the Holocene).

The Quaternary period has been an era of major and often rapid climatic changes, marked by the alternation of cold *glacial stages* and warmer *interglacial stages* (Figure 1.6). Shorter periods of cold conditions are termed

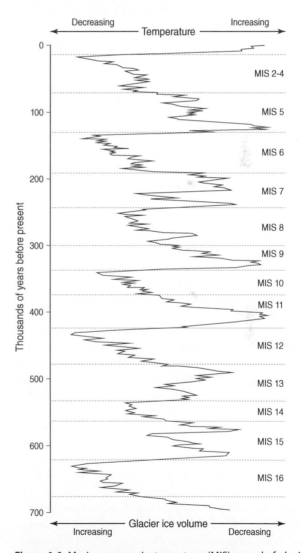

Figure 1.6 Marine oxygen isotope stage (MIS) record of glacial–interglacial oscillations during the past 700 ka. Changes in the $\delta^{18}O{:}\delta^{16}O$ ratio in benthic foraminifera in deep ocean cores record the expansion and contraction of global ice sheets, and thus changes in global temperatures. Dashed horizontal lines mark the main boundaries between glacial stages (even numbers) and interglacial stages (odd numbers).

stades or *stadials*, and shorter periods of warmer climate are referred to as *interstades* or *interstadials*. The Holocene epoch represents the present interglacial, and climatically is probably representative of earlier interglacials if we exclude the effects of recent anthropogenic climate warming. Glacial stages, particularly over the last million years, were characterized by expansion of ice sheets over mid-latitude areas, and by drops in global sea level of up to about 120–130 m below present due to the huge volumes of water locked up in the ice sheets. During the last glacial maximum (LGM) of 26.5–19.0 ka (Clark *et al.*, 2009), glacier ice occupied more than 30% of the present land area of the Earth, compared with about 10% today: the Laurentide Ice Sheet in North America covered much of Canada and extended to latitude 38° N in Ohio; roughly 70% of the present land area of the British Isles was covered by an ice sheet that terminated southwards along a line between South Wales and Yorkshire; and Scandinavia was buried by an ice sheet that extended southwards to Denmark, northern Germany and northern Poland. Climatic cooling during glacial stages was also associated with extension of periglacial conditions and permafrost into areas beyond the ice margin, such as southern England, northern France, the Netherlands, Belgium, central Germany and central Poland, northern states of the USA and extensive parts of Russia, China and Japan. In such areas, evidence for Pleistocene periglacial conditions is locally preserved in the form of relict periglacial landforms, sediments and soil structures (Figure 1.2). As very cold conditions persisted during at least the initial stages of ice-sheet retreat, permafrost and periglacial phenomena also developed in terrain vacated by the retreating ice margins, so that relict periglacial deposits sometimes overlie glacigenic deposits, and periglacial soil structures are developed in glacigenic sediments. Moreover, though ice sheets and permafrost have often been regarded as separate entities, there is growing evidence that permafrost influences the dynamics and erosive behaviour of overlying ice sheets and the resultant sediment-landform assemblages (Waller *et al.*, 2012).

The great majority of relict Pleistocene periglacial features that have been described relate to the last glacial stage, which lasted from ~130 ka until 11.7 ka. This period is known in North America as the *Wisconsinan* stage, and elsewhere as the *Devensian* (Britain), *Weichselian* (NW Europe), *Würmian* (the Alps), *Vistulian* (Poland) and *Valdaian* (European Russia). This was not a period of uniformly cold conditions, but comprised a complex sequence of alternating cold stages and cool temperate interstades. The last global ice-sheet maximum (26.5–19.0 ka) occurred during the *Late Wisconsinan* (*Late Weichselian*) substage of ~31.0–11.7 ka, itself a climatically complex period. Of particular note is the final stadial episode of the Late Weichselian in Europe, the *Younger Dryas Stade* of ~12.9–11.7 ka. This occurred after an interval of cool temperate climate (14.7–12.9 ka), in response to disturbance of oceanic circulation in the North Atlantic Ocean. It involved a brief return to severe cold in NW Europe, with readvance of glaciers in Scandinavia and Scotland, and return of permafrost conditions to unglacierized parts of northern Europe (Isarin, 1997a).

The Holocene epoch (~11.7 ka to the present) has also witnessed systematic climate shifts, albeit of lesser magnitude than those associated with glacial–interglacial transitions. The *Holocene thermal maximum*, sometimes referred to as the Holocene *climatic optimum* or *hypsithermal*, marks the period of highest Holocene global temperatures prior to recent anthropogenic global warming. The Holocene thermal maximum occurred within the period 11.0–7.0 ka, but peaked at different times in different regions, and was most pronounced at high latitudes; Renssen *et al.* (2012) have suggested that the maximum air temperature anomaly relative to pre-industrial levels typically exceeded +2.5 °C north of latitude 60° N. A more recent period of relative cooling, the *Little Ice Age*, occurred during the last millennium. Conventionally, this has been attributed to the period ~1550–1850 AD, but in some areas there is evidence for cooling as early as 1250–1300 AD, and the thermal minimum of the Little Ice Age was reached at different times in different regions. In middle and high latitudes, the associated negative thermal anomaly was of the order of 0.5–2.0 °C, and resulted in widespread advance of glaciers and aggradation of permafrost.

Evidence of periglacial conditions prior to the last glacial stage is sometimes preserved in older Pleistocene stratigraphic sequences that have survived subsequent erosion. These demonstrate that permafrost was present outside ice margins during both the Middle Pleistocene (e.g. Murton *et al.*, 2001a) and the Early Pleistocene (e.g. Kasse, 1993). The oldest record of Pleistocene periglacial conditions is a soil buried under till in northern Missouri. This exhibits soil structures indicative of formation under periglacial conditions and probably former permafrost, and has been dated to ~2.5 Ma (Rovey and Balco, 2010), suggesting that periglacial conditions have intermittently affected mid-latitude environments throughout the entire span of the Pleistocene. Similarly, evidence for former permafrost formation and degradation in central Alaska has been dated to ~2.1 Ma (Beget *et al.*, 2008), suggesting that cold climate conditions have persisted (though not uniformly) in high latitudes throughout much of the Pleistocene.

An important component of our understanding of how periglacial landscapes have evolved in formerly glaciated areas is the concept of *paraglacial landscape modification*. The term *paraglacial* refers to 'non-glacial earth-surface processes, sediment accumulations, landforms, landsystems and landscapes that are directly conditioned by glaciation and deglaciation' (Ballantyne, 2002), and *paraglacial geomorphology* is the study of the way in which glaciated landscapes have adjusted to nonglacial conditions (Ballantyne, 2003, 2013a). The retreat of glacier ice, especially in upland areas, exposes landscapes in an unstable or metastable state. Rock masses stressed by the weight of glacier ice experience decompression, opening of fractures and resultant adjustment in the form of rockfalls, rockslides, rock avalanches and slope deformation. Glacigenic sediments on hillslopes may be reworked by landslides, debris flows, snow avalanches and running water. Valley-floor glacigenic sediments are eroded, transported and redeposited by rivers, forming alluvial fans, alluvial valley-fills, or deltas and bottom deposits in lakes or fjords. In general, paraglacial landscape adjustment and sediment reworking is most rapid during and immediately after deglaciation and declines in importance as rock masses stabilize and glacigenic sediment sources become exhausted. However, enhanced sediment flux within deglaciated catchments may persist for thousands of years, and today still occurs in large catchments that were deglaciated in the Late Pleistocene or Early Holocene. Many depositional landforms in formerly glaciated periglacial environments (such as talus accumulations, debris cones, alluvial fans, floodplain deposits and deltas) are essentially of paraglacial origin, and reflect an unusually rapid period of landscape change in the centuries or millennia following deglaciation, rather than the operation of processes at present-day rates.

To employ relict periglacial features to reconstruct past environments, we need to establish the age of such features. Numerous techniques have been used to date Pleistocene sediments (Walker, 2005; Lowe and Walker, 2015), but three are particularly important in periglacial geomorphology. The first is *radiocarbon dating*. Plant and animal tissues in living organisms contain the same concentration of the unstable isotope ^{14}C (carbon-14) as the atmosphere, and atmospheric concentrations remain roughly constant through time. When a plant or animal dies, the concentration of ^{14}C in the dead tissue reduces at the rate of 1% every 83 years. The rate of decay therefore describes a negative exponential function: 50% of ^{14}C is lost after 5730 years, 75% after 11 460 years, 87.5% after 17 190 years and so on. By measuring the concentration of ^{14}C present in dead organic matter, we can estimate the time of death or burial. Because the concentration of ^{14}C becomes very small with increasing age, radiocarbon dating is usually accurate only for ages up to about 45 ka, though with isotope-enrichment techniques ages of up to 60 ka may be detected. In a periglacial context, radiocarbon dating is usually employed to date organic matter buried under or within sediments such as slope deposits or windblown sand.

Luminescence dating works on the principle that buried sediments are subject to very low levels of natural radiation. This causes release of electrons that become trapped in structural defects within the crystal lattices of minerals. These electrons can be released under controlled laboratory conditions, either by heating (*thermoluminescence (TL)* dating) or by shining a beam of light on to the sample (*optically stimulated luminescence (OSL)* dating). Both procedures result in an emission of light (luminescence) that is proportional in intensity to the concentration of trapped electrons and thus the age of sediment deposition, provided that the luminescence signal was zeroed by exposure to sunlight prior to sediment emplacement. OSL dating has largely replaced TL dating in establishing the age of sediments, as much less prior exposure to sunlight ('bleaching') is required (Aitken, 1998). OSL dating of quartz and feldspar grains has an upper age limit of about 150–200 ka and is primarily employed in a periglacial context to date the deposition of windblown or fluvial sediments (Bateman, 2008).

Finally, *cosmogenic isotope dating*, sometimes termed *cosmogenic radionuclide dating*, measures the concentration of isotopes produced within the lattices of certain minerals as a result of neutron spallation (nucleus fragmentation) due to exposure to secondary cosmic radiation at the Earth's surface (Gosse and Phillips, 2001). The main isotopes used are ^{10}Be (beryllium-10), ^{26}Al (aluminium-26) and ^{36}Cl (chlorine-36). All are radioactive isotopes, with half-lives of ~1360, ~301 and ~705 ka respectively. The concentration of cosmogenic isotopes present within the upper part of an exposed bedrock or boulder surface can be used to determine how long that surface has been exposed to the atmosphere (*surface exposure dating*) and is often used to determine the timing of deglaciation or stripping of overlying sediment cover. Cosmogenic radionuclides have also been used to establish the age of alluvial deposits and landslides, the timing of emergence of *tors* or bedrock protrusions and the age of weathered rock material (*regolith*), as well as to calculate rates of land-surface lowering (e.g. Small *et al.*, 1997, 1999; Anderson, 2002; Phillips *et al.*, 2006; Darmody *et al.*, 2008; Ballantyne *et al.*, 2014).

1.5 The Aims and Organization of this Book

The aim of this book is to provide a comprehensive introduction to periglacial geomorphology, one that incorporates not only the geomorphological and sedimentological processes that operate in present-day periglacial environments, and the landforms, deposits, and sediment structures produced by such processes, but also the identification of relict periglacial features and interpretation of their palaeoenvironmental significance. This poses several challenges. The first is the breadth and diversity of the subject, which necessitates considerable economy of treatment. The second is spatial variation: though the physics of periglacial processes are universal, the conditions under which these operate vary enormously in different periglacial environments. The processes operating arctic tundra lowlands, for example, produce an entirely different assemblage of landforms from those operating in high-relief alpine environments. The third is the need to address both present processes and landforms, and the origins and significance of relict periglacial features.

The book is divided into six parts. The remainder of this introductory section (Chapter 2) provides further background on the contrasting nature of periglacial environments. Part II (Chapters 3–5) is devoted to the characteristics of frozen ground: ground freezing and thawing, permafrost and seasonally frozen ground, and ground ice. This forms the foundation for Part III (Chapters 6–9), which considers the characteristics, origin, and formation of landforms and structures associated with frozen ground, such as thermal contraction cracking of permafrost, ice-cored mounds, *thermokarst* landforms produced by the thaw of ice-rich permafrost, and a range of smaller landforms produced by seasonal freezing and thawing of the ground. Part IV (Chapters 10–12) addresses the weathering or breakdown of rock in periglacial environments, the processes operating on hillslopes, and the wide range of landforms that result from cold-climate slope processes. Part V (Chapters 13–15) considers the operation of fluvial activity, wind action, and coastal processes in periglacial environments, and how these are conditioned by cold climate conditions and frozen ground. Finally, Part VI (Chapters 16 and 17) draws the book to a close by considering how relict periglacial features can inform our understanding of past environmental conditions and the responses of periglacial environments to climate change.

Most of the central chapters (6–15) follow a common format: a brief introduction defining relevant terminology is followed by an account of relevant processes, a description of resulting landforms, deposits, and/or sediment structures, and finally an outline of equivalent relict (Pleistocene) phenomena and their significance. This format allows consideration of relict periglacial features in association with their active counterparts.

The literature on periglacial geomorphology is vast, and any synthesis of the topic has to be highly selective. The main criteria for selection of sources has been scientific importance, innovation, or representativeness, but where possible English-language publications or translations are cited, in acknowledgement of the likelihood that many readers of this book will be monoglot anglophones, or students who have mastered English as the *lingua franca* of scientific research. For those who have not, van Everdingen (1998) has compiled and edited an indispensible multi-language glossary of permafrost and related ground ice terminology, the Rosetta Stone of multilingual periglacial research. Major contributions have been made to our understanding of both periglacial geomorphology and geocryology by Russian, Chinese, Japanese, Scandinavian, French, German, and Polish scientists, with growing contributions by researchers in other European countries and South America. The focus on the English-language literature should not be interpreted as an attempt to promote anglophone hegemony on a subject to which innumerable scientists have contributed from all parts of the globe. The community of periglacial geomorphologists is international in both heritage and membership.

2

Periglacial Environments

2.1 Introduction

As outlined in Chapter 1, the geographical extent of present-day periglacial activity is not easily defined, but here we adopt the view that the present periglacial realm encompasses all unglacierized regions where frozen ground and/or freezing and thawing of the ground significantly influences landform development. The present operation of frost-action processes and the distribution of resultant landforms thus define the extent of the periglacial domain. The driving influence on frost action is climate, which represents the primary control not only on the distribution of permafrost and the depth of seasonal ground freezing and thawing, but also on the frequency of ground-level freeze-thaw events, the depth of winter snowcover, the seasonal availability of liquid water in the upper levels of the ground, the runoff regime of rivers and the propensity for erosion and deposition of sediment by strong winds. At a more local level, however, the effects of climate on ground temperature and geomorphological processes are modulated by vegetation cover and substrate characteristics. This chapter sets the scene for analysis of the processes operating in periglacial environments by outlining the characteristics of periglacial climates and briefly summarizing those of soils and vegetation cover in cold environments.

2.2 Periglacial Climates

No single climatic parameter adequately defines the limit of periglacial climates, though French (2007) suggested that this can be approximated by a mean annual air temperature (MAAT) of +3 °C. This is a useful criterion that encompasses not only the polar, subpolar and high-altitude regions of the Earth, but also areas of shallow seasonal ground freezing. Classification of periglacial climates is inevitably rather arbitrary, as boundaries between climatic zones are gradational, and even within particular areas there may be marked climatic variation relating to such factors as altitude, slope aspect and distance from the coast. French (2007) employed a fourfold

classification of periglacial climates (high arctic, continental interiors, alpine and climates with low annual temperature range) and identified two areas (the Qinghai-Tibet Plateau and Antarctica) that do not fall readily into any of his classes. A similar but modified approach is adopted here. Note that precipitation figures cited here and illustrated in Figures 2.3–2.5 are probably low estimates, particularly for winter snowfall, due to possible undercatch of snow under windy conditions.

2.2.1 Arctic Climates

Serreze and Barry (2014) have provided a comprehensive and accessible account of arctic climates. These can be understood with reference to three key factors. The first is a strongly negative annual radiation budget at the top of the atmosphere (roughly −60 to −120 W m^{-2}) in arctic areas, implying that outward radiative losses exceed inputs. In part, this is due to the fact that north of the Arctic Circle (66.5° N) there is a period of winter darkness that becomes progressively longer with increasing latitude (Figure 2.1), but it also reflects the fact that in summer, solar radiation reaches high latitudes at a shallow angle (i.e. the sun remains low in the sky), so radiative inputs are relatively low despite 24-hour daylight. A second key feature is the development, particularly during autumn, winter and spring, of semipermanent atmospheric high-pressure systems over much of the Arctic. These prevent incursion of frontal systems and associated movement of relatively warm, moist air into northern polar latitudes, with the notable exception of the North Atlantic cyclone track that extends eastwards from a zone of pervasive low pressure (the Icelandic Low) centred southeast of southern Greenland around latitude 60° N, and to a lesser extent the Aleutian Low in the North Pacific. The effect on precipitation totals can be seen in Figure 2.2, in the form of zones of low (<400 mm a^{-1}) precipitation centred on the Canadian Arctic and northern Asia, and a belt of much higher (>600 mm a^{-1}) precipitation crossing from Baffin Island across southern Greenland and Iceland into the Barents and Kara Seas. Finally, heat transfer from the Arctic

Periglacial Geomorphology, First Edition. Colin K. Ballantyne.
© 2018 John Wiley & Sons Ltd. Published 2018 by John Wiley & Sons Ltd.

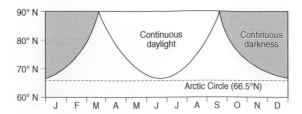

Figure 2.1 Duration of continuous daylight and continuous darkness at latitudes 60–90° N, defined by whether the sun is above or below the horizon.

Ocean tends to reduce temperature extremes in winter in surrounding landmasses, but heat loss to the Arctic Ocean also depresses summer temperatures slightly in coastal locations.

Figure 2.3 depicts climatic summaries for Eureka (79° 59′ N) and Resolute Bay (74° 41′ N), both in the Canadian Arctic Archipelago. In common with most of the high Arctic, winter monthly average temperatures are typically within the range −30 to −40 °C, and summers are short (2–3 months) and cool, with mean July

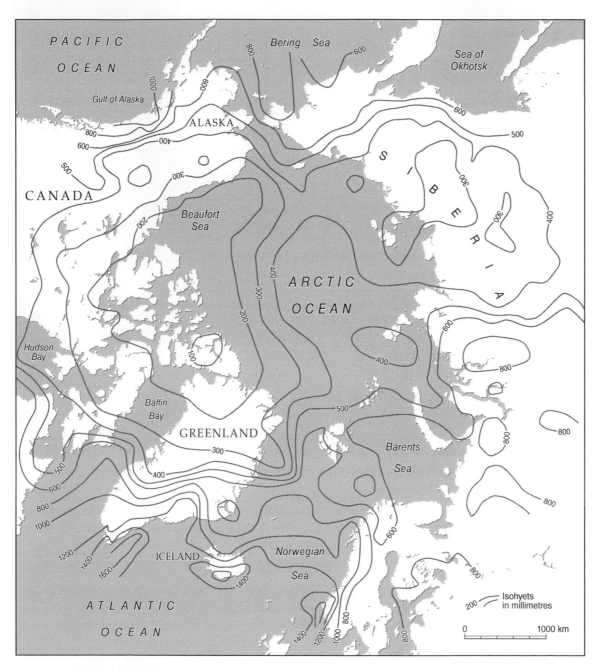

Figure 2.2 Isohyets of mean annual precipitation (MAP) in the Arctic. Isohyet intervals are 100 mm for precipitation <600 mm and 200 mm for precipitation >600 mm. Extensive areas of the high Arctic and arctic or subarctic continental interiors have MAP <400 mm. *Source:* Serreze and Barry (2014). Reproduced with permission of Cambridge University Press.

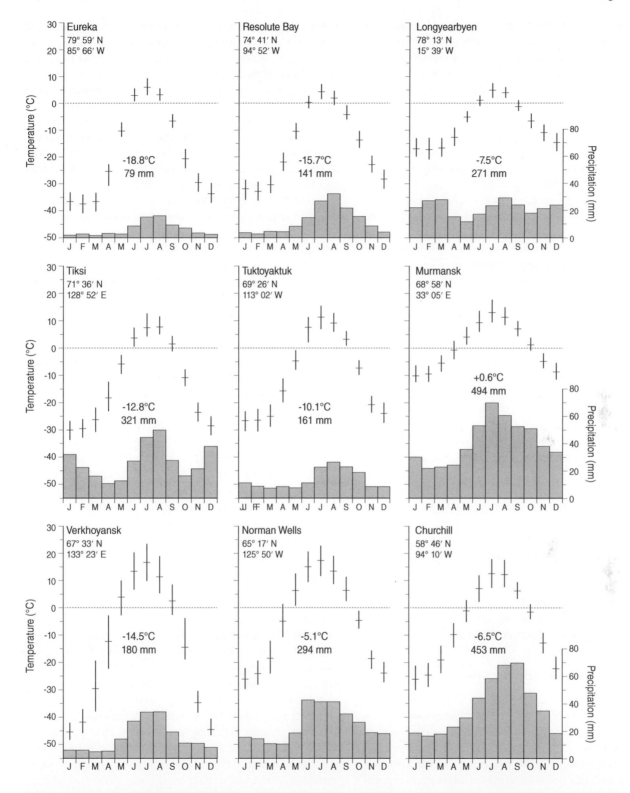

Figure 2.3 Mean monthly temperature and precipitation at high-arctic, low-arctic and continental interior locations near sea level. Horizontal dashes represent mean monthly temperature and vertical lines the range between maximum and minimum mean monthly temperatures. The histograms represent mean monthly precipitation. The figures in the centre of each graph are the mean annual air temperature (MAAT) and mean annual precipitation (MAP).

temperatures of +4 to +6°C, though some summer days are remarkably mild; the author fondly remembers basking in warm (17°C) sunshine one July afternoon at 78°N on Ellesmere Island, and according to one commentator it may be (theoretically) feasible to grow lettuce in benign parts of the Canadian high Arctic. Measured mean annual precipitation (MAP) is low (79 and 141 mm for Eureka and Resolute Bay respectively) and concentrated in the months of June to October, when there is increased cyclonic activity. By contrast, mean monthly winter air temperatures at Longyearbyen on Svalbard (78°13′N) are warmer (–15 to –20°C), and precipitation is moderate throughout the year. Both characteristics reflect the location of Svalbard in the North Atlantic cyclone track, so that relatively warm moist air crosses the archipelago in winter, moderating average temperatures and introducing frontal precipitation; rainfall is rare but not unknown in Longyearbyen in January.

Moving southwards to latitudes around 70°N, Tuktoyaktuk (69°26′N) in the western Canadian arctic experiences a climate similar to that of Resolute Bay, but winter mean monthly temperatures are warmer (–20 to –30°C) and summers are warmer and longer, with four months experiencing above-freezing average air temperatures (Figure 2.3). Tiksi (71°36′N) on the Laptev Sea coast near the mouth of the Lena River has a similar temperature regime but twice as much precipitation, in part reflecting the easternmost reach of the North Atlantic storm belt; the northern Eurasian coast also experiences maximum frontal frequencies in summer, due to differential heating of the Arctic Ocean and land surface. However, Murmansk (68°58′N) near the southern shore of the Barents Sea enjoys an arctic climate that is moderated by open water offshore and incursions of relatively warm maritime airmasses that bring precipitation throughout the year, together with longer, warmer summers. Of all the stations summarized in Figure 2.3, Murmansk alone lacks permafrost by virtue of its positive (+0.6°C) MAAT.

2.2.2 Low-arctic and Subarctic Continental Interiors

In low-arctic and subarctic continental interiors between latitudes 55°N and 70°N, the dominant feature of climate is the extraordinarily wide range of mean monthly temperatures, reflecting the absence of moderating oceanic influences and the predominance of high-pressure systems during the winter months. A zone of high pressure over eastern Eurasia (the Siberian High) builds during the autumn in response to radiative cooling and persists until spring, resulting in the lowest winter temperatures anywhere in the northern hemisphere. Verkhoyansk (67°33′N) experiences a mean January temperature of –48.3°C and extremely limited winter snowfall, but for

five months of the year mean monthly temperatures are above 0°C, peaking at +17.2°C in July. Canadian locations in this zone, such as Norman Wells (65°17′N) and Churchill (58°46′N) have similar, if less extreme, characteristics: very cold winters and moderately warm summers, with 4–5 months averaging > 0°C and mean July temperatures ranging from ~15°C to ~25°C (Figure 2.3). Where permafrost is present, the warm summers permit seasonal thaw to depths of 2–3 m, and where permafrost is absent, seasonal ground freezing can reach similar depths. MAP in subarctic continental interiors ranges from about 180 to 500 mm a^{-1}, with a pronounced summer maximum, in part resulting from convectional rainstorms. However, the relatively high summer temperatures also promote high rates of evapotranspiration, so there is a seasonal soil moisture deficit.

A notable feature of continental high-arctic, low-arctic and subarctic climates is that the air temperature at ground level crosses 0°C comparatively infrequently – typically only about 20–50 times a year. This reflects the steepness of the spring warming and autumn cooling curves (Figure 2.3), so that periods of air temperature fluctuations around 0°C are brief. Summer air temperatures tend to remain consistently above 0°C and winter temperatures consistently below 0°C, except in areas affected by winter cyclonic activity. Moreover, persistence of winter snowcover in spring and the insulating effects of vegetation cover in autumn further reduce the number of freeze–thaw cycles at the ground surface. In consequence, all of these areas are dominated by a single annual cycle of ground freezing and thawing, and short-term ground surface freeze–thaw cycles have very limited impact on periglacial processes in these zones.

2.2.3 Maritime Periglacial Environments

Maritime periglacial environments are those where conditions throughout the year are strongly influenced by the proximity of relatively warm oceanic waters and frequent cyclonic activity. They are characterised by a low annual temperature range and the absence of extreme cold, but also high precipitation, frequent cloudy conditions and strong winds associated with deep depressions, particularly in winter. Permafrost and deep seasonal ground freezing are usually absent, but air freeze–thaw cycles at ground level are frequent, and often associated with the passage of alternating warm and cold airmasses during autumn, winter and spring. The effects of relatively warm offshore ocean waters and frequent cyclonic activity are particularly evident in northwest Norway: Tromsø has an MAP of 1031 mm, an MAAT of +2.8°C and a mean January temperature of –6.5°C, and is thus climatically marginal for periglacial activity despite its high latitude (69°41′N). Much of Iceland experiences similar or slightly warmer conditions, as does the south coast of Alaska: Anchorage (61° 13′ N) has an MAAT of +2.8 °C, a mean January temperature of −11.4 °C and an MAP of 423 mm. Maritime periglacial conditions affect mountains in ocean-proximal locations such as southwest Norway, the Faroe Islands and terrain above ~800 m in Scotland, which experiences MAATs in the range +1.0 °C to +4.0 °C, 2000–4000 mm of annual precipitation, frequent strong winds and only shallow ground freezing, but nonetheless supports a range of active periglacial processes (Ballantyne, 1987a). In the southern hemisphere, as outlined below, maritime periglacial conditions are typical of sub-Antarctic islands such as Kerguelen, South Georgia and high ground in the Falkland Islands.

2.2.4 Antarctica

Glacier ice covers more than 99% of the Antarctic continent, and unglacierized areas are restricted to coastal enclaves and mountains protruding through the ice as nunataks. The ice cover dictates the climate: a surface layer of extremely cold, dense air a few hundreds of metres deep flows radially outwards to form a persistent katabatic airflow that accelerates coastwards and converges in coastal valleys, giving extremely strong winds (up to 200 km h^{-1}) and mean windspeeds of ~15–40 km h^{-1}. The cooling effect of winds blowing from the interior results in mean summer (January) temperatures no higher than −5 °C to 0 °C on the unglacierized margins of

much of the continent, with mean winter (July) temperatures of −20 °C to −30 °C. Permafrost underlies all unglacierized terrain except the coastal fringes of the Antarctic Peninsula (Vieira *et al.*, 2010). The persistence of cold airflow from the interior also means that most unglaciated areas receive very limited snowfall (Figure 2.4), though the incursion of summer weather systems from lower latitudes may generate rain on coastal slopes, particularly on the Antarctic Peninsula. An area of hyper-aridity, the McMurdo Dry Valleys, occupies 4800 km^2 in Victoria Land (77–78° S, 162–164° E). Here MAATs range from −14.8 °C to −30.0 °C (Doran *et al.*, 2002) and a combination of low surface albedo, dry katabatic winds and very low precipitation (snowfall) relative to potential evaporation make this one of the arid parts of planet Earth.

Surrounding the Antarctic continent is the circumpolar trough, a persistent band of low atmospheric pressure centred around 60–65° S. The limited landmasses within this zone experience much higher precipitation and cloud cover, but also milder and less extreme temperatures, than the Antarctic itself. Mean monthly temperatures at sea level on Signy Island in the South Orkney Islands (60° 45′ S), for example, range from −10.2 °C in July to +1.2 °C in February, with an MAAT of −3.5 °C. Amongst the sub-Antarctic islands, South Georgia (54° 17′ S) also has a low annual temperature range, with a sea-level MAAT of +1.7 °C and an MAP of 1395 mm at Grytviken (Figure 2.4), but it is 57% glacierized and supports permafrost under unglacierized high ground. Kerguelen

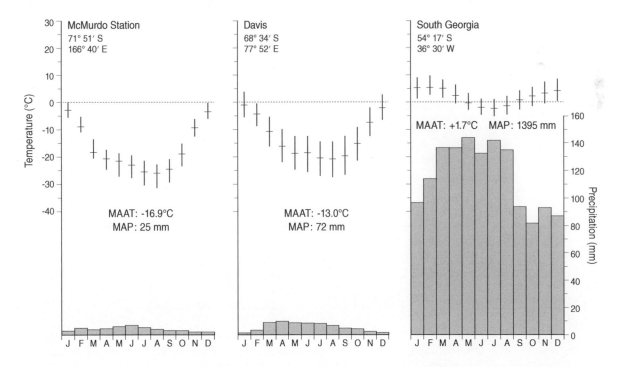

Figure 2.4 Mean monthly temperature and precipitation at Antarctic and sub-Antarctic locations near sea level. Horizontal dashes represent mean monthly temperature, vertical lines the range between maximum and minimum mean monthly temperatures and histograms the mean monthly precipitation.

(49° 15′ S) has a mean monthly sea-level temperature range of +2.1 °C to +8.2 °C, an MAAT of +4.9 °C and an MAP of 730 mm. Kerguelen supports an ice cap over its highest ground, tundra vegetation cover and probably patchy permafrost at high elevations, despite being at a latitude equivalent to that of northern France.

2.2.5 Alpine Periglacial Environments

Alpine periglacial environments are those where altitude and slope aspect (shading) are dominant controls on the annual temperature cycle. The lower limit of alpine periglacial environments broadly coincides with the altitudinal treeline, above which are areas of grasslands, alpine tundra and expanses of bare rock, scree and sparsely vegetated debris cover. In the northern hemisphere, the treeline rises generally southwards, from near sea level (e.g. at 70° N in coastal Norway) to a maximum altitude of over 4000 m at 30–40° N, but is controlled by precipitation and exposure as well as temperature. In most alpine mountains, the treeline occurs at between 2000 m and 4000 m. Barry (2008) has provided a comprehensive account of mountain climates and their controlling factors.

In a stationary atmosphere, temperature declines with altitude according to the environmental lapse rate, which averages ~0.65 °C $(100\,\text{m})^{-1}$, but in practice varies locally, regionally and temporally, and may periodically reverse during temperature inversions caused by katabatic drainage of cold, dense air from high ground into valleys under calm, clear conditions. Figure 2.5 illustrates the difference in temperature regime at two stations on the Niwot Ridge in Colorado. At 3749 m, MAAT is –3.7 °C, mean monthly temperatures range from –10 to –15 °C in

winter and only four months have average temperatures above 0 °C. At 3048 m, MAAT is +1.3 °C, mean monthly winter temperatures are much milder (–5 °C to –10 °C) and six months have average temperatures above 0 °C. Precipitation is high throughout the year, but highest in winter (as snow) and spring, though it exhibits marked interannual variation (Greenland, 1989). On the summit of Sonnblick (3106 m), the highest observatory in the European Alps, the annual temperature cycle is similar to that on Niwot Ridge: MAAT is –4.6 °C, mean monthly winter temperature ranges from –10 °C to –15 °C and three months have average temperatures above 0 °C, though MAP is much higher (2076 mm). Also depicted in Figure 2.5 is the temperature regime on a subarctic mountain summit, Fanaråken (2062 m), at 61°31′ N in the Hurringane Massif of south-central Norway. Mean monthly temperatures here exceed 0 °C in only two months of the year, but the amplitude of the annual temperature cycle is reduced by the passage of relatively warm maritime airmasses during winter.

In general, precipitation tends to increase with altitude on mid-latitude mountains, because airmasses and frontal structures are forced to rise and cool adiabatically, inducing condensation, cloud formation and orographically-enhanced precipitation on windward slopes. Conversely, compression and adiabatic warming of airmasses descending the lee sides of mountains produces relatively warm, dry *föhn* winds and greatly reduced precipitation, so there tends to be a marked precipitation contrast between the windward and lee sides of mountain ranges. This contrast is beautifully illustrated by the Southern Alps in New Zealand, where western windward slopes receive more than 6000 mm annual precipitation but eastern lee slopes

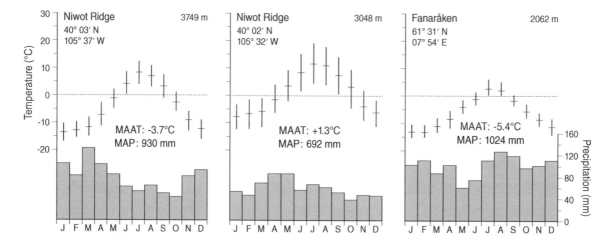

Figure 2.5 Mean monthly temperature and precipitation on the Niwot Ridge, Colorado, USA, a mid-continental location with a large annual temperature range, and in Fanaråken, Norway. Horizontal dashes represent mean monthly temperature, vertical lines represent the range between maximum and minimum mean monthly temperatures and histograms represent the mean monthly precipitation. Note that the precipitation scale is compressed compared with Figures 2.3 and 2.4. *Source:* Niwot Ridge record compiled from data from Greenland (1989).

receive less than 1000 mm. Precipitation is generally greatest in winter on mid-latitude mountains, producing thick seasonal snowcover, but both cyclonic and convectional rainstorms maintain moderate or high precipitation in summer. Mountain areas also experience exceptionally high winds, in part because airflow is forced to rise and accelerate over topographic barriers, but also because mountainous topography funnels airflow through valleys and across cols. The greatest recorded wind speed in the northern hemisphere ($372 \, km \, h^{-1}$) was measured at the summit of Mount Washington (1917 m; 44° 16′ N) in New Hampshire; the mean wind speed at the same location is $83 \, km \, h^{-1}$ in winter and $43 \, km \, h^{-1}$ in summer. On the summit of Sonnblick (3106 m) in the Austrian Alps, wind speeds average $\sim 30 \, km \, h^{-1}$ in winter and $\sim 20 \, km \, h^{-1}$ in summer, but winds exceeding $100 \, km \, h^{-1}$ occur in most months.

In alpine environments, permafrost is usually absent below the treeline but occurs at higher altitudes as a mosaic of patchy permafrost alternating with permafrost-free terrain that grades upslope to discontinuous then, on the highest summits, continuous permafrost, particularly on shaded slopes. However, deep snowcover insulates the ground from winter cooling, so permafrost may be absent even where MAAT is several degrees below the freezing point. Conversely, permafrost may exist at sites where MAAT exceeds 0 °C but cold dense air within coarse bouldery deposits results in a negative ground temperature anomaly (Chapter 4).

In general, alpine environments record a higher frequency of air temperature oscillations across the freezing point because air temperatures remain close to 0 °C for longer periods than is the case at high latitudes or in continental interiors, though the effect on ground temperature freeze–thaw cycles is greatly reduced by seasonal snowcover, which insulates the ground from short-term air temperature fluctuations. At high elevations on tropical or subtropical mountains, the annual temperature range is limited because inputs of solar radiation vary only slightly throughout the year, though permafrost may be present above 4000–5000 m. Nocturnal radiative cooling and diurnal radiative warming at high altitude on low-latitude mountains nevertheless result in frequent 24-hour oscillations of air temperature around the freezing point, causing shallow ground freezing and, where moisture is present, the development of needle ice in the uppermost few centimetres of the soil.

2.2.6 The Qinghai-Tibet Plateau

The Qinghai-Tibet (Tibetan) Plateau averages 4950 m in altitude and occupies an area of approximately 2.5 million km^2 between latitudes 29° N and 37° N. It is bounded to the south by the Himalayan mountain chain and to the north by the Kunlun Shan, and typically consists of a tundra-steppe environment interspersed with mountain ranges and large brackish lakes. The Himalayas act as a barrier to deep penetration by the warm, moist summer south Asian monsoon, though the summer east Asian monsoon affects its eastern reaches. Precipitation is consequently greatest along the southern and eastern parts of the Qinghai-Tibet Plateau (typically $250–500 \, mm \, a^{-1}$) and declines northwards and westwards. Tingri (28° 34′ N, 86° 38′ E; 4348 m) on the southern margin of the plateau has an MAAT of +2.6 °C, a mean January temperature of –6.9 °C, a mean July temperature of +10.5 °C and 271 mm mean annual precipitation. Equivalent figures for Fenghuo Shan on the eastern plateau (34° 20′ N, 97° 52′ E; ~4800 m) are –5.6 °C, –18.2 °C, +4.2 °C and 345 mm respectively. In both locations, almost all precipitation falls between May and September, reflecting monsoonal influence, and winter precipitation is negligible; snowcover is thin and transient, often melting within days under conditions of strong winter sunlight and low humidity. Farther north and northwest, the plateau steppe becomes progressively higher, colder and more arid. Permafrost underlies approximately 1 million km^2 of the plateau (Ran *et al.*, 2012), principally at altitudes >4500 m. Owing to a combination of strong insolation with cold air temperatures, a large number of days experience air freeze–thaw cycles, which are superimposed on an annual temperature cycle of moderate amplitude, with typically 15–25 °C difference between the warmest and coldest months.

2.3 Soils in Periglacial Environments

The term 'soil' is used in two different ways. Engineers and geomorphologists use the term as a loose synonym for a cover of unconsolidated sediment on the land surface, and are primarily concerned with its physical or mechanical properties. Biologists, agronomists and pedologists are more concerned with the textural, hydrological, chemical and biological properties of soil, and its role as a medium for plant growth. In this latter sense, soil is defined as a body of unconsolidated sediment consisting of layers or *horizons* of mineral and/or organic matter that differ in their physical, chemical, mineralogical and/or biological properties from overlying or underlying horizons and from the underlying parent material. The latter may be bedrock or unconsolidated sediment such as till, alluvium or windblown sediment.

The nature of soil is determined by five factors: climate, organisms, topography, parent material and time. Soil characteristics vary not only regionally in response to climate, but also locally, reflecting such influences as slope, aspect, drainage, vegetation cover, organic content and the nature of the underlying substrate. This has made soil classification difficult, and there is still incomplete agreement on terminology. Here, only a very broad sketch of cold-climate soils is provided, following the terminology of the World Reference Base for Soil Resources (Nachtergaele *et al.*, 2000); for more detailed accounts, see Campbell and Claridge (1992), Kimble (2004) and Margesin (2009).

In high-latitude environments, the dominant soils are *cryosols*, predominantly mineral soils formed in the *active layer* (zone of seasonal thawing) above permafrost. Seasonal freezing and thawing dominates pedogenesis, causing downwards translocation of organic matter and both upwards and lateral movement of mineral soil (Bockheim, 2007), though such *cryoturbation* (soil displacement due to freezing and thawing) tends to be most effective in moist, fine-grained soils (Broll *et al.*, 1999). Tarnocai (2004) has proposed a subdivision of cryosols into *turbic cryosols* (strongly cryoturbated, predominantly mineral soils), *static cryosols* (developed in coarse-textured parent materials, with limited or no evidence of cryoturbation) and *organic cryosols* (characterized by accumulation of partly decomposed, acidic plant matter at the surface), but there may be marked differences in cryosol characteristics even within small areas (Lev and King, 1999). Cryosols occur in areas of polar desert, tundra, forest-tundra or boreal forest, and are dominant in Siberia and Yakutia north of 65° N, and across most of Alaska and Canada north of 65–70° N.

Polar deserts also support *leptosols* (thin, stony, azonal soils overlying bedrock) or thicker clast-rich immature soils known as *regosols*. Of more widespread distribution are *histosols*, soils composed of incompletely decomposed plant remains with or without an admixture of sand, silt or clay. These mainly take the form of *peat deposits* or organic-rich silty *muck deposits*. Histosols occur in arctic, subarctic and boreal environments, particularly in poorly-drained areas such as valley floors, swamps and marshlands. They are widely distributed both within and south of the cryosol belt in the northern hemisphere, and dominate much of the southern Hudson Bay lowlands, Mackenzie Delta, northern Finland and parts of western and central Siberia. Also common in arctic and subarctic wetlands are *gleysols*, defined as soils that are saturated (below the water table) within 50 cm of the surface for long periods of time. Prolonged saturation and decaying organic matter result in chemical reduction of iron and manganese, producing a uniform or mottled olive, greyish or dark blue soil with limited horizon development or textural contrasts. Gleysols have a low-arctic to subarctic distribution similar to that of histosols, are dominant in parts of western and central Alaska, and are common in parts of the Yukon, the Mackenzie Delta, the Hudson Bay lowlands and areas of Russia stretching eastwards from the Finnish border to the West Siberian Plain. In general, the percentage cover of gleysols decreases northwards in the northern

hemisphere, and they are of negligible importance in high-arctic areas (Goryachkin *et al.*, 1999). Finally, *podzols* (sometimes called *spodosols*) are soils with a distinctive ash-grey subsurface horizon from which iron, aluminium and organic compounds have been leached to accumulate in an underlying dark illuvial horizon. Podzols predominate in subarctic and particularly boreal forest environments, such as those of northern Québec-Labrador, Scandinavia and northern Russia west of the Urals.

In mountainous areas, the sub-alpine zone of coniferous forest and dwarf shrub tundra also tends to be dominated by podzols (Legros, 1992), though peaty histosols may be present in this zone in areas of high rainfall and restricted drainage. Podzols or immature podzols also underlie many high-alpine grasslands in permafrost-free areas. Cryosols are present in areas of mountain permafrost where active layer depths are <2 m, which corresponds approximately to MAATs < –6 °C (Bockheim, 2015). On high summits where vegetation cover is discontinuous or absent, slopes are steep and surfaces are unstable, immature soils (leptosols and regosols) tend to dominate.

2.4 Vegetation Cover in Periglacial Environments

The most important vegetation boundary in the periglacial realm is the *treeline* or *timberline*. In subarctic or arctic environments, the treeline defines the northern limits of forested terrain, and in alpine environments it defines the altitude of the highest tree-covered slopes. The northern (latitudinal) treeline extends from Alaska to Labrador and from northern Scandinavia to the Bering Strait, and marks the boundary between the two cold biomes of the Earth, the *taiga* or *boreal forest* biome and the *tundra* biome. The term 'treeline' is somewhat of a misnomer in this context, however, as the boundary between the taiga and tundra biomes is usually represented by a transitional zone or *ecotone* that is typically 50–100 km wide and characterized by a mosaic of forest, tundra and wetlands. The southern margin of the ecotone is represented by the northern limit of coniferous forest, and is roughly coincident with the 13 °C mean July air temperature isotherm. Farther north is often a zone of *boreal woodland*, comprising open stands of conifers and denser stands of mixed conifers and deciduous trees (such as birch and poplar), interspersed with moss-covered peatlands in poorly drained sites. Father north still is the *forest tundra* zone, where stands or individual trees (mainly spruce) are interspersed with tundra vegetation comprising mosaics of dwarf birch, alder or willow, ericaceous heathland plants, and wetlands dominated by grasses, sedges and sphagnum moss. Isolated forest stands and gallery forests also extend northwards along major floodplains such as those of the Mackenzie and Lena rivers. The treeline is usually considered to lie at the northern edge of the forest tundra zone, and corresponds approximately with the 10 °C mean July air temperature isotherm; farther north, the growing season is too short to support the growth of mature trees. The southern boundary of the boreal forest biome is equally diffuse, but corresponds roughly with the latitude at which mean monthly air temperatures exceed 10 °C for five months each year.

The taiga zone (Figure 2.6e) exhibits both local homogeneity and regional variety. In cold continental interiors, tree height may be restricted and tree cover discontinuous, with lichens forming the dominant ground cover. In warmer and wetter areas, the boreal forest comprises

Figure 2.6 Vegetation cover in periglacial environments. (a) Polar desert, Svalbard. (b) Polar desert, Ellesmere Island, Canadian Arctic Archipelago. (c) Tundra, Tuktoyaktuk Peninsula. (d) Tundra, Alaska. (e) Wetlands in taiga forest near Churchill, Manitoba. (f) Alpine vegetation: coniferous forest in the valley, alpine grasslands and bare ground above ~2500 m, Savoie, French Alps. *Source:* Courtesy of (a,d) Matthias Siewert, (b) Mark Bateman and (e) Matt Morison.

unbroken stands of tall conifers, with a discontinuous herb and shrub layer and a well-developed moss layer. North American boreal forests are dominated by lodgepole pine and alpine fir in the west, and white spruce, black spruce and balsam fir farther east. Scots pine and Norway spruce dominate the western Eurasian taiga, giving way eastwards to larch. Birch and alder are also common over wide areas.

Within the tundra zone, summer temperatures and particularly the length of the growing season dictate both the extent of vegetation cover and species diversity. In the Canadian Arctic, for example, about 350 species of vascular plants occur in regions close to the treeline, but in the coldest and most northerly regions fewer than 35 species occur. Locally, however, vegetation cover is closely related to favourable sites, notably those that afford protection from the wind or enjoy a supply of water during the growing season. In the most favoured areas near the treeline, tundra vegetation has a threefold structure, with an upper layer of low shrubs or dwarf trees (birch or willow), a middle layer of tussocky grasses and cushion-form herbs, and a ground layer of mosses and lichens. In the most challenging areas, the polar deserts, vegetation is floristically limited and often restricted to a patchy cover of mosses, lichens and occasional tufts of grass. Overall vegetation cover in the tundra zone is usually dominated by grasses and sedges or woody shrubs (Figure 2.6), but the nature of vegetation cover is highly variable and strongly influenced by ground conditions. The most favoured areas support patches of dwarf trees, sedges, alpine meadow grasses, crowberry, mosses and lichens; well-drained ground supports ericaceous woody heath plants; and marshy areas support grasses, sedges and rushes, sphagnum moss, cotton grass, dwarf alder and dwarf willow. A useful five-part bioclimatic subzonation of the arctic tundra biome is outlined in Walker *et al.* (2002).

Alpine environments exhibit altitudinal zonation of vegetation types, though the dominant species present depends on location. A foothill zone of mixed deciduous forest is generally replaced with increasing altitude by a belt of dominantly coniferous montane forest, which in turn gives way to a sub-alpine belt of conifers adapted to cold, snowy winters and a short growing season, such as larch and Norway spruce. At the climatic treeline (roughly delimited by a mean annual air temperature of 3–4°C), the sub-alpine forest belt grades into 'alpine tundra', a heathland cover of dwarf shrubs, such as juniper and rhododendron, which is replaced at still higher altitudes by alpine grassland (Figure 2.6). On high, cold, windswept summits, the vegetation resembles that of polar deserts, consisting of a patchy cover of mosses, lichens, tussocky grass or cushion-form herbs. On the Qinghai-Tibet Plateau, the nature of vegetation cover is strongly controlled by precipitation, which declines from the south and east (where the summer monsoon reaches the fringes of the plateau) to the north and west. On a southeast–northwest transect, the main bioclimatic regions comprise zones dominated in turn by montane forest, steppe grassland then steppe tundra, giving way to semi-desert then desert zones.

2.5 Synthesis

Climate is the primary driver of periglacial processes. Adopting the rule that periglacial climates exist in unglacierized areas where MAAT <3.0°C introduces a wide spectrum of climate types, some with severe winter freezing and a large (>30°C) mean monthly temperature range, others with moderate winter freezing and a limited (<15°C) mean monthly temperature range. High-latitude and continental climates are dominated by the annual air-temperature cycle, which induces a single major ground freezing and thawing event each year, but maritime and alpine climates and much of the Qinghai-Tibet Plateau also experience frequent short-term oscillations of air temperature around the freezing point. Permafrost underlies high-latitude, continental-interior and high-alpine areas, and much of the Qinghai-Tibet Plateau, but may be discontinuous, patchy or absent in low-arctic, subarctic, maritime and low-alpine climatic zones (Chapter 4). Precipitation totals range from extremely low (polar deserts in both hemispheres and parts of the Qinghai-Tibet Plateau) to extremely high (maritime and alpine periglacial environments) and may exhibit a winter maximum (maritime and most alpine environments) or a summer maximum (most arctic areas and continental interiors, and the Qinghai-Tibet Plateau). Almost all periglacial environments experience winter snowcover, though this is limited in the high Arctic and much of Antarctica, and is transient in some maritime periglacial environments, as well as on the Qinghai-Tibet Plateau and on low-latitude mountains due to strong radiative inputs throughout the year. Exceptionally strong winds (>100 km h^{-1}) are characteristic of the coastal fringes of Antarctica, alpine environments and maritime periglacial environments, but all parts of the periglacial domain are susceptible to extreme winds. On a regional scale, climate plays an important role in determining soil type and the predominant role in determining the nature and extent of vegetation cover. The latter responds primarily to the length and warmth of the summer growing season, soil moisture availability and substrate characteristics, all of which produce marked contrasts in the nature and extent of plant cover, ranging from cold deserts through tundra to mature stands of conifers in the boreal or alpine forest zones.

All periglacial environments, irrespective of climatic regime, soil type and vegetation cover, share a common characteristic: they are dominated by freezing of the ground, and the resultant distinctive geomorphological processes and landforms. In the following chapter, we consider the processes that occur during ground freezing and thawing, and in Chapters 4–7 the nature of perennially frozen ground (permafrost) and features associated with the development of bodies of ice within permafrost. A recurrent theme throughout these chapters is not only the influence of climate, but also the ways in which vegetation cover and soil characteristics modulate the response of the ground thermal regime to energy exchange at the surface.

3

Ground Freezing and Thawing

3.1 Introduction

The defining characteristic of periglacial environments is frozen ground, in the form of either perennially frozen ground (permafrost) or ground that freezes and thaws over annual or shorter timescales. In this chapter, we explore the processes that occur as water freezes or thaws in soil or bedrock. We examine first how the temperature regime of the ground is affected by heating and cooling at the ground surface, then how moist soils respond to subzero temperatures by progressive freezing of soil water as temperatures fall below 0 °C. We then consider migration of liquid water within freezing soils and the resultant process of *ice segregation*, whereby discrete bands and lenses of pure ice are formed within soils and porous rock during freezing. The final part of the chapter focuses on *thaw consolidation*, the response of ice-rich soils to thaw, and how this is accompanied by build-up of high positive pore-water pressures that allow saturated soils to deform internally. As we shall see in later chapters, cyclic freezing and thawing of rock and soil is of fundamental importance in explaining a wide range of periglacial phenomena, such as breakdown of rocks, the formation of patterned ground, the slow downslope movement of soil, and soil instability on relatively low-gradient slopes underlain by permafrost.

Cryotic ground is the term applied to ground that has a temperature below 0 °C, irrespective of whether water or ice (or neither) is present. The term *frozen ground* is often used to refer to ground in which ice is present, but it is important to note that the water-to-ice phase transition in soil or rock is dependent not only on temperature, but also on pressure, the size of pores within soil or rock and the presence of soluble impurities in soil water. In the case of fine-grained soils, for example, soil temperature may drop a degree or two below 0 °C before ice nucleation (ice crystal formation) commences, and even then only part of the water freezes. Indeed, one of the most important characteristics of soil freezing is that ice and liquid water often coexist at temperatures below 0 °C, with the proportion of unfrozen water diminishing as the temperature falls. There is

therefore no single 'freezing point' of water contained in soil and rock pores; instead, freezing of liquid water occurs progressively as the ground cools at temperatures below 0 °C (Dash *et al.*, 2006).

3.2 Ground Heating and Cooling

3.2.1 Basic Concepts

Temperature at the surface of the ground is related to air temperature regime, modified by the insulating effects of snowcover and vegetation (Chapter 4). Here we consider first how temperature variation at the ground surface affects the subsurface thermal regime. In general, subsurface temperatures are determined primarily by *conductive heat transfer* from the surface. This means that heat passes down (or up) through soil or rock in the same way as it passes along a metal bar heated at one end. In practice, however, heat may be also gained or lost at particular levels through vapour diffusion or movement of water through soil. Such *nonconductive heat transfer* effects may enhance or diminish the effects of conductive heat transfer, particularly during soil freezing or thawing. Infiltration of meltwater or rainwater may advect heat into frozen soil during thaw, for example, and the upward migration of water to the freezing front as described later in this chapter also involves nonconductive redistribution of heat. It is nonetheless a reasonable assumption that conduction is by far the predominant mode of heat transfer in mineral soils (Kane *et al.*, 2001). In the following account, temperature is usually expressed in °C (degrees Celsius), but in most cases it could equally well be expressed in K (kelvin), which represents temperature above absolute zero (−273.15 °C). For any temperature, an increase of 1 K is identical to an increase of 1 °C, so 0 °C is equal to 273.15 K.

Four concepts are essential to the application of heat conduction theory to ground that is subject to freezing and thawing. The first is *thermal conductivity (k)*, which measures the efficiency with which a substance conducts heat and is given in $W\,m^{-1}\,°C^{-1}$ (watts per metre per degree Celsius) or $W\,m^{-1}\,K^{-1}$ (watts per metre per kelvin,

Periglacial Geomorphology, First Edition. Colin K. Ballantyne.
© 2018 John Wiley & Sons Ltd. Published 2018 by John Wiley & Sons Ltd.

which is dimensionally identical). The thermal conductivity of most dry rocks falls within the range 2.0–10.0 W m^{-1}°C^{-1}, though some shales and marls have thermal conductivities as low as 1.0 W m^{-1}°C^{-1}. That of soils, however, is both moisture- and temperature-dependent. For dry sandy soil, it is as little as ~0.3 W m^{-1}°C^{-1}, increasing to ~2.2 W m^{-1}°C^{-1} on wetting and to ~3.0 W m^{-1}°C^{-1} as the moist soil freezes. Thus heat passes much more readily through rock and ice-rich soil than through dry soil. This is because air has a very low thermal conductivity (~0.025 W m^{-1}°C^{-1}), so dry soils with air-filled pores have much lower thermal conductivity than moist soils with water-filled voids or frozen soils with ice-filled voids. The thermal conductivity of ice at 0°C is four times that of liquid water at 0°C (Table 3.1), which is why ice-rich soil has a higher thermal conductivity than unfrozen moist soil. However, in fine-grained soils, not all water freezes at 0°C, and unfrozen water is present at several degrees below 0°C. This means that the thermal conductivity of frozen fine-grained soil is temperature-dependent (Figure 3.1).

The second key concept is that of *heat capacity*: the temperature change experienced by a body of soil or rock when it is heated or cooled. This can be expressed as *volumetric heat capacity (C)*, expressed in J m^{-3}°C^{-1} (joules per cubic metre per degree Celsius) or J m^{-3}K^{-1} (joules per cubic metre per kelvin), the amount of heat required to change the temperature of 1 m^3 of soil or rock by 1°C, or as *mass heat capacity* (J kg^{-1}°C^{-1} or J kg^{-1}K^{-1}), the amount of heat required to change the temperature of 1 kg of soil or rock by 1°C. For a given input of heat, the resultant temperature change will be greater in material with low heat capacity. Thus, for a specified heat input, the temperature increase of 1 kg of quartz, with a mass heat capacity of 800 J kg^{-1}°C^{-1}, will be more than five times greater than that of 1 kg of water at 0°C, which has a mass heat capacity of 4180 J kg^{-1}°C^{-1}; to raise the temperature of 1 g of quartz by 1°C takes about 0.8 J of heat, whereas to raise the temperature of 1 g of water at 0°C by 1°C takes about 4.2 J. Most dry soils have volumetric heat capacities of 1.0–1.5 MJ m^{-3}°C^{-1} at 0°C (1 MJ = 10^6 J), and water at 0°C has a volumetric heat capacity of 4.18 MJ m^{-3}°C^{-1}. Moist soils therefore have a greater heat capacity than dry soils, and the heat capacity of soils increases approximately linearly with water content (Table 3.1).

The third key concept is that of *thermal diffusivity (κ)*, which is expressed in m^2 s^{-1} (square metres per second) and measures the rate of heat diffusion through a body of soil or rock. Thermal diffusivity is thermal conductivity divided by volumetric heat capacity, thus:

$$\kappa = k/C \tag{3.1}$$

A high value of κ implies a rapid, large change in temperature, and *vice versa*. Thus, quartz, with high

Table 3.1 Thermal properties of soils and constituent materials. *Source*: Williams and Smith (1989). Reproduced with permission of Cambridge University Press. The figures for organic matter and unfrozen soils are approximate.

	Water content (m^{3}m^{-3})	Density (kg m^{-3})	Mass heat capacity (J kg^{-1}K^{-1})	Volumetric heat capacity (10^3 J m^{-3}K^{-1})	Thermal conductivity (W m^{-1}K^{-1})	Thermal diffusivity (10^{-6} m^{2}s^{-1})
Soil constituents						
Quartz	–	2660	800	2128	8.80	4.14
Clay minerals	–	2650	900	2385	2.92	1.22
Organic matter	–	1300	1920	2496	0.25	0.10
Water at 0°C	–	1000	4180	4180	0.56	0.13
Ice at 0°C	–	917	2100	1926	2.24	1.16
Air	–	1.2	1010	1.21	0.025	20.63
Unfrozen soils						
Sandy soil	0.0	1600	800	1280	0.30	0.24
(40% porosity)	0.2	1800	1180	2124	1.80	0.85
	0.4	2000	1480	2960	2.20	0.74
Clay soil	0.0	1600	890	1424	0.25	0.18
	0.2	1800	1250	2250	1.18	0.53
	0.4	2000	1550	3100	1.58	0.51
Peat soil	0.0	300	1930	576	0.06	0.10
	0.4	700	3300	2310	0.29	0.13
	0.8	1100	3650	4015	0.50	0.12

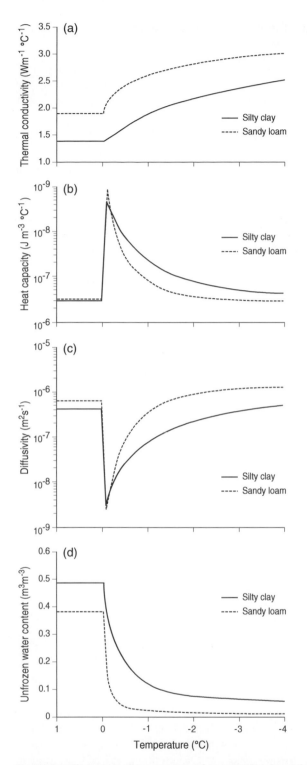

Figure 3.1 Changes in the properties of a moist silty-clay soil and a moist sandy loam within the temperature range +1 to −4 °C. *Source:* Williams and Smith (1989). Reproduced with permission of Cambridge University Press.

thermal conductivity and low heat capacity, has a thermal diffusivity of $4.14 \times 10^{-6}\,\mathrm{m^2\,s^{-1}}$, whereas the thermal diffusivity of water at 0 °C, which has a low thermal conductivity but high heat capacity, is

only $0.13 \times 10^{-6}\,\mathrm{m^2\,s^{-1}}$ (Table 3.1). The case of water is more complicated, however, as differential heating (or cooling) of any liquid or gas results in heat transfer by *convection* (density-driven motion within the fluid) as well as conduction.

The final key concept is that of the *latent heat of fusion* (L_f), which is the heat that must be removed from liquid water at 0 °C to freeze it, or the heat that must be supplied to ice at 0 °C to melt it (sometimes referred to as the latent heat of melting, L_m). During the phase change from water to ice or *vice versa*, this loss or gain of heat involves no change in the temperature of the water or ice. A useful way to envisage latent heat is to consider water with ice floating in it. When heat is slowly added, all the heat energy is used up melting the ice, so there is no change in the temperature of the water until all the ice has been melted. Conversely, during freezing of water, latent heat is released, thereby slowing the freezing process: to cool 1 g of water from 1 °C to 0 °C requires the removal of 4.2 J of heat (4.2 kJ kg^{-1}), but to freeze 1 g of water at 0 °C to form 1 g of ice at 0 °C requires the removal of 334 J (334 kJ kg^{-1}). This is an important effect in considering ground temperature regime in moist soils. During autumn, ground surface temperatures in periglacial environments often drop rapidly, and temperatures in the upper part of the soil fall quickly to 0 °C. Thereafter, however, there is often a prolonged period before soil temperatures drop below 0 °C, because release of latent heat during freezing of soil water counteracts heat loss to the surface. Only when most of the near-surface soil water has frozen do soil temperatures fall below 0 °C in response to surface cooling. This effect, termed the *zero curtain effect* (Outcalt *et al.*, 1990), is most prolonged in soils with high water content. It may also occur during seasonal warming of ice-rich soil, with the soil temperature remaining steady around 0 °C until all ice has melted. The seasonal (autumn freezeback, spring thaw or both) occurrence of a zero curtain effect appears to be linked to mean ground surface temperature and thermal gradient within the freezing or thawing layer, and is not an inevitable consequence of ground freezing or thawing (Putkonen, 2008). The latent heat of fusion and the latent heat of melting nevertheless have a major influence on the thermal diffusivity of freezing and thawing of moisture-rich soils. This is incorporated in the *apparent thermal diffusivity* (κ_a) of soil or rock, which is thermal conductivity divided by the sum of volumetric heat capacity and latent heat capacity, thus:

$$\kappa_a = k/\left(C + L_f\right) \qquad (3.2)$$

Incorporation of latent heat effects therefore markedly lowers the apparent thermal diffusivity within moist or frozen soils or rock.

3.2.2 Conduction of Heat into the Ground

Where temperature at the ground surface remains constant, heat flow vertically downwards from the ground surface (Q_g) is expressed by:

$$Q_g = -k(dT/dz) \tag{3.3}$$

where T is temperature, z is depth and dT/dz denotes the thermal gradient, or change in temperature with depth. The negative sign indicates heat flow in the direction of decreasing temperature. For any layer within the soil with thickness dz and unit area, we can write a heat balance equation, in which the rate of change of heat storage (ΔQ) equals the rate of heat flow into the soil unit (Q_{in}) minus the rate of heat flow out of the soil unit (Q_{out}), thus:

$$\Delta Q = Q_{in} - Q_{out} \tag{3.4}$$

The rate of change of heat storage (ΔQ) in the soil layer can also be expressed by:

$$\Delta Q = C(dT/dt)dz \tag{3.5}$$

where C is volumetric heat capacity, t is time and dT/dt is change in temperature over time for a soil layer with thickness dz. Substituting equations (3.3) and (3.5) into equation (3.4) allows us to express heat balance as:

$$C(dT/dt)dz = k(dT/dz)_{in} - k(dT/dz)_{out} \tag{3.6}$$

This can be rearranged to give the heat conduction equation:

$$\begin{aligned} dT/dt &= k\left[(dT/dz)_{in} - (dT/dz)_{out}\right]/C \cdot dz \\ &= \kappa\left[(dT/dz)_{in} - (dT/dz)_{out}\right]/dz \end{aligned} \tag{3.7}$$

which in differential form is:

$$\delta T/\delta t = k/C\left(\delta^2 T/dz^2\right) = \kappa\left(\delta^2 T/\delta z^2\right) \tag{3.8}$$

However, equations (3.7) and (3.8) assume a constant value of thermal diffusivity (κ). This assumption is unrealistic in practice, because (apparent) soil thermal diffusivity changes with soil properties such as texture and density, as well as with temperature and moisture content, and is strongly affected by freezing and thawing of soil moisture, associated latent heat effects and nonconductive heat transfer. Realistic solutions to these equations can therefore only be achieved by complex modelling approaches that solve simultaneously for small finite differences in temperature and depth (Goodrich, 1982; Lunardini, 1991). A particular difficulty arises in fine-grained soil with small pore spaces. In such soils, not all water freezes at 0 °C, and small changes in temperature below 0 °C can produce marked changes in both apparent heat capacity and apparent thermal diffusivity (Figure 3.1).

3.2.3 Estimating Depth of Freezing and Thawing

Despite the complications just outlined, the depth of freezing (or thawing) of soil can be estimated using the *Stefan solution*. This formula assumes that the latent heat of freezing is the only heat that must be removed when the soil freezes, and that the latent heat (L_f) released by freezing a uniform layer of soil of thickness z in time t is equal to the heat conducted through the overlying frozen layer, thus:

$$L_f(dz/dt) = k_f(v_s/z) \tag{3.9}$$

where k_f is the thermal conductivity of the overlying frozen soil and v_s is the difference between the ground surface temperature and the freezing temperature of soil moisture at depth z_f. Integration of equation (3.9) and solving for depth (z_f) gives:

$$z_f = \sqrt{(2k_f/L_f).\int v_s dt} = \sqrt{(2k_f v_s t)/L_f} = \sqrt{(2k_f I_{sf}/L_f)} \tag{3.10}$$

where $I_{sf}(= v_s \times t)$ is the surface freezing index in degree-days (Andersland and Ladanyi, 2003). Similarly, depth of thaw of a homogeneous frozen soil (z_t) can be approximated by:

$$z_t = \sqrt{(2k_t I_{st}/\rho_s w L_f)} \tag{3.11}$$

where k_t is the thermal conductivity of the overlying thawed soil, I_{st} is the thaw index in degree-days, ρ_s is the dry bulk density of the soil and w is the mass of pore ice in the initially frozen soil, expressed as a decimal proportion of the mass of dry soil. More complex solutions for multi-layered soils with different properties are discussed by Kurylyk (2015). Expressed in SI units, equation (3.10) becomes:

$$z_f = \sqrt{172,800 k_f I_{sf}/L_f} \tag{3.12}$$

(Andersland and Ladanyi, 2003.)

However, because the heat capacity of the soil is neglected in this formula, the Stefan equation tends to overestimate freezing depth, though its performance can be improved through the use of correction factors (Kurylyk and Hayashi, 2016). The *modified Berggren equation* also attempts to rectify this overestimation. This formula assumes that the surface temperature changes from an initial value of v_0 degrees above freezing to v_s ($= I_{sf}/t$) degrees below freezing at the start of the freezing period, and adjusts the Stefan equation by introducing a dimensionless coefficient, λ. Thus, in SI units:

$$z_f = \lambda\sqrt{172,800 k_f I_{sf}/L_f} \tag{3.13}$$

where λ is a function of two dimensionless parameters, α (thermal ratio) and μ (fusion parameter):

$$\alpha = v_0/v_s = v_0 t / I_{sf} \tag{3.14}$$

$$\mu = (C \cdot v_s)/L_v = (C \cdot I_{sf})/L_v \cdot t \tag{3.15}$$

where C is soil volumetric heat capacity ($\mathrm{kJ\,m^{-3}\,^{\circ}C^{-1}}$) and L_v is volumetric latent heat ($\mathrm{kJ\,m^{-3}}$). Graphical solutions for λ as a function of (α, μ) are given in Andersland and Ladanyi (2003).

3.2.4 Temperature Change with Depth

If we consider a situation where the ground surface temperature (T_s) is constant but $<0\,^{\circ}\mathrm{C}$, then temperature increases with depth at a rate that is determined by the thermal conductivity (k) of the ground. If k is unchanging with depth, the temperature at any depth z (T_z) is given by:

$$T_z = T_s + z(\mathrm{d}T/\mathrm{d}z) \tag{3.16}$$

where ($\mathrm{d}T/\mathrm{d}z$) is the *geothermal gradient*, representing steady-state one-dimensional flow of heat to the surface. At temperatures below $0\,^{\circ}\mathrm{C}$, the geothermal gradient ranges from $1\,^{\circ}\mathrm{C}/160\,\mathrm{m}$ ($0.0063\,^{\circ}\mathrm{C\,m^{-1}}$) to $1\,^{\circ}\mathrm{C}/22\,\mathrm{m}$ ($0.045\,^{\circ}\mathrm{C\,m^{-1}}$). If, for example, the mean annual ground surface temperature (MAGST) is $-5\,^{\circ}\mathrm{C}$ and the geothermal gradient is $0.02\,^{\circ}\mathrm{C\,m^{-1}}$, the temperature at $100\,\mathrm{m}$ depth will be approximately $[-5\,^{\circ}\mathrm{C} + (0.02 \times 100)\,^{\circ}\mathrm{C}]$ or $-3\,^{\circ}\mathrm{C}$. Where MAGST is $<0\,^{\circ}\mathrm{C}$, permafrost is normally present (Chapter 4), and the approximate maximum depth of permafrost (z_p) can be estimated by extrapolating the geothermal gradient to the depth at which $T_z = 0\,^{\circ}\mathrm{C}$:

$$z_p = (0\,^{\circ}\mathrm{C} - T_{TOP})/(\mathrm{d}T/\mathrm{d}z) \tag{3.17}$$

where T_{TOP} is the mean annual temperature at the top of the permafrost. For example, where T_{TOP} is $-5\,^{\circ}\mathrm{C}$ and $\mathrm{d}T/\mathrm{d}z$ is $0.02\,^{\circ}\mathrm{C\,m^{-1}}$, z_p is $(5/0.02)\,\mathrm{m}$ or $250\,\mathrm{m}$. As the steepness of the thermal gradient is inversely related to thermal conductivity, for any given value of T_s, ground with high thermal conductivity will be underlain by thicker permafrost than ground with low thermal conductivity. For example, assuming $T_{TOP} = -5\,^{\circ}\mathrm{C}$ and $\mathrm{d}T/\mathrm{d}z = 0.01\,^{\circ}\mathrm{C\,m^{-1}}$, z_p is $(5/0.01)\,\mathrm{m}$ or $500\,\mathrm{m}$, twice as deep as for $\mathrm{d}T/\mathrm{d}z = 0.02\,^{\circ}\mathrm{C\,m^{-1}}$ (Figure 3.2). Variations in permafrost depth are considered in greater detail in Chapter 4.

3.2.5 Temperature Variations with Depth

Ground surface temperatures, of course, are not constant, but undergo periodic fluctuation on annual and diurnal cycles. Here we focus on the annual cycle, which in most periglacial environments dictates the maximum depth of freezing and thawing of the ground. Seasonal temperature variations at the surface are propagated down into the ground, but the amplitude (the difference between the

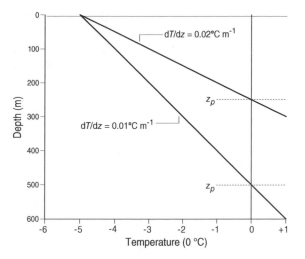

Figure 3.2 Relationship between geothermal gradient (dT/dz) and depth of permafrost (z_p). The upper line is associated with ground exhibiting relatively low thermal conductivity, the lower line with ground exhibiting relatively high thermal conductivity.

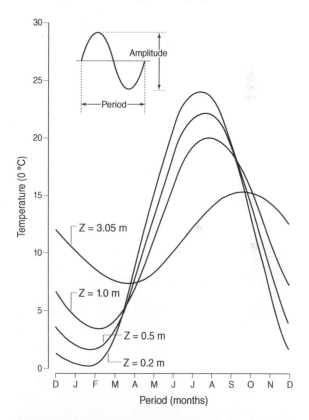

Figure 3.3 Annual temperature variations in unfrozen soil, illustrating diminishing amplitude and increased phase lag with depth (Z). *Source:* Williams and Smith (1989). Reproduced with permission of Cambridge University Press.

highest and lowest temperatures) of seasonal temperature variations diminishes with depth and the phase lag (the delay in response time) increases with depth (Figure 3.3). Eventually, however, a depth is reached at which annual change in temperature is negligible ($<0.1\,^{\circ}\mathrm{C}$).

This is termed the depth of *zero annual amplitude,* and is related to thermal diffusivity. In soil, with relatively low thermal diffusivity (Table 3.1), it occurs at depths of 10–20 m, but in rocks, with higher thermal diffusivity, it often occurs at much greater depths.

For an annual temperature cycle with period p (~365 days), the amplitude of the annual temperature cycle at depth z (A_z) is given by:

$$A_z = A_s \exp\left(-z\sqrt{\pi/\kappa_a p}\right) \qquad (3.18)$$

where A_s is amplitude of the temperature cycle at the surface. The annual range in temperature (T_z) at depth z is given by:

$$T_z = T_a + A_s \exp\left(-z\sqrt{\pi/\kappa_a p}\right) \qquad (3.19)$$

where T_a is mean annual ground surface temperature. If we ignore the geothermal gradient so that temperature is constant with depth, equation (3.19) produces trumpet-shaped curves that define the annual temperature range at any depth (Figure 3.4a); the depth of the 'trumpet' increases as apparent thermal diffusivity (κ_a) increases. The phase lag (t^*) between the temperature cycle at depth z and that at the surface (Figure 3.4b) is given by:

$$t^* = (z/2)\cdot\sqrt{p/(\pi\cdot\kappa_a)} \qquad (3.20)$$

and the temperature ($T_{z,t}$) at depth z at time t (where t is number of days since the surface temperature wave passed through its mean value in spring) is approximated by:

$$T_{z,t} = T_{az} + A_s \exp\left(-z\sqrt{\pi/\kappa_a p}\right)\cdot\sin\left[(2\pi t/p) - z\sqrt{\pi/\kappa_a p}\right]$$
$$(3.21)$$

where T_{az} is the mean temperature at depth z. Equations (3.18) to (3.21) can also be applied to shallow diurnal temperature fluctuations, with $p = 24$ hours. Both cases assume that ground surface temperature regime can be reasonably approximated by a simple sine wave (Figure 3.4b), but this is not always the case. Late-lying snowcover, for example, retards annual ground warming, as does the additional heat required to melt ice in the ground. The pattern of diurnal surface temperature variation may be strongly influenced by direct solar radiation, and thus by slope aspect. In moist soils, release of latent heat (the zero curtain effect) may retard soil cooling at or just below 0 °C until most water is frozen, creating a 'plateau' in the nearsurface temperature curve, though this irregularity tends to disappear with increasing depth.

3.3 Soil Freezing

Many landforms and soil structures that are unique to periglacial environments are related in some way to freezing of moist soil or thawing of ice-rich soil. Pure

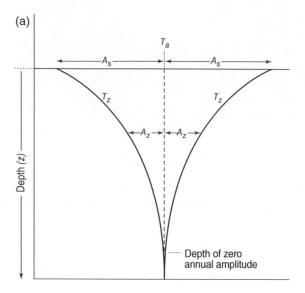

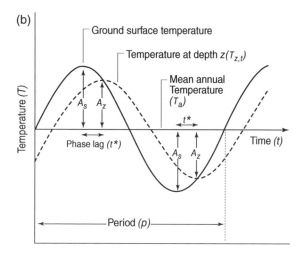

Figure 3.4 (a) Decline in annual temperature range with depth. For an explanation of terms, see text. Note that this diagram ignores the effect of geothermal gradient. (b) Definition of phase lag (t^*), which is the difference in timing between peak ground surface temperature and peak temperature at depth z. For clarity, A_s and A_z are shown as semi-amplitudes (0.5×amplitude). *Source:* Andersland and Ladanyi (2003). Reproduced with permission of Wiley.

bulk (unconstrained) water at atmospheric pressure freezes to become ice at 0 °C or 273.15 K. As the temperature of bulk water is lowered, it dips very slightly below 0 °C, a process referred to as *supercooling.* Such supercooled water is unstable, and further cooling is abruptly interrupted by *ice nucleation,* the formation of stable ice crystals. Ice nucleation releases latent heat, which causes the temperature of the ice to rise to 0 °C, the *equilibrium freezing point* of bulk water under atmospheric pressure (Figure 3.5). In porous media such as soils and rocks, however, there is no single temperature at which all liquid water freezes, but a situation where both liquid water and ice coexist at temperatures below 0 °C, with the proportion of unfrozen water diminishing as temperature falls further below 0 °C. There therefore

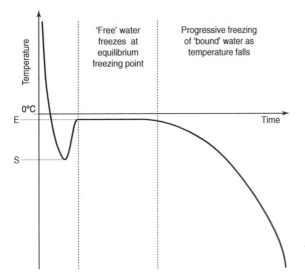

Figure 3.5 Cooling curve for soil water and ice. Initial supercooling of unfrozen water (to point S) is interrupted by ice nucleation. The resultant release of latent heat raises the temperature to the equilibrium freezing point (E), which may be depressed slightly below 0 °C by the presence of dissolved salts. The remaining 'free' or unbound water freezes at the equilibrium freezing point. Thereafter, 'bound' water held in capillary attraction or adsorbed on to mineral surfaces freezes at progressively lower temperatures.

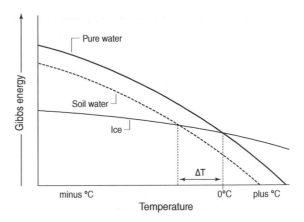

Figure 3.6 Diagrammatic representation of the free energy of ice, pure water and soil water plotted against temperature. Pure water and ice have equal free energies at 0 °C, but the free energy of soil water is reduced by the presence of dissolved salts and other effects, resulting in 'freezing point depression' (ΔT) below 0 °C. In reality, there is no single freezing point of soil water, as diminishing quantities of soil water coexist with ice as the temperature is lowered below 0 °C.

is no single 'freezing point' of soil water, particularly in fine-grained soils, though reduction in the temperature at which ice nucleation occurs is often for convenience termed *freezing point depression*. The melt temperature (T_m) of ice at a given pressure is constant, however, so physicists studying the behaviour of ice in soils and other porous media usually refer to lowering of the melting temperature rather than lowering of the freezing temperature and use the term *premelted water* to refer to water that exists in liquid or liquid-like form at subzero temperatures where $T < T_m$.

The temperature at which water freezes in soils is determined by the Gibbs free energy of soil water relative to that of ice at temperatures below 0 °C. *Gibbs free energy*, also known as *Gibbs energy* or *free energy*, is a parameter that determines whether a particular chemical reaction or phase change is thermodynamically possible at a fixed and constant temperature and pressure. It is thus a measure of chemical potential. In a spontaneous chemical reaction or phase change (one that occurs without a change in temperature or pressure), free energy invariably decreases, so that the change in free energy (ΔG) is negative. Formally, free energy (G) is expressed as:

$$G = H - TS = (E + PV) - TS \qquad (3.22)$$

where H is enthalpy ($= E + PV$), E is the internal energy of a system, P is pressure, V is volume, T is temperature (K) and S is the internal entropy of a system. *Enthalpy* is the heat content of a substance per unit mass, and a change in enthalpy (ΔH) reflects the amount of energy change

in the system during a chemical reaction or change in phase. When water freezes, for example, release of latent heat results in a reduction of enthalpy. *Entropy* is a measure of the disorder of a system. At constant pressure, entropy increases with temperature, but there is a step-change increase in entropy at phase boundaries, such as those between ice and water and between liquid water and water vapour. Gibbs free energy is therefore a measure of the combined effects of changes in enthalpy (ΔH) and entropy (ΔS) in any thermodynamic process, such as the phase change from water to ice. We cannot calculate an absolute value for G, however, but only the change in Gibbs energy (ΔG) implied by a spontaneous reaction, thus:

$$\Delta G = \Delta H - T\Delta S \qquad (3.23)$$

If $\Delta G = 0$, the system is in equilibrium and no net change occurs, but when $\Delta G < 0$, change will occur towards a new equilibrium state with lower free energy.

At the freezing point of bulk pure liquid water under atmospheric pressure (0 °C), the Gibbs energy of ice (G_i) equals that of water (G_w), so the tendency of ice to melt is exactly counterbalanced by the tendency of water to freeze. At temperatures below 0 °C, G_i is theoretically less than G_w (Figure 3.6), but two phases with differing free energies cannot coexist, so all liquid water is spontaneously converted to ice, which has lower free energy. In soil water, however, complications are introduced by three effects, all of which cause freezing point depression and permit the coexistence of premelted water and ice in soil at temperatures below 0 °C: surface tension, the effect of dissolved impurities within soil water and the effects of intermolecular forces acting at the interface between pore ice and the walls of soil pores (*interfacial melting*). Each of these effects is described briefly in the following text; a detailed treatment is provided by Dash *et al.* (2006).

The Scottish physicist William Thomson (Lord Kelvin) observed that small droplets of water freeze at a lower temperature than bulk water, due to surface tension effects that reflect a combination of surface energy and surface curvature; small water droplets with a pronounced surface curvature exhibit higher effective vapour pressure, since the surface area is large in comparison to the volume. This property, often termed the *Gibbs–Thomson effect*, can cause marked freezing-point depression of water dispersed within the pores of fine-grained soils. The magnitude of freezing-point depression is determined by the interfacial curvature of the contact between unfrozen (premelted) water and adjacent ice or soil particles. In fine-grained soils where the gaps between soil particles are smaller, interfacial curvature and thus freezing point depression tend to be greater.

Dissolved impurities lower the freezing point of bulk water because of the lower chemical potential (free energy) of the solvent phase (Figure 3.6). A 3.5% NaCl (sodium chloride) solution, for example, can be cooled to $-2.1\,°C$ before ice nucleation commences, and with increasing concentrations of NaCl the freezing point is further depressed. This property is important as it enables the survival of submarine permafrost under cold saline water that exists at temperatures below $0\,°C$ but does not freeze, and it accounts for the presence of unfrozen saline layers (*cryopegs*) at temperatures below $0\,°C$ within permafrost (Chapter 4). Natural concentrations of dissolved impurities in soil water are usually very weak. However, ice nucleation in soil pores forces dissolved impurities out of the ice (the solubility of most chemical species in ice is negligible) to concentrate them in thin films of liquid water that coat adjacent soil particles. The increased solute concentration within such interfacial films therefore lowers the freezing point of the remaining liquid water, though the process is complicated by other effects acting at ice–grain boundaries (Dash *et al.*, 2006). Moreover, the initial freezing temperature of a saline soil is lower than that of the corresponding solution, and the difference increases with increasing solute concentration (Wan *et al.*, 2015).

Of greater general importance in depressing the freezing point of water, particularly in fine-grained soils, are the effects of *capillarity* and *adsorption*. Both of these effects 'bind' water molecules to adjacent mineral or ice surfaces, reducing the Gibbs free energy of the 'bound' water so that its freezing point is lowered below that of ice at $0\,°C$. *Capillarity* is the result of surface tension: water contained in small spaces, such as soil pores, exhibits both surface *cohesion* (the mutual attractive force that exists between water molecules at the interface between water and air) and *adhesion* to adjacent surfaces. The latter effect reflects the asymmetry of water molecules, which exhibit weak polarity and are attracted by van der Waals forces to regions of opposite charge at solid (mineral grain) surfaces. Capillary action is evident, for example, in the internal cohesion of large water droplets, the rise of water up a narrow tube, the meniscus effect at the top of a water column and the way in which dry soil soaks up water from a reservoir below. Consider a soil that is progressively wetted from below. At first, water entering a soil pore forms a film on adjacent mineral surfaces, and because adhesion is greatest between water molecules nearest mineral surfaces, capillary forces are strong enough to draw water upwards through soil against the pull of gravity. As further water is added, however, the thickness of films of water adjacent to pore walls increases, capillary force is diminished and the water molecules farthest from the pore walls may flow under gravity. Evaporation of soil water near the ground surface therefore causes an increase in capillary force that draws water upwards through the soil. In fine-grained soils with small pore spaces, the proportion of water close to a pore wall is greater than in coarse-grained soils with large pore spaces, so capillary force tends to be greatest in fine-textured soil. As a moist soil freezes, the formation of ice in pores not only reduces the thickness of films of liquid water on pore walls, but also confines the remaining soil water into smaller and smaller spaces. Both effects increase capillary force and thus further reduce the free energy of the remaining water, so that cooling to progressively lower temperatures is required to induce further freezing.

Adsorption is generally considered to involve two forces: 'long-range' van der Waals forces that cause the initial attraction of water molecules to a solid (mineral) surface and strong 'short-range' ionic forces that finalize the settling of additional layers of water molecules on to a solid surface without generating a new chemical species. For water without dissolved impurities, at temperatures below $-1.5\,°C$ most of the unfrozen water remaining in a soil or rock is adsorbed water, though the amount of unfrozen adsorbed water is progressively reduced as temperatures fall. Some adsorbed water may persist in soils at temperatures as low as $-70\,°C$, much lower than occurs under natural circumstances. Figure 3.6 illustrates the effect of dissolved impurities, capillarity and adsorption in depressing the freezing point of water in soil pores.

An important consequence of the effects of capillarity and adsorption in depressing the freezing point of water in soil or rock is that a *free energy gradient* (or *chemical potential gradient*) becomes established. If we consider the situation where cooling below $0\,°C$ is occurring from the surface downwards, so that the ground temperature increases with depth, 'bound' premelted water (at $T < 0\,°C$)

in pores that are largely occupied by ice has lower Gibbs energy than 'free' liquid water in ice-free pores farther down the soil column. As a result, water from deeper within the soil is sucked upwards towards the freezing front. In polar environments where seasonal ground freezing takes place both from the surface downwards and from the top of the permafrost upwards, unfrozen water may be drawn both upwards towards a freezing front descending from the surface and downwards towards a freezing front advancing upwards from the top of the permafrost. Irrespective of the direction of water movement, this effect is referred to as *cryosuction*; it is driven by the difference between ice pressure (P_i) and water pressure (P_w) at the freezing front, and suction potential increases by approximately 1.2 MPa per degree below 0 °C. Cryosuction explains why water tends to drawn towards the freezing front to form bands or lenses of ice within soil, as described below.

Cryosuction is one manifestation of the *matrix potential* or energy status of pore water at any point in a soil, which is usually expressed in terms of *pore-water pressure*. Pore-water pressures are neutral in free water under atmospheric pressure, positive when they exceed atmospheric pressure and negative when they are less than atmospheric pressure. Freezing of moist soil produces strong negative pore-water pressures that induce water migration from parts of the soil where pore-water pressures are greater; water is drawn along a pressure gradient from a zone of higher (neutral or positive) pore-water pressures to a zone of lower (negative) pore-water pressures near the freezing front. Such migration of water towards the freezing front will continue so long as there is sufficient hydraulic connectivity between soil pores (i.e. the pores are not blocked by ice) and there is unfrozen water available in the unfrozen or partly-frozen parts of the soil. As the pressure of water in freezing soils is strongly influenced by its confinement in soil pores and the proximity of mineral surfaces, these effects tend to be limited in coarse-grained soils, where pore spaces are comparatively large and most soil water is 'free' rather than 'bound'. In fine (silt- or clay-rich) soils where pore spaces are small, a much larger proportion of water remains in liquid phase as the temperature falls below 0 °C (Figure 3.1d), and slow freezing of the soil generates strongly negative pore-water pressures.

Pressure may also cause freezing point depression. This is a familiar concept in glaciology, where melting of ice at any depth in a glacier occurs at *pressure melting point*, which is slightly below 0 °C because the ice is subject to above-atmospheric overburden pressures; freezing of water under pressures greater than one atmosphere also occurs at a temperature slightly below 0 °C. The reason for this can be seen in a diagram on which the three phases of water (ice, liquid water and water vapour) are identified on a plot of pressure against

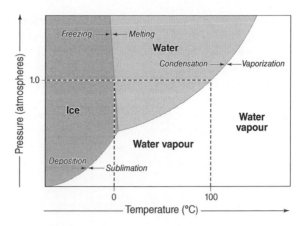

Figure 3.7 Phase diagram for water. The solid lines separating different phases (ice, liquid water and water vapour) are 'coexistence lines' where $\Delta G = 0$; in other words, where the Gibbs energy of ice equals that of liquid water, or where that of liquid water equals that of water vapour. The two coexistence lines meet at the 'triple point' of water where $P = 6.03 \times 10^{-3}$ atmospheres (0.603 kPa) and $T = 0.0098$ °C. The reverse slope of the ice–water coexistence line implies that an increase in pressure causes freezing point depression.

temperature (Figure 3.7). Separating the phases are coexistence lines, which connect all pressure–temperature combinations at which the Gibbs energies of two phases are equal. Unusually (because most liquids near their freezing point freeze under pressure), the coexistence line separating the ice and water phases in Figure 3.7 has a reverse slope, implying that as pressure increases, the freezing point temperature declines. The way in which pressure lowers the freezing point of water is described by a generalized form of the Clapeyron equation:

$$dP/dT = L_f / T\Delta V \qquad (3.24)$$

where dP/dT defines the slope of the coexistence line, L_f is latent heat of fusion, T is absolute temperature (K) and ΔV is the volume change of the phase transition; as water expands by 9% on forming ice, ΔV is a constant, equal to 1.09. The Clapeyron equation holds when both ice and water are equally affected by increased pressure (e.g. the confining pressure imposed by the weight of overlying soil). Under these circumstances, the freezing point of water is only slightly reduced, by 0.074 °C per MPa of pressure. However, as temperature falls further below 0 °C and increasing amounts of liquid water are converted to ice, the difference between ice pressure (P_i) and water pressure (P_w) increases. This difference produces a much larger freezing point 'depression', which can be expressed as:

$$T - T_0 = \left[\left(P_w V_w - P_i V_i \right) T \right] / L_f \qquad (3.25)$$

where T_0 is the normal or 'atmospheric-pressure' freezing point of bulk water at 0 °C or 273.15 K, V_w is the specific volume of water and V_i is the specific volume of ice (specific volume is volume divided by mass, and thus is equal to $1.00\,\mathrm{cm^3 g^{-1}}$ for water and $1.09\,\mathrm{cm^3 g^{-1}}$ for ice). We have seen that as 'bound' water in soil pores progressively freezes, negative pore-water pressures develop, increasing the difference $(P_w V_w - P_i V_i)$ and thus the freezing point depression $(T - T_0)$. Effectively, then, any liquid water remaining in soil pores becomes progressively more difficult to convert to ice as temperature falls, so that minute quantities of unfrozen water persist in fine-grained soils even at the lowest temperatures recorded at the surface of the Earth.

3.4 Ice Segregation in Freezing Soils

A feature that accompanies the freezing of moist soils is the development of millimetre- to centimetre-thick platelets of ice termed *ice lenses* or *segregation ice* that form banded sequences of particle-free ice separated by layers of ice-infiltrated soil (Figure 3.8). Ice lenses form normal to the direction of freezing within the ground, and therefore approximately parallel to the ground surface. The process of ice lens formation during freezing is termed *ice segregation*, and because the formation of ice lenses increases the volume of the soil, it results in *frost heave*, the uplift of the soil surface during freezing. Because the pressures developed by frost heave are sufficiently great to disrupt structures such as roads and building foundations, the processes responsible for ice segregation and the factors that control these processes have been subject to intense study by engineers and physicists.

Historically, a key contribution to understanding ice segregation and frost heave was made by Beskow (1935), who showed that the distribution of ice in freezing soils, whether as ice lenses that grow to exclude soil particles and cause heave or as *pore ice* that simply occupies the pores within the soil, depends on three factors: rate of freezing, overburden pressure and soil texture. He showed that the soils most susceptible to ice segregation and frost heave are those that are sufficiently fine-grained to enable liquid water to be held above the water table by capillary forces under unfrozen conditions, yet sufficiently permeable to enable a supply of water to reach growing ice lenses when freezing occurs. This allowed him to propose a granulometric boundary between *frost-susceptible soils*, in which ice segregation and frost heave is likely to occur, and *non-frost-susceptible soils*, in which ice segregation is unlikely (Figure 3.9). Though this

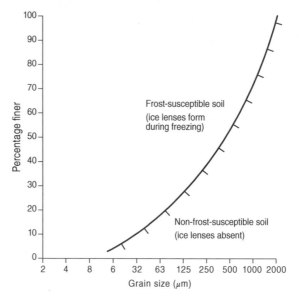

Figure 3.9 Differentiation of frost-susceptible soil (in which ice lenses form during freezing) and non-frost-susceptible soil on the basis of particle size distribution, as proposed by Beskow (1935).

Figure 3.8 Segregated ice lenses in lacustrine silty clay, western Canadian Arctic. The lenses near the top of the permafrost are thickest, and become progressively thinner with increasing depth. The section is about 2 m high. *Source:* Courtesy of Julian Murton.

simplifies the boundary conditions for ice lens formation, it represents a useful approximation. Essentially, the problem of ice segregation in soils breaks down into two parts: (i) the cause of water migration to the freezing front, where it becomes ice; and (ii) the reason why water that freezes within some soils forms discrete ice lenses, instead of simply freezing within soil pores to form pore ice.

3.4.1 Cryosuction

As outlined above, *cryosuction* is the force generated by the establishment of a free energy gradient within a freezing soil, because premelted interfacial water in pores occupied by ice at temperatures below the freezing point has lower free energy than pore water farther down the soil column. Effectively, cryosuction is generated by the same intermolecular forces responsible for the thin premelted films of water that surround soil particles at subzero temperatures; as the temperature falls, the liquid pressure decreases, causing water to flow towards regions of colder temperature. The force generated by cryosuction greatly exceeds that of gravity, so top-down freezing of moist soil drives water from unfrozen soil towards the freezing front. It can be envisaged as 'negative pressure' and is measured in pascals (Pa, where 1 Pa is a force of 1 newton (N) exerted over an area of $1\,m^2$, or $1\,N\,m^{-2}$). However, because the pressures associated with cryosuction are large, they are generally measured in MPa, where $1\,MPa = 10^6\,Pa$. Most explanations of cryosuction employ a form of the Clapeyron relationship that was introduced earlier (equation 3.24), and here we follow the explanation outlined in Davis (2000). If a change is made to any of the three variables that determine the free energy within a freezing soil (temperature, volume and pressure), the Clapeyron relationship describes how the other variables respond to restore thermodynamic equilibrium:

$$\Delta T = \left[\left(V_w \cdot \Delta P_w \right) - \left(V_i \cdot \Delta P_i \right) \cdot T_m \right] / L_f \qquad (3.26)$$

where ΔT is change in absolute temperature T, T_m is the melting point of ice (273.15 K or 0 °C), L_f is the latent heat of fusion, ΔP_w and ΔP_i are changes in the pressure of water and ice respectively, and V_w and V_i are the specific volumes of water and ice. If we assume no change in ice pressure, so that $\Delta P_i = 0$ and $(V_i \cdot \Delta P_i) = 0$, equation (3.26) can be rearranged in terms of change in water pressure:

$$\Delta P_w = \left(\Delta T \cdot L_f \right) / \left(V_w \cdot T_m \right) \qquad (3.27)$$

If we now define P_0 as the pressure on water at T_m, so that $\Delta P_w = (P_0 - P_w)$ for the temperature change $(T_m - T)$, we can derive from equation (3.27) a value for the suction induced by the accompanying temperature change:

$$Suction = \left(P_0 - P_w \right) = \left[\left(T_m - T \right) \cdot L_f \right] / \left(V_w \cdot T_m \right) \qquad (3.28)$$

Since V_w is unity and L_f is constant $(334\,J\,g^{-1})$, suction pressure is entirely temperature-dependent. Setting T at 272.15 K (−1 °C) means that suction $(P_0 - P_w)$ is $(1 \times 334/273.15) = 1.22\,MPa$. Similarly, setting T at 271.15 K (−2 °C) means that suction $(P_0 - P_w)$ is $(2 \times 334/273.15) = 2.44\,MPa$. Thus we can see that, if ice pressure is held constant, suction increases linearly (at $1.22\,MPa\,°C^{-1}$) as temperature decreases.

This description assumes that $\Delta P_i = 0$, but changes in ice pressure result in only small changes in the gradient of the suction–temperature relationship. If both ice and water pressure change equally, the gradient of the line reduces to $\sim 1.15\,MPa\,°C^{-1}$; if P_i changes five times as much as P_w, the gradient is reduced to $\sim 1.11\,MPa\,°C^{-1}$ (Figure 3.10), close to its limiting value of $\sim 1.10\,MPa\,°C^{-1}$. The Clapeyron equation (3.26) therefore explains thermodynamically why lowering the temperature at the freezing plane creates a suction that can draw water towards and into freezing soil, and shows that the magnitude of suction is linearly related to temperature decrease below the melting temperature of ice.

An alternative explanation of cryosuction, sometimes referred to as the *premelting model* (Davis, 2000), uses a slightly different approach. We have seen that within the range of temperatures at which soils freeze, a stable film of supercooled ($T < T_m$) premelted water exists between the ice and soil particles. In the premelting model, the hydrodynamic pressure in a film of premelted water (P_w) is given by:

$$P_w = P_i - \rho_i L_f \left(T_m - T \right) / T_m \qquad (3.29)$$

where ρ_i is the density of ice. This equation implies an increase of water pressure of roughly $1.1\,MPa\,°C^{-1}$, only

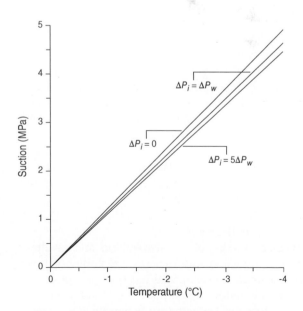

Figure 3.10 Linear relationships between cryosuction and temperature predicted by the Clapeyron equation for $\Delta P_i = 0$, $\Delta P_i = \Delta P_w$ and $\Delta P_i = 5\Delta P_w$.

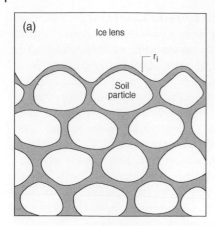
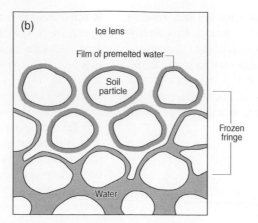

Figure 3.11 (a) Microscopic view of soil particles at the ice–soil interface. The convex-outward radius of the ice between soil particles at the base of the ice lens exceeds the pore radius, so ice cannot penetrate the pore spaces under the ice lens (capillary model). (b) Microscopic view of ice extending down from an ice lens into a frozen fringe.

slightly less than that predicted using equation (3.28). The term $\rho_i L_f (T_m - T)/T_m$ in equation (3.29) is sometimes termed the *thermomolecular pressure*. In the premelting model, it is assumed that the pressure on ice in a freezing soil is initially fixed and equal to the overburden pressure. Fixing P_i at a constant value requires that the thermomolecular pressure increases and P_w decreases as temperature decreases, and that the suction $(P_i - P_w)$ draws water upwards towards the freezing front.

3.4.2 Ice Segregation

The production of multiple layers of discrete surface-parallel ice lenses in freezing soils implies that the locus of lens formation periodically shifts downwards in the soil column so that a new lens begins to form. Despite much experimental and theoretical work, the process is still incompletely understood, and several models of ice lens formation have been proposed (Davis, 2000; Rempel, 2010; Peppin and Style, 2013). Three models of ice segregation are briefly reviewed here. For simplicity, it is assumed (i) that freezing takes place from the surface downwards, though a similar process occurs where freezing takes place upwards from the top of permafrost, and (ii) that the process occurs in frost-susceptible soil, though an essentially similar process may occur in porous bedrock (Walder and Hallet, 1985, 1986; Hallet *et al.*, 1991; Murton *et al.*, 2006).

3.4.2.1 Capillary Theory

The capillary theory of ice segregation and resultant frost heave originated with the classical studies of Taber (1930) and Beskow (1935), and has been elaborated by numerous later authors. A key assumption of this model is that ice does not immediately penetrate soil pores as the temperature drops below T_m. This is because the radius of the convex-outward form of the ice between

soil particles (r_i) exceeds the effective pore radius r_p, so ice cannot invade the pores, the freezing front stalls (Figure 3.11a) and water migrates upwards through the soil to feed a thickening ice lens; release of latent heat during freezing of water at the bottom of the lens also serves to arrest the progression of the freezing front. Eventually, however, a further decrease in temperature and corresponding increase in the pressure difference $P_i - P_w$ at the ice–water interface reduces r_i so that it equals r_p and the ice invades the underlying soil. Formally, the temperature at which ice invades soil pores (T_P) is given by:

$$T_P = T_m \left[1 - 2\gamma_{iw} \big/ \left(\rho_w \cdot L_f \cdot r_p \right) \right] \tag{3.30}$$

where γ_{iw} is the ice–water surface energy and ρ_w is the density of water. In the capillary model, ice lens thickening stops when $T \leq T_p$, at which point ice is assumed to fill the underlying soil pores, blocking them and preventing water from feeding the ice lens. Though this model showed good agreement with early experiments conducted on soils with uniform-sized particles at temperatures close to T_m, it fails to offer an adequate explanation for the initiation of new lenses below the initial lens and yields invalid results for poorly-sorted soils; it has also been argued that the Clapeyron equation on which it is based is valid only for equilibrium phase states, and not for the nonequilibrium states that develop during soil freezing (Peppin and Style, 2013; Ma *et al.*, 2015).

3.4.2.2 Frozen Fringe Models

The limitations of the capillary theory prompted some researchers to propose that ice lens thickening and consequent frost heave can occur when ice has formed a *frozen fringe* under an existing ice lens by growing into soil pores; the frozen fringe therefore separates the bottom of the ice lens from the freezing front (Figure 3.11b). This concept has the advantage that the existence of pore ice below the level of an existing ice lens provides nuclei from which a new lens can develop. It is supported by research demonstrating the existence of films of premelted water at ice–particle interfaces at temperatures below T_P (Dash et al., 2006), as water movement through such films allows the overlying ice lens to continue to thicken until a new lens develops. The frozen fringe concept has been developed by numerous researchers (Peppin and Style, 2013); here, we consider the model of Rempel et al. (2004), who argued that the mechanism for ice lens formation is fundamentally different from that assumed in the capillary model.

Rempel et al. (2004) pointed out that soil and ice grains repel each other across interfacial premelted liquid films as a consequence of the same intermolecular forces that produce such films. Low pressure is thus generated in the films, drawing in surrounding water along thermomolecular pressure gradients, with the rate of lens thickening and frost heave being determined by the mobility of the water through the partially frozen soil. The thermomolecular force (F_T) operating on a single particle embedded in ice that is subject to a uniform temperature gradient ∇T is approximated as:

$$F_T \approx m_s\left(L_f/T_m\right)\nabla T \tag{3.31}$$

where m_s is the mass of ice that could occupy the particle volume. This force is termed *thermodynamic buoyancy* (Rempel et al., 2001), and as equation (3.31) shows, it is proportional to the mass of ice (m_s) that could fit within the volume occupied by the particle and the product of temperature gradient (∇T) with the ratio of the latent heat to bulk melting temperature (L_f/T_m). Rempel et al. (2004) applied this concept to evaluate the net force per unit area supported by all the ice–particle interactions over the connected ice surface from the lower boundary of an existing lens to the base of the fringe. Incorporating this force with the net vertical fluid pressure against the ice surface and balancing both with overburden pressure permits calculation of the vertical gradient in load supported by particle contacts within the frozen fringe and prediction of the depth of initiation of new ice lenses by solving for conditions when this load reduces to zero (Rempel, 2007, 2010; Figure 3.12). This model also permits prediction of responses to particular combinations of soil freezing rate and overburden pressures. As Figure 3.13 shows, Rempel et al. (2004)'s model predicts that at low freezing rate and under low overburden pressure, a single ice lens grows in a stable manner, pushing the soil particles ahead of it, though at higher freezing rates periodic ice lenses form. When overburden pressures are large, no lenses form and no heave occurs. Rempel et al. (2004) found that hysteresis occurs near regime boundaries (dashed lines in Figure 3.13). For example, if soil freezing rate is initially rapid, so that multiple lenses form, but then declines, the soil will not revert to steady lens behaviour until conditions drop below the lower dashed line. Conversely, if freezing rate increases, then a steady lens with no fringe will develop a fringe and periodic lenses will develop at conditions above the upper dashed line.

A key assumption of the Rempel et al. (2004) model and its subsequent elaborations (Rempel, 2007, 2010, 2011) is

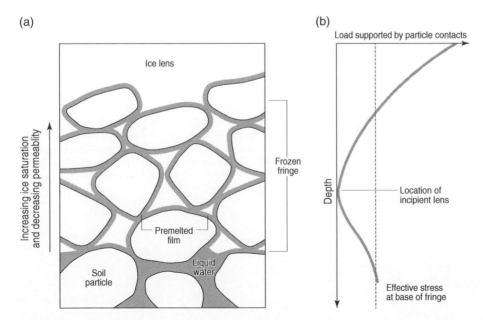

(a)   (b)

Figure 3.12 (a) Ice distribution in the frozen fringe beneath a growing ice lens. (b) Predicted load supported by particle contacts. Above the base of the frozen fringe, the load supported by particle contacts decreases to a minimum because of the large gradients in liquid pressure needed to drive flow through the lower permeabilities caused by increasing upwards ice saturation. At the depth where the minimum load tends towards zero, a new ice lens will form. *Source:* Rempel (2010). Reproduced with permission of Cambridge University Press.

the existence of a frozen fringe, but a review of experimental attempts to demonstrate the presence of a fringe suggests that the evidence is equivocal, and that it is likely that frozen fringes exist only under particular circumstances, depending on such factors as soil texture, rate of freezing and overburden pressure (Peppin and Style, 2013). This consideration has prompted a renewed search for a new model of 'fringe-free' ice lens formation.

3.4.2.3 Geometrical Supercooling and Ice Lens Formation by Crack Extension

Style *et al.* (2011) noted experimental evidence for crack-like behaviour leading to lateral extension of segregation ice in soil during lens initiation. They postulated that lens initiation begins when ice is present within a flaw or void in soil at some distance below the most recent ice lens (Figure 3.14). They employed the Clapeyron equation in the form:

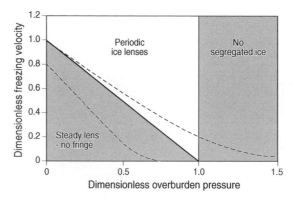

Figure 3.13 Regime diagram displaying the behaviour of a freezing soil as a function of freezing rate and overburden pressure. Dashed lines describe the limits of hysteretic behaviour (see text). *Source:* Rempel *et al.* (2004). Reproduced with permission of Cambridge University Press.

$$P_i - P_D = \rho L_m \left[(T_m - T)/T_m \right] \qquad (3.32)$$

(where P_D is the Darcy pressure of pore water, ρ is the density of the solid or liquid phase, L_m is the latent heat of melting of ice, T_m is the melting temperature of ice and T is the temperature at the flaw) to show that where undercooling of water occurs in the soil, $T_m - T$ is positive and hence the pressure of ice in the flaw (P_i) exceeds the Darcy pressure P_D. As the undercooling ($T_m - T$) increases, P_i increases and the ice in the flaw exerts pressure on the flaw walls. Assuming linear-elastic behaviour and an overburden pressure P_0 at the depth of the flaw, Style *et al.* (2011) showed that the flaw will extend laterally when:

$$P_i = P_0 + \sigma_t \qquad (3.33)$$

where σ_t is the tensile strength of the soil (Figure 3.14a–c). They argued that undercooling of liquid water at temperatures below T_m is caused by *geometrical supercooling*; in essence this occurs because in cohesive soils the temperature T may drop below the temperature for ice lens formation (T_L) without ice lens formation, because of the difficulty of rupturing the soil structure. The condition for ice lens initiation through ice nucleation is given by:

$$T_L - T = (T_m \sigma_t)/\rho L_m \qquad (3.34)$$

implying that when supercooling reaches a critical value, the soil ruptures and a new ice lens is initiated. Style *et al.* (2011) argued that new ice-filled cracks propagate parallel to existing ice lenses because the internal pressure driving crack extension occurs at a constant depth where $T_L - T$ reaches the critical amount required to initiate soil rupture, and also noted that because horizontal cracks have no shearing component,

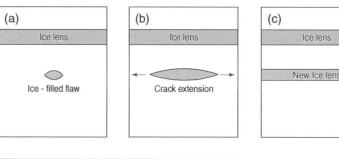

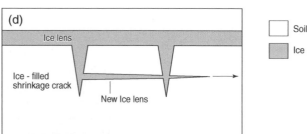

Figure 3.14 Ice lens formation through lateral crack extension. (a)–(c) Crack extension and lens formation from an initial ice-filled flaw (void) in freezing soil. (d) Crack extension from an initial flaw located in the wall of an ice-filled shrinkage crack. *Source:* Reproduced from Style *et al.* (2011) with permission from the American Physical Society.

cracks should propagate along a horizontal plane, as observed in frost-heave experiments.

Style *et al.* (2011) also demonstrated theoretically that formation of a new ice lens both reduces upwards flow of water from the lens and increases upwards flow of water into the lens, allowing it to thicken. Such behaviour implies a surge in heave rate each time a new lens is formed, and that the amplitude of the surge depends on the magnitude of critical supercooling and thus on the fracture resistance of the soil. A key requirement of their theory is the presence of ice in flaws below the level of the most recent ice lens. This may occur within a frozen fringe, but Style *et al.* (2011) showed that flaws located on the walls of ice-filled vertical shrinkage cracks may act as the loci of lens initiation even when a frozen fringe is absent (Figure 3.14d), producing ladder-like (reticulate) patterns of segregated ice, as seen in various laboratory experiments (e.g. Arenson *et al.*, 2008).

The geometrical supercooling model revives aspects of the classical capillary theory, albeit in significantly revised form, and represents a radical alternative to the extensive family of frozen fringe models that has been developed. Numerous questions relating to the exact mechanism of ice lensing in freezing soils and porous bedrocks nevertheless persist (Peppin and Style, 2013). Irrespective of the mechanism of ice lens formation, the expansion of ice within soil pores and lenses, together with the effect of suction, results in the formation of millimetre- to centimetre-sized compressed (consolidated) aggregates of soil. Such aggregates survive subsequent thaw because of the partial irreversibility of the consolidation process, and their size and shape reflect the former spacing and orientation of ice lenses (e.g. Van Vliet-Lanoë *et al.*, 1984; Chapter 9).

From a geomorphological perspective, several aspects of the development of ice lenses in freezing soils and rocks are important. The first is that the resulting frost heave can generate pressures of several MPa that result in *en masse* uplift of the soil. This is important in the generation of patterned ground (Chapter 9) and in causing slow downslope movement of soil through frost creep (Chapter 11). Ice lens development in moist, freezing bedrock has been convincingly shown to split intact rock apart and is therefore an important component of periglacial weathering (Chapter 10). Finally, ice lenses represent a form of *excess ice* in the soil, where the volume of ice exceeds the total pore volume of unfrozen soil. This has major implications for the development of soil structures in the active layer above permafrost and in seasonally frozen ground (Chapter 9), as well as in generating high pore-water pressures during thaw that lead to downslope soil movement through gelifluction and active layer detachment failures (Chapter 11). When ice-rich soils thaw, they give rise to *thaw consolidation*, which we now consider.

3.5 Thaw Consolidation

Thaw consolidation is defined as time-dependent compression of soil during thaw and drainage of pore-water, and results in a reduction in soil volume. It may be considered to have three components: (i) reduction in the volume of ice in the soil as it melts; (ii) reduction in pore sizes as soil grains and grain aggregates reorientate and squeeze closer together during thaw; and (iii) drainage of pore water and excess water produced by thaw of ice lenses.

In uniformly coarse-grained soils where ice lenses or other forms of excess ice are absent, such effects are minimal and little reduction in soil volume occurs. In frost-susceptible fine-grained or poorly sorted soils containing ice lenses, however, thaw consolidation may result not only in a marked reduction in soil volume during thaw, but also in a reduction in soil strength. The latter occurs when the upper layers of soil thaw to form a low-permeability layer that limits the rate of water drainage from the soil. If the rate at which water is liberated from thawing ice exceeds the rate of drainage, some of the stress imposed by the weight of the overlying thawed soil is transferred from grain-to-grain contacts to trapped soil water, raising the pore-water pressure (u) and thus reducing the effective stress (σ'), which 'locks' soil particles together under the weight (W) of the overlying soil, thus:

$$\sigma' = W - u = \gamma z_t - u \qquad (3.35)$$

where γ is the unit weight of thawed soil and z_t is the depth of thawed soil. The loss of soil strength resulting from reduction in effective stress therefore depends on the ratio between the rate of water release from thawing ice in the soil and the rate at which water can escape through the thaw-consolidated soil. This ratio is termed the *thaw consolidation ratio (R)*, expressed as:

$$R = \varphi / 2\sqrt{C_v} \qquad (3.36)$$

In equation (3.36), φ is given by:

$$\varphi = z_t \sqrt{t} \qquad (3.37)$$

where t is time elapsed since the onset of thaw at the ground surface and z_t is the depth of thaw at time t (McRoberts and Morgenstern, 1974). C_v is the coefficient of consolidation and is expressed as:

$$C_v = k^* / \rho_w m_v \qquad (3.38)$$

where k^* is the permeability of the thawed soil, ρ_w is the density of water and m_v is the coefficient of compressibility (Morgenstern and Nixon, 1971). Thus, in a soil with high permeability (k^*), drainage is rapid and excess pore-water pressures are reduced; conversely, both a high value of φ (implying rapid thaw) and a high soil compressibility

(m_v) have the opposite effect. For a soil mass thaw-consolidating under self-weight conditions, the excess pore-water pressure (u) is:

$$u = \sigma' \big/ \left[1 + \left(1/2R^2 \right) \right] \qquad (3.39)$$

where σ' is the effective stress after complete dissipation of excess pore-water pressures (McRoberts and Morgenstern, 1974). This means that a high value of R (approaching or exceeding unity) implies rapid thaw and/or slow escape of water released from melting ice, producing high pore-water pressures. Conversely, a low value of R implies slow thaw penetration and/or rapid soil drainage, and little or no elevation of pore-water pressures. The first situation is likely to arise in fine-grained silt- or clay-rich soil of low permeability containing abundant excess ice (ice lenses) that retards the rate of thaw and yields excess water on melting. The second situation is likely in uniformly coarse-grained sandy soil that contains little or no excess ice. As outlined in Chapter 9, thaw consolidation plays a role in the formation of certain structures in thawing soils, but the key importance of this process is in promoting mass movement on soil-mantled slopes that would otherwise be stable (Chapter 11).

3.6 Synthesis

It is useful to recap the most important points of this chapter, as many of the processes discussed in later chapters are based on the principles outlined here.

First, heat in transferred into and out of the ground primarily by conductive heat transfer, though this may be enhanced or diminished by nonconductive heat transfer, such as advection of heat by water movement in soil. The rate of heat transfer in soil or rock is determined by its thermal diffusivity (thermal conductivity divided by volumetric heat capacity), but this is complicated by latent heat effects, so that both soil freezing and soil thaw may temporarily stall at 0 °C (the zero curtain effect). The amplitude of ground surface temperature cycles diminishes with depth, and subsurface temperature change lags behind that at the surface, with an increase in phase lag with depth.

Second, there is no single temperature at which water freezes in soils or rock, but the proportion of unfrozen water present diminishes as the temperature falls further below 0 °C. Liquid water present in soil or rock at temperatures below 0 °C is termed 'premelted water' and exists because of surface tension effects, the presence of dissolved impurities that depress the freezing point of water and the effects of capillarity and adsorption. Premelted water in frozen soil or rock forms microscopically thin films at the boundary between mineral grains and pore ice, allowing migration of water within frozen ground. Such premelted films have lower free energy than unfrozen pore water, so pore water is driven along a free energy gradient towards the freezing plane by cryosuction, which represents the difference in ice pressure and water pressure at the freezing front. Suction potential increases by about 1.2 MPa per degree below 0 °C, and is greatest in fine-grained soils where a larger proportion of water remains in the liquid phase at temperatures below 0 °C.

Third, freezing of fine-grained soils often results in the formation of lenses of segregated ice, and the consequent increase in soil volume causes powerful frost heave of the ground surface. Such soils are termed 'frost-susceptible'. Segregated ice lenses can also form in some rocks, resulting in fracture. Formation of ice lenses is a consequence of cryosuction, but the reason for ice segregation operating at successively lower levels within a frost-susceptible soil freezing from the surface down is still debated. Traditionally, the capillary theory has been used to explain ice segregation, but other models have gained traction recently. Two of these have been outlined: a frozen fringe model that explains ice lens initiation in terms of a balance between thermodynamic buoyancy, fluid pressure and overburden pressure that recurs at different depths as a soil freezes from the surface downwards; and a model that explains ice lens initiation in terms of supercooling of water and lateral extension of ice-filled cracks from ice-filled flaws in the soil.

During thaw, ice-rich soils experience thaw consolidation in the form of self-weight soil compression that reduces soil volume, particularly through drainage of excess water produced by melting of ice lenses. If soil thaws rapidly and excess water drains slowly through the overlying thaw-consolidated layer, then pore-water pressure rises at the thaw front, potentially to levels that allow soil movement to occur even on low-gradient slopes. This is most likely to occur in fine-grained (silt- or clay-rich) frost-susceptible soils of low permeability, and less likely to affect uniformly coarse-grained sandy soils that contain only pore ice, experience limited (or no) consolidation and drain freely.

The following two chapters extend this discussion. In Chapter 4, we consider the characteristics of permafrost: ground that remains below 0 °C rather than experiencing seasonal freezing and thawing. Chapter 5 is devoted to discussion of the various types of ice that exist in frozen ground.

4

Permafrost

4.1 Introduction

Permafrost is defined as ground that remains below 0 °C for at least two consecutive years, or, more succinctly, as perennially cryotic ground. The term *cryotic* implies a ground temperature below 0 °C, and is used in preference to 'frozen', which suggests the presence of ice. Permafrost is thus defined exclusively by the thermal state of the ground, irrespective of whether or not ice is present, and exists in peat, organic soils, inorganic sediments, rubble and bedrock. Because solar radiation and above-freezing temperatures thaw the uppermost layer of ground during the summer, permafrost does not extend to the ground surface except under special circumstances (for example under perennial snowbeds or cold-based glaciers), but is overlain by the *active layer*, defined as seasonally cryotic ground overlying permafrost, or, where water is present, the nearsurface zone where ground freezes in winter and thaws in summer (Figures 4.1 and 4.2a), though most researchers adhere to the purely thermal definition (Burn, 1998). The term *permafrost table* refers to the upper boundary of permafrost under the active layer.

Though this simple thermal distinction between permafrost and the overlying active layer is adequate for most purposes, it assumes that the depth of seasonal freezing and thawing of the ground remains constant. In reality, annual freezing of the active layer may occasionally fail to reach the permafrost table, or deep thaw during a warm summer may extend into the top layer of permafrost. There is thus a *transition zone* in the uppermost layer of permafrost that alters in status between perennially cryotic and seasonally cryotic over timescales of years to centuries in response to short-term climatic extremes (such as a succession of warm summers), longer-term climate change or change in ground cover. The transition zone therefore represents comparatively short-lived permafrost that occasionally becomes part of the active layer, whereas the underlying cryotic ground represents long-term permafrost. The uppermost part of this transition zone, which experiences the most frequent thaw episodes, is an ice-rich *transient layer* that overlies an even icier *intermediate layer* (Shur *et al.*, 2005; Figure 4.3). The ice-rich transition zone functions as a thermal buffer between the overlying active layer and the underlying long-term permafrost because of the heat required to thaw ice, and is present in both soil and fine-grained porous bedrock; it is discussed in more detail in Chapter 5.

An additional complication is that the continuity of permafrost may be interrupted by layers or zones of unfrozen ground, which are often related to the presence of water bodies at the surface or to subsurface groundwater movement. Such zones of unfrozen ground are referred to as *taliks* and may be noncryotic, where the ground temperature exceeds 0 °C, or cryotic, where the ground temperature is perennially <0 °C but liquid water remains unfrozen because it is saline. The latter is termed a *cryopeg* and is by definition a component of permafrost even though it contains liquid pore water. Where the permafrost base is located within sediments that are saturated with mineralized groundwater, the lowermost parts of the permafrost consist of a cryotic but unfrozen *basal cryopeg*. Various types of talik are illustrated in Figure 4.4.

The term *seasonally frozen ground* (or, more strictly, *seasonally cryotic* ground) is employed to describe ground that undergoes a seasonal cycle of freezing and thawing. It usually refers to the uppermost layer of ground in cold but permafrost-free environments, though some authors also use it to refer to the active layer above permafrost. The depiction of seasonally frozen ground in Figure 4.1b is an oversimplification, as thaw may halt the downwards progression of the freezing front during winter, or renewed freezing from the surface downwards may interrupt spring thaw. Neither permafrost nor seasonal ground freezing is a prerequisite for all frost-action processes, as some superficial periglacial features such as small-scale ground patterning (Chapter 9) and frost creep (Chapter 11) can result from shorter-term shallow ground freezing and thawing.

Periglacial Geomorphology, First Edition. Colin K. Ballantyne.
© 2018 John Wiley & Sons Ltd. Published 2018 by John Wiley & Sons Ltd.

Figure 4.1 Ice wedge in permafrost, Svalbard. The soil above the ice wedge represents the active layer, which freezes and thaws annually. *Source:* Courtesy of Ole Humlum.

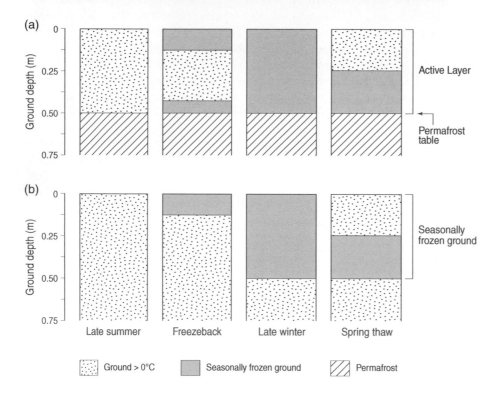

Figure 4.2 Seasonal patterns of ground freezing and thawing: (a) above permafrost; (b) seasonally frozen ground in non-permafrost terrain. In (a), two-sided freezing above 'cold' permafrost is depicted; above 'warm' permafrost, freezing is mainly from the surface downwards only.

4.2 Permafrost Thermal Regime

Permafrost formation is possible where the ground is subject to a negative energy budget at the surface so that mean annual ground surface temperature (MAGST) is below 0 °C. Where this is the case and seasonal frost in the ground survives summer thaw, the freezing front in the ground propagates downwards (*permafrost aggradation*) through heat diffusion. Cooling at the base of the aggrading permafrost is opposed by geothermal heat

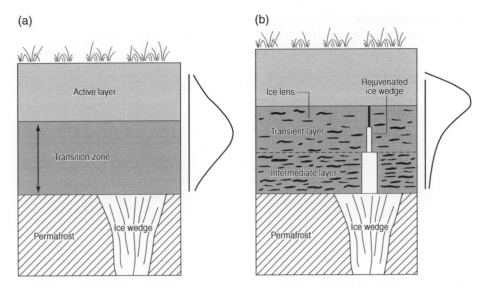

Figure 4.3 Three-layer model of the active layer, transition zone and permafrost. The curve represents the relative probability of annual thaw depth (a) directly following very deep thaw at subdecadal to centennial timescales and (b) with ice enrichment of the transition zone several centuries after the deep thaw event. Ice enrichment of the transition zone in (b) involves formation of ice lenses and upwards growth (rejuvenation) of secondary and tertiary ice wedges. *Source:* Shur *et al.* (2005). Reproduced with permission of Wiley.

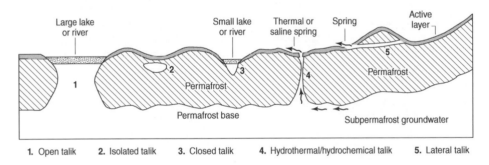

Figure 4.4 Schematic illustration of taliks. Open taliks penetrate the permafrost body completely, whereas closed taliks occupy a depression in the permafrost table, usually below a lake or river; their temperature remains above 0 °C because of heat storage in the surface water body. Isolated taliks are surrounded by frozen ground, and take the form of cryopegs or transient noncryotic taliks. Lateral taliks are overlain and underlain by permafrost. Hydrothermal and hydrochemical taliks are those that are fed by groundwater; the former are noncryotic, and resist freezing because of the heat supplied by flowing groundwater, whereas the latter are cryotic, but freezing is prevented by groundwater salinity.

flux, as well as other potential heat sources such as heat advected by mobile subpermafrost groundwater. Where upwards heat loss exceeds the geothermal heat flux, however, permafrost continues to aggrade downwards until it is in equilibrium with both surface temperature and geothermal heat flux. *Epigenetic permafrost* forms in pre-existing bodies of soil or rock, and aggrades downwards until such thermal equilibrium is achieved; it is therefore younger than the host sediment or bedrock. *Syngenetic permafrost* develops in an accumulating sediment body (for example on a floodplain) and aggrades upwards as the sediment body thickens; it is therefore roughly coeval with the host material. Syngenetic permafrozen sediments often contain plant rootlets and buried organic-rich horizons (palaeosols) that represent

the level of the former active layer prior to burial under accumulating sediment and the consequent rise in the permafrost table. Substantial bodies of Holocene syngenetic permafrost are most widespread on the aggrading floodplains and deltas of major arctic rivers, such as the Lena, Yenisei, Kolyma and Indigirka rivers in northern Siberia and the Mackenzie and Colville rivers in arctic North America, but they rarely exceed a few metres in thickness. Moreover, many thick permafrost bodies contain both epigenetic and syngenetic components (e.g. Shur *et al.*, 2004), and hence are best described as *polygenetic permafrost*.

The *permafrost base* is the maximum depth of permafrost, below which ground temperatures are above 0 °C. As outlined in Chapter 3, the depth of the permafrost base (z_p) can be estimated from the geothermal gradient (dT/dz) and mean annual temperature at the top of the permafrost (T_{TOP}) by the formula:

$$z_p = \left(0°C - T_{TOP}\right)/\left(dT/dz\right) \qquad (4.1)$$

on the assumption that the geothermal gradient is constant (Figure 4.5a). For example, where T_{TOP} is −4 °C and dT/dz is 0.02 °C m^{-1}, z_p is (4/0.02) m or 200 m.

Seasonal variations in the thermal regime of permafrost and the overlying active layer are governed by two further equations introduced in Chapter 3. The amplitude (A_z) of the annual temperature cycle at depth z is:

$$A_z = A_s \exp\left[-z\sqrt{\left(\pi/\kappa_a p\right)}\right] \qquad (4.2)$$

where A_s is the amplitude of the annual temperature cycle at the ground surface, κ_a is apparent thermal diffusivity and p is the cycle period (365 days). Effectively, this means that the amplitude of the annual temperature cycle diminishes nonlinearly with depth (Figure 4.5b) and that the degree of attenuation with depth is determined by the apparent thermal diffusivity of the ground. The depth at which annual temperature fluctuations become negligible (<0.1 °C) is termed the *depth of zero annual amplitude*, which typically ranges from ~10 m to ~25 m, depending on the thermal diffusivity of the ground, and is greatest in bedrock (e.g. Guglielmin *et al.*, 2011a). Moreover, because temperature change at the surface does not propagate instantaneously into the ground, there is a phase lag (t^*) between the surface temperature cycle and that at any depth:

$$t^* = \left(z/2\right)\cdot\sqrt{p/\left(\pi\cdot\kappa_a\right)} \qquad (4.3)$$

so that the maximum and minimum temperatures at, say, 3 m depth may occur 1–2 months after the maximum and minimum temperatures recorded at the ground surface (Figure 4.5c). The maximum and minimum temperature curves depicted in Figure 4.5b therefore do not represent the ground thermal state at any one particular time, as with increasing depth the timing of annual maximum and minimum temperatures becomes increasingly lagged behind temperature change at the ground surface.

The relationship between mean annual temperature at the top of permafrost (T_{TOP}) and mean annual air temperature (MAAT) is complicated by two factors (Throop *et al.*, 2012). The first is the effect of a thermal *buffer layer* of seasonal snowcover, vegetation cover and organic material (plant litter and humus) that creates a temperature difference between MAAT and MAGST. Because snow has very low thermal conductivity, seasonal snowcover tends to insulate the ground surface from winter air cooling. Ground under a snow depth of 0.6–0.8 m can generally be regarded as thermally insulated from the atmosphere. Conversely, during the summer months, vegetation cover shades the ground surface, and dry organic matter (which also has low thermal conductivity) reduces the heat entering the ground. In most environments, the effect of winter snowcover is dominant, so MAGST is warmer than MAAT, the difference in temperature being termed the *surface offset*. The second complication is that because the thermal conductivity of water (0.56 W m^{-1} K^{-1} at 0 °C) is much lower than that of ice (2.24 W m^{-1} K^{-1} at 0 °C), the thermal conductivity

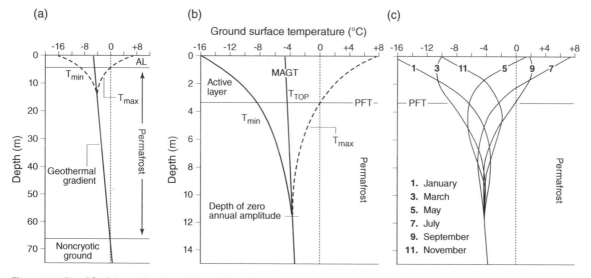

Figure 4.5 Simplified thermal regime of permafrost, showing temperature change with depth. (a) Full depth of permafrost, showing how the permafrost base is defined by the point where the geothermal gradient crosses 0 °C. (b) Expansion of the top part of (a) to show the amplitude of annual temperature change in the active layer and underlying permafrost, the depth of zero annual amplitude and a definition of the active layer as the point where T_{max} crosses 0 °C. (c) Idealized profile of the phase lag in temperature at 2-monthly time intervals, illustrating how temperature at depth lags behind temperature change at the surface. MAGT, mean annual ground temperature; T_{min} and T_{max}, minimum and maximum annual temperature; AL, active layer; PFT, permafrost table; T_{TOP}, mean annual temperature at the top of the permafrost.

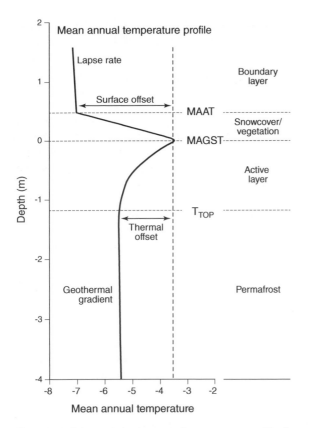

Figure 4.6 Schematic mean annual temperature profile through the active layer and overlying surface boundary layer, showing the relation between air temperature and permafrost temperature. MAAT, mean annual air temperature; MAGST, mean annual ground surface temperature; T_{TOP}, temperature at the top of the permafrost. *Source:* Smith and Riseborough (2002). Reproduced with permission of Wiley.

of thawed soil (k_t) is less than that of frozen soil (k_f), so that heat is more readily conducted out of the active layer when it is frozen than into the active layer when it is thawed. As a result, the mean annual ground temperature (MAGT) becomes progressively colder with increasing depth in the active layer, reaching a minimum at the permafrost table. The resulting temperature difference between MAGST and T_{TOP} is termed the *thermal offset* (Figure 4.6), which increases as the thermal conductivity ratio ($rk = k_t/k_f$) decreases. In mineral soils, *rk* varies from ~0.6 to ~0.9, depending on water content, but in saturated organic soils (peat) it may be as low as 0.3. In bedrock, *rk* approaches unity and there is negligible thermal offset (Smith and Riseborough, 2002).

4.3 Classification of Permafrost

Permafrost has been classified in various ways, notably by its temperature, spatial extent and ice content, though there is incomplete consensus regarding terminology and the boundaries between classes. Measurements of

permafrost temperature range from −24 °C to 0 °C. In the northern hemisphere, measured permafrost temperatures generally range from −15 °C to close to 0 °C (Romanovsky *et al.* 2010a), but Vieira *et al.* (2010) have recorded permafrost temperatures as low as −23.6 °C in the Antarctic. 'Warm' permafrost is that where T_{TOP} is close to 0 °C, or roughly within the range −5 °C to 0 °C; 'cold' permafrost is generally considered to be that where T_{TOP} is −5 °C or lower. However, Throop *et al.* (2012) place the warm–cold permafrost boundary at −1.5 °C, and others have suggested −2 °C or −3 °C (Smith *et al.*, 2010). An important distinction between the two is that whereas the active layer above warm permafrost freezes only from the surface downwards (*one-sided freezing*), that above cold permafrost freezes both downwards from the surface and upwards from the permafrost table (*two-sided freezing* or *bidirectional freezing*).

There is limited consensus on classification of permafrost by spatial extent, with differences between the schemes adopted by Russian and American scientists (Yershov, 1998; Hegginbottom, 2002). The North American literature tends to recognize three or four classes of permafrost on the basis of the percentage of ground underlain by it. Here we follow the scheme employed by Brown *et al.* (2014), which recognizes four classes: (i) *continuous permafrost* (>90% of terrain underlain by permafrost); (ii) *widespread discontinuous permafrost*, often abbreviated to *discontinuous permafrost* (50–90%); (iii) *sporadic discontinuous permafrost*, often abbreviated to *sporadic permafrost* (10–50%); and (iv) *isolated patches* of permafrost (<10%). This and related schemes can be applied to a general increase in the percentage of permafrost underlying the terrain with increasing latitude (sometimes termed *latitudinal permafrost* or *polar permafrost*), or with increasing altitude in midlatitude mountains (*alpine permafrost* or *mountain permafrost*). French (2007) distinguished a further category of high-altitude permafrost (*montane permafrost*) that underlies ~1.8 million square kilometres of high plateau and mountains in Central Asia, mostly in the Qinghai-Tibet Plateau, where it has developed in response to ~3000 m of uplift during the last two million years (Wang and French, 1995a). Additionally, *subsea permafrost* is present below the seafloor on some arctic continental shelves.

In terms of thermal regime, *equilibrium permafrost* is that which is in thermal equilibrium with the present ground surface temperature regime and geothermal heat flux. Conversely, *relict permafrost* is that which formed under colder conditions than now, notably during Pleistocene glacial stages, and which exists both in present-day permafrost environments and in areas where ground surface temperatures are now too high to permit permafrost formation, so that residual permafrost is decoupled from the atmosphere by an overlying layer of noncryotic ground. The term *past permafrost* is applied to terrain, soil or bedrock where there is evidence for the former existence of permafrost even though it is now completely absent (French, 2008; French and Shur, 2010). Extant permafrost is also classified by its ice content. *Ice-bearing permafrost* is simply permafrost that contains ice, whereas *dry permafrost* contains neither water nor ice. *Ice-bonded permafrost* is that in which soil particles are 'cemented' together by intergranular (pore) ice, and *ice-rich permafrost* is that containing excess ice, implying that the ice content exceeds the unfrozen pore volume of the soil, sediment or rock. The term *saline permafrost* is applied to permafrost in which the salinity of pore water causes freezing-point depression so that though ground temperatures are <0 °C, all or part of the water content is unfrozen, as in cryopegs.

4.4 Detection, Mapping and Modelling of Permafrost

Hegginbottom (2002) has reviewed the methodological and cartographic challenges posed in mapping the distribution of permafrost and its characteristics. Traditionally, identification of terrain underlain by permafrost was based on borehole measurements of ground temperatures, the presence of ice in sediment cores, or logging of ice-bearing sediments in exposures. Such site-specific information has been supplemented by mapping of landforms diagnostic of permafrost or permafrost degradation from aerial photographs or high-resolution satellite images. Conspicuous permafrost landforms include thermal contraction polygons (Chapter 6), pingos and palsas (Chapter 7), thaw lakes and other thermokarst features (Chapter 8) and, in mountainous areas, rock glaciers and large, stable ice-cored moraines (Janke, 2005; Fukui *et al.*, 2007; Lilleøren and Etzelmüller, 2011; Lilleøren *et al.*, 2013; Chapter 12). In the discontinuous permafrost zone, remotely-sensed vegetation and topographic data have also been employed to determine near-surface permafrost presence or absence (Nguyen *et al.*, 2009; Panda *et al.*, 2010). Though such approaches remain useful, many areas of permafrost terrain support few diagnostic surface features, and for these areas the presence of permafrost has to be interpolated from sites of known permafrost occurrence or inferred indirectly using air-temperature data from widely spaced climate stations. This approach becomes increasingly uncertain in areas underlain by discontinuous or sporadic permafrost where vegetation cover, substrate characteristics, aspect, elevation and snowcover modulate the surface energy balance so that permafrost distribution is localized. Over the past few decades, much research has been devoted to the development of additional methods for establishing the distribution and characteristics of permafrost, notably through the use of geophysical techniques and through permafrost modelling.

4.4.1 Geophysical Surveys

The application of geophysical methods to the detection of permafrost is reviewed in Hauck and Kneisel (2008), Kneisel *et al.* (2008), Harris *et al.* (2009) and Hauck (2013). Such methods have been employed to determine the depth and lateral extent of permafrost, the active layer and taliks, and they rely on mathematical inversion of data measured at or above the surface to produce models of subsurface properties such as specific resistivity, seismic or electromagnetic wave velocity and permittivity. Geophysical methods differentiate between frozen (ice-bearing) and unfrozen (water-bearing) materials, rather than measuring ground temperature. They suffer from the disadvantage that the inversion modelling is usually poorly constrained (unless supplemented by borehole measurements), so that the resulting model of permafrost extent, depth or characteristics may be ambiguous. They have nonetheless enjoyed considerable success in determining the presence or absence of permafrost at local and regional scales.

A wide range of geophysical methods have been employed in permafrost research (Hauck, 2013). Of these, electrical resistivity tomography (ERT) is particularly suited to permafrost detection (e.g. Fortier *et al.*, 2008b; Lewkowicz *et al.*, 2011), as resistivity increases strongly at the freezing point due to the change from electrically conductive water to nonconductive ice, but this technique requires good galvanic contacts between electrodes and the ground. Electromagnetic induction (EM) techniques avoid this requirement by inducing a varying current in a transmitter loop to measure subsurface electrical conductivity, and thus the presence and character of ice-bearing ground, and these have been employed in the detection of both lowland and mountain permafrost (e.g. Harada *et al.*, 2000; Hauck *et al.*, 2001). This approach has also been extended to airborne electromagnetic (AEM) imaging to provide regional-scale estimates of permafrost distribution at the continuous–discontinuous permafrost boundary (Minsley *et al.*, 2012); these have been combined with remotely-sensed and other data to infer probabilities of permafrost occurrence (Pastick *et al.*, 2013). Ground-penetrating radar (GPR) is an EM technique that employs radio waves, typically in the range 10–1000 MHz, to map subsurface structures. An electromagnetic pulse is emitted by a transmitter and recorded by a receiver after reflection in the subsurface, and strong reflections occur at subsurface boundaries, characterized by contrasting dielectric properties. As the dielectric properties of frozen and unfrozen ground differ markedly, GPR allows high-resolution stratigraphic profiling of permafrost bodies (e.g. Hinkel *et al.*, 2001a). Finally, seismic refraction techniques are based on the travel time of refracted seismic waves: elastic waves generated at the surface travel through the subsurface, are refracted at discontinuities and are recorded at surface geophones. As P-wave velocities in unfrozen sediments (typically 400–1500 m s^{-1}) are much slower than those in ice-bearing permafrost (2000–4000 m s^{-1}), the boundary between the two is recorded as a discontinuity. Most geophysical investigations of permafrost involve 2D transects of limited length, though 3D surveys are feasible. Geophysical techniques are particularly useful in specialized permafrost research, such as the identification of the lower altitudinal limits of permafrost in mountain areas, localized bodies of permafrost, subsurface ice bodies or taliks. Upscaling of geophysical data to inform regional surveys is more problematic, though the use of AEM surveys represents a promising development.

4.4.2 BTS Measurements

In mountainous areas, where permafrost occurrence is strongly influenced by altitude and aspect, the occurrence of permafrost has been inferred from late-winter measurements of the basal temperature of the snow (BTS). This method assumes (i) that deep (>0.8 m) seasonal snowcover insulates the underlying ground from short-term variations in the surface energy balance, so that the BTS reaches a fairly constant equilibrium temperature in late winter or early spring, and (ii) that the BTS value therefore reflects heat flux from the subsurface and serves as an indicator of subsurface thermal conditions, and therefore the probability of permafrost. Early studies using BTS measurements usually assumed that BTS < –3 °C implied probable permafrost occurrence, values in the range –3 to –2 °C implied possible permafrost occurrence and values > –2 °C implied that permafrost was improbable (e.g. Hoezle, 1992; Ishikawa and Hirakawa, 2000). Subsequent research has shown, however, that these assumptions may be invalidated by local conditions, and that ground truthing (excavation of pits in late summer at a sample of BTS sites) is necessary to establish appropriate BTS boundaries for the assessment of permafrost probability. BTS measurements nevertheless provide a useful means of verifying or complementing the results of geophysical surveys (e.g. Gómez *et al.*, 2001; Isaksen *et al.*, 2002; Hauck *et al.*, 2004).

Numerous studies have shown that BTS values are strongly related to potential incoming solar radiation (PISR) and thus slope, aspect and shading (Hoezle, 1992; Julián and Chueca, 2007), or to altitude and thus air temperature lapse rate (Hauck *et al.*, 2004), or both (Lewkowicz and Ednie, 2004; Bonnaventure and Lewkowicz, 2008). Such relationships enable empirical-statistical spatial modelling of predicted BTS values

(and thus permafrost probability) based on PISR and altitude data derived from digital elevation models (DEMs), but the resulting maps of permafrost probability depend to some extent on the treatment of the input BTS data (Lewkowicz and Ednie, 2004). Moreover, Brenning *et al.* (2005) have emphasized the need for stratified sampling to eliminate statistical distortion of results, and the possible limitations imposed by intra-annual and inter-annual variation in measured BTS values (e.g. Imhof *et al.* 2000). An additional challenge in using BTS-calibrated DEMs to model the spatial distribution of permafrost probability in mountain areas is that air-temperature lapse rates are non-uniform (and sometimes reversed) above and below the treeline as a result of winter air-temperature inversions. Lewkowicz and Bonnaventure (2011) have outlined a solution to this problem by modelling the *equivalent elevation* of mountainous terrain, a procedure that takes variable surface lapse rates into account and produces a flattened or inverted DEM-derived surface, but which requires calibration using air-temperature data from multiple locations at different elevations and aspects. Bonnaventure *et al.* (2012) incorporated this approach in a regional permafrost probability model that covers nearly 500 000 km^2 of mountainous terrain in southern Yukon and northern British Columbia. This is based on seven local empirical-statistical permafrost probability models based on winter BTS measurements and summer ground truthing, and on the use of logistic regression to relate predicted BTS values to ground-truthing observations. These models were upscaled by modelling both PISR and equivalent elevation across the entire region and merging the individual local models using a distance-decay function to produce the regional permafrost probability map. Despite its acknowledged uncertainties and inaccuracies, this approach represents a remarkable advance in the use of BTS measurements as the starting point for regional permafrost probability assessment in complex terrain, albeit one that requires a wealth of field data and sophisticated GIS-based modelling.

4.4.3 Modelling of Permafrost Distribution and Temperature

Several numerical models have been proposed for modelling the thermal state of permafrost and its response to perturbation (Riseborough *et al.*, 2008). From the viewpoint of determining permafrost distribution and thermal characteristics, the challenge is to relate T_{TOP} to air-temperature regime, so that permafrost distribution and temperature can be predicted from meteorological data. Nelson and Outcalt (1987) approached this problem through the calculation of a *frost number (Fn)*, defined as:

$$Fn = \sqrt{I_{FA}} \Big/ \left(\sqrt{I_{FA}} + \sqrt{I_{TA}} \right) \tag{4.4}$$

where I_{FA} and I_{TA} are respectively the number of freezing and thawing degree-day sums (°C days) of screen temperature. This index can be modified to incorporate the effect of snowcover, and in this modified form a value of 0.5 is assumed to represent the permafrost limit. Though this approach has been successfully employed to delimit permafrost at continental scales (Nelson and Anisimov, 1993), it is insensitive to local variance in permafrost occurrence (Janke *et al.*, 2012).

An alternative approach is the *TTOP model*, which employs *n-factors* to relate air temperature to ground surface temperature. An n-factor represents the energy balance at the surface as a single dimensionless number that constitutes an empirically-derived transfer function between the air temperature and the ground surface temperature (Klene *et al.*, 2001; Smith and Riseborough, 2002). It is calculated separately for the freezing period (n_f) and the thawing period (n_t), thus:

$$n_f = I_{sf}/I_{FA} \tag{4.5}$$

$$n_t = I_{st}/I_{TA} \tag{4.6}$$

where I_{sf} and I_{st} are respectively the freezing and thawing degree-days sums at the ground surface (°C days) and I_{FA} and I_{TA} represent air (screen) freezing and thawing degree-day sums. The relationship between T_{TOP} and MAAT can be expressed as: T_{TOP} = (MAAT + surface offset + thermal offset), where the surface offset is generally positive and the thermal offset is negative (Figure 4.6). Smith and Riseborough (1996, 2002) have shown that this relationship can be reduced to:

$$T_{TOP} = \left[\left(rk \cdot n_t \cdot I_{TA} \right) - \left(n_f \cdot I_{FA} \right) \right] \tag{4.7}$$

where *rk* is the thermal conductivity ratio ($= k_t/k_f$). It follows that the limiting condition for permafrost occurrence ($T_{TOP} = 0$ °C) is given by:

$$rk \cdot n_t \cdot I_{TA} = n_f \cdot I_{FA} \tag{4.8}$$

Assuming that I_{TA} and I_{FA} can be calculated or interpolated from meteorological data, the key parameters in this model are *rk* (typically 0.6–0.9 for mineral soils, but as low as 0.3 for saturated organic-rich soil) and n_f, which can vary from 0.1 to 1.0 depending on snow depth and MAAT (Figure 4.7); n_t typically varies from ~0.8 to 1.0 and hence is less critical, though exceptionally (on the Qinghai-Tibet Plateau) values as high as 2–3 have been recorded (Lin *et al.*, 2015). Application of the TTOP model to predict permafrost distribution therefore requires an input of empirically-derived n-factors and the assumption or measurement of appropriate thermal conductivity ratios (e.g. Riseborough, 2002; Wright *et al.*, 2003; Juliussen and Humlum, 2007). An interesting conclusion from the application of the TTOP model to Canadian data is that *rk* is the critical variable in determining the southern limit

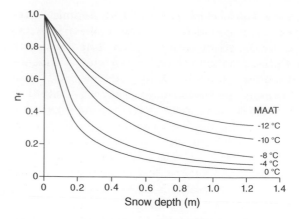

Figure 4.7 Relationship between the n_f factor, snow depth and mean annual air temperature (MAAT). *Source:* Smith and Riseborough (2002), Reproduced with permission of Wiley.

of discontinuous permafrost, whereas snowcover (and hence n_f) is the principal control on its northern limit.

An alternative solution for calculating T_{TOP} based on the amplitude of ground surface temperature (A_s) and MAGST (T_s) is provided by Sazonova and Romanovsky (2003):

$$T_{TOP} = \left[0.5 T_s \left(\left(k_f + k_t \right) + A_s \left(k_t - k_f \right)/\pi \right) \right.$$
$$\left. \times \left(\left(T_s/A_s \right) \left(\arcsin\left(T_s/A_s \right) + \left(1 - \sqrt{\left(T_s^2/A_s^2 \right)} \right) \right) \right) \right] \Big/ k' \tag{4.9}$$

where $k' = k_f$ if the numerator is <0 and $k' = k_t$ if the numerator is >0. This method, however, requires prior calculation of the surface offset and the value of A_s, and assumption or measurement of k_t and k_f. Recently, attention has focused on the use of land surface models (LSMs), which assume various soil parameters and use meteorological data to simulate the distribution and characteristics of permafrost (Chen *et al.*, 2015). For further discussion of both numerical and statistical-empirical modelling of permafrost distribution and temperature, the reader is referred to Riseborough *et al.* (2008).

4.5 Permafrost Distribution

4.5.1 Global Distribution

In the northern hemisphere, the permafrost zone has been estimated to occupy ~22.8 × 10⁶ km², or about 23.9% of the exposed land surface, roughly 70% of which is distributed between latitudes 45° and 67° N (Zhang *et al.*, 1999), with an extensive southern outlier represented by permafrost underlying the Qinghai-Tibet Plateau and the high mountains of Central Asia (Figure 4.8); the total area of permafrost in China is estimated to be ~1.35 × 10⁶ km² (Ran *et al.*, 2012). However, because

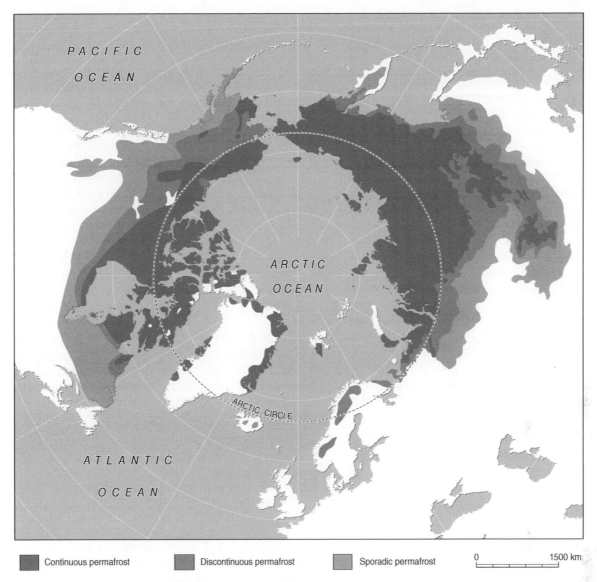

PACIFIC OCEAN

ARCTIC OCEAN

ATLANTIC OCEAN

ARCTIC CIRCLE

■ Continuous permafrost   ■ Discontinuous permafrost   ■ Sporadic permafrost   0 ⊢———⊣ 1500 km

Figure 4.8 Circum-arctic map of permafrost. *Source:* Adapted from Brown *et al.* (2014). Reproduced with permission of the National Snow and Ice Data Center.

much of the northern hemisphere permafrost zone consists of discontinuous or sporadic permafrost alternating with areas of permafrost-free terrain, the figure for exposed land actually underlain by permafrost (the *permafrost area*) is estimated to be markedly lower, within the range $12.2–17.0 \times 10^6 \, km^2$, or 12.8–17.8% of the exposed land area (Zhang *et al.*, 2000). These empirically-based estimates, however, involve considerable extrapolation across remote areas where data are sparse or absent. A simulated estimate based on MAATs and high-resolution global elevation data (Gruber, 2012) suggests that the northern-hemisphere permafrost zone occupies $(21.7 \pm 3.0) \times 10^6 \, km^2$ (19–25% of the total land area) and the northern-hemisphere permafrost area is within the range $12.9–17.7 \times 10^6 \, km^2$, or 13–18% of the total land area. The same study suggests that the global

permafrost area on land in both hemispheres falls within the range $13.2–18.1 \times 10^6 \, km^2$, or $16.0–20.9 \times 10^6 \, km^2$ if subsea permafrost is included. In the southern hemisphere, permafrost underlies a much smaller area of exposed land surface, mainly in mountain areas (the southern Andes and Southern Alps of New Zealand) and the unglacierized parts of Antarctica. According to Vieira *et al.* (2010), permafrost underlies all unglacierized terrain (~50 000 km^2) in Antarctica apart from the coastal fringes of the Antarctic Peninsula and the Antarctic and sub-Antarctic islands. The figures for both hemispheres, however, do not include permafrost under polar ice sheets and ice caps. As extensive parts of the Antarctic Ice Sheet, Greenland Ice Sheet and high-latitude ice caps contain cold-based glacier ice where the basal ice temperature is below pressure melting point, the underlying substrate in such areas constitutes permafrost, though the full extent of such subglacial permafrost is unknown.

The lowest altitude of permafrost rises progressively southwards in the northern hemisphere. In Europe, permafrost extends to sea level in Svalbard (76–80°N), but is limited to altitudes above 700–900 m in northern Scandinavia, 1200–1500 m in southern Norway, 2500–3000 m in the Alps and ~3500 m in the Sierra Nevada of southern Spain (Harris *et al.*, 2009). On Kilimanjaro, the highest mountain in Africa, just 300 km from the equator, thin permafrost occurs above ~4800 m. Similarly, the lower altitudinal limit of permafrost in the North American Cordillera rises from about 1000 m at 60°N to over 3000 m at 33–39°N, and across the Qinghai-Tibet Plateau it rises from ~4200 m in the north to ~4800 m in the south. Despite the high geothermal heat flux associated with active volcanism, permafrost is also present on many high-latitude or high-altitude volcanoes, for example in Iceland (Etzelmüller *et al.*, 2007) and Kamchatka (Abramov *et al.*, 2008), and even on tropical stratovolcanoes (Palacios *et al.*, 2007). Across North America, the boundary between continuous and discontinuous permafrost mainly coincides with the –6°C to –8°C MAAT isotherms, reaching its southernmost extent around Hudson Bay at 55°N (Figure 4.8). In Russia east of the Urals, the continuous–discontinuous permafrost boundary tends to follow the northern margin of boreal forest (taiga), reflecting more effective snow entrapment in forested areas, and thus greater insulation of the ground surface from winter cooling.

Permafrost thickness varies from less than 1 m in peatlands at the southernmost margin of the latitudinal permafrost zone to an estimated ~1500 m in northern Yakutia, in the low-lying plains and river basins bordering the Laptev and East Siberian Seas. This maximum depth reflects an exceptionally long history of permafrost formation, and estimates for the thickness of high-arctic (>70°N) continuous permafrost generally fall within the range 300–700 m, except in coastal locations. On Svalbard, for example, permafrost thickness exceeds 500 m under the mountains of the interior but is less than 100 m near the coast (Humlum *et al.*, 2003). Active-layer thickness in soils and unlithified sediments generally lies within the range 0.2–3.0 m, depending largely on the number of summer degree-days above 0°C and the persistence of snowcover, though the active layer in the coldest unglacierized parts of Antarctica is only a few centimetres thick (Campbell and Claridge, 2006; Bockheim *et al.*, 2007). Conversely, exceptional active-layer depths of up to 5 m have been recorded in sediments near the southern margins of mid-continental permafrost, and active-layer depths reach over 5–10 m in bedrock, which has much higher thermal diffusivity than soil, allowing summer warming to penetrate deeper (Christiansen *et al.*, 2010; Farbrot *et al.*, 2013; Hipp *et al.*, 2014).

4.5.2 Relict Permafrost

Because the depth of permafrost and the overlying active layer are governed by energy exchange processes at the ground surface, and thus ultimately related to MAAT, we might expect that permafrost generally thickens (and the active layer generally thins) with increasing latitude. Though in general terms this is the case (Figure 4.9), the thickness of permafrost in any location is also related to its history. According to Yershov (1998), permafrost may have begun to form in northern Siberia and Yakutia in the early Pleistocene or possibly the late Pliocene. Though these regions experienced intensely cold winter conditions during Pleistocene glacial stages, they were also very arid at such times (Hubberten *et al.*, 2004) and so escaped glaciation. The extreme depth (up to ~1500 m) of permafrost present in these areas therefore represents downwards permafrost aggradation under extremely cold winter conditions over timescales of 10^5–10^6 years. This is an example of the formation of *relict* permafrost

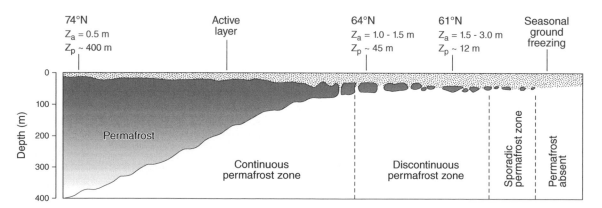

Figure 4.9 Schematic north–south transect of permafrost and active-layer thickness in arctic and subarctic Canada, based on data for Hay River (61°N), Norman Wells (64°N) and Resolute (74°N). Active layer (stippled) is not to scale. z_a, active-layer thickness; z_p, permafrost thickness. *Source:* Adapted from Brown (1970) with permission from University of Toronto Press.

that developed under much colder conditions than those of the present; such relict permafrost is generally polygenetic, with both epigenetic and syngenetic components. In western Siberia, Russian scientists have found evidence for two layers of permafrost: a lower layer of relict permafrost of Pleistocene age and an upper layer of equilibrium permafrost of late Holocene age, separated by a talik that is believed to have formed during the early Holocene thermal optimum (Yershov, 1998). Similarly, stratigraphic evidence from the nonglaciated parts of Yukon and Alaska demonstrates that permafrost in these areas persisted throughout the last interglacial stage, implying that the depth of permafrost (locally >600 m) reflects deep aggradation under Pleistocene cold conditions despite limited top-down thaw during the warmest parts of the last interglacial and the Holocene (Reyes *et al.*, 2010). In the western Canadian arctic, permafrost depths exceed 500 m at sites that remained glacier-free during the last glacial stage, compared with <300 m at sites that were covered by the last ice sheet. Most permafrost that has developed in northern high latitudes since retreat of the last ice sheets is epigenetic, though in areas of postglacial sediment accumulation (such as floodplains, or areas of windblown sediment deposition) epigenetic permafrost may be overlain by a layer of Holocene syngenetic permafrost several metres thick.

Relict permafrost also exists in areas that are now too warm to support permafrost formation. A striking example is the discovery of a body of permafrost at least 93 m thick at a depth of 357 m within sedimentary rocks in northeast Poland, where present MAAT is +7 °C. (Szewczyk and Nawrocki, 2011). This permafrost apparently formed during the last (Late Weichselian) glacial period, when MAGST is estimated to have been approximately −10 °C in northeast Poland, and its survival throughout the Holocene is thought to reflect both very low geothermal heat flow from the underlying rock and insulation from warming surface temperatures by overlying water-saturated sedimentary rocks. Similarly, at Anchorage in Alaska (MAAT +2.2 °C), borehole investigations have identified a body of relict Pleistocene ice-rich permafrost up to 12 m thick a few metres below the ground surface. In this case, the permafrost has survived Holocene warming under a thick insulating layer of moss and peat, which has very low thermal conductivity in summer (Kanevskiy *et al.*, 2013). A tiny (0.134 km^2) body of relict permafrost over 30 m thick reported by Xie *et al.* (2013) near the highest point of the Loess Plateau in China also owes its survival to the insulating properties of an overlying cover of peat.

Subsea permafrost is also relict. The greatest expanses of subsea permafrost underlie arctic continental shelves that escaped Pleistocene glaciation, notably the Beaufort, Laptev and East Siberian shelves. Romanovskii *et al.* (2004) have shown that ice-bearing, continuous permafrost roughly 300–350 m deep extends offshore on the Laptev and East Siberian shelves to water depths of 50–60 m, and that discontinuous permafrost extends to the shelf edge at greater depths. Subsea permafrost on arctic shelves developed during Pleistocene glacial stages, when eustatic sea-level fall (roughly 120–130 m during the last glacial maximum, LGM) exposed shelf surfaces to extremely cold conditions. As the ice sheets melted and rising seas flooded the shelves, subsea permafrost was preserved because of the subzero temperature (−0.7 °C to −3.4 °C; typically around −1.8 °C) of saline seawater at the sea floor, which inhibited top-down warming during successive interglacial stages and during the Holocene. Modelling of the thermal history of subsea permafrost on these shelves suggests that though it may have thawed completely on the outer shelves during interglacial periods, the permafrost underlying the inner shelves at water depths <60 m probably survived the last four glacial–interglacial cycles, a period spanning roughly 400 000 years.

4.5.3 Local Variations in Permafrost Distribution

Throughout the arctic continuous permafrost zone, permafrost is generally ubiquitous except under major rivers that flow throughout the year and under lakes that do not freeze to their beds in winter, so that such water bodies are underlain by closed taliks. In areas of discontinuous and sporadic permafrost, however, permafrost distribution reflects a complex of factors, notably topoclimate (terrain altitude, aspect and shading), snowcover, vegetation cover, the thickness of the surface organic layer, soil texture and soil moisture content, all of which modulate ground surface temperature, creating a mosaic of alternating permafrost and seasonally cryotic ground.

In low-relief areas of discontinuous permafrost, vegetation type, snowcover, organic layer thickness and soil moisture content are the main controls on permafrost present or absence. Analysis of the factors controlling permafrost distribution in the boreal forests of central Yukon by Williams and Burn (1996), for example, demonstrated that organic-layer thickness and soil moisture content account for 95% of the variance in permafrost occurrence; a thick organic layer insulates the underlying ground in summer, limiting the depth of thaw, and high soil moisture content promotes evapotranspiration, which reduces summer ground surface temperature. In central Kamchatka, permafrost occurs under areas of sphagnum moss and creeping pine, but is absent under birch stands, a contrast attributed by Fukui *et al.* (2008b) to the low summer thermal conductivity of sphagnum (0.07–0.19 W m^{-1}K^{-1}) and pine litter (0.06–0.09 W m^{-1}K^{-1}), which insulate the underlying ground from warming and promote permafrost survival under a thin active layer.

Similarly, at the southern boundary of discontinuous permafrost in northern Mongolia, permafrost is largely limited to north-facing forested slopes and flat plains where there is a ground cover of moss and humus (Ishikawa *et al.*, 2005; Etzelmüller *et al.*, 2006; Avirmed *et al.*, 2014). In the sporadic permafrost zone, a thick cover of fibrous peat with low (dry) thermal conductivity has allowed the formation and preservation of patchy permafrost even in areas where MAATs may be as high as 0 °C (Westin and Zuidhoff, 2001; Pissart, 2002). The role of vegetation in dictating permafrost distribution across areas of low relief is also related to winter snowcover. Earlier and thicker snowcover insulates the ground from winter cooling. Because taller vegetation traps drifting snow, ground surface temperatures tend to be higher under taiga forest and tall shrubs than under low shrubs, herbs and bare ground (Morse *et al.*, 2012; Roy-Léveillée *et al.*, 2014), though shading by a forest canopy reduces ground surface temperatures in summer. Permafrost that has formed or persists due to edaphic or vegetation factors in climatically marginal areas has been termed *ecosystem-driven* or *ecosystem-protected* permafrost, as distinct from *climate-driven* permafrost in the continuous permafrost zone (Shur and Jorgensen, 2007).

Alpine permafrost also exhibits marked spatial variability in occurrence, depth and temperature because topography, vegetation cover, snowcover and substrate type modulate energy exchange at the ground surface (Gruber and Haeberli, 2009; Lewkowicz *et al.*, 2012). Reduction in air temperature with increasing altitude represents the primary control on permafrost distribution in mountain areas, but is strongly influenced by aspect; in the northern hemisphere, MAGST may be several degrees higher on steep south-facing than on north-facing slopes (Figure 4.10), so the altitudinal limit of permafrost on the former may be hundreds of metres higher than on the latter (e.g. Gruber *et al.*, 2004b; Noetzli *et al.*, 2007). In the Jotunheimen Massif

of southern Norway, for example, the lower altitudinal limit of permafrost in shaded north-facing rockwalls is ~1200–1300 m, compared with ~1500–1700 m on other aspects (Hipp *et al.*, 2014). Near the lower altitudinal limit of alpine permafrost, variations in the extent, thickness and duration of snowcover result in a patchwork of permafrost (where snowcover is thin) and seasonally-frozen ground under deep snowdrifts (Ishikawa, 2003; Juliussen and Humlum, 2007; Rödder and Kneisel, 2012). Moreover, the assumption that increasing elevation invariably results in lower mean air temperatures is not always satisfied: O'Neill *et al.* (2015) have shown that persistent winter air temperature inversions may result in higher MAATs above the treeline in arctic mountain areas, leading to increased mean annual surface temperatures at higher elevations.

An interesting cause of local variations in permafrost distribution on mountains is the texture of the substrate. Comparative measurements carried out by Harris and Pedersen (1998) of ground temperature in blocky open-work debris and in adjacent areas of fine-grained mineral soil demonstrated that the former experiences much greater seasonal temperature variation and generally lower temperatures throughout the year (Figure 4.11). These authors estimated that in areas of limited snow-lie, MAGTs in blocky debris could be 4.0–7.0 °C lower than in fine-grained mineral soils. A similar contrast was found by Gorbunov *et al.* (2004), who showed that under favourable conditions (limited snowcover, shaded hillslopes and abundant moss cover) MAGTs in blocky debris could be 2.5–4.0 °C lower than MAATs. Such findings imply that equilibrium permafrost may exist within or beneath coarse debris even when MAATs exceed 0 °C.

This phenomenon is evident at the foot of blocky talus (scree) slopes, where permafrost sometimes occurs hundreds of metres below the altitudinal permafrost limit, even though the upper part of the slope is permafrost-free, and under high-level blockfields or blockslopes

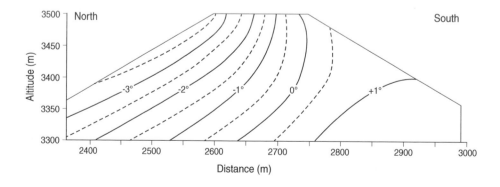

Figure 4.10 Modelled two-dimensional steady-state isotherms of ground temperature distribution in an idealized alpine summit ridge based on borehole temperature data measured on the summit of Stockhorn, Switzerland. Permafrost is absent from the south face but underlies the north face. *Source:* Gruber *et al.* (2004b), Reproduced with permission of Wiley.

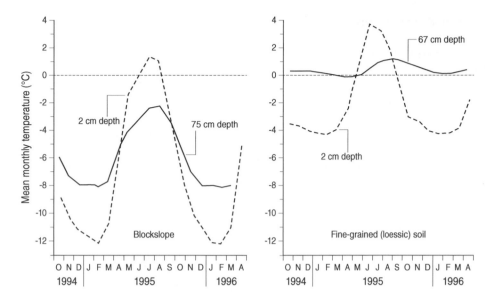

Figure 4.11 Monthly temperature regime measured near the surface and at depth on Plateau Mountain, Alberta, Canada. The blockslope temperatures are not only much lower than those for fine-grained loessic soil, but also display much greater annual amplitude, especially at depth. *Source:* Adapted from Harris and Pedersen (1998). Reproduced with permission of Wiley.

in otherwise permafrost-free terrain. There is limited consensus regarding the explanation of such negative temperature anomalies. One possible explanation, the Balch effect (settling of cold, dense air within inter-clast voids) is rejected by most investigators as it fails to explain seasonal temperature variations at depth (Figure 4.11). Harris and Pedersen (1998) favoured continuous air exchange with the surface during winter (Figure 4.12a) and cooling of subsurface boulders by evaporation or water and sublimation of ice during the summer, but this explanation is valid only where winter snowcover is limited. For slopes covered by thick seasonal snowcover, an alternative explanation is that cold, dense air is advected downslope within the openwork debris cover during winter, displacing relatively warm air that migrates through the debris cover towards the top of the slope (Sawada *et al.*, 2003; Figure 4.12b). An interesting variant of this explanation is the 'chimney effect' advocated by Delaloye and Lambiel (2005) and supported by several subsequent studies (Lambiel and Pieracci, 2008; Phillips *et al.*, 2009; Stiegler *et al.* 2014). This interpretation suggests that during the period of winter snowcover, relatively warm air moves upslope through inter-debris voids under the snowcover, escaping through funnels melted through the snow near the top of the slope and drawing cold air into the lower parts of the slope through the overlying snow (Figure 4.12c). In summer, the movement of air is reversed, with relatively dense cold air moving through the debris layer to the slope foot, limiting the depth of thaw. A different (but possibly complementary) model has been proposed by Juliussen and Humlum (2008), who attributed cooling of

openwork blockfields on Norwegian mountains to conductive heat loss through boulders protruding through the winter snowcover (Figure 4.12d), leading to MAGTs 1.3–2.0 °C lower than in adjacent terrain.

It seems feasible that all these mechanisms operate, but under different circumstances governed by the duration, thickness and permeability of winter snowcover, the length and gradient of the slope and the thickness of openwork blocky debris. Irrespective of mechanism, they create negative thermal anomalies that complicate the relationship of permafrost to surface energy exchange, permitting the formation and survival of permafrost even in areas where MAAT exceeds 0 °C. At the foot of an undercooled scree slope at 990 m altitude in Austria, for example, perennially frozen sediments 5–20 m thick occur below depths of 1–3 m even though MAAT is approximately +4.7 °C (Stiegler *et al.*, 2014).

A related phenomenon in karstic (limestone and dolomite) mountains is the preservation of permafrost in caves in the form of *cave ice*, which can occur even where MAATs are several degrees above freezing (Figure 4.13). The primary process responsible for the preservation of cave ice is density-driven movement of cold air into cave systems in winter. Cold, dense air persists in the caves through the summer months, so that the air temperature within caves remains colder than that outside. Cave ice forms both from freezing of water infiltrating into caves and from diagenesis (firnification) of snow entering caves in winter (Luetscher *et al.*, 2005), and because of latent heat effects may persist for long periods even under conditions of climate warming.

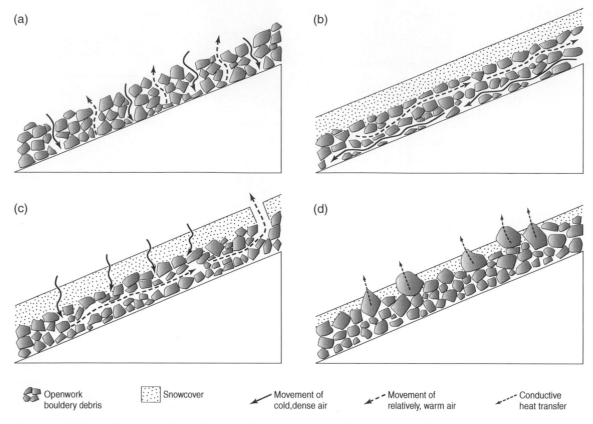

(a)

(b)

(c)

(d)

| Openwork bouldery debris | Snowcover | ⟵ Movement of cold, dense air | ⟵--- Movement of relatively, warm air | ⟵--- Conductive heat transfer |

Figure 4.12 Models for undercooling of openwork bouldery debris. (a) Free exchange of air with atmosphere during winter under snow-free conditions. (b) Downslope advection of cold air and compensatory upslope advection of warm air under snowcover. (c) The 'chimney effect': upslope advection of relatively warm air draws in cold air through overlying snowcover. (d) Cooling by conductive heat transfer through boulders protruding through shallow snowcover.

Figure 4.13 Cave ice in a limestone cave, Picos de Europa, northern Spain. A colour version of this figure appears in the plates section. *Source:* Courtesy of Bernard Hivert.

4.6 Permafrost–glacier Interactions

Glaciologists employ a threefold thermal classification of glaciers, based on whether the ice is at or below pressure-melting point. *Temperate* or *warm-based* glaciers are those in which the ice is at pressure-melting point throughout, apart from a surface layer that experiences seasonal temperature fluctuations; *cold* or *cold-based glaciers* are everywhere below pressure-melting point and frozen to the underlying substrate; and *polythermal glaciers* contain both 'warm' and 'cold' ice, usually being cold-based around their margins but with an up-glacier zone of warm ice. The distribution of permafrost under present-day glaciers and ice sheets is poorly documented, but the occurrence of cold-based ice under parts of the Greenland and Antarctic Ice Sheets and under polar ice caps implies that the underlying substrate is permafrozen. Numerical modelling of Late Pleistocene ice sheets suggests that these were also extensively cold-based (Marshall and Clark, 2002; Kleman and Glasser, 2007), and therefore underlain by widespread permafrost. Subglacial permafrost in most likely to occur (i) in a broad zone (tens to hundreds of kilometres wide) within the margins of ice sheets and ice caps, where the ice cover is relatively thin, (ii) in the interior of large ice masses and at zones of flow divergence where limited strain heat is generated by ice deformation, and (iii) under the margins of polythermal glaciers. Ice-rich permafrost buried by the advance of warm-based glacier ice, however, is likely to thaw because of heat generated by basal sliding and advected into the ground by subglacial meltwater.

Cold-based glaciers move almost exclusively through internal ice deformation, and were traditionally viewed as geomorphologically inactive. This view rested on the assumption that liquid water is absent at the glacier sole, so that basal sliding and subglacial sediment deformation or entrainment do not occur. More recent work, admirably summarised by Waller *et al.* (2012), demonstrates that this is not the case. Because of the presence of nanometre-thick films of premelted water within frozen sediments (especially fine-grained sediments), ice-rich permafrost is never completely frozen, even at temperatures well below 0 °C (Chapter 3). Ice-rich permafrozen sediments may therefore experience deformation or shear under cold-based glacier ice. A useful way of envisaging this is to consider that debris-rich ice at the glacier sole and ice-rich sediment in the upper part of the substrate have similar strength properties, so the zone of ice deformation effectively extends down into the underlying substrate (Figure 4.14).

The style of subglacial sediment deformation is determined by the permafrost temperature, liquid water content and sediment granulometry. Ductile deformation

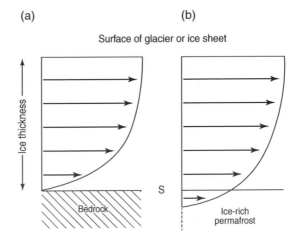

(a)　　　　　　　　　(b)

Surface of glacier or ice sheet

Ice thickness

S

Bedrock

Ice-rich permafrost

Figure 4.14 (a) Velocity profile of a cold-based glacier moving over a rigid (bedrock) bed; movement occurs through internal deformation of the ice. (b) Velocity profile of the same glacier over ice-rich permafrozen sediments; the velocity profile extends into the substrate, causing deformation of the upper part of the permafrost.

is favoured in clay-rich sediments (with relatively high premelted water content) and in warm permafrost (because liquid water content increases with temperature), whereas brittle shear deformation is characteristic of coarse-grained sediments or lower permafrost temperatures. This contrast in response is illustrated by the presence of intact frozen sand bodies termed *sand intraclasts* preserved within subglacially deformed clay-rich sediments. Moreover, the dominant style of deformation may change through time: glacier advance across cold permafrost may trigger an initial phase of brittle shear deformation, succeeded by dominantly ductile deformation as the subglacial permafrost warms (Murton *et al.*, 2004). Ice lenses may provide planes of weakness within deforming subglacial permafrost (Fitzsimons, 2006), or the zone of glacitectonic deformation may extend much deeper into the substrate, allowing *en masse* deformation, shear and entrainment of frozen subglacial sediment above deep-seated planes of weakness such as ice-rich, ice-poor or clay-rich layers. Astakhov *et al.* (1996) have described thicknesses of tens of metres of glacially deformed permafrost in western Siberia, and Murton *et al.* (2004) have reported thicknesses of at least 5–20 m in the western Canadian Arctic. Such thicknesses are much greater than those reported for deformation of unfrozen sediments under warm-based glaciers, because the much more rigid permafrozen sediments have permitted transmission of shear stresses to greater depths. Typical deformation structures include recumbent and S-shaped folds, sediment intraclasts and sand layers, shear structures, angular unconformities (décollement surfaces) and ice clasts quarried from subjacent massive ice layers (Waller *et al.*, 2012).

Permafrost is also implicated in the formation of certain glacial landforms. Prominent amongst these are large multi-crested push moraines, such as those that front many glaciers in Svalbard (Bennett, 2001). The enhanced rigidity of permafrozen sediments allows stress propagation on to the proglacial zone and stacking of successive rafts of frozen sediment at the ice margin during episodes of ice margin advance (Figure 4.15). In areas of shallow permafrost, high pore-water pressures at the base of the permafrost may create a plane of décollement (thrusting) along which displacement of frozen sediment blocks occurs (Boulton, 1999). Melt-out of debris-rich basal ice following retreat of cold-based or polythermal glaciers produces chaotic hummocky moraines, ice-cored sediment cones or 'controlled' (aligned) ablation moraines where melt-out of debris-rich structures imparts linearity to the moraines. Whilst geomorphological and stratigraphic records of mid-latitude ice-sheet glaciation exhibit a bias towards interpretations based on the assumption of warm-based glaciation (Kleman and Glasser, 2007), recent research on glacier–permafrost interactions indicates the need for greater appreciation of the role of subglacial permafrost deformation and sediment entrainment under cold-based glacier ice, and the characteristics of subglacial permafrost deformation structures and permafrost-influenced glacial landforms.

4.7 The Geomorphic Importance of Permafrost

The presence of permafrost profoundly influences the geomorphological processes and resulting landforms of cold environments. Ice-bearing permafrost is effectively impermeable and limits water movement and groundwater storage to the active layer. This ensures that moisture content in the active layer remains generally high during the summer thaw season, promoting the development of distinctive freeze–thaw processes and structures (Chapter 9); however, confinement of groundwater to the active layer also limits baseflow in rivers in the continuous permafrost zone (Chapter 13). Permafrost preserves perennial ice in the form of, for example, ice lenses, massive ground ice and wedge ice (Chapter 5). Such ice bodies not only underlie distinctive landforms, such as thermal contraction crack polygons, pingos and palsas (Chapters 6 and 7), but also, on thawing, create a distinctive family of landforms collectively referred to as *thermokarst* (Chapter 8). Thaw and aggradation of ice-rich permafrost permits reconstruction of past environmental changes through the study of the resulting ice structures (Chapter 5). On coasts and riverbanks, ice-rich permafrost is vulnerable to thermal and mechanical erosion, resulting in rapid bank erosion or coastline recession, and concomitant

Figure 4.15 Multi-crested push moraines at Bergmesterbreen, a polythermal glacier on Svalbard. The rigidity of the frozen sediments has caused stacking of successive rafts of frozen sediment at the glacier margin during periods of ice-margin advance.

release of sediment into rivers and the nearshore zone (Chapters 13 and 15). On slopes, permafrost limits the depth of soil movement through solifluction, yet forms a potential shear surface for active-layer detachment failures (Chapter 11), and ice-rich permafrost may itself deform downslope under gravity or loading, the fundamental process underpinning the formation and movement of rock glaciers (Chapter 12). Finally, some soil structures and landforms associated with permafrost are preserved in permafrost-free mid-latitude areas, providing a valuable record of former periglacial conditions (Chapter 16). All of these properties and effects are considered in greater detail in subsequent chapters, and the effects of climate change on the thermal regime and distribution of permafrost are discussed in Chapter 17.

5

Ground Ice and Cryostratigraphy

5.1 Introduction

The term *ground ice* incorporates all forms of ice that occur within permafrost or seasonally frozen ground. In the latter, as we saw in Chapter 3, ground ice commonly occurs in soil pores or forms discrete lenses or bands of segregated ice. In permafrost, ground ice exists in a large variety of additional forms, such as vertical veins and wedges, massive ice bodies, sills and dykes (Figure 5.1). Structurally, ground ice can be considered as a kind of mineral or a rock that is stable at subsurface temperatures below its melting point: as a mineral, where individual ice crystals or crystal aggregates exist within sediments; as a rock, where it forms a stratigraphically distinct body within permafrozen sediments or bedrock. *Cryostratigraphy* is the study of stratigraphic units in permafrost, based mainly on ground-ice characteristics, aimed at unravelling aspects of its history such as past periods of permafrost degradation and aggradation.

In this chapter, we consider first the genetic classification of ground ice and the structural and textural characteristics of different ground-ice types, before outlining the methods employed to describe and interpret ice bodies exposed in vertical sections and the principles of cryostratigraphy. We then focus on ground-ice distribution in the transition zone below the active layer, the origin and significance of massive bodies of ground ice and the ice complexes (*yedoma*) that underlie vast areas of Siberia and Alaska. The characteristics of particular landforms associated with ground ice are developed in Chapters 6 and 7, and Chapter 8 considers the suite of *thermokarst* landforms and processes that are associated with thaw of ice-rich permafrost. French and Shur (2010) and Murton (2013c) provide authoritative introductions to the origins and characteristics of ground ice and the governing principles of cryostratigraphy.

5.2 Genetic Classification of Ground Ice

In 1972, the Canadian geomorphologist Ross Mackay published a seminal paper entitled 'The World of Underground Ice', in which he introduced a systematic classification of ground ice based on the origin of the water prior to freezing, the principal transfer processes and the resultant ground ice forms. A modified and expanded version of this classification compiled by Murton (2013c) is shown in Figure 5.2. The first-order subdivision of classes is into *intrasedimental ice*, formed when water freezes or ice sublimates within sediments or rock, and *buried ice*, which originates through burial of a body of ice (such as glacier ice or lake ice) under younger sediments. Most ground ice originates either *syngenetically*, implying that the ice forms as a sediment body accumulates, for example on an aggrading floodplain or within accumulating windblown silt (loess) deposits, so the ice is roughly the same age as the sediment it occupies, or *epigenetically*, where the ice forms in an existing stable substrate, so the ice is younger than the ground it occupies. Ice that forms within upwards-aggrading permafrost is sometimes termed *aggradational ice*. As in Mackay's (1972) classification, the various types of ground ice identified in Figure 5.2 are related first to the moisture sources (atmospheric water, surface water, subsurface water and surface ice) and second to the principal processes of moisture transfer (or ice burial, in the case of buried ice).

5.2.1 Pore Ice

Pore ice, sometimes termed *interstitial ice* or 'cement' ice, forms where there is *in situ* freezing of water within soil pores, or where water drawn through the soil to feed a developing ice lens (Chapter 3) freezes within a frozen fringe. In practice, the nature of pore ice depends on the

Periglacial Geomorphology, First Edition. Colin K. Ballantyne.
© 2018 John Wiley & Sons Ltd. Published 2018 by John Wiley & Sons Ltd.

Figure 5.1 Ground ice: ice wedges at the top of the photograph penetrate downwards into ice-rich frozen silts that overlie a body of stratified massive ice. A colour version of this figure appears in the plates section. *Source:* Courtesy of Matthias Siewert.

relative abundance of ice and soil grains, and on the size of the pores in the soil. Yershov (1998) described four categories of what he termed 'ice cement' in frozen soil: 'basal' ice, where soil particles are surrounded by a matrix of ice, so that grain-to-grain contacts are lost; pore ice, which completely fills soil pores but preserves most grain-to-grain contacts; 'film' ice, where ice coats the pore walls but part of the pore space is unfilled; and 'contact' ice, where ice is restricted to intergranular contacts (Figure 5.3). When water freezes in soil pores, the cohesion of the soil increases markedly due to 'cementation' (binding) between soil and ice particles, even though a thin film of premelted water separates the ice from the particles (Williams and Smith, 1989). In consequence, the strength of the frozen soil tends to increase up to the point where ice completely fills pore spaces but does not force intergranular contacts apart. At higher ice contents ('basal' ice), however, stress operating on frozen soil is transmitted through the ice rather than intergranular contacts, and the strength of the frozen soil declines. Of the types of pore ice illustrated in Figure 5.3, only 'basal' ice yields excess water on thawing.

5.2.2 Closed Cavity Ice

Closed cavity ice is sometimes termed *sublimation ice*. The term 'sublimation' strictly refers to a transition of ice to water vapour that does not pass through the liquid phase, but it is often also employed to denote the reverse process as well, and it is in this sense that it is employed here. Sublimation ice forms in open cavities in permafrost, notably thermal contraction cracks, but may also form in closed cavities through water vapour diffusion, forming delicate crystals that coat the cavity walls.

5.2.3 Segregated Ice

As we saw in Chapter 3, segregated ice forms through suction of water to the freezing front, where thin lenses or layers of ice grow, pushing apart the overlying and underlying soil to cause frost heave. In areas of shallow frost penetration, however, ice segregation may take the form of near-surface *needle ice*: long, thin, contiguous ice crystals that develop at or near the ground surface in response to short-term (often nocturnal) ground freezing (Figure 5.4). Such crystals grow perpendicular to the ground surface, often pushing up a thin layer of soil, clasts and organic matter. Needle ice forms through upwards migration of soil water towards a stationary freezing front in the top few centimetres of soil. Critical conditions for its formation are: (i) a rapid drop in air and ground surface temperature, often due to nocturnal clear-sky radiative cooling; (ii) soil moisture availability, to feed ice crystal growth; and (iii) frost-susceptible but permeable soil that permits rapid water migration to the freezing front. Initial ice nucleation occurs as temperatures near the surface drop to between $-1\,°C$ and $-2\,°C$, and crystal growth continues provided that the ground surface temperature remains slightly below $0\,°C$ and there is a continued supply of water so that latent heat released by freezing of water at the base of crystals is sufficient to prevent deeper frost penetration. Needle-ice crystals grow rapidly to lengths of up to 5 cm (and exceptionally to 10 cm) in a single night, and may continue to grow over longer periods if air temperatures remain slightly below $0\,°C$. Where two or more periods of cooling are uninterrupted by thaw, needle ice forms layers, often separated by a very thin soil layer, with each layer representing a cooling period. Because needle ice formation usually occurs through shallow overnight freezing of moist soil, it tends to be most common in maritime subarctic uplands and on mid-latitude mountains, and is the dominant form of ice segregation on high subtropical mountains (Lawler, 1988; Francou and Bertran, 1997; Holness, 2004), but it is usually a short-lived phenomenon.

Ice lenses and ice layers form where subzero ground surface temperatures persist for longer periods, driving the freezing front deeper into seasonally frozen ground, but they also exist within permafrost, particularly

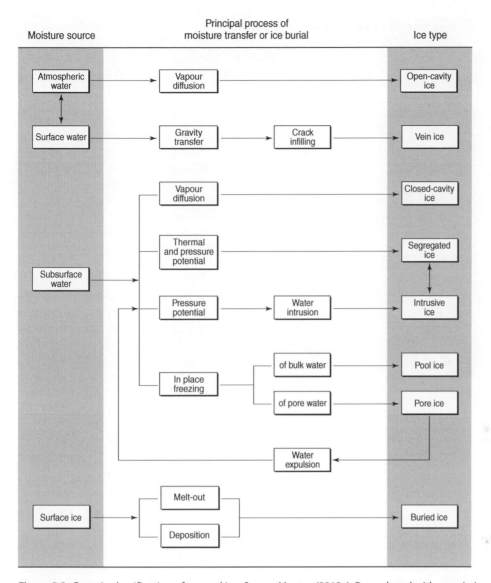

Figure 5.2 Genetic classification of ground ice. *Source:* Murton (2013c). Reproduced with permission of Elsevier.

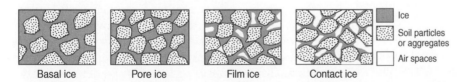

Figure 5.3 Classification of types of pore ice ('ice cement') in frozen soils. Ice separates grain-to-grain contacts in 'basal' ice, but not in 'pore', 'film' or 'contact' ice. *Source:* Yershov (1998). Reproduced with permission of Cambridge University Press.

within the transition zone just below the permafrost table. Because lenses and layers of segregated ice form parallel to the freezing plane, they tend to be roughly parallel to the ground surface, but dipping lenses can form where the freezing plane is tilted, or where the loci of lens development are determined by the structural properties of the host sediment. Under particular combinations of overburden pressure and freezing rate (Chapter 3), successive thin lenses develop in freezing or frozen ground, but when freezing is slow, a much thicker single stable lens may develop (Rempel *et al.*, 2004). Consequently, the dimensions of segregation ice lenses and layers vary greatly: in seasonally frozen ground, they are typically less than 10 mm and sometimes

Figure 5.4 Needle ice crystals, 15 cm long. Segmentation within the crystals indicates more than one period of growth. At the top of the crystals is soil that has been pushed up by crystal growth. The soil at the bottom of the crystals represents the ground in which crystal growth was rooted.

less than 1 mm thick; in permafrost, however, segregated ice occurs in a wide range of sizes, from *micro-lenticular* features (<0.1 mm in thickness and up to a few millimetres long; Bray *et al.*, 2006) to *massive ice* bodies (up to tens of metres thick).

Segregated ice lenses and ice layers typically form in porous, frost-susceptible soil and bedrock, though exceptionally they may develop in sandy sediments under high pore-water pressures (Mackay, 1985). In appearance, they may be transparent or opaque, due to the inclusion of air bubbles, notably tube-shaped bubbles that form normal to the freezing front and are consequently vertical in horizontal lenses. However, because lenses and layers of ice may incorporate mineral particles, clasts and organic matter during freezing of soil, there is a continuum of appearance from pure ice to dirty ice to ice-cemented ground in which individual lenses may be difficult to distinguish. The main crystal axis of segregation ice is usually normal to the plane of the ice lens or layer, and other distinguishing features may include icy coatings on clasts, broken soil peds that match across a lens and air bubbles rising upwards from soil inclusions.

5.2.4 Intrusive Ice

Intrusive ice forms through freezing of water that has been injected under hydrostatic or hydraulic pressure into frozen sediments or bedrock. Freezing of water-saturated, coarse-grained non-frost-susceptible sediments (such as sand, or mixed sand and gravel) results in *pore-water expulsion*. In a closed system from which water cannot escape, this generates hydrostatic pressures capable of fracturing frozen ground, so that water is driven into fractures, where it freezes to form intrusive ice. Hydraulic pressures are generated by groundwater flow under a downslope hydraulic pressure gradient in a confined aquifer beneath or within permafrost, and may also result in hydrofracture of frozen ground and injection of water to form intrusive ice. The stresses generated by water injection and subsequent freezing can result in uplift of the ground surface, particularly in the form of *pingos* (Chapter 7), but such uplift is not generally considered to be a form of frost heave.

Where it forms concordantly with the strata of the host sediment or bedrock, intrusive ice typically creates sills (low-angle, tabular ice bodies) or laccoliths (domed ice bodies); vertical or steeply dipping ice intrusions create dykes, analogous to the forms produced by igneous intrusions. The central parts of ice intrusions often lack incorporated material and hence are pure and fairly transparent, but sediment inclusions may be present at the base of intrusions in the form of clasts, or of layers or tongues of soil particles incorporated in the ice during water injection and freezing. Air bubbles are common in ice intrusions, and may be grouped into concentrations or aligned parallel to the upper margin of ice sills.

Ice dykes studied by Mackay (1989) and Mackay and Dallimore (1992) exhibited several distinguishing features produced by inwards freezing from fracture walls. These included marginal zones of small ice crystals formed during rapid freezing along dyke boundaries, candle-shaped ice crystals and bubble trains orientated at right angles to the opposing dyke walls, a central bubble-rich seam indicating two-sided freezing and a zone of discoloured (brown or yellow) ice produced by solute rejection by inwards-freezing ice and solute concentration in the last water to undergo freezing. It is likely that similar features typify ice sills, though in this case their alignment would represent two-sided freezing from the underlying and overlying permafrost. The characteristics of intrusive ice forming laccoliths under pingos are described in Chapter 7.

5.2.5 Wedge Ice and Vein Ice

Ice wedges develop through repeated thermal contraction cracking in permafrost, and infill of open cracks with water or sublimation ice crystals (Chapter 6). When a thermal contraction crack initially opens in permafrost, water from the surface enters the crack and freezes to form a single narrow, downwards-tapering *ice vein*. Subsequent ground cracking at the same site and freezing of meltwater in the crack produces an adjacent ice vein, and the process is repeated until ultimately an ice wedge develops from lateral increments of vein ice (Figure 5.1). The dominant structural feature of most ice wedges is therefore downwards-converging subvertical ice foliations or laminae, typically 0.5–5.0 mm wide, each of which represents a component ice vein. Adjacent foliations may be concordant or cross-cutting, and may be emphasized by thin bands of mineral or organic matter that have accumulated between the original ice veins. Ice crystals within individual ice veins grow horizontally inwards from crack walls, sometimes meeting at a central zone marked by a concentration of expelled mineral grains or gas bubbles. On the scale of the entire wedge, however, progressive recrystallization of ice results in larger crystals near the wedge margins (where the ice is older) than in the centre. Yershov (1998) also noted that individual ice crystals in syngenetic ice wedges tend to be two to three times larger than those in epigenetic wedges. Air bubbles formed through release of dissolved gas during freezing (or through trapping of air between sublimation ice crystals) are abundant, so pure wedge ice appears white rather than translucent; many wedges, however, contain abundant organic and/or mineral inclusions, giving them a dirty brown or grey appearance.

5.2.6 Dilation Crack Ice

Dilation cracks are formed through the updoming of permafrost by pingo growth (Chapter 7), which generates radial tensile stresses that are relieved by fracturing of the permafrost and overlying active layer. Surface water entering such dilation cracks in permafrost freezes against the crack walls to form dilation crack ice (Mackay, 1985). A characteristic feature of such ice is pronounced vertical banding, with individual bands typically 10–100 mm wide and often distinguished by discoloration due to incorporation of organic impurities. Inwards freezing from crack walls is represented by horizontally aligned ice crystals and bubble trails, and a median bubble-rich vertical seam marking the last water to freeze.

5.2.7 Pool Ice

Pool ice, sometimes called *thermokarst-cave ice* (Shur *et al.*, 2004; Bray *et al.*, 2006) forms through freezing of water in underground cavities and tunnels formed by the melt of other forms of ground ice, particularly those produced by thermal erosion of ice wedges. Pool ice is characteristically pure and translucent, and may be massive or layered. It is frequently characterized by columnar ice crystals, signifying the freezing of bulk water, and where the ice has formed through inwards freezing from cavity or tunnel walls, such ice columns and associated bubble trains radiate outwards from the centre of the cavity (Kotler and Burn, 2000). Sediment deposited on tunnel floors by flowing water or roof collapse may separate pool ice bodies from the truncated tops of ice wedges (Murton and French, 1993b).

5.2.8 Buried ice

Buried ice forms where a surface ice body is buried under a sediment cover and preserved by upwards-aggrading permafrost. Such ice may be of glacial origin, buried under a surface cover of bouldery debris, ablation (melt-out) till or glacifluvial deposits, or of nonglacial origin (lake, sea, river or icing ice, or the icy core of perennial snowbanks), covered by glacifluvial, fluvial, aeolian, colluvial, lacustrine or marine deposits. However, unless the sediment cover exceeds the depth of the active layer, buried ice may thaw within months to decades, depending on the initial thickness of the ice body and that of the overlying sediment. Buried ice is therefore most likely to be preserved where the overlying sediment continues to aggrade rapidly after initial ice burial.

The structural characteristics of buried ice are those of the ice body prior to burial. Thus, buried lake, sea and river ice, all of which represent freezing of bulk water, are characterized by vertically-aligned elongate ice crystals ('candle ice') produced by surface-downwards freezing. Buried snowbank ice tends to be opaque and milky white with granular crystals and vertically orientated air bubbles; sediment (usually of aeolian origin) is either disseminated throughout the ice and/or present in thin bands representing annual accumulation on the former snow surface; slumping of sediment from backwalls and steep bluffs is a one cause of snowbank burial (Christiansen, 1998a). Buried glacier ice consists of polycrystalline ice, in which crystal size generally increases with depth. Ice crystal fabric (the aggregate orientation of crystals) within glaciers may be altered by steady applied stresses, resulting in reorientation of crystals parallel to the direction of former glacier movement, but crystal fabric may vary markedly within glacier ice (Benn and Evans, 2010). In temperate or polythermal glaciers, where basal ice is at pressure melting point, there is often a debris-rich regelation layer produced by melting of ice up-glacier from obstacles, where pressure is high, and refreezing in a zone of reduced pressure down-glacier. Cold-based ice in polar glaciers, however, tends to lack a basal regelation layer.

5.2.9 Transitional, Compound and Modified Ice Types

The genetic classification encapsulated in Figure 5.2 often represents end members of a continuum of ground ice forms that may be contiguous or represent changes in formative conditions. As outlined in Chapter 7, for example, 'pingo ice' incorporates not only the intrusive ice that forms the core of pingos, but also varying components of segregated ice, pore ice, dilation crack ice and even wedge ice (Mackay, 1985). Similarly, poorly drained terrain crossed by thermal contraction crack polygons (Chapter 6) may contain intersecting bodies of pool and wedge ice. Hybrid segregated–intrusive ice forms under conditions of changing water pressure, from the low pressures required for ice segregation to the high hydrostatic or hydraulic pressure required to overcome overburden pressure and form intrusive ice during the formation of massive intrasedimental ice bodies (Mackay, 1989; see below).

Ground ice bodies may also be modified by melting and recrystallization, which tends to produce larger crystals and may alter crystal orientation, or by deformation. The latter may occur through permafrost creep, where ice-rich permafrost undergoes slow downslope deformation (Chapter 11) or unloading of overburden, which can result in up-arching or fracture of massive ice bodies (Mackay, 1985, 1990b; Mackay and Dallimore, 1992). Overriding of ice-rich ground by glaciers and ice sheets causes fracture, folding and erosion, and incorporation of ground ice in the glacier sole; glacier advance over cold permafrost may trigger an initial phase of brittle shear deformation, succeeded by ductile deformation as the permafrost warms (Murton *et al.*, 2004, 2005; Murton, 2005; Waller *et al.*, 2012).

5.3 Description of Ground Ice

The characteristics and properties of ground ice can be described quantitatively in terms of the mass or volume of ice relative to that of the host material, or in terms of *cryostructures* (generally termed *cryotextures* in the Russian literature) that represent the shape of ice bodies and their relationship to the host sediment.

5.3.1 Ice Content

The ice content of frozen ground can be assessed in three ways: by assessing the mass of ice present in a sample relative to that of the host material (*gravimetric ice content (i_G)*); by measuring the volume of ice in a sample relative to that of the whole sample (*volumetric ice content (i_V)*) or by measuring the *excess ice content* (i_E), defined as the volume of ice that exceeds the total pore volume of the ground under unfrozen conditions. All three are generally expressed as percentages, thus:

$$i_G = \left(M_i / M_s \right) \times 100\% \qquad (5.1)$$

$$i_V = \left(V_i / V_{sample} \right) \times 100\% \qquad (5.2)$$

$$i_E = \left(V_W - V_P \right) \times 100\% \qquad (5.3)$$

where M_i and M_s are respectively the mass of ice contained in a sample and the mass of the dry sample, V_i and V_{sample} are the volume of ice in a sample and the volume of the whole sample, and V_W and V_P are the volume of water in a thawed sample and the total volume of pore spaces in the sample. Murton and French (1994), for example, employed volumetric ice content in a fivefold classification of frozen sediment: ice-poor sediment (<25% ice by volume), ice-rich sediment (25–50%), sediment-rich ice (50–75%), sediment-poor ice (75–100%) and pure ice (100%). Excess ice content is a particularly useful measure: ground that contains excess ice is termed *ice-rich* or *ice-supersaturated*, whereas ground containing only pore ice is usually referred to as *ice-cemented* or *ice-bonded*. Excess ice contents of 15–50% are common in both seasonally frozen and permafrozen soils, and ice content may reach 70–80% in permafrost (Pollard and French, 1980; Hodgson and Nixon, 1998; Schirrmeister *et al.*, 2011, 2013). Ice-rich permafrost is termed *thaw-sensitive*, and is susceptible to subsidence on thawing owing to the volume decrease that accompanies the thaw of excess ice and drainage of excess water that such thaw produces.

5.3.2 Cryostructures in Unlithified Sediments

Various schemes have been proposed, particularly by Russian scientists, to describe the structure of ice bodies in soils and rocks (e.g. Yershov, 1998; Melnikov and Spesivtsev, 2000; Jorgenson *et al.*, 2001; Dubikov, 2002; Siegert *et al.*, 2002). In practice, researchers have often adopted elements of various schemes to provide an optimal system for describing the cryostructures they have observed (e.g. Kotler and Burn, 2000; Kanevskiy *et al.*, 2011, 2014). That depicted in Table 5.1 and Figure 5.5 represents a modified version of a scheme originally proposed by Murton and French (1994) and elaborated by Shur and Jorgenson (1998) and Murton (2013c) for cryostructures developed in frozen sediments. An advantage of this scheme is that it permits labelling of units exposed in section, much like the lithofacies codes employed in the description of glacigenic sediments (Benn and Evans, 2010, 365–368). Thus, a planar horizontal lenticular ice body is coded Lhp, and a dirty ice wedge is coded Swd. *Pore cryostructures* typically form in voids between and around mineral grains, particularly in sands and gravels, and *organo-matrix cryostructures* are similar but develop through freezing of peat or peaty soil. *Crustal cryostructures* take the form of a sheath of ice encasing frost-susceptible clasts, typically within silt-rich sediments just below the permafrost table. *Vein cryostructures* are usually vertical or steeply inclined, and typically form in thermal contraction cracks, shrinkage cracks or dilation cracks. Lenses of segregated ice in frozen soil form a

Table 5.1 Cryostructures in frozen ground.

Cryostructure	Primary property or shape	Secondary property	Genetic ice type	Material commonly hosting ice	Typical freezing processes
Pore (P)	Nonvisible (n) Visible (v)		Pore ice	Sand, gravel, porous bedrock	*In situ* freezing of soil or rock pores
Organic-matrix (O)	Nonvisible (n) Visible (v)		Pore ice	Organic-rich soil or peat	*In situ* freezing of peaty soil or peat
Crustal	Entire (e) Partial (p)		Segregated ice	Silty sediment	Ice segregation around clasts
Vein (V)	Vertical (v) Irregular (i)		Segregated ice, wedge ice or dilation crack ice	Various	Freezing in contraction or dilation cracks
Lenticular (L)	Horizontal (h) Inclined (i) Cross-stratified (c) Grouped (g)	Planar (p) Wavy (w)	Segregated ice	Frost-susceptible soil or bedrock	Ice segregation
Bedded or layered (B)	Layer density: Sparse (<5%) (s) Medium (m) Dense (25–50%) (d)	Planar (p) Wavy (w) Curved (c) Ruptured (r)	Segregated ice or intrusive ice (thin ice sills or dikes)	Frost-susceptible soil or bedrock	Ice segregation or injection and freezing of pressurized water
Reticulate (R)	Trapezoidal or prismatic (t) Lattice or blocky (l) Foliated or platy (f)	Unit width: Fine (<5 mm) Medium Coarse (>10 mm)	Segregated ice or intrusive ice	Silt-clay mixtures, especially above massive ice	Freezing of cracks within saturated sediments
Ataxitic (A)	50–75% ice (m) 75–95% ice (d) 95–99% ice (v)	Round (r) Angular (a) Blocky (b)	Segregated, intrusive, buried, wedge or dilation-crack ice	Individual grains, grain aggregates and clasts suspended in ice	Ice segregation, water injection or freezing in contraction or dilation cracks
Solid (S) (>10 cm thick bodies of ice)	Sheet, horizontally stratified (h) Wedge, vertically stratified (w)	Clear (c) Opaque (o) Dirty (d) Porous (p)	Segregated, intrusive, wedge, pool, dilation- crack or buried ice	Massive ice, icy sediments, ice and composite wedges, pingo ice	Ice segregation, injection of water, ice wedge formation, crack infill, freezing of underground pools, burial of surface ice

Source: Adapted from Murton (2013c). Reproduced with permission of Elsevier.

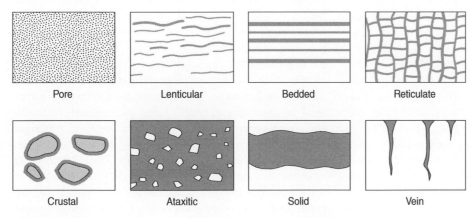

Pore   Lenticular   Bedded   Reticulate

Crustal   Ataxitic   Solid   Vein

Figure 5.5 Schematic illustration of cryostructures, following the scheme in Table 5.1. Ice is shown in black, and the grey objects in the crustal cryostructure are frost-susceptible clasts.

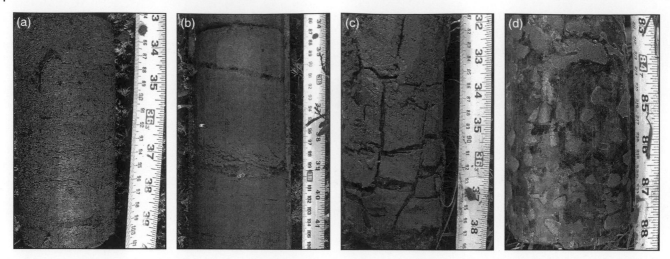

Figure 5.6 Cryostructures in cores extracted from ice-rich permafrost: (a) lenticular; (b) banded; (c) reticulate; (d) ataxitic. A colour version of this figure appears in the plates section. *Source:* Courtesy of Yuri Shur.

discontinuous *lenticular cryostructure* (Figure 5.6a) that can be described in terms of the inclination, thickness, spacing, length and shape of component lenses, and their relationship to one another (parallel, nonparallel, cross-cutting or grouped). Ice lenses typically occur in frost-susceptible soil or bedrock, and their alignment is usually normal to the freezing front at the time of lens formation (Chapter 3), though it may follow structural weaknesses in stratified sediments. In contrast, a *bedded* or *layered cryostructure* (Figure 5.6b) takes the form of continuous thin layers of ice, sediment or both. Bedded cryostructures also form through ice segregation in frost-susceptible soil, particularly in the ice-rich transition zone at the top of permafrost, though they may also represent injection and freezing of pressurized water to form thin sills in permafrost. Rhythmically-organized layered and lenticular cryostructures appear to be typical of syngenetic permafrost that has aggraded upwards into an accumulating sediment body. Research conducted in the CRREL permafrost tunnel in Alaska, for example, has identified three cryofacies characteristic of syngenetic permafrost: (i) a micro-lenticular cryostructure of short ice lenses <0.5 mm thick; (ii) a layered cryostructure comprising repeated layers of ice 2–10 cm thick and 2–5 cm apart, which often coexists with the micro-lenticular cryostructure; and (iii) a lenticular-layered cryostructure comprising ice lenses 0.5–1.5 mm thick and up to 10 mm long that form continuous layers with soil inclusions (Shur *et al.*, 2004; Bray *et al.*, 2006; Kanevskiy *et al.*, 2008).

Reticulate cryostructures (Figure 5.6c) usually take the form of a 3D network of vertical or subvertical ice veins interconnected with horizontal or subhorizontal ice bands that isolate orthorhombic blocks of clay- or silt-rich sediment, though irregular reticular structures lacking systematic orientation of ice also exist. Regular reticulate cryostructures appear to form through a combination of ice segregation and ice infilling of shrinkage cracks or hydraulic fractures. As sediment progressively freezes, moisture migration to the freezing front is thought to cause desiccation and consequent formation of vertical shrinkage cracks, which are then occupied by ice (Mackay, 1974), though in some cases the vertical components may consist of cracks produced by hydrofracture by pressurized water, which then freezes in contact with the crack walls to form ice veins (Mackay and Dallimore, 1992). Irregular or 'chaotic' reticulate cryostructures may also develop through inward freezing of sediments trapped in underground thermo-erosional tunnels within permafrost, and reflect uneven inwards progression of the freezing front under conditions of limited water supply (Fortier *et al.*, 2008a). In some cases, the blocks of sediment may be subdivided by a secondary system of thinner, secondary ice veins.

Ataxitic (or *suspended*) *cryostructures* (Figure 5.6d) are dominated by ice (>50% by volume) in which mineral grains, grain aggregates or clasts are embedded, and are common in the ice-rich transition zone near the top of permafrost. They occur in segregated ice, intrusive ice formed by injection of water into permafrost, vein ice, ice wedges, massive intrasedimental ice bodies and buried ice of glacial origin. Finally, the term *solid cryostructure* is used to denote ice bodies more than 10 cm thick; this may originate in several ways, as segregation ice, through intrusion of pressurized water into permafrost, through the development of ice wedges or composite wedges (Chapter 6) or the icy core of pingos (Chapter 7), through freezing of underground pools of water, through freezing of water in dilation cracks or through ice burial.

As with the genetic classification of ground ice, the classification of cryostructures describes end-members of a continuum of types. Pore, lenticular, bedded, reticulate and ataxitic structures, for example, may coexist in some combination within frozen frost-susceptible soils or rock; thick ice bands (bedded cryostructures) may be transitional to solid horizontally stratified cryostructures; and massive solid ice bodies may support ataxitic cryostructures. Murton (2013c) also distinguishes between *superimposed cryostructures*, which form in initially homogeneous soil or rock, and *inherited cryostructures*, which are partly determined by pre-existing heterogeneity in the host sediment, such as stratification or textural contrasts. Though most cryostructures develop in freezing or frozen ground, they can also develop during periods of thaw, particularly where meltwater from the upper thawed layers of soil percolates downward into the still-frozen soil below, forming pore ice or ice lenses (Jaesche *et al.*, 2003; Harris *et al.* 2008b). Moreover, because liquid water movement can occur within frozen soil, cryostructures may evolve through time, particularly under conditions of changing temperature and/or pressure. A pore cryostructure may evolve into a lenticular cryostructure, for example, or a bedded cryostructure may thicken to form a solid sheet-like cryostructure (Zhestikova, 1982).

5.3.3 Cryostructures in Rock

Relatively little research has been carried out on cryostructures in lithified bedrocks, where the location of ice bodies is often determined primarily by pre-existing cracks (joints), bedding planes, foliation planes and other flaws or zones of weakness. This is evident in a sixfold classification of cryostructures in 'solid and semi-solid' rocks proposed Melnikov and Spesivtsev (2000), which depicts four classes where ice lenses, bands and veins follow pre-existing structural weaknesses, often producing reticulate structures, and two ('fissured' and 'fissured-widened') in which ice fills randomly orientated cracks, in the latter forming an irregular reticulate structure. Similarly, Murton (1996b) described a range of cryostructures in perennially frozen shale, which include lenticular, ataxitic and both regular and irregular reticulate types. The resulting fragmentation of bedrocks by ground ice is termed *brecciation* (Chapter 10) and is common in weakly-cemented sedimentary rocks, such as shale, siltstone and sandstone. Cores extracted from near surface siltstones and shales on Ellesmere Island in arctic Canada, for example, contain on average more than 40% ice by volume (Hodgson and Nixon, 1998). The formation of such excess ice in weak rocks has mainly been attributed to ice segregation (French *et al.*, 1986; Mackay, 1999b), an interpretation supported by laboratory experiments (Murton *et al.*, 2001b, 2006), though Christiansen *et al.* (2005) argued that ice filling joints in coal-bearing siltstone and shale bedrock in an abandoned coal mine on Svalbard represents regelation ice derived from glacial meltwater.

5.3.4 Cryofacies

Geologists studying sequences of sedimentary rocks employ the term *facies* (or *lithofacies*) to characterize the properties of a particular rock (such as its grain size, sedimentary structures and bedding characteristics) that indicate environment of deposition. Russian scientists have extended this practice to describe *permafrost facies* within thick sequences of frozen sediment, and Murton and French (1994) introduced the term *cryofacies* to describe relatively uniform units of frozen ground (in vertical exposures) that are distinguished from adjacent units by volumetric ice content, the nature of sediment and cryostructures (Table 5.2). An assemblage of contiguous cryofacies that forms a genetically distinct unit is referred to as a *cryofacies association*, whereas adjacent cryofacies of uncertain genetic relationship constitute a *cryofacies assemblage*. Identification and differentiation of distinct cryofacies is a key component of cryostratigraphic analysis.

Table 5.2 Cryofacies of frozen sediments. A diamicton is a poorly sorted sediment body, usually involving clasts embedded in finer sediment. Cryostructure coding follows that in column one of Table 5.1.

Cryofacies type	Volumetric ice content	Cryofacies	Code	Dominant cryostructures
Pure ice	100%	Pure ice	PI	L, B, S
Sediment-poor ice	75–100%	Sand-poor ice	SPI	L, B, A, S
		Aggregate-poor ice	API	L, B, A, S
Sediment-rich ice	50–75%	Sand-rich ice	SRI	L, B, A, S
		Aggregate-rich ice	ARI	L, B. A, S
Ice-rich sediment	25–50%	Ice-rich sand	IRS	P, L, B
		Ice-rich mud	IRM	L, B, R
		Ice-rich diamicton	IRD	L, B, R, C
Ice-poor sediment	<25%	Ice-poor mud	IPM	P, L
		Ice-poor sand	IPS	P
		Ice-poor gravel	IPG	P
		Ice-poor diamicton	IPD	P
		Ice-poor peat	IPP	P

Source: Adapted from Murton and French (1994) with permission from the National Research Council of Canada.

5.4 Ice Contacts

The boundaries or *ice contacts* between individual cryo-facies exposed in vertical section may be gradational, where there is a transition between a cryofacies and its neighbours, or sharp, where there is an abrupt contact between two cryofacies. Contacts may be conformable (implying contemporaneity, or uninterrupted formation) or unconformable (implying a time gap between two units). There are three types of inter-cryofacies contacts: freezing contacts, erosional contacts and thaw contacts.

Freezing contacts occur where ice has formed in contact with cryotic ground. At one extreme, these may be sharp and unconformable, for example where the steeply dipping edges of wedge ice are in contact with frozen sediment or where water injected under pressure into icy sediments forms an intrusive ice body. At the other extreme, freezing contacts may be gradational and conformable, showing a progressive downwards sequence: for example, from massive ice to ice-rich mud with numerous ice lenses, frozen mud with fewer ice lenses then frozen blocks of mud enclosed by reticulate cryostructures (Mackay, 1971), or from massive ice to sediment-rich ice then reticulate ice (Harry *et al.*, 1988). Conformable ice contacts generally reflect alignment of ice bodies and intervening sediment bands normal to the direction of heat flow, and are indicative of ice segregation within freezing or frozen sediment.

Erosional contacts represent removal of frozen sediment by erosion, then subsequent sediment accumulation and permafrost aggradation at the same site, producing a discontinuity in the cryostratigraphic record. Erosional contacts are therefore often discordant, unconformable and generally abrupt.

Finally, *thaw contacts*, often termed *thaw unconformities*, mark the position of present or former thawing fronts within permafrost, are often sharply defined and are valuable for interpreting the permafrost history. The base of the present active layer is a thaw contact, sometimes termed a *primary thaw unconformity* (French and Shur, 2010), as are the margins of taliks within permafrost. Older thaw contacts are represented by the base of former, deeper active layers (or former taliks) that have subsequently refrozen during permafrost aggradation (Figure 5.7). Thaw contacts representing a former deeper active layer are termed *palaeo-thaw unconformities*. These truncate the tops of pre-existing ice bodies such as ice wedges, ice veins or massive ice, or may be represented by an abrupt downward change in cryostructure or ice content. They may also be marked by the lower limit of plant roots or rhizomes, or soft-sediment deformation structures (*thermokarst involutions*; Chapter 8), though such structures may terminate some distance above the thaw inconformity (Murton and French, 1993a, 1994).

Figure 5.7 Thaw unconformity truncating banded massive ice, Crumbling Point, NWT, Canada. The zone above the uncorformity represents the palaeo-active layer. *Source:* Courtesy of Julian Murton.

Thaw unconformities marking formerly deeper active layers may reflect localized active-layer deepening due to forest fires, changes in hydrological conditions such as drainage of shallow lakes or river channel abandonment, removal of vegetation or peat cover, or anthropogenic disturbance (e.g. Calmels *et al.*, 2012). Regional-scale thaw unconformities, however, are indicative of climate changes. One example of the latter is a widespread thaw unconformity at ~1–3 m depth in northern Alaska, along the coast of the western Canadian Arctic and in central Yukon (Burn *et al.*, 1986; Burn, 1997), and another is a thaw contact that separates Holocene from Pleistocene permafrost in Siberia (Yershov, 1998). Both are attributable to deepening of the former active layer under the relatively warm conditions of the early Holocene thermal maximum, followed by upwards permafrost aggradation and active-layer thinning as the climate subsequently cooled. The evidence for the regional thaw unconformity exposed in bluffs and ground-ice slumps along the western arctic coast of Canada is summarized by Burn (1997), and includes truncated ice wedges, variations in cryostructure, soft-sediment deformation structures in the palaeo-active layer above the thaw unconformity and colour changes across the thaw contact. Enhanced chemical weathering within the palaeo-active layer has also locally resulted in contrasts in secondary clay mineral composition above and below the thaw unconformity, and at some sites ground ice above the unconformity has also been shown to have a significantly higher oxygen isotope ($\delta^{18}O$) composition than that below, a contrast attributable to strongly negative $\delta^{18}O$ values in Pleistocene permafrost below the unconformity. This isotopic contrast is not always present, however, because of downwards migration and refreezing of Holocene water in the permafrost below the unconformity (Burn and Michel, 1988). Moreover, the true depth of the former active layer may be rather less than that of the thaw unconformity. This disparity reflects several factors, notably (i) formation of excess ice above the unconformity during upwards permafrost aggradation, with consequent substrate thickening as the active layer thinned, (ii) accumulation of sediment from upslope, (iii) peat growth, (iv) incorporation of organic material and (v) soil displacement by ice-wedge growth (Burn, 1997).

A rather different type of thaw contact occurs in parts of the subarctic discontinuous permafrost zone, where noncryotic ground separates a near-surface zone of seasonal frost penetration from the top of relict permafrost. The noncryotic layer is termed a *residual thaw zone* (French and Shur, 2010), and the top of the permafrost represents a transient thaw unconformity.

5.5   Cryostratigraphy

Cryostratigraphy is the study of stratigraphic units within permafrost, based on the description and interpretation of component units and their interrelationships, to establish a sequence of events. Such units are defined primarily by the type, amount, structure and distribution of ground ice present in the host sediments or bedrock as exposed in vertical section or in cores. This information can be used to infer changing thermal and hydrological conditions that have led, for example, to periods of regional or local permafrost thaw, aggradation, erosion or deformation. Studies of the cryostratigraphy of extant permafrost are important in informing the interpretation of structures in mid-latitude sedimentary sequences that were permafrozen during successive Pleistocene glacial stages but experienced thaw during intervening interglacial or interstadial periods.

Conventional stratigraphic analysis of unfrozen sediments or sedimentary rocks employs three general principles to establish a relative sequence of events: (i) the principle of original horizontality, which states that most sediments are deposited as horizontal or near-horizontal layers; (ii) the principle of superposition, which states that layers get younger upwards in an undisturbed sedimentary sequence; and (iii) the principle of cross-cutting relationships, which states that a disrupted sediment sequence is older than the cause of disruption. Where ground ice in permafrost forms epigenetically within a stable substrate, the first of these principles is largely redundant. In cryostratigraphy, where the relative age of cryofacies and individual ice bodies is the primary concern, the principle of superposition is valid where permafrost has aggraded upwards, either as a result of active-layer thinning or due to progressive sediment accumulation or peat growth at the surface. This principle is vitiated, however, where intrusive ice or pool ice forms within pre-existing permafrost and hence is younger than overlying ice-rich ground. In such circumstances, the principle of cross-cutting relationships assumes paramount importance: a sill or dyke of intrusive ice or an ice vein or wedge that disrupts a body of lenticular, layered or reticulate ice must be younger than the disrupted body of ice-rich sediment (Figure 5.8).

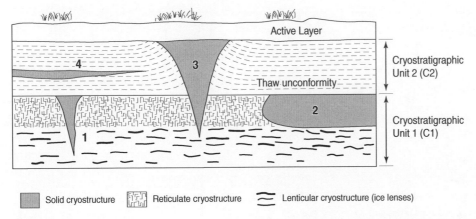

Solid cryostructure   Reticulate cryostructure   Lenticular cryostructure (ice lenses)

Figure 5.8 Cryostratigraphic relationships within a stable substrate. C2 is a palaeo-thaw layer separated from C1 by a thaw unconformity representing the base of a former, deeper active layer. The truncated ice wedge (1) within C1 is younger than the host sediments in C1 but older than the ice in C2. The truncated massive ice body (2) is also younger than the host sediments in C2 if it represents intrasedimental ice, but may be roughly coeval with these sediments if it is buried glacier ice. The upper active ice wedge (3) and intrusive sill ice (4) are younger than the host sediment.

Cryostratigraphic mapping involves logging the cryostructures, cryofacies and ice contacts visible in steep to vertical exposures (such as riverbanks, coastal bluffs or the headscarps of thaw slumps), or in cores extracted from frozen sediments, to infer both the history of the sediments (accumulation, erosion and/or deformation) and the changing thermal and hydrological history of the sediment body, as recorded by ice contacts or changes in cryostructure. Logging of visible structures is often accompanied by laboratory measurements of the ice content of different units, measurements of isotope concentrations (δ^{18}O and δD) of ground ice to differentiate units and the climatic environment under which they formed, and radiocarbon dating of contained organic material (peat, organic-rich soil or plant remains) to constrain the timing of events (e.g. Kotler and Burn, 2000; Kanevskiy *et al.*, 2011).

Many sections exhibit a complex cryostratigraphy. A good example is that mapped by Kanevskiy *et al.* (2008) on the wall of part of the CRREL permafrost tunnel in Alaska, 12–14 m below the ground surface. This depicts two major cryostratigraphic units separated by an erosional contact (Figure 5.9). To the left of the contact, the cryostratigraphy comprises ice-rich silts containing micro-lenticular ice in the form of small subhorizontal ice lenses less than 0.5 mm thick and less than 0.5 mm apart, a cryostructure typical of syngenetic permafrost. The gravimetric moisture content of these deposits is

100–240%, the organic content ranges from 3.3 to 9.5% by weight and radiocarbon ages obtained for peat layers place the age of the sediment at ~35–31 ka. This zone is crossed by seven thin organic layers, and ~0.4–0.6 m below each organic layer is a distinct ice-rich layer; ice veins descend up to 0.5 m from each organic layer. Kanevskiy *et al.* (2008) interpreted this sequence in terms of progressive accumulation of silt and syngenetic permafrost aggradation, with the organic-rich bands representing peat development during periods of slower sediment accumulation, and the vertical distance between each band and the corresponding ice-rich layer representing the former active layer.

The erosional contact marks the margin of a gully that is partly floored by a truncated ice wedge (not shown in Figure 5.9), suggesting that the gully was excavated by thermal erosion acting along the site of the wedge. The sediments filling the gully are mostly ice-poor stratified silts with lenses of sand and numerous inclusions of reworked organic material; the organic content ranges from 7.0 to 22.8% by weight, much greater than in the undisturbed syngenetic permafrost. The cryostratigraphy in the lower part of the gully fill is dominated by pore ice and what Kanevskiy *et al.* (2008) describe as 'latent micro-lenticular' ice, with a lower gravimetric ice content (70–100%) than the undisturbed sediments; that of the upper part of the gully fill is more similar to that of the ungullied permafrost, with a gravimetric

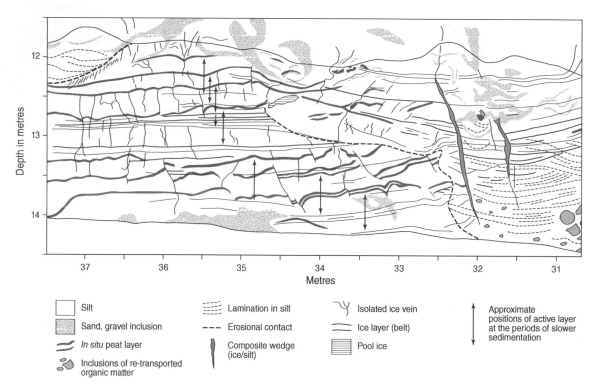

Figure 5.9 Cryostratigraphic record of the wall of part of the CRREL permafrost tunnel: undisturbed syngenetic permafrost is to the left of the erosional contact, and frozen gully infill is right of the contact. *Source:* Reproduced with permission of Kanevskiy *et al.* (2008).

ice content of 110–140%, suggesting a decrease in sedimentation rate during the last stages of gully infilling.

The value of cryostratigraphy in reconstructing both landscape history and permafrost evolution is beautifully illustrated by research carried out at three sites in the Koyukuk and Innoko Flats of west-central Alaska, a low-lying region of lakes, bogs, palsas and peat plateaus underlain by discontinuous permafrost 10–130 m deep. By logging the cryostratigraphy and ground ice content of exposed sections and permafrost cores, Kanevskiy *et al.* (2014) detected three main cryostratigraphic units within elevated peat plateaus. The uppermost unit comprises surface peat, frozen syngenetically or semi-syngenetically as indicated by the dominance of micro-cryostructures and presence of ice-rich bands. The middle unit consists of frozen lacustrine deposits, comprising an upper subunit of organic-rich silt and sedimentary algal peat and a lower subunit with inclusions of silt and terrestrial peat interpreted as representing collapse of lake banks. The lowermost, predominantly epigenetic unit comprises reworked *yedoma* deposits (extremely ice-rich, syngenetically frozen silt containing huge ice wedges), in the form of an upper subunit of silt with inclusions of terrestrial peat and organic material, probably reflecting lacustrine reworking of yedoma soils, and a lower subunit of silt with rare organic inclusions, interpreted as reflecting thaw of yedoma sediments beneath a former lake. The lacustrine organic deposits produced early- to mid-Holocene radiocarbon ages.

Taking as their starting point a low-relief ice-rich yedoma landscape (Figure 5.10a), Kanevskiy *et al.* (2014) employed their cryostratigraphic information to construct a nine-stage model of landscape evolution from the Late Pleistocene to the present. Stages 1–3 represent the Pleistocene–Holocene transition, during which much yedoma was lost by thermokarst subsidence and thermal erosion: water initially ponded at ice-wedge polygon intersections, then deepening and widening of water-filled troughs reached a critical depth (when mean annual water bottom temperatures reached the melting point) at which thawing of yedoma accelerated and thermokarst lakes developed (Figure 5.10b). Under the relatively warm conditions of the early Holocene (stage 4; Figure 5.10c), thaw depths reached a maximum: an open talik developed under thermokarst lakes, lake sediments incorporated sediments from collapsing banks and yedoma survived only in well-drained areas. During the middle to late Holocene (stages 5–8; Figure 5.10d), the cryostratigraphic data suggest that breaching and drainage of lakes occurred, leaving a terrain of ponds, fens and mounds on which peat accumulated, facilitating renewed permafrost aggradation on parts of the former lake bed to form palsas and permafrost plateaus (Chapter 7) under the cooler conditions of the middle Holocene. The present landscape (stage 9; Figure 5.10e)

consists of a mosaic of peat plateaus, thermokarst lakes, bogs, fens and ponds, surmounted by isolated yedoma remnants. As Kanevskiy *et al.* (2014) demonstrated from the accompanying radiocarbon dates, this sequence developed over millennial timescales but was not synchronous across the landscape: some lakes probably expanded as others drained.

The significance of some ground ice types deserves further elaboration. In the next sections, we focus on the nature of ice in the transition zone immediately below the active layer, the characteristics and origins of massive ice bodies within permafrost and the nature of the yedoma or 'ice complex' deposits that underlie vast areas of Siberia and Alaska.

5.6 The Transition Zone

In the introduction to Chapter 4, we saw that the boundary between the active layer and the underlying permafrost is often represented by a *transition zone* comprising an upper ice-rich *transient layer* and a lower, extremely ice-rich *intermediate layer* (Shur *et al.*, 2005; Figure 4.3). The transition zone is best developed in frost-susceptible soils in lowland arctic regions underlain by cold permafrost, where the active layer is seasonally subject to bidirectional ('two-sided') freezing, both downwards from the surface and upwards from the underlying permafrost. It is located immediately below the active layer, and it alters in status between perennially cryotic (permafrost) and seasonally cryotic (active layer) over timescales of years to centuries in response to relatively short-term changes in subsurface thermal conditions, such as a succession of abnormally warm or cold summers, or more prolonged (decadal to centennial) climate changes. The base of the transition zone represents the 'long-term' permafrost table and is marked by a thaw unconformity defined, for example, by the tops (or truncated tops) of primary ice wedges or by changes in cryostructure or excess ice content. Because it is ice-rich, the transition zone acts as an important thermal buffer zone between the active layer and underlying permafrost, as a large heat input is required to thaw the ice (Chapter 3).

The ice in the transition zone represents 'aggradational ice' that forms as permafrost aggrades upwards into the active layer, either epigenetically as the active layer thins in response to colder conditions, or syngenetically as sediment or peat accumulates on the surface (Figure 5.11). Most ice in the transition zone is segregated ice, often containing suspended particles or aggregates of soil or, in frost-susceptible bedrock, detached rock fragments. Lenticular, bedded (layered) and reticulate cryostructures predominate, though ataxitic structures may be present where the ice content is particularly high.

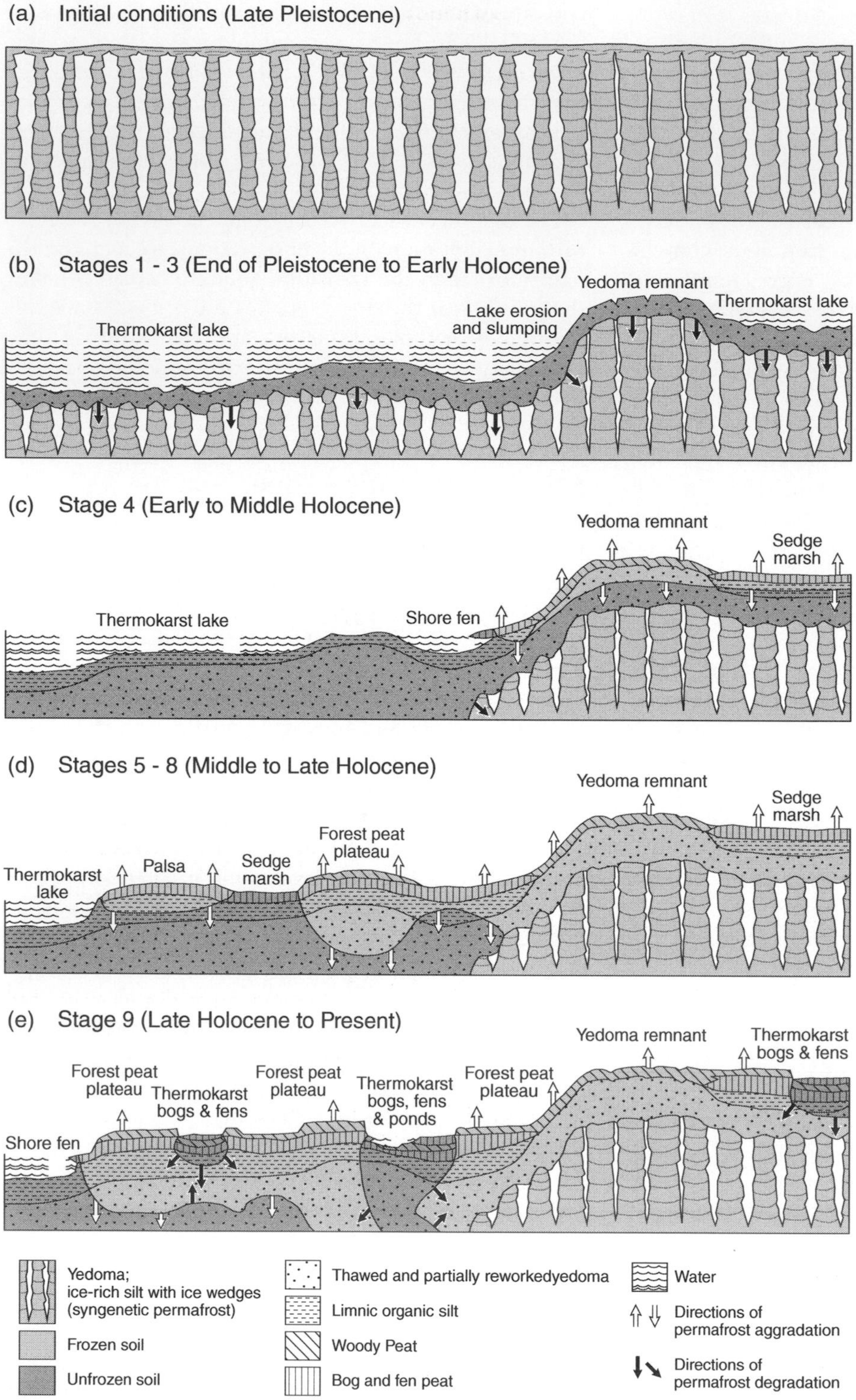

(a) Initial conditions (Late Pleistocene)

(b) Stages 1 - 3 (End of Pleistocene to Early Holocene)

Thermokarst lake

Lake erosion and slumping

Yedoma remnant

Thermokarst lake

(c) Stage 4 (Early to Middle Holocene)

Yedoma remnant

Sedge marsh

Thermokarst lake

Shore fen

(d) Stages 5 - 8 (Middle to Late Holocene)

Yedoma remnant

Sedge marsh

Forest peat plateau

Thermokarst lake

Palsa

Sedge marsh

(e) Stage 9 (Late Holocene to Present)

Yedoma remnant

Thermokarst bogs & fens

Forest peat plateau

Thermokarst bogs & fens

Forest peat plateau

Thermokarst bogs, fens & ponds

Forest peat plateau

Shore fen

Yedoma; ice-rich silt with ice wedges (syngenetic permafrost)

Frozen soil

Unfrozen soil

Thawed and partially reworked yedoma

Limnic organic silt

Woody Peat

Bog and fen peat

Water

Directions of permafrost aggradation

Directions of permafrost degradation

Figure 5.10 Conceptual model of landscape development from the Late Pleistocene to the present in west-central Alaska, based on the cryostratigraphy of frozen sediments and peat plateaus. *Source:* Kanevskiy *et al.* (2014). Reproduced with permission of Wiley.

The high ice content of the transition zone is thought to result from repeated episodes of ice segregation without intervening thaw events. During seasonal freezeback of the active layer above cold permafrost, water is drawn both upwards to feed ice-lens growth as a freezing plane descends downwards from the surface, and downwards to feed ice segregation as a freezing plane rises upwards from the underlying permafrost. In summer, as the active

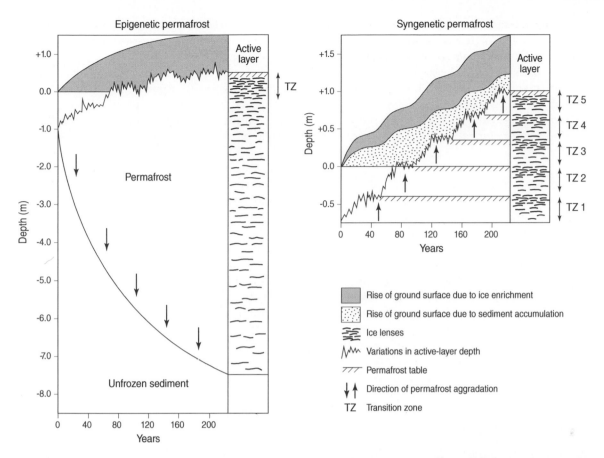

Figure 5.11 Development of the transition zone during epigenetic and syngenetic permafrost aggradation. The latter depicts pulsed sediment accumulation, with concomitant rises in the average position of the permafrost table, producing several transient ice-rich transition zones, as in the accumulation of yedoma deposits. *Source:* French and Shur (2010). Reproduced with permission of Elsevier.

layer thaws, downwards water movement saturates the base of the active layer (Hinkel *et al.*, 2001b) and water is drawn down into the underlying frozen ground by cryosuction, causing renewed ice segregation within the upper layers of permafrost and thus contributing to progressive ice enrichment in the transition zone (Cheng, 1983; Mackay, 1983). This interpretation is supported by field experiments by Burn (1988), who used transparent tubes filled with loam to demonstrate a progressive increase of segregated ice in the uppermost 20 cm of permafrost over a 15-month period, with the ice content increasing at most sites during the summer months. He also calculated from the volumetric ice content of permafrost above a thaw unconformity that the minimum long-term rate of water incorporation into the upper part of the permafrost must have averaged 0.1–0.2 mm a^{-1} since the early Holocene. Further support for ice enrichment of the transition zone by downwards movement of water from the active layer is provided by Burn and Michel (1988), who detected elevated levels of tritium (^{3}H) in ground ice up to 50 cm below the permafrost table. As the atmospheric levels of tritium were enhanced by nuclear weapons testing in the 1950s and 60s, this finding appears to confirm that downwards movement

of meteoric water through the active layer has contributed to recent ice enrichment of the transition zone. The ice content of the transition zone may also be enhanced through the development of ice veins and secondary (rejuvenated) ice wedges (Chapter 6). The higher ice content in the lower part of the transition zone (the intermediate layer) probably reflects its long-term stability: the upper part (the transient layer) is subject to more frequent thaw events, and so the ice in this layer has had less time to accumulate.

The transitional zone is significant cryostratigraphically as a marker horizon limiting the former depth of the active layer. Diagnostic features include its bipartite structure (ice-rich transient layer overlying ice-richer intermediate layer), its overall high excess ice content (typically greater than that of overlying or underlying sediments), its distinctive (particularly reticulate or ataxitic) cryostructures and its location immediately above the tops of primary (or truncated primary) ice wedges (Figure 4.3). The continuous or intermittent accumulation of frost-susceptible sediment at the surface and the consequent upwards aggradation of syngenetic permafrost produce a continuous vertical sequence of ice-rich transition zones (Figure 5.10), resulting in layering in the ice-rich yedoma deposits discussed below. The transient layer is also characterized by solute enrichment, which has been attributed by Lacelle *et al.* (2008) to greater availability of water for geochemical reactions at the base of the active layer, and by Kokelj and Burn (2005) to expulsion of dissolved solutes during freezing. Subsequent upwards permafrost aggradation then sequesters the zone of solute enrichment.

5.7   Massive Ground Ice

The term *massive ice* is usually applied to thick bodies of ice with an ice content exceeding 250%; those with lower ice content are termed *massive icy beds* (van Everdingen, 1998), though these two categories may be difficult to differentiate in stratigraphic sequences where the sediment content varies markedly over short intervals. Massive ice bodies occur within continuous permafrost in the lowland arctic environments of northern Siberia and between the Canadian Arctic Archipelago and Alaska. They usually occur under a cover of Late Pleistocene or Holocene sediments,

though there is borehole evidence for development of massive ice within sedimentary rocks (sandstone and coal) in northeast China (Wang, 1990).

Massive ice forms tabular bodies that range in thickness from a few metres to a few tens of metres, and in horizontal extent from tens of metres to more than a kilometre. It varies markedly in appearance, from pure ice to dirty ice containing mud, sand or, in some cases, clasts up to boulder size, and it may be bubble-free or bubble-rich. Vertical variations in sediment and bubble concentration give many massive ice bodies a layered or foliated appearance that reflects melt-out of sediment-rich bands (Figure 5.12). Massive ice bodies have been interpreted in two main ways: as intrasedimental ice bodies that developed during downwards permafrost aggradation, or as remnants of Pleistocene glacier ice that have been buried by a sediment cover that exceeds the depth of the active layer, and hence are preserved in permafrost. Both explanations are feasible in glaciated lowlands underlain by permafrost, and in some localities both types coexist, even in close proximity (Murton *et al.*, 2005). Because both types of massive ice have a similar appearance, distinguishing between them requires detailed

Figure 5.12 Massive stratified ground ice, Hooper Island, NWT, Canada. The upper parts of the ice body have been deformed as a result of overrunning by glacier ice. *Source:* Courtesy of Julian Murton.

cryostratigraphic, petrological and chemical analyses, though a glacial origin can be excluded in areas that escaped Pleistocene glaciation (Lawson, 1983; Wang, 1990).

5.7.1 Massive Intrasedimental Ice

In a seminal investigation of an exposure of massive ice at Peninsula Point on the western arctic coast of Canada, Mackay and Dallimore (1992) advanced several arguments supporting growth of the ice during downwards permafrost aggradation. The exposure they examined at Peninsula Point exhibited up to 10 m of massive ice overlain by a frozen, frost-susceptible diamicton; a core through the massive ice showed that it was underlain by fine to medium sand that in an unfrozen state would have permitted lateral and upwards flow of groundwater towards aggrading permafrost. Mackay and Dallimore noted that the upper contact between the massive ice and the overlying diamicton was conformable, with sediment bands parallel to the contact and bubble trains indicative of downward freezing originating at the upper contact. They argued that both features support an intrasedimental origin, as a thaw contact (unconformity) might be expected at the top of buried glacier ice. Reticulate ice immediately above the upper contact provided further evidence for downward freezing. Ice dykes penetrating upwards from the massive ice into the diamicton indicated upwards injection of water into frozen ground, implying that the massive ice was younger than the overlying sediments, particularly as similarities in geochemistry and stable isotope values indicated that the ice dykes and massive ice had a common water source. The petrofabric of the massive ice was consistent with an origin by ice segregation, and recycled Tertiary pollen grains recovered from the ice body were derived from the underlying sand, implying that water passing through the sand fed growth of the ice. Moreover, Mackay and Dallimore noted that all of 50 boreholes in the same area had encountered sand at the base of massive ice, and that there was no evidence of sub-ice till, as might be expected if the ice body were glacier ice. Finally, no isotopic discontinuity occurred at the basal contact, implying that the water that fed the massive ice and the water frozen in the underlying sand had the same source.

Mackay and Dallimore (1992) concluded that the massive ice at Peninsula Point grew by downwards permafrost aggradation in an area where groundwater pressure was high enough to push up the overlying frozen ground and open up vertical hydrofracture cracks, now represented by ice dykes, in the overlying frozen diamicton. The term *segregated–intrusive ice* has been widely used to describe such intrasedimental ice bodies: segregated ice is thought to form when the water pressure is relatively low, and intrusive ice when the water pressure is sufficiently high to lift the overlying ice and frozen sediment.

However, as Murton (2013c) has pointed out, ice segregation in such circumstances is not driven by cryosuction of water to the freezing plane, as occurs during freezing of frost-susceptible soil, but by upwards movement of pressurized groundwater to the base of downwards-aggrading permafrost, so that massive ice thickens downwards. The strongly negative (−28‰ to −32‰) $\delta^{18}O$ values obtained for the massive ice and ice in the underlying frozen sand imply that the most likely source of pressurized groundwater was subglacial meltwater, combined with pore water expelled from the underlying sands as they froze. Further support for an intrasedimental origin for the Peninsula Point massive ice was provided by Moorman *et al.* (1998), who found that CO_2 concentrations in gas bubbles within the massive ice are an order of magnitude greater than atmospheric level, consistent with those in soils but much higher than those in glacier ice. Lacelle and Vasil'chuk (2013) have summarized developments in the use of isotope geochemistry to determine the origin and age of ground ice.

The key problem in explaining the formation of intrasedimental massive ice is the source of abundant pressurized water. Both Mackay and Dallimore (1992) and Moorman *et al.* (1998) invoked an explanation proposed by Rampton (1974, 1988). Rampton's model envisages degradation of the uppermost permafrost under an advancing ice sheet (Figure 5.13a), thus opening taliks for subsurface flow of subglacial meltwater that is driven towards the glacier margin by hydraulic pressure. Retreat of the ice-sheet margin and consequent downwards permafrost aggradation outside the retreating ice-sheet margin seals the pressurized groundwater in an semiconfined aquifer, providing abundant pressurized water to feed the growth of the massive ice body (Figure 5.13b). Though this model has been questioned on the grounds that it requires slow retreat of the ice margin to maintain a supply of water and that rupture of overlying permafrost is likely to release pressurized water flowing through taliks (French and Harry, 1990; Mackay and Dallimore, 1992), it is supported in a study by Lacelle *et al.* (2004) of icy sediments in the Richardson Mountains of arctic Canada. Like the massive ice at Peninsula Point, ice in these sediments yielded strongly negative $\delta^{18}O$ and δD values, consistent with a glacial meltwater source, and gas bubbles within the ice produced CO_2 concentrations nine times higher than that typical of glacier ice, but similar to soil CO_2 values. Both the physical and isotopic properties of the icy sediments appear consistent with their origin as segregated–intrusive ice that formed after retreat of the ice-sheet margin, as sediment-laden subglacial meltwater moved through a proglacial talik and froze as the permafrost aggraded downwards.

An intriguing alternative suggested by Rampton (1991) to explain the massive ice in the Tuktoyaktuk Peninsula of western arctic Canada is that such ice may have formed in

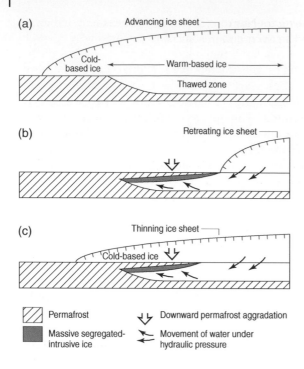

Figure 5.13 Proposed explanations for the growth of massive intrasedimental ice in glaciated lowlands. (a) An ice sheet advances across permafrost, creating a subglacial thawed zone up-ice from the ice-sheet margin. (b) Proglacial downwards permafrost aggradation follows ice margin retreat, and subglacial meltwater under hydraulic pressure moves through a confined aquifer to feed growth of massive intrasedimental ice. (c) Ice-sheet thinning causes permafrost aggradation under cold-based ice, and subglacial meltwater under hydraulic pressure moves through a confined aquifer to feed the growth of subglacial massive intrasedimental ice.

sediments *under* a thinning ice sheet. According to this model, advance of a thick ice sheet across the area resulted in permafrost thaw to depths of 50–100 m or more under warm-based glacier ice. If the ice sheet subsequently thinned and became cold-based, permafrost may theoretically have started to aggrade downwards under the thinning glacier ice. Rampton argues that this scenario would have resulted in subglacial meltwater flowing under hydraulic pressure through a confined aquifer, trapped between the downwards-aggrading subglacial permafrost and the relict permafrost below the thawed layer, and feeding the downwards growth of intrasedimental massive ice (Figure 5.13c). Deformation of the massive ice may be explained by subsequent thickening and readvance of the ice sheet, and it is even possible that the subglacial intrasedimental ice body could have become entrained into the base of the readvancing glacier ice, forming glacially modified or glacially incorporated intrasedimental ice (Murton *et al.*, 2004).

Rampton's models are inapplicable to the formation of intrasedimental massive ice in unglaciated areas.

An alternative possibility is that pore-water expulsion resulting from the freezing of underlying sandy non-frost-susceptible sediments produces sufficient excess water to form massive ice in former intrapermafrost taliks (French and Harry, 1990). In this context, Mackay and Dallimore (1992) noted that of 634 boreholes drilled through massive ice in arctic Canada and western Siberia, 85% showed that the lower contact of the ice body is with sand or sand and gravel, and only 15% found it underlain by clayey sediments. Bearing in mind that some of these massive ice bodies may represent buried glacier ice, this finding suggests that formation of massive intrasedimental ice is favoured by, and possibly dependent on, the presence of a thick, saturated layer of non-frost-susceptible sandy sediment that produces sufficient excess water on freezing (by pore-water expulsion) to nourish the growth of a massive ice layer. An alternative possibility is that high hydraulic pressure is generated in such sediments where an intrapermafrost talik extends from high to low ground, but this situation can only apply where massive ice occurs in valley-floor situations and is inapplicable in areas of low relief.

5.7.2 Buried Glacier Ice

Burial of thinning glaciers and ice sheets under sediment or debris cover is a common occurrence, and in permafrost-free environments results in the formation of hummocky topography as blocks of stagnant ice melt to form enclosed and often water-filled kettle holes. In permafrost environments where the thickness of the supraglacial sediment mantle exceeds the depth of seasonal thaw, stagnant debris-covered glacier ice may become incorporated in permafrost to form buried glacier ice. This is of two types: firn-derived glacier ice, which represents the progressive accumulation of snow on a glacier surface and its subsequent transformation into ice through melting, refreezing, compaction and recrystallization; and basal ice, which has undergone pressure-induced melting and refreezing (regelation) at the glacier sole. Both may be sediment-poor or sediment-rich, foliated or massive, and deformed or undeformed. Regelation ice commonly comprises much smaller crystals than those in firn-derived ice and usually incorporates sediment derived through abrasion and plucking (quarrying) of bedrock, or subglacial sediment or bedrock fragments that are frozen on to the glacier sole (Benn and Evans, 2010). Near glacier margins, debris-rich basal ice may be thrust upwards to englacial or supraglacial positions as a result of compressive flow, particularly in polythermal glaciers where cold-based ice at the glacier margin forces warm-based ice farther upglacier to deform upwards in successive thrust planes and recumbent folds (Hambrey *et al.*, 1996).

Outside mountain areas, the most common source of supraglacial sediment cover is melt-out of sediment that accumulates as ablation till on glacier surfaces. In a study of glacier ice burial near the snout of Stagnation Glacier on Bylot Island in Arctic Canada, Moorman and Michel (2000) counted more than 50 shear planes carrying basal sediments to the glacier surface within 500 m of the glacier terminus, so that the thinning glacier snout was completely covered by supraglacial melt-out till up to 2 m thick. They also observed buried ice extending up to 200 m outside the glacier terminus. Ancient buried glacier ice that was incorporated within permafrost during the retreat of Pleistocene ice sheets occurs in northern Russia (Astakhov *et al.*, 1996; Astakhov and Svendsen, 2002; Svendsen *et al.*, 2004) and western arctic Canada (Murton *et al.*, 2005; Lacelle *et al.*, 2007). Henriksen *et al.* (2003) have interpreted sediment cores from lakes in the Pechora lowland of northern Russia in terms of burial and subsequent thaw of glacier ice that existed within permafrost between ~90 and ~13 ka. Much older buried glacier ice occurs in southern Victoria Land, Antarctica, where dating of tephra within glacigenic deposits implies that underlying buried glacier ice is of Miocene (>8.1 Ma) age (Marchant *et al.*, 2002). In alpine mountain areas where coarse rubbly debris provided by rockfall, debris flows and avalanches has buried small glaciers, buried glacier ice forms the core of *glacigenic rock glaciers*, which continue to move, albeit slowly, through internal deformation of the buried ice (Krainer and Mostler, 2000; Berger *et al.*, 2004; Hausmann *et al.*, 2007; Monnier *et al.*, 2011). These features are considered in more detail in Chapter 12.

Buried glacier ice of Late Pleistocene age may be distinguished from intrasedimental massive ice by its morphological and stratigraphic context, for example where it occurs within moraines (French and Harry, 1990) or is sandwiched between till deposits; it may also exhibit a sharp thaw contact (thaw unconformity) where it meets overlying sediments (Worsley, 1999), in contrast to the gradational upper contact present in intrasedimental ice bodies. The isotopic signature provided by $\delta^{18}O$ and δD in buried glacier ice should approximate that of meteoric water, and the $\delta^{13}C$ values and CO_2 concentrations should be similar to those found in polar glaciers (Moorman *et al.*, 1998). Lacelle *et al.* (2007), for example, employed a range of approaches (cryostratigraphy, petrography, analyses of $\delta^{18}O$ and δD from ice and of the molar concentrations of occluded gases (CO_2, O_2, N_2 and Ar) trapped in ice) to demonstrate that massive ice exposed in a moraine complex in the Ogilvie Mountains of central Yukon represents buried glacial ice preserved in permafrost since the Middle Pleistocene. Utting *et al.* (2016) have demonstrated that analysis of the noble gas (Ar, Kr and Xe) ratios of occluded gases within buried ice provides an additional and potentially powerful tool for discriminating between buried massive ice bodies of different origins.

5.7.3 Massive Ice Hybrids

In some cases, however, intrasedimental massive ice has been modified by glacial over-running or incorporated within glacial deposits. In the Tuktoyaktuk Coastlands of Canada, an area over-run by the Laurentide Ice Sheet, massive ice of intrasedimental or glacial origin is overlain by ice-rich diamicton or sandy silts. The latter exhibit glacitectonic structures indicative of predominantly ductile deformation, and include ice rafts up to 15 m long derived from the underlying massive ice. Murton *et al.* (2004) interpreted these sediments as a subglacially deformed frozen layer (glacitectonite), and the contact with the underlying massive ice as an angular unconformity produced by erosion of the top of the massive ice and incorporation of ice clasts into the deforming subglacial sediments. Elsewhere in the same region, glacially-deformed massive ice and icy sediments exhibit features of both intrasedimental ice and basal glacier ice. The latter exhibits similarity in its ice facies and ice crystal fabrics to basal ice found in extant glaciers and ice-cored moraines, and is separated by an erosional or thaw unconformity from the overlying sediments. It contains bodies of intrasedimental pore ice and segregated ice that existed within Pleistocene sands before glacial over-riding, and is thus a hybrid ice (or icy) body, inferred by Murton *et al.* (2005) to have been preserved by downwards permafrost aggradation under thinning, stagnant basal glacier ice and burial by melt-out till. At North Head in the Tuktoyaktuk Coastlands, two generations of massive ice occur within a single stratigraphic sequence: an older unit of ice that was glacially deformed and eroded beneath the cold-based margin of the Laurentide Ice Sheet, and a younger unit of post-deformation ice that formed during or after deglaciation (Murton, 2005). The former comprises buried glacier ice, massive segregated ice and ice clasts eroded from pre-existing ground ice; the latter contains dykes and sills of intruded ice, massive segregated-intrusive ice, ice wedges and both segregated and pool ice. This sequence was interpreted by Murton in terms of (i) an initial stage of subglacial erosion and deformation of massive intrasedimental ice accompanied by burial of basal glacier ice and emplacement of glacitectonite above the entire glacially-deformed sequence, succeeded by (ii) injection of pressurized groundwater under proglacial permafrost to form the undeformed segregated-intrusive ice, ice dykes and sills. Ice wedges in the sequence represent thermal contraction cracking following deglaciation. Early Holocene warming then resulted in active-layer deepening, truncating the tops of the ice wedges, before mid-Holocene cooling resulted in upwards permafrost aggradation, rejuvenated ice-wedge growth and the formation of pool and segregated ice.

As these examples testify, not all subsurface solid ice bodies fit neatly into the categories of massive intrasedimental ice and buried glacier ice, particularly as the former may have been entrained, deformed or eroded at or below the base of glacier ice, and emplacement of the latter may be succeeded by formation of massive intrasedimental ice. It is interesting to conjecture as to whether subglacially-deformed massive ice or icy sediments should be classified as a form of subglacial deformation till or glacitectonite, or whether they should be regarded as ground ice (permafrost) phenomena that have experienced glacial modification.

5.8 Yedoma

5.8.1 Terminology, Distribution and Characteristics

The Russian word *yedoma* (sometimes transcribed as *edoma*) is widely applied to describe extremely ice-rich sediments that accumulated during the last (Weichselian/Wisconsinan) glacial stage in arctic and subarctic areas that escaped Late Pleistocene glaciation. This vast region is termed *Beringia*, and stretches 3000 km from the Tamyr Peninsula of northern Siberia eastwards across the Bering Strait to encompass much of Alaska and those parts of Yukon that lay beyond the reach of the last

Laurentide Ice sheet. Though 'yedoma' was originally used to describe flat-topped hills underlain by icy sediments, *yedoma*, *yedoma suite* and *yedoma complex* are now employed mainly as stratigraphic terms referring to extremely ice-rich, mainly silty sediments of Late Pleistocene age. The term *ice complex* has also been used to describe such deposits, and in areas of Alaska and Yukon they have been described simply as *muck*, a pejorative descriptor introduced by gold miners who had to remove them to gain access to the underlying auriferous alluvial gravels. An accessible introduction to yedoma is provided by Froese *et al.* (2009), and a useful overview by Schirrmeister *et al.* (2013).

Yedoma deposits are characteristically 10–40 m thick, though locally they may exceed 50 m in depth. In western Beringia (northeast Siberia), they underlie low-relief coastal and lowland plains and broad river valleys (Grosse *et al.*, 2013b); in eastern Beringia (Alaska and Yukon), yedoma occurs in the northern part of the Seward Peninsula and interior Alaska, as well as the Klondike area of Yukon, but it probably has a much wider distribution (Figure 5.14). The total area underlain by yedoma in Siberia and Alaska is thought to exceed one million square kilometres. Because yedoma deposits experienced deep thaw during the early Holocene, in many areas they comprise low, flat-topped residual *yedoma hills* interspersed with thermokarst lakes, drained thermokarst lake basins (*alases*) and *thermo-erosional valleys* incised by rivers into the icy sediments (Grosse *et al.*, 2007; Figure 5.15) and are now exposed along coastal bluffs, undercut riverbanks

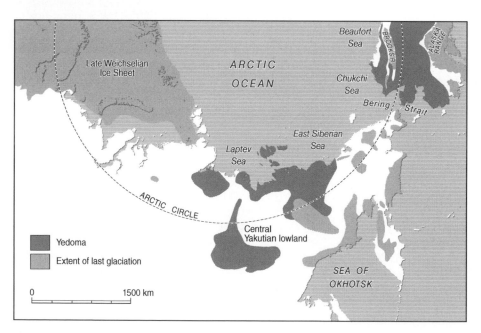

Figure 5.14 Distribution of yedoma deposits in Siberia and Alaska, showing how they occur only outside the limits of the last (MIS 2) Pleistocene ice sheets. *Source:* Murton *et al.* (2015a). Reproduced with permission of Wiley.

and thaw-slump headwalls. Most yedoma deposits are underlain by continuous permafrost hundreds of metres deep, though some extend southwards into the present discontinuous permafrost zone.

The defining characteristics of yedoma deposits are fourfold. First, they are nearly always dominated by massive or faintly stratified silt or silt and fine sand (typically >70%), with a minor clay component and, more locally, coarser sand or gravel. Mean grain size ranges from 10 μm to 300 μm, but the modal grain size is nearly always within the silt to fine sand range (20–100 μm), and some yedoma deposits exhibit bimodal or polymodal grain-size distributions. Second, the sediments are extremely ice-rich, typically seamed with segregated ice in the form of horizontal ice lenses or microlenses that sometimes form a reticulate cryostructure, and by ice-rich bands a few centimetres thick that represent ice enrichment near a former permafrost table and give a layered appearance to yedoma sequences. The gravimetric ice content of such ice-rich sediments commonly ranges from 40% to 100%. Third, yedoma deposits are penetrated by syngenetic ice wedges produced by recurrent thermal contraction cracking of the ground (Chapter 6) over thousands of years as the deposits accumulated. Such wedges often extend the full depth of the yedoma deposits (Figures 5.15 and 5.16) and are typically several metres wide. They often exhibit pronounced 'shoulders', indicating episodic sedimentation and permafrost aggradation, and partial thaw surfaces produced by changes in active-layer depth.

Ice wedges may occupy up to 65% of the overall yedoma sequence (Ulrich *et al.*, 2014), so the total volumetric ice content of yedoma sequences is commonly 60–90%. Finally, yedoma deposits contain abundant organic matter. This may take the form of frozen organic-rich layers or peat lenses within the icy sediments, or of dispersed plant remains in the form of roots, twigs and leaves, and the remains of insects and ostracods. Buried palaeosols within yedoma deposits represent dominantly epigenetic soil formation, whereas dispersed organic material represents syngenetic pedogenesis (Zanina *et al.*, 2011). Grass roots are particularly common in organic-rich layers. The most impressive skeletal remains preserved in yedoma are those of Late Pleistocene herbivorous megafauna, including mammoth, bison and caribou, and occasionally musk ox, camel and woolly rhinoceros, as well as some ice-age predators, notably cave lion, bear and wolf. Less common, if more sensational, are the preserved freeze-dried megafaunal carcasses that are occasionally extracted from thawing yedoma exposures (Shapiro and Cooper, 2003). Because of the high preservation potential of organic remains frozen into yedoma sediments as they accumulated, yedoma deposits are unrivalled as a source of information about the fauna and flora of late Pleistocene arctic and subarctic periglacial environments, and thus of the nature of the climate and climate changes that allowed the dominant species to flourish (Willerslev *et al.*, 2014; Schirrmeister *et al.*, 2016).

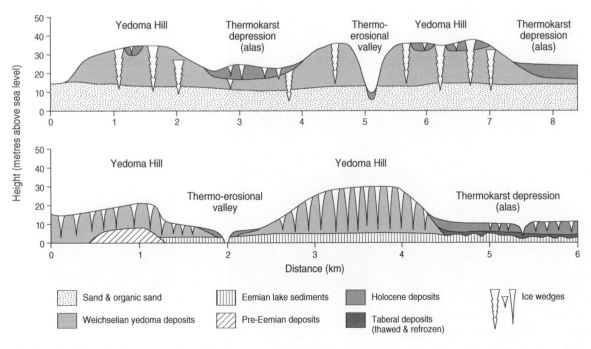

Figure 5.15 Generalized cross-sections of yedoma exposures. Top: Kurungnakh Sise Island, Lena River Delta. Bottom: Oyogos Yar coast, East Siberian Sea. *Source:* Schirrmeister *et al.* (2011). Reproduced with permission of Elsevier.

5.8.2 Age and Environment of Deposition

In Siberia, yedoma deposits are underlain by last (Eemian) interglacial sediments, or by younger alluvial and peat deposits dated to 60–100 ka (Andreev *et al.*, 2002; Wetterich *et al.*, 2008, 2009, 2014), suggesting that accumulation began during the Weichselian Glacial Stage (MIS 4–2). Radiocarbon ages imply initiation of yedoma accumulation before 55 ^{14}C ka on the New Siberian Islands but no earlier than ~27 ^{14}C ka on the western Laptev Sea coast (Schirrmeister *et al.*, 2011). Organic-rich layers within yedoma in the CRREL permafrost tunnel in Alaska (Kanevskiy *et al.*, 2008) and from an exposure on the bank of the Itkillick River in northern Alaska (Kanevskiy *et al.* 2011; Figures 5.16 and 5.17) suggest that most Alaskan yedoma formed over the last ~60 ka, during the Middle and Late Wisconsinan, though

association with dated tephras suggests that yedoma formation in the Klondike area of Yukon began during the Early Wisconsinan (Sanborn *et al.*, 2006). The apparent lack of pre-Weichselian/Wisconsinan yedoma deposits in many areas seems to indicate that most older yedoma deposits did not survive warming during the last (Eemian/Sangamonian) interglacial (MIS 5e). Reyes *et al.* (2010), however, have shown that relict ice wedges survived last-interglacial warming at sites in Alaska and Yukon, and evidence for much older yedoma sequences is present in central Yukon (Froese *et al.*, 2008) and northern Siberia, where cosmogenic [36]Cl dating of wedge ice in a lower yedoma suite implies a Middle Pleistocene age (Gilichinsky *et al.*, 2007).

Analyses of pollen, plant macrofossil, insect, ostracod, testate amoeba and mammal assemblages preserved in yedoma deposits in Siberia suggest that initial sediment

Figure 5.16 Exposures in yedoma sediments. (a) Syngenetic ice wedges in the yedoma deposits at Duvanny Yar, northeast Siberia. (b) Yedoma deposits on the Itkillik River, northern Alaska. Syngenetic ice wedges separate stratified silts. A colour version of (b) appears in the plates section. *Source:* Courtesy of (a) Julian Murton and (b) Mikhail Kanevskiy.

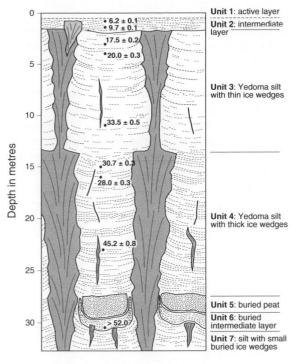

Figure 5.17 Cryostratigraphic units of the Itkillik yedoma exposure (ice wedge widths not to scale) and radiocarbon ages (cal ^{14}C ka) obtained on organic material within the yedoma deposits. Wavy horizontal and dashed lines represent ice lenses or microlenses, which tend to be concentrated in bands within the silts. Shaded areas represent ice wedges. *Source:* Adapted from Kanevskiy *et al.* (2011). Reproduced with permission of Elsevier.

accumulation during the Early Weichselian (MIS 4) took place under very cold, arid conditions with a ground cover of sparse grass-sedge tundra. This was succeeded by development of tundra-steppe vegetation under the less cold and probably less arid environment of the Middle Weichselian (MIS 3), and then a return to extremely cold arid tundra conditions during the Late Weichselian (MIS 2). Strongly negative δ^{18}O and δD values obtained for samples of wedge ice and sediment ice in yedoma indicate intensely cold winter temperatures (Meyer *et al.*, 2002; Wetterich *et al.*, 2011, 2014, 2016). The extremely cold continental conditions of the Weichselian in northern Siberia probably reflected the effect of Eurasian ice sheets in blocking the passage of moisture-bearing airmasses across the area, exposure of large areas of continental shelf to the north by low eustatic sea levels and probably persistent sea-ice cover across the Arctic Ocean (Schirrmeister *et al.*, 2011). Palaeosols associated with Early and Late Wisconsinan full-glacial conditions in yedoma deposits in the Klondike area of Yukon Territory have characteristics analogous to those of both steppe and tundra environments, and also imply arid conditions (Sanborn *et al.*, 2006).

5.8.3   Formation of Yedoma

The evidence provided by exceptionally high syngenetic ice wedges, ice-rich bands marking the former position of the permafrost table, buried palaeosols and youngingupwards radiocarbon ages (Figure 5.17) demonstrates that yedoma sediments progressively accumulated throughout much or all of the last glacial stage, accompanied by upwards permafrost aggradation. The presence of buried palaeosols, ice-rich bands and 'shoulders' on syngenetic ice wedges (Figure 5.16) indicates that rates of sediment accumulation varied throughout this long period. Though some yedoma deposits accumulated continuously throughout the last glacial stage, others exhibit stratigraphic gaps due to intervening episodes of thermal erosion or thaw and thermokarst formation (Schirrmeister *et al.*, 2011).

The nature of the accumulating sediment has excited controversy. Until recently, many Russian researchers viewed yedoma as a dominantly alluvial or lacustrine deposit. Researchers working on yedoma suites in Alaska and Yukon, however, have tended to interpret deposits in terms of primary deposition of aeolian sediment, mainly loess (wind-transported silt), or of loessic deposits reworked by colluvial processes such as slopewash (e.g. Fraser and Burn, 1997; Kotler and Burn, 2000; Muhs *et al.*, 2003; Sanborn *et al.*, 2006), implying that yedoma represents the stratigraphic and genetic equivalent of the loess deposits of mid-latitude North America and Eurasia (Chapter 14). Some Russian researchers have also favoured this view (e.g. Zimov *et al.*, 2006; Zanina *et al.*, 2011), but researchers who have carried out detailed multidisciplinary research on yedoma deposits in northern Siberia have advanced a more complex, polygenetic model (Grosse *et al.*, 2007; Schirrmeister *et al.*, 2011). This model incorporates their observations that: (i) the heavy mineral contents of yedoma sediments differ from area to area, and appear to be related to local sources; (ii) the grain-size distributions of yedoma sediments vary from site to site, as well as in different horizons from a single site; (iii) bimodal and polymodal grain-size distributions suggest polygenetic formation, though an aeolian component is important; and (iv) the sediments are unlikely to represent glacigenic loess (Chapter 14), as they accumulated hundreds of kilometres from the nearest (Eurasian) ice sheets. According to this model, sediment in the form of plant and mineral detritus initially accumulated on extensive snowfields on uplands and cryoplanation terraces, whilst snow-free areas around the snowfield margins experienced comminution of detritus to silt by intense freeze–thaw weathering. Reworking of fine sediment by a combination of meltwater runoff, solifluction and wind is envisaged to have resulted in its progressive accumulation on alluvial fans and foreland plains, accompanied by upwards aggradation of syngenetic permafrost and ice wedges.

According to this interpretation, the yedoma deposits of northern Siberia represent the chronological and stratigraphic equivalent of the mid-latitude Late Pleistocene loess belts, but not their exact genetic equivalent.

This view has been challenged by Murton *et al.* (2015a) in an important multidisciplinary investigation of the Mid to Late Weichselian (MIS 3–2) yedoma at Duvanny Yar (68°38′ N; 159°09′ W), which is considered the Russian stratotype of the 'Yedoma Suite', and has strong affinities with the yedoma exposure investigated by Kanevskiy *et al.* (2011) at Itkillik River in Alaska (Figure 5.16). Murton *et al.* (2015a) convincingly dismissed the possibility of alluvial, lacustrine or polygenetic accumulation of silty deposits at Duvanny Yar on geomorphological, stratigraphic, palaeoenvironmental and palaeoecological grounds, and demonstrated affinities with the properties of primary or reworked loess elsewhere. They explained the characteristic bimodality or multimodality of the yedoma sediments in terms of mixing of grains from various sources: fine silt and clay particles as a background airfall deposit, representing deposition of far-travelled dust carried by high-altitude airstreams; coarse silt as the products of summer dust storms; and fine to medium sand as the product of more localized aeolian transport from floodplains or sandy substrates. Their conceptual model of yedoma accumulation at Duvanny Yar envisaged summer and autumn deposition of loess on an unfrozen, relatively dry surface colonized by grasses, sedges and nongraminoid flowering plants (forbs) that trapped and stabilized the windblown sediments as they accumulated.

Murton *et al.* (2015a) also argued for a broader subcontinental model of extensive aeolian deposition across the Beringian coastlands, in the form of loess, sand dunes and sand sheets deposited by strong winds throughout the region and sourced in part from adjacent exposed continental shelves. They envisaged ice-rich Beringian yedoma as the outcome of Late Pleistocene loess deposition in areas of continuous, cold, ice-rich syngenetic permafrost aggradation, whereas the loess belts of northwest Europe and North America (Chapter 14) represented relatively ice-poor sediment accumulation under fluctuating conditions of cold permafrost, warm permafrost, seasonal ground freezing and permafrost thaw. As a result, the former support huge syngenetic ice wedges that formed continuously over tens of millennia, whilst the latter, now completely thawed, contain evidence for only relatively small ice wedges in the form of ice-wedge pseudomorphs (Chapter 16). This insight elegantly couples our understanding of yedoma with that of cold-climate Pleistocene aeolian deposits across the mid and high latitudes of the northern hemisphere.

Irrespective of origin, the high ice content of yedoma implies that it is extremely thaw-sensitive. This is a cause for concern, because degradation of permafrost within yedoma may induce widespread thermokarst development (Chapter 8), and because terrain underlain by yedoma provides the habitat for numerous arctic and boreal ecosystems that are vulnerable to thaw (Jorgenson and Osterkamp, 2005). Of even greater concern is the fact that yedoma contains a relatively high content of organic material, which on thawing contributes to greenhouse gas emissions (Zimov *et al.*, 2006; Kuhry *et al.*, 2013; Chapter 17).

This chapter has focused on the myriad forms of ground ice that exist in permafrost, but most of these make only a limited contribution to surface morphology, apart from uplift associated with increased substrate volume. The following two chapters describe in greater detail the landforms associated with thermal contraction cracking of permafrost and associated ice-wedge formation, and the development of ice-cored hills (pingos) and other forms of frost mound. The final chapter in this sequence (Chapter 8) is devoted to thermokarst, which develops as a result of thaw of ice-rich permafrost.

6

Thermal Contraction Cracking: Ice Wedges and Related Landforms

6.1 Introduction

One of the most distinctive and widespread features of many high-latitude permafrost environments is polygonal patterned ground defined by a network of shallow troughs or low ridges (Figures 6.1 and 6.2). These remarkable landforms are *frost polygons* (sometimes called *frost-crack polygons* or *thermal contraction crack polygons*) and usually range from ~5 to ~50 m in diameter. Many are orthogonal in plan, but others are pentagonal or hexagonal. They extend across floodplains, deltas, river terraces, drained lakebeds, peaty fens, outwash deposits and till plains, and occur on valley floors, hillslopes and low plateaus. Many have depressed centres that contain ponded water in summer. Others, especially on slopes, have slightly raised centres and depressed margins.

Under each trough lies a wedge-shaped structure, composed of ice, sediment or both, that tapers downward and extends the full length of the trough. Such wedges form through recurrent *thermal contraction cracking* or *frost cracking* of frozen ground. This occurs when winter cooling generates tensile stresses in the permafrost and the overlying frozen active layer. If the tensile stress exceeds the tensile strength of the frozen ground, a thermal contraction crack opens. In areas of abundant snowcover, meltwater entering the crack freezes, forming a narrow *ice vein*. During subsequent winters, cracking is renewed at the same site, adding further ice veins until a slowly thickening *ice wedge* develops. Over thousands of years, ice wedges can grow to widths of several metres, in many areas constituting the dominant form of ground ice. In areas with limited snowcover, however, thermal contraction cracks may become infilled with windblown sediment rather than ice, forming *sand veins*. Lateral accretion of sand veins through recurrent frost cracking at the same site forms *sand wedges*. Wedges comprising both ice and sediment infills are termed *composite wedges*. Collectively, ice wedges, sand wedges and composite wedges are sometimes referred to as *cryogenic wedges*, *frost-fissure wedges* or simply *frost wedges*.

This chapter considers first the process of thermal contraction cracking and polygon evolution, then the growth and characteristics of frost wedges, the modification of wedge structures as permafrost decays and finally the identification and interpretation of relict contraction crack polygons and associated wedge structures. Murton (2013a) provides a concise summary of these topics.

6.2 Thermal Contraction Cracking and Polygon Evolution

As water in soil freezes, the soil expands due to the 9% volume increase as water becomes ice. When most of the water in the soil has frozen, however, further cooling produces net contraction. The amount of contraction is measured by the linear coefficient of thermal expansion or contraction (ψ), given by:

$$\psi = \Delta L / L \cdot \Delta T \qquad (6.1)$$

where L is length, ΔL is change in length and ΔT is change in temperature. For ice within the temperature range for thermal contraction cracking, $\psi \approx 5 \times 10^{-5}\,°C^{-1}$. The value of ψ for frozen sediment varies, depending on its mineralogy, granulometry, organic content, ice and unfrozen water content, degree of saturation and thermal history. For temperatures of $-1\,°C$ to $-10\,°C$, Yershov (1998) cites $\psi = 10^{-2}$ to $10^{-4}\,°C^{-1}$ for frozen clay, 10^{-3} to $10^{-4}\,°C^{-1}$ for frozen sandy silts and silty clays, and 10^{-4} to $10^{-5}\,°C^{-1}$ for frozen sands. Most frozen soils therefore have a much larger coefficient of thermal expansion or contraction than ice. Putting these values in context, $\psi = 10^{-3}\,°C^{-1}$ implies that a 10 m length of frozen ground shortens by 10 cm when cooled by $10\,°C$.

Resistance to thermal contraction produces tensile stress, and cracking occurs if this stress exceeds the tensile strength of the frozen ground. If winter cooling occurs gradually, however, frozen ground responds to tensile stress by creep (internal deformation), which allows the ground to stretch, relaxing the tensile stress. If sufficient creep occurs, tensile stress remains too low to induce cracking. For cracking to occur, therefore, there must be rapid cooling, so that the rate of tensile stress increase exceeds relaxation of stress by creep.

Periglacial Geomorphology, First Edition. Colin K. Ballantyne.
© 2018 John Wiley & Sons Ltd. Published 2018 by John Wiley & Sons Ltd.

Figure 6.1 High-centre frost polygons and thermokarst lakes, central Banks Island, Canadian Arctic Archipelago. Individual polygons form subsets of a larger irregular polygonal net. A colour version of this figure appears in the plates section. *Source:* Courtesy of Tobias Ullmann.

Figure 6.2 (a) High-centre polygons defined by water-filled troughs overlying ice wedges. (b) Flooded low-centre polygons bordered by parallel ridges of sediment pushed up by expansion of underlying ice wedges. *Source:* Courtesy of (a) Matthias Siewert and (b) Julian Murton.

Thermal tensile stress in frozen ground is therefore a function of frozen-ground rheology (creep behaviour), nearsurface ground temperature and rate of cooling.

Our understanding of cracking of frozen ground owes much to Lachenbruch (1962), who developed a nonlinear viscoelastic model of thermally-induced stress that forms the basis of current theory. Lachenbruch's model predicted that frost cracking requires rapid cooling when the temperature at the top of the permafrost is below −15 °C to −20 °C. He showed that crack depth depends on stress distribution at the time of cracking; theoretically, deeper cooling generates deeper cracks, and cracks taper downwards as tensile stress diminishes with depth. Lachenbruch argued that once an ice vein has formed due to freezing of water in an initial crack, such veins form lines of weakness that will be subject to recurrent fracture, allowing growth of an ice wedge (Figure 6.3). He proposed that the size of thermal contraction crack polygons could be explained by stress perturbations induced by a single initial crack, and by the distribution of zones of lower tensile strength in adjacent ground. He also suggested that orthogonal polygons evolve by progressive subdivision, with later cracks forming at right angles to earlier ones, and that random orthogonal crack systems form on isotropic surfaces, and oriented orthogonal systems where initial cracking follows a topographic discontinuity such as a river channel or terrace edge. Non-orthogonal (mainly hexagonal) crack networks he attributed to cracks propagating laterally until they reach a limiting velocity, when they branch at obtuse (~120°) angles, effectively resulting in simultaneous generation of the crack network.

Plug and Werner (2001, 2002) investigated the development of thermal contraction crack networks using numerical modelling that encapsulates fracture initiation and propagation, the influence of open fractures on the adjacent stress field and the 'memory' of past fracture patterns represented by ice wedges. Though their approach been challenged on the grounds that some underlying assumptions are inconsistent with field conditions (Burn, 2004; Plug and Werner, 2008), their models yield several insights into the evolution of crack networks. Amongst the most important are: (i) that thermal contraction crack networks may continue to develop under conditions less severe than those required for network initiation; (ii) that network development is sensitive to extreme temperature events, so that the distribution of frost polygons may not reflect 'average' climate; and (iii) that wedge spacing (and therefore polygon width) reflects extreme winter cooling events (rapidly falling temperatures at low absolute temperatures), which generate the high tensile stresses required to add new fractures to a network.

The importance of rapid cooling at low temperatures in causing frost cracking has been demonstrated in field

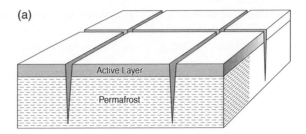

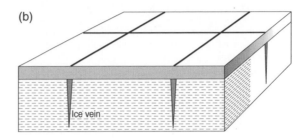

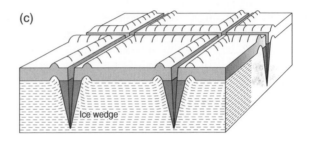

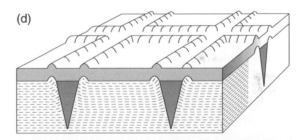

Figure 6.3 Development of ice wedges and frost polygons due to repeated thermal contraction cracking of the ground and infill of cracks by vein ice. (a) Initial cracking (orthogonal), first winter. (b) Formation of ice veins in initial cracks in permafrost. (c) Later cracking of established wedges. Ridges pushed up by volume increase define the margins of low-centred polygons. (d) Incorporation of a new ice vein in the wedge. Wedges progressively widen through time as new ice veins are added.

studies (Table 6.1). Collectively, these show that cracking of ice wedges is associated with T_{TOP} of −15 °C or colder, a steep thermal gradient (≥10 °C m^{-1}) in the upper part of the ground and a rapid decline in air temperatures sustained over periods of 10–100 hours. A strong correspondence between ice-wedge cracking and rapid temperature decline at low temperatures only holds when snowcover is limited, however, as thick snowcover insulates the ground and dampens the effect of the short-term cooling (Mackay, 1993a), and some rapid cooling

Table 6.1 Temperature conditions leading to thermal contraction cracking.

Location	T_a (°C)	T_s (°C)	T_{TOP} (°C)	ΔT_a (°C h^{-1})	ΔT_s (°C d^{-1})	ΔT_{TOP} (°C d^{-1})	$\Delta T/\Delta z$ (°C m^{-1})	References
W Canadian Arctic	−29 to −34	<−20	−18 to −20	−0.1	−0.5 to −0.9	−0.1 to −0.4	−10 to −15	Mackay (1984, 1993a, 1993b)
Northern Québec	−25 to −43	−21.7	−15 to −20	−0.2 to −0.6	−0.9 to −1.7	−0.6	≥ −10	Allard and Kasper (1998)
Bylot Island	−25 to −40	−15 to −29	−13 to −24	−0.2 to −0.9	+0.1 to −1.1	+0.1 to −0.3	−4 to −19	Fortier and Allard (2005)
Svalbard	−21	−15 to −20	−15	n/a	n/a	<−0.3	−9 to −15	Christiansen (2005)

T_a, T_s, T_{TOP}, air temperature, ground surface temperature and temperature at the top of permafrost, respectively, at the time of frost cracking. ΔT_a, ΔT_s, ΔT_{TOP}, cooling rate of air, the ground surface and the top of the permafrost, respectively, before cracking events. $\Delta T/\Delta z$, vertical temperature gradient of the ground.
Source: Adapted from Fortier and Allard (2005) with permission from John Wiley & Sons Ltd.

events may cause cracking within the frozen active layer, but not the underlying permafrost (Matsuoka and Christiansen, 2009; Watanabe *et al.*, 2013). In some cases, atmospheric cooling is transmitted to the top of the permafrost through ridges bordering ice-wedge troughs when the troughs themselves are snow-filled (Christiansen, 2005). The importance of rapid cooling is nicely illustrated by Matsuoka (1999) for a site in Svalbard. During the mild winter of 1990–91, negligible separation of markers on either side of an ice-wedge trough was detected. Conversely, rapid lowering of air temperatures to −35 °C in February 1992 resulted in ground surface temperature falling to −27 °C and the temperature at the top of the permafrost falling to −17 °C. This cooling resulting in opening of the wedge trough by 16 mm, reflecting marked permafrost contraction and probably wedge cracking.

Various characteristics of ice-wedge cracking have been investigated by Mackay (1975, 1984, 1992, 1993b) at sites on Garry Island in the western Canadian arctic. He showed that cracking frequency is variable both spatially and temporally, with 8–42% of wedges cracking at one site in any given year but 22–75% at another, and that cracking frequency is inversely related to snow depth because of the insulating effect of snowcover. He also observed that ice wedges tend to crack along their axes at the same place year after year, producing cracks that average 10 mm in width and taper down to depths of 5 m or more. However, he observed that warming and expansion of the ground in spring cause partial closure of cracks prior to the formation of new ice veins, so that the annual ice increment averages only about 20% of maximum crack width, equivalent to a mean ice-vein width of about 2 mm. He also found that the direction of cracking varies: over an 8-year period, 57% of the wedges that he monitored cracked from the ground surface downwards, while the remainder cracked both upwards and downwards from the wedge top, though the direction of cracking at individual sites was not consistent from year to year. Finally, though there are accounts of gunshot-like reports accompanying crack opening, suggestive of rapid crack propagation along wedge axes, such behaviour is rare; most cracks open incrementally, so that individual crack segments are often offset laterally from one another.

Observations of frost crack initiation on the floors of drained lakes (Mackay, 1999a; Mackay and Burn, 2002) have shown that cracking and ice-vein formation occurred in the first winter after exposure of lake floors to permafrost aggradation. Initial cracks widened to 3–20 cm, much wider than those in mature ice wedges on adjacent ground. Even after partial closure in spring, ice veins up to 30 mm wide developed, and the growth rate (widening) of ice wedges at one site averaged 10–30 mm a^{-1} over the first decade. This extraordinary growth rate probably reflects initially high coefficients of thermal contraction in the newly frozen lake-floor sediments, resulting in annual ground contraction much greater than that characteristic of sites of long-established permafrost. The initial frost cracks were aligned normal to ground contours, then secondary cracks developed parallel to the contour, growing away from the primary cracks and producing a roughly orthogonal network. After 10–20 years, however, cracking gradually ceased due to vegetation colonization, increased snow entrapment and a consequent increase in winter ground surface temperatures, allied to a decrease in the coefficient of thermal contraction of the frozen sediments.

6.3 Ice Veins and Ice Wedges

Ice veins and ice wedges develop when the primary infill of thermal contraction cracks is ice rather than sediment. *Ice vein* refers to ice infilling a single contraction crack; *ice wedge* refers to an ice body constructed from lateral increments of vein ice due to recurrent cracking at the same location (Figures 6.3 and 6.4). Both develop in a wide range of sediments, from ice-rich clays and silts to ice-poor sand and gravel, and even in weathered or

Figure 6.4 (a) Two ice wedges of different widths in the western Canadian Arctic. The wedges are ~2 m long. (b) Conjoint wedges in coarse deltaic deposits, Svalbard. Coalescence of these wedges suggests that they occur near a frost polygon junction. (c) Single ice wedge at Crumbling Point, western Canadian Arctic. (d) Top of an ice wedge exposed in a permafrost tunnel, showing the subvertical laminae that represent progressive addition of ice veins during wedge widening. A colour version of (a) appears in the plates section. *Source:* Courtesy of (a) Matthias Siewert, (b,d) Ole Humlum and (c) Mark Bateman.

fractured bedrock. The main source of ice is snow melt-water, which refreezes in contact with permafrost on entering a frost crack, though hoar ice crystals may grow on crack walls when the air temperature exceeds that of the ground. Conversely, sublimation (the phase change from ice directly to water vapour) may cause loss of ice from open cracks. This effect is most pronounced in polar deserts, where sublimation is favoured by a steep thermal gradient between the air and the ground, low humidity, thin or absent snowcover and strong winds (Raffi *et al.*, 2004; Raffi and Steni, 2011).

Because contraction cracks taper downwards, infilling ice veins also thin downwards. Repeated cracking of ice wedges and the addition of tapering ice veins in the cracks give ice wedges their characteristic triangular shape, as earlier vein-ice increments are pushed aside by the addition of later veins at the centres of the wedges. Such incremental addition of vein ice results in downwards-converging subvertical ice foliation in ice wedges, an effect sometimes emphasized by thin bands of mineral or organic matter that separate constituent ice veins (Figure 6.4). Because ice-wedge growth implies a local

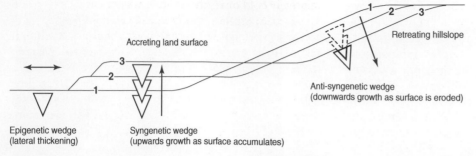

Figure 6.5 Growth of epigenetic, syngenetic and anti-syngenetic ice wedges. Epigenetic ice wedges develop in stable sediment bodies, syngenetic wedges in aggrading sediments and anti-syngenetic wedges under conditions of surface lowering. 1, 2 and 3 represent successive levels of the ground surface. *Source:* Mackay (1990a). Reproduced with permission of Wiley.

increase in nearsurface permafrost volume, ice wedges and adjacent host sediments are subject to deformation. Mackay (1990a) suggested that two forms of deformation occur: (i) upwards shear and deformation of ice and adjacent frozen sediment near ice–sediment boundaries, causing upwards bending or reverse faulting of frozen sediment at wedge margins; and (ii) wedge uplift, due to the lower density of wedge ice compared to the adjacent frozen sediment and lateral squeezing of wedges due to summer expansion of host sediments as the ground warms. Uplift of wedge ice by either or both mechanisms may elevate the top of the wedge into the active layer, causing thaw of the uppermost part of the wedge.

Ice wedges have traditionally been classified as active or inactive, the former being those that are still growing through frost cracking and addition of new ice veins, the latter those in which growth has ceased. This simple classification may be difficult to apply, however, as apparently inactive wedges may be reactivated by infrequent extreme cooling events or changes in nearsurface thermal regime. An alternative classification relates ice-wedge evolution to changes in the land surface (Mackay, 1990a, 1995; Melnikov and Spesivtsev, 2000; Figure 6.5) or climatically-driven changes in the thickness of the active layer (Kasper and Allard, 2001). Three classes of wedge are recognized.

First, *epigenetic ice wedges* grow under approximately steady-state conditions in pre-existing permafrost underlying stable surfaces that experience neither sediment accretion nor sediment removal. Such wedges are therefore younger than the host sediment in which they develop. Recurrent cracking and infilling by ice results in progressive widening but negligible increase in depth or height. These wedges commonly attain widths of 0.5–2.0 m and depths of 2–5 m.

Second, *syngenetic ice wedges* develop where sediment accumulation or peat growth causes upwards aggradation of permafrost. As the permafrost table rises, the top of the growing ice wedge rises accordingly. Where surface aggradation is rapid relative to the frequency of wedge cracking and ice infill, deep narrow wedges form.

Conversely, wide syngenetic ice wedges reflect either gradual surface aggradation and/or relatively rapid wedge growth. Episodic surface accretion results in the formation of nested wedges (Vasil'chuk, 2013; Figure 6.5); continuous surface accretion may produce wedges that broaden downwards in the uppermost part of the permafrost (Morse and Burn, 2013). Syngenetic wedges reach staggering proportions: Yershov (1998) describes syngenetic wedges 50–80 m deep and 8–10 m wide. Such giant wedges, however, are probably restricted to floodplains, deltas and river terraces in northern Siberia, Yakutia and Alaska, in yedoma areas of episodic sediment accumulation that have remained unglaciated over the past 40 ka or longer (Vasil'chuk and Vasil'chuk, 1997; Meyer *et al.*, 2002). Dating of ice in syngenetic wedges in northern Siberia using cosmogenic ^{36}Cl indicates a minimum age of 460 ± 140 ka (Gilichinsky *et al.*, 2007), and the oldest wedges probably exceed 740 ka in age (Froese *et al.*, 2008). In coastal sections in the eastern Siberian Arctic, syngenetic ice wedges dating back to over 50 ka make up over 50% of the outcrop, forming part of an ice complex (yedoma) in which more than 80–90% of the ground is composed of ice (Sher *et al.*, 2005). Similarly, Ulrich *et al.* (2014) employed satellite imagery to derive estimates of 31–63% wedge-ice volume in Late Pleistocene yedoma deposits at four sites in Siberia and Alaska. Holocene syngenetic ice wedges, however, rarely exceed 10 m in depth, and most are much smaller.

Finally, *anti-syngenetic ice wedges* form where there is lowering of the ground surface by erosion or mass movement. Surface lowering results in equivalent lowering of the permafrost table, so the upper parts of the original wedges are truncated by thaw at the base of the descending active layer but new ice veins extend progressively deeper into the substrate (Figure 6.5). Because frost cracking of wedges aligned along hillslopes occurs normal to the surface, the axis of such wedges is also normal to the slope, so that they have a downslope-tilted appearance in cross-section.

Ice-wedge evolution can also be affected by changes in the depth of the active layer caused by changes in air temperature regime, increases or decreases in snow-cover, or the growth or loss of vegetation cover. Active-layer thickening results in truncation of the tops of ice wedges and may be accompanied by reduction or cessation of wedge growth. If the permafrost table subsequently rises and thermal contraction cracking continues or is resumed, *rejuvenated ice wedges* may develop due to survival of vein ice above the truncated wedge surface. Rejuvenated wedges are characterized by the development of one or more nested ice wedges rising from the top of the original wedge (Figure 6.6). Kasper and Allard (2001) showed that the tops of ice wedges in northern Quebec exhibit stepped truncation and rejuvenation structures that relate to alternating episodes of active-layer thickening and thinning. Through radiocarbon dating of organic material near wedge tops, they inferred that Late Holocene active-layer thickening and wedge deactivation were succeeded by active-layer thinning and ice wedge rejuvenation during the Little Ice Age (AD 1450–1910), and again by cooling between 1946 and 1992. Caution is necessary, however, in attributing secondary wedge growth to climatically-induced changes in active-layer depth, as short-term variability in summer thaw depth can also result in the formation of raised ice structures along the axes of ice wedges (Lewkowicz, 1994).

Syngenetic ice wedges can also be employed in palaeo-climatic reconstructions, provided the age of the wedge or the host sediments can be determined by radiocarbon dating of organic inclusions (Lachniet *et al.*, 2012). Oxygen (δ^{18}O) and deuterium (δD) isotope analyses of ice sampled from different generations of Late Pleistocene and Holocene wedges or different parts of a single synge-netic wedge produce isotopic signatures that indicate relatively cold or relatively mild winter conditions (Opel *et al.*, 2011; Streletskaya *et al.*, 2011), but do not readily convert to palaeotemperatures. Vasil'chuk and Vasil'chuk (2009) have identified millennial-scale cooling events (Dansgaard–Oeschger cycles) from the isotope record in Siberian ice wedges, and Meyer *et al.* (2010) have employed isotope analysis of wedge ice to demonstrate severe winter cooling during the Younger Dryas climatic reversal in Alaska.

6.4 Ice-wedge Polygons

Planimetrically, frost polygons can be classified into ran-dom orthogonal systems, orientated orthogonal systems where the pattern of cracking follows some pre-existing terrain discontinuity such as a river channel, and non-orthogonal systems. This simple typology is complicated, however, by the growth of secondary contraction cracks (and eventually ice wedges) that propagate orthogo-nally away from established primary wedges, and of tertiary wedges, which grow away from primary or sec-ondary wedges, subdividing the network into progressively smaller polygons (Figure 6.1). Accumulation of sediments with a high coefficient of thermal contraction also encourages subdivision of the original polygon network. On Bylot Island in the Canadian Arctic Archipelago, for example, Fortier and Allard (2004) found that the primary network of syngenetic ice wedges developed on an out-wash plain had been progressively subdivided by later contraction cracks after accumulation of windblown and organic sediments over the outwash sediments. In general, smaller ice-wedge polygons occur in fine-grained ice-rich soils than in frozen sands, because coefficients of thermal

Figure 6.6 Rejuvenated ice wedge, western Canadian Arctic. The top of the wedge has been truncated by active-layer deepening, but renewed cracking and ice-vein formation following later active-layer thinning have allowed the central part of the wedge to form at a higher level, forming the raised finger of ice near the centre of the wedge. *Source:* Courtesy of Julian Murton.

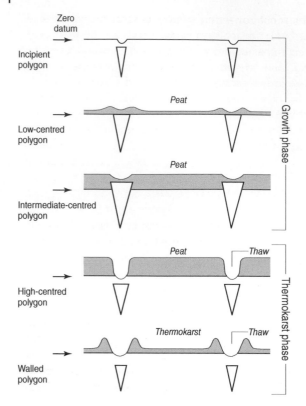

Figure 6.7 Topographic classification of ice-wedge polygons. *Source:* Reproduced from Mackay (2000) under a licence given by Copibec.

contraction are greater for the former, generating higher tensile stresses and thus greater crack density.

Mackay (2000) suggested a topographic classification of polygon morphology. He envisaged that embryonic polygons develop into low-centred polygons through formation of ridges 0.5–1.0 m high flanking wedge troughs, a process that may take several centuries, and that low-centred polygons subsequently evolve into intermediate-centred polygons through peat accumulation in polygon centres (Figure 6.7). High-centred polygons he considered to be an outcome of thaw subsidence of ice-wedge troughs, attributable to rising temperatures, surface disturbance or hydrological changes (Gray and Seppälä, 1991; Steedman *et al.*, 2016), leaving domed areas between the troughs; 'walled' polygons, he interpreted as a degradational phase of polygons with high ice content in the polygon centre. However, polygon evolution may be more complex, interrupted by climatically-driven changes in the growth, collapse and rejuvenation of ridges along polygon margins (De Klerk *et al.*, 2011), and some polygons have an extremely subdued microrelief, with ridges only a few centimetres high or lacking altogether (Morse and Burn, 2013). Mackay's classification, moreover, is based largely on observations made on poorly drained low-lying tundra environments, and may not be universally valid. In parts of the Canadian Arctic

Archipelago, for instance, low-centred polygons are rare, and intermediate- and high-centred polygons predominate (Figure 6.1). High-centred polygons also tend to predominate on elevated or sloping ground, possibly reflecting drainage along troughs (Fortier *et al.*, 2007). On slopes, frost polygons may be obscured by solifluction, which results in infilling of troughs due to slow downslope movement of the active layer. Even on flat ground, polygons may be obscured by vegetation cover, though in such cases their configuration is traceable using ground-penetrating radar (Munroe *et al.*, 2007; Watanabe *et al.*, 2013).

The development of ridges along the edges of ice-wedge troughs has usually been attributed to upwards displacement of frozen sediments to accommodate increases in the volume of growing wedges. Measurements of net displacement of the active layer adjacent to troughs, however, suggest that this explanation is incomplete. Long-term monitoring by Mackay (2000) of the displacement and tilt of rods inserted into the centres and edges of ice-wedge polygons demonstrated slow net transport of the active layer towards polygon troughs. Mackay explained this effect in terms of contrasting contraction and expansion behaviour of frozen ground. During cooling, contraction of frozen ground away from ice-wedge troughs is opposed at polygon centres by similar movement from the opposite sides of polygons. Conversely, expansion of the active layer during spring warming is unopposed at the edges of the ice-wedge troughs, so there is net movement of the active layer towards the troughs. This effect may continue in summer as the underlying permafrost warms and expands, carrying the thawing active layer towards the polygon margins. Net long-term movement of the active layer towards troughs is confirmed by the outwards tilt of spruce trees towards polygon rims (Kokelj and Burn, 2004), and by the deformation characteristics of sediments adjacent to ice wedges (Fortier and Allard, 2004).

The net annual outwards displacement measured by Mackay was 4–6 mm a^{-1} for the nearsurface active layer, equivalent to a 'linear transport coefficient' of $\sim 2.5 \times 10^{-5}\,°C^{-1}$, and this has two important implications. First, long-term movement of mineral sediment and/or peat towards and into ice-wedge troughs encourages localized permafrost aggradation and upwards ice-wedge growth, so that syngenetic (or 'quasi-syngenetic') wedge development may occur even in the absence of peat growth or net sediment accumulation. Second, creep of the active layer and underlying permafrost towards sites of ice-wedge expansion creates a zone of compression and consequent upwards sediment displacement at wedge margins, building the low ridges that flank wedge troughs. Fortier and Allard (2004) have demonstrated upwards movement of sediment at wedge margins of 2.5–4.0 m over the last 3000 years, an average vertical displacement of 0.8–1.2 mm a^{-1}.

6.5 Sand Veins and Sand Wedges

Where thermal contraction cracks fill with mineral particles and/or organic material rather than ice, the resultant structures are termed *primary soil veins* (<10 cm wide at the top) or *primary soil wedges* (>10 cm wide). The adjective 'primary' is used to distinguish such structures from those resulting from 'secondary' infill of sediment following melt of ice veins or wedges. Though there are reports of primary soil veins and wedges infilled with clay-, silt- or gravel-dominated sediment, by far the most common types are *primary sand veins* and *primary sand wedges*, in which the infill is dominated by sand, usually of aeolian origin. For conciseness, these structures are referred to here simply as *sand veins* and *sand wedges*. Murton *et al.* (2000) have provided an excellent account of their characteristics and significance.

Though sand veins and small sand wedges may form in seasonally frozen ground (Seppälä, 1987), almost all documented active examples occur in permafrost, particularly in polar deserts where snowcover and vegetation cover are sparse (Sletten *et al.*, 2003b; Bockheim *et al.*, 2009). They also occur in tundra areas with limited snowcover but local sources of windblown sandy sediment,

but in most low-lying tundra environments they are much less common than ice wedges.

Sand veins described by Murton (1996a) in ice-rich sediments in western arctic Canada take the form of discrete vertical or steeply-inclined layers of massive sand, typically 1–10 mm wide and up to 4 m high, with sharp lateral contacts where they truncate host sediments. Many bifurcate and rejoin, and are both laterally and vertically discontinuous. They occur singly or in collective 'bundles', and each represents the infill, usually by aeolian sand, of a single contraction crack.

Sand wedges exposed in the same area (Figure 6.8) have developed through recurrent thermal contraction cracking and sand-vein infill at the margins of sand-wedge polygons. These wedges are typically 1.0–2.5 m wide and 4–6 m deep, and where they have developed in ice-rich sediments are V-shaped in cross-section, implying that cracking is confined to the wedge. Those penetrating sandy host sediments, however, vary from V-shaped to irregular. Irregular wedges probably develop because the tensile strength of the wedge sediment is similar to that of the host material, so that the locus of cracking shifts periodically across the wedge and adjacent sediments (Murton, 1996a). Such shifts result in veins branching

Figure 6.8 Large epigenetic composite wedge with a predominantly sand infill, Crumbling Point, western Canadian Arctic. Bundles of sand veins can be distinguished within the wedge, which extends downward through a thaw unconformity that truncates massive ice and ice-rich sediments. *Source:* Courtesy of Julian Murton.

from the sides and toes of wedges, interfingering of veins and host sand, cross-cutting of 'bundles' of veins within wedges and development of veins between wedges. Where V-shaped sand wedges form in ice-rich sediments with relatively high tensile strength, the wedge fill may be structureless, but commonly it exhibits vertical or subvertical laminae, sometimes cross-cutting, which represent successive additions of sand veins following cracking events. Where wedges form in sandy host sediments with tensile strength similar to that of the wedge infill, a more complex structure evolves, with cross-cutting 'bundles' of sand veins, irregular lateral offshoots, a splayed toe and inclusions of host sediment. Host strata adjacent to sand wedges are commonly upturned, folded and/or faulted due to lateral compression produced by wedge growth, though Murton (1996a) has also documented downturn of sediment-poor ice adjacent to wedges, probably due to creep generated in the icy host material by the greater density of wedge sand.

Like ice wedges, sand wedges may be epigenetic (developing in stable sediment bodies), syngenetic (developing in aggrading sediments), anti-syngenetic (developing under conditions of surface lowering) or rejuvenated by a period of colder winters following one of relative warmth. Murton and Bateman (2007), for example, have described remarkable near-vertical syngenetic sand infills exceeding 9 m in depth yet only 1–21 cm wide within aeolian sand deposits. The narrowness of these structures implies a delicate balance between cracking frequency and infill, and aeolian sand accumulation. Conversely, at a nearby site, anti-syngenetic sand wedges with wide (≥3.9 m) tops extend down from an erosion surface, and are interpreted as having grown downwards during surface lowering by wind erosion.

Not all sand wedges are produced by sediment infilling thermal contraction cracks. Li *et al.* (2014) have described small (0.5–0.6 m deep) sand wedges formed within a dense polygonal network in the hyper-arid cold environment of the Gobi Desert. These apparently reflect expansion of the saline host sediments due to wetting during rare rainfall events, contraction during subsequent drying, formation of desiccation cracks in existing sand wedges and infill of these cracks by windblown sand, causing episodic lateral growth of the wedges.

6.6 Composite Veins and Composite Wedges

Composite veins and *composite wedges* reflect primary infill of thermal contraction cracks by both ice and mineral sediment. Both ice and soil may fill a single contraction crack, and recurrence of mixed ice and soil infill creates composite wedges composed of alternating ice and soil (sand) veins. At Crumbling Point on the western

Canadian arctic coast, for example, alternating infills of sand and ice are common along the sides of sand wedges penetrating icy sediments (Murton, 1996a). Alternatively, the composition of crack infill may change with a systematic increase or decrease in aeolian sediment supply or snowcover, or both (Figure 6.8). At Crumbling Point, active ice wedges have developed within or beside inactive sand wedges; the latter formed under cold, arid proglacial conditions when aeolian sand was in abundant supply, the former under more humid conditions similar to the present.

6.7 Sand-wedge Polygons

Viewed from above, sand-wedge polygons are indistinguishable from ice-wedge polygons, and they probably evolve in a similar manner. Sletten *et al.* (2003b) have studied the evolution of sand-wedge polygons in the Dry Valleys of Antarctica, one of the most arid periglacial environments on Earth. Measurement of the separation of rods drilled into permafrost on either side of sand-wedge troughs yielded an average widening rate, measured over 39 years, of ~0.6 mm a^{-1}. Slight tilting of the rods inwards towards troughs during this period indicates net movement of the active layer towards the troughs, as described by Mackay (2000) for ice-wedge polygons. As with ice wedges, compression induced at the edges of sand wedges by wedge expansion has resulted in uplift of frozen sediments along wedge margins, forming low ridges on either side of the wedge troughs.

Sletten *et al.* (2003b) suggested that sand-wedge widening drives creep of ice-rich permafrost towards polygon centres, causing upwards bulging. Coupled with net movement of active-layer sediments towards (and into) polygon troughs, this movement produces slow sediment circulation through polygons (Figure 6.9), and over very long timescales allows sand wedges to occupy polygon centres. Sletten *et al.* (2003b) argued that if an average

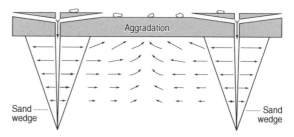

Figure 6.9 Vertical section across a single sand-wedge polygon, showing hypothetical long-term circulation of sediment. Aggradation of sediment between the wedges is offset by loss of surface sediment into thermal contraction cracks, so there is long-term circulation of sediment associated with lateral wedge expansion. *Source:* Sletten *et al.* (2003b). Reproduced with permission of the American Geophysical Union.

widening rate of 0.6 mm a^{-1} is sustained over millennial timescales, it would take 10–100 ka for sand wedges to extend across polygons 10–20 m in diameter, completely replacing the original ice-rich permafrost. This view is supported by their observation that at sites of long-term stability (~10^6 years), sand wedges appears to be continuous from one side of polygons to the other. This outcome is probably very rare, however, as few surfaces remain stable for such a long duration.

The same authors also compared sand-wedge polygons on surfaces of different ages to establish a developmental sequence. Polygons on the youngest (2–4 ka) surfaces are associated with sand wedges 0.5–1.0 m wide underlying shallow curvilinear troughs that intersect at right angles to form irregular tetragons up to 30–50 m across. Polygon networks on older (~12 ka) surfaces comprise 2–3 m wide sand wedges underlying distinct troughs with low raised margins; younger (secondary) troughs intersect the original network, so polygons are smaller, with an increase in equiangular (~120°) triple junctions and thus of five- or six-sided polygons. On the oldest (~1000 ka) surfaces, polygons are mainly pentagonal or hexagonal, 10–20 m wide and entirely underlain by sand-wedge sediments; trough margins rise about 0.8 m above the troughs. These changes in planform appear to reflect long-term adjustment of the crack network to tensile stresses (cf. Plug and Werner, 2001, 2002). Elsewhere in Antarctica, however, Bockheim *et al.* (2009) observed retrogressive development of sand-wedge polygons over very long timescales, with well-developed polygons on surfaces ~117 ka old, moderately-developed polygons on surfaces ~200 ka old and poorly expressed polygons on surfaces 1.0–2.2 Ma in age. They attributed these contrasts to lowering of the coefficient of thermal contraction of permafrost due to very slow sublimation of ice-bonded permafrost.

6.8 Frost Cracking of Seasonally Frozen Ground

So far, this chapter has considered only the effects of thermal contraction cracking of permafrost. In continental climates characterized by severe winter freezing but deep summer thaw, however, thermal contraction cracking of seasonally frozen ground also occurs, both in the active layer above permafrost and in areas of deep seasonal ground freezing outside the permafrost limit (Romanovskii, 1985). During an exceptionally cold winter, a network of seasonal frost cracks extended over parts of a golf course in New Hampshire, USA; crack depths reached 50 cm, and a thin ice vein at least 30 cm long was observed (Washburn *et al.*, 1963). Svensson (1974) described 'earthquake-like shocks' and sharp reports associated with seasonal ground fissuring in

southern Sweden when exceptionally cold polar air passed over the area, and there is evidence of recent thermal cracking in permafrost-free areas of Iceland during exceptionally severe winters. In seasonally frozen ground, however, ice veins melt completely during summer thaw, and often little trace remains of the former crack location.

If an open crack in seasonally frozen ground becomes infilled by mineral sediments, however, a *seasonal soil vein* may survive thaw, and if cracking is repeated at the same site, a *seasonal soil wedge* (sometimes termed a *ground wedge*) may develop. In northern Fennoscandia, for example, seasonal frost cracking has resulted in the formation of irregular polygons up to 12 m wide bounded by shallow troughs, and in winter cracks up to 70 cm deep have been observed. Where seasonal frost cracks have formed repeatedly on the silt floor of a deflation hollow amid sand dunes, *seasonal sand wedges* 35–50 cm deep define small polygons 2–4 m in diameter (Seppälä, 1987). In areas of discontinuous permafrost in northern Quebec, polygons 10–30 m in diameter on the crests of drumlins are defined by shallow troughs underlain by seasonal soil wedges 0.3–1.0 m deep formed by deformation of adjacent soil into thermal contraction cracks in the upper part of the active layer. Charcoal within these wedges suggests that cracking was initiated when fire destroyed the forest cover so that snow-free ground was exposed to severe winter cooling (Jetchick and Allard, 1990). Narrow fissures 0.5–3.0 mm wide and up to 0.9 m deep observed in winter indicate that thermal contraction cracking occurs at present.

Though the seasonal soil wedges or sand wedges described above may be distinguished from frost wedges developed in permafrost by their limited depth, much larger soil wedges are known to develop both in seasonally frozen ground and in the active layer above permafrost (Katasonov, 1973). Melnikov and Spesivtsev (2000) have illustrated seasonal soil wedges up to 2 m deep and 2 m wide developed in a variety of sediments. Such large seasonal soil wedges appear to be restricted to midcontinental subarctic environments characterized by extreme winter cold and deep seasonal freezing of the ground.

6.9 Thaw Modification of Frost Wedges

During permafrost degradation, frost veins or wedges may undergo complex changes in morphology and structure, so that relict (thawed) examples sometimes bear little resemblance to their active counterparts. This process is termed *thaw modification* and reflects deepening of the active layer due to climate change, or to local disruption of nearsurface thermal balance as a result of removal of vegetation cover or peat, exposure of wedges in coastal or riparian bluffs, or thermal erosion by running water. Such changes in nearsurface temperature regime may be progressive, incremental or periodically reversed, further complicating the outcome.

6.9.1 Thaw Modification of Ice Wedges

Replacement of ice in ice wedges by overlying and adjacent sediment results in the formation of *ice-wedge pseudomorphs*, a nomenclature now preferred to 'ice wedge casts', as the latter implies retention of the original wedge morphology. Infill of voids created by melt of ice wedges is nonetheless referred to as *ice-wedge casting* for want of an appropriate synonym. Thaw modification of ice wedges is difficult to study because it operates underground. Murton and French (1993b) overcame this problem by studying the structures associated with thawed and thawing ice wedges exposed in coastal exposures in western arctic Canada. In this situation (which may differ from that of downwards thaw on level terrain), their observations highlighted the importance of thermal erosion by subsurface running water in creating tunnels that were infilled or partly infilled by subsidence or collapse of overlying material, and by inwash of sediment, showing that both vertical and horizontal movements of sediment may contribute to wedge casting. Observations by Bray *et al.* (2006) of structures in the walls of a permafrost tunnel also indicated that thermal erosion of ice wedges by running water is a key element of thaw modification. Murton and French (1993b) emphasized that though ice wedges preferentially develop in ice-rich fine-grained sediments, such sediments are prone to deformation during thaw and thus may not preserve ice-wedge pseudomorphs. By contrast, ice-poor coarse-grained sediments (sands and gravels) often retain their integrity during thaw, and thus may retain a semblance of the original wedge morphology. There is therefore selective preservation of ice-wedge pseudomorphs, which explains why many reported Pleistocene examples occur in sand and gravel deposits.

This point has been confirmed by scaled centrifuge modelling experiments designed to investigate the process and structures associated with thaw modification of ice wedges (Harris and Murton, 2005; Harris *et al.*, 2005). These showed that the ice content and thaw consolidation behaviour of the host sediments exercise a major control on the geometry and size of the resulting pseudomorphs. In sands and sandy silts, thaw consolidation may be largely complete before the wedge melts. These sediments therefore retain much of their pre-thaw integrity, and casting is dominated by tunnel development, normal faulting along wedge walls and slumping of sediment from the tunnel sides and roof. In ice-rich silty clays, however, slow thaw consolidation and high pore-water pressures result in soft-sediment deformation towards voids created by melt of the wedge. In the experiments, such deformation was evident in downwards tilt of marker horizons towards the former wedge axis; in reality, deformation of homogeneous ice-rich sediment could result in modification of the original wedge structure beyond recognition, or in complete disappearance of all trace of the original wedge.

6.9.2 Thaw Modification of Sand Wedges and Composite Wedges

Ice-poor sand wedges undergo limited volume reduction on thaw, and hence might be expected to experience little or no thaw modification. This is true, however, only where sand wedges occupy ice-poor, coarse-grained host sediments that also undergo limited modification during permafrost degradation. Where the host sediment comprises massive ice, or ice-rich silts or clays, the shape of the original sand wedge may be altered by deformation of the host sediment as it thaws. Murton and French (1993b) noted that in response to a past episode of active-layer deepening, the tops of some sand wedges had sagged and spread laterally, and load structures had developed at contacts between the wedge sand and the host material due to deformation of the latter. Sand wedges crossing inclined thaw contacts exhibited folding and shearing, possibly reflecting differential frost heave. It seems likely that complete thaw of sand wedges and adjacent ice-rich host sediment may completely alter the original wedge shape and destroy or modify any internal laminar structure. The effect of thaw on composite wedges probably depends on ice content: sand laminae in ice-poor parts of such wedges may be preserved, but in ice-rich parts all traces of the original structure are likely to be destroyed.

6.10 Frost-Wedge Pseudomorphs and Frost Polygons in Areas of Past Permafrost

6.10.1 Frost-wedge Pseudomorphs

Because frost wedges in thaw-stable sediments have high preservation potential, wedge pseudomorphs represent some of the most important evidence for inferring former permafrost in areas that experienced periglacial conditions during Pleistocene glacial stages (Figures 1.2 and 6.10). In Europe, Asia and North America, frost wedge pseudomorphs are abundant in near-surface sediments beyond the maximum reach of the last Pleistocene ice sheets, providing evidence for permafrost development and severe winter cooling during the last glacial stage (Chapter 16). They also occur inside the limits of the last Pleistocene ice sheets, demonstrating permafrost development under terrain vacated by retreating glacier ice (Gao, 2005; Ewertowski, 2009), and in some

Figure 6.10 (a) Late Wisconsinan ice-wedge pseudomorph in sandy sediments, southwest Ontario, Canada. Note faulting of host sediments along the margin of the wedge. (b) Ice-wedge pseudomorph in coarse gravels, Lincolnshire, eastern England. The infill of clay and sand is derived from a sediment layer that has subsequently been removed by erosion. *Source:* Courtesy of (a) Cunhai Gao and (b) Peter Worsley.

Table 6.2 Typical characteristics of cryogenic wedge pseudomorphs.

Origin	Infill	Infill structures	Adjacent host sediments
Ice wedge	Secondary infill of overlying and adjacent sediments; often massive or deformed	Steeply dipping layers formed by deformation of adjacent sediment; concave-up subsidence structures; faulted sediment blocks; vertically aligned clasts; tunnels or voids	Usually downturned; may be upturned in thaw-stable host sediments; step-faulting common
Primary sand wedge	Primary infill of well-sorted aeolian sand	Vertical or near-vertical lamination or massive (rare); cross-cutting bundles of sand veins	Typically upturned; rarely downturned or undeformed; intruded by apophyses (fingers) of wedge sediment
Composite wedge	Primary infill of aeolian sand and secondary infill of overlying and adjacent sediments	Similar to those of sand and ice wedges, depending on original ice–sand composition.	Upturned or downturned, depending on ice content and host sediment thaw stability

areas during the renewed cooling of the Younger Dryas Stade of ~12.9–11.7 ka (e.g. Isarin, 1997a, 1997b; Liverman *et al.*, 2000; Rémillard *et al.*, 2015). Frost-wedge pseudomorphs in Pleistocene stratigraphic successions also provide evidence for the existence of permafrost during cold stages prior to the last glacial period (e.g. Vandenberghe, 1992b; Kasse, 1993; Murton *et al.*, 2015b).

6.10.2 Identification of Frost-wedge Pseudomorphs

Identification, dating and interpretation of frost-wedge pseudomorphs are not always straightforward.

Identification poses three problems: first, various structures of nonperiglacial origin resemble frost-wedge pseudomorphs exposed in vertical section; second, it is sometimes difficult to distinguish between ice-wedge, sand-wedge and composite-wedge pseudomorphs; and third, it may be difficult to differentiate structures representing thermal cracking of former permafrost from soil wedges formed through thermal cracking of seasonally frozen ground. Some of the criteria for identifying frost-wedge pseudomorphs that represent thermal contraction cracking of former permafrost are outlined in Table 6.2. However, no single criterion is necessarily definitive, all diagnostic characteristics are seldom present in the same exposure and soft-sediment deformation during thaw modification may have altered pseudomorphs so that they no longer retain their original morphology or internal structure. Critical assessment of the structural, stratigraphic, sedimentological and geomorphological context is necessary if true frost-wedge pseudomorphs are to be distinguished from wedge-shaped imposters.

The most convincing test of a frost crack origin for wedge structures is a demonstration that these form part of a polygonal net. Most structures that mimic frost-wedge pseudomorphs are either laterally discontinuous (e.g. root casts, solution pipes, water-escape fissures and load casts), so these alternative origins can be excluded by excavation or employment of ground-penetrating radar (Doolittle and Nelson, 2009). Other wedge-like structures, notably sedimentary dykes, dilation crack infill and fault infill, may be laterally continuous but do not form a polygonal net. Sediment-filled desiccation cracks form a polygonal pattern, but are usually <0.5 m deep, form polygons <2–4 m in width and lack the structural characteristics of frost-wedge pseudomorphs.

Different types of frost-wedge pseudomorphs can usually be distinguished from one another on the basis of diagnostic structural features (Table 6.2), though problems of interpretation may arise when secondary infill comprises aeolian sand, so that the resulting ice-wedge pseudomorphs resemble sand wedges, or where deformation of host sediments has modified or destroyed fill structures. Some older accounts of frost-wedge pseudomorphs require reassessment in light of current understanding of thaw modification processes and the characteristics of different types of pseudomorphs. Many early investigators assumed that frost-wedge pseudomorphs developed from ice wedges only, whereas more recent work based on detailed analysis of wedge structures suggests that sand wedges and composite wedges are strongly (and probably preferentially) represented in the Pleistocene stratigraphic record (Nissen and Mears, 1990; Kolstrup, 2004; Ghysels and Heyse, 2006).

Distinguishing pseudomorphs that represent thermal cracking of permafrost from those of soil wedges formed through thermal cracking of seasonally frozen ground is problematic. Several authors have inferred that seasonal soil wedges are typically <1 m deep and <4 m apart, and thus less than half the size and spacing of nearby permafrost-wedge pseudomorphs (French *et al.*, 2003; Vandenberghe *et al.*, 2004), but Ribolini *et al.* (2014) have described soil wedges reaching depths of 1.5 m. As noted earlier, moreover, seasonal soil wedges up to 2 m deep are illustrated in the Russian literature, and some exhibit structures similar to those of permafrost-wedge pseudomorphs. Caution is therefore necessary in inferring former permafrost from wedge pseudomorphs in continental interiors near the southern limits of Pleistocene permafrost.

6.10.3 Relict Frost Polygons

In mid-latitude areas where the Late Pleistocene ground surface has not been modified by erosion, sediment accumulation or human activity, traces of former thermal contraction cracks may be evident in the form of polygonal *crop marks* that represent the original contraction crack network. Differences between the moisture-retention properties of sediment infilling polygon troughs and those of polygon centres lead to plant growth being enhanced or retarded along polygon boundaries. Such contrasts cannot be detected on the ground but are often clear on aerial or satellite imagery and can be detected by geophysical survey (Doolittle and Nelson, 2009). In North America, numerous polygonal patterns have been detected south of the limit of the last (Late Wisconsinan) ice sheet, for example in Wyoming (Nissen and Mears, 1990), Iowa (Walters, 1994), New Jersey, Delaware and Maryland (Gao, 2014), and these probably reflect thermal contraction cracking of permafrost during the last glacial stage. In Europe, a last glaciation age has also been inferred for polygonal networks outside the Late Weichselian ice-sheet limit, for example in west Jutland, eastern England, France and Flanders (Svensson, 1988; West, 1993; Ghysels and Heyse, 2006; Bertran *et al.*, 2014; Andrieux *et al.*, 2016). Polygonal networks also occur inside the limits of the last Pleistocene ice sheets, indicating the formation of permafrost in such areas after the onset of ice-sheet retreat (Chapter 16).

Exceptionally, relict frost-crack networks of Late Pleistocene age can be distinguished by surface morphology. In northern Norway, Svensson (1988) found well-preserved relict frost-wedge polygons defined by a network of furrows 10–20 cm deep on terraces and raised beaches; the relationship with shoreline chronology suggests that these were last active during the Younger Dryas Stade of ~12.9–11.7 ka. Conversely, excavation has also exposed ancient polygons of much greater age than those described above. Sand-wedge polygons developed in a palaeosol in eastern England, for example, are believed to be more than 430 ka old (Rose *et al.*, 1985; Figure 6.11).

6.10.4 Dating of Frost-wedge Pseudomorphs and Primary Sand Wedges

Frost-wedge pseudomorphs are of limited value in palaeoenvironmental reconstructions if the timing of wedge formation cannot be determined. Luminescence dating techniques provide a means of directly dating sand-wedge and composite-wedge pseudomorphs (e.g. French *et al.*, 2003; Vandenberghe *et al.*, 2004; Buylaert *et al.*, 2009; Guhl *et al.*, 2012; Liu and Lai, 2013; Rémillard *et al.*, 2015), but may yield only a maximal age for wedge formation if bleaching of sediment infill was incomplete (Kolstrup, 2004). Luminescence dating

Figure 6.11 (a) Relict sand-wedge polygon 2.5 m wide exposed by excavation at Newney Green near Chelmsford, England. (b) Small sand wedge in Middle Pleistocene sand and gravel deposits at Broomfield, near Chelmsford, England. Both features are overlain by glacigenic deposits of MIS 12 age, and therefore over 430 000 years old, but the wedge still preserves downwards-tapering laminae representing individual sand veins. *Source:* Courtesy of Jim Rose.

has also been applied to the secondary sediment infill of ice-wedge pseudomorphs, but though this may provide an approximate age for the timing of thaw modification (Owen *et al.*, 1998; Porter *et al.*, 2001), it does not establish the timing of ice-wedge growth and may yield invalid results if infill by host sediments is sampled. Radiocarbon dating usually cannot be applied directly to pseudomorphs, though Wang *et al.* (2003) reported last-glacial-stage radiocarbon ages obtained through radiocarbon dating of eluvial calcium carbonate in relict sand wedges.

Stratigraphic context may also be used to constrain the time of wedge formation, particularly when the age of the host sediment is known (e.g. Briant *et al.*, 2004; Ewertowski, 2009). Epigenetic wedge pseudomorphs represent former wedges that were younger than the host sediments, but older than any overlying sediments; syngenetic pseudomorphs represent wedges that were approximately coeval with adjacent host sediments. Location relative to former glacier limits may also provide useful information. Where frost-wedge pseudomorphs occur in nearsurface sediments or are represented at the surface by a polygon net outside the limits of the last Pleistocene ice sheets, it is often assumed that they represent thermal contraction cracking during the last glacial stage, though in some cases an earlier origin cannot be discounted (Ghysels and Heyse, 2006). Nearsurface frost-wedge pseudomorphs and polygonal nets inside the limits of the last Pleistocene ice sheets imply wedge formation after the onset of ice-sheet retreat.

6.10.5 Palaeoenvironmental Implications

Permafrost-wedge pseudomorphs and relict polygon nets have been widely employed to estimate palaeotemperature, and particularly mean annual air temperature (MAAT) at the time of wedge formation. Such estimates assume that present-day climatic constraints on active wedge development can be applied to relict examples of the same features. Péwé (1966), for example, noted that the distribution of 'active' ice wedges in Alaska coincides approximately with that of continuous permafrost, and thus MAATs of −6 to −8 °C or lower, and argued that frost-wedge pseudomorphs have similar implications. This argument now appears flawed. One reason for this is that it is difficult to draw a meaningful boundary between 'active' and 'inactive' wedges that can be related to MAAT. For example, Mackay (1992) showed there is every gradation between wedges that crack in most years and those that crack very infrequently during extreme winter cooling events. Moreover, Hamilton *et al.* (1983) have shown that ice wedges in Alaska have recently been active under MAATs close to −3.5 °C, and Burn (1990) demonstrated that wedges near Mayo in Yukon Territory (MAAT about −4 °C) must have grown over the previous 30 years. Both studies demonstrate that ice-wedge cracking and growth are not necessarily confined to areas of continuous permafrost with MAATs ≤ −6 °C, and may occur in some areas of discontinuous permafrost.

An additional problem arises in relating the constraints on frost cracking to 'average' climatic conditions. Because the critical thermal tensile stress leading to cracking is a function of several variables (coefficient of thermal contraction, frozen-ground rheology, cooling rate, near-surface winter ground temperature and thermal gradient), it is difficult to relate cracking behaviour or frequency to a single thermal parameter such as MAAT. Thermal contraction and creep behaviour are both affected by ground temperature, sediment texture and ice content, so that, for example, steeper thermal gradients and lower temperatures are required for cracking of fine-grained frozen sands than for ice-rich clayey silts. Moreover, mean annual ground surface temperature (MAGST) is only loosely related to MAAT because of the variable thermal offset of roughly 1–6 °C between the two, and MAAT and MAGST reflect both summer and winter temperatures and give limited indication of the severity or rate of winter ground cooling.

In view of these caveats, numerous authors have cautioned against quantitative association of frost-wedge pseudomorphs with climatic averages (Harry and Gozdzik, 1988; Burn, 1990; Murton and Kolstrup, 2003; Plug and Werner, 2008; Murton, 2013a). These structures, if correctly interpreted, conclusively demonstrate the former presence of permafrost, and in most cases probably imply former continuous permafrost. It is likely that most formed under MAATs lower than about –3 to –4 °C, but this is no more than an approximate upper estimate based on information regarding the distribution of presently 'active' wedges.

The growth of sand wedges, rather than ice wedges, is favoured in areas of continuous permafrost associated with low precipitation, thin or patchy snowcover, limited vegetation cover and transport of aeolian sediment by strong winds. It seems reasonable to infer that similar conditions favoured sand-wedge development in mid-latitudes during Pleistocene glacial stages. More detailed interpretation of the palaeoenvironmental implications of relict sand wedges is difficult. They could represent conditions similar to those of present-day polar deserts, or of steppe-tundra environments where there is a nearby source of exposed sandy sediment such as a floodplain. In some cases, it is likely that Pleistocene sand wedges developed near the margins of former ice sheets, where outwash deposits furnished an abundant supply of sand for entrainment by gusty katabatic winds blowing off the ice. In Europe and North America south of the last ice sheets, ice-wedge, sand-wedge and composite-wedge pseudomorphs of apparently similar age are sometimes found in close proximity (Kolstrup, 1986; Nissen and Mears, 1990). This situation may reflect topographic differences, with sand wedges and composite wedges forming on higher, more exposed terrain that was better drained, relatively snow-free and supported incomplete vegetation cover (Kasse and Vandenberghe, 1998; Ghysels and Heyse, 2006). Alternatively, coexistence of Pleistocene ice and sand wedges could reflect instability in vegetation cover, possibly associated with rapid climate changes, for which there is no appropriate analogue in present high-latitude environments (Murton and Kolstrup, 2003). Only where additional proxy palaeoenvironmental evidence is available is more precise interpretation warranted of the conditions under which relict sand wedges developed (e.g. Huijzer and Vandenberghe, 1998; Kasse *et al.*, 1998).

7

Pingos, Palsas and other Frost Mounds

7.1 Introduction

Though ice exists within permafrost without any resulting surface expression (Chapter 5), in some areas the development of ground ice has caused updoming of overlying sediments, peat or even bedrock, producing a family of landforms sometimes referred to as *frost mounds* or *ground-ice mounds*. The most striking manifestation of this phenomenon is the formation of *pingos*, ice-cored hills that commonly reach heights of 10 m or more. *Palsas* and *lithalsas* are more subdued permafrost mounds, typically 1–7 m high, formed by localized frost heave in frost-susceptible sediments. Other perennial ground ice mounds do not fall readily into either category, and *ephemeral frost mounds* produced during winter freezing but destroyed by summer thaw constitute a related phenomenon. Useful reviews of these features have been published by Gurney (1998, 2001), Pissart (2002) and Ross (2013), but there is still debate concerning classification of frost mounds and incomplete understanding of their genesis.

7.2 Characteristics of Pingos

Pingos are ice-cored hills, usually conical or dome-shaped, that form through the freezing of pressurized groundwater within near-surface permafrost, resulting in the updoming of overlying frozen sediments or even bedrock (Figures 7.1 and 7.2). Most pingos are less than 20 m high, with basal diameters up to about 250 m, but some are much higher or broader. The highest documented pingo in Alaska rises ~54 m above the surrounding plain.

As outlined below, pingo formation requires specific topographic and hydrologic circumstances, and the distribution of active pingos is restricted to locations where these conditions are met. Approximately 1350 pingos occur on the Tuktoyaktuk Peninsula area of the western Canadian arctic, with a further 80–100 located on the adjacent Mackenzie Delta (Mackay, 1998). Roughly 500 more are found in the Yukon, and smaller clusters are scattered throughout other areas of arctic and subarctic Canada. There are over 1200 pingos in northern Alaska (Jones *et al.*, 2012a), and more than 6000 in northern Asia (Grosse and Jones, 2011), as well as clusters in parts of northern Scandinavia, Svalbard, Greenland, Mongolia and the Qinghai-Tibet Plateau. No convincing examples have yet been reported from Antarctica. Grosse and Jones (2011) have estimated that the present global total of active pingos probably exceeds 11 000.

Most pingos fall into one of two types: hydrostatic or 'closed-system' pingos and hydraulic or 'open-system' pingos. For pingo formation, injection of groundwater under high pressure is required to overcome geostatic overburden stress and so allow the growth of an ice core and consequent updoming of overlying frozen ground. Under hydrostatic pingos, high water pressures are generated by expulsion of pore water from saturated coarse-grained sediments as permafrost aggrades around an enclosed talik of unfrozen ground. Hydrostatic pingos most commonly develop on low relief terrain, particularly where drainage of a lake overlying a talik results in inwards advance of permafrost through the underlying saturated sediments, thus isolating the talik as a closed system from which water expelled during freezing of surrounding ground cannot escape (Figure 7.3). Under hydraulic pingos, subpermafrost artesian water pressures are generated by groundwater flowing in a confined subpermafrost or intrapermafrost aquifer from adjacent higher ground. Hydraulic pingos are therefore often located on floodplains or deltas flanked by steep valley-side slopes, on lower hillslopes or on alluvial fans.

Irrespective of origin, all pingos undergo a cycle of growth and degradation over hundreds or thousands of years. The growth phase is marked by progressive uplift of the pingo summit and steepening of pingo sides, often with little change in pingo diameter. But pingo growth contains the seeds of pingo destruction: as the ice core grows, the frozen ground at the summit of the pingo experiences stretching, resulting in the development of dilation (tension) cracks. With continued pingo growth, these widen to form trenches and gullies, thinning the overburden at the pingo summit

Periglacial Geomorphology, First Edition. Colin K. Ballantyne.
© 2018 John Wiley & Sons Ltd. Published 2018 by John Wiley & Sons Ltd.

Figure 7.1 Hydrostatic pingos, Tuktoyaktuk Peninsula, western Canadian Arctic. A colour version of this figure appears in the plates section. *Source:* Courtesy of Emma Pike.

Figure 7.2 Fløtspingo, Reindalen, Svalbard, a hydraulic pingo formed where groundwater flowing under pressure has fed the growth of an ice core within permafrost, updoming the overlying frozen sediments. A colour version of this figure appears in the plates section.

and ultimately exposing the ice core. Progressive thaw of the exposed ice core leads to subsidence and inwards slumping of sediment, widening the summit depression until a pingo in an advanced stage of decay forms a broad depression surrounded by a circular rampart of ice-cored sediment. This cycle of growth and decay is an intrinsic feature of pingo behaviour, and is independent of climate change.

Pingo growth and decay are often accompanied by the development of various ancillary features. Stretching of the flanks of a pingo during growth causes the development of dilation cracks, which sometimes extend on to adjacent ground. An ice-wedge network may develop through thermal contraction cracking on the flanks of pingos. The steep sides of mature pingos are susceptible to various forms of mass movement, including active-layer slides, solifluction and permafrost creep, and debris flows and surface runoff may deposit small fans at the pingo periphery. Springs fed by artesian groundwater or reaching the surface through rupture of frozen ground may generate small-scale frost mounds adjacent to the parent pingo or surface icings composed of water that has frozen on exposure to subzero temperatures at the surface.

7.3 Hydrostatic Pingos

Much of our understanding of hydrostatic pingos stems from the work of the Canadian scientist Ross Mackay. Over 40 years, Mackay carried out a programme of research into pingo behaviour in the Tuktoyaktuk Peninsula area of the western Canadian arctic, and the account in this section is based largely on his findings.

7.3.1 Pingo Formation and Growth

The Tuktoyaktuk Peninsula area is underlain by permafrost 400–700 m deep; mean annual air temperatures (MAATs) range between –10 °C and –12 °C, and mean annual ground surface temperatures (MAGSTs) between –7 °C and –10 °C. The low-lying terrain of the area is pockmarked with thermokarst lakes. The larger and deeper lakes do not freeze to their beds in winter and overlie large taliks of unfrozen water-saturated coarse sandy sediments (Figure 7.3a). Mackay (1988a) has shown that catastrophic lake drainage has been common over the past few thousand years, mainly because rise of lake levels during spring snowmelt results in overflow along ice-wedge cracks and consequent rapid thermal incision of ice-rich sediments to form outlet channels. Between 1950 and 1986, about 65 lakes in the area of pingo concentration drained partially or completely.

Lake drainage results in a drastic change in the thermal regime of sub-lake taliks, allowing permafrost to extend downwards from the surface of the exposed lake floor and inwards from the edges of the talik (Figure 7.3b). When the saturated sands around the margins of the talik freeze, the volume increase associated with the phase change from water to ice results in expulsion of roughly 9% of pore water into the unfrozen zone. The expelled water is slightly saline as a result of solute rejection due to ice crystal growth at the freezing front, so it remains unfrozen even at temperatures of –0.1 °C to –0.5 °C. This water moves under hydrostatic pressure towards areas of thinner nearsurface permafrost, often under residual ponds on the former lake floor, where confining stresses are lower. Freezing of excess water injected at the base of the thinner permafrost results in formation

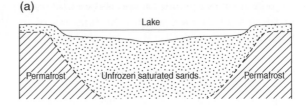

(a)

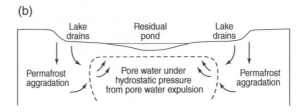

(b)

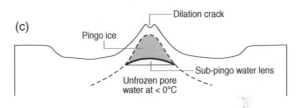

(c)

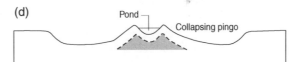

(d)

Figure 7.3 Schematic evolution of a hydrostatic pingo. (a) Development of a suprapermafrost talik of unfrozen saturated sandy sediments beneath a lake. (b) Lake drainage results in downwards and lateral permafrost aggradation into the talik; expulsion of pore water from the aggrading permafrost results in the development of high hydrostatic pressures under a residual shallow pond where permafrost is thinnest. (c) Pore water expulsion into the residual talik leads to the formation of a sub-pingo water lens; progressive downwards freezing results in formation of a thickening ice core and updoming of the overlying frozen sediments, forming a hydrostatic pingo. (d) Exposure of the ice core at the pingo summit causes melt of the core and subsidence of the central part of the pingo. *Source:* Reproduced from Mackay (1998) under a licence given by Copibec.

of a core of ice that forces the overlying frozen ground upwards. Through time, additional pore water expelled from the advancing freezing front migrates to the base of the ice core, where it freezes, causing further uplift of the overlying frozen ground (Figure 7.3c). The nature of the ice forming the pingo core depends on whether the core is in direct contact with the underlying saturated sediment, allowing the formation of segregation ice, or overlies a sub-pingo water lens, in which case massive injection ice develops (Mackay, 1994). Stable isotope studies of ice cores obtained from pingos in Alaska and northwestern Siberia have confirmed the presence of both segregation ice and injection ice in the pingos, and shown that ~50% of ice in the pingos formed during the final stages of permafrost aggradation (Vasil'chuk *et al.*, 2016b).

A sub-pingo water lens forms when the addition of expelled pore water to the base of an ice core exceeds the rate of downwards freezing from the surface. The hydrostatic pressure in this water lens is sufficient not only to support the overburden stress, but also to cause uplift of the ice core and the overlying frozen ground; some pingos are therefore 'floating' on a lens of water (Figure 7.3c). The presence of pressurized water lenses under hydrostatic pingos was spectacularly demonstrated by Mackay (1978), who drilled through permafrost in three growing pingos and the adjacent lake flats, and in every case was rewarded by forceful artesian flow of water gushing out of the drillholes and forming fountains up to 3 m above the surface. This behaviour not only confirms lateral migration of pore water away from the aggrading permafrost to form sub-pingo water lenses, but also demonstrates that permafrost aggradation is capable of producing pore-water pressures that approximate the total geostatic overburden pressure. At one site, drilling revealed a sub-pingo water lens 2.2 m deep.

The presence of water lenses under hydrostatic pingos also explains a phenomenon termed 'pulsating pingos' by Mackay (1977), who showed that the summits of some pingos rise and fall in response to the accumulation and loss of water under the pingo ice core. Addition of expelled pore water to the sub-pingo water lens results in uplift. If the uplift causes rupture of frozen ground at the pingo margin, however, sub-pingo water escapes to the surface, and the pingo temporarily subsides. Mackay (1978) was able to replicate this effect artificially by allowing escape of sub-pingo water through drillholes, causing a pingo to subside by 0.6 m.

In general, pingo growth involves an increase in height but little increase in width, so side slopes progressively steepen and the summit depression gradually expands. Annual growth is therefore greatest at the summit and declines towards the periphery (Figure 7.4). Growth rates vary markedly. Porsild Pingo in the Tuktoyaktuk Peninsula area, for example, is a young landform that developed on the bed of a lake that drained around AD 1900, and reached a height of about 8.5 m by 1972. Summit uplift rates averaged ~16.6 cm a^{-1} from 1972 to 1983 but declined slightly thereafter (Mackay, 1998). Ibyuk Pingo is a much older landform that probably started to grow over a thousand years ago. Despite reaching a height of 49 m, it exhibits continuing summit uplift that averaged ~2.8 cm a^{-1} between 1973 and 1994. Generalizations about growth behaviour are difficult: some of the pingos monitored by Mackay (1998) exhibited negligible growth, others experienced rupture and some showed evidence for alternating summit uplift and subsidence.

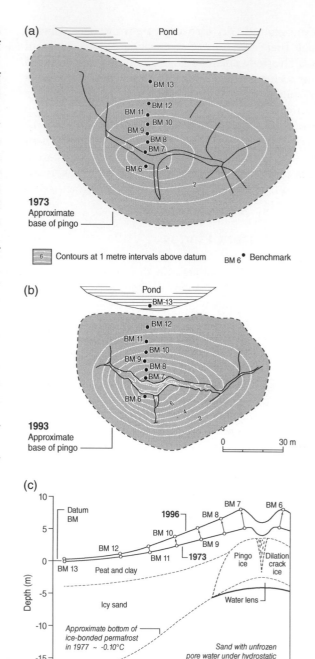

Figure 7.4 (a,b) Contoured maps of a growing hydrostatic pingo surveyed in 1973 and 1993, showing a height increase of about 4 m in 20 years, but without significant change in diameter. (c) Cross-section of the same pingo based on surveys in 1973 and 1996 and drilling in 1977. BM, benchmark. *Source:* Adapted from Mackay (1998) under a licence given by Copibec.

7.3.2 Pingo Ice

Observations of ice exposures in hydrostatic pingos indicate the presence of five types of ice: intrusive ice, segregation ice, pore ice, dilation crack ice and ice-wedge ice (Mackay, 1985). Intrusive (injection) ice forms a massive body of pure ice in the core of a pingo, and probably represents freezing of water at the top of a sub-pingo water lens. Some intrusive ice exhibits alternating banding of clear and bubble-rich ice, the former attributed by Mackay (1990b) to fast freezing during downwards propagation of a cold winter temperature wave, the latter to slower freezing during summer. Segregation ice grows when the pingo core is in direct contact with underlying saturated sediments, and takes the form of ice lenses interlayered with frozen sediment. Unlike intrusive ice, segregation ice may extend deep below the level of the surrounding lake flats. Though segregation ice is normally associated with fine-grained frost-susceptible sediments (Chapter 3), it may also develop in non-frost-susceptible sands where water pressures are high (Mackay, 1979). Depending on the presence or absence of a sub-pingo water lens, therefore, the ice core of a pingo may be composed entirely of intrusive ice, entirely of segregation ice or of varying amounts of both (Mackay, 1994).

Pore ice is pervasive in frozen sediments overlying, surrounding or underlying the pingo core. Dilation ice forms where water entering tension cracks formed by stretching of the frozen ground overlying a growing pingo core freezes on contact with permafrost. It is characterized by vertical banding due to repeated episodes of cracking and freezing of water, and by inclusion of soil, clasts and organic matter that has fallen into open cracks. Ice-wedge ice forms on the pingo overburden via thermal contraction cracking of nearsurface frozen ground (Chapter 6) and tends to be present only on older pingos. All types of ice present in pingos are liable to deformation during pingo growth and to faulting and melting during pingo collapse.

7.3.3 Pingo Degradation and Collapse

Because the sides of pingos steepen during growth (Figure 7.4), pingos are susceptible to sediment loss from their flanks by active-layer sliding, solifluction, debris flow and possibly permafrost creep, all of which reduce the thickness of the overburden mantling the pingo core. The critical trigger of pingo collapse, however, appears to be exposure of the ice core at the pingo summit (Mackay, 1987, 1988b). As a pingo grows, dilation cracks at the summit widen, and sediment at the base of such cracks is spread over an increasing area until the top of the ice core is exposed at the base of a wide, crater-like depression. Melting of the top of the ice core often results in the formation of a shallow pond. Gradual loss of ice from the top of the core steepens the inward-facing slopes around the central depression, causing slumping of sediment towards the centre of the depression. Hydrostatic pingos at an advanced stage of collapse consist of a central, often water-filled depression surrounded by a circular ice-cored ridge (Mackay and Burn, 2011; Figures 7.2d and 7.5).

7.4 Hydraulic Pingos

Hydraulic or 'open-system' pingos occur both in areas underlain by continuous permafrost, for example in Greenland and Svalbard, and in areas of thin or discontinuous permafrost, such as central Alaska. They are of similar dimensions to their hydrostatic cousins. Glacier Pingo in East Greenland is 38 m high, with a diameter of 350 m, and Ny Pingo in Svalbard is 36 m high and 750 m long, but most are much smaller. Hydraulic pingos develop in a range of sediments, including floodplain,

Figure 7.5 Collapsing pingo on the Parry Peninsula, NWT, Canada. *Source:* Courtesy of Chris Burn.

delta, fan or slope deposits and till, but some involve uplift of fractured bedrock (Seppälä, 1988b; Matsuoka *et al.*, 2004). They exist as isolated individuals, small clusters of aligned pingos or sometimes as complex landforms involving several stages of pingo growth and decay located around a common source. A feature of some hydraulic pingos is emergence of a groundwater spring, sometimes perennial, from the summit, flanks or periphery (Figure 7.6). Analysis of the chemistry or isotopic composition of such springwaters has suggested a meteoric origin, implying derivation from precipitation or meltwater that has flowed to the pingo site as subpermafrost or intrapermafrost groundwater (Allen *et al.*, 1976; Gurney, 1998; Yoshikawa *et al.*, 2003), and persistent springflow is often considered diagnostic of a hydraulic origin. Yoshikawa (1998) has shown that groundwater temperature and discharge are critical: springs associated with pingos have low discharge and water temperatures of 3 °C or lower; rapidly discharging or warmer groundwater reaching the surface forms a perennial spring, but without pingo development. It is also notable that springs emerging from some hydraulic pingos are strongly saline, suggesting that liquid water movement through permafrost conduits is possible because of salinity-induced freezing-point depression. Data on growth rates of hydraulic pingos are sparse, though Matsuoka *et al.* (2004) noted that Riverbed Pingo in Svalbard grew from 7.8 m in 1964 to 12.0 m in 1990, a mean growth rate of 17 cm a^{-1}.

The occurrence of hydraulic pingos at slope-foot or valley-floor locations suggests a common genesis through movement of groundwater along a hydraulic pressure gradient to the point of pingo growth (Figure 7.7). This hypothesis, however, raises questions regarding the nature of the subpermafrost or intrapermafrost groundwater flow responsible for feeding pressurized water to feed the ice core. The model outlined in Figure 7.7 may be valid for areas of discontinuous permafrost, where infiltration of meteoric water on permafrost-free upper slopes allows summer recharge of subpermafrost groundwater, and thin permafrost on lower slopes or valley floors permits injection of subpermafrost

Figure 7.6 Surface icing (upslope from snowmobiles), formed where a spring issuing from an open-system pingo has frozen on reaching the surface, Adventdalen, Svalbard.

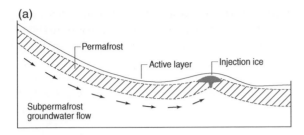

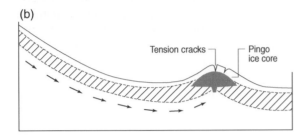

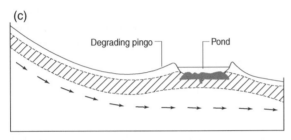

Figure 7.7 Schematic evolution of a hydraulic pingo.
(a) Groundwater moving under hydraulic pressure through a subpermafrost aquifer feeds growth of a core of injection ice under shallow permafrost. (b) Growth of the ice core results in updoming and cracking of the overlying frozen ground. (c) Exposure of the ice core at the surface causes melt of the core and subsidence of the pingo.

groundwater under artesian pressure (Yoshikawa et al., 2003). It appears less feasible, however, in areas of continuous permafrost, where it is more difficult to envisage recharge of groundwater flow, or how pressurized groundwater might reach the surface through permafrost tens or hundreds of metres thick without freezing.

Yoshikawa and Harada (1995) suggested three solutions to this dilemma. The first is that movement of subpermafrost or intrapermafrost groundwater is focused along zones of fractured bedrock on fault lines. Several authors have observed alignment of hydraulic pingos along faults or fracture zones (e.g. Yoshikawa et al., 1996), suggesting preferential flow of subpermafrost groundwater to pingo outlets along corridors of fractured rock. For this to happen, groundwater temperatures must be high enough to maintain an unfrozen routeway to the base of the pingo core. Worsley and Gurney (1996) argued that intrapermafrost flow through continuous permafrost in East Greenland reflects high geothermal heat flux that permits groundwater movement within structural discontinuities, and cited fairly high (2.5–3.5 °C) springwater

temperatures as evidence. This explanation, however, is unlikely to hold for all areas that contain hydraulic pingos.

A second solution, advocated by Liestøl (1977) for some pingos in Svalbard, is that meltwater at the base of polythermal glaciers infiltrates into the underlying unfrozen substrate then migrates under hydraulic pressure beneath the permafrost under the glacier margins to feed a pingo ice core in a nearby area of thin permafrost. This explanation is supported by research carried out Haldorsen et al. (2010), which suggests that springflow through thick permafrost on Svalbard may have been initiated during periods of more extensive glaciation, when unfrozen ground under warm-based glacier ice formed a conduit for meltwater springs that have survived subsequent glacier recession and permafrost aggradation. Numerical modelling by Scheidegger et al. (2012) suggests that spring systems in arctic areas can persist for millennia after exposure by retreat of polythermal glaciers, provided that spring discharge is sufficiently high to prevent downwards freezing of the spring outlet, and a similar explanation may account for the supply of water to hydraulic pingos in areas of thick, continuous permafrost. A final explanation is that some pingos occur on low-lying floodplains, deltas and coastlines that emerged from the sea during the Holocene as a result of glacio-isostatic uplift. Permafrost underlying recently emerged littoral sites is thin compared with that farther upslope or inland, allowing subpermafrost groundwater to penetrate and freeze within saturated fine-grained marine, deltaic or estuarine sediments and form a pingo. Whilst all three explanations may be valid in particular cases, the question of subpermafrost groundwater recharge and penetration of groundwater to the surface through thick permafrost remains unresolved for many hydraulic pingos in areas of continuous permafrost.

The internal structure of hydraulic pingos is also incompletely understood. An exposure through a pingo in Svalbard logged by Yoshikawa (1993) revealed lenses of segregation ice in frozen sediments overlying a massive ice core that he interpreted as injection ice. Drillholes through hydraulic pingos in northern Sweden (Lagerbäck and Rodhe, 1985) and interior Alaska (Yoshikawa et al., 2003) also revealed massive ice cores, suggesting formation by freezing of sub-pingo water lenses or equilibrium between the rate of groundwater supply and that of freezing at the base of the ice core. Several attempts have been made to investigate pingo structure using geophysical techniques (Yoshikawa et al., 2006). Yoshikawa (1993) concluded from resistivity measurements that Riverbank Pingo in Svalbard is underlain by a talik in permafrost 15 m deep. Survey of the same pingo and nearby Innerhytte Pingo using ground-penetrating radar revealed steeply dipping reflectors subparallel to the surface topography (Ross et al., 2005). These could represent reflecting horizons within a massive ice core, or alternating bands of ice and debris. Resistivity measurements carried out on Innerhytte Pingo indicate that it is underlain by either ice-rich frozen bedrock or a core of massive ground ice (Ross *et al.*, 2007). Similar measurements made on two nearshore pingos suggest that their internal structure may be dominated by segregation ice and pockets of massive ice within partly frozen marine muds, though this interpretation may be compromised by low resistivity values due to unfrozen saline pore water. These studies suggest that though geophysical techniques hold promise for pingo investigation, they need to be supplemented by drilling to reduce interpretational uncertainty. Coring of a hydraulic pingo in Mongolia, for example, revealed the presence not only of both intrusive and segregation ice within the pingo core, but also of a lens of sub-pingo water that erupted at the surface under artesian pressure when tapped by the drill (Yoshikawa *et al.*, 2013).

Collapse of hydraulic pingos has received little attention, though the presence of dilation cracks on pingo summits and flanks and of summit craters containing ponds, evidence for active-layer slides on pingo flanks and the existence of residual pingos consisting of ramparts surrounding a collapsed core (e.g. Yoshikawa *et al.*, 2003; Matsuoka *et al.*, 2004) all suggest that the collapse of hydraulic pingos resembles that of hydrostatic pingos, and is similarly triggered by exposure of the ice core due to stretching and thinning of summit overburden as pingos grow.

7.5 Pingo Problems and Problem Pingos

One problem that emerges from the preceding discussion is that of pingo definition and classification. Harris *et al.* (1988) defined a pingo as 'a perennial frost mound consisting of a core of massive ice, produced primarily by injection of water, and covered with soil and vegetation'. But some features interpreted as pingos, particularly those formed in frost-susceptible sediments, may consist mainly of segregation ice (Mackay, 1985, 1994; Ross *et al.*, 2007), so a core of pure injection ice may not be a defining characteristic. Morphologically, size alone is an unsatisfactory criterion; though large (>10 m high) ice-cored hills are almost certainly pingos, smaller or immature forms overlap in dimensions with those of palsas and other frost mounds. Probably the only unifying characteristic of pingos is genetic: the generation of high subpermafrost or intrapermafrost groundwater pressures sufficient to overcome geostatic overburden stress (Ross, 2013).

The distinction between hydrostatic and hydraulic pingos may also be less clear-cut than many accounts suggest. Mackay (1998) noted that some hydraulic pingos occur in areas of coarse-grained alluvial sediments where permafrost is aggrading downwards, suggesting that high sub-pingo pore-water pressures may in some cases

be derived through pore-water expulsion, as within hydrostatic pingos. Moreover, stable isotope analysis of a core recovered from a hydraulic pingo in Mongolia suggested that open-system freezing alternates with semi-closed- or closed-system freezing during pingo growth (Yoshikawa *et al.*, 2013). The presence of perennial springs emanating from hydraulic pingos does, however, indicate that subpermafrost or intrapermafrost groundwater flow along a hydraulic gradient is the main source of pressurized water feeding pingo cores in most cases.

The most intriguing problem in pingo research relates to the subpermafrost plumbing system that feeds the ice core of hydraulic pingos in areas of continuous permafrost. Though the classification of Svalbard pingos outlined by Yoshikawa and Harada (1995) into fracture belt, glacier-fed and nearshore types offers explanations for particular pingos and is supported by evidence based on the chemistry and characteristics of pingo ice and spring water (Yoshikawa, 1993, 1998; Worsley and Gurney, 1996; Matsuoka *et al.*, 2004), several key questions remain unanswered. The most important concern: (i) the source of subpermafrost groundwater recharge; (ii) whether artesian pressure alone is sufficient to promote pingo uplift, or whether uplift essentially reflects freezing pressure; (iii) the nature of unfrozen ground underlying hydraulic pingos (broad taliks or narrow 'dikes' or pipes, as inferred by Yoshikawa (1993)); and (iv) whether hydraulic pingos are underlain by a lens of injected water that feeds growth of a core of injection ice. Such uncertainties can probably only be resolved through survey of the growth behaviour of hydraulic pingos and drilling through pingos and adjacent ground, supplemented by geophysical investigations and chemical or isotopic fingerprinting of water sources.

Some pingos or pingo-like landforms are difficult to explain in terms of the hydrostatic and hydraulic mechanisms just outlined. For example, dome-shaped and elongate pingos occur along low alluvial terraces of the Thomsen River on Banks Island, an area of low relief underlain by continuous permafrost but not associated with drained lake sites. Excavation has revealed cores of both segregation ice and massive ice. Pissart and French (1976) interpreted these landforms as hydrostatic pingos that developed when channel abandonment resulted in permafrost aggradation and pore-water expulsion around the margins of a subchannel talik. Pingo-like ridges over 2 km long in the Sachs River valley on Banks Island may have a similar origin (Pissart and French, 1977). Another example of an unusual pingo is one composed of fractured dolomitic bedrock on low-relief terrain in the northern District of Mackenzie, arctic Canada. St-Onge and Pissart (1990) concluded that it represents a hydrostatic pingo that formed when drainage of a glacial lake triggered permafrost aggradation around a talik, causing development of an ice core within the rock.

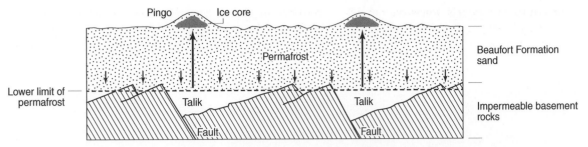

Figure 7.8 Suggested evolution of aligned 'summit pingos' on Prince Patrick Island. As permafrost aggrades downwards, expelled pore water accumulating in subpermafrost taliks is forced upwards and freezes at shallow depth, forming pingos.

Other authors have proposed the existence of distinct pingo types or hydrostatic–hydraulic hybrids. Christiansen (1995) suggested that 'marsh initiated pingos' are represented by eroded ice-cored mounds up to 10 m high on the higher saltmarsh areas of an emergent delta in west Greenland. She envisaged that these formed by freezing of water from adjacent river channels or sediments, with high sub-pingo water pressures presumably generated by permafrost aggradation within unfrozen estuarine mud flats as the sea receded (Gurney and Worsley, 1997). A similar mode of pingo genesis was envisaged by Pissart (1967) to explain estuarine pingos of apparent hydrostatic origin that developed following Holocene marine regression in the Canadian Arctic. Alternatively, however, as the features described by Christiansen (1995) are overlooked by steep valley-side slopes, they may simply represent examples of hydraulic nearshore pingos similar to those in Svalbard described by Yoshikawa and Harada (1995).

Gurney and Worsley (1996) have also challenged the subdivision of pingos into hydrostatic and hydraulic varieties with reference to growing, collapsed and rejuvenated pingos on low relief terrain underlain by continuous permafrost on Banks Island. Some of these occupy drained lake sites and are apparently of hydrostatic origin. Others they attributed to a combination of both hydrostatic and hydraulic processes associated with 'a dynamic system of interlinked taliks' that 'influences groundwater supply to centres of ground ice growth' (Gurney and Worsley, 1996, p. 41). It is difficult, however, to envisage hydraulic development of high sub-pingo groundwater pressures on a low-lying plain, even if these sites do not always conform to the classic drained lake setting of hydrostatic pingos.

Walker et al. (1985) found that in the Prudhoe Bay region of Alaska, steep-sided hydrostatic pingos in drained lake basins are interspersed with much broader mounds with gentle slopes. Drilling of one such mound revealed a core of massive ice. The origin of the broad mounds remains a mystery: they are morphologically distinct from the local hydrostatic pingos; absence of lake basin locations apparently excludes a hydrostatic origin; and there is insufficient relief to explain these

features as hydraulic pingos. Even more intriguing are 'summit pingos' on Prince Patrick Island in the Canadian Arctic. These average 3.3 m in height and 58 m in diameter, and contain ice bodies at shallow depth in continuous permafrost. They are developed along two parallel lines along the main drainage divide, and occur on valley floors, interfluve summits and valley-side slopes. Pissart (1967) suggested that these pingos follow buried faults in underlying impermeable basement rocks. He hypothesized that as permafrost extended downwards through the overlying sands and gravels after deglaciation, groundwater trapped in fault-defined depressions between the descending permafrost and the basement rocks was injected towards the surface to form the pingos (Figure 7.8). These examples suggest that though the hydrostatic and hydraulic explanations for pingo growth are well established, there are numerous variations on these basic themes, and that our understanding of these iconic landforms remains incomplete.

No account of problematic pingos would be complete without mention of the dome-shaped hills that occur on the seabed under the shallow waters of the Beaufort Sea, offshore from the western Canadian arctic coast. These have an average height of ~30 m and were originally interpreted by Shearer et al. (1971) as submarine hydrostatic pingos, formed after marine transgression as cold (<1.0 °C) saline bottom waters displaced freshwater lakes, causing inward freezing of sub-lake taliks and build-up of hydrostatic pressures sufficient to cause updoming of submerged, frozen lakebed sediments. Reinterpretation of these domes as mounds produced by release of decomposing methane gas hydrates into seabed sediments (Paull et al., 2007) now seems a more plausible explanation.

7.6 Segregation Ice Mounds: Palsas, Lithalsas and Related Landforms

The term *palsa* is here employed to describe a peat-covered mound that has grown out of a bog through development of a frozen core of segregated ice, peat and

Figure 7.9 Palsa in Finnish Lapland. *Source:* Courtesy of Matti Seppälä.

mineral sediments (Figure 7.9). A similar mound without peat cover is here referred to as a *lithalsa*, a term proposed by Harris (1993) to replace a host of clumsy synonyms, notably 'mineral palsas' and 'mineral permafrost mounds'. A *peat plateau* is a broad, low, flat-topped area of peat and underlying mineral sediment raised above a surrounding peatland by frost heave, and a *permafrost plateau* is a similar feature composed of mineral sediment and lacking peat cover. All of these features share a common genesis in that they are produced, at least in part, by frost heave due to the formation of ice lenses during downwards aggradation of permafrost in frost-susceptible sediments.

7.7 Palsas

Palsas are peat-covered permafrost mounds, domes or ridges, usually up to about 10 m high and 30 m wide, which rise above permafrost-free bogs and mires (Figure 7.9). They often occur in groups termed *palsa fields*, or in *palsa complexes* where mounds are interspersed with water-filled depressions representing the sites of collapsed palsas (Figure 7.10). Palsas are widely distributed in the circumpolar zone of discontinuous permafrost in Iceland, Fennoscandia, the Kola Peninsula, Siberia, Alaska and subarctic Canada, and are associated with shrub tundra and forest tundra vegetation. They also occur in lower latitudes at high altitudes, for example in the Canadian Rockies, Dovrefjell in south-central Norway and the Daisetsu Mountains in Japan (Brown, 1980; Sone and Takahashi, 1993; Matthews *et al.*, 1997a). Palsa-like mounds have also been reported within the zone of continuous permafrost in eastern Siberia and

Svalbard (Åkerman, 1982), though whether these features are genetically identical to true palsas developed in discontinuous permafrost is debatable (Pissart, 2002).

The southern limit of palsas coincides with MAATs of −1 to 0 °C (Vitt *et al.*, 1994; Pissart, 2002), and the frozen cores of palsas often represent the southernmost appearance of permafrost in the northern hemisphere. The northern limit of palsa formation tends to coincide with MAATs of approximately −7 to −6 °C (e.g. Worsley *et al.*, 1995; Allard and Rousseau, 1999; Vallée and Payette, 2007), and thus with the transition from discontinuous to continuous permafrost. Palsa development is favoured by deep peat cover in wet mire or fen environments where winter snowcover is limited (Seppälä, 2011).

7.7.1 Palsa Structure

The internal structure of palsas holds the key to their formation, and has been investigated by geophysical techniques and through drilling and analysis of core sediments (Fortier *et al.*, 2008b; Kohout *et al.*, 2014). Most palsas consist of a layer of peat of variable thickness overlying frost-susceptible sediments, and contain a permafrost core containing 30–80% by volume of layers of segregation ice, formed by migration of water to the freezing plane as permafrost aggraded downwards. Such ice lenses range from a few millimetres to a few centimetres in thickness, and may be thickest at the base of the permafrost. Segregation ice layers in palsas tend to be dome-shaped rather than horizontal, reflecting either heat loss from the sides as well as the top of a growing palsa or uparching of horizontal ice layers as the palsa grows; the latter explanation is consistent with reports of

Figure 7.10 A complex of palsas and lithalsas growing amid thermokarst ponds in northern Quebec. The rounded mounds in the centre of the photograph lack peat cover (lithalsas), but nearby mounds are peat-covered (palsas). The circular thermokarst ponds are the sites of collapsed palsas and lithalsas. A colour version of this figure appears in the plates section. *Source:* Adapted from Calmels *et al.* (2008). Reproduced with permission of Elsevier.

high-angle fractures and faults in some palsa cores (Fortier *et al.*, 1991; Allard and Rousseau, 1999; Calmels and Allard, 2008). A layer of aggradation ice formed by freezing of water percolating down through the active layer may also occur at the top of the permafrost, particularly where peat cover is thinner than the active layer (An and Allard, 1995; Allard and Rousseau, 1999). Pore ice is also present throughout the frozen core. In most palsas, segregation ice is concentrated in the underlying frozen mineral sediments. As fibrous peat is generally not frost-susceptible (Kujala *et al.*, 2008), ice layers within frozen peat probably represent aggradation ice formed at the base of the palsa as it grows (Seppälä and Kujala, 2009; Seppälä, 2011).

7.7.2 Palsa Growth and Collapse

Palsa growth within a bog reflects seasonal variations in the thermal conductivity of peat. Dry fibrous peat is a good insulator, with low thermal conductivity ($0.07-0.08\,\mathrm{W\,m^{-1}\,K^{-1}}$). The conductivity of saturated peat is much higher ($0.4-0.5\,\mathrm{W\,m^{-1}\,K^{-1}}$), and that of frozen saturated peat is higher still ($1.5-2.3\,\mathrm{W\,m^{-1}\,K^{-1}}$; Kujala *et al.*, 2008). If the surface of an incipient palsa is raised above the surrounding bog and the uppermost peat dries out, the underlying frozen ground becomes insulated from downwards heat flow during the summer and may become permafrost. As the permafrost core aggrades slowly downwards, frost heave pushes the overlying peat upwards, increasing the thickness of the surface layer of dry peat, and thus permitting further downwards permafrost aggradation (Figure 7.11). Insulation of the tops and flanks of palsas by dry peat in summer also explains how palsas can form in marginal areas of discontinuous, sporadic or isolated permafrost where MAATs may be as high as 0 °C (Westin and Zuidhoff, 2001).

Two main hypotheses have been proposed to explain the initial growth of an incipient palsa above the level of

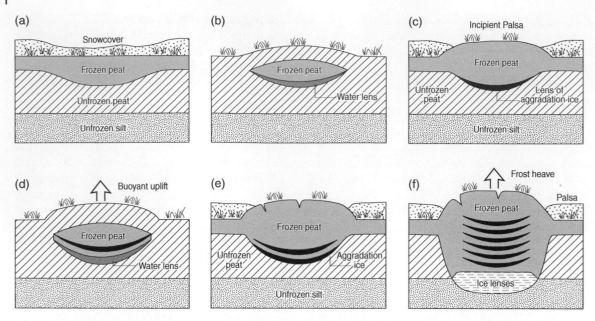

Figure 7.11 Evolution of a palsa through buoyant uplift and frost heave. (a) Deeper winter frost penetration in peat under an area of thinner snowcover. (b) Buoyant uplift of frozen peat during the first summer, leading to development of an underlying water lens. (c) Freezing of the water lens during the next winter, forming a lens of aggradation ice. (d) Renewed buoyant uplift in the following summer, with formation of a second water lens. (e) Freezing of second water lens during the next winter, adding a further layer of aggradation ice. (f) After several years the freezing plane penetrates underlying silts, leading to ice-lens formation and frost heave. *Source:* Adapted from Seppälä (2011). Reproduced with permission of Elsevier.

the surrounding bog or mire. The first is localized removal of winter snowcover by wind, allowing deeper winter freezing under the site of palsa growth. Seppälä (1995a) demonstrated the feasibility of this mechanism by periodically clearing snow from an experimental plot on a mire in Lapland at intervals over three winters. As a result, winter frost penetration was deeper than in the surrounding snow-covered areas, permafrost developed to a depth of 1.1 m and the surface of the cleared plot was heaved 0.3 m above the adjacent mire. As palsas grow, their tops are kept clear of snow by wind, allowing progressively deeper ground freezing, further frost heave and thus continued growth. Conversely, snowdrifts tend to accumulate against the flanks of emerging palsas, inhibiting frost penetration into the surrounding mire (Seppälä, 1994), and thus ensuring a continuing supply of water to feed ice growth in the downward-aggrading permafrost core of the palsas.

An alternative (but compatible) explanation for palsa initiation is that zones of frozen peat, having lower density than the surrounding wetlands, become buoyant and 'float' above the general level of a mire (Outcalt and Nelson, 1984; Nelson *et al.*, 1992). Drying of surface peat on the floating mass then insulates the underlying frozen layer from summer thaw. Seppälä and Kujala (2009) have proposed that a lens of water develops under the buoyant frozen palsa core and freezes during the ensuing winter to form a lens of aggradation ice. During succeeding years, the cycle of buoyant uplift and basal ice aggradation is repeated (Figure 7.11), permitting permafrost to aggrade downwards until it penetrates the underlying mineral substrate, thereby triggering additional palsa uplift through the growth of segregation ice in frost-susceptible sub-peat sediments.

Once palsa growth is initiated, downwards aggradation of permafrost and concomitant formation of segregation ice by water drawn upwards and from the palsa margin by cryosuction continues to heave the palsa upwards until thermal equilibrium between frost penetration and geothermal heat flux is established. Downwards permafrost aggradation slows through time, accounting for the development of thick layers of segregation ice near the base of the permafrost (An and Allard, 1995). By the time a mature palsa has formed, the permafrost core may have reached a depth of up to 20 m, and palsa height (reflecting ice content in the core) may be about one-third of permafrost depth, or up to about 7 m (Allard and Rousseau, 1999).

Palsa growth, however, often triggers potential palsa decay. Wind abrasion may thin the insulating cover of dry peat on palsa summits (Seppälä, 2003), and exposed peat may be eroded by rainsplash and slopewash. Tension and desiccation cracks split the peat cover as a palsa grows, and blocks of peat may slide down the steepening sides to disintegrate in marginal ponds. These effects reduce or remove peat cover, exposing the permafrost core to thaw, subsidence and thermal erosion by ponded water until little is left of the original palsa apart from a shallow pond or thermokarst depression, sometimes encircled by a subdued ridge of sediment and peat. Seppälä (1988a) suggested that a cycle of growth and decay is an intrinsic feature of palsa development, a view supported by numerous descriptions of palsa growth amid the collapsed remnants of former palsas. Others have challenged the universality of this concept, particularly for palsas near the continuous–discontinuous permafrost boundary, where MAATs are sufficiently low to sustain palsa survival after removal of peat cover (Allard and Rousseau, 1999). The stability of permafrost under palsas depends in part on the depth of the organic layer, which insulates the underlying ground during summer, and in part on surface vegetation cover. Jean and Payette (2014b) have shown that permafrost stability is likely to be greater under shrub-covered palsas than under wooded ones, where a heterogeneous vegetation cover promotes snow entrapment and lateral heat transfer. There is nevertheless widespread evidence for palsa degradation due to climate warming and/or changes in snowcover or hydrologic regime near the southern margin of the discontinuous permafrost zone (Zuidhoff and Kolstrup, 2000; Payette *et al.*, 2004; Vallée and Payette, 2007; Thibault and Payette, 2009). In such marginal permafrost environments, palsas are in danger of extinction (Chapter 17).

7.8   Peat Plateaus

Peat plateaus, sometimes termed *palsa plateaus*, are flat-topped areas of peat, often bounded by steep bluffs, that have been raised above the level of surrounding peatlands by frost heave. They are thus genetically related to palsas and often occur next to palsa fields in the discontinuous permafrost zone. Small peat plateaus in northern Quebec averaging 75 by 25 m in extent and 3.3 m in height have been described by Allard *et al.* (1987). Other are much more extensive, particularly near the northern limit of discontinuous permafrost, but most rise less than 2 m above adjacent peatlands. Early work attributed the formation of peat plateaus to uplift driven by the buoyancy of perennially frozen peat in saturated wetlands, or to frost heave due to ice lens growth within the peat (e.g. Outcalt and Nelson, 1984; Harris *et al.*, 1992; Nelson *et al.*, 1992). More recent research, however, has shown that some peat plateaus are frozen to the underlying substrate. In a study of a 1.3 m high peat plateau in the Hudson Bay lowlands, for example, Allard and Rousseau (1999) found that uplift of peat 0.3–0.8 m thick is due to the development of lenses of segregation ice in an underlying silty clay layer 2.2 m thick, with a volumetric ice content of 40–80%. The limited peat uplift at this site appears to reflect the limited thickness of the frost-susceptible silty clay layer.

Uplift of peat plateaus probably reflects establishment of a more uniform negative thermal regime within the peat and underlying sediments than is found in palsas, and frost heave may be limited in large peat plateaus by moisture availability. Like palsas, peat plateaus are susceptible to progressive degradation. Large peat plateaus are often dissected by polygonal desiccation crack networks that focus meltwater runoff and thermal incision, and subsequent backwasting of peat scarps from the resultant thermokarst gullies and depressions isolates individual flat-topped mounds. These have sometimes been referred to as 'degradation palsas', and may resemble true palsas, but their relationship to the parent peat plateau is usually evident. Like palsas, peat plateaus may degrade in response to climatic warming, increased snowcover or increased water levels in surrounding fens (Payette *et al.*, 2004; Vallée and Payette, 2007).

7.9   Lithalsas

Lithalsas are frost-heaved permafrost mounds similar in size and shape to palsas, but lacking peat cover. Like palsas, they have grown through frost heave generated by the development of ice lenses during downwards aggradation of a permafrost core. The morphometry of both active lithalsas and Late Pleistocene relict examples suggests that lithalsas grow outwards from the site of initial uplift as a result of lateral as well as vertical aggradation of the ice-rich permafrost core, ultimately forming mounds up to about 8 m high and 120 m in diameter (Pissart *et al.*, 2011; Wolfe *et al.*, 2014). The genetic similarity of palsas and lithalsas is emphasized by their mutual proximity on mires in the discontinuous permafrost zone (Calmels *et al.*, 2008; Figure 7.10). Some landforms exhibit transitional characteristics, such as palsas with thin peat cover, lithalsas capped by small peat remnants and lithalsas that started growing as palsas but have lost all peat cover (Lagerbäck and Rodhe, 1986; Harris, 1993; Worsley *et al.*, 1995; Matthews *et al.*, 1997a). It is incorrect to conclude, however, that all lithalsas are simply bald or balding palsas, as lithalsas also occur along peat-free river terraces and emergent shorelines, where they appear to have developed *ab initio* without peat cover (Allard *et al.*, 1987; Wolfe *et al.*, 2014). The volumetric ice content of the frozen core of lithalsas exceeds 50%, commonly in the form of reticulate ice lenses that separate soil segments. The cryostratigraphy of some lithalsas is complex, however, and may reflect not only syngenetic differential ice segregation but also epigenetic creep deformation of the initial cryostructure (Iwahana *et al.*, 2012). Sections through two lithalsas investigated by Vasil'chuk *et al.* (2016a) in the Eastern Sayan Mountains of southern Russia exhibited predominantly lenticular, reticulate and layered cryostructure, with up to 60–70% visible ice content; the isotopic composition of constituent ice suggested that the water sources included lake, fen and groundwater as well as precipitation.

Because lithalsas lack an insulating peat cover, they cannot form or survive near the southern limit of discontinuous permafrost, except under special circumstances (Pissart *et al.*, 1998). In northern Quebec, for example, lithalsas are associated with MAATs of –6.7 to –4.3 °C and with cool summers in which the temperature of the warmest month does not exceed 11.5 °C (Pissart, 2000). A survey by Wolfe *et al.* (2014) indicated three prerequisites for lithalsa formation: extensive discontinuous warm permafrost with a thermal regime of –1 to 0 °C and typically MAATs of –4 to –6 °C; a substrate composed of fine-grained, frost-susceptible sediment; and groundwater availability (usually satisfied by proximity to ponds, fens and streams) to feed ice-lens growth. The distribution of lithalsas overlaps that of palsas, but is restricted to the colder, more northerly part of the discontinuous permafrost zone, though lithalsas also occur in high mountain areas; Wünnemann *et al.* (2008) mapped lithalsas at altitudes over 4500 m in the Himalayas. They form on a range of frost-susceptible substrates, including till, alluvium and slope deposits, but particularly lake sediments and marine deposits.

The development of lithalsas in peat-free areas is incompletely understood, though localized vegetation succession, tree colonization or thinning of winter snow-cover by wind may be critical in the initial development of a localized permafrost core (Jean and Payette, 2014a). In some cases, an upper layer of low-density mineral sediments (such as loess) may aid insulation of a developing permafrost core against summer thaw (Pissart

et al., 1998; Saemundsson *et al.*, 2012). Once low incipient lithalsas have formed, development of shrub vegetation cover and an organic soil horizon may further insulate the underlying mineral substrate, promoting further permafrost aggradation, ice segregation and frost heave (Harris, 1993).

Lithalsas that have developed through removal of peat cover from a palsa may be considered landforms at an intermediate stage of the 'palsa cycle'. Calmels *et al.* (2008) reconstructed the history of a 3.4 m-high 'palsa-derived' lithalsa in northern Quebec from borehole evidence. They found that an active layer 1.5 m deep overlies a permafrost core that descends through 8.5 m of clayey silt with abundant ice lenses into underlying sand and bedrock (Figure 7.12). They inferred that the lithalsa probably began to grow as a palsa under conditions colder than those of the present, but subsequently lost its peat cover through erosion. In the years following extraction of the core, settlement affected the top of the mound, ponded water accumulated in a shallow depression and a circular rim ridge formed as the centre subsided. Further shrinkage of the ice-rich permafrost core is likely to cause more subsidence and extension of surface ponding until all that remains is a thermokarst pond surrounded by a ridge of unfrozen sediment, similar to that surrounding nearby ponds (Figure 7.10). As such rim ridges are composed mainly of mineral sediment, they have much greater preservation potential than collapsed palsas. As with palsas and peat plateaus, lithalsas in the discontinuous permafrost zone are sensitive to climate warming; Beck *et al.* (2015), for example, detected a 6% decrease in the area occupied by lithalsas in subarctic Quebec over the period 2004–09.

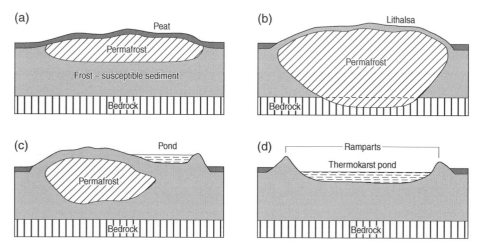

Figure 7.12 Schematic evolution of a lithalsa. (a) Proto-lithalsa, formed by permafrost aggradation and frost heave under an initial peat cover. (b) Mature lithalsa, with permafrost extending several metres into the ground and peat cover removed by erosion. (c) Permafrost degradation, initiation of lithalsa collapse and surface ponding. (d) Collapsed (relict) lithalsa, comprising ramparts of sediment encircling a shallow thermokarst pond. *Source:* Calmels *et al.* (2008). Reproduced with permission of Elsevier.

7.10 Permafrost Plateaus

Permafrost plateaus (lithalsa plateaus) are broad expanses of ground that have been raised a few metres above the surrounding terrain by the development of segregated ice layers in the underlying sediment. They resemble peat plateaus in having well-defined marginal slopes, but carry no peat cover and may be up to 10 m high. Allard *et al.* (1996) reconstructed the evolution of a permafrost plateau developed in thick, fine-grained marine sediments on the eastern shore of Hudson Bay. This landform rises up to 5.6 m above the high tide mark, extends 180 m inland and is 150 m wide. Its surface is flat or gently undulating, except near its inland margin, where thermokarst ponds separate residual mounds. The plateau supports a moss and lichen cover; peat is absent and tree cover is localized.

Drilling showed that the plateau is underlain by four cryostratigraphic layers (Figure 7.13). Allard *et al.* (1996) inferred that the permafrost plateau began to grow around AD 1830 following emergence and exposure of the surface to freezing (present MAAT is about −6 °C). They argued that initial rapid permafrost aggradation produced a layer of reticulated ice. Most plateau uplift, however, has been due to ice-lens formation in the underlying sediment (layer 4 in Figure 7.13). Initiation of thermokarst near the inland margins of the plateau they attributed to snow retention in areas of denser tree and shrub colonization, resulting in deepening of the active layer, thaw of the ice-rich layer of aggradational ice and localized subsidence. Thus whereas the permafrost underlying some parts of the plateau continues to aggrade, that under more densely vegetated areas is experiencing

degradation. Conversely, Chasmer *et al.* (2011) have shown that biomass reduction at the edges of permafrost plateaus in the discontinuous permafrost zone may cause an increase in incident shortwave radiation that augments permafrost thaw and increases surface runoff, leading to a reduction in plateau area.

7.11 Other Permafrost Mounds

Not all permafrost mounds fit readily into the categories described above. Jaworski and Niewiarowski (2012), for example, have described low (0.2–1.3 m-high) *frost peat mounds* on a peat bog in northwest Svalbard. Excavation of one such mound revealed 0.4 m of peat and peaty silt overlying alternating layers of peat and predominantly massive ice, interpreted as injection ice, that extend to 1.5 m depth in continuous permafrost. The presence of massive layers of injection ice suggests that these mounds grew through episodic injection of water under hydrostatic pressure during closed-system freezing, with preservation of ice layers under an insulating cap of peat. Small mounds about 1 m high and 4–10 m in diameter in northern Sweden have been interpreted by Åkerman and Malmström (1986) as hybrid landforms initiated by hydraulic injection of water between a freezing active layer and underlying permafrost or bedrock. They envisaged that the resulting mounds remain above the level of winter snowcover, permitting deeper frost penetration and stimulating further growth due to the formation of segregation ice. Finally, small vegetated mounds called *pounus* occur in mires in discontinuous or sporadic permafrost zones, particularly around the

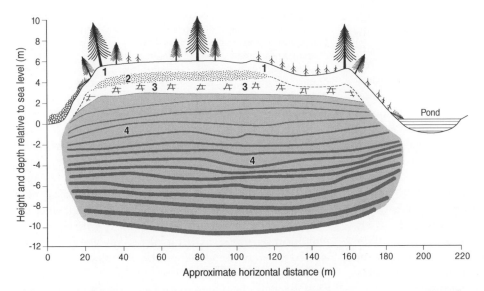

Figure 7.13 Cryostratigraphy of a permafrost plateau developed on silty clay marine sediments on the shore of Hudson Bay. Layer 1 is the active layer; layers 2–4 represent permafrost. Layer 2 contains mainly aggradational ice (50–80% by volume). Layer 3 contains reticulate ice (10–30% by volume). Layer 4 consists of alternating frozen sediment and ice lenses (50–80% by volume), which increase in thickness near the base of the permafrost. *Source:* Allard *et al.* (1996). Reproduced with permission of Wiley.

margins of palsa fens where peat cover is less than 0.6 m thick. Pounus are typically less than 1 m high and 2 m wide, and contain a core of mineral soil (Seppälä, 1988a). The core may freeze and thaw annually, or it may remain frozen through several summers. Seppälä (1998) has documented recent aggradation of permafrost up to 44 cm thick beneath an active layer 55 cm thick in large pounus on a mire in Finnish Lapland. As with palsas, permafrost can form and survive in pounus, as their summits are relatively free of winter snowcover, and as peat cover insulates the core from thaw during the summer.

7.12 Ephemeral Frost Mounds

All of the frost mounds described above have a core of ice-rich permafrost and a life cycle of decades or centuries. By contrast, *ephemeral frost mounds* (or *seasonal frost mounds*) develop in the active layer or, exceptionally, in seasonally frozen ground above bedrock. Ephemeral frost mounds may grow and disappear annually, or they may survive a few years before thawing.

Frost blisters are small ephemeral frost mounds that form during freezeback when a confined aquifer is created between the downward-freezing active layer and the permafrost table. Where groundwater flow through this progressively narrowing conduit is fed by a spring, a hydraulic pressure gradient builds up. If groundwater drainage is blocked by freezing of the active layer to the permafrost table and trapped water cannot escape, the resultant build-up of hydraulic pressure causes uparching of the overlying frozen soil. Freezing of this trapped water lens results in further uplift, forming a frost blister consisting of frozen active-layer sediments overlying a dome-shaped core of injection ice (Pollard and French, 1984, 1985; Pollard, 1988; Figure 7.14). Tension cracks in the updomed frozen sediments may allow escape of pressurized water before freezing of the core is complete. Frost blisters typically achieve heights of 1–4 m, and occasionally up to 8 m, but because the ice core is developed above the permafrost table it may melt completely the following summer, or at most within a few years. Slope-foot frost blister formation is favoured at sites where a perennial spring feeds groundwater flow within a deep active layer composed of sediments with high hydraulic transmissivity, and climatically by long cold winters with limited snowcover. Frost blisters also develop, however, within low-centred syngenetic ice-wedge polygons, where closed-system freezing drives water under hydrostatic pressure towards the polygon centres, and freezing of the resulting water lens causes updoming of overlying frozen sediments. A survey by Morse and Burn (2014) of polygon-centre frost blisters on the Mackenzie Delta showed that though they do not exceed 1 m in height and rarely exceed 10 m in length, they occur in great numbers

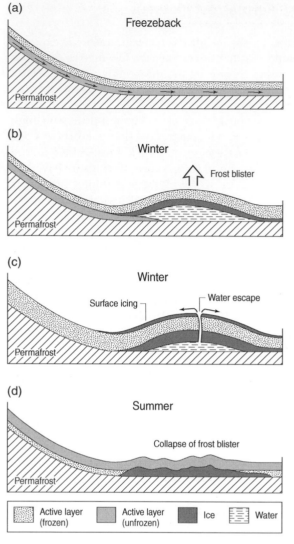

Figure 7.14 Development of a frost blister. (a) Groundwater flow through the active layer during freezeback. (b) Damming of flow as freezing extends to the permafrost table on the valley floor, causing development of high hydraulic pressure and updoming of the overlying frozen ground. (c) Inwards freezing of the water lens, producing further uplift, and escape of water, creating a surface icing. (d) Melt of the ice core during summer, resulting in collapse.

($>$1700 km^{-2}) and are transient but perennial, with individual blisters surviving for up to a decade.

In many cases, however, rupture of the active layer allows pressurized groundwater to escape to the surface without updoming of the overlying frozen ground. In such cases, an *icing blister* composed entirely of ice may form (Figure 7.15). Icing blisters involve freezing of ejected groundwater at the surface, then updoming of the surface icing layer by further injection of pressurized groundwater, which then freezes as a core of injection ice. They also form when pressurized groundwater is injected under a pre-existing surface icing, such as a river icing, and usually melt completely the following summer.

Figure 7.15 Icing blister with summit tension crack, Svalbard. *Source:* Courtesy of Ole Humlum.

Remarkable frost blisters (misleadingly termed 'seasonal pingos') on the Qinghai-Tibet Plateau have been described by Wu *et al.* (2005). These form where subpermafrost groundwater reaches the surface along fault lines and is annually injected under an active layer of frozen sediments to form chains of elongate frost blisters up to 3.5 m high, 30 m long and 15 m wide. Though these thaw and collapse each summer, new frost blisters form in slightly different locations along the faults the following year, posing a hazard to road and rail links in the area. Short-lived frost mounds may also be produced by fragmentation of seasonal ice-cored peat deposits. Hinkel (1988) has described annual formation and thaw of a massive sill of injection ice 1.2 m thick under peat in a flooded marsh in the Brooks Range, Alaska. Thermal erosion by streams has isolated small tabular ice-cored mounds 1 m high, 2–4 m wide and 4–8 m long at the margins of this 'seasonal peat plateau', and though some thaw within a single summer, others have survived for at least four years as a rather unusual type of ephemeral frost mound.

7.13 Relict Permafrost Mounds

As noted above, pingos and lithalsas leave distinct traces on the landscape after thaw, usually in the form of depressions partly or completely encircled by low ramparts of sediment (Figure 7.10). As pingos and lithalsas develop in permafrost environments, relict pingos and lithalsas that formed during the last (Wisconsinan/Weichselian) Glacial Stage can be employed to establish the extent of former permafrost, and to constrain palaeoclimatic parameters at the time of their formation. Conversely, ephemeral frost mounds and peat plateaus have limited preservation potential and are unlikely to be represented in areas of past permafrost. The long-term survival of degraded palsas is debatable. Ponds left by palsa collapse are liable to be infilled by peat or sediment, so traces of former palsas are often rapidly extinguished. Where palsas carry only thin peat cover, however, low annular ramparts of mineral sediment may mark the former site of palsa growth (Seppälä, 1988a). Such palsa remnants are likely to resemble those of relict lithalsas, with which they are closely associated both geographically and climatically (Worsley *et al.*, 1995; Calmels *et al.*, 2008).

Numerous landforms interpreted as relict pingos or lithalsas have been identified both outside and inside the limits of the last Pleistocene ice sheets across northern Europe from Ireland to Russia, as well as in North America and northern Asia (De Gans, 1988; Gurney, 2000; Pissart, 2002, 2003; Kluiving *et al.*, 2010; Makkaveyev *et al.*, 2015). Structures resembling large-scale pingo scars have even been detected on parts of the North Sea Basin that were exposed during the last glacial maximum (LGM) by falling sea level (Long, 1991). Interpretation is often problematic, however, because of the morphological similarity of relict pingos, relict lithalsas and other thermokarst depressions. For this reason, some authors favour the term *ramparted ground ice depressions* to describe the sites of possible former pingos or lithalsas, as this avoids genetic connotations. A further problem is that ramparted depressions can be explained in various other ways, for example as 'rimmed' kettle holes, depressions formed where icebergs were grounded by lake drainage and dead-ice topography. Morphology, topographic context, relationship to glacier limits and the sedimentology and structure of ramparts and depressions need to be considered before inferences can be made regarding the genesis and palaeoclimatic implications of possible relict permafrost mounds.

7.13.1 Relict Pingos

Mackay (1988b) suggested various criteria for the identification of relict pingos or *pingo scars*: rampart volume should approximate that of the depressions from which rampart sediments were derived; rampart deposits should comprise sediments emplaced by active-layer slides, slumping, solifluction, debris flows and running water; and ramparts should exhibit evidence for high-angle normal faults and for sediment-filled casts infilling former dilation-ice cracks. Such information is often difficult to obtain, however, as depressions are often infilled by lake sediments or peat, and exposures through ramparts are rare. Mackay also pointed out that thaw of ice-cored ramparts may leave only very subdued ridges. Topographic context offers more promising criteria. By analogy with active examples, relict hydrostatic pingos should be represented by isolated ramparted depressions occupying former lake floors underlain by coarse sandy sediments, and relict hydraulic pingos should occupy slope-foot, alluvial-fan or valley-floor locations, particularly in permeable sediments fed by groundwater seeps or springs, and should occur individually or in small clusters. Many proposed relict pingos do not meet these criteria.

Some of the most convincing accounts of putative pingo scars concern numerous broad, ramparted depressions, 100–350 m in diameter, located outside the Late Weichselian ice limit in the northern Netherlands (De Gans, 1988; Van der Meulen, 1988). These are located in flat areas of low relief, average 4–5 m in depth and are ringed by low ramparts up to 1.4 m high. Organic infill has been dated to the Late Weichselian, suggesting formation during the last glacial stage as hydrostatic pingos in an area of continuous permafrost (De Gans and Sohl, 1981; De Gans, 1982). There are numerous descriptions of supposed pingo scars elsewhere in northern Europe (De Gans, 1988), but many require reassessment in the light of present understanding of the constraints on pingo formation and alternative explanations. In North America, widespread oval 'prairie mounds' up to 9 m high and 200 m across in southern Saskatchewan, southern Alberta and parts of British Columbia were initially interpreted as pingo scars, but a glacigenic origin now seems more likely both for many of these (Mollard, 2000) and for similar forms in Illinois (Iannicelli, 2003).

Even where there is reasonable evidence favouring a pingo origin for a ramparted depression, establishing the age of such a feature is often difficult and palaeoclimatic inferences are problematic. Mackay (1988b) highlighted the pitfalls associated with using the climates of areas presently occupied by active pingos as a guide to the palaeotemperature implications of pingo scars. Active pingos may have begun to develop under colder conditions than those of the present, and both pingo size and thermal offset need to be taken into account. Isarin (1997b) suggested that hydrostatic pingos imply MAAT $\leq -6\,°C$ and hydraulic pingos MAAT $\leq -4\,°C$. The present distribution of active pingos, however, indicates that the MAAT under which relict pingos developed could have been up to about 5 °C lower than these 'warm-side' limits imply.

7.13.2 Relict Lithalsas

On ground above 500 m on the Hautes Fagnes Plateau in Belgium are numerous closely spaced ramparted depressions. Some are circular or oval and up to 150 m in diameter; some are irregular in plan; others are complex, with cross-cutting rims and hollows suggesting a multigenerational origin. Ramparts are mainly less than 1 m high, but some reach 5 m. Pissart (2000, 2003) has argued that the morphology, location, stratigraphy and clustering of these landforms suggest that they represent relict lithalsas, and he illustrated their close resemblance to degraded lithalsas in the Hudson Bay area (Pissart *et al.*, 2011; Figure 7.10). Dating of peat buried within one rampart demonstrates formation during the Younger Dryas chronozone of 12.9–11.7 ka, the final period of permafrost in northern Europe. In the British Isles, many of the ramparted depressions interpreted as possible pingo scars (Coxon and O'Callaghan, 1987; Ballantyne and Harris, 1994) have similar characteristics, suggesting that these also should be reinterpreted as relict lithalsas (Gurney, 1995, 2000; Worsley *et al.*, 1995; Pissart 2000, 2003), or as landforms transitional between lithalsas and hydraulic pingos (Ross *et al.*, 2011). The characteristics of ramparted depressions in Pennsylvania (Marsh, 1987) also appear consistent with those of relict lithalsas, as do those of Holocene 'pingo-like remnants' in Lapland described by Seppälä (1972).

Pissart (2000, 2002, 2003) has suggested that because active lithalsas occupy a narrow climatic niche, relict lithalsas have excellent potential for palaeoclimatic reconstruction. Active lithalsas in the Hudson Bay lowlands and Great Slave Lake lowlands occupy areas where MAATs are about −4 °C to −6 °C and mean July temperatures fall within the range 9.0–11.5 °C (Wolfe *et al.*, 2014), suggesting that similar conditions pertained in parts of northwest Europe that supported lithalsa growth during the Younger Dryas period. Pissart (2003) tested these temperature ranges against independent reconstructions of Younger Dryas palaeotemperatures based on a range of proxy evidence (Isarin, 1997a, 1997b; Isarin and Bohnke, 1999) and demonstrated concordance between the two. Relict lithalsas therefore appear to be amongst the most fine-tuned periglacial palaeotemperature proxies, possessing both warm-side and cold-side constraints on former temperatures.

Relict pingos and lithalsas, like ice-wedge pseudomorphs, are thermokarst landforms produced by the melt of bodies of ground ice. Other landforms and structures formed by permafrost degradation and associated thaw of ground ice are considered in Chapter 8.

8

Thermokarst

8.1 Introduction

The term *thermokarst* is used to describe landforms, sediments and processes associated with the degradation of ice-rich permafrost. As with many terms in the periglacial lexicon, it is something of an etymological oddity, derived from the apparent similarity between some permafrost-degradation landforms and *karst* landforms developed on limestone and other carbonate rocks. However, thermokarst development is a thermal rather than solutional phenomenon, is independent of underlying lithology and involves a completely different set of processes. Even the resemblance to limestone karst phenomena is incomplete and superficial.

Accounts differ regarding the range of periglacial phenomena that should be considered as thermokarst. Jorgenson (2013) lists 23 different thermokarst landforms, but includes some that reflect short-term thaw of the uppermost part of ice-rich permafrost (active-layer detachment slides, here considered in Chapter 11) and others that reflect thermomechanical erosion operating along riverbanks, coasts and lakeshores (Chapters 13 and 15). Here attention is focused on thermokarst phenomena that generally reflect medium- to long-term permafrost degradation. We have encountered some of these already, in the form of thaw modification of degrading ice wedges (Chapter 6) and the landforms produced through the collapse of frost mounds, such as pingos, palsas and lithalsas (Chapter 7). In the case of pingos in particular, permafrost degradation is autogenic, an inevitable consequence of pingo growth and exposure of the ice core, rather than triggered by extrinsic changes.

Even if some of these features are excluded, however, the range of thermokarst phenomena is large. All share two common features. First, they are developed in ice-rich permafrozen sediments that contain excess ice, so that thaw of ground ice produces a volume of water that exceeds the volume of pore-spaces within the sediment. Thaw is therefore accompanied by volume loss and ground surface depression, sometimes termed *thermokarst subsidence*. The amount of subsidence depends on the volume and distribution of excess ice prior to thaw and on the depth of thaw. For example, if permafrost underlying a 0.5 m-deep active layer contains 50% excess ice, deepening of the active layer to 1.1 m will produce ground surface subsidence of 0.3 m (50% of the 0.6 m increase in active-layer depth). Second, thermokarst phenomena are produced by a change in surface energy balance that alters the thermal equilibrium of nearsurface permafrost, resulting in deepening of the active layer and in some cases formation of a talik. Jorgenson *et al.* (2010) have suggested a fourfold classification of permafrost degradation: (i) transient, involving short-term thaw of the uppermost layer (transient layer) of permafrost during an exceptionally warm summer; (ii) surface, involving longer-term deepening of the active layer without talik development; (iii) intermediate, involving development of a closed talik under the active layer; and (iv) complete, where all permafrost has thawed and an open talik has formed. They also included two additional categories: lateral permafrost degradation, due to bank collapse or thermomechanical erosion by waves or rivers, and internal, caused by subterranean movement of suprapermafrost groundwater through tunnels, often along the margins of ice wedges.

The change in surface energy balance (heat transfer) that initiates permafrost degradation may be caused by either regional changes in climate, such as longer or warmer summers or increased winter snowfall, or a range of more local factors (Table 8.1). Some of the latter are natural, such as soil erosion and forest fire, but others, such as deforestation and stripping of vegetation or peat cover, may be induced by human activity (Lawson, 1986; Nicholas and Hinkel, 1996). By far the most widespread causes of thermokarst development, however, are related to climate shifts that are often amplified by changes in ground surface conditions (Shur and Jorgenson, 2007). Nearsurface permafrost temperatures at high latitudes have increased over the past few decades in response to climate warming (Osterkamp, 2007; Romanovsky *et al.*, 2010a), and numerical models of permafrost response to future climate warming forecast widespread permafrost degradation by the end of the present century (Callaghan *et al.*, 2011). Moreover, global circulation

Periglacial Geomorphology, First Edition. Colin K. Ballantyne.
© 2018 John Wiley & Sons Ltd. Published 2018 by John Wiley & Sons Ltd.

Table 8.1 Factors responsible for initiating or retarding thermokarst.

Factor	Thermokarst initiation (permafrost degradation)	Thermokarst retardation (permafrost stability or aggradation)
Regional factors		
Air temperature	Climatic warming	Climatic cooling
Snowfall	Increased snowfall	Decreased snowfall
	Earlier snowmelt	Later snowmelt
	Earlier snow accumulation	Later snow accumulation
Summer weather	Longer, warmer summers	Shorter, cooler summers
	Increased summer rainfall	Decreased summer rainfall
Local factors		
Vegetation cover	Degradation or removal	Regrowth; tree or shrub growth
	Boreal forest fires	Post-fire tree regrowth
Peat cover	Rupture, erosion or removal	Peat accumulation
	Summer wetting of peat	Summer drying of peat
Snow cover	Local thickening	Local thinning
Hydrology	Lake deepening	Lake drainage
	Formation of drainage channels	Cessation of channel runoff
Overburden	Natural soil erosion	Natural sediment deposition
	Artificial removal of soil	Burial by spoil, gravel pads, etc.
	Compaction	
	Concrete and tarmac surfaces	
Local heat source	Heated buildings, pipelines, etc.	

models of future climate change predict significant increases in precipitation and extreme rainfall events in high latitudes (Walsh *et al.*, 2011; IPCC, 2013). Such changes are likely to promote widespread extension of the area occupied by thermokarst terrain within the compass of a human lifetime (Murton, 2009).

Some thermokarst landforms, particularly thermokarst lakes and drained lake basins on the ice-rich yedoma deposits of the northern Siberian coastal plains and interior Alaska, started to develop during the warming that accompanied the Pleistocene–Holocene transition at ~13–11 ka (Morgenstern *et al.*, 2013; Kanevskiy *et al.*, 2014; Lenz *et al.*, 2016) or during the Early Holocene thermal maximum (Kaufman *et al.*, 2004; Katamura *et al.*, 2006, 2009). Other thermokarst features appear to have been initiated much later, in response to gradual warming or changes in precipitation after about AD 1700 (Jorgenson *et al.*, 2001). Thermokarst is particularly widespread in arctic and subarctic lowlands. In coastal regions inland from the Laptev Sea, for example, 78% of the region is affected by permafrost degradation (Grosse *et al.*, 2006). A detailed inventory of 12 areas of varied surficial geology covering 300 km^2 in arctic Alaska has found that 63% has been affected by thermokarst development (Farquharson *et al.*, 2016a). Thermokarst has also has been recorded on the fringes of Antarctica and

on the high plateaus of China, Tibet and Mongolia, but is localized and less common in alpine permafrost environments.

Here, we consider first the evolution of thermokarst on lowland landscapes in the form of thermokarst lakes, drained lake basins, thermokarst bogs and fens, then lateral erosion of ice-rich permafrost by retrogressive thaw slumps, small-scale thermokarst landforms associated with running water and thermokarst sedimentary structures. The chapter concludes with a brief account of some relict thermokarst features in mid-latitude areas of past permafrost. Useful accounts of the range of permafrost processes and landforms include those by Burn (2013), Jorgenson (2013) and Kokelj and Jorgenson (2013). Jorgenson and Osterkamp (2005) provide a summary of the modes and effects of thermokarst development in the boreal forest zone.

8.2 Thermokarst Lakes and Drained Lake Basins

Globally, one of the most widespread manifestations of thermokarst activity takes the form of *thermokarst lakes* (sometimes termed *thaw lakes*) and the basins formed by lake drainage (Figure 8.1). A thermokarst lake is formed

Figure 8.1 (a) Thermokarst lakes and thermokarst ponds, central Banks Island, arctic Canada. The outlines of submerged frost polygons can be seen on the floor of the central lake, suggesting that it formed recently. (b) Thermokarst lakes flanking a river channel, Mackenzie Delta area, western Canadian Arctic. A colour version of (a) appears in the plates section. *Source:* Courtesy of (a) Tobias Ullmann and (b) Matthias Siewert.

by subsidence of the ground caused by melting of excess ground ice, and occupance of the resulting depression by standing water. If the water body exceeds the depth of winter ice cover, an unfrozen reservoir of water persists throughout the year. Because this water has a high specific heat capacity, it maintains a positive heat flow into the underlying permafrost, causing development of a talik under the lake floor, further thaw of ground ice and consequently continued subsidence and lake deepening. At the same time, erosion of the lake shores results in expansion of lake area. Eventual drainage of the lake leaves an enclosed depression or *drained lake basin* underlain by a talik that is progressively diminished in size by permafrost aggradation. Grosse *et al.* (2013a) provide a useful synthesis of the distribution, formation, morphology, hydrology and drainage of thermokarst lakes.

8.2.1 Characteristics of Thermokarst Lakes

The distribution of thermokarst lakes is closely related to that of arctic, subarctic and boreal lowlands underlain by thick sediment cover and ice-rich permafrost, primarily in areas that were not glaciated during the last glacial maximum (LGM) (in northern and interior Alaska, northwest Canada and central, northern and northeastern Siberia), but also in glaciated lowlands where buried glacier ice became preserved in permafrost. Thermokarst lakes cover nearly 25% of the ice-rich permafrost lowlands across much of arctic Alaska, Canada and Siberia, and locally as much as 40% in some areas, such as parts of the northeastern Siberian coastal lowlands, northern Alaska and the Mackenzie Delta area (Mackay, 1988a; Hinkel *et al.*, 2005; Grosse *et al.*, 2005; Marsh *et al.*, 2009). Grosse *et al.* (2013a) estimated that there are more than 61 000 lakes exceeding $0.1\,km^2$ in lowland permafrost areas with moderate to high ice content, with a total area exceeding $207\,000\,km^2$. Moreover, drained lake basins locally occupy as much as 50–75% of the land area in ice-rich permafrost lowlands, and in many areas coalesced and overlapping basins are present, some of which contain residual lakes or secondary thermokarst lakes that formed after partial or complete drainage of a former lake (Hinkel *et al.*, 2003, 2005; Grosse *et al.*, 2005; Jones *et al.*, 2012b). Collectively, therefore, thermokarst lakes and drained lake basins represent the dominant landforms in such areas. Smaller numbers of thermokarst lakes occur on the Qinghai-Tibet Plateau (Lin *et al.*, 2010; Niu *et al.*, 2011) and in arctic and subarctic peatlands (Sannel and Brown, 2010; Sjöberg *et al.*, 2013). They are rare in alpine environments, except where small lakes have developed due to thaw of buried glacier ice (Kääb and Haeberli, 2001).

Thermokarst lakes range from ~100 m to over 15 km in length, but most are less than $1\,km^2$ in area; those on Richards Island on the western arctic coast of Canada, for example, average $0.33\,km^2$ (Burn, 2002). Smaller water bodies formed by thermokarst subsidence are usually referred to as *thermokarst ponds*. Most thermokarst lakes are circular, oval or elliptical in planform, though D-shaped or rectangular lakes have also been described, and in some areas clusters of lakes share a common long-axis alignment; the factors responsible for producing such *oriented lakes* are considered below. In areas of epigenetic permafrost where the volume of nearsurface excess ice (and thus the potential for subsidence) is limited, thermokarst lakes tend to be shallow, with depths rarely exceeding 5 m. In areas of thick ice-rich syngenetic permafrost such as the yedoma regions of Siberia and interior Alaska, however, thermokarst lakes reach depths

of up to ~25 m, reflecting the large volume loss through sediment compaction that has occurred as a result of thaw of deep, extremely ice-rich sediments. Bathymetric surveys of thermokarst lakes on the north Yukon Plain, an area of shallow ground ice, show that these have depths of 1.6–3.2 m, irrespective of lake area. Conversely, two lakes underlain by deep ground ice on the northern Seward Peninsula have depths of ~13 and ~22 m (West and Plug, 2008).

The bathymetry of thermokarst lakes varies. Those that have formed in epigenetic permafrost with only nearsurface ground ice often have shallow, flat-bottomed basins, whereas small but deep lakes in syngenetic ice-rich permafrost tend to occupy bowl-shaped depressions, and large, deep lakes sometimes occupy asymmetrical basins. Many shallow thermokarst lakes have a deeper central pool flanked by shallow littoral shelves that extend for tens of metres from the shore under less than 1 m of water. From a study of the bathymetry of 28 lakes in Alaska, Hinkel *et al.* (2012) found that broad, shallow littoral shelves tend to be associated with lakes where abundant sandy sediments have been eroded from the margins and deposited around a deeper central pool, whereas lakes developed in ice-rich silty sediments tend to lack littoral shelves because of limited sediment generation by shore erosion.

8.2.2 Formation of Thermokarst Lakes: Initiation

The sequence of events leading to the formation of thermokarst lakes and drained lake basins was illustrated in a classic paper by Czudek and Demek (1970); an adapted form of their model is depicted in Figure 8.2a. This model, however, applies only to extremely ice-rich syngenetic permafrost (yedoma) landscapes, such as those of unglaciated Siberia and interior Alaska, where ice-rich permafrost typically extends to depths of 10–40 m and locally more. In arctic areas underlain by epigenetic permafrost, such as the arctic coastal plain of Yukon, the ice-rich zone represented by ice wedges and segregated ice is typically only a few metres deep (Figure 8.2b), and this restricts the amount of subsidence that can result from thaw of the ice-rich layer, and hence the depth of the resulting thermokarst lake. Moreover, not all arctic lakes evolve initially from thermokarst subsidence, as discussed below.

Thermokarst lakes are initiated by a change in the heat balance at the surface. Radiocarbon dating of organic material from lake basins has shown that, in many arctic, subarctic and boreal environments, widespread thermokarst lake formation was initiated by climate warming and increased precipitation during the Pleistocene–Holocene transition at ~13–11 ka (Romanovskii *et al.*, 2000; Walter *et al.*, 2007; Morgenstern *et al.*, 2013;

Figure 8.2 Schematic evolution of thermokarst lakes and drained lake basins (a) in syngenetic permafrost, where excess ice is tens of metres deep and (b) in epigenetic permafrost, where excess ice extends only a few metres below the surface.

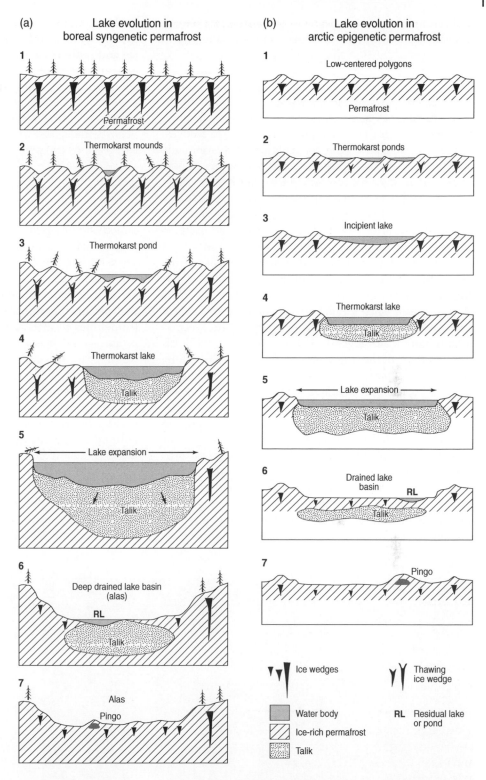

Kanevskiy *et al.*, 2014; Lenz *et al.*, 2016), and locally reached a peak during the early Holocene thermal maximum. In the Yukon Flats of interior Alaska, for example, thermokarst lakes that formed around 12–11 ka have been stable throughout the Holocene (Edwards *et al.*, 2016), and in eastern Siberia much of the landscape of thermokarst lakes and drained lake basins formed prior to 9.0–7.5 ka,

with limited modification thereafter (Romanovskii *et al.* 2004). In other areas, thermokarst lake formation has continued throughout the Holocene, particularly in the form of secondary thermokarst lakes in the drained basins of earlier lakes. Warming and increased precipitation are regionally the most important factors triggering the formation of thermokarst lakes, though terrain disturbance by forest fires, vegetation clearance or construction activity has locally initiated lake development (Burn and Smith, 1990; Lin *et al.*, 2010; Schleusner *et al.*, 2015).

In areas of high-centred ice-wedge polygons underlain by syngenetic permafrost (Figure 8.2a, stages 1–3), thaw of the upper parts of the ice wedges results initially in accumulation of water at polygon intersections, forming *thermokarst pits*, and in thaw of the tops of wedges, forming *thermokarst troughs*. This process eventually results in the formation of hummocky or conical *thermokarst mounds*, also known by the Russian term *baidzherakhi* and, in older literature, *cemetery mounds*, a reflection of their regular spacing. In areas of boreal forest, tilting of trees towards the collapsing margins of thermokarst mounds produces *drunken forests* of toppled and toppling trees that provide evidence of the early stages of uneven ground subsidence. Further subsidence results in coalescence of ponded water to form a shallow thermokarst pond. In poorly drained lowlands of low-centred polygons underlain by epigenetic permafrost (Figure 8.2b, stages 1–3), thermokarst ponds form through thaw of the upper part of ice wedges, breaching of the ridges separating polygons and consequent coalescence of adjacent pools of water. The initial stages of lake development, however, may also simply result from uneven subsidence of ice-rich permafrost, particularly on terrain underlain by massive ground ice.

Standing water at the surface has low albedo, absorbs longwave radiation during the summer months and has relatively high heat storage capacity (Harris, 2002). It therefore imparts heat to the underlying frozen ground, so that temperatures at the water–sediment interface rise several degrees above mean annual air temperature (MAAT). This causes warming and thaw of the underlying permafrost, ground volume loss through melt of excess ice and sediment compaction, and thus gradual subsidence and water deepening. In the initial shallow stages of lake formation, winter ice cover freezes to the bottom, and thaw is consequently slow. When the water body reaches a depth (typically 1–2 m) at which unfrozen water persists under the winter ice cover, thaw is more rapid and thermokarst ponds deepen and expand into lakes.

8.2.3 Thermokarst Lake Growth

Thermokarst lakes grow though both deepening, as underlying ice-rich sediments thaw and consolidate, causing lake basin subsidence, and laterally, through shoreline erosion (Figure 8.2a,b, stages 4 and 5). Once water depth exceeds winter ice thickness, year-round above-freezing temperatures at the lakefloor enhance thawing and initiate the formation of a sublacustrine talik (sometimes termed a *thaw bulb* or *thaw basin*), which progressively deepens until thermal equilibrium is restored. In the central deep pool of a thermokarst lake on the western arctic coast of Canada, for example, Burn (2002) measured average annual water temperatures ranging from +1.5 to +4.8 °C, but under the surrounding shallow littoral shelves (where winter ice freezes to the bottom) he recorded temperatures of −0.2 to −5.0 °C. In summer, heat from the atmosphere is diffused throughout lakes by wind-driven currents and possibly thermally driven convection (Kääb and Haeberli, 2001), and is partly conducted through the sublacustrine talik to the underlying permafrost, causing further thaw of ground ice, basin subsidence and lake deepening. In ice-rich syngenetic

Figure 8.3 Small thermokarst lake, central Banks Island, arctic Canada. Degradation of ice-wedge polygons near the lake margins has formed thermokarst mounds. Fens have colonized the lake margins. A colour version of this figure appears in the plates section. *Source:* Courtesy of Tobias Ullmann.

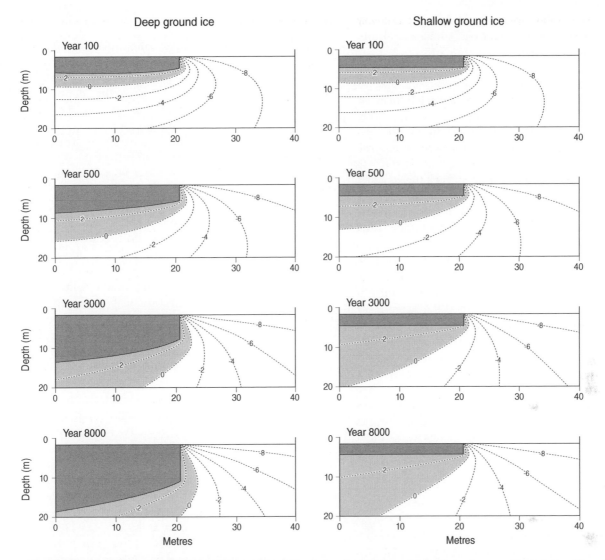

Figure 8.4 Simulated talik development (light grey) and lake deepening (dark grey), based on a nonexpanding lake with an initial depth 2 m, 30% excess ice content in the underlying sediments and an assumed lake-bottom temperature of 3 °C. The model on the left represents an area of deep ground ice (syngenetic permafrost), and that on the right an area of shallow ground ice (epigenetic permafrost). Isotherms are °C. Both models exhibit near-equilibrium conditions after 8000 years. *Source:* Adapted from West and Plug (2008). Reproduced with permission of the American Geophysical Union.

permafrost, where excess ice extends tens of metres below the surface, this process of talik growth and lake deepening continues for millennia. In epigenetic permafrost, where excess ice is confined to the top few metres below the surface, lake deepening stalls at shallow depths (often less than 3 m), but the underlying talik may continue to thicken, eventually reaching much greater depths.

The process of lake and talik deepening has been simulated by West and Plug (2008) using a 2D numerical model of conductive heat transfer, phase change and thaw subsidence of ice-rich sediments under lakes. Their simulations assume an initial lake 2 m deep and ~700 m in diameter with an average lake bottom temperature of 3 °C, underlain by sediments containing 30% excess ice. In areas underlain by deep ground ice

(syngenetic permafrost), their model (Figure 8.4) predicts that lake and talik depth increase approximately as $\sqrt{t}$, where t is time elapsed since the onset of lake-floor thawing, reaching a lake depth of ~22 m after 8000 years, with an underlying talik ~45 m thick. By contrast, in areas underlain by shallow ground ice (epigenetic permafrost), the predicted increase in simulated lake depth over 8000 years is <2 m, but the underlying talik reaches a depth of ~48 m. Effectively, their model predicts that lakes in areas of shallow excess ice achieve steady-state depths within a century, but lakes in areas of deep excess ice continue to deepen until thermal equilibrium is regained, the lake floor reaches sediment lacking excess ice (Morgenstern *et al.*, 2011) or the lake drains. They also showed that increasing the lake bottom temperature markedly increases the rate of lake deepening and talik thickening: in syngenetic permafrost, a ~20 m deep lake forms in ~5000 years with an assumed lake-bottom temperature of 3 °C, but in ~3500 years with a temperature of 4 °C. Their modelling also suggests that an increase in ice content (and therefore thermal diffusivity) in frozen sublacustrine sediments also enhances the rate of lake deepening but may retard the rate of talik development, because of the additional heat required to thaw ice.

Irrespective of the depth of ground ice, all thermokarst lakes undergo lateral expansion as a result of shoreline erosion. In extreme cases, this may involve shoreline retreat of several metres per year, but studies using sequential remotely-sensed imagery suggest that typical rates of lake expansion average 0.3–0.8 m per year (e.g. Jorgenson and Shur, 2007; Jones *et al.*, 2011). Modelling by Plug and West (2009) of heat transfer, thaw subsidence and sediment movement suggests that deep lakes underlain by syngenetic ice-rich permafrost tend to expand more rapidly than shallow lakes in epigenetic permafrost, because steeper bathymetric slopes permit more effective removal of thawed sediment away from lake margins.

Various processes contribute to bank erosion. On many lake shores, the dominant erosional process is summer wave action. In steep ice-rich banks, waves cut a notch or *thermoerosional niche* at or just above water level in the permafrozen ice-rich lakeshore through a combination of ground ice melting and mechanical erosion; the undermined frozen sediments then collapse into the lake, where they thaw and are dispersed by waves. Extension of the sublacustrine talik under lake margins may cause subsidence, bank steepening and slumping (Kokelj *et al.*, 2009), and large *retrogressive thaw slumps* (see below) extending backwards from the shore may expose massive ice bodies and feed mudflows into the lake (Séjourné *et al.*, 2015). In poorly drained lowlands, lake expansion often involves progressive capture of ponds within low-centred ice-wedge polygons (Jones *et al.*, 2011). Ancillary processes aiding lake expansion are ice-push during spring break-up and toppling or slumping of trees, vegetation cover, soil and peat from eroded banks. Parsekian *et al.* (2011) have described an intriguing form of bank recession on the margins of thermokarst lakes involving the generation of floating vegetation mats. They showed that as sublacustrine taliks extend under lake margins, causing thaw and subsidence of lakeshore sediments, the overlying mat of peat and vegetation cover is gently launched into the lake. As the water under floating vegetation mats remains above freezing point in winter, the presence of such mats promotes lateral talik extension and consequent bank collapse. Vegetation mats may extend into lakes by up to 1–2 m per year, so that they eventually occupy up to 4% of the lake area.

8.2.4 Oriented Lakes

Clusters of thermokarst lakes sometimes exhibit similar orientation, with their long axes aligned roughly parallel to one another. In areas of former glaciation, this may be due to lakes occupying depressions between aligned streamlined subglacial bedforms (megaflutes), but some other explanation is required in areas lacking such features. In some areas, preferred lake orientation appears to reflect the direction of the dominant winds that blow across the lakes in summer, setting up a two-cell circulation of currents that focuses erosion of lake shores at both ends, thus causing progressive elongation of the lakes at right angles to dominant summer wind direction. Côté and Burn (2002) tested this hypothesis by calculating the mean orientations of 578 oriented lakes and 145 drained or partly drained lake basins on the Tuktoyaktuk Peninsula of the western arctic coast of Canada and comparing them with predicted lake orientation derived from data on wind directions at two nearby stations. The lakes have a mean orientation of N07° E, and the drained lake basins have a mean orientation of N13° E. The predicted orientation of lakes derived from the (predominantly westerly or easterly) winds ranges from N00° E to N29° E, depending on the dataset employed and the assumed threshold wind velocity. This comparison provides strong support for the hypothesis that oriented thermokarst lakes in some areas are produced by long-axis lengthening due to wind-driven currents, and the similarity in orientation of lakes and drained or partly-drained lake basins suggests that the dominant wind regime has been constant over at least the past few centuries. A Landsat-based study of the orientation of 13 214 thermokarst lakes and 6539 drained lake basins on the Arctic Coastal Plain of Alaska by Hinkel *et al.* (2005) demonstrated that the lakes and drained lake basins share a similar preferred orientation (with long axes aligned between NNE and north) that is constant on emergent terrains of different age, demonstrating that summer wind direction has not changed significantly over the past few millennia.

This explanation, however, is not universally applicable. Preferred lake orientation may in some cases be influenced by the orientation of the primary ice-wedge network during lake formation, uneven distribution of ground ice or preferential retrogressive thaw slump activity (and hence lake extension) on south-facing slopes where insolation enhances ground-ice melting (Grosse *et al.*, 2013a; Séjourné *et al.*, 2015).

8.2.5 Thermokarst Lake Sedimentation

The nature of thermokarst lake sedimentation depends on a wide range of variables, notably the depth and extent of the lake, the composition of the sediment source (the retreating lake shoreline) and the processes responsible for eroding, transporting and depositing sediment. During initial lake-floor subsidence, wood fragments and vegetation remnants are incorporated into lake-floor sediments, and in the boreal forest zone toppled or drowned trees may also be preserved under anaerobic conditions. The complexity of subsequent lake sediments is illustrated in a study by Murton (1996c) of vertical coastal exposures through drained or partly drained lake basins in western arctic Canada. In these exposures, Murton recognized eight distinct sediment facies, six of which reflect deposition in former lakes and two of which formed after lake drainage (Table 8.2). He identified three stratigraphic units: a basal unit, formed during the early development of the lakes (stage 1); a middle unit, representing subsequent lacustrine deposition and sediment modification (stage 2); and an upper unit, representing post-drainage sediment accumulation of peat or aeolian sediment (stage 3).

The stage 1 basal unit represents sediment progradation towards lake centres, and comprises two facies: diamicton, derived from glacial till that caps the sediment around the lake shore, and interpreted as the product of retrogressive thaw slumping that continued to travel towards the lake centre as subaqueous debris flows; and 'impure sand', a poorly-sorted fine sand containing inclusions of peat, mud, clasts and wood or charcoal fragments. The origin of this facies is uncertain. Murton suggested that it may represent episodic bank collapse, subaqueous sediment gravity flows associated with thaw slumping or failure of steep banks or sandy foreset beds. Together, these two facies form a progradational foreset

sequence that advanced episodically towards the centres of lake basins. Syndepositional faulting within basal unit sediments provides evidence of lakefloor subsidence during their accumulation.

The overlying middle unit of stage 2 comprises four aggradational facies. In the deep central basin, organic-rich silt and muddy peat suspended in lake water settled on the former lake floor to form stratified or massive, horizontal or gently dipping beds. Shallow water facies formed on shelves at lake margins included lenses or sheets of pebbly sand and sandy gravel, or a veneer of well-sorted and rippled fine sand (Table 8.2), both of which were attributed by Murton to winnowing of mud and detrital peat from basal unit sediments by currents and wave action. The final shallow-water facies comprises truncated beds of detrital peat up to 2 m thick, and may represent severed sections of the floating vegetation mats described by Parsekian et al. (2011).

Murton (1996c) acknowledged that his depositional model is both idealized and oversimplified, and that some sediment aggradation probably occurred during stage 1 and some progradation of sediment during stage 2. It is also noteworthy that he was working in an area of deep lake basins, and that the sediment sequences within shallow thermokarst lakes may lack the progradational facies he identified. The facies represented in thermokarst lakes are also conditioned by the nature of the terrestrial sediment source: a diamicton facies, for example, is likely to be present only where there is a source of coarse sediment, such as till or gravel, and lakes formed in silty organic-rich frozen sediments may contain only bottom deposits that have settled from suspension.

Table 8.2 Sedimentary facies of deep thermokarst lake basins, Tuktoyaktuk Coastlands.

Environment	Facies	Processes	Facies description
Basal unit	Diamicton	Subaqueous debris flow	Massive silty to sandy clay containing clasts, with peat, wood, charcoal and mollusc inclusions, sand beds ≤15 cm thick and lenses of pebbly sand; often intercalated with impure sand facies
Basal unit	Impure sand	Progradation of littoral shelves	Poorly sorted massive or stratified fine sand containing peat, mud, clasts and wood or charcoal fragments, with foreset beds dipping ≤28° and high-angle syndepositional faults
Central basin	Mud and muddy peat	Settling from suspension	Organic-rich silt to muddy peat, well stratified to massive, horizontal to gently dipping beds up to 30 cm thick; may contain wood or peat fragments, mollusc shells, pebbles and small amounts of sand
Shallow water or beach	Pebbly sand and sandy gravel	Transport by waves and currents	Massive pebbly sand to sandy gravel; fine to coarse sand with clasts up to large cobble size, commonly rounded
Shallow water or beach	Fine sand	Transport by waves and currents	Well-sorted, massive or stratified fine sand with horizontal or gently dipping beds, including planar sand and silt laminae, symmetrical ripples and low-angle cross bedding
Basin margins	Detrital peat	Transport by waves and currents	Massive or stratified accumulation of organic matter ≤2 m thick, horizontal or gently dipping, with abrupt lateral terminations
Drained basin	In situ peat	Peat growth	Massive fibrous peat at surface of basin sediments
Drained basin	Aeolian sand	Wind deposition	Well-sorted massive or stratified fine sand trapped by vegetation cover

Source: Based on Murton (1996c).

Subsequent research has confirmed a general contrast between the organic-rich silty sediments that settle in the centres of lake basins and the predominantly sandy deposits along lake margins (Jorgenson and Shur, 2007; Hinkel *et al.*, 2012). Cores obtained from the floor of Lake El'gene-Kyuele, a 10.5 m-deep lake in northeastern Siberia, show interbedding of silty gyttja (mud formed by partial decay of peat) and layers of fine sand. The former represents settlement of suspended sediment, and the latter have been interpreted as representing pulses of terrigenous sediment delivery to the lake from thaw slumping along the lake margin, driven either by warmer or wetter climatic conditions or by sequential exposure of deep ice wedges aligned parallel to the lake shore (Biskaborn *et al.*, 2013; Schleusner *et al.*, 2015; Figure 8.5). Sediment cores extracted from two deep thermokarst lakes in silt-rich yedoma terrain on the Seward Peninsula, Alaska, however, revealed only silty deposits: sublacustrine taberal silts, a chaotic silt facies formed through collapse of the lake margin, an interbedded organic silt representing reworking of the chaotic silts by wave action and a silt-rich mud that settled from suspension (Farquharson *et al.*, 2016b). The nature of thermokarst lake deposits is therefore

clearly influenced by that of the sediments in which the lakes have developed.

8.2.6 Thermokarst Lake Hydrology and Drainage

For thermokarst lakes without drainage outlets, the annual water balance is dominated by inputs of spring snowmelt and summer rainfall, with a minor contribution from melting ground ice, and by water loss due to summer evaporation (Pohl *et al.*, 2009). Because the snowmelt inputs to such lakes occur in spring, whereas net evaporative losses cumulate throughout the summer, the volume of water stored in such lakes tends to diminish throughout the summer months, causing seasonal shrinkage of lakes with shallow margins (Chen *et al.*, 2013). However, a study of variations in the areas of thermokarst lakes on the Tuktoyaktuk Peninsula of western arctic Canada over the period 1978–2001 has shown that interannual variability in lake area is strongly correlated with cumulative precipitation in the preceding 12 months, implying that variations in precipitation inputs rather than summer temperature (and hence evaporation) are the primary factor influencing interannual changes in lake area (Plug *et al.*, 2008). Over millennial timescales, there is

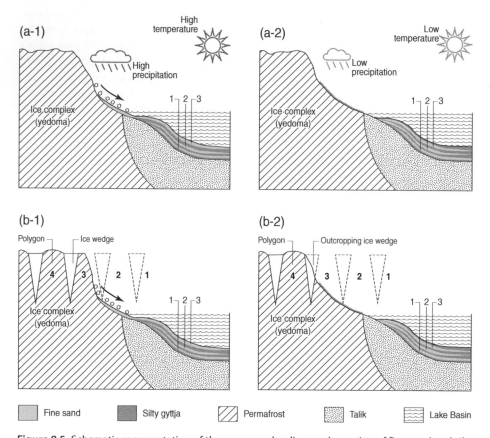

Figure 8.5 Schematic representation of the processes leading to alternation of fine sand and silty gyttja in Lake El'gene-Kyuele, northeast Siberia. Top: Climate-driven increased (a-1) or decreased (a-2) thermoerosion. Bottom: slope destabilization (b-1) and stabilization (b-2) phases driven by recurrent exposure of ice wedges. Both scenarios potentially lead to alternating input of fine sand (a-1 and b-1) and silty gyttja (a-2 and b-2). *Source*: Biskaborn *et al.* (2013). Reproduced with permission of Wiley.

evidence that changes towards a drier or wetter climate may also influence long-term average lake levels (Lauriol *et al.*, 2009; Pestryakova *et al.*, 2012). Large lakes with stable drainage outlets generally maintain a more constant seasonal water volume. In areas of continuous permafrost where water storage is confined to the active layer or deep subpermafrost aquifers, groundwater flux makes a negligible contribution to lake water balance. In the discontinuous permafrost zone, however, where thermokarst lakes are connected by an open talik to subpermafrost groundwater, lakes may lose water by drainage through the sublacustrine talik (Yoshikawa and Hinzman, 2003).

Considered on a millennial timescale, many thermokarst lakes are transient landscape features whose gradual expansion may be abruptly terminated by sudden drainage within a few hours or days. Satellite imagery of thermokarst lake regions almost invariably shows that the area occupied by drained lake basins greatly exceeds that occupied by extant lakes, implying a long history of lake expansion, drainage and renewal, though some thermokarst lakes have apparently persisted since the early Holocene and exhibit no evidence of drainage events (Pestryakova *et al.*, 2012). Drainage may be triggered in various ways: through lateral tapping by rivers, coastal erosion, headward gully expansion, thaw slump development or coalescence with a lower lake; through thaw of an adjacent ice wedge network, creating a drainage pathway; or by formation of an overflow channel that is incised rapidly into ice-rich permafrost through a combination of mechanical erosion and melt of ground ice by the escaping lake water (Mackay, 1988a; Marsh and Neumann, 2001; Marsh *et al.*, 2009). Lake overflow commonly occurs during the snowmelt period, when water inputs are high, but is also known to have occurred during rainstorms or when lake outlets are dammed by snow. Human activity, notably disturbance of tundra surfaces by vehicles and ditching, has also been documented as a cause of lake drainage (Hinkel *et al.*, 2007). In areas of thin or discontinuous permafrost, where sublacustrine taliks extend below the base of the permafrost, lakes may drain completely through the underlying substrate (Yoshikawa and Hinzman, 2003). Drained or partly-drained lake basins formed by subterranean drainage are sometimes termed *thaw sinks*, in which the level of any remaining water body is controlled by groundwater connections (Jorgenson, 2013). In all cases, drainage may be complete, evacuating all standing water from the lake basin, or partial, leaving lakes with lower surface levels or shallow residual lakes or ponds that occupy the lowest parts of the drained lake basin (Figure 8.2a,b, stage 6).

A case study of lake drainage in northern Alaska has been documented by Jones and Arp (2015). In their study area on the arctic coastal plain of Alaska, thermokarst lakes occupy 22.5% of the landscape and drained lake basins account for 61.8%. Sequential remotely-sensed imagery shows that nine lakes >0.1 km^2 in area drained within the period 1955–2004. In July 2014, a lake with an estimated volume of 872 000 m^3 drained catastrophically, probably as a result of bank overtopping caused by a combination of high snow meltwater input and early summer precipitation. Overflow of lake water caused rapid thermoerosion of an ice-wedge network, with formation of a gully 9 m wide, 2 m deep and 70 m long that allowed the lake to drain completely in 36 hours, with 75% of water loss in the first 10 hours, and an observed peak discharge of 25 m^3 s^{-1}.

8.2.7 Drained Lake Basins

The depth of the enclosed depression created by lake drainage reflects that of the former lake, so in areas of epigenetic permafrost, drained lake basins often extend no more than a few metres below the surrounding terrain (Figure 8.2b, stage 6). In yedoma landscapes, however, drainage of primary thermokarst lakes leaves a deep, flat-floored, steep-sided basin called an *alas* or *alas depression*. These can achieve impressive dimensions: those on Kurungnakh Island in east Siberia, for example, are up to 3.5 km long, 3.0 km wide and 30 m deep, and most contain residual lakes, ponds and small streams (Morgenstern *et al.*, 2013). In the former yedoma landscape of west-central Alaska, drainage of deep thermokarst lakes has produced a mosaic of marshes, shallow ponds and low mounds on the former lake floor, providing optimal conditions for peat formation (Kanevskiy *et al.*, 2014; Figure 5.10). Coalescence of individual alases forms steep-sided *thermokarst valleys* that are characterized by abrupt changes in alignment and, initially, a stepped longitudinal profile.

Lake drainage radically alters the thermal conditions on the former lake floor, which is now exposed to prolonged deep winter freezing. As a result, permafrost aggrades into the sublacustrine talik from above, below and all sides, causing it to shrink and eventually disappear. Refrozen talik sediments are sometimes referred to as *taberal deposits* or *taberites*. In some drained lake basins, pore-water expulsion ahead of the advancing freezing front results in the formation of hydrostatic pingos on the floor of the drained lake basin (Figure 8.2, stage 7; Figure 8.6).

A more general consequence of nearsurface permafrost aggradation in drained lake basins is the formation of epigenetic (or, under thickening peat deposits, quasi-syngenetic) ground ice in the highly frost-susceptible organic-rich silty sediments that drape the former lake floor. This mainly takes the form of segregated ice lenses that grow in the transition zone immediately below the permafrost table and the formation and lateral thickening of a new generation of epigenetic ice wedges. Bockheim and Hinkel (2012) have demonstrated that the ground

Figure 8.6 Pingo in a partially drained alas lake basin, central Yakutia. *Source:* Courtesy of Julian Murton.

ice content of nearsurface permafrost in drained thermokarst lake basins initially increases very rapidly, then slows through time. They found that cores retrieved from young (<50 years old) drained lake basins in arctic Alaska contained on average ~20% by volume excess ice, whereas ancient basins that drained several thousand years ago contained on average ~40%. Similarly, the frequency of wedge ice encountered in boreholes also increased progressively from 0% in the youngest basins to 32% in the oldest. As the volume of excess ice increases, the former lake floor experiences upwards heave (Roy-Léveillée and Burn, 2016). In areas where comparatively shallow thermokarst lakes formed in epigenetic permafrost, renewed excess ice formation following lake drainage may elevate the lake basin to, or almost to, the level of the terrain prior to lake formation. In yedoma landscapes, however, the amount of lakefloor elevation is slight relative to the depth of the alas depression formed by drainage of the original lake.

The formation of excess ice within drained lake basins permits the formation and growth of a second generation of thermokarst lakes on former lake floors. In a study of the evolution of a large (7.5 km^2) partly-drained thermokarst lake basin in the yedoma landscape of the East Siberian coastal lowlands, for example, Morgenstern *et al.* (2013) showed that the original lake drained around 5.7 ka, leaving a 20 m deep alas containing two residual lakes. Following drainage, permafrost aggraded in the sublacustrine talik and lake-floor sediments, and a combination of ground ice accumulation, ice-wedge growth, deposition of terrestrial sediments from the alas slopes and peat accumulation resulted in elevation of the alas floor, though the larger residual lake continued to expand laterally, eroding the adjacent alas slope. Both residual lakes then partially drained, exposing terraces under which permafrost started to aggrade, and between the two lakes a small pingo formed. A small secondary thermokarst lake developed on the alas floor some time after ~1.7 ka. This example illustrates the complexity of thermokarst landscape development. The formation and expansion of the original lake represents high-intensity thermokarst evolution. However, even after lake drainage, the alas floor experienced not only the effects of permafrost aggradation, but also localized thermokarst activity associated with the residual lakes and the late-stage development of a secondary thermokarst lake in the taberal deposits.

8.2.8 Cyclic and Alternative Models of Thermokarst Lake Evolution

Because ice-rich permafrost forms in drained lake basins, several authors have proposed that thermokarst lake formation is cyclic, such that several generations of lakes may form, expand then drain at approximately the same location, with renewed development of ice-rich epigenetic permafrost following each successive episode of drainage (Lenz *et al.*, 2016). As Figure 8.2 illustrates, however, this view is tenable only for shallow lakes. For deep lakes that develop in syngenetic permafrost (yedoma), the large volume loss that accompanies the melt of excess ice to produce an alas is irreversible, and secondary thermokarst lakes developed in ice-rich epigenetic permafrost on the alas floor are likely to be much smaller and shallower than the original thermo-karst lake (Morgenstern *et al.*, 2011). Moreover, ice-wedge volumes in drained lake basins are typically much less than in pristine yedoma deposits (Ulrich *et al.*, 2014). Even in the case of shallow lake basins where develop-ment of ice-rich epigenetic permafrost has elevated the former lake floor, development of a later lake at the same site requires some change in the surface thermal regime to trigger renewed thermokarst development, though arguably this may occur through ponding of water in developing second-generation low-centred polygons, without being driven by a change in climate or ground cover (Billings and Peterson, 1980).

Not all thermokarst lakes are initiated by thermokarst subsidence, as the initial water body may have some other origin. Pestryakova *et al.* (2012), for example, noted that shallow oxbow lakes in abandoned river meanders and lakes formed in inter-dune depressions may subse-quently evolve as thermokarst lakes through thaw of the underlying ice-rich permafrost. This theme has been developed by Jorgenson and Shur (2007), who studied lake-basin development on the coastal plain of Alaska. In this area, most terrain contains insufficient ground ice to sustain the development of thermokarst lakes, except in the centres of drained lake basins where organic-rich silty sediments have accumulated. They inferred that the original lakes had formed in the early Holocene through flooding of depressions in an undulating ground sur-face, rather than through thermokarst subsidence, then expanded laterally through wave erosion of lake margins. Following drainage, which mainly occurred in the late Holocene, there has been differential ice aggradation under former lakefloors, with marked uplift of the organic-rich silty sediments underlying basin centres, but limited ice aggradation and hence limited uplift in

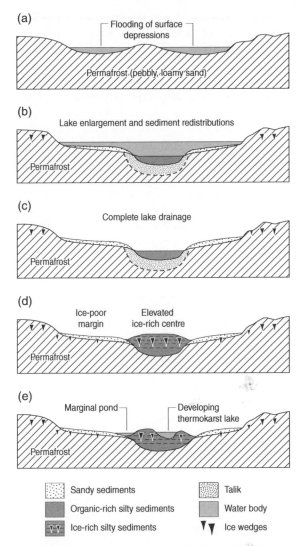

Figure 8.7 Conceptual model for the evolution of lakes and drained basins on the coastal plain of northern Alaska. (a) Flooding of surface depressions. (b) Lake enlargement and talik formation. (c) Lake drainage, leaving silty sediments in the middle of the basin and sandy sediments on marginal shelves. (d) Preferential ground ice formation, elevating the silty sediments. (e) Development of a thermokarst lake on the ice-rich silty sediments. *Source:* Adapted from Jorgensen and Shur (2007). Reproduced with permission of the American Geophysical Union.

the surrounding sandy sediments. As a result, true thermo-karst lakes have subsequently developed in the centres of the drained lake basins, sometimes flanked by smaller shallow water bodies in the relatively low basin margins (Figure 8.7).

Thermokarst lakes may also develop through the thaw of blocks of buried glacier ice. Glacier ice typically becomes buried under sediment cover in three settings: as blocks of stagnant ice covered by proglacial outwash deposits; within ice-cored moraines deposited at the margins of polythermal glaciers and subpolar ice caps or ice sheets; and as a result of the downwasting of debris-covered glaciers in alpine environments (Evans and Twigg, 2002; Dyke and Evans, 2003; Benn and Evans, 2010). In permafrost-free environments, buried ice bodies melt fairly rapidly, forming *kettle holes* (enclosed, water-filled depressions) or hummocky *kame and kettle topography*, comprising enclosed depressions separated by flat-topped, hummocky or conical mounds or ridges (*kames*). In permafrost regions, such ice bodies may survive under a cover of sediment for millennia (Chapter 5), but when the thaw front extends to the ice body, they gradually melt, forming *glacial thermokarst lakes*, which may then evolve in a manner similar to other thermokarst lakes. Chaotic hummocky topography formed by thaw of smaller bodies of buried glacier ice, such as those in ice-cored moraines, is sometimes described as *glacier thermokarst* (Jorgenson, 2013).

8.2.9 Thermokarst Lake Dynamics

Since the 1970s, most arctic and subarctic areas have experienced general climate warming, locally accompanied by a slight increase in precipitation. Such changes might be expected to affect thermokarst lakes in two contrasting ways: by promoting lake initiation and expansion as a result of a warming surface heat budget, but also by increasing the frequency of lake drainage, either by breaching or by water infiltration through an open, through-going sublacustrine talik. Such changes have been monitored over extensive areas through the use of sequential remotely-sensed imagery. In this way, contrasting lake response to recent warming was detected by Smith *et al.* (2005), who tracked the changes in lakes across ~515 000 km^2 of Siberia between 1973 and 1998. During this period, the total number of large lakes >0.4 km^2 in area declined by ~11%, and total lake area declined by ~6%, due partly to complete lake drainage and partly to lake shrinkage. This overall trend, however, masks pronounced spatial differences. In areas underlain by continuous permafrost, lake numbers increased by 4% and total lake area by 12%, mainly due to shoreline erosion. By contrast, in areas of discontinuous, sporadic and isolated permafrost, lakes exhibited a net decline in both number (by 9%, 5% and 6% respectively) and area (by ~13%, 12% and 11%, respectively). This contrast in lake behaviour appears to reflect enhanced lake initiation and expansion in areas of continuous permafrost, where sublacustrine taliks remained sealed, but increased drainage of more southerly lakes, particularly as a result of drainage through open sublacustrine taliks.

Similarly, in a study of shallow closed-depression ponds >0.2 ha in area in Alaska from the 1950s to 2002, Riordan *et al.* (2006) found that those underlain by continuous permafrost on the Arctic Coastal Plain exhibited no net change in area, whereas those in regions underlain by

discontinuous permafrost in the boreal zone decreased in number by 5–54% and in surface water area by 4–31%, a contrast that they interpreted in terms of enhanced drainage of ponds underlain by discontinuous permafrost, or of increased evapotranspiration during longer and warmer summer seasons. Infilling by fen vegetation is also a major factor in lake shrinkage in boreal and low arctic regions, causing lake area to shrink even while the area affected by thermokarst activity is increasing (Parsekian *et al.* 2011; Roach *et al.*, 2011). In three peatland areas underlain by continuous, discontinuous and sporadic permafrost, Sannel and Kuhry (2011) detected limited net change in aggregate lake extent over a 30–50 year period, but significant lake changes in the sporadic permafrost zone, in the form of both lake drainage and infilling by fens, accompanied by widespread initiation of new thermokarst lakes.

Even in areas of continuous permafrost, however, the temporal pattern of lake changes may be complex. Jones *et al.* (2011) employed remotely-sensed imagery from 1950–51, 1978 and 2006–07 to quantify changes in water bodies >0.1 ha on the northern Seward Peninsula, Alaska (Figure 8.8). They found that over the entire observation period, the number of water bodies increased by 10.7%, but the total area of these lakes and ponds decreased by 14.9%. This apparent paradox is largely explained by the formation of remnant ponds following partial drainage

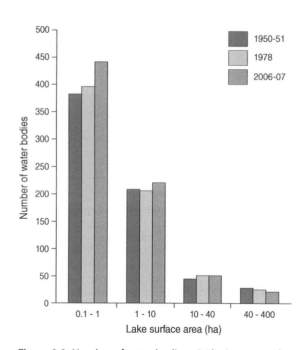

Figure 8.8 Number of water bodies >0.1 ha in area on the northern Seward Peninsula, Alaska, in four size classes (detected on remotely-sensed images) for 1950–51, 1978 and 2006–07. The increase in smaller water bodies is largely attributable to partial drainage of larger lakes, leaving residual small lakes and ponds. *Source:* Jones *et al.* (2011). Reproduced with permission of the American Geophysical Union.

of large lakes, which decreased by 24% in number and 26% in area. This pattern they interpreted in terms of lateral drainage of large lakes, rather than sublacustrine infiltration.

The response of thermokarst lakes to decadal-scale climate changes is therefore complex, depending on a range of factors, including changing precipitation and evapotranspiration, the depth and temperature of the underlying permafrost, groundwater connectivity, substrate permeability (Jepsen *et al.*, 2013), the erodibility of lakeshores, and terrestrialization (the colonization of shallow lake margins by fens; Figure 8.3). Moreover, some authors have sounded a cautionary note regarding the use of sequential remotely-sensed 'snapshots' of lake extent. Plug *et al.* (2008) noted that interannual changes in lake area of up to 4% are possible as a result of variable precipitation inputs in the preceding year, complicating the identification of longer-term trends. A study by Chen *et al.* (2013) of intra-annual variability in lake area within a $4224\,km^2$ area of the Yukon Flats in Alaska showed that total lake area decreased by 42% within a single summer, reflecting a drop in lake surface levels due to evaporation. As this research shows, the time during the summer when remotely-sensed imagery is obtained may significantly influence the interpretation of decadal-scale temporal changes. Accurate assessment of recent temporal changes in thermokarst lake behaviour is important, as expansion of such lakes results in an increase in emission of methane, a potent greenhouse gas (Zimov *et al.*, 1997; Walter *et al.*, 2006; Chapter 17).

8.3 Thermokarst Pits, Bogs and Fens

Permafrost degradation landforms in the form of *thermokarst pits*, *thermokarst bogs* and *thermokarst fens* are particularly widespread in the circum-arctic boreal forest zone, particularly on peatlands, in locations where the ground is underlain by discontinuous 'warm' permafrost that is close to its thawing point (Jorgenson and Osterkamp, 2005). Bogs comprise ombrogenous peat deposits (mainly *Sphagnum* peat) where surface water is derived from precipitation alone, whereas fens are fed by relatively nutrient-rich groundwater or surface flow (Beilman *et al.*, 2001).

Thermokarst pits represent localized subsidence due to uneven thaw of ice-rich permafrost, and are widespread throughout lowland boreal forests. They are typically 1–3 m deep and 2–15 m wide, and are commonly occupied by toppled trees. Standing water fills many pits, promoting the development of an underlying talik, and it is likely that progressive degradation of ice-rich permafrost under and at the margins of individual pits leads ultimately to coalescence of pits and the formation of peat-filled thermokarst bogs, sometimes referred to as *collapse scar bogs*, which are typically 30–150 m in diameter (Racine *et al.*, 1998; Osterkamp *et al.*, 2000; Figure 8.9). Such bogs form roughly circular openings in forests that occupy ice-rich peat plateaus, sediments on abandoned floodplains or lowland loess deposits. They expand at rates of 0.1–$0.5\,m\,a^{-1}$ through lateral heat transfer and consequent thaw of ice-rich permafrost around their margins (Jorgenson, 2013).

Figure 8.9 Thermokarst bog, Alaska. Toppling of trees at the bog margins is due to permafrost degradation. A colour version of this figure appears in the plates section. *Source:* Courtesy of Torre Jorgenson.

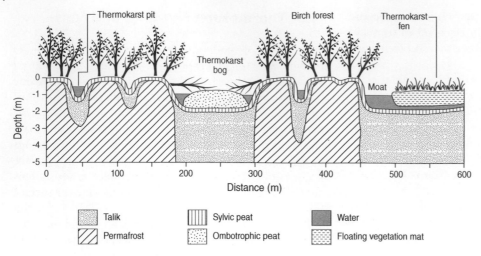

Figure 8.10 Schematic cross-section depicting thermokarst pits, bogs and fens associated with boreal birch forest in the discontinuous permafrost zone (50× vertical exaggeration). *Source:* Adapted from Racine *et al.* (1998).

Thermokarst fens (or *collapse scar fens*) occupy linear depressions, often at the margins of forest stands, where groundwater discharge to surface springs has caused rapid thaw of ice-rich soil, leading typically to 1–3 m of thaw settlement (Figure 8.10). Like thermokarst bogs, they are often associated with thaw of ground ice in lowland loess or silty abandoned floodplains. A common feature of such fens is a 'moat' of slowly flowing water, typically ~1 m deep, at the boundary between the fen and the adjacent forest margin. Heat flow from the fen margin promotes lateral extension of fens through thaw of ice-rich permafrost under adjacent forest stands at rates of up to ~1.0 m a^{-1} (Jorgenson *et al.*, 2001). Farther from the margin, fens often support forb-dominated vegetation supported by a floating mat of roots, rhizomes and peat.

Thermokarst bogs and fens therefore represent two pathways of wetland expansion and succession at the expense of forest stands: an ombotrophic sequence, through the development of pits and *Sphagnum*-dominated bogs within forest stands, and a minerotrophic sequence, involving toppling and drowning of trees adjacent to groundwater-fed fens, with associated development of a forb-dominated vegetation mat. Bogs may be incorporated into expanding fens, increasing the rate of wetland expansion and reduction of forest cover in the discontinuous permafrost zone, particularly for birch stands. A study based on sequential remotely-sensed imagery of an area of 2640 km^2 in the Tanana Flats region of Alaska, where ~42% of the total area is affected by thermokarst, suggests that the area occupied by fens increased by ~29% between 1949 and 1995 (Jorgenson *et al.*, 2001). A more recent investigation of the same area indicates a 7% expansion of thermokarst-generated wetlands at the expense of birch forest over the past 60 years (Lara *et al.*, 2016). Such observations have prompted speculation

that forest ecosystems in some areas underlain by ice-rich discontinuous permafrost may be completely destroyed within the present century (Osterkamp *et al.*, 2000). The onset of thermokarst development on the Tanana Flats has been attributed by Jorgenson *et al.* (2001) to warming in the mid- to late-18th century, but permafrost degradation, thermokarst development and consequent wetland expansion appear to be accelerating in response to permafrost warming over the past few decades (Lara *et al.*, 2016).

8.4 Retrogressive Thaw Slumps

8.4.1 Characteristics of Retrogressive Thaw Slumps

Arguably the most active of all thermokarst landforms, *retrogressive thaw slumps* are rapidly-evolving slope failures resulting from thaw of ice-rich permafrost (Figure 8.11). These landforms have two main morphological components: a steep (vertical or near-vertical) icy headscarp (or headwall) 1–25 m high, and a low-gradient (typically 2–15°) slump floor. These are often separated by a ramp or buttress of sediment that has been detached from the ablating headwall and has accumulated at its foot. The slump floor is occupied by muddy sediments that locally take the form of lobes deposited by debris flows (mudflows) or miniature fans of sediment deposited by meltwater (Figure 8.12). Farther downslope, sediment derived from slumps is delivered directly into rivers, lakes or the sea, or else accumulates within or at the foot of evacuation channels as mudflow tongues and fans. In planform, the headwalls of retrogressive thaw slumps are typically arcuate (U-shaped, C-shaped or horseshoe-shaped) but may be straight or crescentic. In some older accounts, these features are sometimes

Figure 8.11 (a) Large retrogressive thaw slump (megaslump) along the Selawik River, northwest Alaska. Over 700 000 m³ of sediment was evacuated from this slump between its formation in 2004 and its stabilization in 2016. (b) The same slump after stabilization. A colour version of (a) appears in the plates section. *Source:* Courtesy of (a) Benjamin Crosby and (b) Josefine Lenz.

referred to as *ground-ice slumps*, and the scar produced by retrogressive slumping as a *thermocirque* or *thermo-erosional cirque*, but these terms are now considered redundant.

The size of thaw slumps varies greatly, depending on terrain characteristics and their stage of development. Slumps flanking thermokarst lakes in the Mackenzie Delta area extend 50–150 m inland from lakeshores, with

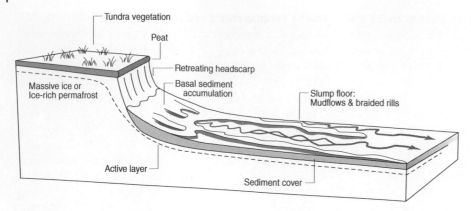

Figure 8.12 Schematic section of the morphological components of a retrogressive thaw slump. Progressive retreat of the headscarp results in extension of the slump floor.

headscarps up to several metres high in massive ground ice, but less in nearsurface aggradational ice (Kokelj *et al.*, 2009). Coastal slumps may be much larger, extending inland up to 650 m from the shoreface and sometimes exceeding 1 km in width (Lantuit *et al.*, 2012a). On the shoulders of valleys in the Richardson Mountains and Peel Plateau in northwest Canada, *megaslumps* up to 52 ha in area with headscarps up to 25 m high have developed in glacial sediments underlain by massive ice bodies (Lacelle *et al.*, 2015); the largest individual megaslumps in this area have delivered up to $10^6\,\mathrm{m}^3$ of sediment to adjacent valley floors (Kokelj *et al.*, 2015). Inactive, stable, vegetation-covered slumps often occur close to their active counterparts; they frequently exhibit step-like discontinuities indicating more than one period of activity or are affected by renewed slumping at present. Such rejuvenated slumps are referred to as *polycyclic thaw slumps*. Of 530 retrogressive thaw slumps documented by Kokelj *et al.* (2009) around thermokarst lakes in the Mackenzie Delta region, 97% are multi-aged, and of 164 slumps identified along the 50 km coastline of Herschel Island in the Beaufort Sea by Lantuit and Pollard (2008), up to 90% are polycyclic.

The sedimentary characteristics of thaw slump deposits in the Tuktoyaktuk Coastlands of western arctic Canada have been described by Murton (2001). These commonly comprise two facies: massive to crudely-sorted diamictons and well-stratified sands or muds. The former represent debris flow deposits and contain randomly dispersed clasts, occasional sand veins and sand lenses, blocks of peat or sod and plant or charcoal fragments. Stratified sands and muds are deposited by meltwater in the form of miniature fans and braided channels. Sheets of diamicton commonly 0.5–4.0 m thick are capped by stratified deposits a few decimetres thick. Both facies may be deformed, eroded or buried by later debris flows.

Retrogressive thaw slumps are restricted in distribution to areas of ice-rich permafrost, in the form of massive ice bodies, sediment-rich ice or ice-supersaturated sediment.

They are common along coasts underlain by yedoma or massive ice, at the margins of thermokarst lakes, along eroding riverine bluffs and on open slopes and valley shoulders. Most research has been carried out on slumps in the Canadian high Arctic, the western Canadian Arctic and Alaska, but they are widespread in parts of arctic Siberia and on the Qinghai-Tibet Plateau of China (Wei *et al.*, 2006) and also occur in Antarctica (Oliva and Ruiz-Fernández, 2015).

8.4.2 Thaw Slump Processes

Thaw slumps are initiated at locations where ice-rich permafrost is exposed at the surface. This commonly occurs on coasts and lakeshores where a combination of wave action and thermal erosion has exposed a scarp of ice-rich sediments or massive ice (Kokelj *et al.*, 2009; Lantuit *et al.*, 2012a), or on riverbanks where fluvial thermoerosion has had a similar effect (Burn and Friele, 1989). Ablation of the exposed ice or icy sediments results in headward retreat (retrogression) of the initial scarp, forming the headscarp and floor of the slump. On open hillslopes, the initiation of retrogressive thaw slumps is more difficult to envisage. Several authors have suggested that this is triggered by *active-layer detachment slides*, which occur in exceptionally warm summers when thaw of ice-rich sediment immediately above or below the permafrost table results in the active layer sliding downslope, locally exposing the underlying ice-rich permafrost (Lacelle *et al.*, 2010; Chapter 11). This explanation is supported by Balser *et al.* (2014), who observed that 16 of 21 new slumps that developed in the Noatak Basin of northwest Alaska between 1997 and 2010 formed within a year following an unusually early and warm thaw season in 2004, conditions that would have favoured active-layer detachment. Some authors have suggested that slumps on hillslopes may also be initiated by thermal erosion of ice wedges, exposing the adjacent ice-rich permafrost, and there are recorded instances of slumps triggered by human activity, as occurred following excavation of permafrost during repairs to the Qinghai–Tibet Highway (Wei *et al.*, 2006).

Once a headscarp has formed, headscarp retreat may be rapid. A three year field-based study of 22 slumps in the lower Mackenzie River basin and Mackenzie Delta by Wang *et al.* (2016) found that annual headscarp retreat rates varied from zero to $15.1 \,\mathrm{m\,a^{-1}}$, with considerable interannual variation at some sites but consistency at others; most values fell within the range 4–$8\,\mathrm{m\,a^{-1}}$. Analyses of the retreat rates of 19 megaslumps in the Richardson Mountains and Peel Plateau areas of northwest Canada using sequential Landsat imagery for the period 1990–2010 suggest that these huge slumps experience remarkable rates of headscarp retreat, averaging 7.2–$26.7\,\mathrm{m\,a^{-1}}$ over the 20-year period, with one year retreat rates in the range 5–$60\,\mathrm{m\,a^{-1}}$ (Lacelle *et al.*, 2015). Wang *et al.* (2016) detected a positive relationship between headscarp retreat rate and headscarp height, which they attributed to the fact that a larger proportion of smaller headscarps is buried under scarp-foot sediment and therefore not exposed to ablation.

Headscarp retreat and the consequent extension of the slump floor are primarily due to ablation of ice in the scarp face, which is caused by incoming shortwave solar radiation and sensible heat inputs (Pufahl and Morgenstern, 1980). Measurements by Grom and Pollard (2008) suggest that during periods of high incoming solar radiation, scarp-face temperatures are slightly elevated relative to those of adjacent undisturbed terrain, accelerating thaw of ground ice in the scarp. Analysis of rates of megaslump headscarp retreat by Lacelle *et al.* (2015) showed that this tends to be greatest on south- and west-facing slopes, where the snowpack melts earliest and radiation inputs are high, but the retreat rates measured by Wang *et al.* (2016) exhibited no relationship with scarp orientation over a three year measurement period.

Thawing of scarp ice has two effects: retrogressive failure (slumping) of the headscarp, exposing a fresh face of massive ice or ice-rich permafrost; and incremental release of saturated sediment, which forms a muddy slurry at the scarpfoot (Figures 8.11 and 8.12). These sediments are then evacuated across the slump floor by meltwater-driven rillwash, streams and small-scale mudflows or debris flows. In megaslumps, sediment is also transported by relatively infrequent deep-seated mass flows triggered by summer rainstorms. At sites where the footslope terminates in the sea, a lake or a river, sediment then enters the water body and is dispersed, locally increasing turbidity and causing major changes in the water chemistry of small thermokarst lakes (Kokelj *et al.*, 2005). Below slumps on open hillslopes or valley shoulders, sediment builds up in gullies and valleys, draining the slope in the form of thick elongate valley-confined sediment tongues that may locally divert streams, triggering

minor slumps along the new stream channels (Kokelj *et al.*, 2015). Pulses of sediment reaching rivers downslope from slumps have major effects on their sediment load, increasing suspended sediment and solute concentrations by several orders of magnitude above those of rivers unaffected by such terrain disturbances (Kokelj *et al.*, 2013).

Headscarp retreat may continue for decades, and is terminated by burial of the retreating headwall under melt-out sediments or by collapse of overhanging soil or peat. Slumps with low headscarps are therefore liable to stabilize more rapidly than those with high headscarps. Headscarp retreat may also cease if it exposes sediment with low ice content. Because slumps retain thicker snowcover in winter and are less shaded by vegetation cover in summer, ground temperatures under slump floors are markedly higher than under adjacent undisturbed terrain, and active-layer depths are consequently greater (Burn, 2000; Kokelj *et al.*, 2009; Figure 8.13). Moreover, slump-floor sediments have increased nutrient status and higher soil pH than undisturbed areas. These factors drive the structure and composition of the vegetation that colonizes immobile slumps, which may differ from that of surrounding tundra or boreal forest (Burn and Friele, 1989; Lantz *et al.*, 2009).

8.4.3 Temporal Changes in Slump Activity

Various authors have employed sequential aerial photographs or satellite images to investigate changes in slump activity over time. Along the coast of Hershel Island,

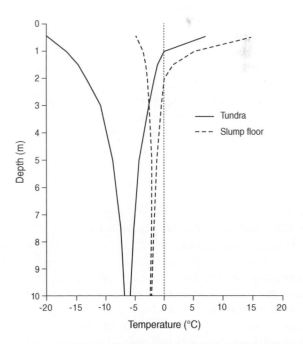

Figure 8.13 Maximum and minimum ground temperatures (2006–07) for undisturbed tundra and a stable vegetated slump surface, Richards Island, NWT, Canada. *Source:* Kokelj *et al.* (2009). Reproduced with permission of Wiley.

northern Yukon Territory, Lantuit and Pollard (2008) found a progressive increase in the number of active slumps, from 46 in 1952 to 75 in 2000, and an accompanying increase in the total area of active slumping, from 185 900 to 484 000 m^2 over the same period. They attributed intensification of slump activity to climatic warming and increased coastal erosion, which had resulted in reactivation of many stabilized slumps. For a sample of 541 slumps bordering thermokarst lakes and ponds in the Mackenzie Delta area, Lantz and Kokelj (2008) measured rates of slump growth and headwall retreat for the periods 1953–70 and 1970–2004. They showed that the mean rate of slump growth in the latter period was about 1.4 times greater than in 1953–70, and that the mean rate of headwall retreat had approximately doubled. These changes they related to climate records for the region, concluding that the marked increase in slump activity between the two periods reflected warming air temperatures.

A more nuanced pattern of change was observed by Lacelle *et al.* (2010) on the Aklavik Plateau of the Richardson Mountains, where retrogressive thaw slumps are located on the shoulders of interfluves. Their comparison of sequential aerial photographs demonstrated an increase in slump initiation from 0.35 new slumps per year in the period 1954–71 to 0.68 per year in 1985–2004, which again may be explained by warming summer temperatures. However, they also found that the number of active mature slumps declined from 46 to 24, a change that they tentatively attributed to burial of headscarps by sediment accumulation, possibly linked to a reduction in precipitation and hence slower removal of slumped sediment from headwalls. There appears to be general consensus, however, that most terrains prone to retrogressive thaw slumping have experienced a significant increase in slump activity over the past few decades, and that such activity is related to enhanced thaw of exposed ice-rich permafrost or massive ice produced by recent climate change, and in particular warming summer air temperatures.

8.5 Small-scale Thermokarst Features: Beaded Streams, Sinkholes and Thermokarst Gullies

Localized degradation of permafrost by running water produces a range of relatively small-scale thermokarst landforms. *Thermokarst water tracks* are shallow parallel or subparallel depressions marking the routeways of suprapermafrost groundwater movement on gentle slopes (Osterkamp *et al.*, 2009), and are broadly equivalent to percolines (seepage lines), which represent zones of concentrated nearsurface seepage in nonpermafrost

environments. In polar desert environments, water tracks are sometimes represented by downslope-aligned bands of vegetated ground separated by bare ground. Those in the hyper-arid environment of the Taylor Valley in Antarctica are enriched in solutes derived from chemical weathering of sediments and dissolution of soil salts (Levy *et al.*, 2012). Thermokarst water tracks are unrelated to ice-wedge polygon networks, but may represent sites of potential or incipient thermokarst gullying.

Most thermokarst landforms associated with surface water flow represent thaw or partial thaw of ice wedges. Small lowland streams crossing moderately sloping ground often follow the troughs above ice wedges, and ponding of water at wedge intersections results in the formation of *beaded streams*, characterized by a of succession shallow (typically 0.5–3.0 m deep) pools linked by narrow channel reaches (Figure 8.14). On the inner coastal plain of northern Alaska, beaded streams form the majority of channel networks, and most initiate from and are fed by lakes; elsewhere they are less common. A remote sensing survey of 445 beaded streams in northern Alaska, Canada and Russia by Arp *et al.* (2015) found that 90% are located on terrain below 210 m altitude and that 50% occupy areas with high ground-ice content, 32% areas of moderate ground-ice content and 18% areas of low ground-ice content, though almost all are located in the continuous permafrost zone. The frequency of pools in a subsample of these streams generally ranges from two to six per 100 m reach, reflecting the size of ice-wedge polygons and thus the frequency of wedge intersections. Beaded streams tend to transition downstream to alluvial rivers as drainage area increases and channel slope declines. Tarbeeva and Surkov (2013) suggested that beaded streams are transient features that become infilled with sediment derived from upstream, but sequential airphoto imagery employed by Arp *et al.* (2015) to investigate changes in a 2.7 km reach over a 64-year period indicated negligible change within that timescale. Their research suggested that though some pools expand through time, others contract due to vegetation encroachment. More remarkably, radiocarbon dating of basal sediments in one deep pool yielded ages of ~13.6–9.0 ka. If valid, these ages suggest that the pool may have been a stable feature in the drainage system throughout the Holocene. This implies that beaded streams are not necessarily indicators of developing thermokarst, even though winter snow accumulation over frozen channels and pools prevents pool water freezing to the bottom, permitting the formation and persistence of shallow taliks under pools.

On flat terrain in the continuous permafrost zone, thaw of the upper parts of ice wedges under standing or flowing water forms *thermokarst pits* at polygon junctions and *thermokarst troughs* between junctions (Jorgenson *et al.*, 2006). These features reflect localized

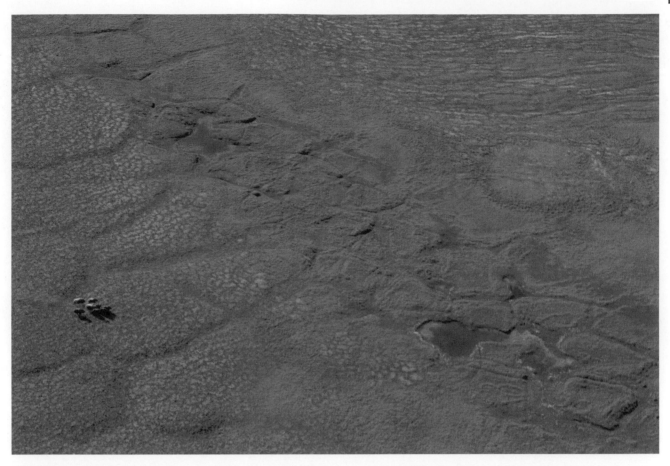

Figure 8.14 Incipient development of a beaded stream along a watercourse in central Banks Island, arctic Canada. Thaw at ice-wedge intersections has created small thermokarst ponds linked by flow along the ice-wedge network. *Source:* Courtesy of Tobias Ullmann.

deepening of the active layer above wedges, and in their initial stages of development involve thaw settlement of typically 1–2 m, or more if water flows through the trough network. If subsidence continues, slumping of adjacent thawed ground into thermokarst pits and troughs results in the formation of flat-topped, conical or hummocky *thermokarst mounds* (*baidzherakhi*) defined by the original polygonal ice-wedge network (Figure 8.3). As described earlier in this chapter, this evolutionary sequence may eventually initiate the formation of thermokarst ponds and lakes, particularly in areas of ice-rich syngenetic permafrost (Figure 8.2).

A particularly destructive form of localized thermokarst involves the formation of *thermokarst gullies*, also known as *thermal erosion gullies* or *thaw gullies*. These features develop where channelized water flow has thawed the underlying ice-rich permafrost and removed the sediment released by melt of ground ice. Such gullies may form as a result of fluvio-thermal erosion of permafrost during exceptional summer rainstorms or due to lake overflow (Marsh *et al.*, 2009; Toniolo *et al.*, 2009; Figure 8.15), but the most impressive examples follow degraded ice-wedge networks. Research by Fortier *et al.* (2007)

on the evolution of a gully system on Bylot Island in the Canadian Arctic Archipelago has shown that gully formation was initiated by the development of *thermokarst sinkholes*, funnel-shaped depressions that occur where water flowing along ice-wedge troughs has entered an open frost crack, thawing the wedge ice and feeding flow of water along subsurface tunnels or pipes following the ice-wedge network and sometimes expanding into adjacent ice-rich permafrost. They observed that sinkholes developed rapidly due to heat exchange from the flowing water into the ice wedge or adjacent permafrost, forming 3–4 m high waterfalls at sinkhole sites within a few days. Most tunnels draining sinkholes formed within ice wedges, reaching depths of 3.5–4.0 m, heights of 0.5–1.7 m and widths of 3.7–9.7 m. The observed tunnels were initially roofed by active-layer sediments, ice-rich silty soil and wedge remnants, but thermal erosion and consequent widening of the tunnels by running water eventually resulted in roof collapse and the development of open gullies that propagated both upstream and downstream, exposing wedge ice and adjacent ice-rich permafrost in gully walls to further thaw. Even when decoupled from the stream channel, the gully walls

Figure 8.15 Thermo-erosional gully, Ellesmere Island, Canadian Arctic Archipelago, which formed within 2 days following overspill of a small glacier-dammed pond.

continued to retreat through thaw slumping, active-layer collapse, slope failure and mudflows, progressively widening the gully floors.

Subsequent research on the same gully system (Godin and Fortier, 2012) showed that the gully continued to extend upslope, reaching 750 m in length, with tributary gullies meeting the trunk gully at former ice-wedge junctions, so that the total area occupied by the main, tributary and relict channels reached ~25 000 m² just a decade after gully initiation. By that time, the areas of most recent gully extension exhibited sinkholes, gully heads and tunnels, whereas the oldest parts of the system were characterized by chaotic thermokarst terrain of drained ice-wedge polygons, thermokarst mounds and wide gullies containing alluvial levées and pools. Gully widening, however, appears to be self-limiting, as thawed soil and the collapsing peat mat eventually insulate gully walls from further thaw. As Godin and Fortier (2012) noted, the rate of gully propagation they observed is amongst the most rapid in the world, testimony to the highly effective heat exchange at the interface between flowing water and ice or ice-rich permafrost. They also noted that hundreds of similar gullies occur on the southwest plain of Bylot Island, demonstrating the effectiveness of the brief period of snowmelt runoff in generating meltwater discharges

that can transform apparently stable polygonal terrain into chaotic thermokarst within a decade.

8.6 Sediment Structures associated with Thermokarst

Structures associated with permafrost degradation include ice-wedge pseudomorphs (Chapter 6) and *thermokarst involutions* (Figure 8.16). The latter are soft-sediment deformation structures that are particularly evident in stratified sediments, or where there is a marked contrast in the composition of superimposed sediment units. Thermokarst involutions form by loading (vertical movement of liquefied soil associated with reversed density gradients), buoyancy, and water escape during the degradation of ice-rich permafrost. These structures were first studied in detail by Murton and French (1993a) in coastal exposures in the Summer Island area of the western Canadian Arctic, where involutions occur above a palaeo-thaw unconformity that represents permafrost degradation in response to warming temperatures at the Pleistocene–Holocene transition (Burn *et al.*, 1986; Burn, 1997; Chapter 5). In these exposures, massive sand overlies a muddy diamicton. Murton and French (1993a) identified within the palaeo-thaw layer several structures that are

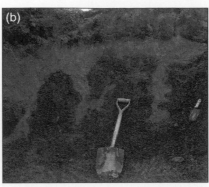

Figure 8.16 Thermokarst involutions within a refrozen palaeo-active layer, Summer Island, western arctic Canada. (a) Pseudo-nodules of sand within diamicton. (b) Load casts of sand between rounded diapirs of diamicton. (c) Ball-and-pillow structures of sand within diamicton. Load casts of sand and diapirs of diamicton occur towards the top of the image. *Source:* Courtesy of Julian Murton.

attributable to thaw of ice-rich permafrost, including: (i) diapirs up to 1.4 m high and 0.7 m wide that mark upwards intrusion of diamicton into the overlying sand; (ii) flat-bottomed or concave-upwards load casts up to 0.9 m wide and a few decimetres deep, where sand has sunk downwards into the underlying diamicton; and (iii) pseudo-nodules and ball-and-pillow structures of sand suspended within the diamicton (Figure 8.16). The pseudo-nodules are typically a few centimetres to a few decimetres in width and thickness, and comprise rounded, horizontally elongated sand bodies that locally form rows; the ball-and-pillow structures have similar dimensions but are less regular, being flat-bottomed, concave-up, acuminate or irregular, and some exhibit pipe-like extensions to underlying sand bodies. Murton and French attributed these structures to establishment of a reverse (i.e. inverted) density gradient during permafrost thaw, with high pore-water pressures generated by thaw consolidation of ice-rich permafrost causing liquefaction or hydroplastic deformation of sediment. In consequence, sand (with a higher bulk density) sank into the underlying diamicton, forming load casts, pseudo-nodules and ball-and-pillow structures, and displaced diamicton ascended into the overlying sand as diapirs. They also observed water-escape structures in slump-floor sediments, comprising upwards-tapering vertical to subvertical alternating streaks of fine sand and muddy sand/sandy mud a few millimetres to more than 10 cm in width.

The interpretation of involutions as the product of permafrost degradation is supported by research demonstrating the establishment of reverse density gradients in thawing soils (Swanson *et al.*, 1999) and scaled centrifuge modelling of rapid thawing of simulated ice-rich permafrost (Harris *et al.*, 2000a). The latter shows that for sand overlying clay, thaw consolidation produced high pore-water pressures and the development of small flame structures, diapirs and load casts at the sand–clay interface. Though the structures described by Murton and French (1993a) are not unique to areas of thermokarst, they are probably widespread in areas subject to present or former degradation of ice-rich permafrost. This is because saturated soil conditions (and associated high pore-water pressures) are likely to develop during thaw consolidation, making soft-sediment deformation extremely likely. This research also has important implications for the study of similar structures in Pleistocene sediments. In particular, the maximum depth of the thermokarst involutions studied at Summer Island terminates above (rather than at) the underlying thaw unconformity, implying that thermokarst involutions cannot be used to retrodict former maximum thaw depth, as inferred by several earlier authors. Murton and French (1993a) also drew an important distinction between thermokarst involutions and *cryoturbation structures* (Chapter 9), which involve mass displacement of sediment due to recurrent freezing and thawing of the ground.

8.7 Relict Thermokarst Phenomena

Landforms representing relict thermokarst in past (Pleistocene) permafrost areas include relict frost polygons that delineate former ice-wedge networks (Chapter 6), pingo scars and ramparted depressions that represent collapsed lithalsas (Chapter 7). Perhaps surprisingly, there appears to be limited published evidence for mid-latitude landforms that represent Pleistocene thermokarst lake basins or retrogressive thaw slumps. In the case of deep thermokarst lakes and associated alas depressions, this absence may in part reflect the lack of Pleistocene analogues for the yedoma deposits in which these features developed. More generally, because these thermokarst landforms are developed in ice-rich permafrost, they may have limited preservation potential once the permafrost has thawed, reducing surrounding terrain to the level of former lake or slump floors, though recognition of these features may be possible from surviving sediment associations (Murton, 1996c, 2001).

Possible examples of the sites of Pleistocene thermokarst lakes occur in the Fenland Basin and Vale of York in eastern England (Ballantyne and Harris, 1994, pp. 80–82). Burton (1987) has described enclosed depressions, generally exceeding 1 km in diameter and up to 7.5 m deep, in the English Fenland. These depressions are drained by shallow outlet channels or are linked by narrow channels. The alluvial terrace gravels surrounding the depressions support polygonal crop marks, marking the sites of former ice wedges, and the margins of the depressions are mantled by solifluction deposits. Burton interpreted these landforms as relict alases analogous to those in Yakutia; subdued arcuate scarps floored by clay deposits at the margins of some depressions he attributed to thaw slumping or thermal erosion. Though these features may indeed represent the sites of former shallow thermokarst lakes, they seem unlikely to be true alas depressions. This is because they have formed in terrace gravels of probable mid-Devensian (mid-Weichselian) age, implying that subsequent permafrost development was epigenetic, so that the depth of former ice-rich permafrost was probably limited (Murton and Ballantyne, 2017). Similar depressions with shallow outlet channels elsewhere in the Fenlands have been described by West (1991), who envisaged that these were formerly occupied by shallow thermokarst lakes analogous to those of western arctic Canada. West also suggested that marked embayments in the Fenlands represent relict retrogressive thaw slumps along the margins of a former glacier-dammed lake. In the chalk lowlands of Norfolk, eastern England, a network of shallow flat-floored depressions 200–300 m in extent enclosed by belts of sand and chalky rubble have been interpreted by West (2015) as further possible sites of former shallow thermokarst lakes.

There appears to be limited evidence for Pleistocene thermokarst lakes in other mid-latitude areas, though Van Huissteden (1990) described possible thermokarst lake deposits relating to mid-Weichselian permafrost degradation in the eastern Netherlands. It is possible that such deposits are much more widespread, but have escaped recognition. A problem with identifying sites of former thermokarst lake basins is that shallow enclosed depressions within Pleistocene sediments may have been formed in a variety of ways (French and Demitroff, 2001), and establishment of a thermokarst-lake origin may therefore be problematic. Arcuate scars representing former retrogressive ground-ice slumps have also rarely been identified in former permafrost environments, apart from the possible examples mentioned above at the margins of former thermokarst lakes. An exception is an account by Brown *et al.* (2015) of arcuate and horseshoe-shaped features up to 187 m wide and 5 m deep located at springlines in the Axe Valley in southwest England, though whether these features represent the scars of translational landslides or progressive retrogressive failure of ice-rich permafrost is uncertain. An intriguing possibility is that some features previously interpreted as nivation hollows in areas of past permafrost may actually represent the scars of former retrogressive ground-ice slumps.

Ice-wedge pseudomorphs represent the most widespread soil structures produced by degradation of ice-rich permafrost (Chapter 6), but thermokarst involutions may also provide evidence of former episodes of permafrost degradation. Structures interpreted as thermokarst involutions have been described at several locations (Figure 8.17). Involutions in the Middle Pleistocene Barham Soil of southeast England, for example, were

Figure 8.17 Thermokarst involutions of last glacial age developed under a stripe pattern, southeast England. The underlying substrate is brecciated chalk. *Source:* Courtesy of Julian Murton.

originally interpreted as active-layer cryoturbations, but Murton *et al.* (1995) demonstrated striking similarities between these features and thermokarst involutions investigated by Murton and French (1993a) on Summer Island in western arctic Canada. The Barham Soil involutions vary in stratigraphic context, but are typically 0.4–0.8 m deep with a horizontal spacing of ~1 m. At one site, they comprise deformed sand wedges and sandy lobate structures separated by flame-like structures composed of material from the underlying palaeosol; at another, clay-enriched lobate structures with narrow necks descend into the underlying sands and gravels. Murton *et al.* (1995) interpreted the Barham Soil involutions in terms of liquefaction of soil in the former thaw layer, downwards movement of denser sediment under temporarily reversed density gradients and compensatory upwards movement of sediment to form diapirs and flame structures. They pointed out that many structures previously interpreted as cryoturbation phenomena (Chapter 9) may actually represent involutions produced by loading associated with degradation of ice-rich permafrost.

Some involuted layers, however, may represent a more complex evolution, involving both permafrost degradation and active-layer cryoturbation (Superson *et al.*, 2010). In Thanet, southeast England, Murton *et al.* (2003) identified a bipartite involuted layer. The lowermost ~0.5 m of this layer contains texturally distinct structures, including downwards-extended lobes of silt or sand, ball-and-pillow structures of silt or sand surrounded by chalk diamicton and upwards-extending diapirs or fingers of chalk fragments or chalk diamicton. By contrast, the upper 0.5–1.5 m of the involuted layer comprises irregular and texturally indistinct involutions. They interpreted this pattern in terms of an initial period of active-layer deepening (OSL-dated to ~21 ka), during which the well-defined structures in the lower part of the involuted layer were formed as a result of loading and consequent soft-sediment deformation. This episode was apparently followed by active-layer thinning (~21–18 ka) and the development of texturally indistinct involutions by cryoturbation operating in the upper part of the involuted layer.

Considerably larger involuted structures, 3–4 m deep, are exposed in gravel pits in the New Jersey Pine Barrens. Flame structures of sand rising 1–3 m into the overlying pebbly diamict and gravels were attributed by French *et al.* (2005) to 'density-controlled mass displacements in the water-saturated thaw zone' during degradation of ice-rich permafrost, and a complex sequence comprising pockets and tongues of downwards-penetrating sand

and tilted and dislocated blocks of sediment was interpreted by the same authors as the product of possible fluvio-thermal erosion of a former sand wedge. The most intriguing features they described are bulbous sediment bodies, 2–4 m high and wide, which penetrate downwards into underlying stratified deposits. The core of these structures comprises a tongue of sand, separated from the host sediments by sandy and pebbly gravel within a sandy matrix. French *et al.* (2005) described these curious features as 'sediment pots' or 'thermokarst kettles' and attributed their formation to thermal erosion of sand-wedge intersections and the mixing, slumping and redeposition of sediment derived from both the wedge and the host sediment. OSL dating suggests that these various features formed before ~60 ka. Broad sediment-filled wedges overlying composite wedge remnants in France have been interpreted as representing former thermokarst gullies that formed as permafrost degraded (Bertran *et al.*, 2014).

Even larger structures at the foot of the northeastern Qinghai-Tibet Plateau have been attributed to permafrost degradation by Vandenberghe *et al.* (2016). At this site, sandy gravels are overlain by alluvial sandy silts, and the deformations comprise a continuous series of regular, symmetrical folds 4.0–4.5 m in amplitude, involving upwards protrusion of the gravels into the overlying silt and downwards sinking of silt into the gravels. OSL dating of the sandy silts suggests that they were deposited during the last glacial stage, and that permafrost degradation may have occurred around 20 ka. Vandenberghe *et al.* (2016) attributed these structures to 'oversaturation and liquefaction' of the sediments 'due to thaw of massive icy lenses within ice-rich permafrost', but acknowledged a problem in explaining the former occurrence of massive ice within the gravels, and also noted that loadcasting of saturated sediments by seismic events may offer an alternative explanation (Van Vliet-Lanoë *et al.*, 2004b).

As these examples indicate, considerable progress has been made in the identification of former periods of thermokarst activity in areas of past permafrost on the basis of involution structures, and luminescence dating of such structures has permitted the approximate timing of permafrost degradation to be identified. As several authors have suggested, numerous structures previously interpreted as the products of active-layer cryoturbation should probably be reinterpreted as thermokarst involutions. Uncertainties persist, however, regarding the interpretation of large involutions and structures for which present-day analogues are imperfectly understood.

9

Seasonally Frozen Ground Phenomena

9.1 Introduction

As outlined in Chapter 4, the term *seasonally frozen ground* refers to ground that undergoes an annual cycle of freezing and thawing. It applies both to the active layer in permafrost environments and the uppermost layer of ground that freezes in winter but thaws in summer in permafrost-free environments. The effects are similar in both cases, except that in areas of cold permafrost freezing takes place both downwards from the surface and upwards from the underlying permafrost, whereas in nonpermafrost areas both freezing and thawing are unidirectional, from the surface downwards. An additional difference is that permafrost impedes downwards drainage of water from the active layer, but downwards drainage is possible in seasonally frozen ground that lacks permafrost or an underlying low-permeability layer such as clay or bedrock.

The physical processes that occur in freezing and thawing soils were introduced in Chapter 3. Several of these have a bearing on the origin of the landforms and soil structures described in this chapter, so it is useful to recapitulate the salient points. The first is that heat transfer into and out of seasonally frozen ground is primarily conductive, though modified by nonconductive effects such as water movement within the soil. Second, there is no single temperature at which all water freezes in soil, but the proportion of unfrozen water present diminishes as temperatures fall below 0 °C. Third, in frost-susceptible soils, pore water is driven by cryosuction towards the freezing front(s), where it forms lenses of segregated ice, causing frost heave of the soil and uplift of the ground surface. Migration of soil water to freezing fronts tends to result in desiccation of unfrozen soil, unless water is replenished by throughflow. Finally, during thaw, ice-rich soils experience thaw consolidation, which inhibits escape of water from the thaw front, generating enhanced pore-water pressures in newly thawed soil. If pore-water pressures at the thaw front are sufficiently elevated, the frictional strength of newly thawed soil is reduced and 'softening' or liquefaction of soil may result.

Here, the effects of recurrent freezing and thawing of the ground are considered in terms of several related consequences: upfreezing of clasts to the ground surface; frost heave of bedrock; the formation of frost-patterned ground; the development of deformation structures (cryoturbations) in soils subject to annual freeze–thaw cycles; and the micromorphology of soils that have experienced repeated freezing and thawing.

9.2 Upfreezing of Clasts

A consequence of seasonal ground freezing in both permafrost and nonpermafrost areas is the upfreezing of stones (clasts) and other objects towards the ground surface. This widespread phenomenon is sometimes termed *frost jacking* or *vertical frost-sorting* and affects not only periglacial environments, but all areas where winter ground freezing occurs, in the form of boulders appearing on the surface of cleared fields or pushing upwards through the tarmac of roads. In periglacial environments it is an important component in the formation of sorted patterned ground, as discussed below, and in the formation of periglacial weathering mantles (blockfields), as considered in the next chapter. One consequence of upfreezing of clasts by repeated freezing of soil is that elongate clasts often tend to assume a vertical or steeply-dipping alignment, a feature that has been widely interpreted as a consequence of upheaving.

Two mechanisms have been proposed to explain upfreezing of clasts: the *frost-push hypothesis* and the *frost-pull hypothesis*. The former rests on the idea that the thermal diffusivity of a buried or partially buried clast is greater than that of the soil in which it is embedded, so that during freezing the 0 °C isotherm passes more rapidly to the base of a clast than to soil at the same level. In frost-susceptible soils, the assumed outcome is formation under the clast of ice that pushes it towards the surface, and that during subsequent thaw the clast is prevented from returning to its original level by sediment movement into the void vacated by the thawing ice (Figure 9.1). One drawback of this explanation is that it requires upwards

Periglacial Geomorphology, First Edition. Colin K. Ballantyne.
© 2018 John Wiley & Sons Ltd. Published 2018 by John Wiley & Sons Ltd.

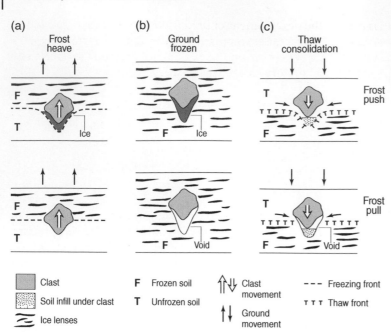

Figure 9.1 *Top:* The frost-push mechanism of clast heave. (a) During freezing, ice forms under a clast, pushing it up. (b) Ice-filled sub-clast void when the soil is completely frozen. (c) During thaw, soil partly infills the void, preventing the clast from regaining its original position. *Bottom:* The frost-pull mechanism of clast heave. (a) During freezing, a clast freezes to surrounding soil; further frost penetration heaves the upper frozen soil and the attached clast. (b) Air-filled void under uplifted clast when the soil is completely frozen. (c) During thaw, soil partly infills the void, preventing the clast from regaining its original position, so there is net upwards movement of the clast over a single freeze–thaw cycle.

movement of clasts through frozen soil; another is that experimental work (Kaplar, 1965; Anderson, 1988) has revealed that upwards movement of clasts is accompanied by the formation of a cavity rather than ice under clasts during freezing.

The frost-pull mechanism is based on the view that as the freezing front descends through the soil, the upper part of a clast freezes to the adjacent frozen soil. As freezing continues deeper into the ground, the upper frozen layers of soil are heaved upwards, carrying the attached clast with them and creating a cavity under the clast. Movement of soil into this void is assumed to prevent resettlement of the clast in its original position (Figure 9.1), so that over recurrent seasonal freeze–thaw cycles there is net migration of the clast towards the surface. The validity of this mechanism has been demonstrated experimentally by Anderson (1988), who monitored the movement of a single clast embedded in frost-susceptible silt. She found that the clast began to move upwards once the freezing isotherm had penetrated about one-third of the way down the clast side, when the adfreeze bond of the frozen soil to the clast was strong enough to support both the weight of the clast and overcome the cohesive bond between the unfrozen soil and the clast. As the freezing front descended deeper into the soil, a cavity opened under the clast, and partial infill of this void by soil prevented the clast from settling in its original position at the end of the thaw cycle. Anderson adduced that the magnitude of clast displacement within a single freeze–thaw cycle depends on the amount of heave of the upper layers of soil, the size, shape and orientation of the clast, and the amount of clast resettlement during thaw.

Experiments by Viklander (1998) provided support for the frost-pull model, but identified the void ratio of the soil (ratio of void space to the volume of solid particles) as a critical parameter; in dense soils with low void ratios, clasts migrated upwards in response to freeze–thaw cycles, but in loosely-packed soils with high void ratios, clasts tended to migrate downwards before being heaved upwards, with some exhibiting net downwards movement over 10 freeze–thaw cycles. Field experiments by Kolstrup and Thyrsted (2010, 2011) involving spheres and orthorhombic blocks of granite embedded in soil at three depths showed that these were heaved upwards during freezing, consistent with the frost-pull mechanism, but returned to their original positions during thaw. Kolstrup and Thyrsted attributed this behaviour to release of the spheres and blocks from the grip of the surrounding thawing soil, so that they settled back into their original positions before sediment moved into the underlying cavity. Some also sank slightly in summer, so that net movement over four years was downwards. An intriguing feature of their research is that granite blocks and spheres exhibited similar initial heave in both frost-susceptible silty soil and (apparently) non-frost-susceptible sandy soil. Kolstrup and Thyrsted tentatively explained the latter in terms of possible 'cryo-vapour-suction', which involves formation of ice lenses in non-frost-susceptible fine sediment through migration of water vapour (rather than liquid water) to the freezing front. Whatever the explanation of this apparently anomalous result, it appears that though the frost-pull mechanism offers a valid explanation for upfreezing of clasts, the conditions under which this operates require further investigation. An interesting aspect of this problem is the behaviour of clast-rich soils such as tills or periglacial weathering mantles (blockfields), where the net migration of clasts during ground freezing and thawing may be conditioned by that of neighbouring and underlying clasts. In clast-rich soils, it seems possible that larger clasts migrate preferentially towards the surface during successive freeze–thaw (heaving–resettling) cycles because the positions they formerly occupied become obstructed by heave-induced jostling of underlying clasts.

9.3 Frost Heave of Bedrock

Though the concept of frost heave is mainly associated with formation of ice lenses in frost-susceptible soil, there are also reports of frost-heaved bedrock in permafrost environments. Frost-heaved bedrock may take the form of isolated uplifted angular blocks, tombstone-like monoliths, dome structures up to 3 m high, structurally-guided upheaved ridges, shallow crater-like depressions or chaotic assemblages of detached blocks (Dionne, 1983; Michaud and Dionne, 1987). Michaud and Dyke (1990) proposed three possible mechanisms to explain these phenomena. First, in some instances it is possible that upwards wedging of joint-bound bedrock blocks occurs as a result of ice segregation within soil-filled joints, but this cannot represent a general explanation as many joints lack an infill of soil. Second, it is possible that blocks are uplifted due to hydraulic pressure in circumstances where freezing from the surface down creates a confined aquifer, but this mechanism can only apply in slope-foot locations where high hydraulic pressures are generated by an elevation differential. Michaud and Dyke concluded that the most plausible general explanation is that water in joints becomes trapped between underlying permafrost and a freezing front descending from the surface, creating a hydraulically closed system. Further freezing of joint water results in a volumetric expansion of 9% associated with the phase change from water to ice, generating high hydrostatic pressures within the joint network. As water is incompressible, these high water pressures are accommodated through upwards heave of individual bedrock blocks or updoming of the surface. On Melville Island in the Canadian Arctic, Dyke (1984) recorded 5 cm of upwards movement of a block during freezing, then subsequent settlement of 3 cm during the ensuing thaw. Dredge (1992) considered that a similar mechanism might have contributed to the formation of incipient blockfields (Chapter 10) in well-jointed limestone in the eastern Canadian Arctic.

Ice segregation nevertheless appears to play an important role in rock breakdown. Field evidence from a range of sites suggests that ice segregation in weak sedimentary rocks such as shale and chalk has been responsible for the development of *brecciated bedrocks* that are seamed by fractures representing the loci of former ice-lens growth (Murton, 1996b), an interpretation supported by laboratory experiments demonstrating ice-lens growth within a block of sound, moist chalk and resultant heave of the block surface (Murton *et al.*, 2006; Chapter 10).

9.4 Patterned Ground: The Embroidery on the Landscape

The term *patterned ground* is used to describe terrain that exhibits regular or irregular surface patterning, most commonly in the form of circles, polygons, irregular networks or stripes. Step-like, oval, lobate and garland patterns have also occasionally been referred to as patterned ground, though in most cases such forms reflect slow downslope movement of soil (Chapter 11). Most investigators have recognized a distinction between *sorted patterned ground*, produced by the alternation of soil and coarse debris at the ground surface, and *nonsorted patterned ground*, defined by microrelief (hummocks, or ridges and furrows) or alternating vegetated and unvegetated ground (Figure 9.2). The distinction between sorted and nonsorted patterns is slightly artificial, however, as nonsorted patterns may grade into sorted forms in locations where clasts are abundant and vegetation cover is incomplete. Although various forms of ground patterning unrelated to frost action occur in non-periglacial environments (e.g. Hallet, 1990; Ahnert, 1994; Wilson, 1995; Wilson and Edwards, 2004), most patterned ground is formed by recurrent freezing and thawing of moist soil. Active periglacial patterned ground features occur not only in high-latitude and alpine environments, but also on mid-latitude mountains and on high subtropical mountains that experience shallow diurnal freeze–thaw cycles. Patterned ground also occurs on recently deglaciated terrain (Ballantyne and Matthews, 1982, 1983; Matthews *et al.*, 1998; Haugland, 2004, 2006), in front of late-lying snowpatches (Kling, 1997), in supraglacial debris covers (Ballantyne, 1979; Kerguillec, 2014), on seasonally flooded ground and under shallow water (Lauriol *et al.*, 1985; Figure 9.3). Frost-generated patterns have even formed near sea level in cool temperate environments such as Newfoundland, Nova Scotia and the Falkland Islands (Leckie and McCann, 1982; Mooers and Glaser, 1989; Wilson and Clark, 1991), and the author has successfully 'grown' miniature frost-sorted patterns in an experimental plot in his garden in Scotland during a single winter (Ballantyne, 1996).

However, not all ground patterns in cold environments are the product of recurrent freezing and thawing of soil. The most conspicuous exceptions are frost polygons (frost-crack polygons) formed through thermal contraction cracking of permafrost, as discussed in Chapter 6.

Figure 9.2 Patterned ground. (a) Active sorted net, northeast Iceland. (b) Active sorted circles (debris islands), arctic Norway. (c) Stone pits, Kalfafjell, Iceland. (d) Sorted stripes, Tinto Hill, Scotland. (e) Earth hummocks (thúfur), Holar Valley, Northern Iceland. (f) Vegetation-defined nonsorted circles (frost boils) 1–2 m in diameter, Howe Island, Alaska. *Source:* Courtesy of (a,c,e) Ole Humlum and (f) Anja Kade.

Another form of patterning occurs in areas of patchy vegetation cover on terrain exposed to strong winds. In such locations, the interplay of wind scour, loosening of soil particles by frost action and selective vegetation growth results in the development of surface patterns in the form of clumps, crescents, ribbons and stripes of vegetation amid unvegetated stony ground (e.g. Ballantyne and Harris, 1994, pp. 260–261; Boelhouwers *et al.*, 2003).

Such *wind-patterned ground* (Chapter 14) is genetically unrelated to patterned ground created by freezing and thawing of the soil.

Most research on frost-induced patterned ground has involved the study of existing features, and the processes responsible for pattern initiation have been inferred from pattern characteristics, internal structure and measurement of soil and clast displacement. There has

Figure 9.3 Sorted net on a seasonally flooded lake margin, Abisko, northern Sweden. *Source:* Courtesy of Matthias Siewert.

also been a tendency for studies to focus on individual pattern types, leading to multiple explanations of pattern origin. A classic review by Washburn (1956), for example, identified 19 mechanisms that were proposed to operate singly or in some combination to initiate or develop patterned ground in cold environments. More recent research, however, suggests that most forms of ground patterning produced by freezing and thawing of moist soil reflect the operation of one or both of two process: differential frost heave and buoyancy-driven soil circulation (Hallet, 2013).

9.5 Patterned Ground Processes

9.5.1 Differential Frost Heave

The concept of differential frost heave has been widely invoked to explain patterned ground formation for over a century (Washburn, 1956, 1979). As we have seen in Chapter 3, frost heave is uplift of soil caused by volume increase due to formation of lenses of segregated ice during soil freezing. *Differential frost heave* is laterally non-uniform frost heave that produces greater soil uplift in some zones than others, resulting in an undulating surface microtopography that may evolve over repeated freeze–thaw cycles into sorted or nonsorted patterns. The kind of pattern that emerges (sorted or nonsorted) appears to reflect initial conditions, notably the abundance, distribution and size of clasts, as well as the presence or absence of vegetation cover. Various explanations have been proposed for the initiation of differential frost heave, notably desiccation cracking of the ground, with more rapid frost penetration under cracks, and lateral variability in soil texture, which promotes greater frost heave in some areas than in others (Goldthwait, 1976;

Washburn, 1979). However, numerical models applying linear stability analysis to the process of soil freezing (Lewis *et al.*, 1993; Kessler *et al.*, 2001; Peterson and Krantz, 2003, 2008; Peterson, 2011) have demonstrated that one-dimensional frost heave is unstable under a range of boundary conditions and tends to develop into differential frost heave, with eventual emergence of a dominant modal wavelength that determines the final pattern spacing. A range of feedback processes, still incompletely understood, is believed to enhance the initial perturbation caused by differential frost heave, permitting the eventual development of both sorted and nonsorted ground patterns.

9.5.2 Buoyancy-induced Soil Circulation

Explanation of both sorted and nonsorted patterns in terms of buoyancy-driven 'convection' cells operating within thawing soils also has a long history, though opinions have differed regarding the cause of soil circulation. The most widely held view is that seasonal thaw of ice-rich soil results in a decrease in soil bulk density with depth, due to thaw consolidation of the upper layers and release of excess water during thaw of ice lenses in the underlying soil. Swanson *et al.* (1999), for example, showed that thawed soils near the permafrost table have a bulk density up to $800\,kg\,m^{-3}$ less than that of the overlying soil. Advocates of soil circulation argue that this unstable density configuration generates 'convection' during thaw of seasonally frozen ground, with upwards movement of relatively low-density saturated soil in the centre of patterned ground features (such as sorted or nonsorted circles) and downwards movement of soil at pattern margins. However, theoretical analysis of thaw consolidation within an ice-rich active layer by Hallet and Waddington (1992) suggested that though the density profile in fine-grained soil immediately above the thaw front may be gravitationally unstable for much of the thaw season, the buoyancy forces generated by density inversion during thaw are insufficient to *initiate* pattern formation. Their analysis also suggested, however, that within existing patterns, buoyant soil near the base of the active layer tends to rise diapirically into the overlying soil. They showed that buoyancy-driven diapirism is likely to be enhanced by thaw of ice lenses concentrated near the base of the active layer (due to upwards freeze-back above cold permafrost), particularly in soils with low-permeability basal layers. It seems likely that though buoyancy-driven soil circulation may not drive pattern initiation, it does produce incremental soil displacements that account, at least in part, for evidence of circulatory soil movement within both sorted and nonsorted patterns developed over permafrost.

9.6 Sorted Patterned Ground

9.6.1 Characteristics

Sorted patterns develop on unvegetated or sparsely vegetated ground where clasts overlie or are embedded within fine soil. On level or gently sloping ground, sorted patterns usually consist of cells or domains of predominantly fine soil surrounded by stony borders that, in the thawed state, may be depressed or raised relative to the fine cells (Figures 9.2a,b and 9.3), though the soil domains are often heaved above the level of surrounding clasts as the ground freezes. The stony borders form regularly-spaced circles or polygons, or an irregular net or labyrinth of clasts separating fine cells of varying sizes and shapes. In some cases, isolated circles of predominantly fine soil termed *debris islands* interrupt an otherwise continuous cover of stony or bouldery debris (Figure 9.2b). Less commonly, *stone pits* develop, in the form of concentrations of clasts surrounded by fine-grained soil (Figure 9.2c). All sorted patterns tend to become elongated along the direction of maximum surface gradient as slope angle increases, grading downslope into stripes consisting of alternating fine soil and clasts (Figure 9.2d).

Pattern widths of individual sorted nets, circles, polygons and stripes range from about 10 cm to 4 m; accounts of larger forms are rare and may reflect accumulation of clasts in the troughs of frost polygons. Pattern width is positively related to the size of clasts that define pattern borders (Goldthwait, 1976; Ballantyne and Matthews, 1982; Feuillet *et al.*, 2011); patterns defined by pebbles rarely exceed 0.5 m in width, while those defined by boulders usually exceed 1–2 m. Within the coarse borders, clast size often diminishes with depth, and platy clasts are often erected (standing on end or on their sides) as a

Figure 9.4 Erected clasts at the margin of a large sorted circle, Mount Bitterness, New Zealand. Clast erection results from lateral pressures exerted on clasts due to expansion of adjacent fine soil during winter freezing.

result of lateral pressures exerted on borders during winter freezing and expansion of fine soil domains (Figure 9.4). Trenches excavated through sorted patterns often reveal clasts occupying V-shaped troughs in finer soil. Some sorted patterns exhibit secondary sorting, with decimetric patterns defined by smaller clasts contained within the fine domain of a larger circle, net or polygon (Warburton, 1990). A particularly striking form of sorted circles occurs in high-arctic permafrost environments, and consists of a gently-domed fine domain, 2–3 m wide, surrounded by a raised annulus of clasts. Beautiful examples occur on raised beaches in Svalbard (Figure 9.5). Excavation of such features has shown that they comprise a central cylindrical plug of soil completely surrounded by coarse gravel, and that both the central plug and the surrounding stone ring extend downwards to the underlying permafrost table (Hallet and Prestrud, 1986; Hallet, 1990, 1998, 2013).

9.6.2 Formation

Most researchers have envisaged the development of sorted patterns in terms of a circulatory model involving upwards and/or outwards movement of sediment in cells, migration of surface and nearsurface material from cells to adjacent borders and compensatory movement

Figure 9.5 Large annular sorted circles, Vardeborgsletta, Svalbard. A colour version of this figure appears in the plates section. *Source:* Courtesy of Ole Humlum.

of sediment towards cell centres at depth. Lateral sorting occurs because sediment moving radially outwards towards the borders is coarser than that moving towards cell centres at depth, so that clast-sized debris accumulates at the cell margins. Measurements of soil and clast displacements on or within large sorted circles developed over permafrost (Hallet and Prestrud, 1986; Washburn, 1989; Hallet, 1998, 2013) are consistent with this general model. Sorted stripes develop on slopes because a gravity-driven downslope component of soil movement is superimposed on soil and clast movement, causing a downslope-orientated helical-coil movement of sediment in fine domains.

There is broad consensus that sorting is initiated by differential frost heave. A long-established view is that displacement of clasts towards pattern margins and of fines towards pattern centres is initiated by uneven penetration of freezing planes into the soil, resulting in upfreezing of clasts perpendicular to the freezing front and migration of fine soil ahead of the freezing front towards zones of slower freezing (Nicholson, 1976; Figure 9.6).

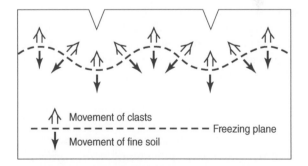

Figure 9.6 Lateral sorting of clasts and fines initiated by inclined freezing planes caused by more rapid freezing under open cracks. Clasts tend to be heaved upwards and towards the zones of more rapid frost penetration, whereas fine soil is driven towards the unfrozen zone under the areas of slower frost penetration.

Once established, this effect should be self-perpetuating, as the freezing front is likely to descend more rapidly below zones of clast concentration than within cells of fine soil, where formation of ice lenses and consequent release of latent heat slows downwards soil freezing. Recurrent upwards and outwards sorting of clasts towards pattern borders during successive cycles of soil freezing and thawing then progressively enhances the textural contrasts between fine domains and stony margins. Furthermore, as soil domains are frost-susceptible and stony margins are not, the former are updomed by frost heave during winter (often above the level of stony margins), so that with the onset of thaw, surface and nearsurface soil and clasts on fine domains migrate radially towards pattern borders.

Various ingenious proposals have been made as to how inclined freezing planes are initiated, but two in particular have traditionally attracted support. The first involves contraction cracking of the ground surface as a result of rapid cooling or desiccation. The resulting contraction cracks act as loci of more rapid freezing-front penetration, so that clasts migrate towards and become concentrated within cracks, forming sorted polygons (Ballantyne and Matthews, 1983; Figure 9.6). The second, termed 'primary frost sorting' by Goldthwait (1976), proposes that randomly-spaced concentrations of finer sediment freeze more slowly than surrounding soil because of preferential development of ice lenses, creating inclined freezing planes. Sorted circles or nets may be formed at first, but interaction of adjacent fine cells may ultimately result in formation of sorted polygons. These concepts have been developed by Van Vliet-Lanoë (1988), who proposed that the formation of contiguous sorted circles and sorted polygons is associated with frost cracking and positive frost-susceptibility gradients (reduction in frost susceptibility with soil depth), whereas isolated or coalescing sorted circles form on soils with negative frost-susceptibility gradients.

A weakness of both these explanations is that, at least in the absence of ground cracking, they fail to explain the regular spacing of many sorted patterns or the controls on pattern dimensions. A solution to this problem was developed by Kessler *et al.* (2001), who proposed that large-scale (2–4 m diameter) sorted circles are self-organized patterns that emerge spontaneously due to differential frost heave, without the need for ground cracking or vertical or lateral variations in soil frost-susceptibility, though such preconditions may arguably influence the initial loci of differential frost heave and determine initial pattern spacing. Their numerical model seeks to explain the formation of sorted circles comprising a central plug of fine soil surrounded by a raised stony ring, such as those in western Spitsbergen (Hallet and Prestrud, 1986; Hallet, 1998, 2013; Kääb *et al.*, 2014; Figure 9.5). This model assumes an initial state comprising an active layer with a cover of stones overlying frost-susceptible soil, a structure that might be produced by upfreezing of clasts and downwash of fine particles. Kessler *et al.* (2001) suggested that slight

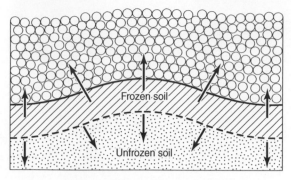

Figure 9.7 Patterned ground initiation by differential frost heave as proposed by Kessler *et al.* (2001). During freezing of the active layer, frost heave drives material away from the freezing front (dashed line) at a small positive perturbation between a layer of soil and an overlying layer of stones. Soil movement is both outwards towards the ground surface and inwards towards unfrozen soil. As soil displacements are not reversed during thaw, this mechanism promotes upwards growth of a plug of soil over recurrent freeze–thaw cycles. *Source:* Kessler *et al.* (2001). Reproduced with permission of the American Geophysical Union.

positive perturbations at the stone–soil interface are incrementally enhanced by movement of soil inwards under the site of the perturbation during annual freezing (Figure 9.7), but that soil movements during thaw are vertical (because of lateral confinement), so that there is a small net increment under the site of the initial perturbation with each annual freeze–thaw cycle. This increment is partly offset by a negative feedback, because a freezing front will tend to penetrate positive soil perturbations before adjacent regions, pushing soil towards the unfrozen regions that surround the perturbation. This negative feedback favours the development of long-wavelength perturbations, achieved through a merger of adjacent perturbations, until wavelength (pattern spacing) stabilizes, limited by soil redistribution rate. Through incremental net soil accretion driven by recurrent annual freeze–thaw cycles, the dominant perturbations form soil plugs, which gradually emerge at the surface through the overlying stony layer. Emergence of soil plugs is followed by updoming and radial expansion of fine soil at the surface, pushing stones to the margins of plugs until all soil from the initial subsurface soil layer is redistributed in fairly regularly-spaced plugs separated by stony rings that extend to the permafrost table (Figure 9.8).

Kessler *et al.* (2001) proposed that extant sorted circles are maintained by differential frost-heave-driven circulation within the soil and stone domains (Figure 9.9). This interpretation is broadly consistent with soil displacements measured on and within sorted circles (Hallet and Prestrud, 1986; Washburn, 1989; Hallet 1998, 2013). Their model replicates the formation of sorted circles with a mean spacing of 3.6 m, representing 750 years of annual freeze–thaw cycles, dimensions and timing that

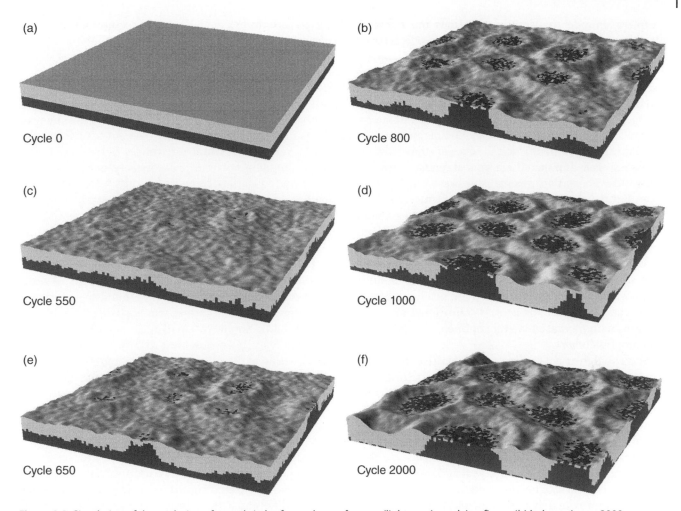

Figure 9.8 Simulation of the evolution of sorted circles from a layer of stones (light grey) overlying fine soil (dark grey) over 2000 freeze–thaw cycles. *Source:* Kessler *et al.* (2001). Reproduced with permission of the American Geophysical Union.

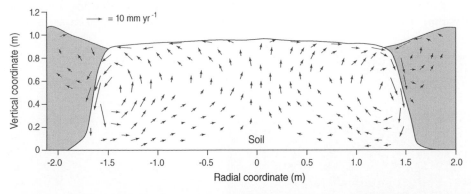

Figure 9.9 Velocity vectors within a sorted circle modelled by Kessler *et al.* (2001). Soil movement is driven by differential frost heave. *Source:* Kessler *et al.* (2001). Reproduced with permission of the American Geophysical Union.

agree closely with field measurements on the sorted circles in Svalbard. Measurement of surface displacement based on terrestrial photogrammetric analysis of three large sorted circles in western Spitsbergen suggests that soil turnover is ~0.1–0.2 $m^3 m^{-2} a^{-1}$, implying complete cycling of soil volume over several centuries (Kääb *et al.*, 2014). Kessler *et al.* (2001) also found that changing the

relative thicknesses of the initial stone and soil layers produced different outcomes, with labyrinthine patterns (sorted nets) forming when the thickness of the soil layer increased to 60% of the active layer depth.

A further contribution to understanding how differential frost heave produces different types of sorted pattern has been made by Kessler and Werner (2003), who developed a numerical model that simulates the development of all common varieties of sorted pattern. Their model is based on the interaction between two feedback mechanisms: (i) the development of inclined freezing planes, with resultant textural contrasts between soil-rich and clast-rich domains; and (ii) lateral squeezing of clast domains during annual freezing and expansion of soil domains. Squeezing and confinement of clast domains are assumed to cause clast movement along the axes of stony borders, producing clast-dominated margins of regular thickness and width. These two mechanisms were employed to drive clast movements over repeated iterations. As three boundary conditions (clast concentration, slope gradient and degree of lateral confinement) were varied, different forms of patterned ground evolved spontaneously. Decreasing stone concentration produced a transition from sorted circles to sorted nets and finally to stone islands (stone pits); increasing gradient produced sorted stripes; and stone islands proved transitional to sorted polygons with increased lateral confinement (decreased soil compressibility).

Differential frost heave also appears to drive the formation of small-scale sorted patterns in nonpermafrost areas of shallow ground freezing, but operates in a different way. Such patterns typically have widths <1.0 m, sorting depths <25 cm and stony domains dominated by clasts <15 cm in length (Figure 9.10). For such small-scale patterns, the dominant sorting mechanism is often needle-ice growth (Pérez, 1992; Matsuoka *et al.*, 2003). Surface or nearsurface needle ice (long, thin ice crystals that grow approximately perpendicular to surface slope) forms on moist soil when the surface temperature drops to between −1 °C and −2 °C, typically reaching lengths of 2–20 cm in response to short-term (often nocturnal) freezing cycles (Chapter 5). The formation of small-scale

sorted patterns by needle-ice heave requires unvegetated, permeable, frost-susceptible soils containing mainly small clasts. During freezing, needle ice grows under surface clasts, raising them above ground level. As needle-ice grows preferentially above patches of fine-textured soil, the layer of upraised clasts forms hummocks and depressions. During thaw, there is radial movement of clasts into depressions, so that over successive freeze–thaw cycles the textural differences between soil and stony domains become progressively enhanced until small sorted patterns appear (Figure 9.11). This may happen quite rapidly. At an experimental site in Scotland, a small sorted net developed in stony frost-susceptible soil as a result of needle-ice growth and differential frost heave in eight freeze–thaw cycles (Ballantyne, 1996), and small sorted stripes formed by a combination of needle-ice growth and downslope movement of sediment on a Scottish mountain were found to re-form on dug-over ground within three winters (Ballantyne, 2001a; Figure 9.2d). Similarly, small-scale sorted circles and polygons have been observed to form on glacier forelands within a decade or two following glacier retreat (Ballantyne and Matthews, 1982, 1983; Haugland, 2004; Feuillet and Mercier, 2012), and laboratory simulations of frost sorting by needle ice confirm that sorted patterns may develop within 20–30 freeze–thaw cycles (Yamagishi and Matsuoka, 2015).

Shallow frost heave may also contribute to the development of small-scale sorted patterns. Matsuoka *et al.* (2003) demonstrated that diurnal freeze–thaw cycles induce differential frost heave within the top 5 cm of soil, particularly if the freezing front does not reach the base of the coarse domain, so that the soil domain rises above the coarse border, promoting outwards movement of soil towards the adjacent coarse border

Figure 9.10 Miniature sorted stripes on basalt regolith, Faroe Islands.

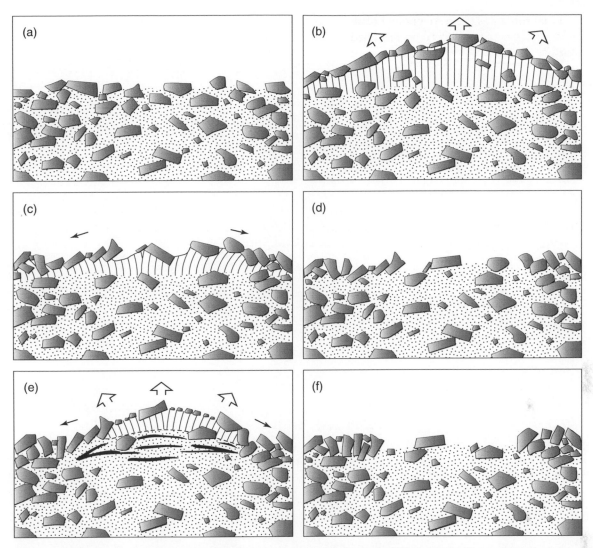

Figure 9.11 Sorted pattern formation by differential needle-ice growth. (a) Clasts embedded in frost-susceptible soil. (b) Differential needle-ice growth causing updoming of surface clasts. (c) Thaw and bending of needle ice, with movement of clasts to dome margins. (d) Complete thaw and incipient lateral sorting. (e) Renewed freezing, with needle-ice growth above, and ice-lens formation within, cells of fine soil; clast movement towards the margins of updomed soil. (f) Miniature sorted net, with clast borders surrounding predominantly fine cells. *Source:* Ballantyne (1996). Reproduced with permission of Wiley.

during thaw. Small sorted patterns nested within larger patterns may reflect the outcome of shallow short-term freeze–thaw cycles operating in ground that also experiences deeper annual freezing and thawing, though the kinetics of such 'secondary sorting' (Warburton, 1990) are unresolved.

A two-dimensional simulation of sorted stripe formation through differential needle-ice growth was developed by Werner and Hallet (1993). Their model assumes preferential growth of needle ice in stone-free zones during each freezing cycle, and calculates the trajectory of iterative clast movements along fall lines dictated partly by slope gradient and partly by the location of unfrozen soil domains. They showed that random spatial variations in soil texture evolve first into irregular stripes of alternating high and low clast concentration, then into

a stable, regularly-spaced stripe pattern with dimensions related, as in field situations, to clast size. Appropriately, their model produced sorted circles and nets when the surface gradient was set at zero.

The simulation models of sorted patterned ground development outlined above are remarkably successful in replicating the characteristics of sorted patterns observed in the field, and represent useful advances towards the development of a general theory of ground patterning by differential frost heave. The possibility remains, however, that thaw-related processes may play some part in the development of large-scale sorted patterns. Hallet and Prestrud (1986) inferred from updoming of soil domains and evidence for subduction of soil at the margins of the remarkable sorted circles in western Svalbard (Figure 9.5) that sediment movement

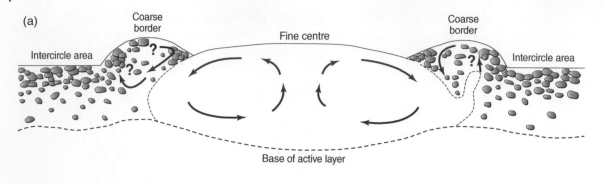

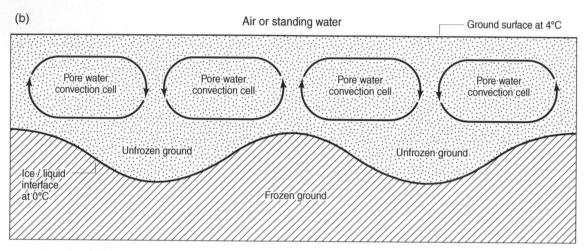

Figure 9.12 (a) Model of sediment displacements within sorted circles proposed by Hallet and Prestrud (1986), reproduced with the permission of Elsevier. (b) Model of free convection of pore water during active layer thawing, with resultant development of corrugations in the permafrost table. *Source:* Krantz (1990). Reproduced with permission of Elsevier.

within the circles could be explained by establishment of a double convection cell in the fine domain of each circle (Figure 9.12a). They attributed such circulation to a decrease in soil bulk density with depth during thaw, but stressed that convective soil movement would occur intermittently during thaw, not simultaneously throughout the full depth of thawed soil. Though the analysis of Kessler *et al.* (2001) suggests that these sorted circles could have formed through differential frost heave alone, it does not exclude the possibility of buoyancy-induced diapiric rise of low-density saturated soil near the base of the active layer into the overlying denser thaw-consolidated soil in the fine domain. If such movements occur, they imply a compensatory inwards movement of soil at depth towards the centre of the fine domain. It is less likely that buoyancy-driven soil movements occur in small sorted patterns in permafrost-free terrain, as these form in the absence of a subsurface layer of ice-rich soil in areas where drainage of soil water is unimpeded.

An alternative theory for the initiation and self-regulation of sorted patterns is based on density-driven movements of pore water rather than soil (Gleason *et al.*, 1986; Krantz, 1990). Because there is a slight increase in the density of liquid water at temperatures up to 4 °C, meltwater at the

thaw front (at 0 °C) is less dense than pore water in the overlying layers of a thawing soil. This model invokes establishment of convection cells during thaw, with descending plumes of relatively dense warmer water and compensatory rise of cooler, less dense water. Downflow of warmer water is envisaged as enhancing thaw at the base of the active layer, so that a pattern of alternating peaks and troughs develops in the permafrost table (Figure 9.12b). Though this model predicts pattern width-to-depth ratios that are consistent with field measurements, the development of subsurface corrugations in the active layer may be a consequence rather than a cause of sorted patterns at the surface. Moreover, the density differences between water at 4 °C and water at 0 °C produce buoyancy effects two orders of magnitude less than those in thawing soils (Hallet and Waddington, 1992). This ingenious solution to the enigma of sorted patterns appears, like many of its forebears, to be redundant.

9.6.3 Environmental and Palaeoenvironmental Significance

Small sorted patterns that form through needle-ice formation or shallow differential frost heave may develop

Figure 9.13 Large, relict vegetated sorted circles at 1050 m in the Grampian Highlands, Scotland. Such sorted circles were formed in the active layer above permafrost during the Late Pleistocene, and have survived intact throughout the Holocene.

in any environment where a rapid drop in air temperatures and overnight ground frost results in ground surface cooling to −1 °C or lower. Such conditions occur in many areas, from temperate mid-latitudes to the higher parts of subtropical mountains (Lawler, 1988). The critical controls on small-scale patterning are unvegetated, moist, frost-susceptible but permeable soils, abundance of small stones and few clasts with dimensions that exceed pattern spacing. Small sorted forms have poor preservation potential, and hence are of limited use in palaeoenvironmental reconstructions.

Relict large-scale sorted patterns defined by bouldery margins have much greater persistence in the landscape (Figure 9.13). Such forms reflect former deep annual ground freezing and thawing. Several authors have proposed that large (≥2 m wide) sorted forms with bouldery margins are indicative of present or former permafrost (Goldthwait, 1976; Washburn, 1980). However, the main function of permafrost in sorted pattern formation is to act as an incompressible aquitard, and there seems to be no reason why large sorted patterns could not form in nonpermafrost areas of deep seasonal ground freezing, providing that the soil overlies an incompressible, impermeable substrate within the maximum annual freezing depth. Reviewing the limited data on the conditions under which active sorted patterns occur, Grab (2002) noted that all sites with active patterns >1.0 m wide are associated with at least localized permafrost, and suggested that patterns >0.8 m wide occur where mean annual air temperature (MAAT) is −1.6 °C or lower. This estimate must be regarded as provisional. Moreover, active sorted circles, polygons and nets 1–2 m wide occur in nonpermafrost areas where edaphic, hydrological or microclimatic conditions are

particularly favourable, such as glacier margins and areas of seasonal waterlogging (Figure 9.3). Relict large-scale (>2 m-wide) sorted patterns can be used as probable indicators of former permafrost only if these circumstances can be ruled out.

9.7 Nonsorted Patterned Ground

Apart from frost polygons formed through thermal contraction cracking of permafrost (Chapter 6), the most common forms of nonsorted periglacial patterned ground comprise circles, nets and stripes defined by alternating vegetated and unvegetated soil, and hummocks and stripes defined by microrelief. There is considerable terminological duplication in the description of nonsorted patterns. Vegetation-defined nonsorted circles have been described as *frost boils*, *mud boils* and *stony earth circles*; for convenience, *frost boil* is adopted here, and vegetation-defined stripe patterns are termed *nonsorted stripes*. Circles or nets defined by microrelief are usually described as *earth hummocks*, and ridge-and-furrow features on slopes are termed *hummock stripes* or *relief stripes*. All such features are 'nonsorted' in the sense that there is no lateral segregation of soil and clast domains, though nonsorted forms may grade into sorted patterns where clasts are abundant. Walker *et al.* (2004) have described a spatial transition from large turf-covered earth hummocks in the southern Arctic, via frost boils in low-arctic and mid-arctic areas, to small earth hummocks in the high Arctic. This continuum suggests that a common suite of processes may explain the formation of different nonsorted forms.

9.7.1 Vegetation-defined Nonsorted Circles (Frost Boils) and Stripes

Frost boils take the form of circular patches of bare or sparsely-vegetated soil surrounded by areas of tundra vegetation cover. They are typically 0.5–3.0 m in diameter, and may occur in isolation, in irregular clusters or as regularly spaced patterns (Figure 9.2f). Frost boils are common in arctic permafrost environments, particularly on poorly drained tundra in well-vegetated mid- and low-arctic areas (Walker *et al.*, 2004), but also occur on well-drained permafrost terrain and in areas of alpine permafrost. When covered by a thin layer of upheaved clasts, such features are sometimes referred to as *stony earth circles*. Zoltai and Tarnocai (1981) identified two types: low-centred frost boils on poorly drained terrain, comprising flat, unvegetated centres surrounded by a raised rim of mineral soil, peat and vegetation; and raised-centre frost boils, which form domed or flat-topped circles of soil that rise above the level of the surrounding tundra vegetation and typically occur in well-drained sites. On gentle slopes, frost boils may form *nonsorted steps*, with flat, unvegetated treads separated by low (15–45 cm) vegetated risers (Shilts, 1978). On steeper slopes, nonsorted vegetation-defined stripes are sometimes present.

Some frost boils support a vegetation mat, but in such cases both the vegetation cover and underlying organic soil layer are much thinner than on adjacent areas. As a result, the ground surrounding bare or thinly-vegetated frost boils is better insulated than the frost boils themselves, which are consequently the sites of both deeper winter freezing and deeper summer thaw (Kade *et al.*, 2006; Kade and Walker, 2008), the latter producing a bowl-shaped depression in the underlying permafrost table. Excavations through frost boils have revealed evidence for subduction of organic matter at frost-boil margins (Walker *et al.*, 2004), and, in some cases, for extrusion of fine sediment near frost-boil centres (Egginton and Dyke, 1982; Ping *et al.*, 2003). Shilts (1978) summarized the characteristics of frost boils ('mudboils') in the Hudson Bay lowlands as comprising four elements: (i) a bowl-shaped ice-rich permafrost table, overlain by (ii) a 'thawed mud substrate' that feeds a diapir, which penetrates (iii) a relatively rigid surface carapace, with individual frost boils being separated by (iv) turf or stone rings underlain by organic soil (Figure 9.14). High concentrations of organic carbon within some frost boils also suggest that organic matter is translocated by soil movement and leaching from the frost boil margins into frost boils at depth (Boike *et al.*, 2008), and radiocarbon dating of organic matter within frost boils indicates that downwards movement of such matter at their margins has occurred at rates of ~1 mm a^{-1} (Dyke and Zoltai, 1980). Study of the concentration of short-lived radionuclides (^{210}Pb and ^{137}Cs) associated with a frost boil in northern Sweden supports the concept of radial movement of surface soil from the centre to the margins and suggests subduction of soil at the margins at rates of a few millimetres per year (Klaminder *et al.*, 2014). During winter, the centres of frost boils experience greater frost heave than the surrounding vegetated ground, suggesting that more rapid frost penetration within frost boils draws water from the surrounding vegetated areas to feed ice-lens growth (Overduin and Kane, 2006; Nicolsky *et al.*, 2008; Watanabe *et al.*, 2012).

Several explanations have been proposed for the origin of frost boils (Van Vliet-Lanoë, 1991), but two have attracted particular support: sediment extrusion (diapirism) and differential frost heave. The sediment-extrusion hypothesis proposes that frost boils result

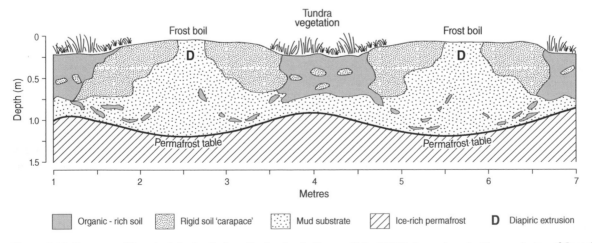

Figure 9.14 Structure of frost boils in the Hudson Bay lowlands. *Source:* Shilts (1978). Reproduced with permission of Canadian Science Publishing.

from expulsion of liquefied mud at the surface during thaw of the active layer. This idea was developed by Shilts (1978), who observed that frost boils in the Hudson Bay lowlands consist of a layer of mud with low liquid limits penetrating an overlying layer of more rigid soil as a diapiric extrusion (Figure 9.14). Shilts postulated that hydrostatic pressures develop within liquefied mud during thaw of the active layer, forcing the mud to the surface. Once a layer of bare soil covers the ground, alteration of the surface thermal regime results in deeper thaw under the developing frost boil, producing a bowl-shaped depression in the permafrost table. During thaw of the active layer, low-density soil with high water content rising upwards in the centre of a frost boil (cf. Swanson *et al.*, 1999) is replaced by gravity-induced soil movement into the underlying depression in the permafrost table, driving circular movement of soil within frost boils (Mackay, 1980). Upwards diapiric movements of saturated soil are most likely to occur in areas of cold permafrost, where upwards freezing from the permafrost table in winter results in ice-rich soil at the base of the active layer, producing excess water and high pore-water pressures during thaw. This process has been observed to produce *mud ejections*, pressurized slurries of fine-grained sediment expelled from the base of the active layer. These tend to occur late in the thaw season during exceptionally warm summers, and often following late summer rainfall events (Holloway *et al.*, 2016). The sediment-extrusion hypothesis of frost-boil formation is supported by observations of freshly exposed soil on frost-boil surfaces, and by measurements of soil movements within frost boils, which have recorded upwards movement of mud of 9–54 cm over a four-year period (Egginton, 1979). This hypothesis is also supported by laboratory experiments designed to investigate the nature of gelifluction (Chapter 11). These have shown that as ice-rich soil thaws, an upwards hydraulic pressure gradient becomes established within the thawed soil, providing a mechanism for extrusion of liquefied mud through overlying thaw-consolidated soil (Harris *et al.*, 2001, 2008a, 2008c). This hypothesis does not, however, explain the regular size and spacing of frost boils in some locations (Figure 9.2f).

An alternative explanation proposes that frost boils emerge from the surrounding vegetation cover due to differential frost heave. Peterson and Krantz (2003) have shown that differential frost heave may develop spontaneously from uniform one-dimensional frost heave; an initial positive perturbation in the freezing soil results in conduction of heat from crest regions to adjacent troughs, increasing frost heave under crests and thus the amplitude of the perturbation, and driving water movement from troughs to feed ice-lens growth under crests. Peterson and Krantz (2008) addressed the subsequent development of frost boils, taking into account the

insulating effects of vegetation cover in intervening areas. They showed that over numerous annual freeze–thaw cycles, a preferred wavelength of differential frost heave gradually develops, thus determining pattern size and spacing. In consequence, the microtopography of the ground surface evolves away from an initial random configuration of frost boils and intervening vegetated troughs towards more regular pattern spacing. They argued that pattern spacing tends to be dictated by the depth at which freezing is most prolonged, in a zone slightly above the base of the active layer. Assumption that this zone lies at 1 m depth yields a theoretical wavelength (pattern spacing) of ~3 m, consistent with the dimensions of large frost boils in the field. Peterson and Krantz (2008) also carried out scaled laboratory experiments involving repeated freezing and thawing of soil 10 cm thick, and reproduced miniature frost boils consistent with development by differential frost heave alone. They suggested that the main role of permafrost is to limit downwards movement of soil water, so that moisture is plentiful during soil freezing.

A different modelling approach employed by Nicolsky *et al.* (2008) to describe soil and water movement within frost boils also emphasized the importance of waterlogged conditions and the role of non-uniform vegetation cover in initiating and maintaining differential frost heave, with movement of both water and soil towards the locus of ice-lens formation under bare or thinly-vegetated frost boils (Figure 9.15). The importance of vegetation in moderating differential frost heave of frost boils has been demonstrated by Kade and Walker (2008), who showed that removal of vegetation from the boils themselves resulted in a 6% increase in active-layer depth and 26% increase in frost heave, whereas addition of a 10 cm thick moss layer to frost boils resulted in a 15% reduction in active-layer depth and a 52% decrease in frost heave.

Though differential frost heave seems to offer an adequate explanation for frost-boil initiation and evolution, the evidence for upwards diapiric movement of soil in some examples suggests that thaw-related soil displacements also contribute to their development, at least in soils with low liquid limits. The two mechanisms are mutually compatible, so it is possible that some frost boils have evolved through differential frost heave alone, and others by diapiric extrusion of mud during thaw, or by the operation of both processes. Nonsorted vegetation-defined stripes are probably produced by similar processes, though with superimposition of a downslope component of soil movement on the circulatory movement of soil implied by the mechanisms discussed above.

9.7.2 Earth Hummocks and Relief Stripes

The term *earth hummocks* describes patterned ground comprising vegetated mounds, typically 0.1–0.6 m high

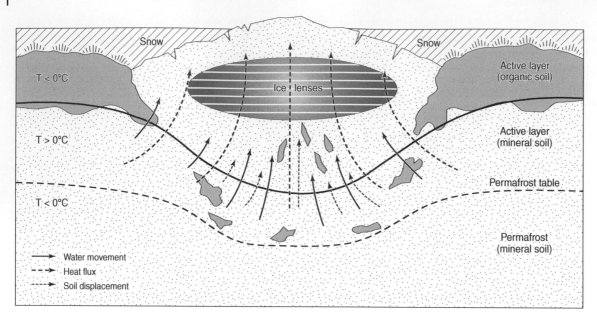

Figure 9.15 Physical processes taking place within a nonsorted circle (frost boil) during freezing. *Source:* Nicolsky *et al.* (2008). Reproduced with permission of the American Geophysical Union.

and 0.5–2.0 m wide, separated by a network of shallow troughs. Such features are also sometimes termed *soil hummocks, tundra hummocks* or *turf hummocks*. The term *mud hummock* is sometimes applied to small unvegetated mounds separated by desiccation cracks or superficial frost cracks, particularly in high-arctic environments where they resemble miniature frost boils and may have a similar genesis (Walker *et al.*, 2004). Earth hummocks are widespread in arctic tundra environments (Tarnocai and Zoltai, 1978; Kojima, 1994; Figure 9.16a), in subarctic permafrost areas south of the treeline (Kokelj *et al.*, 2007), on subarctic, alpine and subalpine mountains (Grab, 1994; Mark, 1994; Scott *et al.*, 2008) and in maritime periglacial environments such as Iceland, where they are known by the collective term *thúfur* and occur in permafrost-free areas (Van Vliet-Lanoë *et al.*, 1998; Figure 9.16b). Small dome-like earth hummocks even occur in areas of shallow (<0.4 m) seasonal ground freezing (Killingbeck and Ballantyne, 2012; Figure 9.16c). Earth hummocks usually occur as 'hummock fields' on flat or gently sloping terrain, and are commonly developed in silt-rich soil containing few or no stones. On gentle slopes, hummocks are sometimes elongate downslope. Where gradients exceed 3–6°, some hummock fields grade into *relief stripes* that form a ridge-and-furrow pattern, usually aligned downslope (Nicholson, 1976; Ballantyne, 1986a; Mark, 1994; Figure 9.16d). This is not always the case, however, as hummocks developed in the shallow active layer above high-arctic permafrost occur on gradients of up to 15° without stripe development; they are sometimes referred to as *slope hummocks* (Lewkowicz, 2011). Earth hummock dimensions and morphology vary widely. Those in

high-arctic lowlands and Icelandic thúfur tend to be knob-shaped, with narrow interhummock troughs and width-to-height ratios of around 2 : 1, whereas hummocks on upland plateaus in the British Isles are typically dome-shaped, separated by broad depressions and have width-to-height ratios of 4 : 1 or more (Killingbeck and Ballantyne, 2012; Figure 9.16).

Excavation of earth hummocks often reveals disruption of underlying soil in the form of updomed soil horizons or diapiric intrusions of fine-grained soil, though hummock structure may be complex and variable (Schunke and Zoltai, 1988; Van Vliet-Lanoë *et al.*, 1998). Where permafrost is present, the permafrost table under hummocks forms depressions similar to those under frost boils. Such depressions reflect shallower thaw under interhummock troughs, which retain snow longer and often contain a thicker insulating cover of vegetation and organic soil than hummock crests. Moreover, not all hummocky microforms in periglacial environments have a core of mineral soil. Several authors have described *boulder-cored hummocks* formed by the updoming of soil above frost-heaved boulders (Harris and Matthews, 1984; Van Vliet-Lanoë and Seppälä, 2002), and hummocky microrelief can develop through uneven accumulation of peat or organic matter (Raup, 1965; Dionne and Gérardin, 1988). Small vegetated mounds called *pounus* occur around the margins of palsa fens in the discontinuous permafrost zone, and reflect localized survival of permafrost under peat cover (Seppälä, 1988a; Chapter 7).

Though it is widely accepted that earth hummocks *sensu stricto* develop through soil displacement induced by annual freezing and thawing of frost-susceptible soil, it is unlikely that all earth hummocks are produced in the

Figure 9.16 (a) Earth hummocks developed over cold permafrost, Ellesmere Island, arctic Canada. (b) Earth hummocks (thúfur) developed in volcanogenic loess, northern Iceland. Permafrost is absent. (c) Earth hummocks at 400 m altitude on Dartmoor, southwest England. (d) Sinuous relief stripes on gentle slopes at 900 m altitude in northwest Scotland.

same way (Grab, 2005). A long-held view is that hummock growth above permafrost is caused by upwards soil displacement due to the generation of cryostatic (essentially hydrostatic) pressure in confined pockets of saturated unfrozen soil during freezeback (Nicholson, 1976; Tarnocai and Zoltai, 1978; Figure 9.17a). However, Mackay and Mackay (1976) demonstrated that confined pockets of unfrozen frost-susceptible soil are likely to be desiccated through migration of pore water to feed ice-lens growth during freezing, so that pore water in unfrozen soil pockets is under tension and cryostatic pressures are not generated.

As an alternative, Mackay (1980) proposed a model for hummock growth that involves very gradual circulatory movement of soil in hummocks, driven by subsurface movement of saturated soil down the bowl-shaped depressions in the underlying permafrost table and by buoyancy-driven rise of mud towards hummock crests (Figure 9.17b). This model appears consistent with evidence for diapirism in some hummocks and an unstable density configuration in thawing soil (Swanson *et al.*, 1999), and with reports of extrusion of liquefied mud at the surface during thaw (Shilts, 1978; Zoltai and

Tarnocai, 1981; Egginton and Dyke, 1982). A rather different model has been proposed by Kokelj *et al.* (2007), who argued that earth hummocks in the subarctic forest zone near Inuvik have been produced by upthrusting of soil due to ice accumulation associated with aggrading permafrost (Figure 9.17c).

None of these models, however, can explain the development of hummocks in permafrost-free terrain, such as the thúfur of Iceland. These features are more akin to hummocks on alpine meadows, and range from subdued forms near snowpatch margins to flat-topped, steep-sided circular forms (Figure 9.16b). There are, moreover, records of Icelandic hummocks reappearing within a few decades in fields cleared for agriculture (Schunke and Zoltai, 1988), which suggests that they develop comparatively rapidly. The internal structure of earth hummocks developed on volcanogenic loess in northern Iceland exhibits upthrusting of soil horizons and tephra layers towards hummock apices. Van Vliet-Lanoë *et al.* (1998) have suggested that these structures can be explained through a three-stage process involving (i) development of a network of shallow thermal or desiccation cracks, (ii) differential frost heave of embryonic hummocks

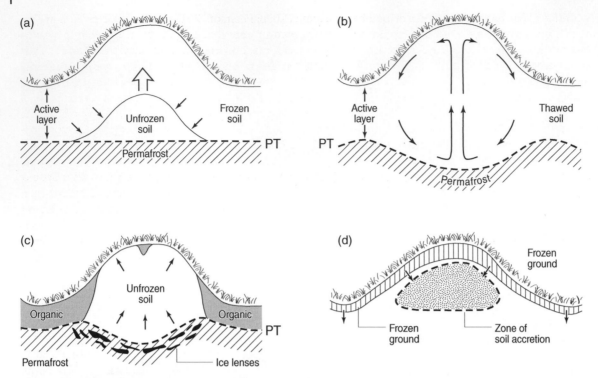

Figure 9.17 Four models of earth-hummock formation. (a) Cryostatic pressure: squeezing upwards of a pocket of saturated unfrozen soil by advancing freezing planes. (b) Soil-circulation model (Mackay, 1980): gradual soil movement down a depression in the permafrost table and buoyancy-driven rise of mud towards hummock crests during thaw. (c) Permafrost aggradation model (Kokelj *et al.*, 2007): ice-lens growth near the top of aggrading permafrost, pushing up the overlying soil. (d) Differential frost heave (Killingbeck and Ballantyne, 2012): migration of silty soil ahead of inclined freezing planes, causing hummock growth over multiple shallow freeze–thaw cycles. PT, permafrost table.

separated by such cracks and (iii) upwards injection of saturated silty soil into dilation cracks during thaw, promoting hummock growth. Van Vliet-Lanoë (1998) has also proposed a more eclectic model that relates hummock formation to initial drainage conditions, frost-susceptibility gradients in the soil and changes in water-table depth. Though this model accounts for contrasting structures observed in earth hummocks in different environments, its generality is uncertain, though it underscores the likelihood that earth hummocks develop in different ways under particular environmental circumstances.

Several recent explanations of hummock genesis have focused on the role of differential frost heave (laterally non-uniform heave of soil during freezing). Differential frost heave is implicit in some of the models described above, but has usually been attributed to lateral variations in initial conditions, such as pre-existing microrelief, ground cracking, soil moisture conditions, soil texture and vegetation cover. However, as we saw in the previous section, numerical models of soil freezing have shown that in moist silty soils, uniform frost heave may develop into laterally non-uniform heave with a characteristic wavelength that determines pattern spacing (Peterson and Krantz, 2003, 2008; Peterson, 2011). Lewis

et al. (1993) calculated that differential frost heave could generate a characteristic longitudinal wavelength compatible with measured hummock spacing. How an initial pattern of alternating swells and depressions develops into mature earth hummocks is less clear. Killingbeck and Ballantyne (2012) have suggested that the initial amplitude of swells and depressions is progressively increased over repeated freeze–thaw cycles by migration of silt particles towards the growing hummock in front of inclined freezing planes (Figure 9.17d), as demonstrated in the experiments of Corte (1966).

Finally, it should be noted that some earth hummocks are formed or modified by non-frost-action processes. Selective entrapment of windblown sediment on tussocks or cushion plants has been invoked to explain the formation of some hummocks (Raup, 1965; Ballantyne, 1986a), and Lewkowicz (2011) has shown that deposition of niveo-eolian sediment on hummock crests contributes to the growth of slope hummocks in high-arctic environments. Slopewash may also enhance hummock development through erosion of interhummock troughs. Given the variability in the morphology of earth hummocks (from gentle domes to large knob-like mounds) and the range of environments in which they occur (from high-arctic tundra to areas of shallow seasonal ground freezing), it is extremely unlikely that all landforms described as 'earth hummocks' have the same origin. The common association between hummocks and frost-susceptible silt-rich (often loessic) soils suggests that differential frost heave is probably implicated in hummock formation in most cases (Tarnocai and Zoltai, 1978; Schunke and Zoltai, 1988; Mark, 1994; Van Vliet-Lanoë *et al.*, 1998; Killingbeck and Ballantyne, 2012), but other processes such as buoyancy-driven soil diapirism, differential soil uplift above aggrading permafrost and soil injection into dilation or desiccation cracks may also be partly or entirely responsible for hummock genesis or evolution.

The development of relief stripes on slopes greater than 3–6° (but only, apparently, in areas of seasonal ground freezing, where permafrost is absent) has not been adequately explained, but seems to imply an overall cellular circulation of soil, on which gradient imposes a gravity-induced downslope vector of soil movement. As noted earlier, not all hummock fields grade into relief stripes as gradient increases; some hummocks on moderate slopes underlain by cold permafrost actually exhibit internal structures indicating slow downslope movement by solifluction or even hummock rotation (Mackay, 1981; Lewkowicz, 2011; Figure 9.18).

9.7.3 Environmental and Palaeoenvironmental Significance

Frost boils and vegetation-defined nonsorted stripes appear to be characteristic of permafrost environments, though theoretically frost boils might develop in nonpermafrost areas of deep seasonal freezing (Peterson and

Figure 9.18 Section excavated through an earth hummock on a slope of 8°, Ellesmere Island, Canadian Arctic Archipelago. Burial of organic soil suggests slow downslope movement of the hummock.

Krantz, 2008; Peterson, 2011). Inactive forms have limited preservation potential on the surface, owing to vegetation recolonization of bare ground. Earth hummocks and relief stripes, on the other hand, may sometimes survive climatic amelioration. Excavation of degraded hummocks and relief stripes on Scottish mountains has revealed well-developed podzols with individual soil horizons following the surface microtopography, demonstrating that these features are inactive and have probably been so for several millennia (Ballantyne, 1986a). Such relict features are, however, not necessarily diagnostic of former permafrost, as both hummocks and relief stripes are known to form on nonpermafrost terrain. In general, however, the internal structures of nonsorted patterned ground may be preserved long after its surface manifestations have vanished (Van Vliet-Lanoë, 1988, 1991, 1998), though the palaeoenvironmental interpretation of such structures is not straightforward.

Perhaps the most enigmatic relict patterned ground features are the irregular nonsorted nets (polygons) and stripes formed on chalkland terrain in England and France (Ballantyne and Harris, 1994; Murton *et al.*, 2003; Bateman *et al.*, 2014). The nets are commonly about 10 m wide and the stripes about 7.5 m wide, and the patterns are defined by sand-filled troughs 2–3 m wide that penetrate up to 1.8 m into the underlying brecciated chalk and separate cells or stripes of chalky diamicton. Luminescence ages obtained for the aeolian sand infill show that these features developed during the last glacial stage, probably over permafrost (as evidenced by the underlying brecciated chalk). Nicholson (1976) suggested that large-scale sediment circulation produced the initial patterns, which were subsequently modified by incorporation and infill of aeolian sand, but the mechanism of sediment circulation remains uncertain, particularly as these features appear to have no exact analogues in present periglacial environments.

9.8 Cryoturbations

9.8.1 Terminology and Characteristics

The term *cryoturbation* is employed in two ways. In the singular, it refers to internal movement or disturbance of soil as a result of repeated, usually seasonal, freezing and thawing. Cryoturbation is therefore broadly equivalent to the concept of *mass displacement* within soil, defined by Washburn (1979, p. 96) as '...the *en* masse local transfer of mobile mineral soil from one place to another within the soil as the result of frost action'. Irrespective of the processes involved, cryoturbation is important in pedogenesis, moving organic matter downwards and thus increasing the storage of organic carbon in the subsoil (Bockheim, 2007; Kaiser *et al.*, 2007). In the plural, the term *cryoturbations* or *cryoturbation structures* is used to describe soil or sediment structures that have developed through recurrent freezing and thawing of the ground, both in the active layer above permafrost and in seasonally frozen ground in nonpermafrost terrain. A related term, *involutions*, is sometimes used to describe such structures, but is descriptive rather than genetic and does not imply a cryogenic origin; involutions can develop in various ways, including loading of saturated sediments by the rapid accumulation of overlying material and internal displacement of sediment due to seismic activity (French, 1986; Van Vliet-Lanoë *et al.*, 2004b). As outlined in Chapter 8, many involutions in past and present permafrost environments probably formed during thaw of ice-rich permafrost (Murton and French, 1993a; Murton *et al.*, 1995; French *et al.* 2005) and are therefore not cryoturbation structures *sensu stricto*. Some authors restrict the use of 'cryoturbations' to soil-deformation structures produced by differential frost heave, but here we follow Vandenberghe (1988, 2013) in applying it to all soil structures that have developed as a consequence of seasonal freezing and thawing of the ground, irrespective of the mechanism involved and the presence or absence of permafrost.

In vertical section, cryoturbations take the form of perturbations within stratified sediments, or vertical interdigitation of two or more sediment layers with contrasting properties, particularly grain size and organic content (Figure 9.19). Common cryoturbation structures representing upwards movement of sediment include diapiric forms, which may be rounded, acuminate (tapering upwards) or irregular, and dyke-like structures representing vertical intrusion of sediment. Downwards movement of sediment is represented by synclinal features, bulbous structures, ball-and-pillow structures comprising an upper-unit sediment body encased within a contrasting lower sediment unit, and dyke- or wedge-shaped features. Some cryoturbations involve regular spacing of structures, others are irregularly spaced and complex, and some conform readily to none of these forms, but consist of irregular folds. Vandenberghe (2013) has proposed a typology of cryoturbation structures (Figure 9.20), which provides a useful basis for discussing formative processes. According to this classification, type 1 cryoturbations form low-amplitude undulations or folds; type 2 cryoturbations are closely-spaced, repetitive, intensely folded forms, generally 0.5–2.0 m in amplitude; type 3 cryoturbations are similar but much smaller; type 4 cryoturbations consist of isolated pocket or drop structures; type 5 cryoturbations involve upwards or downwards displacement of sediment controlled by crack spacing ('dykes'); and type 6 cryoturbations are those exhibiting irregular folding at all scales.

The horizontal expression of cryoturbations is also variable. Many have no obvious microtopographic expression, though some are related to patterned ground features, particularly frost boils and earth hummocks (Shilts, 1978; Van Vliet-Lanoë, 1988, 1991, 1998). In planform, excavated cryoturbations tend to form polygons or nets of varying size defined by contrasting sediment properties (Vandenberghe, 1992a). The form of cryoturbations formed in a volcanic ash layer overlying pumice have been illustrated by Koiwa (2003), who excavated horizontal platforms cutting across an area of cryoturbations at three levels (Figure 9.21) to demonstrate how sinking of the ash into the pumice and compensatory upwards movement of pumice into the ash have produced a nearsurface irregular net of pumice.

Figure 9.19 Cryoturbations in the active layer, Ellesmere Island, Canadian Arctic Archipelago. The active layer is ~40 cm deep.

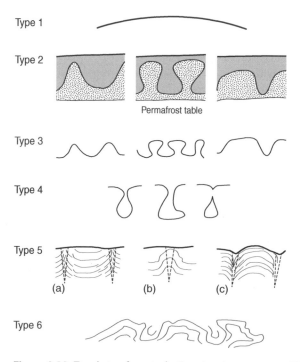

Figure 9.20 Typology of cryoturbation structures proposed by Vandenberghe. *Source:* Vandenberghe (2013). Reproduced with permission of Elsevier.

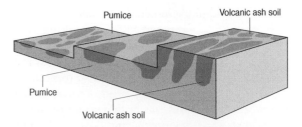

Figure 9.21 Three-dimensional morphology of shallow cryoturbation structures developed in a volcanic ash soil overlying pumice in northeast Japan. *Source:* Adapted from Koiwa (2003). Reproduced with permission of Elsevier.

9.8.2 Formation

Elucidating the processes involved in the development of cryoturbations has proved problematic because of the difficulty of monitoring formation in the field. Most interpretations are based on observations of the structure of involutions exposed in vertical section, supplemented by a limited body of experimental evidence. Three processes, however, are generally acknowledged to (potentially) cause deformation in sediments subject to seasonal freezing and thawing: loading (sometimes termed *load casting*), differential frost heave (sometimes termed *cryostatic heave*) and cryohydrostatic pressure.

In general, sediment deformation by loading requires two conditions: density inversion, where sediment of relatively high density overlies sediment of lower density, and loss of frictional strength in the lower sediment due to liquefaction. A high ice content of soil at the base of the active layer may result in an unstable (inverted) bulk-density profile during thaw (Swanson *et al.*, 1999), though this appears less likely in seasonally frozen ground that is not underlain by permafrost, unless the dry density of nearsurface soil exceeds that of subsurface soil. However, thaw of ice-rich sediment underneath a thaw-consolidated layer may engender undrained conditions, a rise in pore-water pressures and consequent sediment liquefaction, resulting in sinking of overlying sediment into the thawed, liquefied layer and compensatory upwards diapiric intrusion of the latter into the overlying sediment. This process is probably responsible for most cryoturbations of types 2–4 (Figure 9.20), which closely resemble loading structures generated experimentally, and appears to be independent of the nature of sediment contrasts: fine-grained sediments may sink into coarser sediments or *vice versa*. The development of loading structures seems to imply an underlying impermeable layer, such as still-frozen soil or the permafrost table. Vandenberghe (2013) has suggested that the smaller (generally <0.5 m amplitude) cryoturbation structures of types 3 and 4 (Figure 9.20) may represent the lower limit of seasonal thaw, rather than the upper limit of permafrost, particularly in topographic depressions where silts and clays are overlain by sand, so that small-amplitude

loading structures are not necessarily indicative of former permafrost. Under propitious conditions, loading structures may develop rapidly; Swanson *et al.* (1999) have argued that thaw of ice-rich soil at the base of the active layer may promote several decimetres of diapiric movement within a single thaw season.

Not all researchers are convinced regarding the importance of loading during thaw in generating cryoturbations. A notable dissenter is Van Vliet-Lanoë (1991), who considered that the primary process for generating cryoturbations is differential frost penetration. She argued that freezing and ice-lens formation in frost-susceptible soil compress still-unfrozen sediments, and that the resulting cryostatic pressure is responsible for plastic deformation of soil, particularly upwards into desiccation cracks. She stressed that cryoturbations formed in this fashion develop incrementally through repeated freeze–thaw episodes, and that permafrost is not required to allow such cryoturbations to form. Van Vliet-Lanoë also argued that vertical frost-susceptibility gradients are important in determining the outcome of such freezing-induced soil disruption, suggesting that where the surface soil is more frost-susceptible than the underlying soil, injected soil is unable to reach the surface and will form a subsurface diapir; conversely, where the reverse is true, injected sediment may reach the surface to form frost boils or sorted circles. Though the validity of this mechanism as a general explanation of cryoturbations has been questioned (Vandenberghe, 1992b), it offers a feasible explanation of types 1 and 6, and possibly type 5 (Figure 9.20). It receives some support from experiments showing that simple freezing and thawing of saturated frost-susceptible sediment can cause internal deformation (Pissart, 1982; Coutard and Mücher, 1985). A more nuanced picture emerges from a study by Ogino and Matsuoka (2007), who simulated rapid top-down freezing of a two-layer model comprising an upper unit of sandy loam and a lower unit of pumice. They found that ice segregation near the interface of the two units induced upwards movement of coarse pumice grains during freezing, but also that downwards settlement of the (denser) loam occurred during thaw. Such deformation was accentuated during successive freeze–thaw cycles, creating an undulating interface of mounds and depressions, and suggests that frost penetration and thaw settlement may work in tandem to generate cryoturbations in seasonally frozen ground.

An alternative possible mode of cryoturbation is through the generation of high pore-water pressures in pockets of soil that are trapped between a downwards freezing front and an underlying impermeable layer such as the permafrost table. The resulting cryohydrostatic pressure (*sensu* Vandenberghe, 1988, 2013) is thought to expel fluidized soil into zones of lower resistance, in particular injecting it upwards into pre-existing frost or desiccation cracks, so that it reaches the surface to form polygons of sediment injected upwards from farther down the soil profile. Vandenberghe (2013) suggested that type 5 cryoturbations (Figure 9.20) are generated by this process. A possible limitation to this mechanism, as discussed earlier in the context of earth-hummock formation, is that ice segregation during freezing of frost-susceptible soils tends to draw water away from pockets of unfrozen ground, creating negative pore-water pressures (Mackay and Mackay, 1976; Mackay 1980), though it is possible that pore-water expulsion in non-frost-susceptible sandy soil could generate pore-water pressures sufficient to overcome overburden pressures and inject fluidized soil into a crack network.

9.8.3 Interpretation of Relict Cryoturbations

Uncertainties regarding the origin of some cryoturbations and their possible implications for the presence of permafrost (as opposed to deep seasonal freezing of the ground) sometimes make the interpretation of relict Pleistocene examples problematic. Large-amplitude type 2 structures may indicate the former depth of the permafrost table or some other impermeable layer, particularly where downwards-sinking forms are flat-bottomed (Vandenberghe and Van den Broek, 1982). Occasionally, the lower limit of such structures may coincide with the upper surface of sand wedges or ice-wedge pseudomorphs, providing confirmatory evidence of former active-layer depth. A difficulty arises, however, in differentiating many supposed type 2 cryoturbations from thermokarst involutions formed during permafrost thaw, as the inferred process (loading) is identical (Murton and French, 1993a), though in both cases such structures imply former permafrost. Smaller sediment perturbations represented by the other types of cryoturbation illustrated in Figure 9.20 may have formed either in a former active layer or in seasonally frozen ground lacking permafrost, though as Vandenberghe (2013) pointed out, types 3 and 4 may require a subsurface impermeable layer to allow load casting to occur, and the same is true for type 5 structures produced by cryohydrostatic pressure.

An additional problem in the interpretation of some relict involuted structures is that they may reflect two or more periods of soil perturbation. Complex involutions in southeast England, for example, have been interpreted by Murton *et al.* (2003) in terms of a two-stage evolution in which large-amplitude thermokarst involutions formed during an initial phase of permafrost degradation, which was succeeded by permafrost aggradation and the formation of nearsurface cryoturbations in a former active layer. Similarly, deformation structures in sands and loams underlying a Weichselian alluvial terrace in southeast Poland have been interpreted by Superson *et al.* (2010) in terms of a three-stage evolution: (i) irregular structures and broad folds (types 1 and 6 in Figure 9.20) attributed to cryostatic pressure during a period of permafrost aggradation; (ii) small folds and drop-and-pocket structures (types 3 and 4) formed by undrained loading during a later period of waterlogged conditions when the ground was underlain by relatively stable permafrost; and (iii) large thermokarst involutions formed by load casting during subsequent permafrost degradation. As these examples illustrate, some sequences of involuted structures interpreted by earlier researchers as the product of cryoturbation within a former active layer may require reassessment in terms of a more complex evolutionary sequence.

9.9 Pedogenic Effects of Freezing and Thawing

Cryoturbation radically alters the structure of soils in which it occurs, but even in the absence of visible cryoturbation some soils that have been subject to repeated freezing and thawing exhibit small-scale structural features that may persist for millennia, permitting identification of past episodes of former deep seasonal freezing or permafrost occurrence. The most important of these are produced by ice segregation and thaw consolidation acting in moist, frost-susceptible soils, so such microstructures tend to be absent or poorly developed in coarse-grained soils, or in dry soils with limited moisture content. Some microstructures are visible to the naked eye, but others are only detectable in thin sections examined under a microscope. Useful reviews of the effects of freezing and thawing of soil and the microstructures that result have been published by Van Vliet-Lanoë *et al.* (2004a) and Van Vliet-Lanoë (2010).

Amongst the most common features in soils that have experienced seasonal freezing and thawing is a platy or lenticular structure, consisting of small granular aggregates of compressed soil, typically a few millimetres thick and a few centimetres long (Figure 9.22a). These are produced by two processes: vertical compression of soil peds by the formation of lenses of segregated ice, which occupy similar positions during repeated freezing episodes, and desiccation of soil within the peds as moisture is extracted to feed ice-lens growth. In very fine-grained (clay and clay-silt) soils, a prismatic or blocky fabric may develop, reflecting the formation of reticulate segregation ice through a combination of ice lensing parallel to the freezing front and freezing of ice in thin desiccation fissures (Van Vliet-Lanoë *et al.*, 1984; Coutard and Mücher, 1985). Experiments by Pawluk (1988) showed that formation of platy structures by repeated freezing and thawing of a clay-loam till also produces *metavughs*

Figure 9.22 Photomicrographs of structures associated with freezing and thawing of frost-susceptible soil. (a) Lenticular aggregates separated by segregation ice (white). (b) Vesicles (large smooth-walled soil pores). (c) Silt cappings on the upper surfaces of coarse sand grains. *Source:* Courtesy of Brigitte Van Vliet-Lanoë.

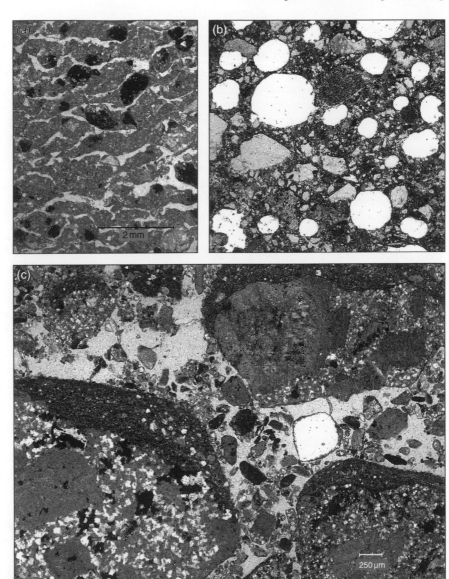

(large connected soil pores) that represent voids left by thawing ice lenses, particularly at the base of clasts or compacted soil aggregates. Not all frost-susceptible soils, however, exhibit a platy structure. Szymanski *et al.* (2015), for example, have shown that it is absent from the active layer under patterned ground in Svalbard, possibly because cryoturbation has prevented the development of a stable platy fabric. Similarly, Harris and Ellis (1980) noted that in solifluctual soils that are subject to gradual downslope movement, a banded pattern of matrix aggregates separated by detritus grains occurs only below the depth of soil movement, suggesting that solifluction tends to modify or destroy lenticular soil structures.

A common feature of seasonally frozen ground is the presence of *vesicles* or air bubbles within frost-susceptible soil. Vesicles are smooth-walled soil pores, commonly round, oval or deformed, that form the dominant voids

in some active-layer soils (Szymanski *et al.*, 2015; Figure 9.22b). Freeze–thaw experiments suggest that vesicles represent escape of air during thaw consolidation of ice-rich soil, when high pore-water pressures induce soil liquefaction or thixotropic conditions, expelling air from the consolidating soil (Van Vliet-Lanoë *et al.*, 1984; Coutard and Mücher, 1985), though it is possible that they result from release of air from soil water during freezing (Van Vliet-Lanoë, 2010).

A further common effect of seasonal freezing and thawing is the downwards translocation of clay and silt particles within the soil. In part, this probably reflects downwards migration of fine soil ahead of the advancing freezing front (Corte, 1966), but it is also attributable to illuviation (downwash) of fines by percolating meltwater during thaw. These fines are intercepted by the upper surfaces of clasts and platy soil aggregates (Figure 9.22c).

They form coatings of clay, silt and fine sand, sometimes termed *cutans*, on the tops of stones and coarse sand grains; the occurrence of such coatings on the sides or lower surfaces of grains probably implies subsequent grain rotation due to upfreezing, cryoturbation or solifluction. On platy aggregates, they result in a distinct fining-upwards pattern, increasing the density of individual aggregates and contributing to their stability and persistence. In permafrost areas, downwards translocation of fines stops at the permafrost table, sometimes forming lenticular patterns of silt enrichment, confusingly referred to as 'silt droplets' in some older accounts.

Lenticular peds (or blocky peds in clay-rich soils), vesicles and silt cappings represent the most common diagnostic characteristics of moist frost-susceptible soils that have experienced recurrent freezing and thawing. Less common features include a distinct banded fabric produced by alternation of compacted platy aggregates and silts washed into the voids vacated by thawing ice lenses (Harris and Ellis, 1980), vertically-aligned sand grains and rock fragments, and chemical precipitates. Vogt and Larqué (2002), for example, have interpreted calcite, gypsum and bassanite precipitates as the products of seasonal soil freezing and thawing of the ground, though all three depend on the mineralogy of the parent materials. Microscopic Fe-Mn nodules detected in active-layer soils on Svalbard have been attributed by Szymanski *et al.* (2015) to reduction of iron and manganese under waterlogged (anaerobic) conditions in summer and to immobilization due to freezing-induced desiccation of the soil in winter.

The various soil cryostructures described above are of particular interest in that, unless disturbed by cryoturbation, bioturbation or human activity, they form stable features that persist for millennia, and hence provide evidence of former (Pleistocene) deep seasonal freezing and thawing of the ground. Laboratory simulations have shown that platy fabrics, vesicles and silt cappings all form under conditions of deep seasonal freezing, and do not require the presence of permafrost (Van Vliet-Lanoë *et al.*, 1984; Coutard and Mücher, 1985; Pawluk, 1988). In northern Italy, for example, laminations attributable to ice lensing, small cryoturbation structures and silt cappings on clasts are largely restricted to the topmost metre of the soil, and appear to represent former deep seasonal freezing rather than permafrost (Cremaschi and Van Vliet-Lanoë, 1990). Such features nevertheless provide diagnostic criteria for distinguishing periglacial from non-periglacial deposits. Detailed study of the micromorphology of a thick (45 m) exposure of poorly-sorted diamictons in coastal cliffs in Wales allowed Harris (1998) to differentiate a lower unit of apparent periglacial origin (containing abundant silt cappings on sand and gravel particles) from an overlying thick intermediate unit lacking cryogenic microstructures and

a topmost unit containing diffuse grain coatings and lenticular aggregates. He interpreted both the lowermost and the uppermost units as solifluction deposits, but the intermediate unit he interpreted as the product of paraglacial reworking of till by debris flows. Elsewhere, however, the stratigraphic context of cryogenic microstructures in soils demonstrates formation in a former active layer above permafrost, for example where they occur above fragipans (see below) or sediments containing ice-wedge pseudomorphs (Van Vliet and Langohr, 1981; Ballantyne and Harris, 1994, pp. 110–112; Van Vliet-Lanoë, 1998). The stability and persistence of cryogenic microstructures is testified by their presence in the Barham palaeosol in eastern England, which incorporates platy aggregates, banded fabrics and silt cappings on sand grains (Rose *et al.*, 1985). As the Barham palaeosol is believed to be of Middle Pleistocene (MIS12) age, these features represent the combined effects of ice segregation and particle translocation that occurred in a former active layer over 430 000 years ago but have survived intact to the present.

9.10 Fragipans

In many regions of past permafrost, a dense, indurated but uncemented horizon exists within the soil, often at depths of ~0.4–1.0 m. Such indurated horizons are referred to as *fragipans*. Though the properties of fragipans vary, they are typically ~15 cm to ~1 m thick and often characterized by a coarse prismatic, columnar or blocky structure. Fragipans tend to have high bulk density compared with that of overlying soil, and relatively low clay content. Their upper surfaces tend to be fairly sharply defined, but their lower boundaries are typically gradational. Fragipan formation is incompletely understood (Bockheim and Hartemink, 2013), but some have been interpreted as having formed either at the top of upward-aggrading permafrost or through ice segregation at the base of a former active layer.

The permafrost hypothesis of fragipan formation was developed by Van Vliet and Langohr (1981) with reference to fragipans developed in loessic and glacigenic silty soils in Belgium. These fragipans exhibit a compact platy or lenticular structure within a polygonal system of vertical fissures, accumulation of translocated clay on their upper surfaces and a greater bulk density than the overlying soil (Figure 9.23). Van Vliet and Langohr (1981) attributed the lenticular structure of fragipans to ice segregation during permafrost aggradation, with the formation of fissures being due to development of a reticulate cryostructure, implying that the tops of fragipans represent the former permafrost table. This interpretation suggests that fragipans represent the ice-rich transition zone at the top of former permafrost (Shur *et al.*, 2005; Chapter 5).

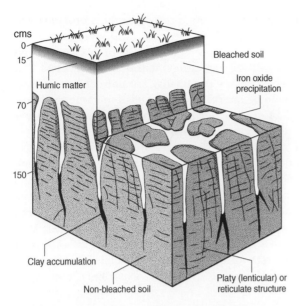

Figure 9.23 Schematic representation of a fragipan in an imperfectly drained silty soil. *Source:* Adapted from Van Vliet and Langohr (1981). Reproduced with permission of Elsevier.

An alternative interpretation is that an indurated horizon may develop at the bottom of the active layer, attributable partly to accumulation of illuviated clay and silt particles washed down from higher in the soil profile, but primarily to intense ice segregation due to refreezing of percolating meltwater late in the summer thaw period, as described by Mackay (1983). In both cases, induration is thought to reflect compression of soil between growing ice lenses and compaction (reduction in pore spaces) during thaw consolidation.

Several authors have supported this interpretation, and suggested that fragipans are not only indicators of past permafrost, but also provide an indication of the depth of the former active layer. Payton (1992), for example, analysed lowland fragipans in northeast England that exhibit compact lenticular structures and polygonal fissuring, and attributed these to formation in association with permafrost during the Younger Dryas Stade of ~12.9–11.7 ka, the final period of permafrost conditions in the British Isles. Similarly, Harris (1991) described an indurated layer at depths of 0.8–1.5 m on the Anglesey coast of Wales, characterized by a strong lenticular structure and silt cappings on grains, and overlain by small cryoturbation structures. He inferred that this probably represents permafrost that developed after the retreat of the last ice sheet, and possibly during the Younger Dryas Stade. A study by French *et al.* (2009) of fragipans within soils on the coastal plain of the eastern United States yielded more equivocal conclusions. The fragipans in this area comprise compacted soil at 1–2 m depth, apparently penetrated by seasonal frost cracks, and exhibit platy structures at the bottom of the compacted layer, suggesting to these authors that the lower parts of the fragipans approximate the depth of the former palaeo-permafrost table. However, they conceded that these fragipans could represent older (possibly non-periglacial) low-permeability horizons that acted as a 'surrogate' for the permafrost table, such that water-saturated sediments were trapped between the base of seasonal frost penetration and the fragipan, promoting the development of an overlying argillic horizon of relatively high bulk density.

The uncertainties inherent in this study highlight the problems of employing fragipans as indicators of the presence of former permafrost. Fragipans occur in parts of the world that have never experienced periglacial conditions, and have been attributed, for example, to the effects of seismic ground motions in saturated soil (e.g. Certini *et al.*, 2007) or lithological discontinuities within soils (e.g. Aide and Marshaus, 2002). In loess deposits, fragipan formation may occur through hydroconsolidation, a form of internal collapse caused by a combination of loading (the accumulation of overlying loess) and wetting. Collapse appears to be favoured by progressive post-depositional accumulation of clay mineral material at particle contacts, and results in the formation of a dense, tightly-packed layer that subsequently dries and contracts, forming a polygonal system of vertical cracks (Smalley and Markovic, 2014; Smalley *et al.*, 2016). As these examples illustrate, interpretation of fragipans in terms of former permafrost and the depth of the former permafrost table has to be approached with caution, except where there is confirmatory evidence in the form of ice-wedge pseudomorphs, relict sand wedges or relict lenticular or reticulate cryostructures indicative of former ice segregation.

9.11 Synthesis

Repeated freezing and thawing of soils, particularly moist frost-susceptible soils, is responsible for producing a wide range of nearsurface structural features, some of which are manifest at the surface as patterned ground. Upfreezing of clasts is widespread, and is most readily attributed to the operation of frost pull, the adfreezing of clasts to adjacent freezing ground and subsequent upwards frost heave of both, though the operation of the process in clast-rich diamictons is poorly understood. Heaving of bedrock has been explained in terms of hydrostatic pressures generated by downwards freezing above permafrost. Explanation of sorted and nonsorted patterned ground features has traditionally been based on the concept of multiple origins, but numerical models complemented by field and laboratory experiments suggest that most patterns are initiated by differential frost heave. The subsequent evolution of some patterned ground features, however, probably involves additional mechanisms, of which buoyancy-induced soil movement during thaw and gravity-induced movements of both surface and subsurface soil are probably the most important. The nature of patterning (sorted or nonsorted) appears to be mainly determined by initial boundary conditions, such as abundance of clasts, soil texture, moisture content and the presence or absence of vegetation cover. Some patterned ground features, such as earth hummocks and frost boils, are related to the development of underlying cryoturbation structures, but the formation of some cryoturbations remains contentious: some authors have favoured density inversion and undrained loading during thaw as the primary cause of most cryoturbations, whereas others have argued that generation of cryostatic pressure (differential frost heave) or cryohydrostatic pressure within freezing soil is more important. It seems likely that both sets of processes operate, but produce rather different types of soil deformation.

There is greater consensus regarding the origin of small-scale structures within seasonally freezing ground (platy or prismatic structures, vesicles and silt or clay coatings on clasts and soil aggregates), but it is now evident that indurated horizons (fragipans) are polygenetic features that should be interpreted in terms of marking the lower limit of a former active layer only in cases where there is supporting cryostructural or other evidence.

The effects of seasonal freezing and thawing of the ground are explored further in the next two chapters, which deal with the weathering of rock in periglacial environments and with solifluction, the slow downslope movement of soil produced by ground freezing and thawing. Chapter 16, the penultimate chapter, describes how some of the features discussed above have informed our understanding of past periglacial conditions in mid-latitude regions that now experience temperate climates.

10

Rock Weathering and Associated Landforms

10.1 Introduction

The term *weathering* describes the breakdown of rock and mineral particles at the Earth's surface. *Physical weathering* is the disintegration of rocks as a result of mechanical stress; *chemical weathering* is the progressive decomposition of minerals within a rock as a result of chemical reactions; and *biotic weathering* is the term given to weathering processes that are stimulated or enhanced by organisms such as fungi and lichens. Physical and chemical weathering processes often operate in combination. Fracturing of rock enhances access for water, the essential ingredient of all chemical weathering processes, and exposes fresh rock surfaces to chemical attack (Hoch *et al.*, 1999). Chemical alteration of constituent minerals weakens rocks so that they are more likely to fracture when stressed. Though geomorphologists have often attempted to isolate particular weathering processes for investigation in the field or the laboratory, rock breakdown is a complex phenomenon that usually involves several different processes operating over long timescales, during which the ambient environment may have undergone marked change. To complicate matters further, rocks vary enormously in mineral composition, nature of intergranular bonding, tensile strength, porosity, capillarity, permeability and structural integrity.

Traditionally, rock breakdown in cold environments has been attributed mainly to mechanical weathering processes, and in particular to *freeze–thaw weathering* caused by stresses generated by the 9% volumetric expansion of water when it freezes in joints, cracks or pores within rock. Freeze–thaw weathering has been commonly assumed to produce 'angular, frost-shattered detritus' (Figure 10.1). Chemical weathering has often been considered to be negligible, because low temperatures retard chemical reactions. Both these views require reassessment in the light of evidence suggesting that a range of physical, chemical and biotic weathering processes contribute to rock breakdown in periglacial environments (Hall *et al.*, 2002; Thorn *et al.*, 2011).

Before we consider such processes, the intrinsic variability of different rock types merits emphasis. The strength of intergranular bonding determines the tensile strength of intact rocks, and hence their sensitivity to fracture or disaggregation, but varies greatly between rock types. In igneous and metamorphic rocks, in which crystals are 'welded' together, such bonds are usually very strong, but in sedimentary rocks, in which grains are cemented together, bonds are often much weaker. A second important consideration is rock porosity, which determines the amount of water a rock can hold. In igneous rocks, porosity is often <2%, but in sedimentary rocks it commonly ranges from 5 to 30%. For any particular rock type, tensile strength tends to decline as porosity increases. Porosity also influences permeability: the readiness with which rock absorbs water. In granite, for example, permeability is about $10^{-12} \, m \, s^{-1}$, but most sedimentary rocks have permeabilities of around $10^{-7} \, m \, s^{-1}$ or more. Finally, structures such as microcracks, foliation, bedding planes and concentrations of finer or coarser grains or of particular minerals may form zones of mechanical weakness (Nicholson and Nicholson, 2000), and may facilitate or impede pore-water movement. Variability in rock properties means that the effectiveness of particular weathering processes is constrained by rock type. In particular, it is useful to draw a distinction between weathering of resistant rocks with low porosity and high tensile strength (such as granite, basalt and quartzite), in which breakdown is often concentrated along pre-existing joints and microcracks, and that of porous sedimentary rocks of low tensile strength, in which cracking, fracture and disaggregation may be initiated within intact rock (Matsuoka, 2001b; Matsuoka and Murton, 2008).

10.2 Physical Weathering Processes

Investigation of physical weathering processes in cold environments has involved three approaches: laboratory experiments, monitoring of rock behaviour in the field

Periglacial Geomorphology, First Edition. Colin K. Ballantyne.
© 2018 John Wiley & Sons Ltd. Published 2018 by John Wiley & Sons Ltd.

Figure 10.1 Blockfield composed of angular 'frost-shattered' schist debris at 1900 m on Mount Bitterness, South Island, New Zealand. Appearances can be deceptive: some clasts are edge-rounded by granular disaggregation, and fine sediment is present just below the surface layer of debris.

and numerical modelling. Much early laboratory research was based on subjecting samples of intact rock to freeze–thaw cycles and evaluating the effect of temperature regime, moisture availability and rock properties on the rate of rock breakdown (Lautridou and Ozouf, 1982). Many such experiments, however, involved unrealistic procedures (such as very rapid freezing and thawing, or immersion in water), and few attempted to discriminate the *processes* responsible for rock breakdown. More recent laboratory research (e.g. Matsuoka, 1990a; Hallet *et al.*, 1991; Prick, 1995, 1997; Murton *et al.*, 2006; Jia *et al.*, 2015) has been more successful in elucidating weathering constraints and processes, though still based mainly on the breakdown of intact samples of porous sedimentary rock. Field monitoring of weathering conditions and processes has benefited from the introduction of data loggers to record rock temperatures, rock moisture conditions and rock strain (e.g. Matsuoka, 2001a, 2008; Sass, 2004, 2005a; Hall *et al.*, 2008; Hasler *et al.*, 2012), providing data concerning microclimatic conditions at the rock surface and the timing and causes of crack enlargement. Finally, numerical modelling (e.g. Walder

and Hallet 1985, 1986; Anderson, 1998; Hales and Roering, 2007; Anderson *et al.*, 2013) has played a vital role in generating testable predictions about weathering processes. Recent developments are reviewed by Matsuoka and Murton (2008) and Murton (2013b).

10.2.1 Frost Weathering

The term *frost weathering* refers to all processes associated with rock breakdown through the freezing of water. Lautridou (1988) identified four frost-weathering effects: (i) *frost wedging*, due to freezing of water in pre-existing cracks; (ii) *frost splitting*, the fracture of intact rock; (iii) *frost scaling*, the detachment of flakes from the rock surface; and (4) *granular disaggregation* (or *granular disintegration*), the detachment of grains or grain aggregates from a rock surface (Figure 10.2). The term *macrogelivation* is sometimes employed to describe breakdown of rocks by frost wedging or frost splitting, producing clast-sized debris, and *microgelivation* has been used to refer to breakdown of intact rock by frost weathering to produce small rock fragments, grain aggregates and

Figure 10.2 Four effects of frost action on rock. (a) Frost wedging, the opening of joints due to expansion of joint water on freezing. (b) Frost splitting of a boulder, Hornsund, Svalbard. (c) Scaling (exfoliation) of thin flakes of rock, Svalbard; the ice-filled polygonal cracks may represent cracking due to thermal stress. (d) Rounding of the surfaces of a granite tor due to granular disaggregation. *Source:* Courtesy of (a) Ole Humlum, (b) Zuzanna Swirad and (d) Adrian Hall.

individual grains, collectively referred to as *grus*. These terms thus reflect the size of frost-weathering products, not the processes involved (Hall and Thorn, 2011). Frost weathering is believed to operate in three ways: by volumetric expansion, ice segregation and hydrofracture. These are discussed below, together with other physical weathering processes that contribute to rock disintegration in cold environments.

10.2.1.1 Volumetric Expansion

The *volumetric expansion* theory of frost weathering rests on the premise that the 9% expansion of water as it freezes generates stresses sufficient to break down rock by widening existing cracks to produce clast-sized debris, or microcracks and rock pores, detaching grains and flakes from sound rock. In a closed system where water completely fills voids, the stress exerted by

freezing increases as the temperature falls, reaching a maximum of 207 MPa at $-22\,^{\circ}\text{C}$, an order of magnitude greater than the tensile strength of the strongest rocks. Although this process has long been assumed to dominate frost weathering, doubts have arisen regarding its efficacy. Three considerations are pertinent. First, for volumetric expansion to be effective, rocks need to be saturated, otherwise stresses generated by freezing will be reduced through expansion of ice into air pockets. Second, freezing must occur rapidly in order to create a closed system from which water cannot escape, otherwise freezing stresses will be relieved by pore-water expulsion. Finally, a small reduction of temperature below $0\,^{\circ}\text{C}$ has no effect, as water in rocks tends freeze at temperatures below $0\,^{\circ}\text{C}$ (Matsuoka, 2001b; Hall, 2004, 2007). Several authors have questioned the likelihood of all three conditions being met, prompting a search for other explanations of rock breakdown in cold environments.

Despite these caveats, volumetric expansion probably operates near the surface of rocks where freezing is rapid and pores and microcracks are water-filled or nearly so, resulting in granular disaggregation and small-scale flaking (Matsuoka and Murton, 2008). Laboratory experiments by Jia *et al.* (2015) have shown that rapid freezing of sandstone causes rock deformation even if samples are only partially saturated, with progressive reduction of compressive and tensile strength and an increase in porosity over successive freeze–thaw cycles. These experiments indicate that recurrent freezing and thawing of rocks, even under subcritical saturation, induces rock fatigue (internal weakening) due to progressive destruction of intergranular bonds. At the grain scale, freezing of water in microcracks and within gas–liquid inclusions has also been implicated in the comminution of quartz sand (Konishchev and Rogov, 1993) and the fracture of quartz grains (Etlicher and Lautridou, 1999; French and Guglielmin, 2000; Woronko and Hoch, 2011; Woronko and Pisarska-Jamrozy, 2016). Volumetric expansion of water upon freezing in cracks (frost wedging) offers the most plausible mechanism for joint widening (Figure 10.2a) and for the detachment of boulders from bedrock. It is also widely accepted to be a major cause of rockfall from cliffs (Matsuoka, 1990b; Matsuoka and Sakai, 1999; Sass, 2005b; Chapter 12). Matsuoka (2001a) showed that joint widening in the Japanese Alps is associated with both short-term freeze–thaw cycles in autumn (Figure 10.3) and refreezing of meltwater entering joints at the beginning of the spring thaw. Similarly, at rockwall sites in the European Alps he measured slight (<0.1 mm) opening of cracks in response to short-term freeze–thaw cycles, more pronounced opening (0.1–0.5 mm) in association with seasonal freezing and occasional marked crack dilation (>0.5 mm) in response to refreezing of meltwater in cracks during seasonal thaw (Matsuoka, 2008). These studies showed that repeated frost wedging produces irreversible crack enlargement and block detachment, confirming that volumetric expansion operates in fractures to cause bedrock breakdown.

10.2.1.2 Ice Segregation

The formation of segregation ice in bedrock resembles the process of ice segregation in frost-susceptible soils. As outlined in Chapter 3, water migrates to the freezing front during freezing of frost-susceptible soils, feeding the growth of lenses of segregation ice. The *ice segregation theory of frost weathering* maintains that a similar process operates in moist, porous, frost-susceptible rocks: water drawn to the freezing front within such rocks forms ice lenses, and ice-lens growth is believed to generate sufficient stress to initiate and widen microcracks, thus fracturing the rock. As ice-lens formation

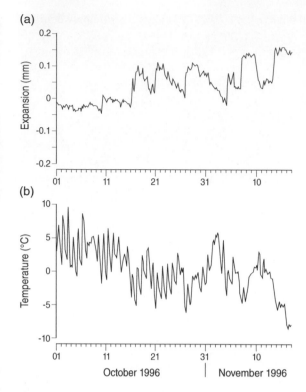

Figure 10.3 Measurements of (a) crack widening and (b) temperature at the crack surface at a site in the Japanese Alps, showing the association between freezing events and crack-widening events during seasonal freezeback. *Source:* Adapted from Matsuoka (2001a). Reproduced with permission of Wiley.

occurs normal to the direction of cooling, fracturing by ice segregation produces a system of subparallel cracks within the rock.

A theoretical model developed by Walder and Hallet (1985, 1986) shows that this process can occur independently of freeze–thaw cycling and suggests that crack expansion through ice segregation is most effective within the range −4 °C to −15 °C; at higher temperatures, ice-segregation pressures may be too low to sustain crack growth, and at lower temperatures water migration to sites of ice-lens formation is inhibited. Experimental evidence provides support for pore-water migration in freezing rock (Prick *et al.*, 1993; Prick 1995), and laboratory and field observations (Akagawa and Fukuda, 1991; Mackay, 1999b) indicate the efficacy of ice segregation as a mechanism for fracturing sound rock, but suggest that effective temperatures for ice-segregation fracturing of some lithologies may be slightly higher than −4 °C (Matsuoka, 1990a). Hallet *et al.* (1991) measured acoustic emissions produced by microfracturing within a sandstone block subjected to unidirectional freezing and thawing, and showed that 90% occurred between −3 °C and −6 °C. They noted that pressures up to 30 MPa could theoretically be generated by ice segregation, roughly twice the tensile strength of the strongest rocks.

Murton *et al.* (2001b, 2006) have convincingly demonstrated the effectiveness of ice segregation as a mechanism of rock breakdown. They kept the lower halves of large blocks of unsaturated chalk at temperatures below 0 °C (simulating permafrost) and subjected the upper halves to temperature cycling around 0 °C, simulating annual freeze–thaw of the active layer. Measurements of the resultant heave produced evidence for an initial phase of limited rock dilation attributable to microcrack development, followed by much more rapid heave due to fracturing and macrocrack formation. Rock-surface heave during both freezing and thawing cycles was observed, incompatible with fracture by volumetric expansion but consistent with sustained ice segregation in parts of the blocks that remained frozen during thaw. Fractures formed during the experiments contained segregated ice and were aligned parallel with the freezing front. Those produced by unidirectional (top-down) freezing were concentrated close to the rock surface, reflecting upwards migration of water towards the descending freezing plane, but fractures produced by bidirectional freezing (downwards from the surface, upwards from the 'permafrost table') were concentrated in the lower part of the 'active layer' and the uppermost part of the 'permafrost', confirming that ice segregation and crack development occurred in portions of the blocks that remained below 0 °C (Figure 10.4). Significantly, the fracture pattern produced by two-sided freezing resembles that of ice-rich fractured limestones, shales and sandstones in arctic permafrost environments and of brecciated chalk in areas that experienced permafrost conditions during Pleistocene cold stages (Murton, 1996b).

Similar laboratory experiments have demonstrated that rock fracture due to ice segregation is not limited to porous sedimentary lithologies. Duca *et al.* (2014) used acoustic emissions to record microcracking within a sample of intact gneiss in the range −2.7 °C to −0.5 °C and showed that progressive propagation of microcracks generated a continuous macrocrack within the sample. Furthermore, the environmental conditions favouring ice segregation (slow freezing or sustained subzero temperatures in moist but unsaturated rock) are common in nature (Anderson, 1998). An important corollary is that rock fracture can occur *within* frozen ground, provided that unfrozen water is present and temperatures are not too low to impede water migration to sites of ice segregation. This implies that rock fracture near the base of the active layer and just below the permafrost table may occur during the summer months, when such conditions are most likely to be met.

10.2.1.3 Hydrofracture

The rock-freezing model of Walder and Hallet (1985, 1986) identified a special case where rapid cooling of saturated rock inwards from all sides results in expulsion

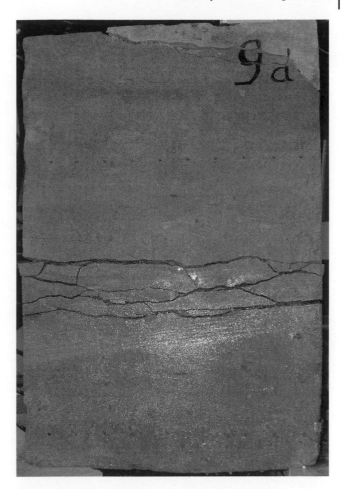

Figure 10.4 Experimental fracture of a chalk block by two-sided freezing and ice segregation. The block is 45 cm high. *Source:* Courtesy of Julian Murton.

of pore water ahead of the freezing plane. If pore water expelled into the unfrozen part of the saturated rock cannot escape, the resulting elevated pore-water pressures may lead to *hydrofracture* of the rock, causing it to split apart. Though pore-water expulsion in rock has been observed in laboratory experiments (Akagawa and Fukuda, 1991; Hallet *et al.*, 1991), the critical requirements (rapid cooling, development of a closed system and near-saturation) seem likely to limit this process in intact rock. Michaud *et al.* (1989) have described possible evidence for hydrofracture in the form of cavities in basalt bedrock, surrounded by angular fragments ejected by explosive rock disintegration, and attributed this phenomenon to sudden release of internal strain energy generated by rapid freezing of saturated rock and associated build-up of water pressure. Such violent hydrofracture of rock appears to be rare, but rapid freezing of water in a pre-existing crack or joint that has been sealed by ice may generate very high pressures (up to 10 MPa) in water trapped beyond the freezing front (Davidson and Nye, 1985; Sass, 2004), and may represent a mechanism for crack extension, joint widening and block displacement.

10.2.2 Hydration Shattering

The process of *hydration shattering* has been advocated by some researchers (White, 1976) as a possible mechanism of rock breakdown in cold environments. Hydration shattering is generated by the pressure exerted on the sides of pores by water adsorbed on to clay particles. Because the surfaces of clay particles have a slight negative charge, adsorbed water molecules become 'ordered' or aligned by polarity: the positive poles of water molecules are attracted to clay surfaces and their negative poles project away from these surfaces. Layers of water molecules adsorbed on to clay surfaces are rigid, resist freezing and are capable of exerting pressure against pore walls. The presence of like (negative) charges at the 'free' sides of clay particles on opposite pore walls means that a repulsive force is generated across rock pores. In large pores, this force is negligible. In small (<5 μm) pores, however, the pore space may be dominated by adsorbed water, generating tensile stress between pore walls. As temperatures fall below 4 °C, expansion of adsorbed water on pore walls increases tensile stress, and increased humidity caused by cooling towards and below 0 °C results in thickening of adsorbed water layers and enhanced alignment or 'ordering' of adsorbed water molecules, increasing the repulsive forces between pore walls. Recurrent changes of rock temperature around the freezing point thus produce adsorption–desorption cycles that alternately increase and relax internal stress, potentially producing rock fatigue and eventual fracture.

Laboratory experiments by Mugridge and Young (1983) and Fahey and Dagesse (1984) suggest that hydration shattering may be effective in the fracture or comminution of sorption-sensitive argillaceous and fine-grained carbonate rocks subject to repeated temperature fluctuations around 0 °C, but it is difficult to separate such effects from those of frost weathering. Hydration shattering may also play some role in the breakdown of chemically weathered rocks containing secondary clay minerals. On present evidence, it appears that hydration shattering plays a limited or negligible role in the breakdown of non-argillaceous rocks.

10.2.3 Thermal Stress

On heating, rock tends to expand by an amount determined by its coefficient of linear expansion (ψ), which commonly lies in the range 10^{-5}–$10^{-6} °C^{-1}$ at temperatures close to 0 °C. On cooling, rock contracts by a similar amount. Thus, a 10 m long sandstone slab with $\psi \approx 10^{-5} °C^{-1}$ expands by 1 mm when heated slowly and uniformly by 10 °C, and contracts by the same amount when cooled by 10 °C. Under natural circumstances, heating and cooling of rock surfaces is often rapid, so

the surface temperature changes faster than that of the underlying rock, producing a steep thermal gradient in the outermost few centimetres of rock. The tendency for the rock to expand or contract during heating and cooling is thus resisted, generating compressive stresses at the surface during heating and tensile stresses during cooling. As the tensile strength of rock is much lower than its compressive strength, rapid cooling is more likely to cause fracture than rapid heating. Rapid cooling events occur in high-latitude and high-altitude environments when air temperatures are below 0 °C but solar radiation has raised rock-surface temperatures well above 0 °C. Lewkowicz (2001), for example, reported rock-surface temperatures up to 31 °C higher than air temperatures at 80° N on Ellesmere Island. If solar radiation is abruptly reduced by cloud cover, the surface layers of rock chill rapidly, creating tensile stress in the outermost parts of the rock.

Thermally-induced stresses created by impeded expansion and contraction of rock surfaces are thought to cause fracture in two ways. Recurrent low-magnitude thermal stress events may slowly weaken rock, causing *thermal stress fatigue* that ultimately results in fracture. Conversely, a single high-magnitude thermal stress event may trigger sudden tensile failure, a form of breakdown termed *thermal shock* (Hall, 1999). In practice, thermal shock is likely to affect rocks already weakened by thermal stress fatigue.

Though thermal weathering has been mainly associated with rock breakdown in hot deserts (McKay *et al.*, 2009), the concept has been embraced by researchers working in cold arid regions such as Antarctica, where lack of moisture inhibits frost weathering (Hall, 1997a, 1998a, 1999; Hall and André, 2001). The case for thermal stress as a weathering agent in such environments rests largely on measurements of rock-surface and subsurface temperatures. Such measurements show that during the summer months, rock surfaces frequently experience temperature changes $\geq 2 °C\ min^{-1}$ during short-lived warming and cooling events. Citing Yatsu (1988), Hall and co-workers accepted $2 °C\ min^{-1}$ as the threshold for thermal shock, but the validity of this assumption has been convincingly challenged (Boelhouwers and Jonsson, 2013). The operation of thermal stress in cold, arid environments may nevertheless be represented by centimetre-scale polygonal crack patterns on boulder surfaces (Figure 10.2c), apparently consistent with tensile fracture by thermal stress (Hall, 1999), though a review by Hall and Thorn (2014) suggests that an *orthogonal* crack system may be the 'one true expression' of thermal shock in rock. Modelling of the fragmentation of boulders in the hyper-arid environment of the McMurdo Dry Valleys in Antarctica provides possible evidence for thermal weathering, as large boulders that experience marked internal thermal gradients appear to have released small clasts much more frequently than the resulting (approximately isothermal) clasts have themselves fragmented (Putkonen et al., 2014).

The possible role of thermal stress in causing granular disaggregation has been evaluated by Hall and André (2003), who investigated temperature change at grain scale on rock surfaces in Antarctica. At this scale, temperature changes $\geq 2\,°C\,min^{-1}$ are frequent, with unidirectional thermal changes of nearly $10\,°C\,min^{-1}$ and rapid bidirectional (cooling plus warming) changes of up to $22\,°C\,min^{-1}$ being recorded. Further grain-scale temperature measurements obtained from Antarctica by Hall et al. (2008) highlight the complexity of differential heating of individual minerals, suggesting that changing intergranular stress due to rapid thermal changes may cause granular disaggregation, though the link between rapid surface temperature change and grain detachment remains to be demonstrated.

10.2.4 Salt Weathering

Salt efflorescences (surface accumulations) and subflorescences (salt accumulations in cracks and voids) have been widely documented on rocks in arid polar environments (Campbell and Claridge, 1987; Gore et al., 1996), suggesting that *salt weathering* of rock occurs in such areas. At least 30 different types of salt have been recorded in the cold deserts of Antarctica (Keys and Williams, 1981), of which the most common include sodium chloride (halite), sodium sulphate (thenardite and mirabilite), calcium sulphate (gypsum) and magnesium sulphate (epsomite and hexahydrite). Such salts are either of meteoric origin or produced by chemical weathering of rocks and soils. Salt is believed to cause rock breakdown in several ways, notably through growth of salt crystals within pores or microcracks, swelling of anhydrous salts by absorption of water (salt hydration) and thermal expansion and contraction of salt crystals (Ruedrich and Siegesmund, 2007).

Stresses generated by salt crystal growth represent the most important mechanism of salt weathering. Salt crystal growth occurs in rock pores or microcracks due to supersaturation of saline water through evaporation or cooling, potentially generating crystallization pressures sufficient to fracture or disaggregate rock (Goudie and Viles, 1997). Steiger (2005), for example, calculated that crystal growth from a 105% sodium chloride solution can theoretically generate a pressure of 15 MPa, which exceeds the strength of most rocks. The effectiveness of salt crystallization as a weathering agent depends on the degree of supersaturation and the location (surface or subsurface) at which it occurs, which in turn reflect salt type and evaporation rate (Rodriguez-Navarro and Doehne, 1999). Salt hydration and consequent expansion may also generate stress within rock. At 0 °C and >60%

humidity, for example, the rapid transition of thenardite (Na_2SO_4) to mirabilite (the hydrated form of sodium sulphate, $Na_2SO_4 \cdot 10H_2O$) is accompanied by a volume change of about 300%. For salts with a hydration phase, therefore, hydration rather than crystallization pressure may be the dominant cause of weathering. Finally, some salt crystals also have moderately high coefficients of expansion (that of halite is roughly four times that of quartz), suggesting that rocks containing salt crystals may be more susceptible to thermal stress. Irrespective of the mechanism involved, experimental evidence indicates that salt weathering is most effective in salt-rich sedimentary rocks with high porosity, high capillarity and low tensile strength. Because salt weathering operates at the grain and microcrack scale, its main effects are disaggregation and rounding of rock surfaces, flaking or exfoliation, and small-scale fracture.

Salt weathering has been promoted as an important agent of granular disaggregation and flaking in cold, arid environments, where lack of moisture inhibits frost weathering but the dryness of the air encourages evaporation. Weathering phenomena developed on granite in southeast Ellesmere Island and on the Sør Rondane Mountains of Antarctica include rounded rock outcrops, mushroom-shaped pedestal rocks, *tafoni* (deep hollows in rock surfaces), *weathering pits* (pan-shaped depressions on outcrop surfaces) and accumulations of grus (granular rock debris), features attributed by Watts (1983) and Matsuoka (1995) to very prolonged weathering by salt crystallization (Figure 10.5). Salt weathering has been associated with the formation of tafoni and related cavernous weathering features in Antarctica by several other investigators, though biological weathering by lichens, abrasion by windborne grains and thermal stress may play some role in the development of such features (Campbell and Claridge, 1987; André and Hall, 2005; Guglielmin et al., 2005; Hall and André, 2006; Strini et al., 2008; Figure10.5c). On the Sør Rondane Mountains, where weathering of rock has occurred for over a million years, biotite gneiss bedrock has been reduced by salt weathering to sulphate-rich silty soil 30–40 cm thick (Matsuoka et al., 2006). In the same area, small cubes of very porous welded tuff soaked in halite and thenardite solutions disintegrated in just 5 years, though cubes soaked in pure water or gypsum solution exhibited little damage (Matsuoka et al., 1996). Collectively, the available evidence suggests that salt crystallization and possibly other forms of salt weathering represent a dominant weathering process under very arid cold conditions, but are probably of secondary importance in wetter periglacial environments.

Laboratory studies of the effects of combined 'frost-and-salt' weathering (e.g. Jerwood et al., 1990; Williams and Robinson, 1991, 2001) suggest that some salts enhance frost weathering of porous sedimentary rock

Figure 10.5 (a) Tafoni in sandstone, Ellesmere Island, Canadian Arctic Archipelago. (b) Weathering pits on a granite tor, Cairngorm Mountains, Scotland. (c) Advanced cavernous weathering of a boulder, Dry Valleys, Antarctica. *Source:* Courtesy of (b) Adrian Hall and (c) Peyman Zawar-Reza.

but that others have negligible effect or even reduce rock breakdown. To some extent, this contrast in response may reflect differences in experimental protocols, and it is questionable whether experiments involving rapid freezing of cut-rock samples that have been soaked or immersed in saturated solutions of a particular salt replicate natural circumstances. Under such circumstances, a mixture of different salts is normally present, leading to ionic interactions, so that both solubility and crystallization pressures differ markedly from those of the individual constituent salts. Some salt combinations seem to enhance rock breakdown, whereas others reduce weathering effects (Robinson and Williams, 2000), but such behaviour is incompletely understood in the context of combined frost-and-salt weathering (Williams and Robinson, 2001). Moreover, most saline solutions freeze at temperatures slightly below 0 °C, implying that the weathering effectiveness of a mild freezing regime may be reduced by the presence of salt. For example, sodium chloride has a eutectic temperature of −21.2 °C, implying that at higher temperatures some unsaturated sodium chloride solution persists in liquid form in association with ice. Such behaviour may enhance ice segregation by facilitating water migration through rocks at low temperatures, but this remains to be tested.

10.2.5 Stress Release and Tectonic Stress

Most rocks originate deep underground, under the weight of overlying rock. When such rocks are exposed at the surface, this overburden stress is released, allowing slight expansion of nearsurface rock. In most environments, such *stress release* is very gradual, reflecting removal of overburden over millions of years, and is partly relieved by gradual rock creep. In areas formerly occupied by Pleistocene ice sheets, however, unloading of rock masses from under the weight of the overlying ice occurred over a few millennia, resulting in release of strain energy induced by ice loading, and comparatively rapid 'rebound' (dilation) of glacially-loaded rock masses. As most rocks have limited elasticity, rapid deglacial unloading results in rock fracture and propagation of the internal joint network. Such paraglacial stress release is thought to be a cause of rock-slope failure and enhanced rockfall activity in recently deglaciated terrain (Chapter 12). The most important effect of stress release in formerly glaciated periglacial environments is probably the formation of joints that may be exploited by frost wedging and hydrofracture, and of microcracks that focus rock breakdown by ice segregation or salt crystallization. Joints and microcracks formed by tectonic stress (Amadei and Stephansson, 1997) have a similar function.

10.3 Chemical Weathering Processes

Rocks are composed of minerals, usually in crystalline form, and all crystals have a unique structure known as a crystal lattice. Decomposition of minerals like quartz, feldspar and olivine by chemical weathering causes lattice breakdown: chemical bonds in the original lattice are broken and new chemical bonds are formed. Such changes occur when minerals come into contact with percolating rainwater (which is usually slightly acidic), soil water and groundwater (which are often more acidic), and organic acids produced by plant decay or secreted by organisms such as lichens. Chemical weathering is important because it transforms relatively insoluble minerals into comparatively soluble substances that are removed in solution, and because it reduces the strength of minerals and intergranular bonds, making rocks more vulnerable to mechanical breakdown.

The chemical reactions involved in crystal lattice breakdown are incompletely understood, but a small number of reaction types dominate most chemical weathering processes. *Oxidation* is a reaction between minerals and oxygen. It particularly affects ferromagnesian minerals, forming rust-coloured *weathering rinds* (thin, oxidized bands at rock surfaces). *Reduction* is the opposite: the removal of oxygen ions from minerals, usually in anaeorobic conditions below the water table. *Hydrolysis* involves a chemical reaction between water (or weak acids) and silicate minerals such as feldspars. It involves dissociation of water molecules into hydrogen (H^+) ions and hydroxyl (OH^-) ions, and substitution of the cations (positively charged ions) of calcium (Ca^{2+}), magnesium (Mg^{2+}), sodium (Na^+) and potassium (K^+) in the crystal lattice of minerals with H^+ ions. The cations thus released recombine with hydroxyl (OH^-) ions to form hydroxides, such as calcium hydroxide ($Ca(OH)_2$) and sodium hydroxide (NaOH), which are relatively soluble and readily removed in solution. Within the rock, cation replacement results in the formation of *secondary clay minerals*, such as kaolinite and illite, that are mechanically weaker than the original mineral, making the rock less resistant to physical weathering. *Carbonation* is the reaction of calcium carbonate (calcite) and magnesium carbonate with weak carbonic acid formed through the combination of water with carbon dioxide. This reaction produces calcium bicarbonate and magnesium bicarbonate, which are readily removed in solution, and underlies the so-called 'solution' of carbonate rocks such as limestone.

Solution (or *dissolution*) occurs when all or part of a mineral becomes dissolved in water percolating though rock or sediment. Most minerals are only very slightly soluble in water: crystalline quartz (silica, SiO_2), for example, has a saturated solubility of $12\,mg\,l^{-1}$. However, chemical reactions such as oxidation, hydrolysis and carbonation create products that are much more soluble than the original minerals. The readiness with which any particular ion enters solution is termed its *mobility*, and it depends on the form in which the ion occurs. For example, the sodium (Na^+) ion is much more mobile in the form of sodium chloride (NaCl) than in sodium hydroxide (NaOH). The mobility of individual ions also varies with temperature and the acidity of soil water or groundwater.

The rate at which chemical weathering occurs reflects climatic, chemical and mineralogical controls. Precipitation regime dictates the propensity for *leaching*, the removal of dissolved solids by surface and subsurface runoff. Temperature controls the rate at which chemical reactions occur. Chemical weathering ceases at temperatures below the freezing point; above the freezing point, reaction rate tends to increase as temperature increases. The chemical environment within rocks and soil also influences weathering rates. A particularly important consideration is pH, which is a negative logarithmic measure of hydrogen ion (H^+) concentration, and thus of the acidity of percolating water. Rainwater and snow meltwater are usually slightly acidic (pH typically 5–7), but the pH of soil water is often further reduced by addition of organic acids produced by decay of plant matter. At low pH (4–5), the mobility of cations is greatly increased and acidity enhances hydrolysis and carbonation. The effect of pH, however, varies from mineral to mineral; silica, for example, becomes more soluble as pH increases. Another important chemical influence on weathering rate is *chelation*, which occurs when a mineral cation bonds with an organic acid, particularly fulvic acid produced by chemical breakdown of humus in soils. Chelation greatly increases the mobility of certain mineral constituents, particularly silica, iron and aluminium.

Though the rates at which minerals are chemically altered in a leaching environment are determined partly by the chemistry of that environment, it is possible to identify a rough scale of susceptibility to chemical weathering amongst the common minerals. Those that crystallized under high temperatures tend to be most vulnerable to chemical decomposition. Thus, olivine and pyroxene are often least stable; feldspar, biotite, albite and muscovite usually have intermediate stability; and quartz is often the most stable. Differential chemical weathering of minerals tends to weaken rock structure, so that unweathered minerals are more readily detached by physical weathering. Only where very pure calcareous rocks (chalk, limestone and dolomite) are decomposed by carbonation is near-total removal of weathering products in solution accomplished, leaving a residue of insoluble impurities. In many rocks, however, chemical weathering alters the original (primary) minerals into new (secondary) minerals that are chemically more stable. The most important secondary minerals are clay minerals such as illite and kaolinite, and amorphous (noncrystalline) hydrous oxides of iron, aluminium, silica and titanium. These are chemically inert, but mechanically weaker than the minerals from which they were derived.

10.3.1 Chemical Weathering in Cold Environments

Though *a priori* considerations (low temperatures, prolonged ground freezing, slow humic decay and, in arid areas, sparse vegetation and lack of a leaching environment) suggest that chemical weathering is likely to be limited in periglacial environments, there has long been evidence to the contrary. Rapp (1960) presented data suggesting that solutional loss may be the principal agent of denudation in the mountains of northern Sweden, and the importance of solution in cold environments was confirmed by measurements of dissolved calcium and magnesium in springs, rills and rivers draining limestone terrain in the Canadian High Arctic (Cogley, 1972; Woo and Marsh, 1977) and Rocky Mountains (Ford, 1971). Tedrow and Krug (1982) showed that slopewash deposits below dolomite ridges in the Arctic include a range of secondary clay minerals, as well as carbonate precipitates indicative of solution of rock farther upslope. Dixon *et al.* (1984) identified several chemical weathering processes operating on a nunatak in Alaska, including solutional grain-size reduction of grus, loss of mobile cations and alteration of biotite to vermiculite.

Evidence for chemical weathering of rock and soil abounds in cold environments, in the form of salt efflorescences, weathering rinds, rock coatings and rock surface pitting, but direct measurement of chemical weathering processes is difficult. Recent studies have approached this problem in a number of ways, particularly (i) hydrochemical studies of solute concentrations in runoff, (ii) investigation of clay mineral content in regolith, (iii) measurement of differential weathering of minerals, and (iv) investigation of the effects of weathering on rock surfaces.

Numerous hydrochemical studies have been carried out of solute concentrations in surface waters in cold environments (e.g. Caine, 1992b; Lewkowicz and Wolfe, 1994; Meiklejohn and Hall, 1997; Campbell *et al.*, 2001, 2002; Keller *et al.*, 2007; Krawczyk and Pettersson, 2007;

Beylich *et al.*, 2004, 2005; Beylich and Laute, 2012). Some studies focus on a calculation of the average basin denudation rate implied by removal of dissolved solids, others on spatial or temporal variability in solute concentration or solute load, and a few on hydrochemical budgets (Buttle and Fraser, 1992). All show that solute removal constitutes a significant component of sediment transport by streams, but net rates of solute removal rarely exceed $10\,t\,km^{-2}\,a^{-1}$, except on carbonate rocks or where chemical weathering is enhanced by unusual levels of acidity (Thorn *et al.*, 2001). Such rates are low in comparison with those measured in humid temperate environments, in part due to the restricted runoff season in cold environments. Rapid snowmelt and rainfall tend to dilute solute concentrations, which are usually highest in springs, where percolating groundwater or soil water has been in prolonged contact with mineral surfaces and has achieved near-saturation with respect to particular ions (e.g. Darmody *et al.*, 2000; Beylich *et al.*, 2005). Measured solute concentrations in many areas are therefore mainly a reflection of water movement through soil, rather than through rock. Thus although the available hydrochemical data indicate the operation of widespread (if low-intensity) chemical weathering in cold environments, they throw little light on the role of chemical activity in rock breakdown, or on the nature of weathering processes. Enhanced solute concentrations, however, occur near the base of the active layer and in the underlying transient layer in permafrost environments. These have been attributed by Kokelj and Burn (2005) to expulsion of dissolved solutes from freezing water during active-layer freeze-back, though Lacelle *et al.* (2008) have emphasized the role of greater water availability above the permafrost table in promoting geochemical aqueous reactions.

Studies of the secondary mineral content of *in situ* periglacial regolith (weathered rock debris) provide a more direct way of assessing the long-term effects of chemical weathering, but are largely restricted to investigation of blockfields that escaped glacial erosion during the Late Pleistocene, and possibly for much longer (Ballantyne, 1998a; Allen *et al.*, 2001; Marquette *et al.*, 2004; Paasche *et al.*, 2006; Goodfellow *et al.*, 2009, 2014b; Hopkinson and Ballantyne, 2014). Such studies have detected a wide range of clay minerals indicative of significant chemical alteration, but these have probably formed over very long timescales under a range of climatic conditions, and may not be representative of present-day weathering processes.

Differential weathering of minerals exposed on bedrock surfaces permits assessment of relative weathering rates, particularly where a protruding mineral, such as glacially polished vein quartz, has experienced negligible lowering since deglaciation. For glacially-abraded rock surfaces in northern Sweden, André (2002) showed that postglacial surface lowering has averaged $\sim0.2\,mm\,ka^{-1}$ on granite and metaquartzite, ~1.0 mm ka^{-1} on biotite-rich crystalline rocks and ~5 mm ka^{-1} on carbonate sedimentary rocks. Similarly, for ice-scoured metamorphic rocks on the Hardangervidda in Norway, Nicholson (2009) calculated rates of Holocene surface lowering of 0.05–2.20 mm ka^{-1}, with a mean of 0.55 mm ka^{-1}, and glacially-abraded outcrops of pyroxene-granulite gneiss in Jotunheimen, Norway, have yielded average Holocene lowering rates of 4.8 ± 1.0 mm ka^{-1} (Matthews and Owen, 2011). Biochemical weathering may be largely responsible for differential weathering of the crystalline rocks at these sites, but it is possible that physical weathering (granular disaggregation) has also contributed to rock-surface lowering. Differential chemical weathering can also be detected at the scale of individual grains. McCarroll (1990) found that on exposed gneiss bedrock in Jotunheimen, feldspar minerals protrude on average 0.24 mm above adjacent pyroxenes, but that on rock exhumed from under soil cover, the pyroxenes rise on average 0.43 mm above the feldspars. McCarroll attributed the preferential weathering of iron-rich pyroxene on exposed rock to chelation by organic acids released from lichens. Such acids are unlikely to have affected the exhumed rock, which would have been buffered from the effects of chelation by the pyroxene-rich soil, and here weathering of feldspars (probably by hydrolysis) has dominated. Minimum rates of surface loss inferred from differential weathering of pyroxenes and feldspars are very low: roughly 0.03 mm ka^{-1} for the exposed rock and 0.05 mm ka^{-1} for the buried rock.

Studies of the effects of chemical weathering on rock surfaces using electron microprobe analyses of chemical composition, derivation of chemical weathering indices (Darmody et al., 2005; Goodfellow et al., 2009) and electron microscopy to investigate the effects of chemical alteration within rock offer promising methods for establishing relationships between chemical alteration and rock decomposition, but have rarely been employed in a periglacial context. An exception is a study by Dixon et al. (2002) of weathering rinds and rock coatings in Karkevagge, northern Sweden. This research demonstrates that rinds on schists in this area are the products of a range of processes, including iron accumulation and selective oxidation, hydration and dissolution of certain mineral components (Ca, K, Si, Mg), leading to increased porosity, grain disaggregation, microfracture and cementation by precipitates. Dixon et al. (2002) also recorded a range of accretionary rock coatings, including iron films, silica glazes, alumina glazes, sulphate crusts and gypsum efflorescences. Other studies in the same area (Dixon et al., 2001; Thorn et al., 2002) attempted to determine rates of weathering by placing polished rock discs or pebble-sized rock fragments in different surface settings or at varying depths in soils and measuring mass loss over several years. Such experiments demonstrate more

rapid mass loss in carbonate rocks compared with granites, and show some relationships with vegetation cover and pH, but provide limited information on processes or long-term weathering rates, particularly since chemical weathering rates change through time as more mobile constituents are removed (Keller et al., 2007), or as chemically-altered rocks become more susceptible to mechanical disintegration.

The studies described above formed part of a research campaign that employed a wide range of approaches to investigate chemical weathering processes, effects and rates in Karkevagge. This programme included measurements of dissolved solids in surface waters (Darmody et al., 2000; Campbell et al., 2001, 2002) and of mineral alteration in soils (Allen et al., 2001), and studies of weathering rinds, rock coatings and efflorescences (Darmody et al., 2002; Dixon et al., 2002) and of weathering of the tops of large boulders, pebbles at the soil surface and polished rock discs (Dixon et al., 2001; Darmody et al., 2005; Thorn et al., 2002). Initial results were summarized by Thorn et al. (2001), and a fuller summary of findings has been placed in the context of chemical weathering of mid-latitude alpine environments by Dixon and Thorn (2005) and Thorn et al. (2011). Two features emerged from this research. First, it highlights the importance of mineral composition in determining weathering rates. The schists of Karkevagge contain local concentrations of finely disseminated pyrite (FeS$_2$), which has oxidized to produce weak sulphuric acid (H$_2$SO$_4$), thereby lowering the pH of the soil and water and facilitating various reactions: weathering of muscovite to illite; release of silica, aluminium and potassium into groundwater; oxidation of iron to produce rock coatings; breakdown of calcite, releasing calcium into groundwater; and precipitation of gypsum (CaSO$_4$) on rock surfaces. The presence of pyrite thus greatly (if locally) enhances weathering and the release of dissolved solids above levels expected for this environment. This influences concentrations of total dissolved solids measured in the river draining the valley (39–52 mg l^{-1}) and the denudation rate based on solute concentrations, estimated by Darmody et al. (2000) to be 19.2 t km^{-2} a^{-1}. For comparison, the equivalent figures for the nearby Latnjavagge drainage basin are 12–18 mg l^{-1} and 4.9 t km^{-2} a^{-1} respectively (Beylich et al., 2004, 2005; Beylich, 2011), and are probably more representative. The second general outcome of this programme is that it underscores the difficulty of integrating weathering studies using different approaches. Though it is difficult to argue with the general conclusion that 'chemical weathering must be viewed as a potentially important contributor to landscape development in cold regions' (Thorn et al., 2001), it remains difficult to assess the relative contributions of chemical weathering and physical weathering to rock breakdown in cold environments.

10.4 Biotic Weathering Processes

Most research on biological weathering in cold environments concerns the role of rock-colonizing micro-organisms such as algae, lichens and fungi. The limited data available suggest that under favourable circumstances these may enhance chemical alteration of minerals (*biochemical weathering*) and even detach grains and flakes of rock (*biomechanical weathering*), making rock surfaces more vulnerable to other weathering and erosive processes.

Lichens are widespread on rock surfaces in periglacial environments. Several types are thought to be responsible for superficial weathering of rock, though lichen–rock interactions remain poorly understood (Ascaso *et al.*, 1990). Some epilithic (surface) lichens secrete acids that enhance chemical alteration of surface and nearsurface minerals (McCarroll, 1990; Arocena *et al.*, 2003), and endolithic lichens growing within rock may also cause biomechanical breakdown. A study of the effects of the endolithic lichen *Licidea auriculata* on gabbro in a subarctic mountain environment (McCarroll and Viles, 1995) demonstrated that this lichen colonizes rapidly, affecting almost all boulders on moraine crests within 240 years. It penetrates up to a millimetre into rock, detaching miniature flakes and reducing the hardness of rock surfaces. By measuring the volume of rock detached from affected surfaces, McCarroll and Viles calculated a minimum rate of rock removal of $0.0012\,\mathrm{mm\,a^{-1}}$; extrapolated over a millennial timescale ($1.2\,\mathrm{mm\,ka^{-1}}$), this represents a significant contribution to rock breakdown. Conversely, precipitation of hematite in rock voids as a result of the biochemical action of epilithic lichens has been shown by Guglielmin *et al.* (2011b) to contribute to case hardening of the outer surfaces of tafoni in Antarctica. Given the ubiquitous nature of lichens in cold environments, further study of their biochemical and biomechanical effects seems warranted.

The same is true for algae and fungi. Observations by Hall and Otte (1990) suggest that chasmolithic (crack-dwelling) algae play a significant role in the breakdown of granite on nunataks rising above the Juneau Icefield in Alaska. These authors suggested that alternating hydration (expansion) and dehydration (contraction) of the mucilage polymer sheath of algae causes opening of microcracks and detachment of surficial rock flakes, and that this process may result in the loss of several hundred grams of rock per square metre each year. Fungi have been shown by Etienne (2002) and Etienne and Dupont (2002) to be implicated in the formation of weathering rinds on basaltic boulders in Iceland. These authors showed that organic acids released by fungi enhance rind formation through oxidation of ferromagnesian minerals, etching of crystals, dissolution of minerals and causing an increase of rind porosity. Microcracks develop at the boundary between the rind and underlying sound rock, and are opened up by ice-lens growth or colonization by other organisms, causing detachment of rind fragments 1–3 mm thick (Figure 10.6). Arocena *et al.* (2003) observed that weathering of granitic rocks on the Qinghai-Tibet Plateau has been enhanced by growth of fungal hyphae along cleavage planes in chloritic minerals.

10.5 Weathering Processes in Periglacial Environments

Given the complexity of weathering processes and the diversity of approaches employed in their study, generalization concerning the efficacy of different processes is difficult. Several points nevertheless deserve emphasis. The first is that there is overwhelming field and experimental evidence that frost weathering makes a major contribution to rock breakdown in all periglacial environments where water is available for freezing in rock. Ice segregation causes fracture, flaking and brecciation in porous sedimentary rocks, though whether such effects extend to low-porosity resistant lithologies remains to be demonstrated. Volumetric expansion of water on freezing remains the most plausible mechanism

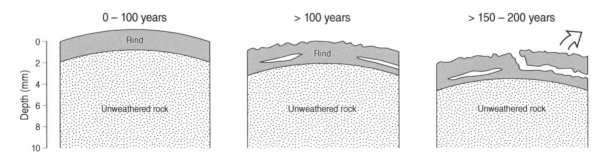

Figure 10.6 Detachment of weathering rinds in basalt. (a) Development of a weathering rind ~2 mm thick. (b) Frost weathering removes the outer part of the rind. Surface roughening increases and microcracks form within the rind. (c) Microcrack enlargement by ice-lens growth, fungi, endolithic lichens or algae leads to rock flaking. *Source:* Adapted from Etienne (2002). Reproduced with permission of Elsevier.

for joint enlargement and block detachment, and almost certainly causes granular disaggregation and consequent edge-rounding of outcrops and boulders, though the processes responsible for such rounding probably extend beyond frost weathering alone. The role of hydrofracture remains contentious: the evidence base is slight, but this process may yet prove important in crack extension and joint enlargement.

Second, other physical weathering processes operating in cold environments are probably restricted by lithology or climate. Hydration shattering appears unlikely to have a significant role in the breakdown of nonargillaceous rocks. Present evidence suggests that though salt weathering may be predominant in causing granular disaggregation of rock in arid, polar environments, and possibly in coastal regions, elsewhere it is probably of secondary importance. Despite powerful advocacy, the case for thermal stress as a major weathering agent in periglacial environments remains to be conclusively demonstrated.

Third, there is clear evidence that chemical weathering is important in cold climates, at least where precipitation is sufficient to create a leaching environment with resultant removal of dissolved solids in runoff. Measurements of solute concentrations in springs and rivers show that chemical weathering makes a significant and sometimes dominant contribution to sediment removal from drainage basins (Chapter 13), but the role of chemical weathering in rock breakdown in periglacial environments remains obscure. The few studies that have addressed this topic suggest that the main role of chemical weathering may be to reduce rock strength through mineral alteration and increased rock porosity, so that rocks are more vulnerable to physical disintegration. It is also difficult to provide a realistic assessment of the role of biotic weathering, though the handful of pertinent studies suggest that this may play a more important role in 'preparing' rock surfaces for mechanical breakdown (granular disaggegation or flaking) than has hitherto been acknowledged.

The remainder of this chapter considers landforms produced by cold-climate weathering processes. We begin with an account of landforms that have developed through the operation of solution processes on carbonate rocks in arctic, subarctic and alpine environments.

10.6 Cold-climate Karst

The term *karst* is used to describe the distinctive landforms in areas where solution of carbonate rocks (such as limestone and dolomite) is a dominant process. As noted earlier, the 'solution' process involves reaction of calcium and magnesium carbonates with weak carbonic acid to produce calcium and magnesium bicarbonates, which are readily removed in solution. Apart from surface and nearsurface solution microforms, collectively referred to as *karren*, the defining characteristic of karst landscapes is the development of an efficient underground drainage system (Ford and Williams, 2007). Surface waters enter underground drainage through a variety of routeways or sinkholes, notably *dolines*, which are hollows formed by solution or collapse of a subterranean chamber, and *poljes*, which are large, enclosed, flat-floored depressions subject to seasonal or occasional flooding. Subsurface water flow may be diffuse, through pores and fissures, or may follow a cave system formed by solutional widening of the joint network, emerging as springs (resurgences). Over time, solutional widening and collapse of subterranean drainage routeways may isolate residual rock remnants as *karst pinnacles* or *karst towers*. Additionally, in limestone areas where surface rivers have incised rapidly into bedrock, *fluviokarst gorges* are formed. Dry valleys lacking surface drainage are a further common feature of karst topography.

Limestone and dolomite underlie large swathes of the Canadian Arctic Archipelago, but here karst effects are limited to surface phenomena, such as small solution pits, karren and solutionally widened joints, because continuous permafrost usually restricts solution to the surface and active layer, though groundwater flow within carbonate rocks has been reported for an area of continuous permafrost in northern Yukon (Utting *et al.*, 2012). Limestone caves in unglaciated parts of northern Yukon appear to have developed prior to the Quaternary (Lauriol *et al.*, 1997a). Farther south, in areas of thinner or discontinuous permafrost, signs of underground drainage become evident. Numerous closed depressions, ephemeral lakes and karst springs occur in dolomite and gypsum near the continuous–discontinuous permafrost boundary west of Great Bear Lake at latitude 64–69° N (Van Everdingen, 1981; Hamilton and Ford, 2002), and a remarkable assemblage of karst features occurs on Akpatok Island (60° N). These include dry valleys, sinkholes, dolines and poljes, but resurgences are rare, suggesting that surface waters descend through vertical conduits within permafrost to emerge as submarine springs (Lauriol and Gray, 1990; Gray and Lauriol, 1993). It is difficult to envisage how such underground drainage routeways could have been initiated within permafrost under present conditions, suggesting that subterranean drainage is inherited from a time when permafrost was absent, possibly under warm-based glacier ice (Ford, 1987, 1996). The most remarkable karst landscape in subarctic Canada, however, is the Nahanni karst in the Mackenzie Mountains (62–63° N), an area of discontinuous permafrost. This is thought to have escaped glaciation for at least 125 ka, and comprises a *karst labyrinth* of gorges, dolines and isolated towers (Ford, 1973; Brook and Ford, 1978). The highest

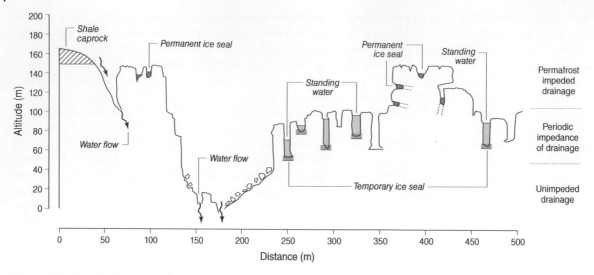

Figure 10.7 Model of karst morphology in rugged terrain underlain by discontinuous permafrost. *Source:* Adapted from Ford (1987). Reproduced with permission of Wiley.

parts lie within the zone of permafrost, and here caves and sinkholes are blocked with ice. Low ground represented by poljes and the floors of sinkholes and gorges displays unimpeded drainage, implying that permafrost is absent. The intervening zone is one where ice seals the bottom of dolines and sinkholes, until the weight of accumulating water causes rupture and rapid underground drainage (Figure 10.7).

Remarkably, limestone dolines containing active sinkholes occur at 78°N, below the Holocene marine limit, near the west coast of Svalbard (Salvigsen and Elgersma, 1985), and imply free drainage to the base of the permafrost. It seems likely that underground drainage at this site was initiated as the sea receded, prior to local permafrost development, and that geothermally-heated springs have since maintained an open talik. Large-scale karst landforms also occur on the carbonate rocks of the middle Lena River Basin in central Siberia (59–61°N), on a plateau underlain by continuous and discontinuous permafrost. These features include underground-draining lakes on watershed locations, funnel- and bowl-shaped sinkhole depressions, dry riverbeds and karst pinnacles. Resurgences associated with taliks on lower slopes feed the Lena River and its tributaries, and solutional microforms are numerous on limestone outcrops. Spektor and Spektor (2009) suggested that underground drainage through permafrost in this area is maintained by heat transfer from summer rainfall infiltrating into the rock and heat released by condensation of water vapour in karst hollows. The maintenance of unfrozen water conduits within continuous permafrost nevertheless remains an unresolved problem (Chapter 7), and the underground drainage networks in such areas are probably inherited from a time when permafrost was discontinuous or absent (Ford, 1987, 1996; Haldorsen *et al.*, 2010).

Such problems of karst development do not arise in areas of discontinuous or sporadic permafrost, where water movement is possible in open taliks. In alpine mountains, karst phenomena are widespread on carbonate rocks (Figure 10.8). Karst features on Castleguard Mountain at the margin of the Columbia Icefield, for example, include not only karren on outcrops at all altitudes, but also vertical shaft-like or funnel-shaped sinkholes at altitudes over 2000 m and numerous springs at lower altitudes, though the famous Castleguard cave system (which extends several kilometres under the icefield) is mainly a relict feature (Ford, 1983). In alpine and subalpine environments there is also evidence that frost weathering plays an important role in widening the mouths of caves developed in carbonate bedrocks through detachment of grains and small clasts from the cave roofs and walls (Oberender and Plan, 2015).

In sum, though only superficial karst phenomena are present in high-arctic locations underlain by deep, continuous permafrost, there is abundant evidence in the northern hemisphere for a wide range of karst phenomena related to underground drainage in areas of thin or discontinuous permafrost, as well as in alpine mountains. These stand as testimony to the efficiency of limestone and dolomite solution processes in cold environments, traditionally viewed as the most challenging for karst landform development.

10.7 Tors

A *tor* is a residual mass of bedrock that rises above its surroundings, is isolated by free faces on all sides and has been formed by differential weathering of bedrock

Figure 10.8 Alpine karst features at 2470 m in the French Alps. (a) Small dolines (sinkholes) formed by roof collapse of an underlying cave system. (b) Solution runnels (karren) formed along a rock joint.

and removal of weathered products from the surrounding ground. Tors occur in all climates and on a wide range of lithologies, but are particularly well developed on granitic rocks. They vary in height from a few metres to more than 40 m. Some, particularly on granites, are rounded, with 'woolsack'-shaped boulders; some are tabular; and others present a shattered, angular appearance (Figure 10.9). In present and former periglacial environments, tors commonly occur on plateaus, where *summit tors* crown the high ground, *scarp-edge tors* stand proud of cliffs or shoulders and *valley-side tors* rise from rubble-mantled slopes. They are absent from areas that experienced severe glacial erosion during the last glacial stage, implying that tor emergence has occurred over timescales greater than the 9–16 ka since the downwastage of the last Pleistocene ice sheets.

Tors 'grow' above the surrounding terrain because the rate of weathering and erosion of the surrounding regolith-covered ground exceeds that of tor surfaces. Such differential weathering often reflects lower joint density within tors than under adjacent terrain, but it may also be due to localized differences in mineral composition or bedrock structure (Goodfellow *et al.*, 2014a). Although differential weathering leading to tor emergence in mid- and high-latitude locations has traditionally been attributed to frost action, it is now evident that a range of weathering processes is implicated. Salt weathering appears to have been important in the development of tors in polar deserts (Selby, 1972; Watts, 1983; Matsuoka, 1995), and elsewhere there is

evidence that chemical weathering has contributed to tor formation (Fahey, 1981; Darmody *et al.*, 2008). The presence of tafoni and weathering pits on many tors confirms that processes other than frost weathering have been active in tor formation. Once a tor surface has emerged from the surrounding regolith, however, tor growth may be enhanced by more rapid weathering of rock buried under the regolith (which retains water) than of the exposed tor surface (which sheds water). Equally important is the comminution and stripping of regolith surrounding tors (Bjornson and Lauriol, 2001; Ballantyne, 2010), regolith removal being accomplished by solifluction, wash, wind erosion, removal of chemically-weathered products in solution and, in some cases, limited glacial erosion.

Some granite tors in cold environments appear to have existed throughout much of the Quaternary. A vertical cosmogenic nuclide age profile obtained for a granite tor on the Cairngorm plateau in Scotland indicates a tor emergence rate of 31 mm ka^{-1}; taking into account periods of glacier ice cover, this rate implies that the largest tors on the plateau (Figure 10.9a) began to emerge up to 1.3 million years ago (Phillips *et al.*, 2006), and is consistent with exposure age data that indicate tor surface ages >600 ka. Similarly, granite tors in northern Finland appear to have been in existence for nearly 1.0 Ma (Darmody *et al.*, 2008). Smaller tors and those developed on less resistant bedrock are likely to be younger. Cosmogenic ^{10}Be dating of samples from the tops of granite tors on Dartmoor in southwest England has yielded apparent exposure ages

Figure 10.9 (a) Granite tor, Cairngorm Mountains, Scotland. (b) Angular schist tor, Old Man Range, South Island, New Zealand. (c, d) Tors in interior Alaska. A colour version of (a) appears in the plates section. *Source:* Courtesy of (b) Barry Fahey and (c,d) Matthias Siewert.

that cluster around 36–50 ka, suggesting Late Pleistocene tor emergence (Gunnell *et al.*, 2013), but may be partly compromised by burial under glacier ice during the last glacial maximum (LGM) and rapid rates of tor surface erosion.

Though some tors in former periglacial environments have developed in unglaciated terrain (Figure 10.9b–d), many others survived successive Quaternary glaciations under a cover of cold-based glacier ice that was frozen to the underlying substrate. In northern Scandinavia and on the Cairngorm Mountains of Scotland, for example, erratic boulders occur on or near tors, implying tor burial under ice during at least one period of glaciation (André, 2004; Hall and Phillips, 2006). Cosmogenic nuclide exposure ages obtained for tors in Fennoscandia and Scotland confirm tor preservation under cold-based glacier ice during multiple glacial cycles (Stroeven *et al.*, 2002; Phillips *et al.*, 2006; Darmody *et al.*, 2008). Briner *et al.* (2003) provided conclusive evidence of the survival of small tors on Baffin Island under the last Laurentide Ice Sheet by showing that though tor surfaces record

exposure ages >60 ka, erratics dumped on tors by the downwasting ice sheet give exposure ages of 17 ka or younger.

Syenite and granite tors that were formerly buried by glacier ice exhibit a variety of forms, ranging from essentially intact examples, through glacially-modified tors, to low-lying slabs or plinths from which the original tor superstructure has been removed by glacial erosion (André, 2004; Hall and Phillips, 2006; Figure 10.10). Tors exhibiting different degrees of glacial modification sometimes occur in close proximity, suggesting that the thermal regime of the overlying ice was highly variable, with 'cold patches' where the ice was frozen to the underlying substrate alternating with zones of wet-based sliding ice. It is ironic that plateau tors, once thought to be diagnostic of nunataks that escaped burial by the Pleistocene ice sheets, have now been shown to provide potential not only for constraining the minimum (rather than the maximum) altitude of former ice sheets, but also for throwing light on the thermal regime of these long-vanished glaciers (Hall and Sugden, 2007).

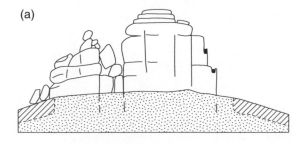

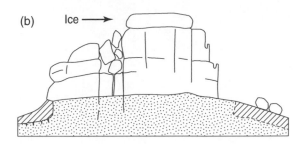

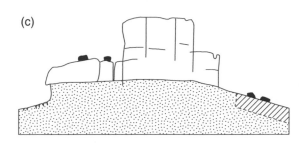

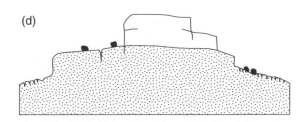

Figure 10.10 Model of progressive glacial modification of granite tors. (a) Unmodified tor, with loose and toppled blocks, and weathering pits on the top and flanks. (b) Tor exhibiting slight glacial modification, with loss of fragile superstructure and removal of exposed blocks and regolith (oblique shading) from the stoss side. (c) More advanced glacial modification, characterized by a residual monolith, displacement of blocks and erratics (black) resting on tor surfaces. (d) Residual tor, comprising a tor plinth rising above glacially-abraded bedrock, with evidence of glacial plucking on the lee side. *Source:* Hall and Phillips (2006). Reproduced with permission of Wiley.

10.8 Blockfields and Related Periglacial Regolith Covers

10.8.1 Classification and Characteristics

The most conspicuous manifestation of rock weathering in present and former periglacial environments is a cover of bouldery *regolith* (weathered rock debris) that mantles the underlying bedrock (Figure 10.1). Such regolith covers are common on mid- and high-latitude mountains, and are sometimes referred to as *felsenmeer* or *mountain-top detritus*. They are particularly well represented on plateaus that have escaped significant glacial erosion, particularly in Scandinavia, Scotland, Svalbard, Greenland, arctic Canada, Labrador and Antarctica (Goodfellow, 2007; Rea, 2013). The most distinctive examples take the form of *blockfields*, a term that implies a cover of coarse openwork bouldery debris with no fine sediment at the surface (Figure 10.11a). On many lithologies, however, the regolith forms a stony *diamicton*, comprising clasts embedded in a matrix or infill of fine sediment (Figure 10.11b). In reality, 'true' blockfields and diamict regoliths form end-members of a continuum of weathering response, so to avoid terminological complexity the term *blockfield* is here employed to refer to all bouldery regolith covers in present or former cold environments, and those lacking fine sediment at the surface are hereafter termed *openwork blockfields*. The latter tend to have formed on resistant but well-jointed igneous and metamorphic rocks that have weathered to produce predominantly boulder-sized debris, whereas diamictic blockfields tend to be typical of weaker rocks that have broken down to produce a mixture of clasts and abundant fine sediment.

It is also useful to discriminate between autochthonous (*in situ*) blockfields and allochthonous (externally derived) block deposits. *Autochthonous blockfields* develop through weathering of bedrock with negligible downslope displacement of weathered debris, and so exhibit abrupt changes in clast lithology at geological contacts, preserve veins or dykes as lines of surface boulders and show a downwards transition through fractured rock to sound rock (Figure 10.12). *Allochthonous block deposits* take various forms, including *blockslopes* or *debris-mantled slopes* of bouldery rubble that has migrated downslope to cover hillslopes, and *blockstreams*, where an allochthonous openwork boulder deposit has accumulated on valley floors.

10.8.2 Autochthonous Blockfields

Most reported depths for autochthonous blockfields vary from a few decimetres to about 1.5 m, though some excavations have achieved depths of 3 m without reaching bedrock. Structural characteristics vary, but excavations often reveal a downwards transition from an openwork, clast-supported or matrix-supported bouldery upper layer to an underlying zone of fractured bedrock, sometimes with an intervening layer dominated by fine sediment (Rea *et al.*, 1996; Ballantyne, 1998a; Paasche *et al.*, 2006; Goodfellow *et al.*, 2009; Figure 10.12). Exposed boulder surfaces may exhibit rounding by granular

Figure 10.11 (a) Pit excavated in an openwork quartzite blockfield on An Teallach, a mountain in northwest Scotland. Exposed boulder surfaces are slightly rounded by granular disaggregation, and fine sediment underlies the openwork layer. (b) Pit excavated in a sandstone blockfield on the same mountain.

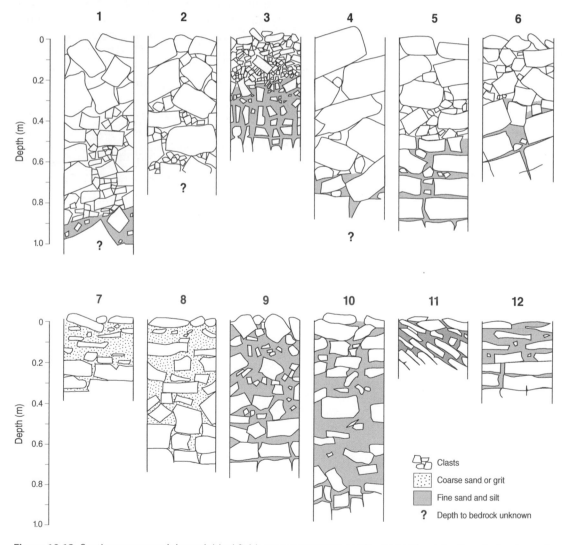

Figure 10.12 Sections excavated through blockfields on mountains in northwest Scotland. Sections 1–6 represent openwork blockfields on quartzite (1,2), granulite (3), microgranite (4) and gneiss (5,6). Sections 7 and 8 are sandy diamictons on arkosic sandstones. Sections 9–12 represent frost-susceptible silt-rich diamictons developed on granulite (9,10) and mica-schist (11,12). *Source:* Ballantyne (1998a). Reproduced with permission of Wiley.

disaggregation, though subsurface clasts are typically angular. The size and shape of the largest clasts are often related to the density of joints and fractures in the underlying bedrock, and mean clast size sometimes declines with depth. Tors or rock outcrops protrude through many blockfields, and some blockfields support ground patterning in the form of relict ice-wedge polygons or large-scale sorted circles and nets. Ballantyne (1998a) classified autochthonous blockfields in Scotland into three types: (i) openwork blockfields, with an infill of fine (<2 mm) sediment at some depth below the surface, typically developed on well-jointed resistant lithologies (Figure 10.11a); (ii) sandy diamictons, in which clasts are embedded in a non-frost-susceptible cohesionless sandy matrix, typically developed on sandstones and some granites (Figure 10.11b); and (iii) silt-rich frost-susceptible diamictons, which often show evidence of vertical frost sorting of clasts, typically developed on schists. The three categories are transitional, reflecting the relative abundance of fine sediment, the silt content and the degree of vertical sorting of clasts (Figure 10.13). Secondary clay minerals are often present in the fine-sediment fraction of blockfields (Rea *et al.*, 1996; Ballantyne, 1998a; Marquette *et al.*, 2004; Paasche *et al.*, 2006; Goodfellow *et al.*, 2009). Though clay is usually a very minor component of blockfield granulometry, the presence of secondary clay minerals such as gibbsite and kaolinite implies some degree of chemical weathering.

10.8.3 Age and Origin: The Blockfield Enigma

An intriguing feature of autochthonous blockfields is that almost all are apparently relict, even in present-day

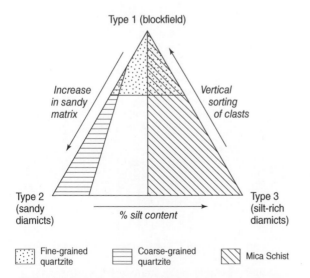

Figure 10.13 Relationships between different types of blockfield on mountains in northwest Scotland, showing the range of regolith types characteristic of fine-grained quartzite, coarse-grained quartzite and schist. *Source:* Adapted from Ballantyne (1998a). Reproduced with permission of Wiley.

permafrost environments. Surface boulders are commonly covered by mosses and lichens, and sometimes by thin soils or peat, and a contrast between the rounding of exposed boulders and the angularity of subsurface boulders suggests prolonged surface stability. Thin blockfields have developed through frost wedging of fissile limestones in the Canadian Arctic during the Holocene (Dredge, 1992), but in many permafrost areas glacially-abraded rock outcrops have experienced little modification since the retreat of the last ice sheets, and some still retain erosional microforms such as striae and chattermarks. Though disruption of ice-scoured bedrock by frost heave of joint-bound blocks occurs in some permafrost areas (Dionne, 1983; Dyke, 1984; Michaud and Dionne, 1987; Michaud and Dyke, 1990), these effects are localized and have not produced regolith covers analogous to the blockfields on mid- and high-latitude plateaus.

Two theories have been advanced to explain blockfield evolution. The first is that blockfields are inherited from chemically-weathered regolith (saprolite) that initially formed under warmer and wetter conditions during the Neogene but has been modified by frost action (particularly upfreezing of boulders) under periglacial conditions during the Pleistocene (e.g. Nesje, 1989; Rea *et al.*, 1996; Marquette *et al.*, 2004; Whalley *et al.*, 2004; Fjellanger *et al.*, 2006; Paasche *et al.*, 2006). Four arguments have been advanced to justify this interpretation: (i) blockfields often occur on palaeosurfaces of (apparent) Neogene age; (ii) secondary clay minerals such as gibbsite and kaolinite indicate advanced chemical weathering; (iii) comminution of fine sediment to silt also suggests chemical weathering; and (iv) the lack of evidence for blockfield formation on glacially-abraded bedrock during the Holocene suggests that recent frost action has been ineffective in producing blockfields, even on plateaus underlain by permafrost. Others, however, have favoured a Pleistocene age and periglacial origin for plateau blockfields, arguing that frost weathering has been the dominant agent of blockfield formation (Ballantyne, 1998a), an interpretation suggested by the characteristic angularity of subsurface clasts and by the absence of rounded corestones, chemically-rotted boulders or saprolite indicative of advanced chemical weathering. This view has been supported by Goodfellow *et al.* (2009, 2014b) and Hopkinson and Ballantyne (2014), who showed that the fine matrix within blockfields in northern Sweden and Scotland is essentially a product of physical weathering, being characterized by low clay content, minor elemental losses, limited development of secondary clay minerals, mixing of primary and secondary clay minerals, and absence of chemical etching of primary minerals. A wider survey of periglacial regolith covers by Goodfellow (2012) showed that these are distinguished by low clay:silt ratios, and that the presence of secondary clay minerals cannot be considered diagnostic of a pre-Quaternary origin. Conversely, the structural and matrix characteristics of the blockfields investigated by Ballantyne (1998a), Goodfellow *et al.* (2009) and Hopkinson and Ballantyne (2014) suggest that the blockfields they examined (and probably most others) formed through exploitation of stress-release joints by frost wedging, and that the fine matrix is predominantly a product of granular disaggregation.

It is possible to reconcile these conflicting viewpoints by considering the dynamic nature of plateau surfaces. The progressive emergence of tors during the Pleistocene (Phillips *et al.*, 2006; Darmody *et al.*, 2008) and rates of alpine bedrock erosion, summit lowering and regolith production calculated using cosmogenic isotopes (Small *et al.*, 1997, 1999; Goodfellow *et al.*, 2014b) suggest that Neogene plateau surfaces have been lowered by a few tens of metres during the Pleistocene through progressive regolith formation and removal. Such surface lowering implies that regolith covers have undergone continuous slow renewal, as losses through surface erosion are offset by lowering of the weathering front at the regolith–rockhead contact. Survival of a pristine Neogene regolith cover is therefore extremely unlikely, though formation of blockfields from rock modified by Neogene weathering remains plausible. If surface lowering during the Pleistocene has progressed below the maximum depth of Neogene weathering, then a periglacial blockfield dominated by physical weathering products is the likely outcome. However, unless a plateau surface has experienced complete regolith removal by glacial erosion at some time during the Pleistocene, the present blockfield cover is inevitably inherited from earlier, vanished regolith covers, and traces of a Neogene ancestry may or may not be present (Ballantyne, 2010; Figure 10.14). As Goodfellow *et al.* (2009) pointed out, blockfield development is likely to occur at the base of the active layer, where water is abundant and underlying bedrock is subjected to an annual freeze–thaw cycle or fracture by ice segregation. It follows that apparently relict blockfields in permafrost environments may still be undergoing subsurface weathering by frost action and other agents. This concept also helps to explain the lack of Holocene blockfield development on glacially-abraded bedrock, where there is no inherited fracture network and where water runs off at the surface instead of concentrating (and freezing) at the base of a regolith cover.

10.8.4 Blockfields and Glaciation: Periglacial Trimlines

Plateau blockfields often extend to the edge of glacial troughs, or are restricted to high ground above the altitudinal limit of landforms produced by glacial erosion during the last glacial maximum (LGM) of 26.5–19.0 ka. In some areas, the boundary between the two is defined by a *periglacial trimline* that delimits the approximate upper level to which glacial erosion has removed or 'trimmed' the regolith cover (Figure 10.15). Such trimlines have formed in two ways. Some, for example in the

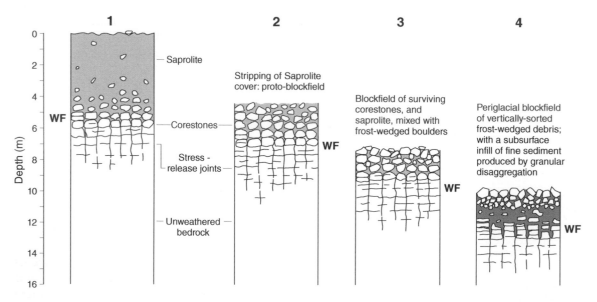

Figure 10.14 Model of blockfield evolution through progressive lowering of plateau surfaces during the Quaternary. WF, weathering front, at which bedrock weathering is focused. (1) Pre-Quaternary weathering profile, comprising saprolite (chemically-weathered regolith) and rounded corestones. (2) Stripping of regolith, leaving a lag of corestones with residual saprolite. (3) Initiation of Quaternary frost weathering: modification of remnant pre-Quaternary regolith and frost heave of surviving corestones and frost-weathered boulders. (4) Mature periglacial blockfield, formed by frost weathering at the weathering front and frost heave of boulders during successive Quaternary cold stages. *Source:* Ballantyne (2010). Reproduced with permission of Wiley.

Figure 10.15 Periglacial trimline on Mount Hoffman, Yosemite, USA. The trimline marks the boundary between a lower zone of glacially-scoured bedrock and the summit blockfield, and represents the maximum altitude of the Tioga Glaciation icefield (last glacial stage). The blockfield zone remained above the icefield as a palaeonunatak.

Southern Alps of New Zealand or Yosemite in California, are *ice-marginal trimlines*, representing the maximum altitude of glacier ice at the LGM, with blockfields surviving on nunataks that remained above the ice (Ballantyne, 2013b). In many other areas, however, blockfields exhibit clear evidence of having survived glaciation under a cover of cold-based glacier ice. Such evidence takes the form of erratics and 'perched' glacially-transported boulders resting on blockfields (Figure 10.16), meltwater channels incised into the regolith cover or ablation moraines dumped on the blockfield surface (Dredge, 2000; Marquette *et al.*, 2004; Fjellanger *et al.*, 2006). The juxtaposition of landforms indicative of vigorous glacial erosion at lower altitudes and blockfields preserved under ice on adjacent plateaus is explained by the thermal regime of former ice sheets. On low ground, and particularly within glacial troughs, wet-based, sliding ice at pressure melting point has resulted in removal of pre-existing regolith and widespread subglacial abrasion and quarrying of bedrock. Conversely, on high ground, cold-based glacier ice below pressure melting point has moved by internal deformation only, without basal sliding. This is because the adhesive strength of the ice–substrate interface exceeds basal shear stress when the ice is below its pressure melting point, so that a cover of cold-based ice within an ice sheet tends to preserve rather than erode underlying blockfields (Kleman and Glasser, 2007; Fabel *et al.*, 2012). In such cases, *englacial trimlines* mark the thermal boundary between cold- and warm-based ice within the former ice sheet. Mature plateau blockfields are therefore indicative of terrain that escaped erosion by glacier ice at the LGM, and possibly throughout the Pleistocene, either on former nunataks or under a protective cover of cold-based ice.

10.8.5 Allochthonous Block Deposits: Blockslopes and Blockstreams

Allochthonous block deposits occur as *blockslopes* (or *debris-mantled slopes*), comprising a mantle of coarse debris on hillsides with gradients up to about 35°, and *blockstreams*, which consist of a gently-sloping cover of coarse openwork rubble that occupies valley floors and tends to be elongated along the valley axis. Blockslopes are distinguished from talus slopes (Chapter 12) in that the debris mantle is not emplaced by rockfall, but has

Figure 10.16 Glacially-transported erratic boulder resting on a granite blockfield on Grytøya, arctic Norway.

migrated downslope through some combination of frost creep, gelifluction, debris flow and possibly permafrost creep. Blockslopes often merge upslope with autochthonous blockfields, and are sometimes contiguous downslope with valley-floor blockstreams. Some blockslopes appear to have developed through downslope movement of debris since the LGM (Ballantyne and Harris, 1994, pp. 176–178), but others are much older. Barrows *et al.* (2004), for example, showed that the exposure age of boulders on a relict blockslope in Tasmania increases with distance from the parent scarp from ~22 ka to ~498 ka, suggesting slow and probably intermittent downslope movement of boulders over half a million years.

Blockstreams have attracted much more attention, on account of their striking appearance (Figure 10.17). Some famous examples, such as the Blue Rocks, Hickory Run and Devil's Racecourse blockstreams in the Appalachian Highlands comprise fields of massive openwork boulders, some several metres in length, that are tightly wedged together (Potter and Moss, 1968; Clark and Ciolkosz, 1988). Blockstreams exhibit considerable morphological diversity: most consist of expanses of openwork boulders, but some exhibit pits, elongated depressions, longitudinal furrows and boulder steps, ridges and lobes, features that may represent melt-out or flow structures similar to those on relict rock glaciers. Blockstreams are typically composed of well-jointed igneous and metamorphic rocks such as basalt and quartzite. The openwork boulder layer is commonly 1–3 m thick, and inverse grading is common, with the largest boulders at the surface. Two features of such coarse blockstreams is that they are manifestly relict, and that all documented examples occur outside the limits of

Late Pleistocene glaciation. Active, mobile debris mantles or *kurums* occur in arctic Russia (Romanovskii and Tyurin, 1986) and at high altitude in the Kunlun Mountains of China (Harris *et al.*, 1998), but these are of limited thickness (0.3–0.4 and 0.15 m, respectively) and appear to be no more than distant cousins to the thick bouldery relict blockstreams described elsewhere.

Superb examples of blockstreams occur on the Falkland Islands (52° S) in the South Atlantic. Pleistocene glaciation of these islands was limited to cirques, so most of the terrain was exposed to periglacial conditions during successive Quaternary cold stages. The term 'stone runs' has been applied both to valley-floor blockstreams and to large relict sorted stripes that extend upslope to bedrock scarps or plateau blockfields and merge downslope into the blockstreams. The latter are typically 1–2 km long and several hundred metres wide; the largest, quaintly named Princes Street after the main shopping thoroughfare in Edinburgh, is 4 km long and about 400 m wide. The Falklands blockstreams are composed of an openwork cover of large (typically 0.5–1.5 m long) subangular quartzite boulders (Figure 10.17) that often exhibit conspicuous inverse grading and overlie a stony diamict. Several theories have been proposed to explain these deposits, but most researchers have accepted that first advocated by Andersson (1906), who proposed that the blockstreams developed under periglacial conditions though frost weathering of bedrock, downslope transport of the resultant rubble by solifluction, vertical frost sorting of debris and immobilization through eluviation of fine sediment. However, noting evidence for advanced chemical weathering of component boulders and the underlying diamict, André *et al.* (2008) have suggested a more complex origin.

Figure 10.17 Quartzite blockstream on the Falkland Islands. The boulders are up to 2 m long. *Source:* Courtesy of Jim Hansom.

This involves regolith formation by chemical weathering during the Neogene, late Neogene stripping and downslope movement of regolith, soil formation under temperate conditions and finally downslope movement, accumulation and frost-sorting of debris during Pleistocene cold stages. Optically-stimulated luminescence ages obtained by Hansom *et al.* (2008) on the diamict sediments underlying Falkland blockstreams range from >54 ka to ~16 ka, suggesting that block movement was active until the end of the last glacial stage. However, cosmogenic [10]Be exposure ages obtained for boulder surfaces on two blockstreams by Wilson *et al.* (2008) range from ~182 ka to ~731 ka, demonstrating that boulder accumulation and movement extended back to the Middle Pleistocene. Boulders on the striped blockslopes that 'feed' the blockstreams yielded generally younger [10]Be exposure ages (~42 ka to ~398 ka), consistent with more recent derivation from upslope sources. The Falkland blockstreams are thus ancient features that achieved their present extent over several Pleistocene cold stages, and possibly longer.

Relict blockstreams similar to (but generally smaller than) those in the Falkland Islands occur in various areas that formerly experienced periglacial conditions, for example in the Appalachian Mountains (Clark and Ciolkosz, 1988; Whittecar and Ryter, 1992; Park Nelson *et al.*, 2007), the Lesotho Highlands in South Africa (Boelhouwers *et al.*, 2002), the Snowy Mountains of Australia (Caine and Jennings, 1968) and the Ligurian Alps in Italy (Firpo *et al.*, 2006). Not all are necessarily as old as the Falkland blockstreams; six boulders on a basalt blockstream in the Snowy Mountains, for example, yielded cosmogenic [36]Cl exposure ages within the range ~17 ka to ~23 ka, suggesting that this blockstream may be no older than (or was last active during) the LGM (Barrows *et al.*, 2004). Most blockstreams have been attributed to release of boulders from bedrock by frost wedging and subsequent downslope and downvalley movement of coarse bouldery debris under periglacial conditions, but the nature of such movement remains speculative. One school of thought attributes boulder movement to solifluction within a matrix of fine sediment, subsequently removed by wash; others favour frost creep as a mechanism; others still have suggested sliding failure due to generation of high pore-water pressures during thaw of the underlying fine sediments (Wilson, 2013). An important consideration is that mean annual temperatures within openwork bouldery debris are often a few degrees lower than those in adjacent soils or ambient mean annual air temperatures (MAATs)

(Harris and Pedersen, 1998; Gorbunov *et al.*, 2004; Juliussen and Humlum, 2008; Chapter 4), so that deep seasonal freezing or permafrost may have been locally present under blockstreams during periods of activity. Extension of freezing into the underlying stony diamict enhances the possibility that movement (by frost creep, gelifluction or even permafrost creep) was essentially generated below the openwork boulder layer, with the latter being carried passively downslope or downvalley on underlying mobile sediments. Inverse grading has usually been attributed to frost heave, though in a mobile openwork deposit this may reflect differential movement ('jostling') of boulders: once a large boulder reaches the surface, it stays there because voids are too small to allow its descent. Given the uncertainties attending the origin of relict blockstreams and related allochthonous block deposits, it is premature to conjecture their palaeoclimatic significance, though calculations based on the elevation of Appalachian examples (Park Nelson *et al.*, 2007) suggest that these formed under conditions conducive to the development of permafrost.

10.9 Brecciated Bedrocks

Though there is an extensive literature devoted to such impressive weathering products as blockfields and blockstreams, much less attention has been given to the effects of cold-climate weathering on relatively weak sedimentary rocks. This anomaly may reflect the fact that resulting regolith is rather unremarkable in appearance, or is often buried under alluvial, colluvial or aeolian deposits. An exception is a study by Murton (1996b) of brecciated (fragmented) chalk exposed in sea cliffs in southern England (Figure 10.18). Nearsurface brecciation of sedimentary rocks in former periglacial environments has usually been attributed to the operation of frost weathering in seasonally frozen ground or the active layer. Murton, however, noted that experimental evidence (e.g. Lautridou and Ozouf, 1982) shows that repeated freezing and thawing of chalk results in breakdown into granular debris, not brecciation. Moreover, he observed a close resemblance between the brecciated chalk of southeast England and brecciated ice-rich

Figure 10.18 Brecciated chalk bedrock partly overlain by loessic deposits, Sussex, southern England. The brecciation becomes progressively coarser with depth, probably reflecting a former decrease in ground ice with depth during conditions of Pleistocene permafrost. *Source:* Courtesy of Julian Murton.

limestones, sandstones and shales in present-day permafrost environments (e.g. French *et al.*, 1986). Pursuing the analogy, he argued that brecciation of the chalk had resulted from the formation of segregated ice lenses in permafrost immediately below the level of the former permafrost table. Subsequent experimental work (Murton *et al.*, 2006) appears to vindicate this interpretation. An important consequence of this work is that brecciated chalk and other nearsurface frost-susceptible brecciated bedrocks provide evidence of former permafrost, and may thus contribute to palaeoenvironmental reconstructions. Murton and Lautridou (2003) have suggested that formerly ice-rich brecciated bedrocks in the English Channel coastlands were probably vulnerable to rapid erosion by running water, resulting in rapid valley development under permafrost conditions. These ideas have potential applicability in all former permafrost areas underlain by frost-susceptible bedrock.

Figure 4.13 Cave ice in a limestone cave, Picos de Europa, northern Spain. *Source:* Courtesy of Bernard Hivert.

Figure 5.1 Ground ice: ice wedges at the top of the photograph penetrate downwards into ice-rich frozen silts that overlie a body of stratified massive ice. *Source:* Courtesy of Matthias Siewert.

Periglacial Geomorphology, First Edition. Colin K. Ballantyne.
© 2018 John Wiley & Sons Ltd. Published 2018 by John Wiley & Sons Ltd.

Figure 5.6 Cryostructures in cores extracted from ice-rich permafrost: (a) lenticular; (b) bedded; (c) reticulate; (d) ataxitic. *Source:* Courtesy of Yuri Shur.

Figure 5.16b Yedoma deposits on the Itkillik River, northern Alaska. Syngenetic ice wedges separate stratified silts. *Source:* Courtesy of Mikhail Kanevskiy.

Figure 6.1 High-centre frost polygons and thermokarst lakes, central Banks Island, Canadian Arctic Archipelago. Individual polygons form subsets of a larger irregular polygonal net. *Source:* Courtesy of Tobias Ullmann.

Figure 6.4a Two ice wedges of different widths in the western Canadian Arctic. The wedges are ~2 m long. *Source:* Courtesy of Matthias Siewert.

Figure 7.1 Hydrostatic pingos, Tuktoyaktuk Peninsula, western Canadian Arctic. *Source:* Courtesy of Emma Pike.

Figure 7.2 Fløtspingo, Reindalen, Svalbard, a hydraulic pingo formed where groundwater flowing under pressure has fed the growth of an ice core within permafrost, updoming the overlying frozen sediments.

Figure 7.10 A complex of palsas and lithalsas growing amid thermokarst ponds in northern Quebec. The rounded mounds in the centre of the photograph lack peat cover (lithalsas), but nearby mounds are peat-covered (palsas). The circular thermokarst ponds are the sites of collapsed palsas and lithalsas. *Source:* Adapted from Calmels *et al.* (2008). Reproduced with permission of Elsevier.

Figure 8.1a Thermokarst lakes and thermokarst ponds, central Banks Island, arctic Canada. The outlines of submerged frost polygons can be seen on the floor of the central lake, suggesting that it formed recently. *Source:* Courtesy of Tobias Ullmann.

Figure 8.3 Small thermokarst lake, central Banks Island, arctic Canada. Degradation of ice-wedge polygons near the lake margins has formed thermokarst mounds. Fens have colonized the lake margins. *Source:* Courtesy of Tobias Ullmann.

Figure 8.9 Thermokarst bog, Alaska. Toppling of trees at the bog margins is due to permafrost degradation. *Source:* Courtesy of Torre Jorgenson.

Figure 8.11a Large retrogressive thaw slump (megaslump) along the Selawik River, northwest Alaska. Over 700 000 m³ of sediment was evacuated from this slump between its formation in 2004 and its stabilization in 2016. *Source:* Courtesy of Benjamin Crosby.

Figure 9.5 Large annular sorted circles, Vardeborgsletta, Svalbard. *Source:* Courtesy of Ole Humlum.

Figure 10.9a Granite tor, Cairngorm Mountains, Scotland.

Figure 11.7a Solifluction sheet advancing over bedrock, southern French Alps. *Source:* Courtesy of Pascal Bertran.

Figure 11.14 Two recent active-layer failures on the Fosheim Peninsula, Ellesmere Island, Canadian Arctic Archipelago. An older vegetated failure separates the two. The tops of ice wedges are visible in the polygons exposed by the recent failure on the left. *Source:* Courtesy of Antoni Lewkowicz.

Figure 12.1a Coalescing talus cones below rock gullies, Svalbard. The lower-gradient cone is a debris cone fed by debris flows and snow avalanches. *Source:* Courtesy of Ole Humlum.

Figure 12.7 Downslope transport by debris flows of quartzite debris across darker sandstones, Beinn Eighe, Scotland. The levées of some individual flow tracks are clearly evident.

Figure 12.9c Glacigenic rock glaciers, Holar Valleys, Iceland. Note residual glacier ice at the headwall. *Source:* Courtesy of Ole Humlum.

Figure 13.9c Braided river channels on a periglacial sandur, central Banks Island. *Source:* Courtesy of Tobias Ullman.

Figure 14.10 Late Weichselian aeolian sand facies in northwestern Europe. (a) Facies 1, of Late Pleniglacial age. Massive bedding and alternating sand and loamy sand beds with synsedimentary deformations. The black band separates the deformed OCS I from the undeformed OCS II. (b) Facies 2, of Late Pleniglacial age (OCS II). Alternating bedding of fine sand (lighter) and loamy fine sand to sandy loam (darker), with crinkly lamination and small-scale injection features caused by deposition on a wet surface. (c) Facies 3, of Younger Dryas age (YCS II); horizontal and low-angle cross-bedding in fine to medium sand deposited on a dry surface in source-proximal dunes. (d) Facies 4, of Younger Dryas age (YCS II), showing large-scale dune slipface cross-bedding in fine to medium sand in source-proximal dunes. *Source:* Courtesy of Kees Kasse.

Figure 15.3 Ice-rich yedoma sediments exposed by coastal erosion, Muostakh Island, southern Laptev Sea. The large syngenetic ice wedges are thought to be of Late Pleistocene age. Coastal retreat is ~50 m a^{-1} in this area. *Source:* Courtesy of Hans Hubberten.

Figure 15.4 Retrogressive thaw slumps on the southeast coast of Herschel Island, Beaufort Sea Coast, arctic Canada. The large slump extends ~500 m inland and is ~400 m wide; the headwall reaches a height of ~25 m. *Source:* Courtesy of Hugues Lantuit.

11

Periglacial Mass Movement and Hillslope Evolution

11.1 Introduction

Hillslopes in periglacial environments differ from those elsewhere on Earth not so much in their morphology as in the distinctive processes that affect their evolution. This chapter considers some of the processes that operate on soil-mantled slopes with gradients of up to about 30–35°. Chapter 12 is devoted to processes and landforms associated with steeper slopes, particularly cliffs and cliff-foot accumulations of rockfall debris. Slope processes associated with running water are considered in Chapter 13.

The downslope movement of soil or rock without the aid of running water is termed *mass movement*, and a striking feature of periglacial environments is that mass movement sometimes occurs even on very gentle slopes. The most pervasive mass-movement process operating in cold environments is *solifluction*, the slow downslope movement of soil due to cyclic freezing and thawing of the ground. Soils in permafrost environments may also experience *permafrost creep*, the gradual downslope movement of nearsurface ice-rich permafrost, and *active layer failures*, which involve relatively rapid sliding or flow of active-layer soils over the permafrost table. Here we consider first the nature and effects of these mass-movement processes, then the processes and landforms associated with late-lying and perennial snowcover (*nivation*) and the development of stepped hillslope profiles (*cryoplanation*). The chapter concludes with a survey of slope forms characteristic of periglacial environments.

11.2 Solifluction Processes

The term *solifluction* encompasses four different components of mass movement, all of which result from cyclic freezing and thawing of soils (Matsuoka, 2001c; Harris, 2013; Figure 11.1). Of these, *needle-ice creep* is the most superficial, taking the form of elevation of a thin layer of soil and clasts on the tips of long, thin crystals of ice that form at the soil surface, and its deposition farther downslope as the crystals melt. *Frost creep* is similar, but operates to greater depth, and involves downslope movement of soil resulting from repeated cycles of volumetric expansion (frost heave) during freezing and contraction (settling) during thaw. *Gelifluction* is the slow gravity-induced shear deformation of soil associated with the generation of high pore-water pressures during thaw of ice-rich soil. *Plug-like deformation* is the somewhat clumsy term used to describe slow downslope movement of the active layer resulting from thaw of ice-rich soil in the zone immediately overlying the permafrost table. These four processes often produce distinct vertical velocity profiles (Figure 11.1), but frequently operate in some combination.

In most early studies, surface soil movement due to solifluction was recorded by measurement or survey of painted lines or stones, or of pegs set in the soil, and subsurface displacement was assessed by inserting flexible or segmented tubes vertically into the soil and measuring deformation through later excavation. More recent research has involved burial of electronic sensors linked to digital loggers, often with simultaneous measurements of surface heave, air, surface and subsurface temperature and soil moisture status (e.g. Yamada *et al.*, 2000; Jaesche *et al.*, 2003; Kinnard and Lewkowicz, 2005; Matsuoka, 2005, 2014), and sophisticated instrumentation has been developed to enable simultaneous monitoring of soil volume change, shear strain at various depths, pore-water pressures, thermal regime and snowcover (Harris *et al.*, 2007, 2011). Surface velocities are usually expressed as $mm\,a^{-1}$, and volumetric velocities as $cm^3\,cm^{-1}\,a^{-1}$. The latter represents the volume of soil that moves through a centimetre-wide across-slope transect in a year, and is calculated from the form of the vertical velocity profile. Most measurements of solifluction rates are limited to a few years, but several datasets incorporating more than a decade of annual or cumulative measurements have been published, some of which permit annual variability in movement to be linked to possible climatic controls (Price, 1990; Smith, 1992; Åkerman, 1996, 2005; Gorbunov and Seversky, 1999; Jaesche *et al.*, 2002; Ridefelt *et al*, 2009; Matsuoka, 2010; Ballantyne, 2013c).

Periglacial Geomorphology, First Edition. Colin K. Ballantyne.
© 2018 John Wiley & Sons Ltd. Published 2018 by John Wiley & Sons Ltd.

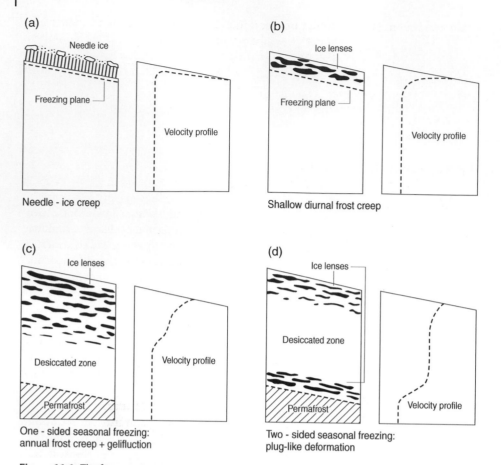

Figure 11.1 The four components of solifluction, defined by freezing regime. The left-hand diagram in each case illustrates ice distribution during diurnal (a,b) and annual (c,d) freezing, and the right-hand diagram illustrates the form of the resultant velocity profile. One-sided seasonal freezing (c) also causes annual frost creep and gelifluction even when permafrost is absent.

Solifluction processes generally operate only in frost-susceptible soils in which ice lenses develop during freezing, so it is useful to review some principles regarding freezing and thawing of moist, frost-susceptible soils. Freezing behaviour reflects the amplitude and duration of freezing cycles, the presence or absence of permafrost, soil granulometry and soil moisture content. During diurnal or other short-term freeze–thaw cycles, the freezing plane may remain stationary within a few centimetres of the surface (Figure 11.1a), allowing upwards migration of water to feed needle ice. More prolonged or more severe freezing, however, allows the freezing plane to descend deeper into the ground, forming ice lenses at shallow depth (Figure 11.1b). During annual freezing of seasonally frozen ground or of the active layer above permafrost, this process is continued to greater depth (Figure 11.1c), though ice lens formation tends to diminish with increasing depth because migration of water to feed ice-lens growth depletes moisture in the underlying unfrozen soil, leaving a desiccated zone beneath the zone of ice-lens development. Finally, where the active layer is underlain by cold permafrost, soil freezing occurs both from the top down in response to subzero air temperatures, and also slowly upwards from the permafrost table. In this situation, ice lenses develop both near the surface and near the base of the active layer, often leaving an intervening zone of desiccated soil (Figure 11.1d).

In all cases, the formation of excess ice in the soil results in a volumetric increase that causes it to heave approximately perpendicular to the soil surface, though in the case of needle-ice formation such heave is limited to a very thin layer of soil and clasts, which rapidly collapses during thaw. Thaw of soils containing ice lenses is more complex. Such soils undergo *thaw consolidation*, caused by closure of voids previously occupied by ice lenses and expulsion of the water contained in such voids (Chapter 3). Thaw consolidation results in reduction of soil volume and consequent settling of the soil under its own weight. However, as the upper layers of soil thaw, they form a low-permeability layer that limits the rate at which excess water produced by melting ice lenses can escape. If the rate of water release due to thaw exceeds the rate of water escape, the stress imposed by the weight of the overlying soil is transferred from interparticle contacts to the trapped soil water, raising pore-water pressures (u) and thus reducing the effective stress (σ_n'), which 'locks' soil particles together under the weight (W) of the overlying soil, thus:

$$\sigma'_n = W\cos\alpha - u \qquad (11.1)$$

where α is surface gradient. Reduction of effective stress lowers the shear strength of the soil, so that as pore-water pressures rise, the soil at the thaw front becomes soft and readily deformed, permitting downslope soil movement even on gentle gradients. It is this process that causes gelifluction and contributes to plug-like deformation, the former reflecting thaw consolidation as ice lenses melt in the upper parts of the soil, the latter melting of ice lenses at the base of the active layer and in the transient layer at the top of cold permafrost.

Generation of high pore-water pressures in thawing ice-rich soils thus depends on the ratio between the rate at which water is produced by thaw of soil ice and the rate at which water can escape through the overlying thaw-consolidated soil. As outlined in Chapter 3, this ratio is termed the *thaw consolidation ratio (R)*, given by:

$$R = \varphi/2\sqrt{C_v} \qquad (11.2)$$

where C_v is the coefficient of consolidation (which ranges from about $10^{-1}\,\mathrm{cm^2\,s^{-1}}$ for sandy soils to about $10^{-5}\,\mathrm{cm^2\,s^{-1}}$ for clayey soils) and φ is given by:

$$\varphi = z_t/\sqrt{t} \qquad (11.3)$$

where t is time elapsed since the onset of thaw at the ground surface and z_t is the depth of thaw at time t (McRoberts and Morgenstern, 1974). A high value of R (approaching or exceeding 1.0) indicates rapid thaw and/or slow escape of water released from melting ice,

conditions conducive to the generation of high pore-water pressures; a low value of R indicates slow thaw penetration and/or relatively rapid soil drainage, so that high pore-water pressures do not develop and soil movement by gelifluction is unlikely.

11.2.1 Needle-ice Creep

The term *needle ice* refers to long, thin ice crystals that develop at or near the ground surface in response to short-term (often overnight) ground freezing (Figure 11.2; Chapter 5). Such crystals grow perpendicular to the ground surface, often pushing up a thin layer of soil and stones. Needle ice forms through upwards migration of soil water towards a stationary freezing front in the top few centimetres of soil. Critical conditions for its formation are: (i) a rapid drop in air temperature, often due to nocturnal clear-sky radiative cooling; (ii) soil moisture availability, to feed ice-crystal growth; and (iii) frost-susceptible but permeable soils, to permit rapid water migration to the freezing plane. Needle-ice crystals grow rapidly to lengths of up to 3 cm (and exceptionally to 10 cm) in a single night, and may continue to grow if subzero air temperatures persist. As ground surface temperatures rise above 0 °C, ablation occurs at both the top and the base of the crystals, causing them to bend and topple as they melt.

Because needle ice forms in response to short-lived freeze–thaw cycles associated with overnight freezing, it is of limited importance in high-latitude environments where the annual freeze–thaw cycle predominates, but of widespread occurrence in alpine, maritime periglacial and subtropical high mountain environments, where

Figure 11.2 Needle ice crystals 12 cm long that have pushed up a layer of soil and small stones. Segmentation in the crystals indicates two periods of crystal growth uninterrupted by thaw. When the crystals melt, they deposit the uplifted soil and stones downslope from their original positions.

diurnal freeze–thaw cycles are frequent (Lawler, 1988). It is a common winter occurrence even at low altitudes in maritime mid-latitude environments, where an overnight frost can result in extensive needle-ice development.

Needle-ice creep occurs because the ice crystals heave a thin layer of soil and clasts perpendicular to the slope, but on melting deposit this sediment farther downslope. If settlement is vertical, then the amount of downslope movement (D_c) for a single freeze–thaw (heaving–settling) cycle is:

$$D_c = h \tan \alpha \qquad (11.4)$$

where h is the amount of heave and α is slope angle. However, because ice needles bend and topple in a downslope direction during thaw, they tend to deposit soil and clasts farther downslope than equation (11.4) predicts, and clasts released from thawing ice needles may roll even farther downslope, enhancing creep rates. Matsuoka (1998a) found that for slopes dominated by needle-ice creep in the Japanese Alps, cumulative creep rate over a year (D_{cum}) was related to cumulative frost heave over the same period (h_{cum}), as approximately:

$$D_{cum} = 3h_{cum} \tan^2 \alpha \qquad (11.5)$$

reflecting downslope bending of melting needle ice and rolling of released clasts. As equation (11.5) suggests, gradient is a primary control on rates of creep, along with soil texture, soil moisture availability and the frequency of short-lived freeze–thaw cycles (Holness, 2004). On unvegetated slopes affected by frequent freeze–thaw cycles, rates of surface movement can be surprisingly rapid. Pérez (1992) measured average surface creep rates of 29–95 mm a^{-1} at high altitude in the Andes, with some clasts moving up to 214 mm a^{-1}, and Francou and Bertran (1997) measured creep rates of up to 1000 mm a^{-1} at 5200 m altitude in the Bolivian Andes. Both sites experience frequent, shallow freeze–thaw cycles, and though other processes contribute to clast movement, needle-ice creep predominates. Data provided by five years of time-lapse photography and multisensor monitoring showed that unvegetated stony soil on a 21° slope on Mount Ainodake in the Japanese Alps moved downslope at an average rate of 271±79 mm a^{-1} in response to 40–70 diurnal freeze–thaw cycles per year, mainly in spring and autumn (Matsuoka, 2014). Mackay and Mathews (1974) found that on areas of sorted stripes at 1900 m altitude in British Columbia, needle-ice creep moved clasts downslope at an average rate of 150 mm a^{-1}, and fine soil at an average rate of 350 mm a^{-1}. On Marion Island in the maritime Antarctic, Holness (2004) measured rates of surface movement averaging 532 mm a^{-1} on fine-textured soils, 161 mm a^{-1} in stony areas and 26 mm a^{-1} in the coarsest debris. Needle-ice creep is active even on unvegetated hillslopes in the British Isles: Caine (1963)

recorded median rates of 70–254 mm a^{-1} on slopes of 9–19° in the English Lake District, and Ballantyne (2001a) found that clasts on a 23° slope in Scotland had moved on average 376 mm in one year.

Movement of debris by needle-ice creep is, however, confined to a thin surface layer of soil and stones, and most movement is confined to the uppermost 50 mm of the soil, effectively the thickness of the uplifted soil layer. The rapid rates of shallow solifluction recorded on mid-latitude mountains and in high-altitude subtropical mountain environments largely reflect needle-ice creep of a superficial layer of debris or soil. Slow downslope movement of soil at greater depths at such sites is mainly a result of the additional effects of frost creep, caused by formation of ice lenses during periods of deeper frost penetration (Matsuoka, 1998a, 2014). Velocity profiles for sites dominated by needle-ice creep consequently exhibit rapid decline with depth, with the uppermost soil moving much faster than the underlying layers (Figure 11.1a). Volumetrically, therefore, needle-ice creep may be less effective than other components of solifluction. Mackay and Mathews (1974) calculated that volumetric velocities at a site in British Columbia amounted to ~13 cm^3 cm^{-1} a^{-1}, and Matsuoka (1998a) calculated values of ~55 cm^3 cm^{-1} a^{-1} for a 29° slope but only ~3 cm^3 cm^{-1} a^{-1} for a 14° slope at high-level sites in the Japanese Alps.

11.2.2 Frost Creep

Frost creep is downslope movement of soil resulting from frost heave of soil perpendicular to the slope and contraction of heaved soil in a more vertical direction during thaw (Figure 11.3). During successive freeze–thaw cycles, a target on the surface therefore describes a ratchet-like trajectory. Frost creep may occur in saturated non-frost-susceptible soil, but the amount of heave is limited to that produced by freezing of pore water, and is consequently slight (<3% of unfrozen soil volume). In frost-susceptible soils, the formation of ice lenses permits much greater volumetric expansion, more pronounced frost heave and thus much more rapid frost creep (Figure 11.1). The theoretical maximum surface displacement associated with a single cycle of heaving and settling (often termed *potential frost creep*) is given by equation (11.4), which assumes vertical settling of the soil. However, surface tension effects during thaw consolidation produce apparent cohesion between soil particles, so that soil tends to settle at an angle between vertical and perpendicular to the slope. This effect is termed the *retrograde component* of soil movement (b), so the actual amount of surface creep (D_f) associated with a single frost-heave cycle is:

$$D_f = b \cdot h \tan \alpha \qquad (11.6)$$

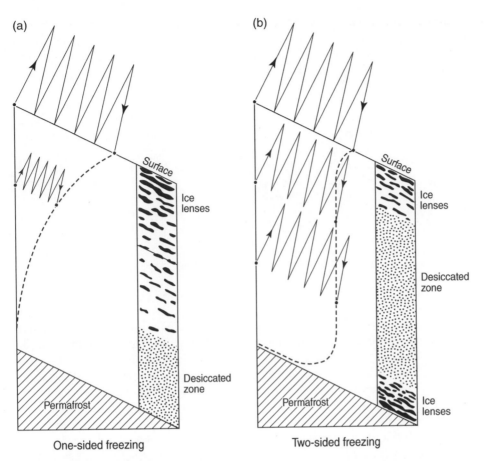

Figure 11.3 Schematic frost-creep trajectories of soil particles associated with five annual cycles of (a) one-sided and (b) two-sided ground freezing (heave) and thawing (settling). The amount of heave and resettling has been exaggerated for clarity. The dashed line represents the resultant vertical velocity profile, and the columns indicate concentration of ice lenses in the frozen soil. In (a), the amount of heave diminishes downwards, reflecting decline in excess ice volume with increasing depth. In (b), most heave results from ice segregation near the base of the active layer, producing an S-shaped velocity profile. One-sided freezing in non-permafrost terrain produces a velocity profile identical to (a).

where $0 < b < 1$. As suggested by equation (11.6), creep rates are determined by gradient (α), frequency of freeze–thaw cycles and amount of heave (h), which reflects the freezing rate (slow freezing favours ice-lens development), the frost-susceptibility of the soil, the depth of freezing and the availability of soil water to feed ice-lens growth. Matsuoka (1998b) incorporated some of these variables in a model of annual surface creep velocity (v_s):

$$v_s = x \cdot n \cdot z_f \left(silt + clay\right)\tan^2 \alpha \qquad (11.7)$$

where x is an empirical constant, n is the number of frost-heave cycles, z_f is the thickness of fine soil in the top 15 cm of the ground and ($silt + clay$) is the percentage of silt plus clay in the fine fraction of the soil. Applied to data from high-level sites in the Japanese Alps where diurnal heaving cycles predominate, this model explained 83% of the observed variance in surface velocity.

Frost creep occurs in response to both shallow freeze–thaw cycles and deep seasonal freezing. It probably reaches its maximum effectiveness on the unvegetated upper slopes of subarctic and mid-latitude mountains, where shallow but intense diurnal freeze–thaw cycles are frequent in spring and autumn (Coutard *et al.*, 1996; Matsuoka, 1998a, 2005, 2014). In such areas, surface velocities of 5–50 mm a^{-1} are typical, though the effectiveness of freeze–thaw cycles in generating frost creep is strongly dependent on moisture availability (Sato *et al.*, 1997). Velocity profiles are typically concave downslope (Figure 11.4), mainly reflecting a decrease in the frequency of freeze–thaw cycles with depth. Volumetric velocities reported by Matsuoka (1998a) for high-level sites in the Japanese Alps were 25 cm^3 cm^{-1} a^{-1} for a 29° slope and 14 cm^3 cm^{-1} a^{-1} for a 12° slope, and thus fairly similar in magnitude to those calculated for needle-ice creep at the same sites. Longer-term measurements reported by Matsuoka (2005) for high-altitude sites in the Japanese Alps ($\alpha = 14°$) and Swiss Alps ($\alpha = 10°$) yielded rather lower volumetric velocities, averaging 7.2 and 5.3 cm^3 cm^{-1} a^{-1} respectively. Shallow frost creep produced by diurnal freeze–thaw cycles at midsummer also dominates mass-movement on the Sør Rondane Mountains in Antarctica,

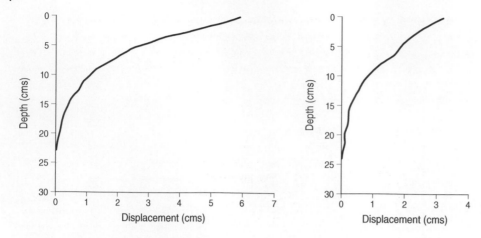

Figure 11.4 Deformation of strain probes by frost creep over 6-year periods on alpine slope crests in the Japanese Alps (left) and Swiss Alps (right), showing the characteristic concave-downward reduction in creep velocity. *Source:* Matsuoka (2005). Reproduced with permission of Wiley.

where Matsuoka and Moriwaki (1992) recorded surface creep velocities of up to 15 mm a^{-1} despite extreme aridity and very limited thaw depth.

In alpine environments at locations where a cover of coarse debris or persistent snowcover limits or precludes diurnal frost heave, frost creep is dominated by heave and settling due to deep annual freezing and thawing of the ground (Matsuoka *et al.*, 1997; Yamada *et al.*, 2000). The annual freeze–thaw cycle is also dominant in generating frost creep in most subarctic and arctic environments (Harris *et al.*, 2011). In areas of seasonally frozen ground or warm permafrost, both freezing and thawing are unidirectional from the surface downwards, producing heave during autumn freezeback and settling during spring thaw. Because ice lenses tend to be concentrated near the soil surface during annual freezing, the amount of heave declines nonlinearly with depth, and consequently creep rate declines in a similar fashion (Figure 11.3a). Though heave is most pronounced during autumn freezeback, it may continue during winter due to freezing of unfrozen water that remains in fine-grained frozen soils at temperatures below 0 °C. A second minor heave episode may also occur during thaw, due to downwards percolation of meltwater from the thawed upper layers of the soil into the still-frozen soil below, causing renewed ice segregation and slight but detectable heave (Jaesche *et al.*, 2003; Harris *et al.*, 2008b).

During seasonal freezeback in areas of cold permafrost, freezing planes descend both down from the surface and upwards from the permafrost table. Soil water migrates both upwards to feed ice-lens growth near the surface and downwards to form ice lenses at the base of the active layer, resulting in formation of an intervening zone of stiff, desiccated soil. Heave of the active layer in these circumstances therefore reflects volumetric expansion at both the top and the base of the active layer, and

in some cases the latter predominates (Mackay, 1981). During thaw, water from the uppermost thawed soil percolates downwards into the underlying still-frozen desiccated zone and refreezes near the base of the active layer (Mackay, 1983), causing slight summer frost heave. When the ice-rich soil at the base of the active layer thaws in late summer, the overlying soil settles *en masse*, contributing to plug-like deformation of the entire active layer (Figures 11.1d and 11.3b), as described below.

11.2.3 Gelifluction

Measurements of soil movement during thaw of frost-susceptible soils often show that displacement exceeds that attributable to frost creep alone, implying that some additional form of strain deformation occurs during thaw of ice-rich soil. Washburn (1967) introduced the term *gelifluction* to explain this additional movement, which he attributed to slow 'saturated soil flow' caused by an increase in the moisture content in thawing ground to the point where the liquid limits of the soil are exceeded. This explanation was initially widely accepted, but more recent research suggests that gelifluction represents elasto-plastic deformation of soil 'softened' by high pore-water pressures rather than slow viscous flow.

Understanding of the nature of gelifluction has been transformed by laboratory experiments carried out both at full scale (by freezing and thawing an artificial slope composed of soil of predetermined properties) and through scaled modelling of a tilted soil sample in a geotechnical centrifuge. The latter method allows scaling down of the thickness of the soil to 10% or less of its prototype thickness through the application of a compensatory increase in gravitational acceleration. Both approaches enjoy advantages over field-based experiments, including the ability to control environmental parameters (freezing rate, freezing temperature and moisture supply), determination of boundary conditions (soil texture and gradient) and continuous monitoring of soil temperature, frost heave, thaw consolidation, pore-water pressure and soil movement. Early experiments, summarized by Harris (1996), investigated the distribution of shear strain with depth, the effect of soil granulometry and the relative contributions of frost creep and gelifluction. Subsequent simulations, both at full scale (Harris *et al.*, 1995, 1997, 2008a; Harris and Davies, 2000) and using a geotechnical centrifuge (Harris *et al.*, 2000b, 2001, 2003, 2008c; Kern-Luetschg and Harris, 2008) focused on the timing and trajectory of soil movement during freezing and thawing, and the effects of changing pore-water pressures at the thaw front.

These studies demonstrated that one-sided freezing and thawing of frost-susceptible soils results in a surface displacement trajectory with a clearly identifiable gelifluction component (Figure 11.1c). They also showed that gelifluction occurs only during thaw consolidation of the upper part of the soil profile; thawing of deeper layers causes thaw consolidation but little or no downslope displacement. This finding is consistent with field measurements, which suggest that on slopes that experience one-sided freezing, gelifluction is confined to the uppermost 0.5–0.6 m of soil, even when ice lenses occur at greater depths (Matsuoka, 2001c; Matsumoto *et al.*, 2010). Harris and Davies (2000) explained this phenomenon in terms of progressive soil consolidation and associated reduction in void ratio during both soil freezing (due to negative pore-water pressures associated with ice segregation) and thaw, as soil at depth consolidates under the weight of overlying soil. There is consequently a progressive reduction in void ratio, and thus water content, as the thaw front penetrates deeper into the soil, so moisture content immediately behind the descending thaw front decreases with increasing depth, and undrained shear strength increases downwards to the point at which gelifluction terminates.

The nature of gelifluction in the uppermost few decimetres of thawing soil reflects changes in pore-water pressure as the thaw front descends. Both full-scale simulations and centrifuge modelling have shown that the downwards passage of the thaw front is associated first with negative pore-water pressures (tension) during the phase change from ice to water, followed by a rapid rise in pore-water pressures to positive and generally above-hydrostatic values as thaw is completed. This change reflects the transition from an ice-supported soil containing unfrozen water under cryosuction (negative pore-water pressures) to a completely unfrozen soil in which consolidation is associated with expulsion of excess pore water. As the thaw front

moves downwards through the soil, the period of maximum pore-water pressures occurs later at progressively lower depths, generating an upwards hydraulic gradient through the thawed soil profile and seepage of meltwater towards the surface. Both the raised pore-water pressures associated with thaw consolidation and the resultant upwards seepage pressures reduce the frictional strength of the soil by transferring some self-weight stress from interparticle contacts to pore water released by melting ice. This reduction in strength 'softens' the soil so that it deforms in a downslope direction under its own weight.

Gelifluction therefore represents pre-failure strain deformation induced by raised pore-water pressures and upwards seepage pressures as the thaw front descends through the upper layers of a thawing, ice-rich soil. As noted above, such deformation was traditionally assumed to be a form of viscous flow, but Harris *et al.* (2003) demonstrated that centrifuge modelling at different scales produces similar displacement rates when compared with prototype values, implying that gelifluction cannot represent viscosity-controlled flow, and must be interpreted as a form of elasto-plastic strain, though the recoverable (elastic) component of strain is much smaller than the permanent (plastic) component. In summary, therefore, gelifluction is gravity-induced elasto-plastic shear deformation of a frictional soil caused by the generation of pore-water pressures to high but subcritical levels at the thaw front. Field measurements appear to vindicate this interpretation. Jaesche *et al.* (2003) demonstrated a close relationship between unfrozen water content and soil deformation, and strain-probe measurements reported by Kinnard and Lewkowicz (2005) showed that gelifluction occurs as a series of discrete, abrupt displacements near the descending thaw plane, which they interpreted as indicating localized micro-shearing along ice-lens boundaries.

Laboratory simulations have also shown that gelifluction operates even in poorly-sorted clast-rich diamictons (Calmels and Coutard, 2000) and that the velocity profile produced by gelifluction is dependent on soil granulometry. In silty soils, a concave-downslope velocity profile similar to that produced by frost creep is typical, but small changes in cohesion due to increased clay content tend to shift the locus of maximum strain farther down the soil profile, producing a convexo-concave profile (Harris *et al.*, 2008c; Figure 11.1c). A convex-downslope profile may also result in situations where the root mat restricts soil movement near the surface. For a given slope and freezing depth, surface velocities are a direct function of frost heave and thaw settlement, implying that soil properties and ice content represent the main controls on deformation rate (Harris and Smith, 2003). For a given heaving ratio and resultant thaw settlement, however, surface velocities are strongly correlated with the sine of the slope angle (α), which determines shearing stress (= W sin α, where W is the weight of overlying soil).

11.2.4 Plug-like Deformation

In areas of cold permafrost where there is two-sided freezing of the active layer, solifluction often incorporates seasonal slow downslope movement of the entire active layer in late summer. Such *plug-like deformation* is a form of shear deformation that accompanies thaw of ice-rich soil at the base of the active layer (Egginton and French, 1985; Matsuoka and Hirakawa, 2000) or the underlying ice-rich transient layer (Lewkowicz and Clarke, 1998; Harris *et al.*, 2011). Laboratory experiments simulating solifluction associated with two-sided freezing above cold permafrost have demonstrated that thaw of ice lenses at the base of the active layer generates above-hydrostatic pore-water pressures and upwards seepage pressures; shear strain is concentrated in this ice-rich zone, with the overlying soil being transported passively downslope on the deforming layer (Harris *et al.*, 2008a; Kern-Luetschg and Harris, 2008; Kern-Luetschg *et al.*, 2008). The nature of basal soil deformation may represent a form of gelifluction, involving plastic deformation of soil 'softened' by high pore-water pressures, though it is possible that localized shear occurs, particularly along ice lenses, but without development of a continuous shear plane (Lewkowicz and Clarke, 1998; Harris *et al.*, 2011). A feature of plug-like deformation is that though surface velocities may be lower than those reported for frost creep and gelifluction, volumetric velocities tend to be greater, as movement involves the full thickness of the active layer. At a site in Svalbard where gelifluction and plug-like deformation operate within an active layer 110 cm deep, Matsuoka and Hirakawa (2000) recorded a volumetric velocity of ~200 $cm^3 cm^{-1} a^{-1}$, which may represent the upper limit for solifluction in any environment. Volumetric displacements of 20–80 $cm^3 cm^{-1} a^{-1}$ have been reported from sites elsewhere in the Arctic, and are probably more representative of areas of continuous permafrost where active layer depths do not exceed 0.5–0.6 m (Mackay, 1981; Lewkowicz and Clarke, 1998). However, plug-like deformation may not occur every year, as it is dependent on the seasonal thaw front reaching the ice-rich soil at the base of the active layer. Field measurements in Svalbard by Harris *et al.* (2011) demonstrated the development of above-hydrostatic pore-water pressures and consequent basal soil shearing in years of relatively deep seasonal thaw, but much lower pore-water pressures and absence of basal shearing when summer thaw failed to reach the transient layer.

11.2.5 Rates of Movement

Rates of solifluction have been reported from a wide range of periglacial environments (Table 11.1). In an overview of published data, Matsuoka (2001c) detected relationships between three measures of activity (surface velocity (V_s), volumetric velocity (V_v) and maximum depth of movement) and mean annual air temperature (MAAT; Figure 11.5). Surface velocities rarely exceed 50 mm a^{-1} in areas of continuous permafrost where MAAT is < –6 °C, but often exceed 100 mm a^{-1} at sites where MAAT falls within the range –3 °C to +5 °C (Figure 11.5a). Many of the highest values have been measured at mid- to low-latitude high-altitude sites that experience frequent diurnal freeze–thaw events, resulting in rapid but shallow needle-ice creep and frost creep. Conversely, the maximum depth of movement is greatest (often >0.5 m) in the continuous permafrost zone, but tends to decline at MAATs above about –5 °C (Figure 11.5c). Plug-like deformation has been recorded at several of the sites with MAATs < –6 °C, but needle-ice creep, frost creep and shallow gelifluction operating in the uppermost 40 cm of the soil dominate non-permafrost sites with MAATs ≥ –3 °C. Volumetric velocity is conditioned by both surface velocity and depth of movement, and is greatest where MAATs range from –8 °C to –2 °C (Figure 11.5b), suggesting that solifluction is most effective as an agent of mass movement in areas of warm permafrost or deep seasonal ground freezing.

Matsuoka (2001c) also examined the effect of depth of annual freezing and thawing on the same three variables. For non-permafrost sites, surface velocity exhibits no consistent relationship with annual freezing depth, because frequent shallow freeze–thaw cycles can result in high surface velocities due to needle-ice creep and superficial frost creep. Volumetric velocities, however, tend to increase with annual freezing depth, reaching a maximum of about 50 $cm^3 cm^{-1} a^{-1}$. In permafrost areas, surface velocities tend to increase only slightly with increasing active-layer depth, but volumetric velocities exhibit a marked increase, reflecting a greater depth of mobile soil. Matsuoka concluded that the key climatic factors controlling rates and depths of solifluction movement are frequency of freeze–thaw cycles and depth of ice-lens formation. Both factors are, however, modulated by local circumstances, notably moisture availability, subsurface thermal regime, soil texture and vegetation cover (Ridefelt *et al.*, 2011). A further control is gradient, but though this is important in regulating local movement rates, global surface velocities are only weakly correlated with slope angle. Matsuoka (2001c) nevertheless noted that, apart from sites dominated by diurnal freeze–thaw, surface velocities tend to be limited by the relation $V_s = 100 \tan \alpha$. He found no consistent relationship between global volumetric velocities and gradient.

Table 11.1 Measured rates of solifluction.

Location	MAAT (°C)	Frost type	Slope gradient (°)	Surface velocity (mm a⁻¹)	Depth of movement (cm)
Polar environments					
Ellesmere Island, arctic Canada	−19.0	CPF	5–9	17–31	60
Sør Rondane, Antarctica	−18.5	CPF	11–20	10	12
Melville Island, arctic Canada	−16.5	CPF	4–5	16	>65
Cornwallis Island, arctic Canada	−16.5	CPF	7	30	–
Banks Island, arctic Canada	−14.5	CPF	2–8	6	–
Garry Island, arctic Canada	−11.0	CPF	3–7	5–7	60
NE Greenland	−9.7	CPF	10–14	9–37	–
Svalbard	−8.0	CPF	6–31	30	51–90
Svalbard	−7.0	CPF	7–15	22	110
SW Yukon	−7.0	CPF	14–18	13	52
Svalbard	−5.0	CPF	10	30–51	–
Svalbard	−4.9	CPF	2–36	9–68	15–48
Subpolar environments					
Schefferville, Québec	−5.0	WPF	8	90	65
Kebnekaise, N Sweden	−4.0	WPF	7–25	19	39
Abisko Mountains, N Sweden	−4.0	WPF	5–25	8–52	–
Iceland	−4.0	SF	7–10	9	20
Okstinan, northern Norway	0.0	SF	5–17	21	30
Mid-latitude mountain environments					
Northern Tibetan Plateau	−6.0	WPF	15–25	3–30	–
West Tianshan, Kazakhstan	−5.0	WPF	5–25	65	40–80
Daisetsu Mountains, Japan	−5.0	WPF	12–27	30	
Colorado Front Range, USA	−3.0	WPF	6–18	2–17	–
Swiss Alps	−3.0	WPF/SF	7–12	13–34	21–41
Canadian Rockies	−2.0	SF	21	5	33
Hohe Tauern, Austria	−2.0	SF	10–20	210	50
Southern Japanese Alps	−2.0	SF	14–30	169	20
Coast Mountains, Canada	−2.0	SF	10–15	250	8
Olympic Mountains, USA	−1.0	SF	12–24	28	–
Northern Japanese Alps	−1.0	SF	11–30	260	22
Swiss Alps	−0.8	SF	~25	41	38
French Alps	0.0	SF	2–34	5–160	–
Central Otago, New Zealand	0.0	SF	7–14	4	–
Fannich Mountains, Scotland	+3.5	SF	14–16	9	39
Kitakami Mountains, N Japan	+4.2	SF	10	250	12
Nikko, central Japan	+6.6	SF	28–41	10	15
Nagano, central Japan	+6.7	SF	22	34	32
High-altitude tropical mountain environments					
Bolivian Andes	−1.0	DF	19–28	1000	18
Venezuelan Andes	+1.0	DF	18–25	163	3–5
Venezuelan Andes	+1.0	DF	19–25	60	5

MAAT, mean annual air temperature; CPF, cold permafrost; WPF, warm permafrost; SF, seasonal freezing; DF, diurnal freeze–thaw.
Most velocities are average values for several sites.
Source: Matsuoka (2001c). Adapted with permission of Elsevier.

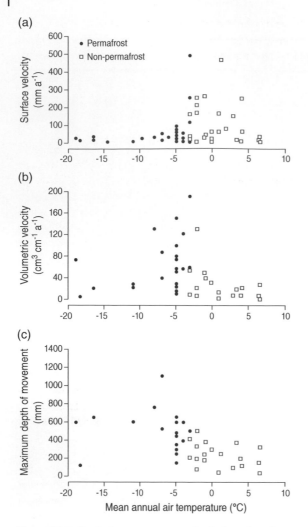

Figure 11.5 Downslope movement of soil plotted against mean annual air temperature (MAAT). (a) Surface velocity; (b) volumetric velocity; (c) maximum depth of movement. *Source:* Matsuoka (2001c). Reproduced with permission of Elsevier.

11.3 Solifluction Landforms

11.3.1 Solifluction Terraces and Lobes

Though soils undergoing slow downslope movement by solifluction may form smooth, featureless slopes (Egginton and French, 1985; Lewkowicz and Clarke, 1998; Matsuoka and Hirakawa, 2000; Ridefelt *et al.*, 2010, 2011), in many areas soil movement has produced *solifluction terraces* or *solifluction lobes* defined by steep frontal risers that mark the downslope extent of a more gently sloping tread. Solifluction terraces extend across the slope approximately parallel to the contour for distances of tens to hundreds of metres. Solifluction lobes are lobate in planform and rarely exceed 25 m in width. Most authors have followed Benedict (1970) in identifying four basic forms: turf-banked terraces, turf-banked lobes, stone-banked terraces and stone-banked lobes (Figure 11.6). *Turf-banked terraces* and *turf-banked lobes* are composed of unsorted soil and clasts, terminate downslope at a vegetated riser and often support partial or complete vegetation cover on the tread. In Benedict's definition, *stone-banked terraces* and *stone-banked lobes* consist of an upper layer of clasts overlying relatively stone-free subsoil and bounded downslope by a bouldery riser, but the term has also been applied to lobes and terraces where a bouldery riser is backed by a tread of vegetated soil. Stone-banked terraces and lobes occur mainly on higher slopes where there is a surface cover of coarse debris, and turf-banked terraces and lobes tend to form in soils containing abundant fine sediment, particularly on mid-slope or slope-foot locations. The term *solifluction sheet* is here used to denote soils that have moved uniformly downslope without riser development except at the downslope terminus (Figure 11.7).

Figure 11.6 (a) Turf-banked lobe on Beinn a' Bha'ach Ard, Scottish Highlands. (b) Stone-banked lobe advancing over bedrock, southern French Alps. *Source:* (b) Courtesy of Pascal Bertran.

Figure 11.7 (a) Solifluction sheet advancing over bedrock, southern French Alps. (b) Terminus of solifluction sheet in western Spitsbergen. A colour version of (a) appears in the plates section. *Source:* Courtesy of (a) Pascal Bertran and (b) Ole Humlum.

All of these forms are transitional to one another. Depending on the abundance of coarse debris and the degree of vertical frost-sorting of clasts, terrace and lobe surfaces may support a continuous cover of organic soil and vegetation, isolated clasts embedded in a matrix of fines, a superficial cover of clasts or a layer of openwork clasts several decimetres thick. On convex hillcrests, terrace or lobe treads often merge upslope into blockfields or solifluction sheets. As gradient increases downslope, however, terrace risers tend to become increasingly crenulate in planform, in response to localized variations in rates of movement; many features referred to as solifluction lobes are actually the lobate downslope extensions of solifluction terraces, marking zones of accelerated movement. On mid-slope locations, lobe-fronted terraces sometimes override one another, isolating individual lobes from their neighbours. Thick lobes and terraces also occupy convex footslopes, reflecting deceleration of mobile soil as gradient decreases.

The key feature defining solifluction terraces and lobes is the frontal riser. Though there appears to be general agreement that risers represent retardation of soil movement, the literature on solifluction is remarkably reticent on how they develop. One possibility is that movement of soil is non-uniform in a downslope direction. If an area of soil is moving more rapidly than soil immediately downslope, compressive stresses at the boundary between the two may cause soil thickening and eventual over-riding of the soil downslope, forming a riser. This seems feasible on concave slopes where gradient diminishes downslope, but less so on convex upper slopes where there is progressive downslope steepening. Alternatively, riser formation may relate to the vertical velocity distribution. All forms of solifluction apart from plug-like deformation produce a reduction in velocity with depth, implying that the upper layers move faster than those beneath (Figure 11.1). Local retardation of surface movement by vegetation, roots or interlocking boulders may result in localized soil thickening and incipient riser development (Price, 1974). It is notable that solifluction sheets (lacking risers) are best developed in areas of sparse vegetation cover. Once an incipient riser has formed, the more rapid movement of nearsurface soil will tend to steepen it, giving rise to steep, bulging or even overhanging risers, particularly on turf-banked

terraces and lobes. It is less certain how riser generation is propagated across the slope for tens or hundreds of metres in the case of solifluction terraces.

Typically, both lobes and terraces are fronted by risers 0.1–2.0 m high (Table 11.2), though exceptionally risers up to 6 m high have been reported. Matsuoka (2001c) has proposed a typology to explain differences in lobe riser height in terms of the depth of soil subject to freezing and thawing. In environments that experience a shallow diurnal freeze–thaw regime, superficial frost creep produces thin lobes with low (<0.2 m) risers, often associated with small sorted stripes on unvegetated ground. Such miniature lobes are characteristic of subtropical high mountains (Francou and Bertran, 1997), marginal periglacial environments (Vieira *et al.*, 2003) and areas of shallow diurnal frost creep in Antarctica (Matsuoka and Moriwaki, 1992). They also develop on the upper slopes of mid- and high-latitude mountains where the depth of frost-susceptible soil is limited (Matsuoka *et al.*, 2005; Ridefelt and Boelhouwers, 2006). Conversely, in areas of deep seasonal freezing or warm permafrost, thicker terraces and lobes with risers 0.3–2.0 m high are common on mid-slope or slope-foot locations, reflecting the greater depth of the annual freeze–thaw cycle. Matsuoka suggested that on slopes underlain by cold permafrost, plug-like deformation produces thick lobes and mobile soil hummocks (e.g. Mackay, 1981; Gorbunov and Seversky, 1999), though in arid high-arctic environments terraces and lobes are often absent and smooth solifluction sheets predominate. In general, deeper soil movements produce thicker lobes, though as surface velocities usually exceed the rate of lobe advance, risers may build up to thicknesses that exceed the depth of movement.

Though solifluction lobes vary greatly in size (Table 11.2), analyses of lobe dimensions have revealed some interesting relationships. Within particular areas, riser height, tread width and tread length all tend to be

Table 11.2 Mean morphometric characteristics of solifluction lobes.

Location	Sample size	Riser height (m)	Riser angle (°)	Tread length (m)	Tread width (m)	Slope angle (°)	References
Mount Kosciusko, Australia	29	0.66	39	6.7	9.3	12	Costin *et al.* (1967)
Jotunheimen, Norway	20	1.00	42	7.3	8.8	18	Matthews *et al.* (1986b)
Southern Canadian Rockies	58	0.41	–	2.6	3.9	19	Smith (1987)
Outpost Mountain, Yukon	85	0.81	65	11.2	20.9	13	Hugenholtz and Lewkowicz (2002)
Abisko Region, northern Sweden	81	1.00	39	10.0	13.0	11	Ridefelt and Boelhouwers (2006)
Sierra Nevada, southern Spain	195	0.56	–	5.3	4.2	10	Oliva *et al.* (2009)

positively correlated (Hugenholtz and Lewkowicz, 2002; Matsuoka *et al.*, 2005; Oliva *et al.*, 2009), suggesting that overall lobe dimensions increase over time as solifluction of soil from upslope feeds lobe growth. This scenario is supported by radiocarbon dates obtained on organic material buried by lobe movement, which indicate that long-term rates of frontal advance (generally $1-10 \, mm \, a^{-1}$) are markedly slower than rates of surface movement on lobe treads (Benedict, 1976; Liu *et al.* 1995; Matsuoka, 2001c), implying progressive lobe thickening over long timescales. Various authors have proposed that lobe growth is terminated by rupture of the riser, release of sediment as a mudflow and resultant subsidence of the tread. This argument is supported by accounts of breached or collapsed lobes (Matthews *et al.*, 1986b; Hugenholtz and Lewkowicz, 2002; Ridefelt and Boelhouwers, 2006) and of accumulation of mudflow sediments on lobes farther downslope (Matsuoka *et al.*, 2005; Matsuoka, 2010). The author witnessed such collapse when an attempt to trench a solifluction lobe during the thaw season led to a torrent of mud and stones being ejected violently from the lobe front, carrying away his spade, lunchbox and field assistant. Such observations appear to confirm that lobes and terraces result from the slow accumulation of solifluction soil behind relatively immobile risers until the front ruptures, with subsequent slow rebuilding of the lobe over long timescales (Kinnard and Lewkowicz, 2006). This model is supported by evidence from the Fannich Mountains in Scotland, where mean surface velocities measured on solifluction lobes are low ($7.8-10.6 \, mm \, a^{-1}$; Ballantyne, 2013c), but radiocarbon ages obtained for a soil buried by lobe advance indicate instantaneous burial over a distance of at least 3 m (Ballantyne, 1986b). The mode of lobe advance is contentious. Some authors have favoured slow 'caterpillar-tread' advance, with gradual overturning and burial of the riser (e.g. Benedict, 1970; Smith, 1987; Elliott, 1996), but others have supported more episodic movement by basal shear or rupture (e.g. Matthews *et al.*, 1986b; Elliott and Worsley, 1999; Kinnard and Lewkowicz, 2006). It seems likely that both regular and rapid modes of advance operate, with periods of gradual advance being interrupted by infrequent episodes of lobe failure and shallow mudflow activity.

Analyses of lobe dimensions and distribution have also revealed links to environmental controls, particularly moisture availability, altitude and vegetation cover (Ridefelt *et al.*, 2010). Studies on mountains in Yukon suggest that a minimum threshold snow depth is a prerequisite for lobe development, with the highest density of lobes occupying northerly aspects associated with late snowcover (Hugenholtz and Lewkowicz, 2002). In northern Sweden, solifluction lobes diminish in size eastwards, reflecting a decrease in precipitation, and the largest turf-banked lobes are associated with high soil moisture content due to deep and prolonged snowcover (Ridefelt and Boelhouwers, 2006). In the Sierra Nevada of southern Spain, high-altitude solifluction lobes display very limited evidence of activity under the present semi-arid climate, and appear to have developed under wetter or snowier conditions in the past (Oliva *et al.*, 2009, 2011).

The internal structure of solifluction landforms is variable. In his classic account of solifluction in the Colorado Front Range, Benedict (1970) found that turf-banked lobes consist of a loamy diamict with a platy structure overlain by a vegetated surface humus layer and underlain by one or more buried organic layers (Figure 11.8a). Stone-banked lobes in the same area consist of an upper layer of openwork cobbles and boulders underlain by a loamy diamict with a platy structure and few clasts, sometimes underlain in turn by a buried organic-rich soil horizon (Figure 11.8b). In both cases, the long axes of clasts under the tread have a strong preferred downslope orientation due to alignment within a deforming soil matrix, though clasts under risers often exhibit transverse alignment parallel to riser orientation. The platy structure observed by Benedict and others reflects compression of soil between ice lenses during freezing, and tends to be best preserved in immobile soil below the maximum depth of gelifluction (Harris and Davies, 2000). Subsequent excavations in a variety of locations have revealed additional features. Elliott (1996) described an inverted podzolic soil under a turf-banked lobe, providing conclusive evidence for overturning of the riser during movement, and Matsuoka and Hirakawa (2000) have also illustrated overturning and burial of soil layers at lobe fronts. Matthews *et al.* (1986b) found colluvial layers indicative of upslope lobe collapse within an excavated lobe, and Kinnard and Lewkowicz (2006) reported planar discontinuities near lobe margins and burial of vegetation downslope by shallow colluvial layers indicative of frontal rupture. Lobes excavated by Oliva *et al.* (2011) exhibited alternating mineral and organic layers, the former relating to periods of solifluction activity and the latter to periods of stability. Micromorphological analyses of soil samples from solifluction lobes show several features: downslope alignment of detrital grains within the zone of mobile soil; small-scale banding, probably due to accumulation of translocated matrix material on the upper surfaces of coarser soil layers compacted by ice lens formation; and ephemeral vesicles in upper soil layers, representing gas bubbles expelled during seasonal freezing of soil water (Harris, 1985).

In environments characterized by shallow solifluction, stratified slope deposits may form through downslope movement of thin (<0.3 m thick) sheets, terraces or lobes composed of an upper coarse pebble layer over a lower layer of fine sediment. Burial of treads by successive thin sheets or lobes advancing downslope produces a distinct

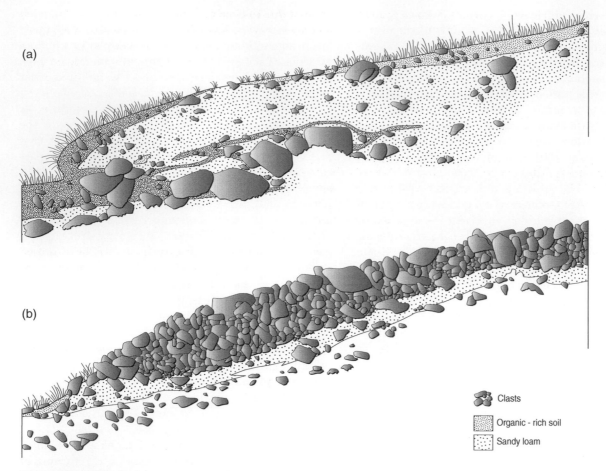

Figure 11.8 Longitudinal sections excavated through (a) a turf-banked lobe and (b) a stone-banked lobe, Niwot Ridge, Colorado Front Range. *Source:* Adapted from Benedict (1970) with permission from the Regents of the University of Colorado.

stratigraphy of alternating layers of openwork pebbles and fine sediment (Figure 11.9). Such stratified deposits occur at high altitudes on subtropical mountains where a diurnal freeze–thaw regime prevails (Francou, 1990). A similar mechanism probably explains alternating matrix- and clast-rich layers within stone-banked lobes in alpine and subantarctic environments (Bertran *et al.*, 1995; Boelhouwers *et al.*, 2003), and may be responsible for the formation of thick stratified slope deposits (*grèzes litées*) during Pleistocene cold periods.

Several studies have employed radiocarbon dating of organic soil buried by lobe advance to establish the timing of lobe initiation, estimate long-term rates of advance or identify periods of accelerated movement or enhanced activity (e.g. Benedict 1970; Smith, 1987; Elliott and Worsley, 1999; Matthews *et al.*, 1993, 2005; Kinnard and Lewkowicz, 2006; Oliva *et al.*, 2011). Such studies show that rates of frontal advance tend to be slow (1–$10\,\mathrm{mm\,a^{-1}}$), and several have identified periods of more rapid lobe movement. Interpretation of accelerated movement is problematic, however, as several possible climate variables are involved, and individual lobes may have experienced

episodic rupture and collapse. Only where concordant radiocarbon chronologies are available from several sites can the palaeoclimatic implications of enhanced solifluction lobe movement be assessed. Most authors have interpreted former periods of accelerated movement in terms of either climatic deterioration (particularly increased depth of freezing in permafrost-free areas) or increased moisture availability.

11.3.2 Ploughing Boulders and Braking Blocks

Ploughing boulders are boulders that have moved downslope more rapidly than the adjacent soil, forming a furrow upslope (Figure 11.10), and often pushing up a mound of soil or turf downslope. Recent movement is indicated by the presence of a deep niche, often extending to the base of the boulder, between the upslope edge of the boulder and the furrow. Ploughing boulders are common on vegetated slopes on mid-latitude mountains (e.g. Allison and Davies, 1996; Ballantyne, 2001b; Berthling *et al.*, 2001a, 2001b; Hall *et al.*, 2001; Grab *et al.*, 2008) but appear to be rare at high latitudes and may

Figure 11.9 Stratified slope deposits formed through successive burial of the stony treads of shallow stone-banked lobes under lobes advancing downslope at 5000 m altitude in the Bolivian Andes. *Source:* Courtesy of Pascal Bertran.

Figure 11.10 Small ploughing boulder in the Fannich Mountains, northwest Scotland, showing the deep furrow formed by downslope movement of the boulder through the soil.

be restricted to areas of seasonal ground freezing or warm permafrost. More locally, they occupy a wide range of slopes (1–38°), and several authors have noted an association with active solifluction lobes and with frost-susceptible soils with low plastic and liquid limits. Evidence of 'ploughing' is usually restricted to larger boulders, suggesting that movement faster than the surrounding soil requires boulders to exceed a critical mass, though there appears to be no upper limit to boulder size; Reid and Nesje (1988) have described a giant ploughing boulder in Norway with a volume of $137 \, m^3$ located at the foot of a furrow 42 m long. Several studies have shown that ploughing boulders have rotated during movement, so that their long axes tend to be aligned downslope, and morphometric analyses have shown that furrow length and width are often positively correlated with boulder size (Ballantyne, 2001b; Hall *et al.*, 2001; Grab *et al.*, 2008).

Measured rates of boulder movement range from a few millimetres to a few centimetres per year (Table 11.3),

with displacement being restricted to the periods of freezing and thawing (Berthling *et al.*, 2001a, 2001b) and being greatest during thaw (Gorbunov, 1991). Most authors have accepted that downslope movement of ploughing boulders is due to frost creep (seasonal heave and settling of boulders) and/or gelifluction, and excavations revealing excess ice underneath boulders (Gorbunov, 1991; Ballantyne, 2001b) support this explanation. The downslope advance of boulders at rates exceeding that of the adjacent soil has been explained by Ballantyne (2001b) in terms of the higher thermal conductivity (diffusivity) of boulders in comparison with that of the surrounding soil. He proposed that during seasonal freezing, the freezing front descends more rapidly through the boulder, causing ice formation at the base of the boulder and heave of the boulder perpendicular to the slope. Conversely, during seasonal thaw, the thaw front reaches the base of the boulder whilst the surrounding soil is still frozen, resulting in elevation of pore-water pressures under the boulder to above-hydrostatic levels. Movement is initiated by thaw of the soil immediately downslope, and takes the form of both vertical settlement of the boulder and sliding of the boulder over softened or liquefied soil, pushing up a mound of soil at its leading edge. Such movement permits trapped water to escape, allowing stability to be regained. The relative contributions to boulder movement of basal sliding and differential thaw settlement remain unknown.

Table 11.3 Measured rates of ploughing boulder movement.

Area	Period (years)	Sample size	Altitude (m)	Slope (°)	Average rate of movement (mm a^{-1})		
					Min.	Mean	Max.
British mountains							
Central Grampians, Scotland	2	7	790–880	8–32	1.5	3.1	7.0
Lochnagar, NE Scotland	3–4	12	800–980	8–28	0.3	2.8	8.7
Fannich Mountains, Scotland	2–20	7	790–820	13–25	3.6	12.6	30.3
Pennines, northern England	10	5	685–820	5–22	0.4	23.2	64.0
Other areas							
Tatra Mountains, Poland	1–8	30	1130–1920	15–34	1.4	13.0	32.5
Hardangervidda, Norway	5	27	1290–1360	~11	3.6	8.4	25.2
Rock and Pillar Range, New Zealand	34	8	1450	12–33	6.2	11.4	18.8

Source: Data from Ballantyne (2001b), Berthling *et al.* (2001a) and Grab *et al.* (2008).

Figure 11.11 Relict stone-banked lobe composed of granite boulders, Mourne Mountains, Northern Ireland. Such boulder lobes were probably last active under permafrost conditions during the Younger Dryas Stade (12.9–11.7 ka). *Source*: Courtesy of Peter Wilson.

The antitheses of ploughing boulders are *braking blocks*: large boulders so deeply embedded in the soil (or in permafrost) that they remain stationary or relatively immobile as the surrounding soil moves around them, sometimes accumulating on their upslope side. These features have attracted only passing mention (Washburn, 1979, p. 224), but have been employed in Antarctica to estimate rates of downslope regolith movement (Putkonen *et al.*, 2012).

11.4 Pleistocene Solifluction Landforms and Slope Deposits

Though small-scale solifluction features have limited preservation potential, large, clast-dominated relict solifluction terraces and lobes that were formed during Pleistocene cold stages are still preserved as periglacial relicts. In Scotland and Ireland, for example, massive stone-banked terraces and lobes with risers 1.0–3.0 m high occur on granite mountains (Figure 11.11), and smaller relict terraces and lobes with risers 0.5–1.0 m high occur on schist and quartzite mountains (Ballantyne and Harris, 1994; Cunningham and Wilson, 2004). The lower limit of relict solifluction lobes coincides locally with the upper limit of glaciers that formed during the Younger Dryas Stade (~12.9–11.7 ka), suggesting that these relict forms were active at that time. Similar relict stone-banked terraces in the Hex River Mountains of South Africa have been described by Boelhouwers (1999), who attributed their formation to seasonal frost creep during the last glacial maximum (LGM).

Evidence for solifluction during Pleistocene cold stages is also widespread on lowland areas that remained outside

Figure 11.12 Gelifluctate ('head') deposits on the Gower Peninsula, south Wales. (a) Massive, crudely-stratified gelifluctate overlying stratified slope deposits. (b) Fabric of gelifluctate at the same site, consisting of angular clasts embedded in a matrix of fines. The ruler is 30 cm long. *Source:* Courtesy of Rick Shakesby.

the reach of successive ice sheets or were exposed to periglacial conditions as the last ice sheets retreated. In formerly glaciated areas, frost-susceptible glacigenic deposits have been extensively modified by solifluction, most notably producing smooth *solifluced till sheets*. More locally, solifluction has also modified glacial landforms such as drumlins and moraines, reworking glacigenic sediment in a downslope direction and sometimes burying organic deposits. Solifluced till sheets retain many attributes of the parent till, such as striated, facetted and erratic clasts, but the long axes of clasts tend to be aligned downslope, and structures indicative of frost action and solifluction (vertical sorting of clasts, platy soil structure, colluvial lenses or layers) may also be present.

In areas that remained outside the limits of Pleistocene glaciation, such as southern England and northern France, periglacial slope deposits are derived from local regolith, though sometimes with a windblown silt component. Harris (1987, 2013) and Hutchinson (1991) identified two classes: granular deposits derived from non-argillaceous bedrocks such as sandstone, granite and limestone, and clay-rich deposits derived from argillaceous bedrocks. Low-angle shear planes within clay-rich periglacial slope deposits suggest that these moved downslope as active-layer detachment failures (see below), but granular slope deposits are generally attributed to former solifluction. Such deposits in England are often referred to as *head deposits* ('head' is an archaic term originally used by quarrymen to describe unconsolidated deposits overlying bedrock), but are more generally described as *gelifluctate*. Typically, Pleistocene gelifluctates

comprise locally-derived angular clasts embedded in a frost-susceptible silt-rich matrix; clasts exhibit low-angle dips approximately parallel to the slope, and clast long axes have a strong preferred downslope orientation, consistent with downslope transport within a deforming soil (Figure 11.12). Occasional slope-parallel layers or lenses of well-sorted silt or sand intercalated within gelifluctates reflect former slopewash episodes or reworking of aeolian silts, and some gelifluctates exhibit crude stratification, particularly those composed of platy clasts derived from schist or sedimentary bedrocks. Fractured bedrock at the base of the deposit may be overturned in a downslope direction (Harris, 1987; Mears, 1997).

Gelifluctates often mantle smooth convexo-concave hillslopes, thickening downslope into a valley-fill deposit several metres deep that represents progressive accumulation of sediment during Pleistocene cold stages. In upland areas, such accumulations have often been incised by rivers, producing valley-floor *solifluction benches* characterized by steep frontal bluffs backed by a more gentle tread. Outside the limits of Pleistocene glaciation, such valley-floor gelifluctate accumulations are composed of locally derived sediments, but inside former glacial limits, solifluced till often predominates (Ballantyne and Harris, 1994; Harrison, 2002). In the Central Highlands of Germany, layered gelifluctates overlie granite saprolites and deeply weathered basalts (Völkel *et al.*, 2001), and in coastal locations around the English Channel, diamictic gelifluctates of various ages mantle relict cliffs and bury raised shorelines (Harris, 1987; Bates *et al.*, 2003). In the chalklands of southern England and the Paris Basin, a gelifluctate composed of chalk and flint clasts embedded in silty or loamy calcareous mud (in England termed *coombe rock*) forms a thin mantle over valley sides but reaches a thickness of several metres to a few tens of metres on the floors of dry valleys and on raised marine platforms (Laignel *et al.*, 2003). The periglacial origin of coombe rock in England is attested by the presence of the remains of mammoth and woolly rhinoceros within the chalky gelifluctate.

Solifluction during Pleistocene cold stages is also implicated in the formation of certain types of *stratified slope deposits*. Various processes produce stratified hillslope deposits, including debris flow, slopewash and dry grainflows (van Steijn *et al.*, 1995; Chapter 12), but all of these appear inadequate to explain the remarkable deposits referred as *grèzes litées* ('bedded grits'). These comprise alternating matrix-rich and openwork beds typically 10–15 cm thick, have gentle to moderate dips (usually <25°) and exhibit lateral and longitudinal continuity over tens of metres (Figure 11.13). Openwork beds are typically composed of fine to medium gravel, and may exhibit normal (fining upwards) or inverse (coarsening upwards) sorting, or normal sorting superimposed on inverse sorting within a single bed. The intervening matrix-rich beds are frost-susceptible and often exhibit platy structures typical of those produced by compression of soil peds by ice-lens formation.

Classic examples of grèzes litées occur in the Charentes region of western France. These were originally interpreted as representing alternating episodes of slopewash and solifluction, but features characteristic of slopewash are absent, and it is difficult to envisage this combination of processes producing bedding of such striking regularity, rhythmicity and lateral continuity.

Bertran *et al.* (1992) and Ozouf *et al.* (1995) have argued that the grèzes litées of Charentes (Figure 11.13) represent successive over-riding of shallow, stone-banked solifluction sheets, analogous to that observed in the high Andes by Francou (1990), with frost sorting of openwork layers producing inverse sorting, and overturning of risers producing superimposed normal sorting. This interpretation may also explain similar stratified slope deposits in northeast France (Laurain *et al.*, 1995), England and Wales (Ballantyne and Harris, 1994), the foothills of the Pyrenees (Garcia-Ruiz *et al.*, 2001) and the southern Carpathians (Urdea, 1995), though the complex stratigraphy of some sites in these areas suggests the localized operation of debris flow and sheetwash in addition to solifluction. A similar explanation may also account for relict stratified slope deposits in central Pennsylvania (Gardner *et al.*, 1991) and on the Qinghai-Tibet Plateau (Liu *et al.*, 1999). This model implies shallow solifluction (needle-ice creep, frost creep and possibly gelifluction in the underlying fine soil) operating in response to frequent shallow freeze–thaw cycles. Preconditions for the formation of grèzes litées by downslope movement of shallow gravel-covered sheets of debris are absence of vegetation cover and an upslope source of friable bedrock that weathers to form small clasts, granules and frost-susceptible soil. As the downslope advance of thicker stone-banked lobes in response to deep seasonal freezing and thawing of the ground may also produce similar, if less regular and less continuous, stratified slope deposits (Bertran *et al.*, 1995; van Steijn *et al.*, 1995; van Steijn, 2011), caution is required in drawing palaeoenvironmental inferences from such deposits.

Figure 11.13 Stratified slope deposits at Verteuil, western France, consisting of alternating matrix-rich and openwork fine gravel beds 5–15 cm thick. Compare with Figure 11.9. *Source:* Courtesy of Pascal Bertran.

11.5   Active-layer Failures

Solifluction represents a form of slow mass movement at subcritical stress levels. In contrast, localized rapid downslope movement of the active layer also occurs on gentle to moderate slopes underlain by permafrost. Such rapid mass movements are termed *active-layer failures*, and take the form of translational (surface-parallel) landslides that usually involve the full depth of the active layer. They may be triggered by intense rainstorms, deepening of the active layer by forest fires, erosion of the footslope by rivers or the sea, or by artificial excavation. The most common trigger of active-layer failure, however, is rapid thaw of ice at the base of the active layer or in the transient layer at the top of permafrost as a result of exceptionally warm conditions in late summer. The resulting soil movements take two forms: *active-layer detachment slides*, where blocks of intact soil move downslope over permafrost on a slope-parallel shear plane or a thin layer of liquefied sediment; and *flowslides* (or *skinflows*), where the sliding mass loses coherence and continues to move downslope as a viscous flow (McRoberts and Morgenstern, 1974; Harris *et al.*, 2008c) though many active-layer failures exhibit hybrid characteristics. The nature of movement depends partly on moisture content and partly on soil texture: cohesionless sandy or silty soils with low liquid limits tend to form flowslides, whereas clay-rich soils with relatively high liquid limits tend to move as translational slides.

Active-layer failures occur in a wide range of environments underlain by ice-rich permafrost (McRoberts and Morgenstern, 1974; Leibman, 1995; Lewkowicz and Harris, 2005a; Niu *et al.*, 2005, 2016). Most comprise an arcuate (convex-upslope) headscarp, a slide track and a mass of displaced sediment farther downslope. The transported sediment may consist of an intact block or blocks of soil, rafted downslope on a thin layer of liquefied sediment, or a complex of compressional ridges that form a lobate runout deposit (Figure 11.14), or both. Lewkowicz and Harris (2005b) identified three types: (i) compact failures, involving limited displacement of an intact block of soil, often at a footslope bordering an incised floodplain; (ii) elongated failures, which may be initiated anywhere on the slope and sometimes extend the full slope length, and are characterized by a bare slide track, fragmented residual blocks of soil and a toe zone of deformed sediments that exhibit internal folds, shears and diapiric intrusions (Lewkowicz, 2007); and (iii) complex failures, which represent successive movements over periods of up to a few days, and involve retrogressive failure of the headscarp after the initial sliding event. The dimensions of individual failures vary greatly, from compact forms less than 30 m in width and length to elongate or complex failures up to 100 m wide and 700 m long. Active-layer detachments occur on low to moderate gradients, often <20° and sometimes as low as 4–7°.

Much of our understanding of the behaviour of active-layer detachment failures derives from study of those on the Fosheim Peninsula (80° N) on Ellesmere Island in the Canadian Arctic Archipelago (Harris and Lewkowicz, 1993, 2000; Lewkowicz and Harris, 2005a, 2005b; Lewkowicz, 2007). Slopes in this area are underlain by continuous permafrost, active-layer depths are typically 0.5–0.75 m and soils are dominantly clays or silts with

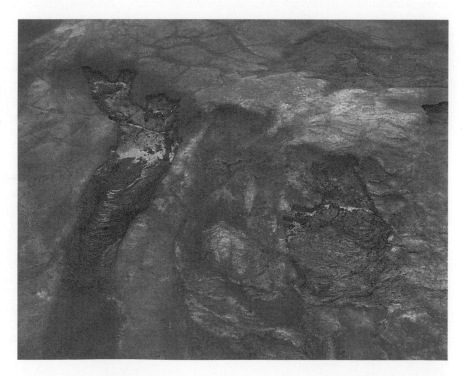

Figure 11.14 Two recent active-layer failures on the Fosheim Peninsula, Ellesmere Island, Canadian Arctic Archipelago. An older vegetated failure separates the two. The tops of ice wedges are visible in the polygons exposed by the recent failure on the left. A colour version of this figure appears in the plates section. *Source:* Courtesy of Antoni Lewkowicz.

low to medium plasticity. Over 500 detachment failures have been mapped at three sites with a total area of $18.4 \, km^2$, and Lewkowicz (1992) estimated that failure frequency between 1950 and 1990 averaged 35 events per decade within an area of $5.3 \, km^2$. This average, however, obscures the intensified nature of sliding in particular years: 75 failures were observed within the same area in the summer of 1988 alone, and other years with particularly warm summers have also witnessed multiple failure events (Lewkowicz and Harris, 2005a). Lewkowicz (2007) monitored two contrasting failures that occurred after a week of high air temperatures in August 2005. The smaller involved a block of soil $940 \, m^2$ in area that detached from a low scarp and moved ~5 m over a gradient of 12–15° before stabilizing about 4 hours after initial failure. The larger occurred on a gradient of 16–19° near the slope crest. The advancing front mobilized soil in its path, pushing up numerous compression ridges, and attained an eventual length of 330 m over 3–5 days, with the toe of the slide extending over a slope of 3–7°. The headscarps of both failures retreated after initial failure, with detached headscarp blocks sliding down the slide track.

The cause of active-layer failure on the Fosheim Peninsula is rapid melting of ice-rich soil at the base of the active layer or at the top of permafrost in late summer. Two-sided freezing results in the formation of a layer of segregation ice at the base of the active layer. If thaw of this ice takes place more rapidly than water can escape through the overlying thaw-consolidated soil, high pore-water pressures build up, reducing the effective normal stress within the soil and thus its frictional strength. Lewkowicz and Harris (2005b) assessed the sensitivity of slopes affected by active-layer failure using the planar infinite-slope stability model:

$$F_s = \frac{c' + z(\gamma - m\gamma_w)\cos^2\alpha \tan\phi'}{z\gamma \sin\alpha \cos\alpha} \tag{11.8}$$

where F_s is the factor of safety (=1 at the point of failure; values >1 indicate stability), c' is effective cohesion, z is the depth of the slip surface, γ is unit weight of soil, γ_w is unit weight of water, m is the ratio of the height of the water table above the slip surface to the depth of the slip surface (z), α is the slope gradient and ϕ' is the effective angle of friction. Rapid melting of excess ice at the base of the active layer causes an increase in m, potentially to artesian values ($m > 1$) if excess water is produced faster than it can escape through the soil (Harris and Lewkowicz, 2000). From equation (11.8), it can be seen that an increase in m results in a decrease in the factor of safety (F_s). Two other critical variables are effective cohesion (c') and the effective angle of friction (ϕ'). As the soils at failure sites on the Fosheim Peninsula yielded $F_s > 1$ for peak (intact) strength, Lewkowicz and

Harris (2005b) argued that prior to failure, shear strength had been reduced to near-residual values (with $c' = 0$ and a reduction in ϕ') by the development of small, progressive shear strains caused by gelifluction (plug-like deformation) at the base of the active layer. Scaled centrifuge modelling (Harris *et al.*, 2008c) has confirmed that the stability of thawing slopes is sensitive to fairly small changes in soil properties, particularly clay content. Even assuming residual strength conditions, however, it is difficult to account for the movement of active-layer slides across low-gradient footslopes, except in terms of dynamic loading by the moving mass and exceptionally low undrained shear strengths produced by high pore-water pressures (Lewkowicz, 2007).

Harris and Lewkowicz (1993) showed that mobile soil blocks in active-layer failures retain their integrity because two-sided freezing results in cryodesiccation of the central part of the active layer, forming a strongly structured soil mass that resists deformation. Conversely, they also demonstrated that the ice content at the base of the active layer tends to exceed the liquid limits of the soil, allowing the development of a layer of liquefied sediment a few millimetres to a few centimetres thick, on which intact soil blocks glide downslope.

Not all active-layer detachment failures are triggered by rapid late-summer thaw. Niu *et al.* (2005) observed active-layer failure on a slope of 6–7° initiated by excavation with subsequent headscarp retreat for over a decade, and some failures have been triggered by exceptionally intense summer rainstorm events (Cogley and McCann, 1976). In subarctic environments, forest fires cause widespread active-layer failure by removing surface peat and other organic material, allowing penetration of the thaw plane into the underlying ice-rich permafrost. In such cases, failure may occur weeks or months after the fire (Lewkowicz and Harris, 2005a).

Active-layer detachment failures are significant for several reasons. In areas where they are endemic, they dominate mass transfer of sediment on slopes (Table 11.4), and if summer temperatures in arctic environments increase as a result of climate change, the frequency of activity in failure-prone sites is likely to increase (Chapter 17). Permafrost degradation occurs under the unvegetated slide track, sometimes generating secondary mudflows, and the track depression forms a site of enhanced snow accumulation and consequent meltwater runoff and soil erosion by surface wash (Kokelj and Lewkowicz, 1998; Lewkowicz and Kokelj, 2002). Finally, past active-layer failures are indicative of slopes in a state of critical conditional stability, and are therefore particularly sensitive to any form of engineering development that may tilt the balance towards renewed failure.

The subdued landforms produced by active-layer failures have limited long-term preservation potential. However, shallow, low-angle planar shear surfaces revealed in

Table 11.4 Rates of mass transfer (10^3 t m km^{-2} a^{-1}) by active-layer failures in the Mackenzie Valley (65°N), northwest Canada, and on the Fosheim Peninsula (80°N), Ellesmere Island.

	Mackenzie Valley		Fosheim Peninsula	
	Site KP 182 (0.37 km^2)	Black Top Creek (3.6 km^2)	Hot Weather Creek (0.82 km^2)	Big Slide Creek (7.6 km^2)
Downslope	6.7–19.0	40–84	5.9–12.0	17–38
Vertical	2.2–5.9	8.3–17.0	1.6–3.4	3.5–7.8
Horizontal	6.3–18.0	39–83	5.6–11.0	17–37

Rates for site KP 182 assume a recurrence interval for fire causing detachment failure of 100–200 years. Rates for the Fosheim Peninsula sites assume that detachment failures remain visible for 50–100 years. The unit area calculations are based on valley slope area only. The unit of mass transfer (10^3 t m km^{-2} a^{-1}) is thousands of metric tons (10^3 t) multiplied by travel distance in metres (m) per square kilometre per year.
Source: Data from Lewkowicz and Harris (2005a).

excavations and borehole cores provide evidence for Pleistocene active-layer failures in clay-rich soils and argillaceous rocks in mid-latitude areas that were formerly underlain by permafrost. In southern and central England, for example, shallow shear planes have been detected under gentle to moderate slopes underlain by various argillaceous lithologies (Hutchinson, 1991; Ballantyne and Harris, 1994). These slopes are now stable, suggesting that such shear planes represent failure of a former active layer over ice-rich permafrost. Bertran and Fabre (2005) have identified similar evidence for former active-layer sliding on a low-gradient (~3°) slope in southwest France. At this site, failure occurred within a 0.5 m thick clay layer overlain by colluvium. They demonstrated that failure within the clay must have occurred under undrained conditions produced by thawing of excess ice, and inferred that the slide represented active-layer detachment over a former permafrost table, thus providing evidence for southwards extension of permafrost as far as southwest France (45°N) during the coldest part of the last glacial stage. Active-layer detachment slides triggered by permafrost degradation may also have contributed in some areas to the accumulation of Pleistocene valley fill deposits (Harrison *et al.*, 2010) and relict periglacial slope deposits (Pawelec and Ludwikowska-Kedzia, 2016).

11.6 Permafrost Creep

Permafrost creep is the slow downslope deformation of ice-rich permafrost under the influence of gravity, or, more specifically, the shear stress imposed by the weight of overlying sediment and ice. This shear stress increases linearly with overburden weight (and hence with depth) and nonlinearly with increasing gradient. The creep process is complex (Yershov, 1998, pp. 145–156). In part it involves deformation of ice bodies within sediment due to small-scale shearing and recrystallization, similar to the creep of glacier ice. In ice-rich soil, however, an

additional factor is the presence of thin films of unfrozen water that surround soil particles, facilitating deformation under gravity-imposed stresses (Johnston *et al.*, 1981; Chapter 3). In engineering terms, permafrost creep represents a form of long-term secondary creep, as deformation neither attenuates through time nor accelerates due to progressive weakening of the soil–ice structure under constant stress. Permafrost creep has important engineering implications, as it can affect the long-term stability of structures (Andersland and Ladanyi, 2003; Figure 11.15). In a geomorphological context, it assumes its most prominent manifestation in the movement of *rock glaciers*, accumulations of predominantly coarse debris that move slowly downslope through the deformation of subsurface ice and/or ice-rich sediments (Chapter 12).

Two factors affect the propensity for permafrost creep and resulting creep rate. Temperature is critical: in 'warm' permafrost, at temperatures only a few degrees below 0 °C, unfrozen water content is high and ice-crystal boundaries lose their rigidity. Conversely, in 'cold' permafrost unfrozen water content is low and ice bodies are more resistant to internal deformation. Huang *et al.* (1986) demonstrated an exponential relationship between creep rate and temperature, reflecting an increase in unfrozen water content with increasing temperature. Soil texture is also important, in that clay- and silt-rich soils are more likely to contain high concentrations of excess (segregation) ice, and also to have lower frictional strength than coarse-textured soils.

The importance of temperature and surface gradient in regulating rates of permafrost creep is evident from measurements made at sites with contrasting ground thermal regimes. On gentle (2–6°) slopes on Melville Island (75°N), where mean annual ground temperature (MAGT) is about −16 °C, Bennett and French (1990) measured downslope displacements averaging 1.1 mm a^{-1} at 35 cm depth (just below the permafrost table) and 0.4 mm a^{-1} at 65 cm depth, with no detectable movement at 95 cm depth. Movement was greatest in late summer

Figure 11.15 Tilted wooden piles representing the foundations of a building in Longyearbyen, Svalbard, that was destroyed during the Second World War. Since then, the piles have been slowly tilted downslope as a result of permafrost creep and/or solifluction.

and early autumn, when temperatures in the upper part of the permafrost reached their seasonal maxima. The measured mean volumetric transport rate of sediment due to permafrost creep at this location was $3.6\,cm^3\,cm^{-1}\,a^{-1}$, markedly less than the mean ($20.3\,cm^3\,cm^{-1}\,a^{-1}$) for solifluction of the overlying active layer (Bennett and French, 1991). In contrast, on slopes of 9–25° underlain by warm permafrost (MAGT –3 °C) on the Tibet Plateau, Wang and French (1995b) measured downslope permafrost creep rates averaging $4.4\,mm\,a^{-1}$ at 1.6 m depth, $2.8\,mm\,a^{-1}$ at 2.2 m depth and $1.6\,mm\,a^{-1}$ at 2.8 m depth, though their results exhibited marked variability between sites. Creep rates of 2.5–$3.0\,mm\,a^{-1}$ have been recorded in clay soils on slopes of 15–31° underlain by warm permafrost (MAGT –2 °C) in the Mackenzie Valley, with strain being most pronounced in ice-rich zones within the soil, and locally detectable to depths of up to 40 m (Savigny and Morgenstern, 1986). The most rapid rates hitherto reported relate to alpine permafrost on a glacier foreland in the Swiss Alps, where comparison of sequential aerial photographs indicated surface movement of up to $0.5\,m\,a^{-1}$ (Kääb and Kneisel, 2006), though it is unclear if this movement reflects permafrost creep alone, or whether solifluction of the active layer played some role in ground displacement at this site.

A distinctive type of permafrost creep was recorded by Dallimore *et al.* (1996a), who monitored the deformation of massive ground ice overlain by 2–5 m of till using inclinometer readings from boreholes. Almost all the deformation occurred within the ice body, with the overlying till being carried passively downslope. One of the two measurement sites exhibited irregular strain

patterns attributable to topographic complexity. At the other, strain was concentrated near the bottom of the ice, producing a displacement profile similar to that produced by internal deformation of glacier ice (Figure 11.16), and consistent with a model of deep-seated creep in massive ground ice developed by Ladanyi *et al.* (1995). Averaged net downslope displacement rates were small (3–$4\,mm\,a^{-1}$), and both sites experienced marked seasonal upslope movements in late winter and early summer, probably reflecting seasonal thermal contraction.

Unequivocal evidence for permafrost creep in mid-latitude environments during Pleistocene cold stages is difficult to establish, but it has been implicated by several authors as a possible mechanism for large-scale slope movements where competent caprocks overlying argillaceous strata have undergone flexing and lateral migration due to deformation of the underlying clay. This phenomenon is termed *cambering*, and usually occurs on the flanks of valleys that have been cut through the caprock into the underlying clays (Figure 11.17). Bending and lateral extension of the caprock are accompanied by the development of deep, contour-parallel fractures termed *gulls*, which separate coherent blocks of rock, and by tilting of the latter towards the valley axis. Often associated with cambering is *valley bulging*, uparching of clay bedrocks along the valley floor to form a broad anticline with the fold axis running roughly parallel to the valley axis. In England, for example, cambering and valley bulging are associated with almost all clay bedrocks (Hutchinson, 1991; Ballantyne and Harris, 1994; Hutchinson and Coope, 2002). Parks (1991) proposed a general model of cambering and valley bulging,

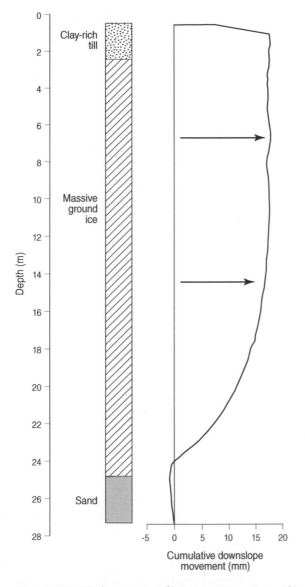

Figure 11.16 Displacement profile representing creep within massive ground ice over seven years; strain is largely confined to the lowermost third of the ice. *Source:* Adapted from Dallimore *et al.* (1996a). Reproduced with permission of Wiley.

additional possibility that bottom-upwards thaw of ice-rich clay may have promoted displacement of thawed clay under the overlying still-frozen ground, enhancing the amplitude of valley bulging. Whilst Parks' model offers a feasible explanation of cambering and valley bulging, the extent to which these structures are attributable to permafrost creep or to deformation of clay bedrocks during thaw is uncertain. It is likely that cambering and valley bulging represent the outcome of multiple cycles of permafrost aggradation and degradation during Pleistocene glacial and interglacial stages.

11.7 Nivation

The term *nivation* denotes a suite of weathering and slope processes that are enhanced by the presence of late-lying or perennial snowcover. 'Snow' in this context usually implies compact firn (density $400–800\,kg\,m^{-3}$), sometimes containing ice layers produced by refreezing of meltwater (Kawashima *et al.*, 1993). Processes traditionally associated with sites of prolonged or perennial snowcover include intensive freeze–thaw activity, enhanced chemical weathering, slopewash and accelerated solifluction (Thorn, 1988; Thorn and Hall, 2002). Transport of debris entrained in creeping or sliding snow may also occur, but appears to be limited mainly to movement of clasts over bedrock surfaces (Jennings and Costin, 1978).

Measurements of nivation processes in a range of periglacial environments suggest that the most important function of most late-lying or perennial snowbeds is sustained production of meltwater throughout the thaw season, so that the underlying substrate experiences prolonged wetting and the pronival zone immediately downslope is often saturated, generating overland flow in the form of rillwash (Thorn, 1976, 1979; Ballantyne, 1985; Hall, 1985; Nyberg, 1991; Caine, 1992a; Kariya, 2002). Vegetation is sparse or absent in areas covered by late-lying snow, so surface wash often removes fine sediment from the pronival zone, leaving a gravel lag deposit or a *boulder pavement* of larger clasts (Hara and Thorn, 1982), and depositing silt, sand and fine gravel farther downslope. Saturation of soil by meltwater may also promote accelerated solifluction downslope from snowpatches (Kariya, 2002; Thorn and Hall, 2002). There is less certainty about the effects of protracted snowcover on weathering processes. Studies based on weathering-rind thickness, boulder surface hardness and clay mineralogy suggest that chemical weathering of soil and rock surfaces is enhanced by prolonged wetting under melting snow (Thorn, 1975; Ballantyne *et al.*, 1989; Thorn *et al.*, 1989; Hall, 1993), but the effect of prolonged snowcover on physical weathering processes remains contentious. Berrisford (1991) found that fractures and

arguing that downwards aggradation of permafrost into clays would have been accompanied by ice segregation and frost heave, which would have stressed the overlying caprocks and initiated permafrost creep in the clay (Figure 11.17b,c) and thus caused valleyward extension of the caprock and opening of gulls between adjacent caprock blocks. He also envisaged that during later permafrost degradation, thaw consolidation of low-permeability clay would have been accompanied by a reduction in shear strength, causing relatively rapid deformation under the stresses imposed by the weight of the overlying caprock (Figure 11.17d), settling and tilting of caprock blocks, and viscous flow of saturated clay towards the valley axis. Hutchinson (1991) noted the

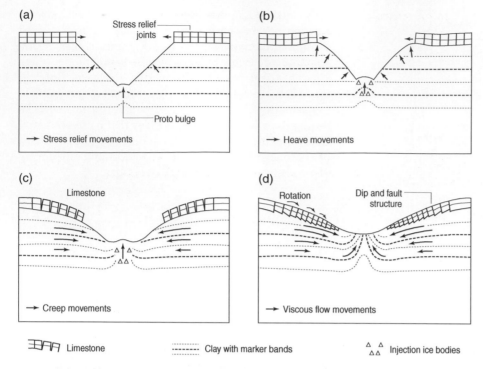

Figure 11.17 Stages in the development of cambering and valley bulging. (a) Valley incision into clay bedrock results in the formation of stress-release joints in the caprock and in the development of a proto-bulge on the valley floor. (b) Frost heave increases the size of the bulge and favours joint opening in the caprock. (c) Permafrost creep within ice-rich clay produces cambering of valley sides, detachment of caprock blocks and increased valley bulging. (d) Thaw of excess ice within ice-rich clays causes viscous flow of clay strata towards the valley axis, enhanced cambering and settling, and rotation of caprock blocks. *Source:* Adapted from Parks (1991).

flaking are much more common on boulders in areas of late-lying snow than in neighbouring sites lacking late snowcover, though the processes responsible for this contrast are uncertain. Other studies suggest that weathering rates may actually be reduced by late-lying snowcover (Caine, 1979; Benedict, 1993).

Though most research on nivation relates to mountain environments, some of the most distinctive nivational effects occur in arctic permafrost environments where perennial beds of firn and ice overlie continuous permafrost. At such sites, groundwater within the active layer upslope enters the uppermost layers of firn and flows over underlying ice to emerge at the foot of the firn field (Ballantyne, 1978a), feeding pronival rillwash throughout the summer. In other cases, most pronival runoff is generated by snowmelt (Lewkowicz and Young, 1990). Other aspects of arctic nivation are embodied in a model by Christiansen (1998a), based on observations in northeast Greenland. This model (Figure 11.18) incorporates sediment entrainment and deposition by pronival runoff, and adds three additional elements: sediment input through niveo-aeolian deposition (Chapter 14), small sediment flows on and within snowpatches, and retrogressive thaw slumping of sediment at snowpatch backwalls. Such slumping is apparently triggered by high pore-water pressures generated as groundwater from upslope is forced to the surface, and has the effect of progressively enlarging the hollows occupied by late-lying and perennial firn.

Nivation was once considered to be responsible for the formation of small cirque-like hollows (Washburn, 1979), but reappraisal of the effectiveness of nivation processes has led to the view that their main role is to modify existing landforms, such as valley heads or the risers of solifluction terraces (Thorn, 1988). Terms such as *nivation hollow*, *nivation niche* and *nivation bench* have been employed to describe landforms formed or modified by nivation processes (e.g. Christiansen, 1998a; Palacios *et al.*, 2003; Kariya, 2005; Figure 11.19), though there is sometimes limited morphological affinity between the features described by different authors. Common characteristics of nivation landforms include a well-defined break of slope between the backwall and the wash zone, a crescentic or scalloped backslope and small fans of washed sediment downslope. Relict nivation hollows or niches have been identified by several authors in areas that lay outside the limits of the last Pleistocene ice sheets (e.g. Gullentops *et al.*, 1993; Christiansen, 1996), and Christiansen (1998b) has shown that nivation activity in northeast Greenland

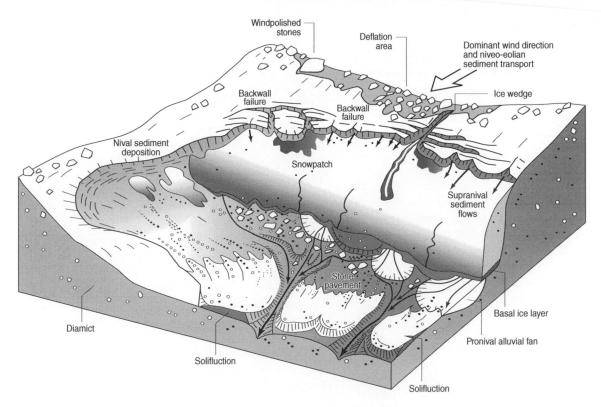

Figure 11.18 High arctic nivation model proposed by Christiansen (1998a) on the basis of observations in northeast Greenland. The model depicts the situation in late summer, with a perennial snowpatch occupying a nivation hollow that is being progressively enlarged by retrogressive failure of the backwall. *Source:* Adapted from Christiansen (1998a). Reproduced with permission of Wiley.

Figure 11.19 Small nivation hollow at 1120 m altitude on Ben Alder, Scottish Highlands. A late-lying snowpatch feeds meltwater to a flooded, partly vegetated wash zone.

was enhanced by climatic deterioration during the period AD 1250–1690 in response to intensified snowdrift and snowmelt runoff.

11.8 Cryoplanation

The term *cryoplanation* has been used in several ways. In its broadest sense, it implies long-term landscape evolution through scarp retreat in cold environments (Figure 11.20). More usually, the term is used to describe the development under cold conditions of broad, low-gradient bedrock benches separated by scarps or backslopes. Such *cryoplanation terraces* tend to occupy hillslopes with low to moderate gradients (usually between 6° and 25°), often imparting a step-like profile to the slope. Low-gradient bevelled summits surmounting flights of cryoplanation terraces are sometimes described as *cryoplanation surfaces*. Like 'nivation', 'cryoplanation' is also often employed as a collective term to designate a suite of formative processes, notably frost weathering, chemical weathering, solifluction and slopewash. Indeed, nivation is usually considered an integral component of cryoplanation, and the range of processes invoked to explain the formation of cryoplanation landforms is similar to that used to explain nivation landforms (Hall, 1998b; Thorn and Hall, 2002). Thus, the difference between features described as 'nivation benches' and 'cryoplanation terraces' may be one of size or maturity, not genesis (Margold *et al.*, 2011). An additional difficulty is that definitions of cryoplanation terraces and surfaces are usually based on morphology and location, and are so broad as to include all bench-like features on hillslopes in periglacial environments, irrespective of size and origin. There has consequently has been somewhat uncritical acceptance of 'cryoplanation' as the inevitable mode of origin of hillslope benches and summit flats, with insufficient consideration of alternative origins. The most convincing examples of 'true' cryoplanation terraces are those where erosional benches cut across the bedrock structure and form a flight of alternating treads and risers.

11.8.1 Cryoplanation Terraces

Step-like flights of bedrock terraces occur both in cold, arid continental interiors and in former periglacial areas such as central Europe (Washburn, 1979). Particularly striking examples occur in northern Yukon (Lauriol and Godbout, 1988; Lauriol, 1990). Relict cryoplanation terraces have also been identified in England (Gerrard, 1988), and possible examples occur in the Appalachians south of the limit of Pleistocene glaciation (Clark and Hedges, 1992). Miniature 'cryoplanation' benches in Antarctica have been described by Hall (1997b), and possible cryoplanation terraces occur at high altitude in the central Andes (Grosso and Corte, 1991). Reported dimensions vary widely. According to Priesnitz (1988) and Czudek (1995), terrace treads range in width from a few metres to more than a kilometre, though most are tens or a few hundreds of metres wide; terrace lengths (across slope) vary from tens of metres to a few kilometres; and backslopes (risers) are typically 1–20 m high. Tread gradients usually fall within the range 1–10°, and riser gradients within 15–40°, though some terraces are backed by steep rock scarps. Risers are usually composed of bedrock or coarse rubble. Treads are mantled by regolith or colluvium 0.5–2.0 m deep, comprising clasts embedded in fine sediment, though some support openwork blockfield debris. Several accounts refer to thermal contraction cracks, large-scale sorted patterned ground, solifluction lobes, loess or rills on tread surfaces; most report that terraces are vegetation-covered and apparently inactive. Cryoplanation terraces appear to be preferentially developed on well-jointed rocks that are susceptible to breakdown by frost wedging: Lauriol (1990) observed that in northern Yukon they occur on argillite, quartzite, limestone, dolomite and sandstone, but are absent on granite and shale. Some terraces exhibit structural control, but others cut across bedding planes and geological contacts. Tors occur on some cryoplanation features (Czudek, 1995), suggesting that long-term surface lowering has played a role in terrace evolution (Anderson, 2002).

With the possible exception of small 'cryoplanation' features in the Arctic and Antarctic (Hall, 1997b), almost

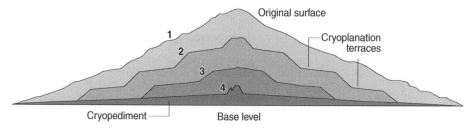

Figure 11.20 Idealized model of long-term landscape evolution by cryoplanation. Though some landscapes in cold continental interiors exhibit stepped surfaces, the extent of pre-Quaternary land-surface inheritance is uncertain. It is unlikely that the full evolutionary sequence depicted has ever reached completion.

all reported examples of cryoplanation terraces appear to be largely inactive, so the mode of formation has been inferred from present morphology. The most widely invoked explanation of terrace formation involves scarp retreat through nivation operating at the foot of risers and transport of weathered products across terrace treads by solifluction and running water (Czudek and Demek, 1971; Reger and Péwé, 1976; Lauriol, 1990; Czudek, 1995). A number of ancillary processes may also operate, such as comminution of sediment on treads by weathering, removal of fine sediment by wind erosion and solutional denudation (Lauriol *et al.*, 1997b; Lamirande *et al.*, 1999). Reger and Péwé (1976) argued that permafrost is essential for terrace formation, forming a 'base level' for erosion. This idea has found little support, though permafrost may play a role by confining water movement to the active layer, thereby enhancing sediment movement across treads by wash and solifluction. Altitudinal coincidence between cirques and cryoplanation terraces in Alaska appears to support a 'nivational' origin for the latter, as both cirque glaciers and perennial snowbeds imply proximity to the regional snowline in locations where conditions are marginal for glaciation (Nelson, 1989).

These explanations fail to account for the loss of sediment from terrace treads, however, unless tread sediment is transported over the backslope of subjacent terraces, which seems unlikely (Hall, 1998b). Alternatively, it is possible that all tread sediment is ultimately comminuted to fine particles and removed by wind or in solution, implying terrace development over very long timescales. Moreover, studies of nivation processes, and particularly of rock weathering (Chapter 10), suggest that scarp retreat through retrogressive 'nivation sapping' of bedrock is also likely to have operated very slowly, again suggesting that mature cryoplanation terraces must be very old. It is notable that well-developed cryoplanation terraces are invariably located outside the limits of Late Pleistocene glaciation, often in areas that remained unaffected by glacial erosion throughout the Pleistocene (Reger and Péwé, 1976; Lauriol and Godbout, 1988; Lauriol, 1990; Clark and Hedges, 1992; Czudek, 1995), suggesting that large cryoplanation terraces represent prolonged slope evolution. Büdel (1982) even suggested that the origin of cryoplanation features and cryoplanation surfaces lies in erosion surfaces formed during the Neogene but modified by periglacial processes during the Quaternary. This viewpoint is logical in the sense that hillslopes undergo continuous slow evolution, so cryoplanation terraces are likely to have evolved from slope irregularities formed under pre-Quaternary conditions in areas that escaped subsequent modification by glacial erosion (French, 2016). Irrespective of the degree of inheritance and mode of formation, it appears that the evolution of flights of mature cryoplanation terraces has involved the slow operation of a range of denudational processes over timescales of 10^5–10^6 years or longer.

No account of cryoplanation would be complete without mention of the 'Giant Steps' of Bug Creek in the Richardson Mountains of Canada. This vast natural staircase is excavated in sandstone and ascends a gentle (6°) slope in twelve vast subhorizontal terraces averaging 285 m in width, separated by rubble-covered scarps averaging 12 m in height. These features differ from most cryoplanation terraces in that the distal parts of treads support blockfields and 'block barricades' of weathered boulders that average 0.7 m in height and act as a retaining barrier, impeding valleyward movement of fine tread sediment. Sinkholes in this zone and springs emerging below backslopes suggest subterranean drainage despite the presence of continuous permafrost. Lauriol *et al.* (2006) suggested that these remarkable landforms represent modification by periglacial processes of pre-Pleistocene landforms, but their evolution, like that of cryoplanation terraces elsewhere, remains enigmatic.

11.8.2 Cryopediments

Often associated with cryoplanation terraces are *cryopediments*, low-gradient footslopes that descend from a mountain edge, merge downslope with alluvial deposits, and were formed or modified under periglacial conditions (Figure 11.20). As with cryoplanation terraces, many of the most impressive examples occur in cold, dry unglaciated continental interiors such as those of Yukon, Alaska and Siberia (Priesnitz, 1988). Relict cryopediments of Pleistocene age have also been identified in unglaciated parts of western and central Europe (e.g. Gullentops *et al.*, 1993; Vandenberghe and Czudek, 2008), and French (1973) has described small cryopediments in chalk areas of southern England. All of these areas escaped Pleistocene glaciation, implying either that cryopediments represent pre-Pleistocene landforms modified by periglacial processes or that they formed *ab initio* during successive Pleistocene cold stages.

Investigation of cryopediments in the Barn Mountains of northern Yukon led French and Harry (1992) to favour the first of these interpretations. The cryopediments that fringe mountains in Yukon are particularly impressive, descending gracefully from mountain scarps to adjacent valley floors over distances of up to a few kilometres (French and Harry, 1992; Fried *et al.*, 1993). Long profiles are straight or gently concave-upwards. Gradients typically average about 3°, but may reach up to 12° at the upslope end and as little as 1° at the pediment foot. Most meet the backslope at a marked break of slope, which may be covered with coarse rubble derived from upslope, and form a single uninterrupted surface, though younger, lower pediments are sometimes nested within older ones.

They cut across bedding planes, faults and other geological structures, tend to be best developed on weak sedimentary rocks and are underlain by continuous permafrost. Sediment cover rarely exceeds 1–2 m in depth, confirming an essentially erosional origin. Morphologically, the Yukon cryopediments appear to differ only superficially from pediments in hot arid environments. Development of the latter is usually attributed to prolonged erosion by channelized runoff and sheetflow under circumstances in which sediment transport capacity exceeds sediment supply, so that there is limited sediment storage on pediment surfaces. French and Harry (1992) observed that there is no evidence for such activity on pediments in the Barn Mountains, and inferred that these are essentially relict pre-Pleistocene landforms that were episodically modified during periods of the Pleistocene when higher precipitation favoured frost-wedging of backslopes and transport of colluvium across pediment surfaces by solifluction.

Researchers working on relict cryopediments in Europe, however, have argued that they were extensively developed during Pleistocene cold stages. This argument was developed by Vandenberghe and Czudek (2008), who compared the characteristics of cryopediments in Moravia with those in the lowlands of Belgium and the Netherlands. The former extend 250–1500 m from a pronounced backslope at gradients of 1–3°, cutting across weak Tertiary sediments. The latter are wider and very gently inclined (0.03–0.05°), lack a backslope, cut across unconsolidated Tertiary and Early Pleistocene sediments and are mantled by a thin (0.5–2.0 m thick) cover of Late Pleistocene aeolian sand. In both areas, some pediments have more than one level, implying the development of later, lower pediments at the expense of earlier surfaces. In Moravia, a Pleistocene or Plio-Pleistocene age is demonstrated by relationships with river terraces, and in the catchment of the Mark River on the Netherlands–Belgium border, relationships with overlying aeolian sand cover suggest that pediments were developing as recently as the LGM. Vandenberghe and Czudek (2008) attributed cryopediment formation in both areas to rapid snowmelt runoff over frozen ground, with vertical lowering of pediment surfaces dominating over backwall retreat.

The two explanations of cryopediments outlined above are not mutually exclusive. It appears that cryopediment development may have involved periglacial modification of pre-Quaternary pediments in unglaciated terrain, or initiation and development of cryopediments during Quaternary cold stages, at least on weak substrates. Like blockfields, tors, cryoplanation surfaces and cryoplanation terraces, cryopediments form members of a family of landforms that have developed under cold conditions over very long timescales in areas that escaped glacial modification, in some cases retaining characteristics of pre-Quaternary ancestry.

11.9 Slope Form and Slope Evolution

With the exception of stepped slopes that have evolved through the formation of cryoplanation terraces, hillslopes in present or former periglacial environments differ little in form from those elsewhere (Pawelec, 2011). In part, this is because slopes underlain by resistant bedrocks evolve very slowly over millions of years, so present-day hillslopes are largely inherited from the slopes that existed during the Neogene or earlier (André, 2003), or reflect glacial erosion or tectonic activity during the Quaternary. Following French (2007, pp. 216–224), we can identify four slope forms that are characteristic of periglacial environments (Figure 11.21). However, given the complexities introduced by Neogene inheritance, Quaternary climate change, sea-level change, glaciation, tectonic activity and variability in rock resistance, an infinite variety of slope forms are possible, and the four profiles illustrated are no more than idealized end-members of a continuum of possible forms.

Slopes dominated by bedrock cliffs (Figure 11.21a) are usually developed in massive resistant bedrocks with high compressive strength. They form through river incision, coastal erosion or steepening of pre-existing slopes by glacial erosion. Debris from cliffs accumulates at the foot to form a talus slope, characterized by an upper rectilinear slope at 34–38° and a basal concavity. A wide range of processes affect talus accumulations, notably debris flow, debris transport by snow avalanches and permafrost creep, which results in the formation of talus rock glaciers (Chapter 12).

Rectilinear debris-mantled slopes are straight slopes, usually inclined at angles of 28–38°, characterized by a thin veneer of debris overlying bedrock, though occasional outcrops of resistant bedrock may interrupt the debris mantle, or more resistant strata may crop out as low cliffs (Figures 11.21b and 11.22). Such slopes are characteristic of upland areas of fairly uniform geology that have not been steepened by glacial erosion. The restricted depth of the debris mantle implies that production of debris by weathering of the underlying bedrock is balanced by mass loss at the surface. Development of rectilinear debris-mantled slopes appears to indicate prolonged hillslope evolution during which the bedrock slope has adopted an overall gradient at which production and removal of debris are roughly in balance. Outstanding examples occur in Antarctica (Augustinus and Selby, 1990) and on sedimentary rocks in Svalbard, where smooth slope profiles are occasionally interrupted by low cliffs of more resistant bedrock. Selby (1971) argued that a combination of salt and frost weathering has been responsible for producing and comminuting the debris cover on such slopes in Antarctica, with removal of fine material by wind, but French and Guglielmin (1999) considered mass movement to be of greater importance

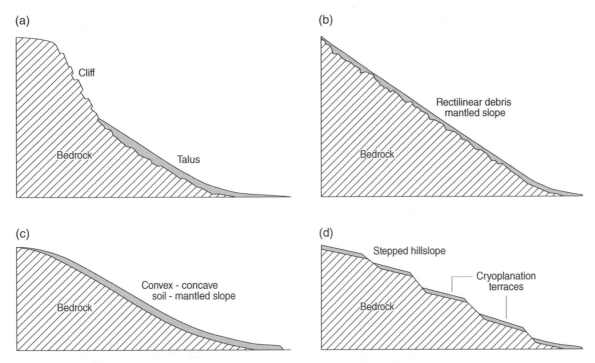

Figure 11.21 Slope forms in periglacial environments: (a) cliff and talus accumulation; (b) rectilinear debris-mantled slope; (c) convexo-concave slope characteristic of lowlands underlain by low-resistance bedrock; (d) stepped slope profile due to the development of cryoplanation terraces.

Figure 11.22 Rectilinear debris-mantled slopes, north Yukon, Canada. *Source:* Courtesy of Matthias Siewert.

in debris removal. Microrelief in the form of solifluction landforms, sorted stripes or debris-flow tracks
may be present.

Convexo-concave slopes (Figure 11.21c) dominate
present and former periglacial environments of low to
moderate relief. They are usually developed on less
resistant rocks, completely covered by soil or debris,
and rarely exceed 30°, though bands of more resistant
rock may crop out, forming crags or valley-side tors.
In periglacial environments, convexo-concave slopes are
usually dominated by solifluction, and may support a
microrelief of solifluction lobes, solifluction terraces
and/or sorted or nonsorted stripes. Nivation activity
related to late-lying or perennial firn patches may result
in the formation of localized breaks of slope with a
steeper backwall and gently inclined foreslope.

Stepped slopes (Figure 11.21d) are those on which
cryoplanation terraces have developed, and are characterized by alternation of gently sloping (1–10°) treads
and steeper (15–40°) risers up to 20 m high. Though
stepped slopes are cut in bedrock, treads support a
shallow (0.5–2.0 m) cover of colluvium or gelifluctate,
and tread microrelief may include large-scale sorted
patterned ground, solifluction lobes and rills. Stepped
slopes appear to be preferentially developed on well-
jointed lithologies, and are indicative of prolonged
hillslope evolution, uninterrupted by glacial modification, throughout much of the Quaternary. Low-gradient
cryopediments are also characteristic of areas that escaped
glacial modification during the Quaternary, and may
represent inherited pediments of Neogene age (French
and Harry, 1992) or, on weak substrates, low-gradient
slopes that developed during Quaternary cold stages
(Vandenberghe and Czudek, 2008).

Kirkby (1984, 1987) devised a numerical simulation
of slope development under periglacial conditions. He
assumed that regolith undergoes landsliding at gradients
greater than 26.5° and solifluction at lower angles, and
used empirical data on rates of solifluction to calibrate
rates of sediment movement. Figure 11.23a shows
Kirkby's model simulation for a 100 m-high slope with
an initial gradient of 70° and debris removal from the
slope foot. Under these assumptions, the model yields a
progressive decline in gradient and development of a
convexity at the crest. If no debris removal is assumed,
the model predicts the development of a basal concavity as
gelifluctate accumulates at the slope foot (Figure 11.23b)
and the evolution of a convexo-concave slope similar to
those in present and former periglacial environments.
The more general implication of Kirkby's model appears

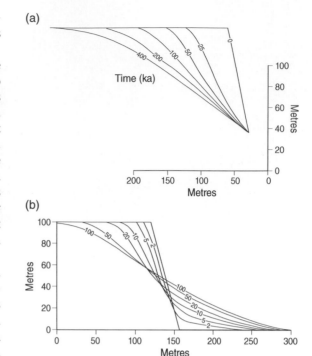

Figure 11.23 Models of periglacial slope evolution: (a) with basal
removal of debris; (b) with basal sediment accumulation. *Source:*
Kirby (1984). Reproduced with permission of Schweizerbart
Science Publishers.

to be that in areas of weak bedrock, initially steep slopes
may have evolved into much gentler convexo-concave
slopes within the last million years.

Another numerical model pertinent to periglacial
environments is that developed by Anderson (2002) to
explain the evolution of low-gradient, blockfield-covered
summit flats in the Wind River Range of Wyoming.
Anderson assumed a regolith production rate, determined
by cosmogenic isotope concentrations in blockfield
regolith and the underlying bedrock, of $14\,mm\,ka^{-1}$
(Small *et al.*, 1999), and that regolith production was
balanced by frost creep of blockfield debris towards
glacial troughs at the plateau edge. Intriguingly, his model
suggests that the plateau surface could have evolved from
progressive lowering and flattening of interfluves within
the past few million years. Applied to an initially irregular hillslope, this model may help to account for the
development of cryoplanation terraces, not as features
that develop through headwards sapping, but as the
outcome of progressive slope decline operating on terrace
treads. The difficulty, as with all attempts to model
hillslope evolution, is that the initial form of such slopes
is unknown, and probably unknowable.

12

Talus Slopes and Related Landforms

12.1 Introduction

Amongst the most widespread landforms in mountain environments are aprons of rockfall debris or *talus* that cover the lower parts of steep rock slopes (Figure 12.1). The slopes formed by such debris accumulations are known as *talus slopes*. Rockfall is the main process responsible for talus accumulation, but in periglacial environments talus slopes have often been modified by other processes, particularly snow avalanches, debris flow, permafrost creep and debris movement over snow and ice. Here we consider first the characteristics of rockfall talus accumulations, then examine the ways in which these are affected by other processes to produce a range of distinctive landforms.

12.2 Rockfall Talus

12.2.1 Characteristics

Talus slopes take three forms: talus sheets, where rockfall has been fairly uniform across slopes; talus cones, which result from concentration of rockfall debris downslope of gullies; and coalescing talus cones, formed by the lateral merging of individual cones (Figures 12.1 and 12.2). Geophysical profiling of talus accumulations in the Alps and Svalbard indicates maximum sediment thicknesses of 3–45 m overlying bedrock (Hoffmann and Schrott, 2002; Otto and Sass, 2006; Sass, 2006, 2007; Siewert *et al.*, 2012), though in some areas an intervening layer of till may separate talus deposits from the underlying bedrock.

Talus accumulations often appear to be composed exclusively of boulders, but exposures usually reveal a shallow surface layer of coarse openwork debris overlying a clast- or matrix-supported diamicton. The presence of fine sediment within talus accumulations implies that they have formed not only by accumulation of rockfall debris, but also by inwash of sand and silt grains released from source rockwalls by granular disaggregation. In permafrost areas, talus sediments beneath the active layer in talus deposits frequently contain ground ice; the

ice content in three periglacial taluses investigated by Scapozza *et al.* (2015) in the Swiss Alps, for example, range from 20% to 50%. The surfaces of rockfall-dominated talus slopes exhibit a general downslope increase in clast size. Such *fall sorting* has been explained in terms of the greater kinetic energy of larger clasts in motion, or as a consequence of the higher frictional losses experienced by small clasts travelling over rough surfaces. Once established, fall sorting is self-perpetuating, as individual clasts tend to accumulate amongst others of similar calibre. Particle shape also influences downslope sorting, as equidimensional clasts tend to roll or bounce farther downslope than platy clasts. Across-slope variations in clast size have been explained in terms of variable joint density in the rockwall source and by localized sliding and dry avalanching (grainflow) of talus debris.

12.2.2 Models of Talus Accumulation

In long profile, unmodified rockfall talus slopes comprise an upper straight slope and a basal concavity, though the latter may be absent where the slope foot is subject to erosion by rivers or coastal wave action. The straight slope usually has a gradient of 33–38°. Traditionally, this gradient was interpreted as the *angle of repose* of talus debris, or the gradient at which debris comes to rest after avalanching down the slope. This interpretation implies that debris accumulating near the slope crest periodically reaches an unstable angle and is redistributed downslope by debris avalanches (dry grainflows). However, Statham (1976) argued that the upper slope gradient is less than would be expected from redistribution of debris by grainflow, and pointed out that this process explains neither the basal concavity nor fall sorting. He suggested that the long profile of talus slopes reflects the travel distance of individual clasts falling on to the slope, together with downslope displacement of clasts by rockfall impact. Statham argued that clasts falling on to talus surfaces possess kinetic energy, and can therefore travel some distance down gradients lower than the theoretical repose angle. He considered that the basal concavity forms at an early stage in talus accumulation, when cliff

Periglacial Geomorphology, First Edition. Colin K. Ballantyne.
© 2018 John Wiley & Sons Ltd. Published 2018 by John Wiley & Sons Ltd.

Figure 12.1 Talus accumulations below a cliff of sedimentary rocks, Svalbard. (a) Coalescing talus cones below rock gullies. The lower-gradient cone is a debris cone fed by debris flows and snow avalanches. (b) Coalescing talus cones. The bench at the foot of the talus on the left is a talus rock glacier, formed by deformation of ice-rich permafrost within talus debris. A colour version of (a) appears in the plates section. *Source:* Courtesy of Ole Humlum.

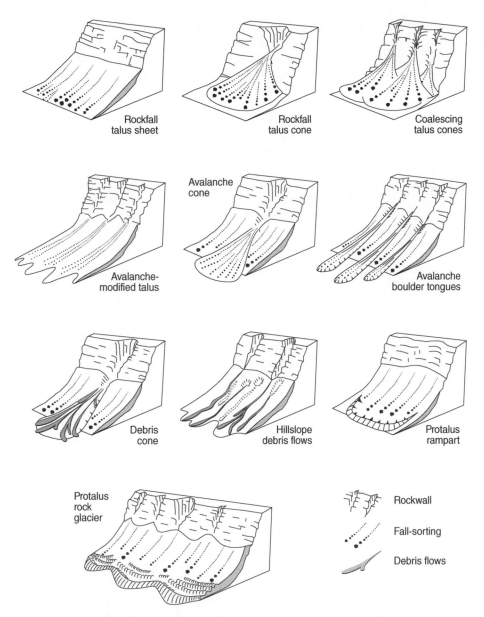

Figure 12.2 Types of talus slopes and related landforms. *Source:* Reproduced from Ballantyne and Harris (1994) with permission from Cambridge University Press.

height (and thus the kinetic energy of falling clasts) is high. Using a simple laboratory model, he also showed that as the upper straight slope increases in height, progressively fewer clasts have sufficient kinetic energy to reach the base of the slope, so that the length of the basal concavity diminishes as the upper straight slope grows (Kirkby and Statham, 1975). This is broadly in accord with research by Francou and Manté (1990), who found that talus at the foot of rockwalls in the Alps is characterized by a straight upper slope and a basal concavity separated by a point of inflection at a gradient of ~33°. They concluded that the upper slope is kept at a fairly constant gradient by the combined action of rockfall deposition and progressive downslope movement of

debris, whereas the basal concavity reflects rockfall deposition alone (Figure 12.3). A more detailed variant on this model has been proposed by Sanders *et al.* (2009) and Sanders (2010) based on the stratigraphy of talus in the Alps. Their research demonstrated that an immature, unsorted, low-gradient rockfall talus accumulated first, with subsequent build-up of a steep proximal slope dominated by fall-sorted rockfall inputs and episodic downslope redistribution of debris by grainflows. They showed that the proximal talus accumulations appear to have progressively prograded over the gentler distal slopes or concave debris fans as a result of inputs of debris deposited by large rockfalls and sediment transport from the proximal slope by debris flows and

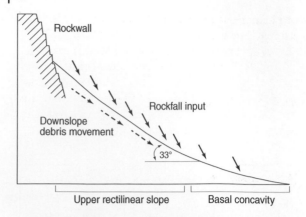

Figure 12.3 Model of talus development proposed by Francou and Manté (1990). The upper rectilinear slope experiences both rockfall input and downslope movement of debris. The basal concavity is a zone of rockfall input without significant downslope movement of debris. *Source:* Francou and Manté (1990). Reproduced with permission of Wiley.

other processes. Bithell *et al.* (2014) performed simulation experiments that offer a different perspective, suggesting that the initial stages of talus accumulation are characterized by large-scale debris avalanching events, replaced later by intermittent redistribution of debris by shallow debris sliding (grainflow) and infrequent disturbances that may extend the full depth of the deposit. Their model suggests that progressive build-up of debris on the upper slope triggers instability, mobilizing much of the talus accumulation. Talus accumulations thus evolve dynamically through time, though their evolution requires consideration not only of rockfall inputs, but also of the processes responsible for transporting debris downslope and their morphological implications.

12.2.3 Rockfall in Periglacial Environments

The term *rockfall* is generally employed to describe the fall of relatively small ($<100\,m^3$) bodies of rock debris that are released from cliffs by block detachment, toppling or sliding along joints, though some authors extend the term to include much larger slope failures involving up to $\sim 10^6\,m^3$ of rock. The impressive size of talus accumulations in alpine environments has led many commentators to conclude that rockfall is particularly pronounced under cold conditions, due to release of debris from cliffs by frost wedging operating within joints (Chapter 10). Several approaches have been developed for assessment of rockfall activity and associated weathering and retreat of rockwalls (Krautblatter and Dikau, 2007; Luckman, 2013). Average frequencies recorded for rockwalls at sites in the in the Rocky Mountains fall within the range 0.49–0.70 rockfalls per hour (Luckman, 1976; Gardner, 1983a). On Svalbard, Åkerman (1984) recorded an average of 1.5 rockfalls per hour on west-facing slopes and 1.1 per hour on east-facing ones.

These figures illustrate the high frequency of rockfall events in some periglacial environments, but take no account of the magnitude of individual falls or the area of contributing rockwalls, and represent only spring and summer rockfall activity.

A better method of evaluating rockfall activity is to measure the volume of debris deposited below a rock face of known area over a given time period, and to express the result in terms of average *rockwall retreat rate*. One approach is to measure the volume of debris resting on spring snowcover on talus slopes, but this has the disadvantage that summer rockfall is not included. Another approach involves establishing a grid of sampling surfaces across a talus slope and periodically measuring the volume of debris deposited on each surface, but this method suffers from the drawback that only a small proportion of the talus surface is sampled. A third approach has involved painting sections of rockwalls then estimating volumetric losses from fresh rock exposures, but this involves sampling only a small part of the cliff face. A problem with all these methods is that the sampling period is usually brief, so that the representativeness of the results is uncertain (Luckman, 2013). Some researchers have used lichenometry to assess spatial or temporal patterns of rockfall (Luckman and Fiske, 1995; McCarroll *et al.*, 1998, 2001; Sass, 2010), though this approach may be compromised by boulder remobilization and by burial of earlier rockfall debris by later rockfalls. More recently, terrestrial laser scanning and sequential digital terrain models based on aerial photography and airborne LiDAR (Light Detection and Ranging) have been employed to evaluate loss of rock and ice from alpine cliffs (Fischer *et al.*, 2011; Kenner *et al.*, 2011; Müller *et al.*, 2014).

Rockwall retreat rate over longer timescales may be assessed by estimating talus volume below a rockwall of known area. If the duration of talus accumulation (time since deglaciation) is known, then the average rate of postglacial rockwall retreat can be calculated as:

$$R_r = V_{talus} / \left(A_r \cdot t \right) \tag{12.1}$$

where R_r is rockwall retreat rate (mm a^{-1} or m ka^{-1}), V_{talus} is the volume of the talus excluding void spaces (m^3), A_r is the area of the contributing rockwall (m^2) and t is time in years since deglaciation. Talus depth is often difficult to estimate, however, though it may be determined by geophysical methods (Hoffmann and Schrott, 2002; Otto and Sass, 2006; Sass, 2006, 2007). The average postglacial rockwall retreat rate, however, is unlikely to be representative of present rockfall activity. This is because rockwall instability, and hence rockfall release, tends to be greatest in the centuries following rockwall exposure from under glacier ice and to diminish thereafter. Enhanced rockfall in the period following deglaciation probably reflects extension and dilation of the nearsurface joint network due to paraglacial stress release, possibly enhanced by frost-wedging operating within rock joints (Ballantyne, 2002; Hinchliffe and Ballantyne, 1999, 2009; Sanders *et al.*, 2009) and seismic activity triggered by glacio-isostatic uplift (Ballantyne *et al.*, 2014). In some mountain areas, past rockfall activity may also have been enhanced by increased frost-wedging of rock joints during periods of Holocene climatic deterioration, such as the Little Ice Age of the 16th–19th centuries AD (McCarroll *et al.*, 1998, 2001), and there is convincing evidence that the frequency of rockfalls of all magnitudes has increased over the past few decades as a result of warming and

thaw of permafrost in high-alpine rockwalls (e.g. Sass, 2005b; Rabatel *et al.*, 2008; Huggel *et al.*, 2010; Ravanel and Deline, 2010; Fischer *et al.*, 2012). Rockwall retreat rate is therefore likely to have varied markedly through time.

Table 12.1 summarizes estimates of rockwall retreat rate in arctic and alpine environments. Because of the uncertainties outlined above, these figures should be treated with caution. It is nonetheless notable that many 'arctic' retreat-rate estimates are markedly lower than those for alpine areas. This difference may partly reflect contrasts in freeze–thaw activity, with arctic sites experiencing a dominant annual freeze–thaw cycle whereas

Table 12.1 Measured rockwall retreat rates in periglacial environments.

Location	Lithology	Rockwall retreat ($m\,ka^{-1}$ or $mm\,a^{-1}$)		
		Minimum	Mean	Maximum
Arctic environments, present day[a]				
1. Yukon, Canada	Syenite, diabase	0.00	–	0.02
2. Yukon, Canada	Dolomite, quartzite	0.02	0.07	0.17
3. Spitsbergen	Limestone, sandstone	0.02	–	0.20
4. Northern Sweden	Micaschist	0.04	–	0.15
5. Franz Josef Land	Basalt	0.05	–	0.07
6. Northern Finland	Metasediments	0.07	0.18	0.60
Alpine environments, present day[a]				
7. Bavarian Alps	Limestone	0.01	0.10	0.40
8. Bavarian Alps	Limestone	0.01	–	0.17
9. Sierra Nevada, USA	Various	0.02	0.28	1.22
10. Bavarian Alps	Limestone	0.05	–	0.10
11. French Alps	Dolomitic Limestone	0.05	–	0.50
12. French Alps	Various	0.05	–	2.50
13. Canadian Rockies	Limestone	0.06	–	0.26
14. Polish Tatra Mts	Granite	–	0.70	–
15. Colorado Front Range	Various	–	0.76	–
16. Polish Tatra Mts	Limestone	0.10	0.84	3.00
17. Swiss Alps	Schist	0.13	–	0.36
18. Austrian Alps	Gneiss, Schist	0.41	0.85	1.46
Arctic environments, Holocene[b]				
19. Spitsbergen	Amphibolite	0.03	0.07	0.11
20. West Greenland	Basalt	0.03	1.04	4.17
21. Northern Finland	Micaschist, various	0.04	0.36	0.94
22. West Greenland	Basaltic breccia	0.05	–	2.40
23. Spitsbergen	Quartzite	0.10	0.70	1.58
24. Ellesmere Island	Dolomite, limestone	0.30	–	1.30
25. Spitsbergen	Sandstone, shales	0.33	–	1.96
26. Spitsbergen	Limestone, sandstone	0.34	–	0.50

(Continued)

Table 12.1 (Continued)

Location	Lithology	Rockwall retreat (m ka⁻¹ or mm a⁻¹)		
		Minimum	Mean	Maximum
Alpine environments, Holocene[b]				
27. Austrian Alps	Schist, gneiss	0.02	–	0.52
28. Sierra Nevada, USA	Various	0.02	–	1.22
29. Bavarian Alps	Limestone	0.06	0.10	0.37
30. Yosemite, USA	Granite	–	0.22	–
31. Blanca Mts, Colorado	Various	0.05	0.42	0.82
32. Bavarian Alps	Limestone	0.10	0.50	1.00
33. Nanga Parbat, Kashmir	Various	0.10	–	7.00
34. Swiss Alps	Schist, gneiss	0.12	–	3.10
35. Rocky Mts, USA	Various	0.30	–	4.60
36. Austrian Alps	Schist, gneiss	0.70	–	1.00
37. Austrian Alps	Gneiss, schist	0.42	0.64	0.89
38. Swiss Alps	Granite, gneiss	0.64	2.51	3.97
39. Swiss Alps	Various	0.80	–	1.50

[a] 'Present-day' rates are based on measurements of rockfall accumulation on talus or lichenometry.
[b] 'Holocene' rates are based on volume of talus accumulation since deglaciation.
Source: Based on data in Sass and Wollny (2001), Hoffmann and Schrott (2002), Curry and Morris (2004), Sass (2007, 2010), Moore *et al.* (2009), Siewert *et al.* (2012) and Götz *et al.* (2013).

alpine cliffs are subject to numerous diurnal freeze–thaw cycles. The main reason, however, may be the greater instability of high alpine cliffs, particularly in areas of large tectonic stress. Estimates of Holocene rockwall retreat rates in arctic environments tend to be greater than those based on recent debris accumulation, implying more rapid cliff retreat in the past, possibly due to the influence of infrequent high-magnitude rockfall events. In alpine environments, there is greater overlap between present retreat rates and mean Holocene rates, suggesting that some alpine rockwalls have remained unstable since deglaciation, or that recent rates have locally been enhanced by degradation of permafrost underlying steep rockwalls (Chapter 17).

Rockwall retreat rate is also conditioned by lithology and joint density. Olyphant (1983) demonstrated that fracture density has been the dominant factor determining the degree of talus development in Colorado, and studies by André (1993) in Svalbard showed that fissured rockwalls of quartzite and mica-schist yield much more debris than massive sandstones and conglomerates. Douglas *et al.* (1991) have shown that the density of microcracks in otherwise intact rock also controls the rate of debris release, and Moore *et al.* (2009) have demonstrated a negative exponential decrease in rockwall retreat rate with increasing rock mass strength. Krautblatter *et al.* (2012a) noted that rates of rockfall and rock-slope failure from carbonate rockwalls in the Reintal valley in the Alps are

markedly higher than those estimated for other alpine environments, and suggested that carbonate dissolution along potential sliding planes favours both low- and high-magnitude release of rock from cliffs. Such considerations help to explain the wide within-environment variability in rockfall rates evident in Table 12.1.

Rockfall in periglacial environments has been widely attributed to frost wedging, the prising open of cracks and joints by ice and subsequent detachment of debris from cliffs as joint ice thaws. Frost wedging is, however, only one of several processes operating on cliffs, and rockfall may also be triggered by stress release, earthquakes, thermomechanical stress, progressive failure along fractures, extreme rainfall events (Krautblatter and Moser, 2009) or build-up of hydrostatic pressure within joint networks. The case for frost wedging as a primary cause of rockfall in cold environments rests partly on data relating to rockfall timing. Rockfall inventories for Baffin Island (Church *et al.*, 1979) and some alpine areas (Luckman, 1976; Francou, 1988) indicate maximum rockfall activity during spring, as ice in joints thaws. At high elevations in the Alps, most rockfall occurs during the summer (Ravanel *et al.*, 2010). On Mount Ainodake in Japan, Matsuoka and Sakai (1999) found that rockfall rate peaked 5–15 days after meltout of the cliff face, when seasonal thaw penetration reached an estimated depth of 1 m, but was only rarely associated with diurnal freeze–thaw cycles or precipitation events.

Other rockfall inventories emphasise diurnal variations in rockfall activity. Diurnal freeze–thaw cycles are frequent on alpine cliffs and penetrate to depths of 50 cm (Matsuoka, 1994); climbers scaling alpine cliffs are acutely aware that warming of the rock face after a night of below-freezing temperatures may trigger a cascade of boulders. In the Rocky Mountains, rockfall frequency peaks at mid-day, when sunlight reaches frozen cliffs (Gardner, 1983a), and a similar effect is evident on Svalbard (Åkerman, 1984). In both areas, however, rockfalls also coincide with summer rainstorms, implying that build-up of hydrostatic pressure in joints also releases rockfall debris. At high altitudes in the Nepal Himalaya, seasonal variation in rockfall activity is evident on north-facing slopes that remain frozen throughout the winter, but cliffs facing other aspects release rockfall throughout the year in response to diurnal freeze–thaw cycles (Regmi and Watanabe, 2009). Analysis of the temperature conditions associated with small- to medium-scale rockfall events at high altitudes in the Alps suggests that most rockfalls occurred during brief periods of exceptional high temperatures, a pattern interpreted by Luethi *et al.* (2015) in terms of rock-slope destabilization due to advective thaw, or possibly stress redistribution caused by large temperature variations.

The importance of freeze–thaw events in rockfall release is supported by a study of the distribution of talus accumulations in the Southern Alps of New Zealand, which suggests that most taluses are located immediately below the zone of maximum potential frost cracking (Hales and Roering, 2005). Evidence for freeze–thaw-generated rockfall also comes from monitoring of frost wedging in bedrock cracks (Matsuoka, 2008), which demonstrates slight crack opening during diurnal freeze–thaw cycles, greater opening (0.1–0.5 mm) during seasonal freeze-back and pronounced crack expansion (>0.5 mm) due to refreezing of meltwater in cracks during seasonal thaw, potentially triggering subsequent rock detachment.

Measurements of rock temperatures and joint dilation associated with steep bedrock slopes underlain by permafrost, however, suggest a more complex response that involves not only crack opening due to a combination of frost action and thermomechanical forcing during periods of freezeback, but also shearing and expansion along joints during rock-face warming (Hasler *et al.*, 2012). The latter effect reflects reduction in shear strength at rock–rock contacts, rock–ice interfaces or ice-cemented infills within joints due to conductive or advective warming of the rock (Hasler *et al.*, 2011), possibly combined with an increase in cleft-water pressures associated with melt of snow and ice. It offers an explanation for exceptional rockfall activity in the European Alps in the hot summer of 2003 (Gruber *et al.*, 2004a) and for evidence of enhanced rockfall following periods of exceptionally warm weather (Allen and Huggel, 2013).

Warming and thaw of permafrost (ice-filled fractures) is also implicated as a cause of large (10^4–10^7 m^3) rockslides or rock avalanches originating from rockwalls underlain by permafrost (Fischer *et al.*, 2012; Huggel *et al.*, 2012). Warming and thaw of ice in ice-bonded rock reduces rock mass strength (Davies *et al.*, 2001), and there are reports of massive ice exposed in the detachment zones of recent rockslides, suggesting that rising temperatures in ice-bonded rock masses triggered failure. The mechanics of rock-slope failure associated with warming and thaw of ice-bonded rocks are complex, involving *inter alia* ductile deformation of ice within joints, failure of rock–ice contacts, changes in hydrostatic pressure and fracture of internal rock bridges. A rock–ice mechanical model developed by Krautblatter *et al.* (2013) suggests that warming affects ice-bonded permafrost rockwalls at two temporal scales: (i) slow, deep subcritical destabilization, due primarily to rock-mechanical changes operating over timescales of months to millennia; and (ii) rapid deformation along existing shear planes, reflecting accelerated ice creep, rock–ice detachment and reduction in total friction. The former may partly explain a delay in the timing of major rock-slope failures following episodes of rapid Lateglacial or Holocene warming (McColl, 2012; Ballantyne *et al.*, 2014), the latter the relationship between exceptionally warm summer temperatures and enhanced rockslide or rockfall frequency in the European Alps. Slow subcritical rockslide deformation may also exhibit seasonal patterns, with more rapid deformation following the spring release of snow meltwater, which releases latent heat on refreezing at depth, and warming of ice in fractures to nearly 0 °C. Conversely, cold-air ventilation within joint networks during autumn and winter lowers ice temperatures, temporarily reducing rockslide deformation (Blikra and Christiansen, 2014).

12.2.4 Talus Shift

The term *talus shift* describes movement of debris down talus surfaces, irrespective of the process involved. On talus slopes monitored by Gardner (1979) in the Rocky Mountains, some boulders remained stationary over several years, but others exhibited movement of up to 70 m. Average downslope displacements ranged from 13 to 88 cm a^{-1}, with large spatial and temporal variations. Measurements elsewhere confirm the highly variable and episodic nature of talus shift (Luckman, 1988), but suggest that the rates recorded by Gardner are unusually high. In the Colorado Rockies, for example, Caine (1986) recorded average displacements of generally less than 3 cm a^{-1}. Such contrasts probably reflect differences in the processes involved. Gardner (1979) inferred that the main causes of boulder movement were rockfall impact, snow avalanches, debris flows, surface runoff and slope failure. In contrast, debris flow and avalanche activity were of limited importance at the sites monitored by Caine (1986), except below gullies. Rates of talus shift tend to diminish downslope, reflecting a downslope increase in boulder size and decreased frequency of rockfall impact.

Though rockfall impact, dry grainflows, debris flow and snow avalanches are the main agents of talus shift on slopes below rockwalls, a different set of processes influences debris transfer down taluses where the headwall has been almost buried by talus accumulation. At such a site in the Andes, Pérez (1993) found that the displacement of coarse debris averaged $3.7\,\text{cm}\,\text{a}^{-1}$, but smaller clasts had moved downslope at an average rate of $15.2\,\text{cm}\,\text{a}^{-1}$ and areas of gravelly sand recorded an average shift of $22.5\,\text{cm}\,\text{a}^{-1}$. He concluded that though rockfall impact contributes to sediment movement, finer debris also experiences superficial creep through diurnal needle-ice creep and displacement by small-scale debris slides and flows.

12.2.5 Stratified Talus Deposits

Exposures within talus accumulations often reveal crude stratification in the form of stacked inversely-graded beds, layers of contrasting particle size, alternating beds or lenses of openwork and diamict facies, or some combination of these. Such structures are referred to as *stratified talus deposits* or *stratified scree*, and are produced by several processes (Van Steijn *et al.*, 1995, 2002; Garcia-Ruiz *et al.*, 2001; Sass, 2006; Sass and Krautblatter, 2007; Van Steijn, 2011; De Vet and Cammeraat, 2012). Rockfall impact or slope oversteepening may trigger *dry grainflows* that move downslope through multiple particle collisions that maintain dispersive stresses within the mobile debris. During movement, kinematic sieving results in inverse (coarsening upward) grading, and over-riding of the main mass by finer debris in the final stages of movement produces crude stratification (Figure 12.4a). Hétu *et al.* (1994) identified a variant of this process termed *frost-coated clast flows*, in which debris movement triggered by rockfall impact is facilitated by a thin ice coating on clast surfaces, which reduces interparticle friction. Frost-coated clast flows on talus at Gaspésie (Quebec) follow tracks 1–3 m wide and up to 500 m long, with well-developed marginal levées and coarse distal lobes. Successive over-riding of flows creates stratification, with finer beds representing channel deposits and coarse beds representing levées (Figure 12.4c). Recurrent debris flows (see below) may also produce lenticular stratification within talus due to alternation of diamict facies (from channel sides and the inner parts of levées) with coarse openwork debris (representing channel deposits or the outer parts of levées; Nieuwenhuijzen and Van Steijn, 1990).

Slow downslope movement of stone-banked solifluction lobes (Chapter 11) may also result in the formation

of stratified slope deposits, particularly in high mountain areas (Francou, 1990; Bertran *et al.*, 1995). In such cases, the stony fronts of mobile sheets and lobes are buried by matrix-rich sediment as debris advances downslope (Figure 12.4b). The result is lenticular or continuous stratification, with a coarse inverse-graded surface layer overlying a matrix-rich unit then a further normally-graded openwork layer. Stratification in talus may also be produced by remobilization of grainflow deposits by slopewash (Van Steijn and Hétu, 1997), burial of grainflow or rockfall deposits by debris flows (Harris and Prick, 2000) or the alternation of mass-movement processes such as dry grainflow or debris flow with episodes of niveo-aeolian sediment deposition (Hétu, 1991, 1995; Figure 12.4d).

Although stratified talus deposits have often been assumed to reflect modification of talus under periglacial conditions, dry grainflow, debris flow and slopewash operate under a wide range of climatic conditions. Solifluction, frost-coated clast flows and niveo-aeolian deposition imply cold but not extreme conditions. Interpretation of the environmental significance of stratified talus deposits is, moreover, complicated by convergence in the sedimentological signatures of different stratification processes, so that sedimentological analyses may not always produce unambiguous conclusions regarding former stratification processes or environmental conditions (Van Steijn *et al.*, 1995, 2002).

12.2.6 Relict Talus Accumulations in Former Periglacial Environments

In mid-latitude areas that experienced periglacial conditions during the Pleistocene, relict talus accumulations are widespread at the foot of cliffs. Such relict taluses often support vegetation cover and exhibit modification by gullying and debris flows. In formerly glaciated areas, relict taluses appear to have formed mainly under Lateglacial cold conditions, with limited accumulation of debris under the milder conditions of the Holocene (Hétu and Gray, 2000). In Scotland, for example, present rockfall delivery appears to be one to two orders of magnitude lower than rockfall rates during the Younger Dryas Stade of ~12.9–11.7 ka (Hinchliffe and Ballantyne, 1999, 2009). During the Younger Dryas Stade, frost wedging of rockfall debris was probably favoured by a combination of strong insolation with much colder air temperatures than at present, though stress release within glacially-steepened rockwalls probably also enhanced Lateglacial rockfall activity. André (1997) has shown that stress release following recent deglaciation in Svalbard has been roughly an order of magnitude more effective than frost wedging in promoting rockfall from cliffs. It is possible that many relict talus slopes are primarily the products of paraglacial stress release operating within

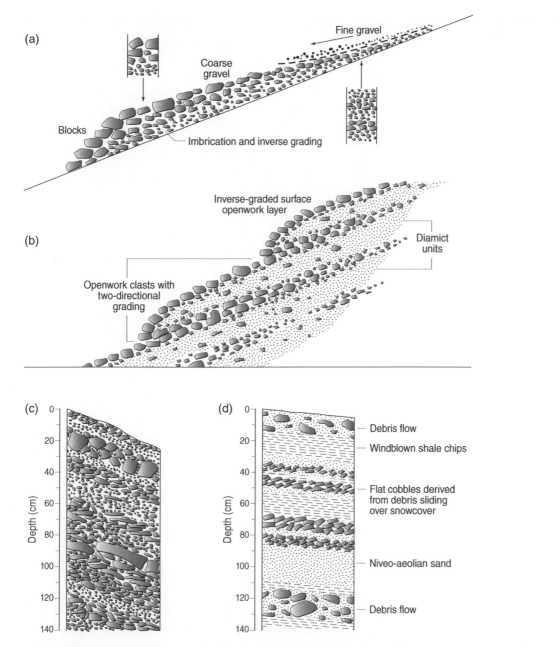

Figure 12.4 Mechanisms of talus stratification. (a) Dry grainflow. (b) Downslope advance of stone-banked solifluction lobes. (c) Frost-coated clast flows (logged section). (d) Alternation of sediment movement and niveo-aeolian deposition (logged section). *Source:* (a), (c) and (d) from Van Steijn *et al.* (1995). Reproduced with permission of Wiley.

rockwalls, together with major rockfalls attributable to thaw of permafrost ice in joints, and that the role of frost wedging in releasing debris from cliffs after deglaciation was secondary.

Studies of relict talus accumulations in upland Britain have shown that ~30% by weight of constituent sediment consists of fine (<2 mm) material produced by rockwall weathering, implying that nearly a third of rockwall retreat since deglaciation reflects granular disaggregation rather than clast detachment (Hinchliffe *et al.*, 1998; Curry and

Black, 2003). These studies have also shown that the upper parts of relict taluses are characterized by stacked debris-flow deposits, implying extensive reworking of talus slopes by recurrent debris flows after the main period of rockfall accumulation had ceased. Exposures within lithified calcareous talus deposits of Late Pleistocene age in the Alps also provide evidence for extensive reworking of rockfall debris by mass movement, in the form of debris-flow deposits intercalated with openwork breccias representing dry grainflows (Sanders, 2010).

12.3 The Geomorphic Role of Snow Avalanches

In many mountain areas, snow avalanches play a major role in modifying the form of talus slopes and in transporting debris across talus surfaces. Here we consider first the characteristics of snow avalanches and then their effects on the landscape.

12.3.1 Characteristics of Snow Avalanches

The factors causing snow avalanche release are complex, but three principal types can be identified: loading of the snowpack by snowfall (*direct action avalanches*), development of structural discontinuities within the snowpack (*delayed action avalanches*) and loss of cohesion within snow, usually as a result of thaw (*wet snow avalanches*). Direct and delayed-action avalanches usually occur during winter and wet-snow avalanches during spring and early summer, though wet slab avalanches may be triggered in winter by positive air temperatures and rainstorms (Eckerstorfer and Christiansen, 2012). Avalanches vary in size from miniature *sluffs* to large avalanches incorporating 10^5–10^6 m^3 of snow. Velocities vary greatly, but are often 20–40 m s^{-1} for dry snow avalanches and 10–30 m s^{-1} for wet snow avalanches. The shear stress at the base of most avalanches tends to be low (0.5–5.0 kPa), but impact pressures are of the order of 10–1000 kPa.

Avalanche size and frequency reflect climatic and topographic controls (Butler and Walsh, 1990; Butler *et al.*, 1992; Laute and Beylich, 2014): snow avalanching is most frequent in mountains that experience abundant snowfall and development of a thick seasonal snowpack, and on slopes where snow accumulation is favoured by topography and dominant wind direction. Optimal gradients for snowpack accumulation and release fall within the range 25–55°, and especially 35–45°. The main sites of avalanche initiation are open slopes, gullies, cliff ledges and lee-side slopes where drifting snow accumulates. Although some avalanches, such as plateau-edge cornice falls, entrain weathered bedrock (Eckerstorfer *et al.*, 2013), most sediment-rich avalanches are full-depth avalanches initiated at the ground surface, wet-snow avalanches that cross snow-free ground in spring or avalanches that erode underlying snowcover as they move, though even when an avalanche crosses exposed loose debris it may have limited erosional impact. The process of debris entrainment is poorly understood. Wet, cohesive snow avalanches entrain debris by 'snowballing', whereas less cohesive wet snow avalanches erode by scouring underlying debris, but only where flow is sufficiently concentrated (Jomelli and Bertran, 2001). The erosional potential of avalanches following tracks below the tree line is usually reduced by vegetation cover.

12.3.2 Avalanche Modification of Talus Slopes

One major effect of snow avalanches is erosion of debris from upper talus slopes and its redeposition farther downslope (Luckman, 1977; Figure 12.2). Avalanches striking or originating on the upper parts of talus slopes may strip away surface clasts, exposing underlying finer debris or bedrock. The effects on talus form are an overall reduction in gradient on the upper slope and formation of a long basal concavity that may even extend a short distance up the opposite slope (Luckman, 1978; Jomelli and Francou, 2000). Downslope redistribution of debris tends to destroy fall sorting. Avalanche-transported clasts are typically angular, reflecting collisions during transport, and are frequently deposited on top of boulders (Figure 12.5a,b). Such *perched clasts*, and associated drapes of soil across boulder surfaces, reflect deposition from ablating snow. Also diagnostic of avalanche-modified talus are *debris tails*, which are ridges of debris up to 1 m high and 5–15 m long that taper downslope in the lee of large boulders and reflect scouring by avalanches of the adjacent talus surface.

12.3.3 Other Aspects of Snow-Avalanche Erosion and Deposition

Avalanches may also contribute to the formation of *avalanche chutes*, which are gutter- or funnel-shaped gullies, often with a U-shaped cross-profile, that are incised into upper rockwalls (Figure 12.5c). Accumulation of wind-blown snow at the crests of avalanche chutes forms deep overhanging snow cornices in winter. Eckerstorfer *et al.* (2013) have shown that in sedimentary rocks on Svalbard, conditions under cornices are highly favourable for frost weathering of the underlying rock by ice segregation. Loosened debris is incorporated into the base of the cornice as it accumulates, plucked from the plateau edge as the cornice tilts away from the slope crest, then transported on to the talus below when the cornice collapses. Rapp (1959) suggested that avalanche-transported debris also abrades the walls and floors of chutes and detaches rock on impact. He noted that avalanche chutes on upper slopes appear too deep to have formed during the Holocene alone, and suggested that upper slopes deeply incised by avalanche chutes may represent the flanks of nunataks that remained above the last Pleistocene ice sheets; alternatively, avalanche chutes may have been preserved under cold-based glacier ice with limited erosive potential during successive episodes of Pleistocene ice-sheet glaciation.

High-velocity, laterally-confined snow avalanches generate large impact stresses, and can excavate *snow avalanche impact pits* at sites where avalanche tracks terminate abruptly at the valley floor. The debris thrown up by avalanche impact often forms an asymmetric ridge

Figure 12.5 (a) Snow avalanche debris on a talus slope, Longyeardalen, Svalbard. (b) Snow avalanche debris after deposition. Note the loose, angular, 'perched' debris emplaced by ablating snow. (c) Snow-filled avalanche chutes feeding snow avalanches that have descended over talus, Larsbreen, Svalbard. (d) Avalanche impact rampart, Jostedalen, Norway. Wet-snow avalanches occur annually down the slope right of the river, and the force of impact has ejected rounded boulders from the riverbed on to the opposite bank, forming the rampart. *Source:* Courtesy of (a,c) Markus Eckerstorfer, (b) Hanne Christiansen and (d) Ole Humlum.

or *avalanche impact rampart* on the far side of the pit (Figure 12.5d). Avalanche impact ramparts tend to have steep proximal slopes and gentle distal slopes. They usually rise less than 5 m above the surrounding terrain, though ramparts up to 9 m high occur in the Canadian Rockies (Smith *et al.*, 1994). In New Zealand, individual pits range from 200 to 50 000 m^2 in area (Fitzharris and Owens, 1984). Corner (1980) classified avalanche impact features into *impact tongues*, where avalanches have excavated debris from a stream bed and spread it as an apron or rampart across the far bank, *impact pits*, which consist of water-filled depressions with similar distal debris accumulations, and *impact pools*, where avalanches entering a water body have excavated a basin bounded distally by a partly submerged ridge of debris. The most common impact landforms are probably associated with avalanches impacting stream channels and spreading an apron of ejected material across the far bank (Figure 12.5d), sometimes in the form of a radial splay of debris that extends beyond the rampart (Owen *et al.*, 2006). In Jostedalen, Norway, Matthews and McCarroll

(1994) found that 50–60% of rampart debris had been excavated from the floor of a river channel, the remainder consisting of clasts eroded from the avalanche track. Similarly, Smith *et al.* (1994) calculated that the volume of debris in avalanche impact ramparts in the Canadian Rockies exceeds that excavated from riverine impact pools, implying an additional component of avalanche-transported debris. The large impact ramparts in the Rocky Mountains apparently represent deposition of copious volumes of debris by infrequent high-magnitude impact events (Johnson and Smith, 2010). Those investigated in Jostedalen receive frequent but smaller additions of debris (Matthews *et al.*, 2015).

The most widespread landforms produced by avalanche deposition are *avalanche boulder tongues*, which extend across valley floors beyond the foot of talus slopes. At their most distinctive, they form *roadbank tongues*: embankments of avalanche debris that rise several metres above the valley floor and are markedly concave downslope but convex, flat-topped or asymmetrical in cross-section (Figure 12.6). In the Canadian Rockies, the

Figure 12.6 Small avalanche boulder tongue (roadbank tongue) flanked by debris-flow tracks, Southern Alps, New Zealand.

largest examples are up to 200 m wide, 700 m long and 15–25 m thick. Roadbank tongues reflect reworking of talus by avalanches that follow the same track year after year. In contrast, *avalanche fan tongues* are steeper, fan or splay-shaped veneers of debris deposited by less laterally-constrained avalanches; many fan tongues represent reworking of talus debris by both avalanching snow and debris flows. Luckman (1977) distinguished between fan tongues and *avalanche runout fans* that reflect deposition of debris by avalanches on valley floors below the treeline. The latter resemble alluvial fans but display evidence of avalanche activity, such as perched boulders, angular debris, soil drapes and damaged vegetation (Butler *et al.*, 1992; Ballantyne, 1995).

12.3.4 The Significance of Snow-Avalanche Activity

The importance of snow avalanching as a geomorphic agent is conditioned by both climatic and topographic controls. In some mountain areas, talus slopes are largely unaffected because avalanche activity is restricted by limited snowfall and the steepness of some cliffs, which inhibits snow accumulation. Avalanches are also of very localized geomorphic importance in wetter but milder environments such as the Scottish Highlands, where winter thaw limits snow accumulation (Ballantyne, 1989; Luckman, 1992). In mountains that receive heavy snowfall, however, the volumes of sediment transported by snow avalanches may be substantial. In the Torlesse Range

of New Zealand, roughly 40 000 m^3 of debris accumulated in 23 years at a site of recurrent avalanche activity (Ackroyd, 1986). Perhaps more typically, Luckman (1988) found that up to 50 m^3 of debris is transported by individual avalanches in the Canadian Rockies, and he measured avalanche debris accumulation rates averaging 0.01–5.6 mm a^{-1} over 13 years. Similar accumulation rates have been recorded in other high mountain areas (Gardner, 1983b; Bell *et al.*, 1990). André (1990a) recorded debris accumulation rates of 0.04–8.3 mm a^{-1} on Svalbard, with 'dirty' avalanches depositing up to 40 mm of sediment in a single year. As most avalanche debris represents reworking of talus rather than erosion of bedrock, however, rockfall *input* into talus debris systems generally exceeds that of avalanching snow (Luckman, 1988).

12.3.5 Relict Avalanche Landforms and Deposits

There are few accounts of relict avalanche landforms, possibly because many Pleistocene avalanche deposits have been reworked or buried by later slope processes. In the Scottish Highlands, relict roadbank tongues 30–50 m wide and 5–7 m thick may originally have been deposited during the Lateglacial period, though the survival of fragile debris tails suggests modification by more recent avalanche activity (Luckman, 1992). Debris cones below rockwall gullies in former periglacial environments may have accumulated partly by snow-avalanche deposition, even though the surfaces of such cones have been reworked by debris flows. In sections exposed within debris cones ('colluvial fans') in western Norway, Blikra and Nemec (1998) identified snow-avalanche facies characterized by abnormally large clast sizes relative to bed thickness, uneven or discontinuous bed geometries, predominantly openwork textures, coarse normal grading and disorganized clast fabric due to settling of debris during snowmelt. Radiocarbon dating of intercalated organic soils showed that avalanche deposition was pronounced during the Younger Dryas Stade and during episodes of Late Holocene climatic deterioration, suggesting that snow-avalanche facies have potential for identifying episodes of climatic deterioration in Quaternary stratigraphies (Blikra and Selvik, 1998). Bertran and Jomelli (2000), however, have described snowflow facies in the Alps that comprise an upper coarse openwork layer overlying a clast-supported diamicton. They suggested that some of the sedimentological features attributed by Blikra and Nemec (1998) to snow avalanche facies may also occur within rockfall debris and possibly debris-flow deposits, implying that identification of snow avalanche facies in exposures may be problematic.

12.4 Debris-flow Activity

The term *debris flow* describes the rapid downslope flow of poorly sorted debris mixed with water, and is also used to describe the tracks followed by individual flows. Debris flows typically move at average velocities of a few metres per second. Movement is distinct from fluvial transport in that water content is usually only 10–40% by weight and the sediment–water mixture undergoes flow *en masse*. Two types of debris flow are widespread on steep slopes: *hillslope flows* on open slopes, and *valley-confined flows*, which originate in bedrock gullies and are channelled for part of their length along the gully floor. In periglacial environments, the main effect of debris flows is usually reworking of talus, though flows also occur on slopes mantled by frost-weathered regolith or glacigenic sediments. In particular, debris flows often rework large volumes of glacigenic deposits on recently deglaciated terrain (Ballantyne and Benn, 1994a; Curry, 1999). In general, the runout distance of hillslope flows increases with the volume of mobilized sediment, though laboratory experiments suggest that small increases in water volume augment runout distance but that increases in clay content diminish it (Hürlimann *et al.*, 2015).

12.4.1 Characteristics of Debris Flows

Debris flows follow long, narrow tracks that emanate from upper-slope gullies and are continued downslope by parallel levées (marginal ridges) that terminate in one or more lobes of bouldery debris (Figure 12.7). Gullies eroded in unconsolidated deposits on upper slopes usually decline in width and depth downslope, and levées characteristically diminish in height towards the slope foot. The size of flow tracks is related to the volume of the flow, the abundance and coarseness of debris and the length of the runout slope. In the Alps, for example, Van Steijn *et al.* (1988) recorded flow lengths of 240–570 m and widths of 3–30 m on mean gradients of 19–28°. Levées, track deposits and lobes comprise heterogeneous matrix-rich sediment, frequently clast-supported, though the outer parts of levées may consist of openwork clasts. Individual flow deposits often exhibit inverse grading, coarsening both upwards and outwards as a result of dispersive pressures during flow. Where hillslope flows cross talus slopes, percolation of water and fine sediment through the underlying talus debris results in sieve deposition of clasts in the form of flat-topped lobes (Shakesby and Matthews, 2002). In some cases, thin, low-gradient sediment splays are deposited in the footslope zone. Such slope-foot deposits locally exhibit a stratigraphic sequence consisting of cobble-rich diamicton overlain in turn by a pebble-rich diamicton, pebbly sand lenses, massive silty sand or sandy silt, and thin spreads of laminated sands and silts (Matthews *et al.*, 1999).

Debris flows erode sediment from the upper parts of talus slopes and deposit it near or beyond the talus foot, ultimately lowering mean talus gradient and producing a long basal concavity. Repeated flows from the same source deposit moderate-gradient (12–25°) *debris cone*s that extend beyond the talus foot (Figures 12.1a and 12.2), and large valley-confined flows may also contribute to the accumulation of *alluvial fans* with gentler (<12°) gradients. Like snow avalanches, debris flows tend to destroy fall sorting on talus by depositing small clasts and fine sediment on lower slopes.

12.4.2 Debris-flow Processes

Debris flows occur when increasing pore-water pressure within sediments reduces shearing resistance, leading to slope failure and flow. Shearing resistance is expressed as:

$$SR = c' + (W \cos \alpha - u) \tan \phi' \qquad (12.2)$$

where c' is the effective cohesion of the debris (usually negligible in talus), W is the weight of debris at a potential failure plane, α is the surface gradient, u is pore-water pressure and ϕ' is the effective friction angle of the sediment. As pore-water pressure increases, the frictional strength of the debris $[(W \cos \alpha - u) \tan \phi']$ is reduced. Failure occurs when shearing resistance falls below $(W \sin \alpha)$, the shearing force generated by the downslope component of debris weight.

Figure 12.7 Downslope transport by debris flows of quartzite debris across darker sandstones, Beinn Eighe, Scotland. The levées of some individual flow tracks are clearly evident. A colour version of this figure appears in the plates section.

Many debris flows in periglacial environments originate as shallow translational (slope-parallel) landslides, typically 0.8–1.5 m thick, on steep (25–45°) slopes (Zimmermann and Haeberli, 1992; Hürlimann *et al.*, 2015). On slopes underlain by permafrost, failure may involve the full depth of the thawed layer. The transition from sliding to flow involves progressive liquifaction and remoulding of the sliding mass, with the addition of further water increasing intergranular distances and decreasing the strength of the moving debris. Other modes of debris-flow initiation also occur: flood torrents in gullies may be transformed into debris flows by the addition of debris from the channel bed and sides, and impulsive loading of saturated bed sediments may initiate debris flow (Bovis and Dagg, 1992). Van Steijn *et al.* (1988) invoked failure of debris dams in gullies during rainstorms as an important process of debris-flow generation on talus slopes in the French Alps.

Though rapid snowmelt or melting ground ice may initiate debris flows (DeGraff, 1993; Harris and Gustafson, 1993; Ballantyne and Benn, 1994a), most are triggered by intense rainstorms. Various attempts have been made to define threshold conditions of rainfall intensity and duration for the onset of sliding failure (Guzzetti *et al.*, 2008), but these offer only approximate guidelines

as propensity for failure is also influenced by antecedent soil-moisture conditions (Church and Miles, 1987), gradient, and the depth and texture of the soil. Watanabe (1985), for example, estimated that debris flows in the Japanese Alps occur when precipitation exceeds 120 mm in 24 hours, but widespread debris-flow activity occurred in Longyeardalen, Svalbard, after only 31 mm of rain in 12 hours (Larsson, 1982). This difference reflects permafrost at shallow depth under the Svalbard talus slopes, as permafrost impedes downwards drainage of water and thus allows rapid build-up of pore-water pressures in the active layer. In the European Alps, debris flows during the snowmelt season may be triggered by modest rainstorms (<20 mm in 24 hours) because of the additional contribution of snow meltwater (Schneuwly-Bollschweiler and Stoffel, 2012). Kotarba *et al.* (1987) illustrated the influence of antecedent soil-moisture conditions by observing that rainfall intensities of 20 mm h^{-1} trigger debris flows in the Tatra Mountains after a period of wet weather, whereas intensities of 40 mm h^{-1} may fail to generate flows if the ground is initially dry. In 1987, more than 600 debris flows occurred within three areas of the Swiss Alps during two separate rainstorm events. The first involved 100–180 mm of precipitation, but rainfall intensities seldom exceeded 12 mm h^{-1}.

The second period of debris-flow activity followed a total of 120–230 mm of rainfall, during which intensities reached 40 mm h^{-1} (Zimmermann and Haeberli, 1992).

The nature of debris flow is contentious (Coussot and Meunier, 1996; Iverson, 1997). Some authors have advocated a form of Bingham flow, in which a raft of debris is carried downslope by laminar flow acting along the sides and base of the flow track. In such 'muddy debris flows', the mixture of fine sediment and water forms an interstitial fluid that lubricates grain motion. Others have viewed movement as a product of dispersive pressures generated by interparticle collisions within a heterogeneous flowing mass ('granular' or 'cohesionless' debris flows; Takahashi, 1991). Though the two processes may be transitional, the latter seems more applicable to flow of mixtures of debris and water down steep slopes. The operation of dispersive pressures is supported by observations of movement of boulders to the surface, front and sides of flows, forming marginal levées and terminal lobes. Flows tend to consist of rapid surges separated by periods of zero or slow flow. The form of sediment runout reflects flow viscosity, which is highly sensitive to small changes in water content. Low-viscosity flows with high water content tend to form elongate tongues of debris (Figure 12.7), whereas high-viscosity flows with low water content may spread out over footslopes as broad fans or lobes. Surface and distal debris-flow deposits may be modified by dewatering (fluid escape), hyperconcentrated (fluvial) transport and slopewash during the final stages of debris emplacement (Blikra and Nemec, 1998; Matthews *et al.*, 1999).

12.4.3 Debris Flows as Agents of Sediment Transport

Under propitious conditions (hillslope gradients of 28–40°; abundant loose sediment; frequent intense rainstorms), debris flow represents one of the most effective agents of sediment transfer in cold mountain environments. Rapp (1960) calculated that debris flows in Kärkevagge (northern Sweden) accounted for 46% of the total volume of debris moved by rapid mass-movement processes. Similarly, Kotarba and Strömquist (1984) found that debris flow is volumetrically the most important of all slope processes in the Polish Tatra Mountains. The pattern of activity differs from area to area, depending on the volume of sediment transported and the frequency of debris-flow activity. In general, individual flows in high-latitude environments tend to be smaller than those in mid-latitude alpine mountains. A rainstorm in Longyeardalen on Svalbard in 1972 triggered multiple debris flows, but the total sediment transported amounted to only ~7000 m^3 within a 6.8 km^2 area (Larsson, 1982); the small size of individual flows during this event reflects restriction of debris entrainment to

the active layer. In Swedish Lapland, the volume of sediment transported by individual flows documented by Nyberg (1985) ranged from <10 to >10 000 m^3. Of the 600 documented flows triggered by exceptional rainstorms in the Swiss Alps in 1987, more than 100 involved >1000 m^3 of sediment, 20 had volumes in excess of 10 000 m^3 and three had volumes greater than 100 000 m^3 of debris (Rickenmann and Zimmermann, 1993); in general, the magnitude of individual debris flows in any particular area is inversely related to frequency (Van Steijn, 2002; Figure 12.8). The largest debris flows tend to originate in major rock gullies, and most flows starting on open hillslopes carry less than 500 m^3 of sediment.

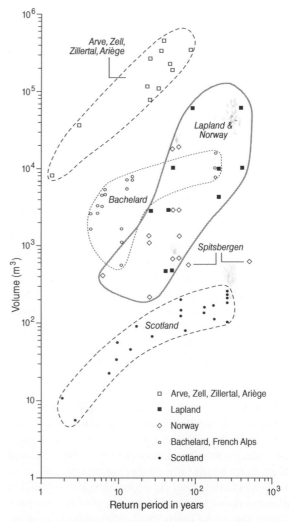

Figure 12.8 Magnitude–frequency relationship for debris flows on European mountains. Arve, Zell and Zillertal are located in the central Alps, Ariège is in the Pyrenees and Bachelard is in the French Alps. *Source:* Van Steijn (2002). Reproduced with permission of Wiley.

The frequency of debris-flow events also tends to be greater in alpine mountains than in arctic or subarctic environments. André (1990b) estimated that the return period for major events in Svalbard is 80–500 years, and Rapp and Nyberg (1981) suggested that 50–400 years is typical for Swedish Lapland. In contrast, the recurrence interval of major events is 5–45 years in the French Alps (Nieuwenhuijzen and Van Steijn, 1990; Van Steijn, 1996), and a return period of 6–13 years has been recorded over the last century on a debris cone in the southern Alps (Strunk, 1992). Such contrasts probably reflect the much greater frequency of extreme rainfall events in many mid-latitude mountains. The available data on flow magnitude and frequency (Figure 12.8) clearly indicate that debris flows tend to be both larger and more frequent in most alpine mountains than in high latitudes.

12.4.4 Relict Debris-Flow Landforms and Deposits

A conspicuous legacy of debris-flow activity in formerly glaciated mountains takes the form of debris cones and alluvial fans, often vegetation-covered, deposited along the flanks of glacial troughs. These essentially paraglacial landforms result from entrainment and redeposition of glacigenic deposits following deglaciation, and they ceased to accumulate when sediment supply upslope became exhausted (Ballantyne, 1995, 2002; Owen and Sharma, 1998). Various processes may contribute to cone and fan accumulation, but the dominant agent of sediment reworking in most areas has been recurrent debris-flow activity. The scale of relict paraglacial cones and fans varies widely, depending on local relief and the volume of sediment available for reworking, from small cones 100 m wide to thick fans 5 km wide (Owen, 1989). Some of the largest fans reflect catastrophic debris-flow activity. Derbyshire and Owen (1990) showed that some paraglacial fans in the Karakoram Mountains comprise only four debris-flow units, each 5–20 m thick, interbedded with fluvial and glacifluvial sediments. Such huge debris-flow deposits represent remobilization of vast quantities of debris during or after deglaciation.

There is also evidence that enhanced debris-flow activity has coincided with periods of Holocene climatic deterioration. Radiocarbon dating of periods of enhanced sediment accumulation in lakes in northern Scandinavia and the Polish Tatra Mountains suggests that past intensification of debris-flow and other rapid mass-movement processes corresponds to periods of colder climate such as Little Ice Age of the 16th–19th centuries AD (Kotarba, 1992). In Norway, radiocarbon-dated records of debris-flow deposition suggest enhanced activity due to Late Holocene climatic deterioration (Matthews *et al.*, 1997b; Sletten *et al.*, 2003a), and on Svalbard, André (1990b)

found lichenometric evidence for enhanced activity within the last 40 years, about 80 years ago, around 600 years ago and possibly around 1800 years ago. Local variations in the timing of periods of enhanced activity are to be expected, as flows are usually triggered by extreme rainstorms rather than general climatic deterioration. Such extreme events may reduce the threshold for subsequent activity by triggering a cycle of gully incision and vegetation removal, thereby prolonging the period of enhanced debris-flow activity.

Debris-flow deposits are frequently encountered in Quaternary sediment sequences in vertical exposures through debris cones, alluvial fans, relict talus accumulations and valley fills. Small-scale hillslope flows are characterized by lenticular structures visible in transverse sections, localized inverse grading and alternating clast- and matrix-supported diamict elements, with openwork structures representing the distal parts of levées (Nieuwenhuijzen and Van Steijn, 1990; Van Steijn, *et al.*, 1995). Laterally-extensive debris-flow units representing major depositional events take the form of massive or crudely-stratified diamictons a few centimetres to a few metres thick, with clasts of various sizes embedded in a sand- or silt-rich matrix. Individual flow units often resemble glacial till, but can be distinguished by the preferred downslope orientation of clasts and by structural and lithofacies characteristics, such as inverse clast grading, slope-parallel lenses or stringers of gravel, and thin beds of sand or granules (Curry and Ballantyne, 1999). Superimposition of successive flow units imparts crude stratification to debris-flow sequences, with contacts defined by discontinuities or by thin beds of silt or sand that reflect surface wash following flow immobilization. In many debris cones and alluvial fans, debris-flow facies are interbedded with fluvial deposits, or with snow-avalanche facies (Blikra and Nemec, 1998).

12.5 Rock Glaciers

Amongst the most impressive landforms in cold mountain environments are rock glaciers (Figure 12.9). Useful sources regarding these fascinating landforms include a collection of papers edited by Giardino *et al.* (1987), monographs by Haeberli (1985) and Barsch (1996), and reviews by Whalley and Martin (1992), Clark *et al.* (1998), Burger *et al.* (1999), Haeberli *et al.* (2006), Springman *et al.* (2012) and Kääb (2013).

12.5.1 Definition and Classification

A *rock glacier* is a thick lobate or lingulate (tongue-shaped) mass of debris that has moved slowly downslope through the deformation of subsurface ice and/or ice-rich

Figure 12.9 (a) Muragl rock glacier, Engadin, Switzerland, showing conspicuous tranverse ridges in its lower part. (b) Talus rock glaciers, Vardeborgsletta, Svalbard. (c) Glacigenic rock glaciers, Holar Valleys, Iceland. Note residual glacier ice at the headwall. (d) Glacigenic rock glacier, Reindalen, Svalbard. A colour version of (c) appears in the plates section. *Sources:* Courtesy of (a) Regula Frauenfelder and (b,c) Ole Humlum.

sediments. Rock glaciers have been classified by morphology (lobate, tongue-shaped, spatulate), location (valley-wall, valley-floor), constituent debris (talus or morainic debris) and activity (active, inactive, relict), but most debate focuses on the glacial or nonglacial (permafrost) origin of internal ice. Some authors question the existence of buried glacier ice within rock glaciers, asserting that the term 'rock glacier' should be restricted to landforms produced by the creep of ice-supersaturated mountain permafrost, and that landforms reflecting the deformation of buried glacier ice should be termed *debris-covered glaciers* (e.g. Haeberli, 1985; Barsch, 1996). Conversely, Whalley and Martin (1992) defend the notion that 'true' rock glaciers are glacially-derived, and suggest that talus-foot landforms resulting from the creep of ice-rich permafrost should be termed *protalus lobes*. A third view, consistent with the definition proposed above, is that rock glaciers may be of either glacial or nonglacial origin, and represent a continuum of ice-deformation permafrost landforms, the transition between the glacial and nonglacial types reflecting the relative inputs of ice and debris (e.g. Humlum, 1996, 1998a, 2000; Clark *et al.*, 1998; Berthling, 2011).

To some extent, this controversy is terminological, depending on the breadth of definition applied to the term 'rock glacier' and whether a genetic or generic definition is preferred. It is nonetheless useful to recognize a distinction between *talus rock glaciers* or *talus-derived rock glaciers*, which have developed through the deformation of ice-supersaturated permafrost within talus to form bench-like, lobate or lingulate extensions of the lower parts of talus slopes, and *glacigenic rock glaciers* or *glacier-derived rock glaciers*, which have formed through the deformation of a core of residual glacier ice under a cover of bouldery supraglacial debris (Monnier and Kinnard, 2015; Figures 12.9 and 12.10), though in certain circumstances hybrid rock glaciers containing both ground ice of permafrost origin

(a)

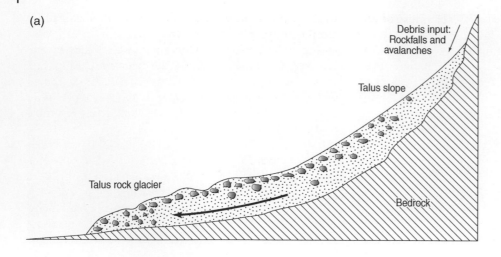

(b)

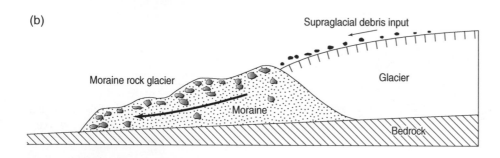

(c)

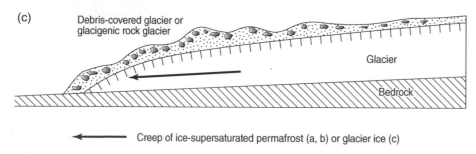

Creep of ice-supersaturated permafrost (a, b) or glacier ice (c)

Figure 12.10 Genetic classification of rock glaciers: (a) talus rock glacier; (b) moraine rock glacier; (c) glacigenic rock glacier. Some authors prefer to refer to glacigenic rock glaciers as 'debris-covered glaciers'.

and ice of glacial origin also occur (Ribolini *et al.*, 2010). Some authors also identify an intermediate category of *morainic rock glaciers* (or *debris rock glaciers*), which have developed though the deformation of ice-supersaturated permafrost within moraines (Barsch, 1996; Figure 12.10b). Others have suggested a rockslide origin for some rock glaciers, but though rockslide runout deposits sometimes mimic rock glaciers, landforms produced by processes other than the deformation of buried ice or ice-rich sediment cannot be considered true rock glaciers. Geophysical investigations of some talus rock glaciers nevertheless suggest that the dominant mode of debris supply has been infrequent high-magnitude rockfall events (Isaksen *et al.*, 2000; Degenhardt, 2009).

12.5.2 Rock-glacier Distribution

The distribution of active talus rock glaciers is dictated by climate, topography and debris supply. They occur in all high mountain ranges presently underlain by permafrost, as well as in polar environments (e.g. Berthling *et al.*, 1998, 2000; Isaksen *et al.*, 2000), and have traditionally been considered most widespread in areas where snowfall is insufficient to sustain glacier ice. A survey of rock-glacier distribution in the Italian Alps, for example, indicates that rock-glacier activity is positively correlated with increasing elevation (decreasing temperature), but negatively correlated with precipitation (Scotti *et al.*, 2013). Humlum (1998a), however, showed that that both talus rock glaciers and glacigenic rock glaciers also occur in moderately humid periglacial climates with cool summers. As permafrost is required for rock glacier development, few occur where mean annual air temperature (MAAT) exceeds $-2\,°C$ and the majority occur where MAAT $< -6\,°C$. Under favourable circumstances, rock glaciers exist in great numbers: there are more than 500 in Svalbard, more than 600 in the southwest Alps and more than 300 in the southern Carpathians; Scotti et al. (2013) identified over 600 in the central Italian Alps, Krainer and Ribis (2012) recognized 3145 in the Austrian Alps and Falaschi et al. (2014) recorded nearly 500 in the Andes of Argentina. Rock glaciers also occur in the peripheral regions of Antarctica, particularly on the Antarctic Peninsula and adjacent islands (Serrano and López-Martínez, 2000).

At a local level, the distribution of rock glaciers reflects an abundant supply of coarse debris, and thus lithological and structural controls (Etzelmüller and Frauenfelder, 2009; Angillieri, 2010). In the southwestern Alps, rock glaciers coincide with massive fractured cliffs of granite, limestone and sandstone, but rarely occur on schists, whereas in the central Pyrenees most occupy a small outcrop of crystalline rock (Evin, 1987; Chueca, 1992). In the Brooks Range, Alaska, they are best developed below cliffs of massive sedimentary rocks interbedded with thin-bedded pelitic rocks (Calkin et al., 1987). Various authors have observed that the location of rock glaciers often coincides with that of rockwalls with a high joint density. On Disko Island in West Greenland, the average spatial density of rock glaciers is ~24 per $100\,km^2$, reflecting rapid weathering of headwalls composed of well-jointed basalt (Humlum, 2000). In the Swiss Alps, Ikeda and Matsuoka (2006) distinguished 'bouldery' from 'pebbly' talus rock glaciers, the former nourished by rockfall from crystalline rocks and massive limestones, the latter by debris from rockwalls underlain by shales and platy limestones.

12.5.3 Rock-glacier Morphology

Talus rock glaciers are often lobate (width > length), valley-wall forms, whereas glacigenic rock glaciers are often tongue-shaped (length > width), valley-floor landforms that sometimes extend upvalley to a source area of exposed glacier ice. However, morphology and location do not always allow discrimination between the two types. Talus rock glaciers, particularly in cirques, are sometimes markedly lingulate, and small lobate rock glaciers may develop from ice-cored lateral moraines, yet are manifestly not talus-derived. Lobate rock glaciers may also develop through creep of ice-rich permafrost under blockslopes (Gordon and Ballantyne, 2006). The dimensions of rock glaciers vary greatly: reported lengths range from 10 m to 2000 m and occasionally more, and

widths from 40 m to >1000 m. Most talus rock glaciers are 50–300 m long and several hundred metres broad, and laterally coalescing talus rock glaciers may extend for several kilometres along the talus foot. Immature examples and the 'pebbly' talus rock glaciers described by Ikeda and Matsuoka (2006) form ramp-like bulges on or at the foot of talus accumulations, and may resemble protalus ramparts, as outlined below.

Active rock glaciers have steep fronts that typically rise 15–70 m above adjacent terrain (Figure 12.9), though frontal heights of up to 150 m have been reported. Mean frontal gradients generally fall within the range 35–45°, implying that most active rock-glacier termini are maintained above a stable 'angle of repose' by constant advance. Rock glaciers tend to be completely covered by bouldery debris except at the terminus (where finer sediment may be present) and, in the case of glacigenic rock glaciers, at the head, where glacier ice may be exposed. Many exhibit a surface relief of concentric arcuate transverse ridges and depressions, with the former rising up to 10 m above the latter (Figure 12.9a,c). Longitudinal ridges, meandering furrows, pits and marginal ridges may also be present, and small natural cairns and ice-walled melt ponds sometimes occur on the lower parts of glacigenic rock glaciers. Transverse ridges and depressions appear to reflect longitudinal compression in an outer zone of decelerating flow, and have been attributed by Frehner et al. (2015) to buckle folding of a viscous upper boulder layer overlying a less viscous ice layer subject to layer-parallel compression, with preferential growth of a dominant wavelength that determines a fairly regular spacing of ridges and depressions. Longitudinal ridges and depressions appear to be characteristic of acceleration and extension, particularly in the headward zone. Transverse ridges develop over millennia, and are transported passively on the surfaces of rock glaciers by deformation of underlying ice (Kääb and Weber, 2004). Boulders on the surfaces of rock glaciers are commonly aligned down-flow, though preferred orientation may be locally dictated by surface microrelief.

12.5.4 Rock-glacier Structure

Information concerning the internal structure of rock glaciers has been obtained through shallow excavations, investigation of natural exposures, extraction of cores and various types of geophysical survey. All rock glaciers comprise two basic units: a surface mantle of debris 1–5 m thick and a core of glacier ice or ice-supersaturated permafrost (Figure 12.11). The debris mantle typically comprises clasts 0.2–5.0 m long. It is transported passively on the deforming core and is broadly equivalent to the active layer of seasonal freezing and thawing, effectively acting as a thermal filter that protects the frozen core from melting (Humlum, 1997). On talus rock glaciers,

Figure 12.11 Glacier ice core exposed in a glacigenic rock glacier, Holar Valleys, Iceland. *Source:* Courtesy of Ole Humlum.

the surface boulder mantle may reflect frost heave of boulders and downwash of finer debris, though on glacigenic rock glaciers it represents melt-out of englacial debris (Elconin and LaChapelle, 1997; Ackert, 1998). Coarsening-upwards of the debris mantle on all types of rock glacier is probably caused by gradual kinematic sieving: jostling of boulders during movement so that the largest reach and remain at the surface.

The case for a glacier ice core within some rock glaciers rests on several lines of evidence, including (i) continuity between some rock glaciers and glacier ice upvalley (Figure 12.9c), (ii) upvalley merging of rock-glacier margins with lateral moraines, (iii) historical evidence demonstrating evolution from an ice glacier, (iv) geophysical investigations, and (v) the characteristics of massive ice observed in exposures (e.g. Krainer and Mostler, 2000; Berger *et al.*, 2004; Hausmann *et al.*, 2007). Observations by several authors (Humlum, 1996; Elconin and LaChapelle, 1997; Monnier *et al.*, 2011) have shown that in many tongue-shaped rock glaciers, massive ice occurs immediately under the surface debris mantle (Figure 12.11), and that ice crystal size, ice foliation, bubble shape and debris organization are consistent with a glacier-ice origin. Geochemical data have also been employed to confirm a glacier-ice core (Steig *et al.*, 1998). Indeed, as argued by Ackert (1998), differentiation of glacigenic rock glaciers from debris-covered glaciers may simply reflect the relative size of the ice accumulation area: when this is small, debris is deposited directly on the ablation zone, and the rock-ice mass behaves as a glacigenic rock glacier; when it is large, the body behaves more like an ice glacier, albeit one in which ablation is inhibited by the debris mantle.

Our knowledge of the internal structure of talus rock glaciers derives mainly from geophysical surveys, sometimes complemented by borehole evidence (e.g. Berthling *et al.*, 2000; Isaksen *et al.*, 2000; Ikeda, 2006; Fukui *et al.*, 2008a; Degenhardt, 2009; Monnier *et al.*, 2008; Leopold *et al.*, 2011). The Murtèl–Corvatsch rock glacier in the Swiss Alps has been studied in particular detail (Haeberli *et al.*, 1998, 2006; Springman *et al.*, 2012). Here a boulder mantle 3 m thick is underlain successively by 17 m of massive ice, 10 m of ice-supersatured silty sand and 20 m of coarse debris in an ice matrix (Vonder Mühll and Klingelé, 1994). Two rock glaciers investigated by Hausmann *et al.* (2012) in the Ötztal Alps, Austria, comprise a 4–6 m thick surface debris layer overlying 20–30 m of ice-rich permafrost, then 10–15 m of ice-free sediments. Boreholes or exposures in other talus rock glaciers have revealed rock–ice mixtures, stratified debris-laden ice, massive ice and alternating layers of ice and rock; ice typically represents 40–90% of the total volume (Haeberli *et al.*, 2006), but its origin remains uncertain. Haeberli and Vonder Mühll (1996) noted that the variability of electrical resistivity within ice-supersaturated rock-glacier permafrost implies multiple ice origins, including freezing of sub- and supraperma-frost groundwater, burial of surface ice or firn patches by debris and possibly burial of stagnant glacier ice. In some alpine talus accumulations, permafrost underlies the lower part of the talus but upper slopes are permafrost-free (Lambiel and Pieracci, 2008; Scapozza *et al.*, 2011), lending support for the view that meteoric water percolating through talus may freeze near the talus foot and contribute to the development of an ice core. Humlum *et al.* (2007) have demonstrated that debris-rich snow avalanches

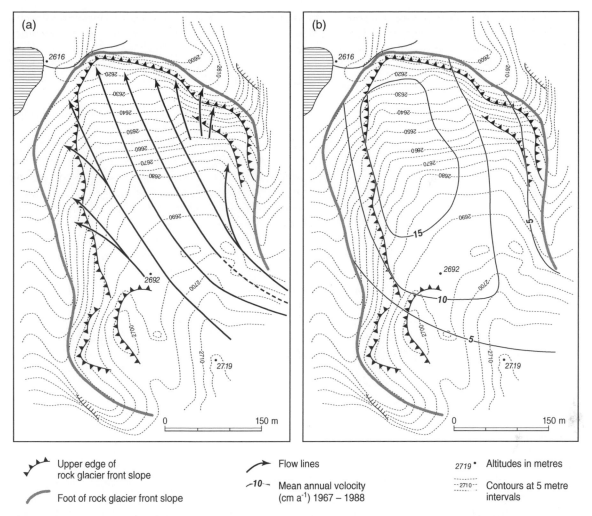

⤳	Upper edge of rock glacier front slope
⟋	Foot of rock glacier front slope
➤	Flow lines
⟋10⟋	Mean annual velocity (cm a⁻¹) 1967 – 1988
2719 •	Altitudes in metres
⋯2710⋯	Contours at 5 metre intervals

Figure 12.12 Vectors (a) and rates (b) of surface movement of the Macun rock glacier, Swiss Alps. *Source:* Barsch and Zick (1991). Reproduced with permission of Schweizerbart Science Publishers.

also make a major contribution to both the debris and ice content of some rock glaciers. Understanding the genesis of ice-supersaturated permafrost in talus rock glaciers nevertheless remains a challenge. The presence of a frozen debris-rich layer at the base of rock glaciers is attributed by Haeberli *et al.* (1998), Humlum (2000) and Kääb and Reichmuth (2005) to over-riding of boulders that have tumbled down the front of the rock glacier and then been buried as it advanced.

12.5.5 Rates and Processes of Rock-Glacier Movement

Measured rates of movement on the surfaces of active rock glaciers generally range from a few centimetres to about two metres per year, and are thus one to two orders of magnitude less than those of most alpine glaciers. In general, 'cold' polar rock glaciers move more slowly than 'warm' alpine rock glaciers. Surface displacement rates of 5–70 cm a⁻¹ have been recorded for the former,

but a much wider range of velocities and rates of surface change has been measured on the latter (Kääb *et al.*, 2002; Bollmann *et al.*, 2015). Annual rates of movement vary markedly in response to short-term climatic forcing (Serrano *et al.*, 2006, 2010; Kääb *et al.*, 2007). Mean annual velocities along the axis of the Laurichard rock glacier in the French Alps between 1986 and 2006, for example, ranged from 0.39 to 1.44 m a⁻¹ and are positively correlated with variations in ground temperature. As with ice glaciers, rates of surface movement generally diminish towards the margins, and vectors of horizontal displacement tend to diverge towards the terminus (Figure 12.12). Surface velocities are greater (sometimes much greater) than rates of frontal advance, implying reduction in average velocity with depth. Strain-rate variations often imply extending flow near the head of some rock glaciers and compressive flow towards the snout, consistent with the notion that transverse ridges are associated with compressive flow. The low velocities of rock glaciers imply formation over several millennia

(Sloan and Dyke, 1998; Konrad *et al.*, 1999). A radiocarbon age obtained for moss fragments recovered from the Murtèl–Corvatsch rock glacier, for example, indicates a minimum age of about 2000 years, and a probable age spanning most or all of the Holocene (Haeberli *et al.*, 1999).

The high ice content of both talus and glacigenic rock glaciers has prompted suggestions that rock-glacier movement may be modelled as secondary or 'steady-state' creep of ice under low to moderate driving stresses, according to Glen's flow law:

$$d\varepsilon/dt = B\tau^n \tag{12.3}$$

where $d\varepsilon/dt$ is strain rate, B is a temperature-dependent constant, n is a quantity that lies between 1.9 and 4.2 (Arenson and Springman, 2005; $n = 3$ is often assumed for pure ice) and τ is shear stress, given by:

$$\tau = \rho g z \tan \alpha \tag{12.4}$$

where ρ is rock glacier density, g is gravitational acceleration, z is depth and α is mean surface gradient. Though this model is a useful first approximation, variations in debris content in the frozen core cause variations in apparent viscosity, concentrating strain deformation at particular depths. Wagner (1992), for example, found that 75% of the surface movement of the Murtèl–Corvatsch rock glacier reflects deformation between 28 and 30 m depth, with negligible movement of the frozen coarse debris that forms the lowermost 30–50 m of the rock glacier. Subsequent borehole deformation measurements made on talus rock glaciers in the Swiss Alps have confirmed the presence of distinct shear zones where horizontal and vertical differential displacements are concentrated (Arenson *et al.*, 2002).

Though there is general agreement that movement is primarily caused by secondary creep of subsurface ice under shear stress, the possible role of basal sliding remains contentious. Basal sliding seems unlikely for talus rock glaciers where the basal layers consist of frozen coarse debris with high yield strength or where permafrost extends into the underlying bedrock, though geodetic measurements of the velocity of a talus rock glacier in the Austrian Alps suggest that sliding may be implicated in recent acceleration of its snout area (Hartl *et al.*, 2016). Sliding appears much more likely to contribute to movement in glacigenic rock glaciers, particularly in alpine environments (Krainer and Mostler, 2000). For a rock glacier in Wyoming that has undergone up to 25 m of horizontal displacement in 30 years, Potter *et al.* (1998) inferred that 50% of the surface displacement might be attributable to basal sliding.

The importance of rock glaciers in terms of debris transport is difficult to establish. Various attempts have been made to derive debris transport rates or headwall retreat rates from rock-glacier volumes, but these require assumptions to be made regarding the duration of rock-glacier movement, rock-glacier thickness and percentage debris content. Barsch and Jakob (1998) calculated average rates of horizontal sediment transport of $7.1 \times 10^6\, t\,a^{-1}\,km^{-2}$ for three talus rock glaciers in the Khumbu Himalaya, but conceded that this figure represents no more than an order-of-magnitude estimate, and Gärtner-Roer (2012) calculated sediment transfer rates of $0.24–1.10 \times 10^6\, t\,a^{-1}$ for two rock glaciers in the Swiss Alps. From the volumes of two well-researched rock glaciers in the Swiss Alps, Barsch (1996) estimated an average Holocene rockwall retreat rate of $1.5–3.4\,m\,ka^{-1}$, and extrapolation of the same approach to other Alpine rock glaciers yielded an average rate of $0.8–1.5\,m\,ka^{-1}$. Humlum (2000) assessed the rockwall retreat rates implied by the volumes of talus rock glaciers and glacigenic rock glaciers in West Greenland. His results suggest that average rockwall retreat rates of $2–6\,m\,ka^{-1}$ are associated with the former and of $5–15\,m\,ka^{-1}$ with the latter. These rates are exceptionally high (cf. Table 12.1), probably reflecting the role of rock glacier movement in evacuating rockfall debris from the base of rockwalls and thus preventing progressive headwall burial under accumulating talus.

12.5.6 Rock Glaciers and Climate

Several studies have indicated that active rock glaciers occupy a distinctive climatic niche that is defined partly by temperature and partly by precipitation. There is widespread agreement that active rock glaciers occur only where MAAT $< -2\,°C$, allowing the survival of perennial ice in the rock glacier core. Mean annual temperatures within the cores of talus rock glaciers in the Alps range from $-1\,°C$ to $-3\,°C$ (Barsch, 1996). In mountains with high snowfall, however, the equilibrium line altitudes (ELAs) of ice glaciers may extend below the lower limit of permafrost, and potential sites of rock-glacier formation are likely to be occupied by glacier ice. In mountains where snowfall is light or moderate, ice glacier ELAs occur well above the lower limit of permafrost, and rock glaciers may be widespread if topography and debris supply are favourable. A model by Haeberli (1983) that identifies the climatic constraints on talus rock glaciers in the Alps (Figure 12.13) has been shown to be reasonably robust in constraining the distribution of similar rock glaciers in other mountain areas (Brazier *et al.*, 1998). Studies by Humlum (1998a) of the altitudes of glaciers and rock glaciers at high latitudes, however, reveal only slight climatic differences between the two; at the rock-glacier sites, MAAT tends to be only very slightly higher, and mean annual precipitation slightly lower, than at sites occupied by ice glaciers. Humlum found that local topoclimatic factors are paramount: ice glaciers occur where snow inputs are high relative to debris inputs, and rock glaciers where debris inputs are high relative to snow inputs.

Because talus rock glaciers are associated with at least sporadic permafrost, the lower limit of active talus rock glaciers can be used to identify the approximate lower boundary of permafrost occurrence (Jakob, 1992; Lambiel and Reynaud, 2001; Lilleøren and Etzelmüller, 2011). Rock glaciers at the climatic threshold for survival may be vulnerable to slight climatic variations and alternate between active and inactive states (Refsnider and Brugger, 2007). However, rock glacier behaviour is controlled by ice content, ice temperature, internal structure, topography and debris input, so identification

of a distinctive climate signal from rock glacier response is potentially compromised.

12.5.7 Relict Rock Glaciers

Relict rock glaciers no longer contain a core of glacier ice or ice-supersaturated permafrost. Those that formed under Pleistocene periglacial conditions occur not only in alpine mountains at lower altitudes than their active counterparts (Aoyama, 2005; Janke, 2007; Colucci *et al.*, 2016), but also in areas where permafrost and active rock glaciers no longer exist, such as northwest Greece (Hughes *et al.*, 2003) and southwest Poland (Zurawek, 2002). They retain some of the features of active rock glaciers, notably a surface mantle of coarse debris and sometimes subdued longitudinal and transverse ridges. They are thinner, however, due to melt-out of subsurface ice, and their margins have usually degraded to a stable angle of ≤35° (Figure 12.14). They may be difficult to distinguish from rockslide runout deposits (Ballantyne *et al.*, 2009), and palaeoclimatic inferences based on their distribution require caution.

Where relict rock glaciers can be positively identified, they offer considerable potential for palaeoenvironmental reconstruction. They imply past permafrost and probably a former MAAT < −2 °C, as well as a copious debris supply (Paasche *et al.*, 2007). In alpine environments where rock glaciers are active at present, the presence of relict rock glaciers at lower elevations offers a means of establishing the change in altitude of the lower limit of permafrost, and thus MAAT depression below present values (Frauenfelder *et al.*, 2001; Moran *et al.*, 2016).

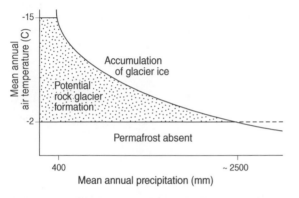

Figure 12.13 Haeberli's model of climatic constraints on talus rock-glacier formation. The stippled area represents possible combinations of mean annual air temperature (MAAT) and mean annual precipitation that permit the development of talus rock glaciers. *Source:* Reproduced from Haeberli (1983) with permission from the United States National Academy of Sciences.

Figure 12.14 Relict rock glacier on Mt Olympus, South Island, New Zealand. Subdued longitudinal and transverse ridges reflect melt-out of internal ice.

12.6 Pronival (Protalus) Ramparts

The term *protalus rampart* has been widely used to denote ridges or ramps of debris formed by accumulation of sediment at the foot of perennial firn or ice beds. Though many such ramparts occur near the foot of talus slopes, others have developed in mid-slope locations or downslope of rock slabs, prompting Shakesby (1997) to advocate the term *pronival rampart* ('snow-front rampart') to encompass all firn-foot debris ridges irrespective of location; the term *protalus rampart* is appropriate only for pronival ramparts in talus-foot locations.

12.6.1 Characteristics of Pronival Ramparts

Documented examples of active pronival ramparts range from ridges less than a metre high to ramps with distal slopes up to 30 m high (Figure 12.15a). Most are less than 300 m long. Ramparts may be arcuate, straight or sinuous, and comprise one or more ridges. Distal slopes are generally steep (26–43°) and sometimes represent the angle of rest of cascading debris, but reported proximal slope gradients are variable (0–44°). The bulk of many pronival ramparts consists of a clast- or matrix-supported diamicton, sometimes veneered with open-work boulders (Ballantyne, 1987b; Figure 12.15b). Ramparts are often described as accumulating at the foot of 'snowbeds', but these usually comprise seasonal snow-cover underlain by perennial ice or firn with densities of 600–900 kg m^{-3} (Kawashima *et al.*, 1993; Shakesby *et al.*, 1999). Formation of pronival ramparts at the foot of seasonal snowpatches seems less likely, as the lower margins of seasonal snowbeds recede during snowmelt. Seasonal snowcover may also be too soft to allow particle transport by rolling or bouncing, though Pancza (1998) has described movement of clasts down late-lying seasonal snow to feed pronival ramparts in the Alps.

12.6.2 Rampart Formation

Rampart formation has traditionally been attributed to the accumulation of clasts that have fallen from cliffs and rolled, bounced or slid to the foot of a firn field, but several commentators have cast doubt on the efficacy of this mechanism. However, field experiments by Pérez (1988) and Hedding *et al.* (2010) have demonstrated that most clasts released above or near the top of a firn field reached ramparts at the foot of the firn, with some being arrested by the ramparts and others being deposited on or beyond their distal slopes. Several other mechanisms also contribute to transport of debris down firn fields and help explain the presence of fine sediment in many ramparts. Ono and Watanabe (1986) demonstrated that fines may be deposited on

Figure 12.15 (a) Arcuate protalus rampart, arctic Norway. The crest is 115 m long, and the distal slope is 15–20 m high. (b) Proximal slope and crest of the same rampart. The coarse angular debris on the proximal slope represents rockfall debris arrested at the rampart. (c) Miniature pronival ramparts formed by snow-push, Romsdalsalpane, Norway. *Source:* Courtesy of John Matthews.

ramparts by supranival debris flow, and sediment-rich snow avalanches have been shown to drape debris over the crests and distal slopes of ramparts (Ballantyne, 1987b; Matthews *et al.*, 2011). The presence of fine sediments in some ramparts may also be explained by reworking of glacigenic deposits from upslope (Harris, 1986), possibly by snow avalanches. Other suggested secondary transport mechanisms include the gradual descent of debris that has previously come to rest on firn, probably by snowcreep, and transport of debris by supranival wash, or by subnival solifluction or debris flow (Shakesby *et al.*, 1995). Rampart formation by 'bulldozing' of till by basal sliding and possibly creep of firn fields, in a manner analogous to the formation of push moraines, has been proposed by Shakesby *et al.* (1999). Though only small (<1.2 m-high) ramparts have been observed to form in this way (Figure 12.15c), it is possible that larger features may be built by repeated push events.

Rampart evolution is poorly understood. Ballantyne and Kirkbride (1986) envisaged rampart growth during thickening of a stationary firn field, with progressive migration of the rampart crest away from the talus foot (Figure 12.16). This model implies that rampart growth is self-limiting, as continued thickening of the firn increases basal shear stress, ultimately transforming the firn field into a small glacier, with consequent destruction or modification of the rampart. Numerical modelling indicates that firn movement becomes significant when the distance between the rampart crest and the talus edge is 30–70 m, implying that debris ridges at greater distances from the talus foot are probably of different origin (Ballantyne and Benn, 1994b). Though this model is supported by field research (Hall and Meiklejohn, 1997), it is valid only for ramparts located at the talus foot below steep firn fields, and is unlikely to be applicable in all cases (Hedding *et al.*, 2010).

The relationship between protalus ramparts and talus rock glaciers is contentious. One view is that all protalus ramparts represent embryonic rock glaciers (Barsch, 1996), but this conflicts with observations of debris

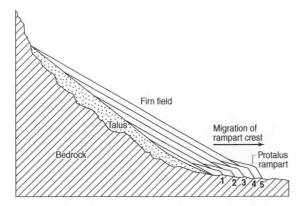

Figure 12.16 Model of rampart development at the foot of a progressively thickening firn field. *Source:* Ballantyne and Kirkbride (1986). Reproduced with permission of Wiley.

transport down firn fields and the accumulation of such debris at the foot of the firn, particularly in non-permafrost areas. An alternative is that the accumulation of some ramparts begins only *after* an incipient protalus rock glacier has developed at the foot of a firn field to trap cascading debris. It is also possible that some protalus ramparts evolve into rock glaciers through burial of ice under accumulating debris, or that ramparts in permafrost areas undergo internal deformation and metamorphose into rock glaciers on reaching a critical thickness (Pancza, 1998). A different perspective, developed by Shakesby *et al.* (1987) and Luckman and Fiske (1995), is that pronival ramparts and talus rock glaciers represent the outcomes of entirely different evolutionary sequences.

Given the range of possibilities, it is not surprising that it may be difficult to distinguish protalus ramparts and embryonic talus rock glaciers in the field. In particular, incipient rock glaciers consisting of a single bench-like ramp or ridge that runs along the foot of a talus slope closely resemble protalus ramparts. There are numerous accounts of situations where supposed 'protalus ramparts' occur adjacent to talus rock glaciers, and a simpler explanation of many such features is that they are actually embryonic talus rock glaciers, genetically identical to others nearby but at a less advanced stage of development.

12.6.3 Relict Pronival Ramparts

Shakesby (1997) has highlighted the difficulty of differentiating relict pronival ramparts from moraines or avalanche ramparts, and some features identified as relict protalus ramparts are almost certainly rockslide runout deposits or antiscarps produced by gravitational slope deformation. Though attempts have been made to formulate diagnostic criteria for rampart identification (Ballantyne and Kirkbride, 1986; Hedding and Sumner, 2013), the range of circumstances under which ramparts form impedes formulation of definitive criteria. Some guidelines can be identified: pronival ramparts are normally located on or at the foot of talus slopes or below rockwalls, and rarely extend more than 300 m across the slope; distal slopes usually exceed 25–35°, but proximal slopes may be gentler; and where debris is derived exclusively from cliffs, clasts are mainly angular and of the same lithology as the rockwall source. Protalus ramparts probably occur only within 70 m of the talus foot. These guidelines, however, do not always exclude other origins for scarp-foot or talus-foot ridges.

The formation of a pronival rampart requires the development of a perennial firn field, implying a delicate balance between winter snow accumulation and summer melting. This property has been explored by

Ballantyne and Kirkbride (1986), who found that the altitudes of relict ramparts of Younger Dryas age rise eastwards across the Scottish Highlands, closely paralleling the ELAs of contemporaneous glaciers. This pattern they interpreted in terms of an eastwards decline in snowfall, arguing that heavy snowfall would be required to balance snowmelt at low altitudes in the west, but only limited snowfall would be needed to replenish firn fields at higher altitudes in the east. Holocene protalus ramparts have also been employed as indicators of former environmental conditions. Butler (1988) has suggested that protalus ramparts and talus rock glaciers that formed in Idaho during three episodes of Neoglacial climatic deterioration developed at progressively higher altitudes, and inferred that the three episodes were of diminishing climatic severity. Derivation of palaeoclimatic inferences, however, requires that pronival ramparts are correctly identified, and this has not always been the case (Shakesby, 1997). Relict ramparts have also been employed to calculate former rockfall rates by dividing the rampart volume by the area of the contributing rockwall. Using this approach, Ballantyne and Kirkbride (1987) calculated that the volumes of relict protalus ramparts in upland Britain imply rockwall retreat of 1.1–1.5 m during the Younger Dryas Stade, equivalent to a rockwall retreat rate of ~1.4–4.0 m ka^{-1}, depending on how long perennial firn fields persisted on the slopes above the ramparts.

12.7 Synthesis

Talus slopes and related landforms constitute a complex debris transport system. Though rockfall is normally the primary input, several processes modify the form and sedimentology of talus accumulations. Figure 12.17 summarizes these processes and associated geomorphological responses, but simplifies the range of possible relationships between modifying processes. Many taluses have intermediate morphological and sedimentological characteristics (Jomelli and Francou, 2000), and do not fall readily into classifications such as those outlined in Figures 12.2 and 12.17.

The environmental or palaeoenvironmental significance of talus landforms and deposits varies. Unmodified rockfall talus and debris flow-related landforms and deposits develop under both periglacial and non-periglacial conditions. The environmental implications of stratified talus deposits are variable, often implying cold but not extreme conditions. Avalanche landforms suggest the build-up of a thick seasonal snowpack. Pronival ramparts require the survival of perennial ice and firn, and thus climatic conditions at the threshold for the development of glacier ice. Talus rock glaciers imply at least sporadic permafrost, and probably MAAT < −2°C, but as their development is also controlled by debris input, the absence of talus rock glaciers does not necessarily imply absence of permafrost. The development of glacigenic rock glaciers is conditioned partly by debris supply and partly by glacier mass balance.

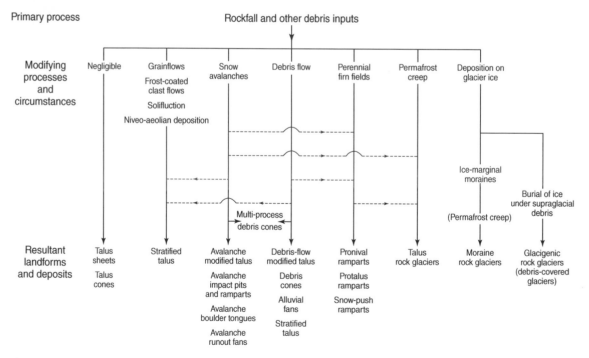

Figure 12.17 The talus debris-transfer system: processes and geomorphological response. Relationships between processes are simplified. Many talus landforms are composite.

The talus debris transport system represents an interrupted sediment cascade (Müller *et al.*, 2014): debris on the talus surface may be stored *in situ* for long periods, moved slowly downslope by rockfall impacts or permafrost creep, or transferred rapidly by snow avalanches or debris flows to the slope foot, where it may be arrested by protalus ramparts or evacuated by talus rock glaciers. Relict taluses on which rockfall input is negligible may experience net erosion, usually by debris flows, which redeposit sediment in debris cones at the foot of gullies. In some instances, however, debris reaching the foot of a talus may be removed by fluvial or coastal erosion, processes that are considered in the following chapters.

13

Fluvial Processes and Landforms

13.1 Introduction

Despite the distinctive geomorphological effects of ground freezing, the broad outlines of many periglacial landscapes differ little from those of other environments. In part, this similarity reflects landscape evolution over very long timescales. Arctic and subarctic landscapes evolved under much warmer conditions than now during the Neogene Period (~23.0–2.58 Ma), and have experienced only limited modification under periglacial conditions during the Quaternary (André, 2003; French, 2016), though some have been extensively transformed by Pleistocene glacial erosion and deposition. Another reason for the broad similarity of landscapes across different climatic zones is the important role played by fluvial processes. Many high-latitude periglacial landscapes are dominated by fluvial landforms. Gullies, gorges, V-shaped valleys and intersecting spurs dominate in the higher parts of periglacial catchments, and the lower depositional reaches support broad floodplains crossed by braided or meandering rivers, often flanked by river terraces or alluvial fans, that terminate in deltas at the coast (Figure 13.1).

The notion of high-latitude environments as regions of marked fluvial activity may appear counterintuitive. Many arctic areas receive less than 400 mm mean annual precipitation, and runoff often ceases for 6–9 months (or longer) each year. To understand why fluvial activity represents one of the most important geomorphological processes in periglacial environments, we need to consider first the nature of water storage and release in cold (particularly permafrost) areas, then the geomorphological consequences.

13.2 Periglacial Hydrology

All periglacial environments share a number of distinctive hydrological characteristics. The most important are: (i) reduction or cessation of runoff during winter; (ii) dominance of snow and ice as seasonal storage components and sources of summer runoff; and (iii) presence of frozen ground, which impedes downwards movement of groundwater. The relative importance of these elements depends on climate, relief, the presence or absence of permafrost and the depth of seasonal freezing or thawing. Here we consider first the hydrological characteristics of high arctic permafrost environments, which represent periglacial hydrology in its most distinctive form, and then how these are modified in other periglacial environments. A comprehensive account of the hydrological characteristics of permafrost environments is given by Woo (2012).

13.2.1 Active-layer Dynamics in High-Arctic Environments

The stream runoff regime in high-arctic environments is strongly linked to hillslope runoff characteristics, which are in turn controlled by the thermal and hydrological dynamics of the active layer. These have been summarized by Woo (1986) in terms of three seasonal 'snapshots' relating to winter freezing, spring snowmelt and summer thaw (Figure 13.2).

During autumn and winter freezing (Figure 13.2a), the air is much colder than the ground, producing a strong negative heat flux. Two-sided freezing of the active layer occurs, and the formation of segregation ice at both freezing fronts results in vertical movements of water through the soil to feed growing ice lenses near the top and base of the active layer. Where such water is not replenished by lateral suprapermafrost groundwater flow, a zone of desiccated soil develops between two ice-rich zones; where there is a lateral input of groundwater, ice lenses may develop throughout the active layer during freezing.

The spring thaw (Figure 13.2b) is initially dominated by melt and sublimation of snowcover due to incoming radiative energy and the onset of above-zero air temperatures. As the snowpack is initially below freezing point, meltwater percolating downwards through the snow refreezes at depth, forming basal ice layers and releasing latent heat that warms the surrounding snowpack until it is 'ripe' or isothermal at 0 °C (Quinton and Marsh, 1998).

Periglacial Geomorphology, First Edition. Colin K. Ballantyne.
© 2018 John Wiley & Sons Ltd. Published 2018 by John Wiley & Sons Ltd.

Figure 13.1 Fluvial landforms on Svalbard. A tributary stream draining a V-shaped valley runs over a large alluvial fan to join a braided river on the valley-floor floodplain. *Source:* Courtesy of Matthias Siewert.

Meltwater also infiltrates the upper parts of the active layer, and on refreezing convects heat into the soil. Such refreezing meltwater may also seal soil pores, stopping further infiltration and thus promoting ponding and surface runoff as snowmelt continues.

Following the removal of snowcover (Figure 13.2c), the thaw front descends at a rate proportional to the square root of time elapsed since the onset of ground thaw, thus:

$$z_t = b\sqrt{t} \qquad (13.1)$$

where z_t is the depth of thaw in metres at t days after the onset of ground thawing. The value of b depends on the substrate, ranging from 0.06 in clay soil to 0.15 in dry gravel. The descent of the thaw front through ice-rich soil is actually much slower, however, because of the additional heat input required to melt ice (Woo and Xia, 1996). The thawed zone of the active layer stores both meltwater and rainfall inputs. Runoff can occur as groundwater flow in the saturated zone above still-frozen ground, throughflow in the unsaturated zone above the water table and overland flow (surface runoff) when the water table reaches the surface due to saturation of

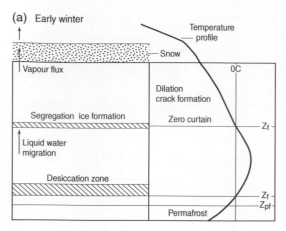

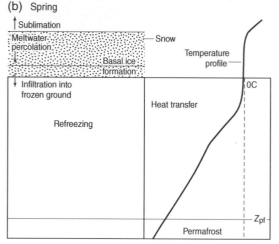

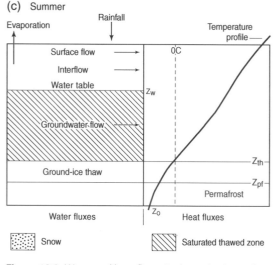

Figure 13.2 Water and heat fluxes in the active layer above cold permafrost during (a) early winter, (b) spring and (c) summer. *Zf,* depth of freezing fronts during early winter freezeback; *Zpf,* permafrost table; *Zth,* depth of summer thaw; *Zw,* depth of summer water table. *Source:* Woo (1986). Reproduced with permission of Taylor & Francis.

the thawed zone. Evaporation may also result in significant water loss, especially just after snowmelt, when abundant water is available at or near the ground surface and energy inputs are high.

13.2.2 Hillslope Hydrology in High-Arctic Environments

In the spring, hillslope runoff begins when much of the snowpack has 'ripened' to 0°C and refreezing of water near the base of the snow and the top of the active layer limits further vertical percolation of meltwater. When snowmelt begins, the storage capacity of thawed ground is limited because the frost table is at or close to the ground surface. As a result, initial downslope delivery of meltwater takes the form of drainage within the snowpack, oversnow flow or slush flow. As snowcover is reduced, there is a period of abundant surface flow over still-frozen snow-free ground (Figure 13.3). Such surface runoff, though brief, usually accounts for a much greater proportion of annual hillslope runoff than subsurface flow through thawed soil, as the latter predominates only after most of the seasonal snowpack has melted (Woo and Steer, 1983; Figure 13.4). Shallow soil pipes at the boundary between an upper organic soil horizon and lower mineral horizon may also be routeways for runoff during snowmelt, particularly in subarctic areas (Carey and Woo, 2000).

Rapid runoff by overland flow or shallow pipeflow ensures that there is little delay in meltwater reaching the stream channel network. After the end of the main snowmelt event, however, the increasing storage capacity of the deepening active layer results in a marked decline in surface runoff. Subsurface flow dominates thereafter, reducing the rate of water delivery to the stream channel. At this stage, late-lying or perennial snowbeds assume increasing importance as sources of hillslope runoff (Lewkowicz and Young, 1990), though summer rainstorms may generate subsurface runoff peaks and even reactivate overland flow if the storage capacity of the active layer is exceeded (Figure 13.4). During freezeback, hydrologic activity on slopes is limited to snow accumulation, water migration to freezing fronts and sometimes throughflow in the active layer below the descending freezing front. Such throughflow can result in a build-up of hydraulic pressure under a cap of frozen ground. If this ruptures, the released water freezes at the surface, forming a sheet-like mass of ice that is referred to as an *icing*, *Aufeis* (German) or *naled* (Russian). Hydraulic pressure generated within a freezing active layer may also cause updoming of the ground, forming a seasonal frost mound (Chapter 7).

13.2.3 River Runoff in High-Arctic Environments

The very largest arctic rivers, such as the Ob, Lena and Yenisei Rivers in Siberia, and the Mackenzie and Yukon Rivers in North America, have their headwaters south of the limits of continuous permafrost and flow throughout the winter. The discharge of the Lena River near Yakutsk (62° N), for example, is typically 900–1500 $m^3 s^{-1}$ under a 2 m thick ice cover by the end of the winter, but reaches up

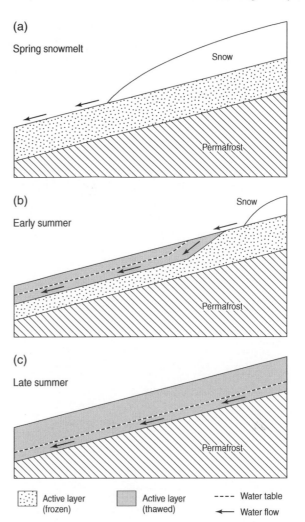

Figure 13.3 Slope runoff. (a) In spring, snowmelt-fed surface runoff over the frozen active layer predominates. (b) By early summer, surface runoff from residual snowbeds sinks to join subsurface groundwater flow sourced by thaw of ice in the active layer in a saturated zone below a shallow water table. (c) By late summer, groundwater flow is confined to a thin saturated zone above the permafrost table, unless reinvigorated by rainfall.

to 100 000 $m^3 s^{-1}$ during snowmelt and ice break-up in June. Break-up begins in May south of Yakutsk, then extends downriver. During break-up, ice jams and logjams may raise the water level by up to 8–10 m, causing extensive inundation of the floodplain (Gautier and Costard, 2000).

Smaller rivers whose catchments lie entirely within the high Arctic cease to flow during the winter months. At the end of winter, many arctic river channels are largely ice-free but choked with snow. During the first days of snowmelt and hillslope runoff, meltwater tends to accumulate as a saturated layer at the base of the river channel snowpack, so the onset of channel runoff occurs shortly after that of hillslope runoff (Figure 13.4). Snow in river channels retards flow until build-up of water allows streams to overtop or tunnel through snow dams and to carve down through snow to the channel floor (Woo

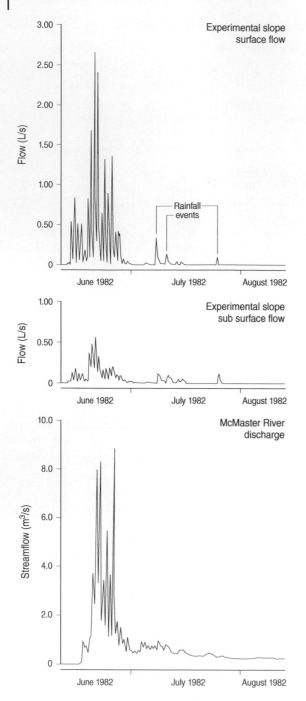

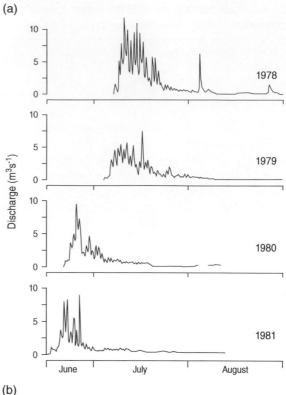

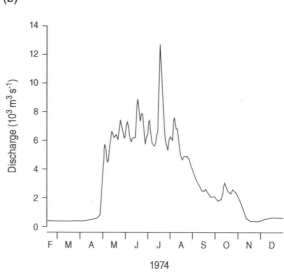

Figure 13.4 Hillslope surface flow, hillslope subsurface flow and river discharge for 1982, McMaster River catchment, Cornwallis Island, arctic Canada (74° N). The onset of flow on slopes precedes that in the channel due to accumulation of water behind snow dams in the channel. *Source:* Reproduced from Woo and Steer (1983) with permission of the National Research Council of Canada.

Figure 13.5 (a) Runoff regime of McMaster River, which drains a 33 km² catchment on Cornwallis Island (74° N) in the Canadian Arctic Archipelago, 1978–81, showing interannual variations in the timing and magnitude of the annual snowmelt flood. The two discharge spikes after the end of the snowmelt flood in 1978 represent rainfall events. (b) Runoff regime of the Liard River, a tributary of the Mackenzie River, February–December 1974. The Liard River drains a 275 000 km² subarctic catchment (57° 30′– 61° 20′ N) with its headwaters in the northern Rocky Mountains. The abrupt start of the snowmelt and ice break-up period is evident, but high discharge is maintained throughout the summer by melt of snow and ice in the higher parts of the catchment, and by summer rainfall. The maximum discharge represents exceptionally high rainfall in the middle and lower catchment. *Source:* Adapted from Woo (2012). Reproduced with permission of Ming-ko Woo.

and Sauriol, 1980; Kane *et al.*, 1991). Under clear-sky conditions, snowmelt discharge then builds rapidly to a peak. In high arctic rivers, the snowmelt flood (*nival flood*) usually occurs within 2–4 weeks of the summer solstice, and so radiative inputs are high. Conspicuous diurnal discharge fluctuations (Figures 13.4 and 13.5) are characteristic of small and medium-sized rivers, and reflect rapid snowmelt response to diurnal changes in solar radiation and temperature inputs; in large basins, however, these may be dampened by snowmelt inputs reaching the trunk stream at different times.

Unless cool, cloudy weather intervenes, the snowmelt flood typically lasts 10–20 days in small- and medium-sized catchments and is terminated by reduction in meltwater inputs due to depletion of snowcover. In the arctic, up to 90% of water discharge and sediment transport occurs within this brief period (Priesnitz and Schunke, 2002; Beylich and Ginz, 2004; Forbes and Lamoureux, 2005; McDonald and Lamoureux, 2009; Figures 13.4, 13.5 and 13.6a). After the snowmelt flood, streamflow is sustained mainly by suprapermafrost groundwater flow, itself fed by melt of residual snowbeds and of ground ice in the active layer. Because of the limited storage capacity of the shallow active layer, however, high-arctic streams exhibit a 'flashy' response to summer rainstorms (Dugan *et al.*, 2009), but these are infrequent. Groundwater inputs diminish as temperatures fall, so the channels of tributary streams tend to dry up before the onset of winter freezeback (Figure 13.6b). The insulating effects of snow accumulation in large river channels may allow the survival of suprapermafrost taliks under channels, and these sometimes persist throughout the winter (Arcone *et al.*, 1998; Rivkin, 1998). Exceptionally, groundwater flow in continuous permafrost may also follow conduits that have remained open within the permafrost since deglaciation (Haldorsen *et al.*, 2010), or conduits in karstified carbonate rocks (Utting *et al.*, 2012). Subchannel taliks may form routeways of water movement during winter, and water under hydraulic pressure in subchannel taliks sometimes emerges at the surface and freezes, forming sheet-like masses of layered ice called *river icings*.

13.2.4 Water Balance in High-Arctic Environments

In its simplest form, the annual water balance of high-arctic river basins is expressed as:

$$\Delta s = Pn - (ET + Q) \tag{13.2}$$

where Δs is change in catchment storage, Pn is precipitation, ET is evapotranspiration and Q is runoff. The annual water balance of McMaster River basin (74° N) on Cornwallis Island in the Canadian Arctic for 1975–82 (Woo, 1983) is shown in Table 13.1. A number of points emerge. First, summer rainfall accounts on average for only 21% of water inputs, highlighting the predominance of snowfall in catchment water input. Second, for the period 1975–81, the runoff ratio (Q/Pn) averaged >0.8, implying that most water inputs leave the basin as river discharge rather than evaporative losses. Finally, net changes in annual storage tend to be small, reflecting the limited storage capacity of the shallow (0.3–0.7 m deep) active layer. Measured water balances elsewhere in the Arctic tend to be similar, but show regional variations (Woo and Young, 1997). For three nested catchments in northern Alaska, for example, water balance averages for 1993–97 indicate that summer rainfall accounted for a much higher (56–67%) proportion of precipitation inputs, and that runoff ratios were lower (0.50–0.65), though still much higher than the global mean runoff ratio of ~0.36 (Kane *et al.*, 1998; Lilly *et al.*, 1998).

13.2.5 River Runoff Regimes in Other Periglacial Environments

The hydrologic regime summarized in this section is typical of arid and semi-arid high-arctic environments, and represents periglacial hydrology in its most distinctive form.

Figure 13.6 (a) Schei River, Ellesmere Island (78° N) near the end of the snowmelt flood. (b) A tributary of the Schei River in early August. Almost all snow in the catchment has melted and the streambed has dried up.

Table 13.1 Annual water balance of McMaster River Basin, arctic Canada, 1975–81. Water balance calculated from September to September.

Year	Precipitation (mm)	Rainfall (mm)	Runoff (mm)	Evaporation (mm)	Change in storage (mm)
1975–76	191	28	161	31	−1
1976–77	120	23	155	31	−37
1977–78	273	61	213	38	+23
1978–79	178	42	143	30	+5
1979–80	165	37	130	51	−16
1980–81	215	67	148	47	+21
Mean	190	43	158	38	−1

Source: Reproduced from Woo (1983) with permission from the American Association of Geographers.

This dominantly *nival* (snow-fed) regime nonetheless represents only one of several that characterize periglacial environments (Woo *et al.*, 2008a; Woo, 2012). Church (1974) identified four categories of arctic and subarctic river regime, namely the *arctic nival* regime already outlined, a *subarctic nival* regime, a *proglacial* regime and a *muskeg* (wetland) regime. Glacier-free mid-latitude mountain catchments exhibit an *alpine nival* regime.

The subarctic nival regime is characteristic of rivers in the discontinuous permafrost zone. Such rivers experience pronounced snowmelt floods, but high flows due to summer rainstorms are more frequent and may generate higher discharges than is the case with the arctic nival regime. Peak water levels may be caused by backwater effects of ice dams during spring break-up. Groundwater contributions to summer runoff are greater due to the lack of continuous permafrost and the greater storage capacity of the active layer. Flow may cease completely in winter, though large rivers with headwaters at lower latitudes may flow throughout the year, fed by baseflow from unfrozen ground.

The proglacial regime describes the runoff characteristics of rivers that are fed partly by melt of glacier ice. Flow commences in a manner similar to that of the arctic nival regime, though the snowmelt flood tends to be much more prolonged, as reserves of snow and ice in glacierized catchments persist throughout the melt season. Rainfall produces a flashy runoff response and may generate peak flows, as may catastrophic floods (*jokulhlaups*) produced by the drainage of glacier-dammed lakes.

The muskeg or wetland regime is characteristic of extensive areas of low relief with poor drainage, such as the Hudson Bay lowlands or those of western Siberia, but smaller areas of wetland also occur in most lowland arctic areas, and even in polar deserts. Such wetlands tend to increase water storage and reduce flood peaks, leading to a more subdued response to both snowmelt and rainstorm-generated runoff, though studies of the hydrologic characteristics of arctic, subarctic and boreal wetlands

(Glenn and Woo, 1997; Young and Woo, 2000; Woo and Young, 2006) indicate hydrologic regimes that are fairly similar to arctic and subarctic nival regimes. Wetland river runoff declines or ceases in winter, is dominated by snowmelt floods in summer and exhibits a flashy response to summer precipitation, as the ground is already close to saturation. The wetland regime differs from the arctic and subarctic nival regimes mainly in that: (i) extensive lowland areas become flooded during the snowmelt period; (ii) evaporation is higher, due to increased surface water exposure; and (iii) annual runoff ratios (Q/P) are lower than those in arctic nival basins (Woo *et al.*, 2008b). In some subarctic and high-boreal wetlands, active-layer storage may exceed snowmelt as a source of runoff generation (Gibson *et al.*, 1993). Woo (1986) also identified a *spring-fed regime* that is often associated with carbonate rocks in the discontinuous permafrost zone. In contrast to the other types described above, streams with a spring-fed regime tend to have relatively stable discharges, as the primary source of water supply is groundwater (Haldorsen *et al.*, 2010; Utting *et al.*, 2012).

Surface runoff in alpine areas ceases or declines during winter in the higher parts of mountain catchments. As in the arctic, snowmelt driven by radiant energy influx is usually the dominant input to stream runoff, though intensive summer rainstorms are more frequent than in high latitudes, and often produce annual flood peaks (Kattelmann, 1991; Höfner, 1995). Extreme floods may result from rain-on-snow events, when a combination of intensive rainfall and rapid snowmelt generates exceptional discharges. In most alpine mountains, the spring snowpack is thicker and is distributed over a much greater altitudinal range than in high latitudes, and snowmelt runoff is consequently much more prolonged (up to 3 months) as snowmelt begins in the lower parts of alpine basins several weeks before the onset of melt at high altitudes. Runoff ratios for alpine permafrost catchments are high, typically in the range 0.6–0.75, reflecting rapid runoff of snowmelt over frozen ground (Yang *et al.*, 1991).

An unusual aspect of some mountain rivers is that coarse bed materials allow infiltration of flow into the bed in the upper part of the catchment, and exfiltration and resumption of channel flow in the lower catchment (Woo *et al.*, 1994).

The runoff regimes summarized in this section represent a continuous spectrum of hydrological behaviour. Marsh and Woo (1981), for example, showed that arctic basins in which late-lying snowcover is extensive at higher altitudes exhibit a runoff regime similar to that of proglacial rivers, and for similar reasons alpine runoff regimes may resemble proglacial regimes. The common rudiments of all periglacial runoff regimes are prolonged winter accumulation of snow, and its subsequent release during a period of intense snowmelt, which produces a runoff regime that is remarkably efficient in translating water inputs into geomorphologically effective flood events (Woo, 2012).

13.3   Slopewash

The term *slopewash* refers to the downslope transfer of sediment by flowing water on hillslopes. Slopewash can be subdivided into *surface wash* and *subsurface wash* components. Both may involve the transport of both particulate and dissolved solids, though subsurface movement of particulate solids is usually negligible.

13.3.1   Surface Wash

Particulate transport by surface wash occurs when the drag exerted by flowing water exceeds entrainment resistance. Surface wash may take the form of sheetwash, when flow is unconcentrated, or rillwash, where flow is focused in rills a few centimetres deep, though in practice the two are transitional. On vegetated ground, flow is channelled between earth hummocks or vegetation tussocks, and in the Arctic it may follow indistinct 'water tracks' of concentrated runoff that follow the maximum gradient. Observations of vigorous overland flow during snowmelt (Figure 13.7) have prompted some authors to suggest that surface wash is a highly effective form of particulate sediment transfer, but field measurements suggest that this is only the case under favourable circumstances. For Svalbard, Jahn (1961) inferred a maximum denudation rate of $6.7\,\mathrm{mm\,ka^{-1}}$ at the most active sites at the downslope margins of late-lying snowbeds, but at other sites the denudation rate was much less. Lewkowicz (1983) calculated that denudation due to transport of particulate sediment by surface wash on Banks Island in arctic Canada ranged from 0.1 to $2.6\,\mathrm{mm\,ka^{-1}}$. In Kärkevagge, northern Sweden, Strömquist (1985) recorded maximum transport equivalent to a denudation rate of approximately $0.5–1.0\,\mathrm{mm\,ka^{-1}}$. On Devon Island, arctic Canada, Wilkinson and Bunting (1975) found that average sediment yield was only about 65 g per metre of slope width per year.

Figure 13.7 Rillwash over high-arctic slopes, central Banks Island, Canadian Arctic Archipelago. *Source:* Courtesy of Tobias Ullman.

Collectively, these studies suggest that despite the apparent erosive capability of turbulent overland flow during spring snowmelt, surface wash of particulate sediments in high-arctic permafrost terrain is low compared with rates recorded in some non-permafrost environments. This anomaly can be explained by the limited availability of erodible sediment. Only where overland flow has access to readily detachable silt, sand or fine gravel does particulate transport achieve high concentrations. Potential sources of erodible sediment include ground disturbed by frost heave or desiccation, soil exposed by active-layer failures and windblown particles trapped within the snowpack. Surface wash that does not have access to erodible sediment sources usually transports a negligible particulate load irrespective of runoff velocity and turbulence. The role of terrain disturbance in enhancing slopewash yield has been illustrated by Lewkowicz and Kokelj (2002), who showed that particulate transport on terrain exposed by active-layer failures on Ellesmere Island produced much higher mean sediment yields ($1200 \, g \, m^{-2} \, a^{-1}$) than undisturbed terrain ($<75 \, g \, m^{-2} \, a^{-1}$). They also demonstrated that particulate sediment yield tends to increase as median grain size falls, at least over the range $250 \, \mu m$ (medium sand) to $6 \, \mu m$ (fine silt), and that a negative correlation exists between sediment yield and percentage vegetation cover. The latter relationship explains the high levels of particulate transport associated with late-lying snowbeds, which on melting expose unvegetated ground.

In alpine environments, measured soil loss by surface wash is greater, but very variable. On alpine terrain in Colorado, Bovis and Thorn (1981) estimated an area-weighted slopewash denudation rate of $0.1 \, mm \, a^{-1}$ ($100 \, mm \, ka^{-1}$), much higher than any reported for high-latitude areas. Estimated average denudation for tundra meadows was $0.01 \, mm \, a^{-1}$, for dry tundra $0.1 \, mm \, a^{-1}$ and for late-lying snowbeds $1.0 \, mm \, a^{-1}$. Rainsplash was inferred to be the main agent of soil detachment outside areas dominated by melt of late-lying snowpatches, which occupied only 3% of the study area but contributed ~50% of particulate sediment yield. It therefore appears that exposure of bare soil by shrinkage of late-lying snow tends to result in particularly effective slopewash transport (Christiansen, 1998a). As nival meltwater often percolates into thawed ground a short distance downslope, however, the transport paths of entrained particles are short, leading to the accumulation of fine-grained slopewash sediments immediately downslope from sites of late-lying or perennial snowbeds (Chapter 11).

Solute transport in surface wash refers to the movement of dissolved solids by overland flow. In general, solute concentration exhibits a nonlinear negative relationship with discharge, due to dilution effects: fast-flowing runoff waters fed by snowmelt or rainfall have less opportunity to achieve ionic saturation than slow-moving waters fed mainly by groundwater. An increase in solute concentrations with distance downslope from melting snow has been reported by several authors and provides evidence of solutional denudation by surface wash, though lack of data makes it difficult to generalize on the magnitude of this process. For rillwash over carbonate terrain in arctic Canada, Cogley (1972) measured solute concentrations of $70–130 \, mg \, l^{-1}$. Such concentrations indicate significant solutional denudation but are modest in comparison with those measured on mid-latitude carbonate terrain. A study by Lewkowicz (1983) of solute concentrations in surface wash on noncarbonate terrain on Banks Island produced variable results, with estimated solutional denudation at four experimental plots ranging from 2 to $74 \, mm \, ka^{-1}$. These values are 8–30 times greater than particulate transport at the same sites, highlighting the dominance of solute transport at sites not affected by terrain disturbance. On disturbed terrain, however, particulate transport may greatly exceed solutional transport (Lewkowicz and Kokelj, 2002). Conversely, the formation of salt efflorescences at sites of permafrost degradation, particularly in marine sediments, may temporarily raise concentrations of total dissolved solids (TDS) in surface wash by an order of magnitude or more (Kokelj and Lewkowicz, 1998).

13.3.2 Subsurface Wash

Subsurface wash of particles is usually negligible, as soil pore spaces tend to be smaller than most soil particles. Exceptions occur where fine sediment is washed through a framework of clasts, such as in blockfields or talus, or where near-surface pipes occur in the soil. Small, discontinuous soil pipes (natural tunnels) occur in the active layer in subarctic environments (Gibson et al., 1993; Carey and Woo, 2000), but no data exist for suprapermafrost pipeflow transport. Seppälä (1997) has described much larger pipes developed along ice wedges in northern Quebec. Such tunnels act as conduits for particulate sediment entrainment and transport, and often represent sites of thermokarst gully initiation (Fortier et al., 2007; Godin and Fortier, 2012; Chapter 8).

Subsurface transport of dissolved solids is of much greater importance. Groundwater flow in the active layer is usually weakly acidic and is in continuous contact with mineral soil, and consequently achieves higher concentrations of dissolved solids than fast-moving surface runoff. For permafrost environments, however, the data regarding concentrations of dissolved solids in soil water are limited. On Ellesmere Island in arctic Canada, Woo and Marsh (1977) found that the total hardness (calcium plus magnesium concentration) of soil water in the active layer rose from 120 to 290 ppm as the runoff season progressed, and demonstrated the importance of biogenic

CO_2 in raising the levels of solute concentrations. On non-carbonate terrain on Banks Island, Lewkowicz and French (1982) also detected a rise in active-layer solute concentrations through the melt season (from 200–240 to 400–500 $mg\,l^{-1}$) due to progressive increase in soil water residence times. They estimated an associated denudation rate of 3.5–29.0 $mm\,ka^{-1}$, which implies that transport of dissolved solids by subsurface wash equals or exceeds particulate sediment removal by surface wash in the same area.

In his assessment of denudation in the low-arctic mountainous catchment of Karkevagge in northern Sweden, Rapp (1960) concluded that solute transport represents the dominant mode of removal of rock materials. Later work in the same valley supports his assessment, reporting a solutional yield of >46 $t\,km^{-2}a^{-1}$ (Campbell *et al.*, 2002), roughly equivalent to a denudation rate of >17 $mm\,ka^{-1}$. This is probably exceptional, however, reflecting the unusual mineralogy of the local schists. In the nearby Latnjavagge catchment, measurements of TDS in slopewash produced very low values (<10 $mg\,l^{-1}$), and small creeks mostly yielded TDS averages of < 40 $mg\,l^{-1}$ (Beylich *et al.*, 2004); the overall solute yield of this catchment is 4.9 $t\,km^{-2}a^{-1}$ (Beylich, 2011), equivalent to a bedrock denudation rate of ~1.8 $mm\,ka^{-1}$. In the Green Lakes Valley, a high-alpine catchment in Colorado, Caine (1992b) recorded solute yields of 5–20 $t\,km^{-2}a^{-1}$, roughly equivalent to a denudation rate of 2–8 $mm\,ka^{-1}$. It is difficult to judge the wider representativeness of such results, however, or of the extent to which solute yields diminish with latitude and increasing brevity of summer groundwater flow. The low yield of solutes in most arctic rivers (see below) suggests that in many catchments sediment transport is dominated by removal of non-dissolved sediments.

13.4 Slushflows

Slushflows are rapid movements of water-saturated snow during seasonal snowmelt. They occur in many arctic and subarctic areas (Rapp, 1985; Hestnes, 1998; Larocque *et al.*, 2001), but less often in alpine environments, though accounts of slushflows in Kyrgyzstan (Elder and Kattelmann, 1993) and the Pyrenees (Furdada *et al.*, 1999) suggest that they may be more common in high alpine valleys than has hitherto been appreciated.

Slushflows tend to follow stream courses, but also occur on open slopes. They are caused by an increase in water in the snowpack that continues until failure occurs as a result of increasing pore-water pressures and weakening of intergranular bonds (Gude and Scherer, 1998). As saturation of the snowpack is rare on steep slopes, slushflow release tends to occur on valley floors or hillslopes with gradients of 5–20°. Snow saturation is favoured by rapid thaw, though rainfall may also contribute to slushflow initiation. The geomorphic effects of slushflows sometimes resemble those of snow avalanches (Chapter 12). Erosional effects include abrasion and entrainment of bedrock, scouring of unconsolidated deposits, formation of debris tails, excavation of impact pits and modification of stream channels. Slushflow deposits include perched debris and drapes of soil over boulders, and they often form a veneer of sediment on steam banks (Nyberg, 1989; Clark and Seppälä, 1988; Barsch *et al.*, 1993). Slushflow deposits may also form small boulder tongues along stream courses. The most distinctive slushflow landforms are large streamlined *slushflow fans* or *whaleback fans*, which resemble alluvial fans but are distinguished by the absence of channels, a surface cover of loose, poorly sorted debris and a reversed proximal slope (Nyberg, 1989).

Rapp (1960) considered that slushflows have the greatest geomorphic capacity of all avalanche types in northern Scandinavia. In northern Sweden, postglacial slushflow deposits reach thicknesses of 4–5 m, implying an average accumulation rate of 0.5–0.6 $m\,ka^{-1}$ (Nyberg, 1987). Though slushflows in some areas occur annually or every few years, there is also evidence for infrequent catastrophic slushflows or *slush torrents*. André (1990b) used lichenometry to establish the age of boulder tongues on slushflow fans in Svalbard, and concluded that they represent mobilization of large amounts of debris (1300–7000 m^3) by catastrophic slush torrents with return intervals of ~500 years. Barsch *et al.* (1993) estimated that a single slush torrent on Svalbard in 1992 reworked 6000 metric tons of sediment. In 1953, a slush torrent destroyed a hospital in Longyearbyen, Svalbard, causing three fatalities.

13.5 Sediment Transport in Periglacial Rivers

The sources of sediment transported by rivers in periglacial environments are highly variable. In any area, the dominant sources are determined by a complex interaction of relief, climate, lithology, vegetation cover, slope processes and access to stores of unconsolidated sediment such as till, glacifluvial deposits and alluvial fans (Beylich, 2011). In lowland arctic environments, a large component of nondissolved sediment is derived from reworking of channel and bar deposits and through riverbank erosion; in arctic environments of moderate relief, slopewash, slushflows, solifluction, retrogressive thaw slumps, active-layer detachment slides and thermokarst development in headwaters may be important in delivering sediment to streams (Bowden *et al.*, 2008; Lamoureux and Lafrenière, 2009; Lewis *et al.*, 2012;

Kokelj *et al.*, 2013; Lamoureux *et al.*, 2014; Favaro and Lamoureux, 2015); and in alpine environments, rockfall, snow avalanches and debris flows may contribute intermittently to fluvial sediment transport. The dominant source of sediment in many periglacial rivers, however, consists of unconsolidated sediments derived from glacigenic deposits exposed by the retreat of glacier ice. Reworking of such deposits implies that present rates of fluvial sediment transport in formerly glaciated areas are often elevated well above background denudation levels (Church and Slaymaker, 1989; Ballantyne, 2002; Church, 2002; Mercier, 2011).

Rivers transport sediment in three ways: in solution, in suspension and as bedload. Dissolved and suspended sediment loads are calculated by multiplying water discharge by sediment concentration. The concentration of dissolved solids in streams tends to drop as discharge increases (Figure 13.8). This is because at low flow, rivers are fed mainly by slow-moving groundwater that contains relatively high concentrations of solutes, particularly during warm summers that lead to active-layer deepening (Lafrenière and Lamoureux, 2013). During high-flow events, however, a much greater proportion of runoff is provided by quickflow (overland flow), which tends to contain a much lower concentration of dissolved solids. The concentration of suspended sediments tends to rise with increasing discharge (Figure 13.8), reflecting in part an increase in stream power, and partly access of rising streamwaters to sources of fine sediment (McDonald and Lamoureux, 2009). This relationship is, however, subject to sediment availability, and rivers generally carry suspended sediment concentrations well

below the maximum possible for a given water discharge. The relationship between bedload and discharge is more complex. Often, bedload transport is negligible until a discharge threshold is exceeded, at which point the streambed is mobilized, greatly increasing the amount of sediment being transported.

The relationship between sediment concentration and water discharge can be obtained by determining the concentration of dissolved and suspended sediment for samples taken at different water discharges (Figure 13.8). If a continuous stream discharge record is available for the entire runoff season, the computed sediment–water discharge relationship can be used to calculate the approximate total annual sediment yield. To allow catchments of different sizes to be compared, the total sediment yield is divided by catchment area to calculate specific sediment yield ($t\,km^{-2}\,a^{-1}$), which can be converted to a denudation rate ($mm\,ka^{-1}$) by dividing by bedrock density. For solute yield, this procedure gives reasonably reliable results. The measured yield of suspended sediments, however, can be distorted by infrequent extreme events such as exceptional rainstorms (Dugan *et al.*, 2009; Lewis *et al.*, 2012), and hence short-term records must be treated with caution. The greatest uncertainty lies in estimation of bedload yield. This is usually achieved by employment of a bedload transport formula that expresses specific bed transport rate ($kg\,m^{-1}\,s^{-1}$) in terms of such variables as clast size in the stream bed, stream power (a function of water discharge, slope and channel width), stream depth and the specific gravity of water and transported sediment. Several different approaches have been used to estimate bedload transport in sand-bed and gravel-bed

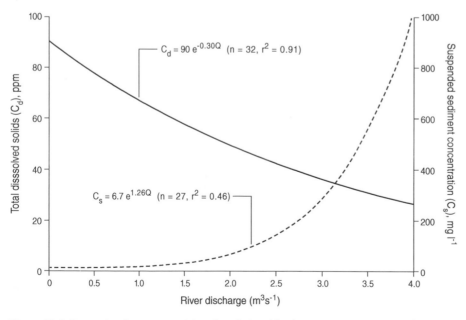

Figure 13.8 Regression lines summarizing the relationships between concentration of total dissolved solids (TDS) and concentration of suspended sediment with river discharge (Q) for a 29 km² catchment partly underlain by carbonate rocks, Ellesmere Island, Canadian Arctic Archipelago (78° N).

rivers, and all yield different results. As it is usually impossible to check the accuracy of the particular formula employed, most bedload transport estimates should be regarded as order-of-magnitude approximations (Martin, 2003).

Given these caveats, data on specific sediment transport yields in periglacial environments must be treated with caution. Those for rivers draining a range of catchment sizes in arctic North America (Table 13.2) nonetheless suggest some generalizations. First, transport of suspended sediment and bedload is usually much greater for glacierized than for glacier-free catchments. This is to be expected in view of the high load of suspended sediment in meltwater draining warm-based and polythermal glaciers (Hallet *et al.*, 1996; Hodgkins *et al.*, 2003; Benn and Evans, 2010), as well as fluvial reworking of unvegetated till and outwash deposits in the proglacial zone. Second, the sediment yield attributable to transport of dissolved solids is generally low, and probably exceeds that of suspended sediment only on carbonate terrain (McCann and Cogley, 1974) or where rivers have limited access to sources of fine sediment. Third, bedload transport probably dominates in most glacierized basins, but its importance in glacier-free basins is difficult to assess because of the limited data. The occurrence of braided gravel-bed rivers in polar deserts with limited vegetation cover suggests significant bedload transport in trunk streams during floods, though in low-arctic and subarctic lowlands crossed by meandering or anastomosing rivers, suspended sediment transport may exceed bedload transport (Gautier and Costard, 2000; Huisink *et al.*, 2002). Data on sediment yield in low-arctic and subarctic catchments (Table 13.3) suggest that very low to moderate flux of dissolved solids and suspended sediment is typical for non-carbonate terrain in such environments (Beylich, 2011).

In steep alpine streams, suspended sediment and bedload movement during snowmelt floods tend to dominate the sediment transport regime, though solute yield is by no means negligible (Orwin *et al.*, 2010a). In a high-alpine catchment in the Alps, Veit and Höfner (1993) found that ~95% of particulate sediment transport is associated with the snowmelt runoff flood. They calculated an annual sediment yield of $88 \, t \, km^{-2} a^{-1}$ for the entire ($1.3 \, km^2$) catchment, but noted the strong influence of vegetation cover on sediment flux: a subcatchment in alpine tundra produced a sediment yield of $8.2 \, t \, km^{-2} a^{-1}$, but one in the sparsely vegetated subnival zone yielded $251 \, t \, km^{-2} a^{-1}$. In the latter zone, slopewash related to snowmelt was a major contributor of suspended sediment. Elsewhere, reworking of glacigenic sediment by debris flows has been identified as a dominant source of sediment in steep mountain streams.

The effectiveness of periglacial rivers as agents of sediment transport can be gauged by comparing estimated total sediment yields with those of other environments. Such comparisons suggest that in largely or entirely nonglacierized catchments, sediment yields are relatively low; available estimates fall well below the $130 \, t \, km^{-2} a^{-1}$ proposed as an average for North American rivers (Table 13.4). Conversely, those for glacierized catchments generally lie well above this average (Table 13.2). The North American average, however, reflects human disturbance in many catchments. Excluding this influence, it seems likely that the sediment yields from nonglacierized arctic and subarctic catchments fall within a range similar to that for undisturbed catchments in mid-latitude environments. However, data summarized by Gordeev (2006) suggest that the total suspended sediment flux of large rivers draining into the Arctic Ocean is very low ($\sim 227 \times 10^6 \, t \, a^{-1}$), and represents only about 1% of the global figure. Gordeev suggested that this low output represents a combination of factors, including thin regolith cover, low precipitation, extensive permafrost, low winter flow regime, trapping of sediment in lakes, reservoirs and wetlands, and a low level of anthropogenic terrain disturbance.

The effectiveness of sediment transfer by rivers in small catchments can also be compared with that of slope processes in the same catchments, or by monitoring 'source-to-sink' movements of sediment within catchments (Orwin *et al.*, 2010b; Favaro and Lamoureux, 2015). Beylich (2011) has summarized relevant data for four small ($7-23 \, km^2$) catchments in subarctic and low-arctic environments in Iceland and Lapland. His results indicate that mass transfer by rivers over periods of 8–13 years tends to exceed total mass transfer on hillslopes, though the dominant mode of fluvial transfer (solution or suspension) varies from site to site (Table 13.3).

13.6 Bank and Channel Erosion

Like rivers elsewhere, those in cold environments derive part of their sediment load from bed scour and bank erosion, particularly during floods. Two factors affecting the efficacy of bank and channel erosion are, however, particularly characteristic of periglacial environments: the presence of seasonal snow and ice in channels, and the operation of *thermal erosion*, the meltout of frozen bank sediments in permafrost areas.

13.6.1 The Role of Snow and Ice

Snow, river ice and icings in channels may either limit or enhance erosion. Snow accumulation can protect riverbeds and banks against the early stages of snowmelt floods (Woo and Sauriol, 1980; Priesnitz, 1990; Kane *et al.*, 1991). Similarly, *bottom-fast ice* that freezes on to riverbeds, banks and bars during winter may protect them

Table 13.2 Specific sediment yields and implied bedrock denudation rates for some periglacial rivers in arctic North America.

River and location	Basin area (km²)	Year	Specific sediment yield (t km⁻² a⁻¹)				Bedrock denudation rate (mm ka⁻¹)[a]			
			Solutes	Suspended sediment	Bedload	Total	Solutes	Suspended sediment	Bedload	Total
Nonglacierized basins										
Heather Creek, Ellesmere Island	6	1991	1.8	1.0	–	–	0.7	0.4	–	–
Whitebear Creek, Bathurst Island	8	1976	3.6	22.8	–	–	1.4	8.6	–	–
Andy Creek, Yukon	13	Mean	19.0	71.0	–	–	7.2	26.8	–	–
Douglas Creek, Yukon	35	Mean	52.0	328.0	–	–	19.6	123.8	–	–
Snowbird Creek, Bathurst Island	61	1976	4.4	25.6	2.5	32.5	1.7	9.7	0.9	12.3
Mecham River, Cornwallis Island	95	1970	–	22.1	–	–	–	8.3	–	–
		1971	21.1	12.7	2.9	36.7	8.0	4.8	1.1	13.8
Hot Weather Creek, Ellesmere Island	155	1990	4.7	17.3	–	–	1.8	6.5	–	–
		1991	1.8	1.7	–	–	0.7	0.6	–	–
Colville River, Alaska	53,000	1962	1.1	5.8			0.4	2.2		
Glacierized basins										
Decade River, Baffin Island	13	1965	–	154.0	–	–	–	58.1	–	–
Schei River, Ellesmere Island	91	1973	–	150.0	–	–	–	56.6	–	–
Upper South River, Baffin Island	90	1967	4.0	48.8	1114.0	1166.8	1.5	18.4	420.4	440.3
		1968	1.9	0.8	29.0	31.7	0.7	0.3	10.9	12.0
Middle River, Baffin Island	106	1967	1.8	65.5	570.0	637.3	0.7	24.7	215.1	240.5
North River, Baffin Island	189	1967	2.0	86.5	–	–	0.8	32.6	–	–
Lewis River, Baffin Island	205	1963	3.2	232.5	1049.0	1284.7	1.2	87.7	395.8	474.8
		1964	2.1	68.1	236.0	306.2	0.8	25.7	89.1	115.5
		1965	2.4	126.9	482.0	611.3	0.9	47.9	181.9	230.7

Where appropriate, data have been aggregated or averaged to yield best-estimate figures. Dash (–) implies no data available.
[a] Bedrock denudation rate assumes a bedrock density of 2.65 t m⁻³.
Source: Data from Church (1972), Lewkowicz and Wolfe (1994) and Priesnitz and Schunke (2002).

Table 13.3 Specific sediment yields and implied bedrock denudation rates for some low-arctic and subarctic catchments.

Catchment	Area (km²)	Years	Sediment yield (t km⁻² a⁻¹)		Denudation rate (mm ka⁻¹)ᵃ	
			Dissolved solids	Suspended sediment	Dissolved solids	Suspended sediment
Eastern Iceland						
Hrafndalur	7	2001–09	29.0	19.0	10.9	7.2
Austdalur	23	1996–2009	8.0	42.0	3.0	15.8
Lapland						
Latnjavagge	9	1999–2009	4.9	2.4	1.8	0.9
Kidisjoki	18	2001–09	3.1	0.25	1.2	0.1

ᵃ Bedrock denudation rate assumes a bedrock density of 2.65 t m⁻³. In practice, most dissolved solids and suspended sediments are derived from sediment cover.
Source: Data from Beylich (2011).

Table 13.4 Estimated total annual sediment yields for some periglacial catchments.

River and location	Sediment yield (t km⁻² a⁻¹)	Implied denudation rate (mm ka⁻¹)ᵃ
Snowbird Creek, Bathurst Island	33	12
Mecham River, Cornwallis Island	37	14
Colville River, Alaska	82	31
Yenesei River, Siberia	5	2
Ob River, Siberia	7	3
North American average	130	49

ᵃ Implied denudation rate refers to equivalent bedrock denudation, assuming a bedrock density of 2.65 t m⁻³.
Source: Data from Table 13.2 and Clark (1988).

against scour during snowmelt runoff. Conversely, breaching of temporary snow dams may generate potentially erosive floods, and saturation of snow may initiate erosive slushflows. Large rivers that flow throughout the winter support a winter ice cover up to 2 m thick. During spring break-up, tabular floes of fractured ice may form ice jams that trigger flooding, causing erosion of channel banks above normal flood levels and the formation of floodplain gullies and scour holes (Smith and Pearce, 2002). Along the channels of major rivers, rapidly moving ice floes may erode sediments from riverbanks and the channel bed, and push sediments (including boulders) along riverbanks. Alluvial channels subject to frequent ice scour during ice-jamming events sometimes develop a distinctive stepped two-level channel cross-sectional morphology, with the upper step reflecting erosion by both ice floes and flood events during the ice-free period (Boucher *et al.*, 2009).

River ice may also play a role in the development of *boulder-paved channels*. These are broad, shallow channels floored by closely packed boulders that form a fairly level 'pavement', very unlike the irregular boulder beds of most mountain streams. According to Davies *et al.* (1990), the boulders probably represent a lag deposit washed out from bouldery till, and have been squeezed into the channel bed by the weight of winter ice in the channel. They attributed the close packing of the boulders to frost heave and thaw consolidation of underlying fine sediments.

Periglacial rivers in which winter runoff is sustained by baseflow or outflows from wetlands or lakes often develop *river icings* due to the rupture of surface ice cover under hydraulic pressure (Pollard and Van Everdingen, 1992; Hu and Pollard, 1997). This results in overflow of released water, which freezes rapidly under subzero air temperatures. Overflow events may occur several times during a single winter, producing sheet-like masses of layered ice several metres thick across channels and adjacent floodplains. The largest icings, representing outflow from several sources, may extend several kilometres down-channel, and river icings covering tens of square kilometres have been reported in Siberia, Mongolia, Alaska and northern Canada. Large river icings not only store winter baseflow, thereby extending the duration of meltwater runoff, but may also protect the channel from scour during the annual snowmelt flood. Conversely, however, thick icings may divert flood runoff, extending fluvial erosion laterally across floodplains and adjacent low-lying terrain, and encouraging channel migration.

13.6.2 Thermal Erosion

Ice-rich permafrost also plays an ambivalent role in dictating rates of bank erosion. During the period of snowmelt runoff, when water temperatures are low, ice tends to bond bank sediments, restricting bank erosion to small-scale collapse of thawing sediment. Later in the runoff season, when river temperatures rise, heat convected by flowing water at the base of channel walls allows rivers to undercut adjacent riverbanks, forming a *thermo-erosional niche*. Costard *et al.* (2003, 2014) showed that the most rapid undercutting and thermal erosion of ice-cemented riverbanks results from a combination of high water temperatures and mechanical erosion, particularly during periods of flood discharge when water levels are elevated and thus in contact with ice-rich sediments. Numerical models of fluvial thermal erosion (Randriamazaoro *et al.*, 2007; Dupeyrat *et al.*, 2011) show that high water temperature, high discharge and relatively high ice temperature promote rapid thermomechanical undercutting of frozen riverbanks, and that ice content in cohesionless sediments controls rate of erosion. Ultimately, undercutting causes large blocks of bank sediment to collapse into the channel, with ice wedges often forming the 'hinges' of bank collapse. Collapsed blocks of ice-rich sediment rapidly disintegrate within the river channel due to thaw and fluvial removal of released sediment. On the Colville River delta in Alaska, bank retreat due to undercutting and collapse locally reaches $12\,\mathrm{m\,a^{-1}}$, and on the Mackenzie River delta, $180\,\mathrm{m}$ of bank retreat in 19 years has been recorded. On reaches of the Lena River in Siberia, bank retreat of 19–$24\,\mathrm{m\,a^{-1}}$ has been documented, but such rapid erosion is very localized (Gautier *et al.*, 2003).

Thermal erosion of riparian bluffs is particularly effective in ice-rich yedoma sediments. Kanevskiy *et al.* (2016) have documented the rates of retreat of a $680\,\mathrm{m}$ long, $35\,\mathrm{m}$ high yedoma bluff on the outside of a meander of the Itkillik River in northern Alaska, where bank retreat is dominated by a combination of thermo-erosional undercutting and collapse along ice wedges, combined with thaw of frozen soils on exposed bluffs. Retreat rates average $\sim11\,\mathrm{m\,a^{-1}}$, though the most actively eroding parts of the bluff are retreating roughly twice as fast. These retreat rates imply that almost $70\,000\,\mathrm{t}$ of sediment are removed annually from the bluff.

13.7 River Channels

A wide range of channel types and channel patterns occur in cold environments (Vandenberghe and Woo, 2002). In mountain areas, predominantly bedrock channels drain steep terrain, often feeding braided or meandering river systems in trunk valleys. In lowland arctic and subarctic environments, rivers occupy meandering, braided or anastomosing channel systems (Figure 13.9). A *meandering* river is a single-channel stream in which the sinuosity (ratio of channel length to straight-line distance) exceeds ~1.5. A *braided* river has a network of interconnected channels divided by a system of depositional *braid bars* that may be flooded during periods of high discharge. An *anastomosing* (or *anabranching*) river occupies a network of channels separated by more permanent vegetated islands that are rarely (if ever) overtopped during floods. The three types are transitional. Bars may be exposed in meandering streams during low flow, the main channels within braided systems often meander, and many braided streams incorporate semipermanent vegetated islands as well as transient bars. In general, alluvial channel morphology is determined by the balance between available energy (determined by stream discharge and gradient) and the amount and coarseness of sediment load. Sediment availability is often strongly influenced by vegetation cover. Vandenberghe (2001) has suggested a general typology for periglacial channel patterns based on these factors (Figure 13.10).

13.7.1 Bedrock Channels

In areas of steep relief, periglacial rivers characteristically occupy narrow, single-thread channels. The long profile of such streams is often irregular, reflecting variations in the resistance of the underlying bedrock, so that steep bedrock-incised reaches alternate with gentler alluvial (depositional) ones. Channel planform is straight or sinuous in bedrock reaches, and gently meandering or braided in alluvial reaches. In bedrock reaches, the channel floor is often occupied by poorly sorted, tightly wedged boulders. Alluvial fans are deposited where steep mountain streams join a floodplain.

A study of the characteristics of a bedrock-channelled mountain stream in Jostedalen, Norway, has revealed several distinctive characteristics (McEwen and Matthews, 1998). Sediment transport is dominated by episodic bedload movement controlled by the frequency of competent floods. Coarse sediment is supplied to the river by snow avalanches and by frost wedging of bedrock in the channel floor and walls. As a result, clasts in the streambed show no evidence of downstream fining, and even in distal channel reaches most clasts are angular, unlike those in other alluvial environments.

13.7.2 Meandering and Anastomosing Channels

Meandering and anastomosing channels are widespread in lowland arctic and subarctic environments (particularly wetlands) that support a continuous cover of tundra or taiga vegetation (Figure 13.9a,b). Many rivers in northern Russia have meandering or anastomosing planforms despite a nival runoff regime and the presence of permafrost (Gautier and Costard, 2000; Huisink *et al.*, 2002). Stable channel systems in such areas are favoured by low stream gradients and vegetation cover. The latter not only limits sediment supply by slopewash, but also tends

Figure 13.9 (a) Meandering river, north Yukon. (b) Anastomosing river channels, northern Yakutia. (c) Braided river channels on a periglacial sandur, central Banks Island. A colour version of (c) appears in the plates section. *Source:* Courtesy of (a,b) Matthias Siewert and (c) Tobias Ullman.

Ground Thermal regime	Stream power: sediment supply	Vegetation absent	Patchy vegetation cover	Continuous vegetation cover
Continuous or discontinuous permafrost	Low ↕ High	Braided channels	Braided or meandering channels / Braided channels	Meandering channels / Anastomosing (anabranching) channels
Discontinuous permafrost or deep seasonal ground freezing	Low ↕ High			Meandering channels / Anastomosing (anabranching) channels / Braided channels

Figure 13.10 Schematic representation of periglacial alluvial river types as a function of vegetation cover, ground freezing conditions and sediment supply. *Source:* Adapted from Vandenberghe (2001). Reproduced with permission of Elsevier.

to bind channel banks, so that channel migration is often limited to erosion on the outside of meander bends and *avulsion* or channel switching is rare. Stream load is usually dominated by transport of suspended sediment rather than coarse bedload.

Some arctic rivers have channel forms transitional between braided and meandering. Such channels occur when a braided system develops only during the snowmelt flood, and during the rest of the runoff season flow is largely confined to a single meandering channel (Vandenberghe, 2001). Transitional channels also develop as a river adapts to a change in discharge regime or sediment supply, and may thus represent a transient response to environmental change (Vandenberghe and Woo, 2002). On some arctic rivers, such as the Thomsen River on Banks Island, meandering reaches alternate with braided reaches where the banks are less confined or channel gradients are steeper.

13.7.3 Braided Channels

The development of high-energy braided rivers with unstable multi-thread channels that are prone to frequent avulsion is favoured by steep channel gradients, erodible banks, abundant sediment supply, high flood discharges and dominantly bedload transport. Such conditions pertain in polar deserts, where vegetation cover is limited and runoff is dominated by the rapid melt of snow and ice over frozen ground, but braided rivers occur in all periglacial environments where stream power is high and sediment is abundant (Figure 13.9c). Braided systems also occur in some alpine areas where trunk streams flow over broad floodplains. The braid bars that divide channels are initiated during periods of high discharge when both bed scour and deposition occur. At such times, spiral secondary flows develop. In zones of flow convergence, bed velocities are increased, causing channel scour. Where flows diverge, a reduction in bed velocity promotes deposition and the development of braid bars that break the flow into a network of channels as discharge decreases and water level declines. Gravel-bed braided streams are dominated by lozenge- or diamond-shaped longitudinal or diagonal mid-channel bars and point bars attached to channel banks, whereas sand-bed braided streams tend to be dominated by lingulate (tongue-shaped) bars.

13.8 Alluvial Landforms in Periglacial Environments

13.8.1 Sandar

The Icelandic term *sandur* (plural *sandar*) describes braided floodplains of dominantly coarse sediments deposited by proglacial rivers, but has been extended to include similar floodplains deposited by rivers with a nival runoff regime in periglacial environments (*periglacial sandar*; Figure 13.9c). Sandar are occupied by rapidly shifting, generally aggrading braided rivers, with high competence and high bedload transport rates. The surfaces of sandar are characterized by rapidly changing sites of local erosion and deposition that reflect fluctuations in water discharge due to variations in the rate of glacier melt or snowmelt. Sandur gradients tend to decline downstream, and whereas the upper and middle parts of many sandar are characterized by a few dominant channels, lower reaches often support a network of wide, shallow channels that are occupied even at low to moderate flow. Clast size decreases downstream, reflecting declining competence, but also exhibits marked lateral variations (Ballantyne, 1978b). On aggrading sandar, net sediment deposition occurs during the waning stages of high flood events, sometimes revealing new channels and bars, whilst sheet deposits laid down by very high floods veneer areas more distant from the main channels. Release of sediment from river icings on sandur surfaces may produce a number of distinct minor features, including thin sediment drapes, cones of dumped sediment, linear ridges formed by melt-out of sediment within icing channels and subsidence structures (Bennett *et al.*, 1998). In general, sandur deposits tend to be sedimentologically immature, and dominated by poorly sorted gravels with normal imbrication. In vertical exposures, they are represented by massive or crudely bedded gravel, and the distinction between individual beds often reflects differences in matrix properties.

Many sandar in formerly glaciated arctic and alpine environments are paraglacial deposits, in the sense that the sediment budget is dominated by fluvial reworking of till or outwash deposits. If the rate of sediment supply diminishes as such sediment sources become depleted, sandur degradation succeeds aggradation, and stream incision results in the formation of terraces along sandur margins (Ballantyne, 2002; Church, 2002). Glacio-isostatic uplift may also result in a falling base level, promoting incision of earlier sandur deposits and prolonging the period of paraglacial sediment reworking (Church and Ryder, 1972).

13.8.2 Alluvial Fans

In common with other landscapes, arctic and alpine environments support alluvial fans at sites where tributary valleys meet low-gradient floodplains (Figure 13.11). Such fans may achieve impressive dimensions: one described by Catto (1993) at the margin of the Aklavik Range near the Mackenzie delta is 3.9 km long and 2.2 km wide, and declines in gradient from 8.5° at the fan head to <1° at its distal reaches. Catto concluded that fan

Figure 13.11 Oblique aerial photograph of alluvial fans, Svalbard. The parallel levées on fan surfaces indicate debris-flow deposition on fan surfaces. *Source:* Courtesy of Matthias Siewert.

formation in this semi-arid permafrost environment reflects a combination of fluvial sedimentation, deposition by snowmelt-induced debris flows and movement of sediment by solifluction; channel development is confined to the fan head, where the feeder channel is incised by up to 3 m.

In alpine mountains, alluvial fans often represent an important form of postglacial sediment storage on valley floors. Many such fans are relict paraglacial features, formed by entrainment and redeposition of glacigenic sediments after the retreat of Pleistocene glaciers (Church and Ryder, 1972; Ballantyne, 2002). They are often compositionally complex, consisting of stratified and sorted fluvial gravels interbedded with massive diamictons emplaced by debris flows (Eyles and Kocsis, 1988).

13.8.3 Deltas

Where periglacial rivers enter lakes or the sea, they deposit prograding deltas similar to those of warmer environments, but often with a number of distinctive characteristics (Walker, 1998). The largest periglacial deltas are those deposited on Arctic Ocean shelves by major northwards-flowing rivers such as the Lena and Mackenzie. These are characterized by a complex network of shifting distributary channels (over 800 on the Lena Delta) flowing through a low-relief landscape studded by shallow thermokarst lakes, extensive backwater wetlands and peat-covered islands. The Mackenzie Delta comprises a patchwork of channels with a total area of 1744 km^2, wetlands covering 1614 km^2, floodplains covering 6446 km^2 and nearly 50 000 lakes covering 3331 km^2 (Emmerton *et al.*, 2007). The annual snowmelt flood and associated break-up of channel ice up to 2 m thick represent the dominant hydrologic and geomorphic

events on these distinctive landscapes. Ice jams re-route floodwaters, which scour out new channels (Prowse and Lalonde, 1996), and as the annual flood subsides, vast quantities of fine sediment are deposited over the delta surface. Aggradation of permafrost under channel levées and overbank deposits may result in hydrological isolation of channels, lakes and alluvial flats until renewed channel scour during ice break-up redirects flow, causing lake drainage and channel migration.

The Colville River Delta in Alaska provides an excellent illustration of the dynamics of a large delta in a permafrost area (Walker and Hudson, 2003). The Colville River drains a catchment of ~53 000 km^2 that extends from the Brooks Range to the Arctic Ocean. Its delta has an area of 650 km^2, and is traversed by 34 major distributaries with both braided and meandering channel forms. Channel banks comprise both fine and coarse alluvial sediment with widespread peat cover, and ice wedges are commonly exposed. The intervening terrain includes areas of sand dunes, shallow lakes, wetlands and mudflats.

During winter, flow ceases entirely and a saltwater wedge extends upstream under the river ice for a distance of at least 60 km. In spring, the initial flow takes place beneath the river ice and over bottomfast ice and adjacent bars and banks. Break-up begins when surface ice begins to fracture and move, and progresses downriver from the delta apex. The annual snowmelt flood peaks over a few days in early June, reaching a maximum discharge of up to 6000 m^3 s^{-1}. Floodwaters typically rise 3–5 m, inundating up to 60% of the delta surface. More than half the annual water discharge of the Colville River is expended within about 20 days following the onset of flooding, and 62% of the annual sediment load (estimated to average ~5.8×10^6 t a^{-1}) is transported within 2 weeks of break-up. During the flood, flow around ice jams causes bed scour and channel migration, and leaves debris draped over adjacent mudflats, sandbars and tundra surfaces. Even after the flood peak has subsided, localized riverbank erosion persists, partly through scour of bank sediments by floating ice and partly by thermo-erosional undercutting and resultant collapse of blocks of frozen sediment into distributary streams.

13.9 Valley Form

13.9.1 V-shaped Valleys and Rock Gorges

In areas of moderate to high relief, many periglacial rivers occupy steep, V-shaped tributary valleys or gorges (canyons) with steep rocky sidewalls (Figure 13.12). Parts of the Canadian Arctic Archipelago, for example, consist of low plateaus deeply entrenched by fluvial incision with evidence of recent channel deepening, and tributary valleys flanking sandar on Svalbard are characteristically steep and rock-cut. The upper reaches of many alpine streams also flow through steep-sided valleys and gorges produced by fluvial incision.

Though fluvial entrenchment can often be attributed to a drop in base level (sea level) due to glacio-isostatic or tectonic uplift, the mechanism of valley-floor incision in such areas may be distinctly periglacial. McEwen and Matthews (1998) have illustrated the importance of frost wedging in detaching bedrock from the channel floor

Figure 13.12 Gorge excavated in sandstone, Ellesmere Island, NWT, arctic Canada.

and adjacent rock banks of a river in Norway, and it seems likely that this process is of wider importance in enabling periglacial rivers to cut down through bedrock. For rock gorges along a Norwegian alpine stream, McEwen *et al.* (2002) calculated a minimum Holocene channel incision rate of 0.15–$0.39\,\text{mm}\,\text{a}^{-1}$, which they attributed mainly to frost wedging acting on the channel floor and removal of detached debris during floods. Much greater rates of bedrock channel incision occur in the Himalayas, where channel downcutting through rock is estimated to be 9–$12\,\text{mm}\,\text{a}^{-1}$ (Leland *et al.*, 1998). Whipple *et al.* (2000) have shown that river diversion in Alaska has resulted in bedrock-channel incision of 10–$100\,\text{mm}\,\text{a}^{-1}$ since AD 1912, though the extent to which this reflects frost wedging of rock as opposed to bedload impact and hydrodynamic 'plucking' acting along rock joints is uncertain. The role of freeze–thaw weathering in bedrock channels deserves further investigation, and may help to explain the evolution of bedrock gorges in cold environments.

13.9.2 Valley Asymmetry

A feature of some shallow river valleys in arctic permafrost environments is *valley asymmetry*, the tendency of valleys to adopt a skewed cross-profile, with one valley-side slope generally steeper than the other. It should be emphasized, however, that many arctic valleys do not exhibit systematic asymmetry, and the extent to which valley asymmetry is a characteristic or exceptional phenomenon remains to be resolved. Periglacial valley asymmetry is generally attributed to microclimatic differences on opposing slopes, and tends to be most pronounced where valleys are cut in weak bedrock or unconsolidated sediments. Most accounts of valley asymmetry in arctic areas describe slopes with northerly aspects as being systematically steeper than those with southerly aspects. In such cases, asymmetry has usually been attributed to greater insolation intensity on south-facing slopes, which is inferred to cause the development of a deeper active layer and to promote more effective solifluction, resulting in more rapid slope decline on south-facing than on north-facing slopes.

In some arctic areas, however, valley sides with northerly aspects are gentler than those facing west, east or even south. On Banks Island in arctic Canada, French (1971) observed systematic asymmetry, with steep southwest-facing slopes and gentle northeast-facing slopes. This he attributed to the effect of prevailing westerly winds, which cause preferential snow accumulation (and thus enhanced gelifluction and nivation activity) on northeast-facing slopes, and enhanced evaporation (and thus soil drying) on southwest-facing ones. He concluded that streams have consequently migrated laterally northeastwards, undercutting and steepening southwest-facing slopes through mechanical and thermal erosion. Kennedy and Melton (1972) have shown that a downvalley reversal in asymmetry may occur within individual catchments, reflecting changing microclimatic circumstances and thus the effectiveness of particular hillslope processes. They concluded that no single directional asymmetry characterizes river valleys in permafrost areas, suggesting that valley asymmetry in mid-latitude areas that experienced periglacial conditions during the Pleistocene cannot be unambiguously interpreted in terms of formative processes or climatic controls.

13.10 Pleistocene Periglacial Rivers

Beyond the southern margins of the ice sheets that invaded mid-latitude areas during the Pleistocene, the forerunners of rivers such as the Mississippi and Rhine flowed under conditions that varied between temperate interglacial and full-glacial or *pleniglacial* (Vandenberghe *et al.*, 2014; Lindgren *et al.*, 2016; Chapter 16). During cold periods, the hydrologic regimes of mid-latitude rivers and the supply and calibre of transported sediment varied greatly compared to those of the present. Under pleniglacial conditions, many major rivers were fed in part by glacial meltwater, and some, like the Thames in England, were diverted southwards to their present courses by the most extensive Pleistocene ice sheets. Those rivers that were not nourished by glacial meltwater were, like arctic rivers at present, fed by a dominantly nival regime, and flowed through steppe-tundra landscapes that were underlain by permafrost or experienced deep seasonal ground freezing. Falling sea level also influenced fluvial dynamics. At the last pleniglacial, for example, the Thames and Rhine appear to have been confluent in the southern North Sea Basin, an area now completely submerged, but then a low-lying plain.

The history of some Pleistocene rivers in Europe has been reconstructed from the distribution, altitude and sedimentology of terraces that flank present floodplains. Such terraces represent former floodplains that were abandoned during later episodes of river incision. Exposures cut through terraces allow reconstruction of the changes that have affected river systems through stratigraphic and sedimentological analyses of fluvial and aeolian deposits, intervening buried soil horizons and periglacial structures such as ice-wedge pseudo-morphs and involutions. Initially, researchers drew a simple distinction between cold periods characterized by aggrading braided streams dominated by coarse bedload transport, and temperate periods characterized by river incision, reversion to single-thread meandering channels and transport of most sediment in suspension. More recent work, however, has shown that fluvial

response to Quaternary climate changes was rather more complicated.

The balance between fluvial aggradation and incision is essentially determined by that between sediment supply and river transport capacity, the former being strongly influenced by the nature and extent of vegetation cover, and the latter by hydrologic regime and particularly flood events. The magnitude and frequency of floods reflect not only direct climatic inputs such as precipitation and snowmelt runoff, but also the presence or absence of permafrost and the nature of vegetation cover. Vandenberghe (2003, 2008) has proposed that incision mainly occurred during periods of climatic transition, and that floodplain stability or aggradation was more typical of periods of relative climatic stability, both cold and temperate, though not all such changes may be preserved in the stratigraphic record. Incision during temperate-to-cold climatic transitions probably reflected an increase in flood runoff relative to sediment supply, whereas incision during cold-to-temperate transitions probably resulted from a reduction in sediment supply due to increasing vegetation cover. This proposition is broadly supported by studies of fluvial sediment sequences in northwest and central Europe (e.g. Huisink, 2000; Antoine *et al.*, 2003b; Kasse *et al.*, 2003; Bridgland and Westaway, 2008; Stemerdink *et al.*, 2010), and is illustrated in a general model of climate-driven fluvial terrace development proposed by Maddy *et al.* (2001) (Figure 13.13). This model has, however, been challenged by Lewin and Gibbard (2010), who concluded that river terraces in southeast England represent bedrock incision and subsequent aggradation of coarse sediments under predominantly cold conditions. Murton and Belshaw (2011) developed this argument, concluding that river incision and planation of underlying frost-susceptible bedrock in southern England occurred under very cold, arid permafrost conditions, when thermal erosion by rivers of limited discharge resulted in the formation of broad, low-gradient erosion surfaces across the underlying frost-susceptible bedrock, with subsequent floodplain aggradation under wetter periglacial conditions being caused by enhanced transport of coarse sediment to the valley floor by solifluction.

Channel patterns and modes of sediment deposition in European rivers nonetheless exhibit contrasts between prolonged periods of temperate and periglacial climate. During the former, single-thread meandering channels were (and are) the norm, with sedimentation dominated by point-bar accretion, accumulation of fine-grained sediments in meander cutoffs and overbank deposition of fine-grained organic-rich sediments during floods. The nature of fluvial systems under periglacial conditions varied, depending in part on relief and on the calibre of sediment supply. Van Huissteden *et al.* (2001) have shown that during the Middle Weichselian in Europe, a

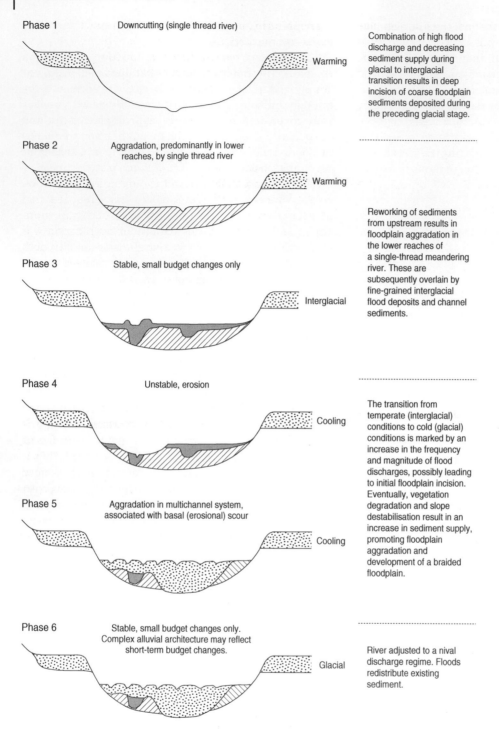

Phase 1 Downcutting (single thread river) Warming

Combination of high flood discharge and decreasing sediment supply during glacial to interglacial transition results in deep incision of coarse floodplain sediments deposited during the preceding glacial stage.

Phase 2 Aggradation, predominantly in lower reaches, by single thread river Warming

Phase 3 Stable, small budget changes only Interglacial

Reworking of sediments from upstream results in floodplain aggradation in the lower reaches of a single-thread meandering river. These are subsequently overlain by fine-grained interglacial flood deposits and channel sediments.

Phase 4 Unstable, erosion Cooling

Phase 5 Aggradation in multichannel system, associated with basal (erosional) scour Cooling

The transition from temperate (interglacial) conditions to cold (glacial) conditions is marked by an increase in the frequency and magnitude of flood discharges, possibly leading to initial floodplain incision. Eventually, vegetation degradation and slope destabilisation result in an increase in sediment supply, promoting floodplain aggradation and development of a braided floodplain.

Phase 6 Stable, small budget changes only. Complex alluvial architecture may reflect short-term budget changes. Glacial

River adjusted to a nival discharge regime. Floods redistribute existing sediment.

Figure 13.13 Model of floodplain and terrace development during a single glacial–interglacial–glacial cycle. *Source:* Adapted from Maddy *et al.* (2001). Reproduced with permission of Elsevier.

nival runoff regime operated and braided gravel rivers drained high-relief areas. Those in the European lowlands, however, were typically aggrading, anastomosing sand-bed rivers flowing across marshy alluvial plains, similar to those that cross Siberian wetlands today. During the last pleniglacial, the style of fluvial sedimen-

tation was nevertheless sensitive to change; throughout much of lowland northern Europe, the onset of colder conditions and development of continuous permafrost after ~28 ka was represented by a period of widespread channel incision that marked the transition from dominantly anastomosing channels to braided river systems

(Mol *et al.*, 2000). In lowland England, the last plenigla-cial appears to have coincided with marked aridity and a consequent decrease in fluvial activity and abandon-ment of braided floodplains (Briant *et al.*, 2004), together with incision due to thermal erosion and a reduction in sediment supply (Murton and Belshaw, 2011). Research based on an optically stimulated luminescence (OSL)-dated sedimentary record for alluvial fan deposits in northwest Germany places the onset of fan deposition at the warm-to-cold transition at 29 ka, with strong fan progradation around 24 ka, followed by evidence for runoff reduction and deposition of overlying aeolian sand under more arid conditions during the late Pleniglacial (Meinsen *et al.*, 2014).

Changes in former river behaviour are often detectable only for major climatic shifts. Evidence for short-term climatic changes within the last pleniglacial are rarely evident in fluvial sedimentary archives, probably because of the absence of marked changes in precipitation inputs and vegetation cover, or else because brief changes in sedimentation pattern have not been preserved in the stratigraphic record (Van Huissteden *et al.*, 2001; Kasse *et al.*, 2003; Briant *et al.*, 2004; Lewin and Gibbard, 2010). The large climatic fluctuations of the Late Weichselian Lateglacial in Europe, however, are locally recorded by distinct changes in fluvial sediment sequences. In low-land England, the Lateglacial (Windermere) Interstade of ~14.7–12.9 ka was characterized by the development of meandering rivers that deposited gravelly and sandy silts with high organic content, reflecting a reduction in sediment supply due to establishment of stable veg-etation cover. Under renewed periglacial conditions during the Younger Dryas Stade (~12.9–11.7 ka), channel erosion and braided river sedimentation predominated. In southern England, rapid climatic amelioration and re-establishment of vegetation cover in the early Holocene resulted in a gradual return to meandering channels, accompanied by infilling of secondary channels and overbank sedimentation of organic-rich fine sediment (Brown *et al.*, 1994). In central Germany, a rapid increase in tree cover around 14.6 ka was accompanied by a change from minerogenic to predominantly organic alluvial sedimentation, and a decline in tree cover due to climatic deterioration after 12.9 ka coincided with renewed clastic sedimentation, though such changes probably affected only tributary catchments (Mausbacher *et al.*, 2001). Some larger European rivers exhibited limited change in behaviour during the Younger Dryas Stade. Moreover, changes in channel pattern during the Lateglacial–Holocene climatic transition were some-times complex. The lower Rhine, for example, changed twice from a braided to a meandering channel pattern, and gradual adoption of a single-thread channel extended over a long timespan (Erkins *et al.*, 2011).

The shifts in fluvial behaviour outlined above are exem-plified by the Vecht River in the Netherlands (Huisink, 2000). Cooling around 42 ka was followed by incision and an increase in stream energy, but a braided system did not develop until after ~31 ka, when establishment of con-tinuous permafrost and a reduction in vegetation cover increased sediment supply. Increasing aridity towards the end of the late pleniglacial reduced the energy of the fluvial system, and there is evidence for widespread aeolian deposition at this time. Lateglacial warming at ~14.7 ka was accompanied by the development of a low-energy meandering river, but renewed cooling during the Younger Dryas Stade (~12.9–11.7 ka) resulted in more open vegetation cover, increased sediment supply and a final return to a high-energy braided system. The river resumed a meandering planform as warming restored a dense vegetation cover in the early Holocene.

13.11 Synthesis

Except in areas of intense glacial modification, periglacial landscapes are often dominated by fluvial landforms that differ only in detail from those of warmer climates. Periglacial environments nonetheless experience very distinctive hydrological regimes, characterized by cessa-tion or reduction in runoff during winter, storage of winter precipitation as snowcover and its release as a nival flood in spring, and, in permafrost areas, very limited groundwater storage. Sediment is supplied to streams by bank erosion, slopewash over frozen ground, slushflows, solifluction, debris flows and other mass-movement processes. Measured rates of sediment transport are highly variable, but indicate that the majority of sediment is transported in suspension or as bedload, and that sediment transport is greater for gla-cierized than glacier-free catchments. Despite the brev-ity of the runoff season, sediment yields appear to be of the same order of magnitude of those of undisturbed mid-latitude catchments.

Vegetation cover plays a critical role in determining sediment supply, and thus the channel planform of allu-vial rivers, which ranges from braided to anastomosing and meandering. Features unique to periglacial rivers include thermo-erosional niches, icings and ice-scour features relating to break-up of river ice. Bedrock channels may be deepened by frost wedging in winter, and some arctic and subarctic rivers in low-relief terrain flow in distinctive asymmetrical valleys. Periglacial sandar form where high-energy braided rivers transport abundant coarse sediment, and periglacial deltas are characterized by extensive break-up flooding associated with bed scour and channel migration. Pleistocene per-iglacial rivers displayed marked changes in behaviour (aggradation or incision) and channel pattern (braided, anastomosing or meandering) associated with changes in the balance between flood discharges and sediment supply that accompanied major climate shifts and resultant changes in hydrologic regime and vegetation cover.

Hillslope processes and fluvial transport, however, are not the only agents of sediment transport operating in periglacial environments. The operation of another major agent, wind action, is explored in the next chapter.

14

Wind Action

14.1 Introduction

There is widespread evidence that wind action played an important role in sediment erosion, transportation and deposition during Pleistocene cold periods, draping extensive areas of mid-latitude lowlands in thick covers of windblown sand and silt. Conversely, in many present-day periglacial environments, the importance of wind as a geomorphic agent is more localized. The key to this apparent paradox lies in the availability of suitable sediment at the ground surface, and in particular the role of vegetation cover in protecting soil from wind erosion. During Pleistocene glacial periods, powerful, gusty winds scoured sparsely-vegetated terrain exposed by ice-sheet retreat or falling sea level, and swept across floodplain sediments deposited by annual floods. Vast amounts of sediment transported by Pleistocene winds accumulated across lowland areas as *sand sheets* or *coversands*, locally forming undulating *dunefields*, whilst more far-travelled windblown silt accumulated to form *loess* deposits up to a few hundred metres thick. Present-day *aeolian* or *eolian* (wind-related) activity in periglacial environments is more restricted, and tends to be significant only in areas where vegetation cover is limited, such as polar deserts, high plateaus, floodplains, deltas and partly-vegetated Pleistocene aeolian deposits.

This chapter first outlines the basic processes of aeolian entrainment, transportation and deposition, then explores aspects of wind erosion and deposition in present-day cold environments, before focusing on the important role played by wind in shaping the landforms and deposits of mid-latitude landscapes during Pleistocene glacial stages. Seppälä (2004), Brookfield (2011) and Wolfe (2013) provide useful overviews of the role of wind in periglacial environments.

14.2 Aeolian Processes

Wind erosion involves two distinct processes: *deflation*, the entrainment of soil particles, and *abrasion*, which involves detachment of grains from rock or soil as a result of the impact of windblown sand, silt or snow particles. Entrainment of soil particles occurs when the shear stress exerted by wind overcomes resistance to particle movement. Shear stress is a function of velocity, turbulence and surface roughness; shearing resistance reflects particle weight, cohesion and packing. Two critical wind-shear velocities may be identified for particle entrainment: the fluid threshold, which is determined by the magnitude of drag and lift forces associated with the passage of wind over and around a particle; and the impact threshold, which relates to the effects of impacting particles. For cohesionless sand-sized particles (0.06–2.0 mm), both threshold velocities increase with increasing grain diameter, but the impact threshold for any given grain diameter is lower than the fluid threshold (Figure 14.1). This means that once movement of particles has been initiated at the higher (fluid) threshold velocity, it can be sustained even at the lower (impact) theshold velocity. For particles finer than about 0.06 mm, however, fluid and impact threshold velocities tend to be inversely related to particle size, due to the increase in interparticle cohesion as particle size decreases. In general, particles 0.04–0.4 mm in size (coarse silt to medium sand) require the lowest velocities of critical wind shear, and are thus most readily entrained. The relationships depicted in Figure 14.1, however, hold only for dry, loosely-packed well-sorted sediments such as dune sands; where there is a mixture of sediment sizes or sediments are tightly packed, the threshold wind velocities tend to be higher.

Particle size is also the dominant determinant of mode of aeolian transport. Silt- and clay-sized particles (<0.06 mm) have low settling velocities relative to turbulence velocities, and can therefore be transported in *suspension*, remaining within turbulent airflows for several days and travelling long distances. Such suspended fine particles are collectively referred to as *dust*. Most dust particles coarser than ~0.02 mm travel close to the ground, and are vulnerable to trapping by vegetation; finer particles may be lofted several kilometres into the atmosphere, and may therefore travel much farther before deposition. The grain size of dust deposits thus

Periglacial Geomorphology, First Edition. Colin K. Ballantyne.
© 2018 John Wiley & Sons Ltd. Published 2018 by John Wiley & Sons Ltd.

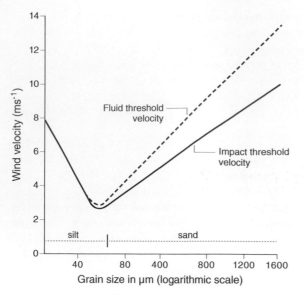

Figure 14.1 Approximate fluid and impact thresholds for quartz grains of different sizes, plotted against wind velocity two metres above the ground surface.

tends to decrease downwind. Sand particles up to about 1 mm in diameter mainly travel in a series of discrete steps by *saltation* or 'bouncing'. Most saltating grains travel no higher than about 20 cm above the surface, unless travelling over hard surfaces such as bedrock, frozen soil or ice. Coarse sand particles (>0.5 mm) may also travel by *creep* (*reptation*), or rolling over the surface, and are likely to travel only short distances from the point of entrainment (Nickling and McKenna Neuman, 2009).

Deposition occurs when wind velocities drop below the threshold for particle movement. This may reflect a reduction in wind shear or turbulence, but may also occur where particles are blown on to more sheltered locations (such as lee slopes or the downwind slopes of sand dunes), become trapped amongst vegetation or adhere to wet surfaces. Accretion of creeping or saltating grains is referred to as *tractional deposition*, whereas accumulation of suspended grains is termed *grainfall deposition*. Dry avalanching of previously-deposited grains down the steep prograding slip faces of sand dunes is sometimes termed *grainflow deposition*.

14.3 Wind Erosion in Present Periglacial Environments

14.3.1 Environmental Controls on Wind Erosion

Many cold environments experience strong winds, especially during winter at high latitudes, in coastal and proglacial settings, and on mountain ranges. At Port Martin in Antarctica, 24-hour average velocities exceeding $46\,\mathrm{m\,s^{-1}}$ ($165\,\mathrm{km\,h^{-1}}$) have been recorded, and gusts exceeding $50\,\mathrm{m\,s^{-1}}$ ($180\,\mathrm{km\,h^{-1}}$) have been measured at

Pangnirtung Pass on Baffin Island. Very strong, persistent winds are characteristic of katabatic airflow in proglacial environments (Nickling and Brazel, 1985; McKenna Neuman, 1990a) and over mountain ranges, where regional airflow is concentrated and accelerated. The mean winter windspeed at the summit of Mount Washington (1915 m) in New Hampshire, for example, is $23\,\mathrm{m\,s^{-1}}$ ($83\,\mathrm{km\,h^{-1}}$), the mean summer windspeed is $12\,\mathrm{m\,s^{-1}}$ ($43\,\mathrm{km\,h^{-1}}$), and a gust of $103\,\mathrm{m\,s^{-1}}$ ($371\,\mathrm{km\,h^{-1}}$) has been recorded. The summit of Cairn Gorm (1245 m) in Scotland experiences annual maximum 24-hour windspeeds of $26–40\,\mathrm{m\,s^{-1}}$ ($93–145\,\mathrm{km\,h^{-1}}$) and annual maximum gusts of $49–76\,\mathrm{m\,s^{-1}}$ ($177–275\,\mathrm{km\,h^{-1}}$). Moreover, as the drag force exerted on particles is proportional to the density of air, which increases as temperature declines, wind is theoretically a more effective agent of deflation in cold environments than in hot deserts (Pye and Tsoar, 2009). Wind-tunnel experiments have shown that sand transport rates may be up to 70% greater at temperatures of $-40\,°C$ than at temperatures of $+40\,°C$, owing to the difference in air density (McKenna Neuman, 2004).

The rather localized wind erosion evident in many present periglacial environments is thus not a reflection of low windspeeds, but is 'surface limited' by the availability of entrainable fine sediment. McKenna Neuman (1993) identified five factors that control sediment availability in cold environments: vegetation cover, development of surface lag deposits, surface wetness, snowcover and ice. Vegetation cover reduces near-surface wind shear and protects soil particles from both deflation and abrasion, so that in many forested and tundra landscapes wind erosion is negligible. Regionally, extensive expanses of bare ground occur only in polar deserts. More locally, unvegetated terrain occurs on sandar and other unvegetated floodplains, seasonally flooded deltas, beaches, glacier forelands, windswept moraine crests, Pleistocene dunefields and sand sheets, and high plateaus. Even where vegetation cover is thin or absent, however, wind erosion may be limited by the formation of a protective gravel lag deposit (Mackay and Burn, 2005). On a recently deglaciated till sheet in Iceland, for example, Boulton and Dent (1974) observed that the cover of surface clasts increased from 30–40% one year after exposure to approximately 90% after four years, due to loss of fines by deflation. Wind erosion may thus be self-limiting in situations where removal of fines creates a clast-armoured *deflation surface* that inhibits further erosion (Figure 14.2). Surface moisture also limits entrainment of sand or silt grains from unvegetated terrain; even low moisture contents drastically limit deflation (Wang *et al.*, 2014), and capillary rise of soil moisture in the active layer in permafrost environments may ensure that soils remain moist at shallow depth throughout the summer. Binding of soil particles by salts precipitated at the ground surface may also substantially reduce wind erosion.

Figure 14.2 Wind erosion in periglacial environments. (a) Deflation surface carpeted by lag gravels, central Iceland. The boulders were deposited by the last ice sheet. (b) Residual 'island' of windblown sand ~0.4 m thick on a plateau deflation surface, Vestidalur, Faroe Islands. *Source:* Courtesy of Ole Humlum.

Snowcover also protects underlying soil from deflation, but strong winds strip snow from exposed sites such as ridges, moraine crests and palsas, so that these are exposed to deflation and wind abrasion in winter. The widespread occurrence of *niveo-aeolian* deposits composed of mixed sand and snow (see below) provides evidence of wind erosion and aeolian transport during winter, and in some arctic and subarctic areas winter aeolian transport exceeds that of the snow-free summer months (McKenna Neuman, 1990a; Bélanger and Filion, 1991).

The role of ice in inhibiting or promoting wind erosion of soils is complex. Though ice forms a strong bond between surface soil grains, sublimation of pore ice releases particles that are readily entrained by wind, and exposed frozen soils are very susceptible to abrasion by wind-driven particles (McKenna Neuman, 1989, 1990b; Van Dijk and Law, 2003). Moreover, the hardness of frozen surfaces permits grain saltation at relatively low wind velocities. During violent storms, even pebbles may be blown over icy surfaces, and saltation of coarse granules to heights of 4 m above frozen surfaces has been recorded (McKenna Neuman, 1990a).

14.3.2 Erosional Landforms

The most extensive landforms created by wind erosion are *deflation surfaces* (or *deflation plains*) from which surface fines have been winnowed away by deflation and abrasion, leaving a surface armoured by clasts and carpeted by lag gravels (Figure 14.2). Deflation surfaces form on diamict substrates, such as till and regolith, in which clasts are embedded in a matrix of fines. In periglacial environments, they are most common in polar deserts, notably those that experience strong katabatic winds from adjacent ice caps or ice sheets, such as the unglacierized fringes of Antarctica and Greenland (Campbell and Claridge, 1987; French and Guglielmin, 1999), and are locally associated with dunefields. Deflation surfaces also occur on windswept mountain ridges and plateaus. In Iceland, erosion of overlying fragile andosols (volcanogenic loess deposits) has exposed widespread deflation surfaces. Arnalds (2000) has estimated that a combination of overgrazing, loosening of sediment by needle ice and removal of sediment by wind erosion, rainsplash and wash has resulted in the formation of 15 000–20 000 km^2 of desert terrain since settlement of Iceland 1150 years ago. Current rates of sediment loss are estimated at $2–3\,Mt\,a^{-1}$, and were probably much higher within the past millennium. Similar rapid rates of accelerated soil erosion by wind have been identified as a contributory cause of abandonment of Norse settlements in Greenland (Jacobsen, 1987).

Deflation surfaces are not always devoid of vegetation. Remnant vegetated 'islands' of vegetation-covered soil or aeolian sediment occur in some areas on mountain plateaus (Figure 14.2b) and are widespread along the fringes of the cold deserts of Iceland. On the stony soils of some deflation surfaces, scattered dwarf shrubs and cushion plants may survive, locally creating *wind-patterned ground* in the form of clumps, crescents and ribbons of vegetation amid sterile stony ground (Boelhouwers *et al.*, 2003).

In partly vegetated Pleistocene or Holocene sand seas and dunefields, wind erosion produces *blowouts*, which take the form of elongate or basin-shaped depressions surrounded by eroding sand scarps. Blowouts may be excavated by deflation entirely within aeolian sand deposits, or may extend to the underlying substrate. Most are a few metres deep, though 10–15 m deep ones occur, for example in the Great Kobuk sand dunes of Alaska. Blowouts are initiated when protective vegetation cover on aeolian deposits is removed or stressed as a result of climate change, a rise or fall of the water table or overgrazing; in subarctic forest tundra, fire may also initiate periods of aeolian activity (Filion *et al.*, 1991; Matthews and Seppälä, 2014). Many arctic and subarctic dunefields of Pleistocene or Early Holocene origin exhibit evidence of subsequent intermittent blowout activity. For example, dunefields formed in Finnish Lapland (68°N) through aeolian reworking of glacigenic sediment in the Early Holocene appear to have stabilized by ~7.0 ka, but have experienced episodic wind erosion since ~4.8 ka due to burning of pine or birch cover and changes in water-table depth, so that now only small remnants of the original dunefield remain (Seppälä, 1995b). Active blowouts in this area are protected from winter wind erosion by snowcover, but experience removal of a 2–3 cm-thick layer of sand as a result of deflation each summer.

Wind abrasion may also erode rock surfaces, particularly pebbles, boulders and low outcrops that protrude above the general level of the surrounding terrain. The collective effects of wind abrasion include, depending on rock composition, surface roughening, frosting or smoothing (*windpolishing*), the formation of surface microforms (miniature grooves, pits and fluting, usually aligned parallel to dominant wind direction) and facetting (the development of planar or slightly convex surfaces, usually aligned normal to dominant wind direction). Such features have traditionally been attributed to 'sandblasting', or abrasion by saltating sand particles within a few decimetres of the ground surface, but there is convincing evidence that suspended silt and ice particles

also contribute to wind abrasion of rock surfaces (Whitney and Splettstoesser, 1982; Schlyter, 1994). The hardness of ice particles increases as temperature falls: at −44°C, ice particles have a hardness similar to that of quartz. Experiments by Dietrich (1977) have demonstrated that even at higher temperatures (−10°C to −25°C), windblown ice particles can abrade common minerals, though some field observations have failed to detect evidence of significant abrasion by windblown ice (McKenna Neuman, 1990a). Whereas saltating sand particles mainly abrade low-lying rock or clast surfaces facing upwind, suspended silt and ice particles appear to be capable of abrading the higher parts of outcrops and boulders, as well as lee, lateral and even undercut faces.

The most widespread manifestation of wind erosion of rock takes the form of *ventifacts*, boulders or exposed bedrock surfaces that have been shaped, worn, facetted, indented, grooved or polished by the abrasive action of windblown particles (Knight, 2008; Figure 14.3). Ventifacts occur in the cold, arid polar desert landscapes of northeast Greenland and Antarctica, but seem to be much less common in arctic tundra environments, probably because vegetation cover inhibits aeolian entrainment of sand grains. In the Taylor Valley, Antarctica, for example, 60% of surface clasts exhibit wind-abrasion features, reflecting the effects of strong winter katabatic winds blowing downvalley from the adjacent ice sheet (Gillies *et al.*, 2009). Long-term observations on ventifacts by Mackay and Burn (2005) suggest that abrasion of resistant rock by wind-driven particles is an extremely slow process, operating over millennial timescales; French and Guglielmin (1999) have estimated minimum long-term abrasion losses of $1–3\,mm\,ka^{-1}$ for ventifacts in Antarctica. Much greater mean rates of rock surface loss (0.015 and $0.022\,mm\,a^{-1}$) were measured on rock

Figure 14.3 Ventifact (windpolished boulder) resting on a deflation surface, central Iceland. The boulder is ~0.4 m high and exhibits pitting, fluting and polishing by aeolian abrasion. *Source:* Courtesy of Ole Humlum.

surfaces in East Antarctica by Spate *et al.* (1995), but these included the effects of flaking by salt weathering as well as wind abrasion. In many parts of Antarctica, bedrock surfaces exhibit both large-scale cavernous weathering forms (*tafoni*) and smaller-scale pitting (*honeycomb weathering*). These appear to be primarily due to salt weathering of bedrock (Chapter 10), but it is likely that their formation has been aided by wind erosion (French and Guglielmin, 1999). Hall (1989) proposed that rock weathering in polar deserts may be aided by packing of windblown sand into rock cracks, and the pressure exerted on such granular infill by wind and wind-driven particles. He also noted that lodgement of aeolian sediment in cracks opened by other weathering processes may prevent subsequent closure. The efficacy of these mechanisms remains unsubstantiated.

Ventifacts in former periglacial environments have been interpreted as indicative of cold, dry, windy conditions with sparse vegetation cover (Demitroff, 2016) and are locally present both within and at the base of Pleistocene coversand deposits (Hoare *et al.*, 2002). Because ventifacts and other windpolished surfaces preserve evidence of dominant wind directions, surviving examples in mid-latitude areas have been used to reconstruct Late Pleistocene palaeowind directions (e.g. Schlyter, 1995; Christiansen and Svensson, 1998;

Christiansen, 2004). The interpretation of such evidence may be problematic, however, due to the difficulties in determining the age of windpolishing and in establishing the significance of results in terms of former atmospheric circulation patterns (Vandenberghe *et al.*, 1999). In arid areas, former wind direction may also be determined by the alignment of *yardangs*, streamlined hills carved from bedrock by wind abrasion and deflation. Few examples have been described in present periglacial environments, but streamlined hills ~150 m high in the Pannonian Basin of Hungary have been interpreted by Sebe *et al.* (2011) as *mega-yardangs* within an aeolian landsystem formed by strong northwesterly and northerly winds under cold arid conditions during the Pleistocene.

14.4 Aeolian Deposits in Present Periglacial Environments

14.4.1 Sandy Aeolian Deposits

Sandy aeolian deposits in cold environments form both dunefields and sand sheets, though the two often occur together. The term *dunefield* describes a cluster or area of *sand dunes* (mounds, hills or ridges formed by drifting sand; Figure 14.4), whereas *sand sheet* refers to a level or

Figure 14.4 Dune slipface at the margin of the Great Kobuk dunefield, northwest Alaska. *Source:* Courtesy of Eduard Koster.

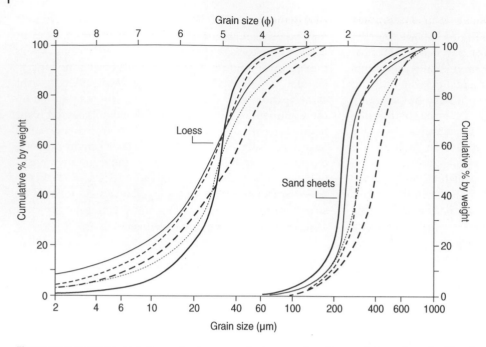

Figure 14.5 Typical cumulative grain-size curves for loess and aeolian sand-sheet deposits. The absence of overlap is because loess is produced by accumulation of airborne dust, whereas aeolian sand is a tractional deposit. Most aeolian sand deposits are better sorted (steeper curves) than loess.

gently undulating aeolian sand accumulation. The critical sedimentological difference between dunes and sand sheets is that the latter lack *slip faces*, which are formed by sand cascading down the steep lee sides of dunes as grain-flow. The term *coversand* is sometimes used to refer to relict sand sheets of Pleistocene age, particularly in Europe, and *sand sea* refers to an extensive area of aeolian sand deposits, whether in the form of dunefields or sand sheets, or both (Pye and Tsoar, 2009). In general, cold-climate aeolian sand deposits are moderately to well sorted, with median grain sizes in the range 100–500 μm (Figure 14.5).

Koster (1988) has summarized the known distribution of areas of partly-active dunefields and sand sheets in arctic and subarctic North America and Scandinavia (Table 14.1). The global distribution of active cold-climate aeolian deposits is probably much greater. Dunefields and sand sheets also occur in Antarctica, Greenland (Christiansen *et al.*, 1999) and Iceland (Arnalds *et al.*, 2001, 2012), but the full extent of active cold-climate aeolian sand deposits elsewhere remains undetermined. Most areas of partly-active aeolian sand deposits initially accumulated during the Pleistocene, in areas that remained unglacierized during the last glacial stage, or formed through aeolian reworking of glacigenic deposits soon after retreat of the last ice sheets. At the last glacial maximum (LGM), for example, extensive deposits of aeolian sand accumulated in most unglacierized lowland areas of Alaska (Lea and Waythomas, 1990), and many of these exhibit localized aeolian reworking at present (Koster and Dijkmans, 1988; Galloway and Carter, 1994).

The scattered dunefields of northern Scandinavia formed through aeolian reworking of glacigenic sediments in the Early Holocene, but have since been intermittently reactivated; only small remnants are active at present (Seppälä, 1995b). Much present-day aeolian sand deposition in arctic and subarctic areas therefore represents reworking of sand deposits of Late Pleistocene or Early Holocene age. The main exceptions occur near active floodplains, outwash plains or other sources of unvegetated fine sediment, such as beaches and deltas.

14.4.2 Dunefields

The morphological and sedimentological characteristics of cold-climate dunefields (Figure 14.4) resemble those of hot arid environments, though rates of dune migration tend to be slower owing to seasonal snowcover and freezing of moisture within dunes. Studies of dune migration in the hyper-arid Victoria Valley in Antarctica, for example, have shown that the rate of downwind dune migration averages ~1.5 m a^{-1}, significantly less than rates reported for hot deserts (Bourke *et al.*, 2009). The dominant landforms in most active cold-climate dunefields are *transverse dunes*, with wavy crests aligned normal to formative wind direction, and *parabolic dunes*, which are crescentic or U-shaped in plan form, with the extremities pointing upwind. The latter usually form on vegetated surfaces downwind of a local sand source. Concave-downwind *barchan dunes* or *barchanoid dunes* are less common, and occur mainly in areas where vegetation

Table 14.1 Arctic and subarctic dunefields and sand sheets, Scandinavia and North America.

Location	Latitude	Extent	Age	Dominant landforms
Scandinavia				
Northern Sweden	68°N	2	LG–Holocene	Parabolic + transverse dunes
Northern Finland	68°N	2	LG–Holocene	Parabolic dunes
Alaska				
Big Delta	65°N	2	WG and older	Parabolic dunes
Yukon Flats	66°N	2	WG and older	Sand sheets
Tanana River Valley	65°N	4	WG and older	Parabolic and longitudinal dunes
Kuskokwim Valley	63°N	3	WG and older	Parabolic dunes
Koyukuk Valley	66°N	3	LW–Holocene	Sand sheets; parabolic, transverse and longitudinal dunes
Kobuk Valley	67°N	2	LW–Holocene	Sand sheets; parabolic, transverse, longitudinal and barchanoid dunes
Arctic Coastal Plain	71°N	4	LW–Holocene	Parabolic and longitudinal dunes
Subarctic Canada				
Labrador	54°N	2	LG–Holocene	Parabolic and transverse dunes
East Hudson Bay	57°N	2	LG–Holocene	Parabolic dunes
Yukon Territory	62°N	1	–	Sand sheets
Arctic Canada				
Banks Island	73°N	1	Holocene	Sand sheets
SE Baffin Island	67°N	1	LW–Holocene	Sand sheets and dunes

Category 1: <100 km^2; 2: 100–1000 km^2; 3: 1000–10 000 km^2; 4: >10 000 km^2.
WG, Weichselian/Wisconsinan (last) Glacial Stage; LW, Late Weichselian/Wisconsinan; LG, Lateglacial.
Source: Based on data in Koster (1988).

cover is sparse or absent, such as the Dry Valleys of Antarctica and the Kobuk Valley in Alaska. All of these types are asymmetrical, with steep (30–34°) downwind slip faces and gentler upwind slopes. *Longitudinal dunes* with straight or wavy crests aligned subparallel to dominant winds also occur in some dunefields, and are sometimes superimposed on transverse forms (Galloway and Carter, 1994). Other types occur more locally, including complex and compound dunes, cliff-top dunes (Bégin *et al.*, 1995), bowl-shaped blowout dunes formed by deflation of pre-existing aeolian deposits (Seppälä, 1995b), climbing and falling dunes (Seppälä, 1993) and the slipfaceless *whaleback dunes* found in Victoria Valley in Antarctica (Bristow *et al.*, 2010). Cold-climate dunes vary in size from miniature *phytogenic dunes* less than 1 m high, which form due to trapping of sand by vegetation, to mature parabolic dunes several tens of metres high, found in the partly active dunefields of Alaska.

The limited information on the sedimentology of cold-climate dunes suggests that even those with a frozen core (due to permafrost aggradation) exhibit internal structures similar to those of their counterparts in hot deserts, though thermal contraction cracking during periods of stability may result in the development of sand veins or sand wedges. Only the *denivation* (ablation) structures associated with niveo-aeolian deposits (see below) appear to be distinctive

to cold-climate dunes, though alternating coarse and fine sand strata in some vegetated cold-climate dunes have been interpreted as representing cyclic aeolian and niveo-aeolian deposition (Ruz and Allard, 1995). Parabolic, transverse and barchanoid dunes generally exhibit steeply dipping (32–34°) foreset laminae, produced by grainflow and grainfall deposition on downwind slip faces, and topset beds, often composed of climbing ripple structures, on upwind faces and dune crests. However, alternating episodes of erosion and deposition, together with changes in dominant wind direction, wind strength and sediment supply, may introduce cross-bedded structures within dunes (Pye and Tsoar, 2009), so that internal stratigraphy is often complex.

14.4.3 Sand Sheets

The accumulation in some areas of gently undulating sand sheets rather than dunefields is incompletely understood, but probably reflects either limited sediment supply or trapping of windblown grains by vegetation. Sand sheets rather than dunefields are particularly characteristic of sites where wind has reworked fine-grained fluvial sediments from unvegetated floodplains. As river discharges decline following the annual nival or glacial flood, extensive areas of fine-grained sediment are exposed on bars and other low-lying surfaces. When these sediments dry out, they are vulnerable to aeolian entrainment and redeposition across adjacent terrain. In such situations, aeolian deposits often exhibit sedimentological changes with distance downwind from the floodplain source. Dijkmans and Törnqvist (1991) identified four main facies associated with aeolian reworking of fine sediments exposed on proglacial sandar in Greenland. On and adjacent to the floodplain, they identified *fluvio-aeolian deposits* (see below) comprising horizontally-interbedded aeolian sand and fluvial silt deposits. On sparsely vegetated terrain near to the floodplain, they found *proximal aeolian sand sheets* composed of horizontal or undulating laminated fine to medium sand layers representing tractional deposition, interbedded with coarse-grained deflational lag deposits. Farther from the floodplain on partly vegetated terrain, they identified *distal aeolian sand sheets* dominated by horizontally-bedded, fine to medium sand interbedded with vegetation debris layers, reflecting deposition on vegetated surfaces. Finally, adjacent valley-side slopes and low bedrock ridges supported *aeolian silt deposits* comprising massive microlaminated silts, formed by grainfall deposition on to vegetated surfaces. This four-part scheme has been elaborated by Willemse *et al.* (2003), who in addition identified *interdune/deflation facies* representing eroded areas within sand sheets, *phytogenic dune facies* associated with localized sand deposition amongst vegetation clumps and *distal back-dune/peat mire facies* formed by concurrent grainfall and peat accumulation in waterlogged depressions (Table 14.2).

In some active sand sheets, however, fewer facies are represented. Those investigated by McKenna Neuman and Gilbert (1986) on Baffin Island are dominated by proximal aeolian facies comprising interstatified deflation surfaces (thin planar beds of coarse sand, granules and small pebbles) and thin deposits of medium sand. Elsewhere, particularly in mountain environments, sand sheets formed of aeolian and niveo-aeolian sediments are massive and poorly sorted, reflecting short transport paths and the mixture of grains of different sizes during deposition from melting snowcover.

Some of the most active sand sheets occur in Iceland, where 22% of the land area consists of sandy deserts composed of windblown sediments derived from outwash plains and volcanic eruptions, and where typical aeolian transport rates are of the order of $100-1000\,\mathrm{kg\,m^{-1}\,a^{-1}}$ (Arnalds *et al.*, 2012). Mountney and Russell (2004) have documented the characteristics of extensive sand-sheet deposits in northeast Iceland. Though primarily derived from outwash deposits, these are also nourished by volcanoclastic sediments derived from periodic eruptions of the Askja volcano. The upwind margin of the sand sheet is characterized by surface lag deposits, ventifacted boulders and sand deposition in the lee of obstacles. This zone forms the main source of sand in the sand sheet proper, which occupies an area of $100\,\mathrm{km^2}$ and consists of a continuous cover of poorly or moderately sorted sand up to 10 m thick. Surface features comprise ripple forms and longitudinal *zibars* (low-amplitude bedforms) 60–80 m apart and up to 1.5 m high. Downwind of the

Table 14.2 Lithofacies characteristics of riparian aeolian deposits, west Greenland. Facies are ordered downwind from the active floodplain.

Facies	Location	Primary structures	Main processes
Proximal aeolian (fluvio-aeolian) facies	On or close to active floodplain	Parallel-laminated medium- to coarse-grained alternating fluvial and aeolian sands	Deposition of fluvial sediments; deflation; tractional deposition of aeolian sand
Interdune and deflation facies	Within sand sheets	Thin planar-laminated medium sands and ripple strata truncated by coarse-grained beds; thin beds of cross-laminated sands	Localized erosion and deposition; cross-laminated sands represent deposition by nival meltwater
Phytogenic dune facies	Within localized dense vegetation	Undulating laminae of fine to medium sand, with high-angle grainfall and grainflow laminae	Deposition amongst vegetation clumps
Aeolian facies for partly vegetated surfaces	Sites with 5–25% vegetation cover	Undulating beds of fine–medium and medium–coarse sand; massive, faintly laminated coarse sand beds; coarse-grained truncation surfaces	Grainfall, tractional deposition and selective erosion
Aeolian facies for stabilized surfaces	Sites with continuous vegetation cover	Thick (1–6 m) deposits of regularly layered organic-rich laminae and organic-poor sands	Episodic deposition of aeolian sands and silts over peaty horizons
Distal back-dune and peat mire facies	Back-dune waterlogged depressions	Layered peat (thin bands of alternating aeolian silt and peat) or massive peat-loess sequences	Infall of aeolian silt and concurrent or episodic peat accumulation
Aeolian silt facies	Upland areas farthest from the active floodplain	Massive mottled silts up to 1 m thick. Micro-laminae of alternating fine and coarse silt layers	Infall of aeolian silt

Source: Based on Willemse *et al.* (2003).

main sand sheet, aeolian deposits become increasingly sporadic around the flanks of the volcano; here, the dominant features are aeolian megaripples composed of low-density pumice granules and pebbles, reflecting the volcanogenic origin of some aeolian sediment in this region.

The relationship between sand sheets and permafrost has been studied by Pissart *et al.* (1977), who examined sections through stabilized sand sheets up to 7 m thick on low terraces adjacent to an active floodplain on Banks Island, arctic Canada. These deposits exhibit horizontal stratification of aeolian beds up to 20 mm thick, intercalated with peat layers. Individual beds are moderately sorted, with typical median grain sizes of 195–380 μm, but overall granulometry ranges from silt-sized grainfall deposits to coarse sand and granules on buried deflation surfaces. Individual beds are laterally continuous for several metres, but locally interrupted by ice wedges and ice veins; horizontal lenses of segregation ice are also present. The stratigraphic characteristics of this site imply (i) that sand accumulation was gradual, and alternated with periods of stability and vegetation growth, and (ii) that sand-sheet accumulation was accompanied by slow upwards permafrost aggradation, with associated development of ice lenses and syngenetic ice wedges and ice veins.

In cold arid environments underlain by continuous permafrost, aeolian sand is also important in the formation of cryogenic sand wedges and sand veins (Murton *et al.*, 2000; Chapter 6). This may occur in two main ways. Primary infill occurs when aeolian sand rather than meltwater fills open thermal contraction cracks, causing the development of sand veins or sand wedges without formation of vein ice or wedge ice. Such primary cryogenic sand wedges and sand veins occur widely in polar deserts and more locally in tundra environments where there is a nearby source of windblown sand. Secondary infill occurs where aeolian sands overlie ice veins or ice wedges, and subsequent melt of ice results in infill of contraction cracks by sand derived from the overlying sand deposits.

14.4.4 Fluvio-aeolian Deposits

The term *fluvio-aeolian* designates sediment accumulations of mixed fluvial and aeolian origin and was introduced by Good and Bryant (1985) to describe sand sheets that are subject to both aeolian processes and ephemeral streamflow in the Sachs River Valley on Banks Island. At this site, annual snowmelt results in meltwater flow across frozen aeolian deposits in networks of shallow channels. After flow has ceased, wind scour removes the finer grains from sand-sheet surfaces, leaving planar deflation surfaces covered by lag deposits of granules or pebbles, and reworked sand accumulates downwind as thin, low-angle sheets. Sections through the sand sheets reveal ripple-laminated or horizontally-laminated sand layers about 10 cm thick, locally interrupted by scour-and-fill channel structures. Good and Bryant attributed alternating coarse (lag) and fine (depositional) layers in the aeolian sediments to cyclic deflation and accumulation, suggesting that the former occurs when surfaces dry out, and proposed that this sediment association may represent an analogue for evenly laminated Late Pleistocene aeolian sands described by Ruegg (1983).

Recent fluvio-aeolian deposits also occur on and beside proglacial sandar in west Greenland. Those described by Dijkmans and Törnqvist (1991) comprise horizontal aeolian sands interbedded with inverse-graded fluvial silts, whereas those investigated by Willemse *et al.* (2003) are characterized by parallel-laminated, medium-coarse sand of alternating fluvial and aeolian origin, with thin accumulations of aeolian sand infilling shallow abandoned channels. These authors classified such fluvio-aeolian deposits as a particular facies of aeolian sand sheets (Table 4.2), rather than as a distinctive category of cold-climate aeolian deposit.

14.4.5 Niveo-aeolian Deposits

In areas of seasonal snowcover where winds transport both fine sediment and drifting snow, distinctive *niveo-aeolian* deposits are formed. Such deposits may represent either simultaneous deposition of mixed aeolian sediment and snow, or deposition of windblown sediment on pre-existing snowcover (Figure 14.6). Most niveo-aeolian deposits are annual, in the sense that underlying or interstratified snow melts completely during the summer. Occasionally, however, buried snow may persist under thick niveo-aeolian deposits for one or more summers, and perennial deposits of interstratified sediment and snow occur under a protective cover of windblown sediment under exceptionally cold, dry conditions in Antarctica (Ayling and McGowan, 2006). Annual niveo-aeolian deposits occur in most areas of active cold-climate aeolian activity, including polar deserts, arctic and subarctic dunefields and sand sheets (Koster and Dijkmans, 1988; Dijkmans, 1990), floodplains (McKenna Neuman, 1990a; Lewkowicz and Young, 1991) littoral sand dunes (Ruz and Allard, 1995), alpine plateaus (Ahlbrandt and Andrews, 1978) and mid-latitude maritime mountains (Ballantyne and Whittington, 1987). Niveo-aeolian deposits are also an important component of sediment transfer at some nivation sites (Christiansen, 1998a). In all of these locations, sediment is entrained from unvegetated surfaces during months when snowcover is incomplete, or during winter when wind strips snowcover from the crests of low ridges, dunes and hummocks. In some areas, niveo-aeolian deposition is the dominant form of aeolian sediment accumulation (Bélanger and Filion, 1991).

Figure 14.6 Niveo-aeolian deposits of interbedded sand and snow layers overlying dune slipface deposits, Great Kobuk dunefield, northwest Alaska. *Source:* Courtesy of Eduard Koster.

Most niveo-aeolian sediments are dominated by medium–coarse sands (0.2–2.0 mm), implying mainly tractional deposition of saltating and creeping grains. Transport distances of sand-sized grains are generally short (<1 km). Fresh deposits reveal concentrations of sediment at the top and base of the snowpack, layers of mixed snow and sediment (implying simultaneous deposition), and occasional sediment-rich layers separated by clean snow (Figure 14.6). Because of mixing of grains of different sizes during ablation of underlying snow, niveo-aeolian deposits tend to be poorly or moderately sorted, but laminae reflecting either primary depositional structures or superimposition of sediment layers during snowmelt may sometimes be preserved.

During ablation of underlying and interstatified snow, niveo-aeolian sediment becomes concentrated as a thickening layer at the top of the snowpack. Further snowmelt often results in the formation of *denivation* (ablation) structures within the sediment. On planar sand sheets, denivation structures are usually inconspicuous, though tension cracks occur on slopes and small

cavities may develop within the sediments as the underlying snow melts. The most pronounced denivation structures are associated with lee slopes, particularly the slip faces of dunes, where annual niveo-aeolian deposits often reach thicknesses of 0.1–0.4 m. On high-altitude dunes in Colorado, Ahlbrandt and Andrews (1978) noted that downslope sliding and ablation of buried snow results in the development of tensional structures in supranival sands on the upper parts of slip faces, and in compressional structures farther downslope. The tensional structures they observed included stairstep faults, stretched laminae and breccia; compressional features included folds, warps and overturned beds, with localized overturning and oversteepening of laminae. Denivation features associated with slip faces in the Great Kobuk sand dunes of Alaska include snow-cored ramparts up to 4 m high, contorted bedding due to snow sliding, sinkholes and snow-cored hummocks formed by differential ablation of underlying snow, miniature snow-cored ridges, slope-foot meltwater fans and tension cracks that develop in wet sand cover due to differential snow ablation or downslope movement of underlying snow (Koster and Dijkmans, 1988). Dimpled or pitted surfaces are also characteristic of some freshly deposited niveo-aeolian sands.

Dijkmans (1990) has described the sedimentological signature of denivation structures developed on the slip faces of small phytogenic dunes on a riparian sand sheet in west Greenland. The main features include faulting of slip-face strata due to ablation of underlying snow, chaotic bedding produced by slumping, rotated breccias and concave beds formed by sediment subsidence, and normal faults and tension cracks due to downslope movement of buried snow. Dijkmans noted that though denivation phenomena on sand sheets tend to be destroyed by summer deflation as soon as the surface dries out, deformation structures developed in dune-foreset cross-bedding may be preserved under prograding slip-face beds. The low preservation potential of most denivation structures, however, means that niveo-aeolian facies are often difficult to identify in Pleistocene coversands and dunefields.

14.4.6 Loessic (Silt) Deposits and Silt–Sand Intergrades

The definition of *loess* adopted here follows that proposed by Pye (1995): a terrestrial clastic deposit, mainly of silt-sized particles, that formed through the accumulation of windborne dust. Though the modal grain size of loess deposits varies greatly (depending on wind energy, the size of source sediments and distance from source), the great majority of loess deposits contain >50% silt; those with >20% sand are referred to as *sandy loess*, and those with >20% clay as *clayey loess*. Typical loess deposits are fine-skewed, with a tail of progressively finer size grades that usually extends into the clay fraction (Figure 14.5). In most loess deposits, quartz grains are dominant, the remainder being composed of feldspars, carbonates, heavy minerals, clay minerals and sometimes shards of volcanic glass; mineralogical composition is determined by that of the sediment source. Loess deposits are generally loosely packed, with high porosity, and occur both *in situ* as gently undulating *loess sheets* and as *reworked loess* that has been transported downslope by slopewash or solifluction.

Though silt-sized particles may originate in a number of ways, including rock weathering, fluvial abrasion and wind abrasion, the most important process responsible for the production of silt in most cold environments is glacial abrasion of bedrock. Glacigenic silt is not only exposed in till sheets, but also forms a major part of the sediment load of proglacial rivers. As annual flood discharges diminish, silt is draped over the bars and abandoned channels of proglacial sandar, forming a copious and annually replenished source of fine sediment for aeolian transport. Not all loess, however, is of glacigenic origin. Extensive loess covers in Central Asia, Northern China and central South America probably reflect aeolian reworking of silt particles produced by rapid weathering of adjacent high, recently uplifted mountain ranges, and many hot deserts are surrounded by a marginal zone of loess derived from weathering of desert surfaces, sometimes termed *peri-desert loess* or *desert loess*.

The dust particles that form loess are transported in suspension within turbulent airflows, mostly within 100 m of the surface, though finer dust may be lofted to altitudes of several kilometres. Deposition may reflect gravitational settling of particles, the formation of airborne aggregates through wetting or electrostatic attraction, downwash by rainfall, downwards turbulent diffusion or advection of dust-laden air towards the surface. Because smaller particles remain in suspension at lower turbulent velocities, the coarsest dust tends to be deposited closest to the sediment source, with progressive downwind fining. Critically, however, particles may be rapidly re-entrained unless they settle on moist ground or vegetation cover, both of which form sediment traps for dust infall.

Though cold-climate loess deposits of Pleistocene age cover vast tracts of mid-latitude terrain, areas of significant active loess accumulation are limited in most present periglacial environments, though extensive loess deposits are present in lowland areas of Alaska (Figure 14.7). The principal high-latitude sites of loess deposition are terraces, slopes and low plateaus bordering braided floodplains. In Alaska, proximal loess deposits near sandar have accumulated at rates of 0.2–$2.0\,\mathrm{m\,ka^{-1}}$ during the Holocene (Muhs *et al.*, 2003). At sandar sites in west Greenland, aeolian silt deposits form the most distal (downwind) facies of recent aeolian deposition (Dijkmans and Törnqvist, 1991; Willemse *et al.*, 2003), but these deposits differ from typical Pleistocene loesses in that they are less than 1 m thick, massive, microlaminated

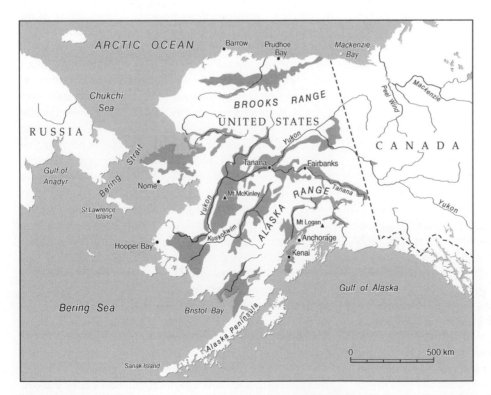

Figure 14.7 Distribution of loess deposits in Alaska. *Source:* Adapted from Muhs *et al.* (2003). Reproduced with permission of Elsevier.

and compact, with rather coarse median grain sizes due to relative proximity to sediment sources. In the Matanuska Valley of southern Alaska, Holocene loess thickness, sand content and sand-plus-coarse-silt content decrease within ~40 km of the outwash source area, but fine silt content increases along the same transect (Muhs *et al.*, 2004). Alternation of loess and palaeosol horizons at this location shows that loess deposition has been episodic, or at least that during some periods loess accumulation has been reduced, allowing soil formation to take place. Similarly, Late Holocene loess accumulation in low-centred ice-wedge polygons on Bylot Island has been interrupted by periods of bryophyte growth, a pattern that Fortier *et al.* (2006) attributed to shifts in summer wind regime and active-layer moisture conditions.

Far-travelled aeolian dust at high latitudes appears to be insufficiently abundant at present to cause significant accumulation of distal loess deposits, though it may contribute to silt enrichment in *loessic soils*, in which dust infall forms a significant proportion of the parent material. Similarly, dust infall occurs in cold mountain environments (see below) and contributes to the formation of silt-rich montane soils. There appears to be no exact present-day periglacial analogue for the Pleistocene distal loess deposits that mantle many mid-latitude terrains, even though deposition of aeolian dust entrained from braided floodplains is locally significant in some temperate mid-latitude areas (Eden and Hammond, 2003).

In areas where sand sheets grade laterally into loess sheets, transitional deposits termed *silt–sand intergrades* may occur. Such deposits exhibit subhorizontal beds of fine to medium sand that alternate with beds of very fine sand to silt. The former dominate in proximal locations, grading laterally (upwind) into distal sand-sheet deposits, but with increasing distance from the sediment source, beds of very fine sand and silt become increasingly dominant, grading downwind into proximal loess (massive silt) deposits. Contacts between the finer and coarser beds may be sharp or normally graded. Lea (1990) suggested that the fine-to-medium sand strata represent tractional deposits formed by particle saltation and creep, whereas the very fine sand-to-silt layers represent short-term suspension and turbulence-modified saltation. Schwan (1986) interpreted sand-silt intergrades as the product of seasonal contrasts in wind velocity, with coarser beds deposited in winter and finer beds in summer. Lea (1990), however, argued that this interpretation is inconsistent with large interannual and intraseasonal variations in windspeed in cold environments, and pointed out that normally graded contacts imply deposition of coarse–fine couplets in a single event. An intriguing feature of silt–sand intergrades is that they imply that the finer beds are not re-entrained by the high winds that deposit the overlying coarse beds, despite bombardment by saltating grains. Preservation of the fine beds may be explained by ground freezing or by wetting (and thus increased interparticle cohesion) through capillary rise of groundwater or rainfall; alternatively, they may be protected by vegetation cover that grows upwards through the accumulating deposit, the development of salt crusts or snowcover. The last possibility suggests that some silt–sand intergrades represent alternation of aeolian deposition of fine beds and niveo-aeolian deposition of coarser layers.

14.4.7 Montane Aeolian Systems

Exceptionally strong and turbulent winds are generated by concentration and acceleration of airflow over mountain barriers, and consequently many high plateaus are dominated by sparsely vegetated deflation surfaces from which fines have been removed (Figure 14.8a). Thorn and Darmody (1985a), for example, found that deflation surfaces at 3460 m in the Colorado Front Range are dominated by particles ≥4 mm in diameter, implying removal of finer sediment by wind. On the deflation surface of An Teallach in Scotland, almost all particles <6 mm have been removed. At the same site, sand grains eroded from the plateau have been deposited downwind as poorly sorted sand sheets up to 4 m thick that began to accumulate in the early Early Holocene (Ballantyne and Morrocco, 2006).

On vegetated plateaus and ridges, wind erosion exploits openings in the vegetation cover to form *deflation scars*, small patches of lag gravels surrounded by miniature soil scarps. On windward slopes and ridges in the Tatra Mountains, Izmailow (1984) recorded rates of scarp retreat averaging ~22 mm a^{-1}. On windswept plateaus dominated by dwarf shrub vegetation, slow downwind migration of the vegetation cover may produce distinctive wind-patterned ground in the form of vegetation clumps, crescents and stripes (Figure 14.8b), though the origin of such patterns is enigmatic. On slopes, the equivalent landforms are *turf-banked terraces*, step-like landforms with steep vegetated risers and gently sloping, sparsely vegetated treads. Such terraces appear to have formed where downslope movement of debris by frost creep has been impeded by the formation of vegetation stripes across the slope. It is not known, however, whether the vegetation stripes formed first, or whether they developed on sheltered risers in response to terrace formation (Ballantyne and Harris, 1994, pp. 261–267).

Though exposed plateaus often form the source of aeolian and niveo-aeolian deposits on sheltered lee slopes, some plateaus support extensive accumulations of windblown sediment derived from adjacent cliffs. Particles dislodged from rockwalls during storms are blown upwards in an accelerating near-vertical airflow, then settle through lower-velocity airflow on to plateau surfaces, where they are trapped by snow or vegetation

Figure 14.8 (a) Deflation surface and remnant deposit of windblown sand, Ward Hill, Orkney Islands, Scotland. (b) Wind stripes developed normal to dominant wind direction on a plateau in northern Scotland. The pole is 1 m long. (c) Pit excavated in poorly sorted, massive plateau-top aeolian sediments on the summit of The Storr, Isle of Skye, Scotland. The pit is 2.9 m deep.

cover (Hétu, 1992). Such plateau-top aeolian deposits are particularly common on maritime mountains, such as those of the Faroe Islands. The summit plateau of The Storr (719 m) in Scotland is mantled by aeolian sediment up to 2.9 m thick (Figure 14.8c) that has accumulated at rates of 0.1–0.6 m ka^{-1} since the adjacent basalt cliff was exposed by a rockslide ~6100 years ago (Ballantyne, 1998b). Such plateau-top aeolian deposits are massive and poorly sorted: samples from those on The Storr contain 11–40% by weight silt, 32–64% sand and 4–36% fine gravel; the largest clast recovered was 109 mm long.

In addition to aeolian reworking of sediments derived locally from exposed plateaus or cliff faces, cold mountain environments also experience slow accumulation of far-travelled aeolian dust. For the alpine zone of the Colorado Front Range, Thorn and Darmody (1985b) have shown that aeolian infall is dominated by poorly sorted silt, with mean particle sizes ranging from 6 to 108 μm. Short-term measurements indicate average annual dust influx of 14 g m^{-2} for the Green Lakes Valley in the Colorado Rocky Mountains (Caine, 1986), 3 g m^{-2} for the Wind River Mountains in Wyoming (Dahms and Rawlins, 1996) and 11 g m^{-2} for the Coast Mountains of British Columbia (Owens and Slaymaker, 1997), though Izmailow (1984) reported a much larger range of values

(1–265 g m^{-2}) for the Tatra Mountains. Exceptionally high values of 102–614 g m^{-2} reported by McGowan *et al*. (1996) for the Lake Tekapo area in New Zealand reflect aeolian reworking of silt from nearby outwash plains. Even though most mountain areas experience only modest infall of aeolian dust, accumulation of such sediment throughout the Holocene has played an important role in enriching mountain soils with silt-sized sediment, enhancing the frost susceptibility of such soils and influencing their pedological development (Muhs and Benedict, 2006).

14.4.8 The Qinghai-Tibet Plateau

Aeolian deposits in the form of dunefields, sand sheets and loess deposits are widespread across the arid or seasonally arid areas of the Qinghai-Tibet Plateau, and many are actively accumulating at present. The principal sediment sources are lake beds, outwash plains, floodplains and pre-existing aeolian deposits, and many deposits appear to be locally derived (IJmker *et al*., 2012), though across this vast area there are regional differences in both the pattern and the timing of sediment distribution. Lehmkuhl *et al*. (2000) noted that Pleistocene loess in southeast Tibet tends to be concentrated below 4000 m altitude and Holocene sandy loess at higher altitudes, whereas near Lhasa climbing dunes and sand sheets occur on the windward sides of mountains above ~5300 m, with loess deposits at lower altitudes (Sun *et al.*, 2007). OSL dating of aeolian deposits in the southern Qinghai-Tibet Plateau suggests that episodic deposition has occurred over the past 100 ka, with possible enhanced sediment accumulation at 83–79 ka, 45 ka, 33 ka, 21–16 ka, 8–7 ka and 3 ka (Lai *et al.*, 2009), but in the northeastern Qinghai-Tibet Plateau the major period of sand and loess accumulation occurred during the Early Holocene (10.5–7.0 ka), and was attributed by Stauch *et al.* (2012) to trapping of aeolian sediments by vegetation cover under wetter conditions associated with the strengthening of the Asian summer monsoon. Several studies have also identified enhanced aeolian accumulation of both sand and sandy loess during the Late Holocene and under present conditions. This may reflect the onset of colder and drier conditions (IJmker *et al.*, 2012), but it is possible that recent enhancement of aeolian sediment transport represents overgrazing-induced vegetation degradation and consequent reworking of pre-existing aeolian deposits (Kaiser *et al.*, 2009b).

14.5   Quaternary Aeolian Deposits

During Pleistocene glacial stages, the aeolian processes that presently operate rather locally in high-latitude environments were active on a vast scale, draping extensive mid-latitude areas in a deep mantle of wind-borne sediment. In areas proximal to sediment sources, this took the form of aeolian sand deposits (coversands and dunefields), whilst more far-travelled windborn silts formed extensive and thick deposits of loess. The global extent of Pleistocene aeolian sediments probably exceeds 10 million km^2. On many mid-latitude plains, cold-climate aeolian deposits represent the dominant form of periglacial sediment accumulation. Pleistocene aeolian deposits also accumulated in arctic and subarctic environments that escaped glaciation, or were exposed to wind action following the retreat of the last Pleistocene ice sheets.

Differentiation of cold-climate aeolian deposits from those that accumulated under temperate interglacial conditions has been based on several criteria, including: (i) the presence of structures related to seasonal ground freezing or permafrost, such as involutions, frost cracks and ice-wedge pseudomorphs; (ii) inclusion of cold-climate pollen, insect and molluscan assemblages; (iii) mineralogical evidence for derivation from coeval glacigenic deposits; (iv) the presence of denivation structures indicative of niveo-aeolian deposition; (v) conformable stratigraphic relationships with underlying till or proglacial outwash deposits; and

(vi) dating evidence (mainly radiocarbon and OSL dating) that demonstrates accumulation during cold periods. In most areas, the most recent accumulation of cold-climate aeolian deposits occurred during the last glacial stage, but similar episodes of extensive aeolian deposition certainly accompanied earlier episodes of Pleistocene ice-sheet expansion, and some loess deposits are known to have accumulated since the beginning of the Pleistocene, ~2.58 million years ago.

14.5.1   Pleistocene Coversands and Dunefields

In northern Europe, a belt of Pleistocene coversands and dunefields extends almost continuously from the Netherlands through Germany, Denmark, Poland and Belarus to Russia (Figure 14.9a). The northern limit of the north European sand belt is broadly coincident with (but overlaps) the southern limit of the last Fennoscandian ice sheet. In the northern Great Plains of North America, similar deposits are also widespread, mainly south of the limit of the last Laurentide Ice Sheet, occupying extensive areas of North and South Dakota, Minnesota, Nebraska and Kansas, and extending southwards into Wyoming and Colorado, with northern outliers in Alberta and Saskatchewan (Figure 14.9b). Extensive areas of Pleistocene windblown sand deposits also occupy inland valleys of Alaska (mostly outside the limits of the last Pleistocene glaciers) and the Alaskan coastal plain (Lea and Waythomas, 1990). Smaller sand sheets and dunefields developed following ice-sheet retreat in both continents, for example in northern Scandinavia (Seppälä, 1995b) and parts of the western Canadian Arctic (Murton *et al.*, 1997; Bateman and Murton, 2006). Pleistocene coversands also occur in southwest France (Sitzia *et al.*, 2015) and in England, though in comparison with the continental sand seas these are thin and patchy (Bateman, 1995). Discontinuous coversands occur on the Drakensberg foothills of South Africa (Telfer *et al.*, 2014). The full extent of Pleistocene cold-climate aeolian sand sheets in continental interiors elsewhere is uncertain.

Most documented Pleistocene aeolian sand deposits form coversands rather than dunefields, though in the European sand belt Weichselian coversands are progressively replaced eastwards by dunefields of similar age, possibly reflecting increased aridity. Moreover, some areas have witnessed extensive reworking of Pleistocene aeolian sands into dunefields under the temperate conditions of the Holocene (e.g. Mann *et al.*, 2002). Unmodified Pleistocene coversands generally take the form of flat to gently undulating sand sheets with low (<5 m) relief, often several metres deep, but thicker where aeolian sand has infilled valleys. They incorporate not only aeolian sand deposits, but also fluvio-aeolian and niveo-aeolian sediments, though differentiation of facies is sometimes problematic.

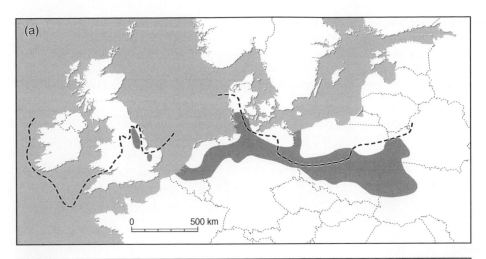

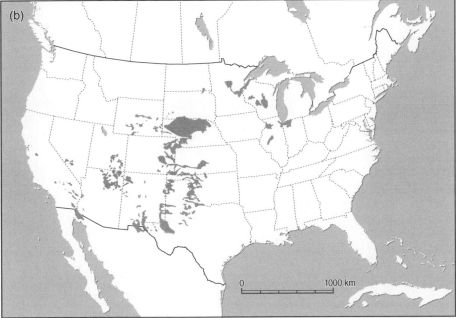

Figure 14.9 (a) Approximate distribution of major cold-climate dunefields and sand sheets (shaded zones) in Europe, from the British Isles to Poland; their continuation beyond eastern Poland is not shown. The dashed line is the approximate southern limit of the last (Late Weichselian) ice sheet. *Source:* Koster (1988). Reproduced with permission of Wiley. (b) Distribution of major cold-climate dunefields and sand sheets in the coterminous United States. *Source:* Adapted from Muhs (2004). Reproduced with permission of Elsevier.

14.5.2 Pleistocene Coversands and Dunefields of the North European Plain

Particularly detailed information on Pleistocene aeolian sand deposits exists for the belt of low-lying terrain that extends from Belgium and the Netherlands through Germany to Poland (Figure 14.9). In these areas, most aeolian deposits that have been investigated relate to the last glacial period, though at some localities these overlie earlier deposits. Four Late Weichselian (~31.0–11.7 ka) chronostratigraphic units have been identified in this area: Older Coversands I and II (OCS I and II) and Younger Coversands I and II (YCS I and II) (Koster, 1988; Bateman and Van Huissteden, 1999; Kasse *et al.*, 2007; Vandenberghe *et al.*, 2013).

Kasse (2002) identified four characteristic facies of Weichselian aeolian sand in northwestern Europe: (1) fluvio-aeolian deposits; (2) horizontally bedded sand-sheet deposits; (3) laminated or low-angle cross-bedded sand-sheet deposits; and (4) dune slipface deposits (Table 14.3). Facies 2 and 3 are by far the most widespread, mantling both valleys and interfluves, and form the characteristic facies of most European coversands (Figure 14.10). Widespread deposition of these facies across the North European Plain reflects the abundance of sediment sources (fluvial and deltaic sediments from the exposed North Sea Basin, together with glacigenic deposits south of the Late Weichselian ice-sheet margin), absence of major topographic barriers to aeolian transport,

Table 14.3 Facies types and characteristics of Pleistocene aeolian sands in Northern Europe.

Facies	Location	Primary structures	Inferred processes
1. Fluvio-aeolian deposits	River valleys and former river valleys	Faintly-bedded or massive silty sand containing lenses of coarse sand, small-scale current ripple laminae and clayey-silty drapes	Aeolian sands partly reworked by shallow runoff and frequently disturbed by freeze–thaw processes and wedge formation
2. Horizontally-bedded sand-sheet deposits	Widespread in valleys and on interfluves	Alternating beds of fine sand and loamy very fine sand	Sand layers: traction load deposited on dry surfaces Loamy layers: suspended particles deposited on wet surfaces
3. Laminated or cross-bedded sand deposits	Widespread in valleys and on interfluves	Horizontal lamination and low-angle cross-bedding; often overlies facies 2	Deposition of planar beds or small wind ripples on dry surfaces
4. Dune slipface deposits	Rare in western Europe; increases eastwards	Large-scale cross-bedding	Grainflow deposition on the lee side of sand dunes

Source: Based on facies descriptions in Kasse (2002).

Figure 14.10 Late Weichselian aeolian sand facies in northwestern Europe. (a) Facies 1, of Late Pleniglacial age. Massive bedding and alternating sand and loamy sand beds with synsedimentary deformations. The black band separates the deformed OCS I from the undeformed OCS II. (b) Facies 2, of Late Pleniglacial age (OCS II). Alternating bedding of fine sand (lighter) and loamy fine sand to sandy loam (darker), with crinkly lamination and small-scale injection features caused by deposition on a wet surface. (c) Facies 3, of Younger Dryas age (YCS II); horizontal and low-angle cross-bedding in fine to medium sand deposited on a dry surface in source-proximal dunes. (d) Facies 4, of Younger Dryas age (YCS II), showing large-scale dune slipface cross-bedding in fine to medium sand in source-proximal dunes. A colour version of this figure appears in the plates section. *Source:* Courtesy of Kees Kasse.

sparseness of vegetation cover and high wind energy. Where facies 2 and 3 occur together, the latter usually overlies the former, possibly reflecting progressive drying out of depositional surfaces.

The available dating evidence suggests that deposition of aeolian sands in northern Europe occurred in three main phases. The earliest, the Late Pleniglacial phase (~26–18 ka), coincided broadly with the LGM and widespread permafrost, and was in many areas initially characterized by reworking of aeolian sands by surface runoff to form the fluvio-aeolian deposits of facies 1. In numerous localities, these deposits are overlain or truncated by the Beuningen Gravel Bed, which represents a period of deflation and constitutes a distinct stratigraphic marker in northwestern Europe. The second phase (~16–13 ka) was the most important aeolian phase in many areas, characterized by widespread deposition of sand-sheet deposits of facies 2 and 3 by dominantly northwesterly or westerly winds (Kasse, 1997; Kasse et al., 2007; Vandenberghe et al., 2013; Meinsen et al., 2014). During this phase, dune deposits (facies 4) also developed in central Europe, and low late-stage dunes developed farther west in response to increasing vegetation colonization. Aeolian activity was reduced in some areas by development of forest cover after ~14 ka. Finally, in western and north-central Europe, the Lateglacial period, and particularly the Younger Dryas Stade of ~12.9–11.7 ka, witnessed renewed aeolian activity, which in some areas extended into the Early Holocene (Kaiser et al., 2009a; Zielinski et al., 2015). Most coversands in England were deposited at this time (Bateman, 1995, 1998; Bateman et al., 2000; Baker et al., 2013), but in Poland Younger Dryas aeolian activity is less evident, possibly reflecting persistence of forest cover (Manikowska, 1991). Sand-sheet deposits of facies 3 are most common, with more local representation of facies 2 at the base of low dunes and rare representation of facies 4. Westerly or southwesterly winds appear to have been dominant during this period (Isarin et al., 1997).

Kasse (2002) suggested that there is no simple correspondence either between phases of aeolian activity and stadial–interstadial oscillations or between phases of deposition and particular aeolian sand facies. The most active aeolian phases appear to correspond with periods of increasing aridity at the end of cold periods (the Late Pleniglacial and Younger Dryas Stade), and to extend into ensuing temperate periods, possibly reflecting a lag in vegetation colonization of shifting sand deposits.

14.5.3 Pleistocene Loess Deposits

Muhs (2013) provides an excellent overview of the distribution, properties and significance of Pleistocene loess deposits. These are typically massive or weakly laminated, highly porous and commonly buff in colour, but may be brown, yellow, red or grey. Like loess deposits in present periglacial environments (Figure 14.5), they are generally poorly sorted and dominated by silt, though mean grain size may range from fine (8–16 μm) to coarse (32–63 μm) silt, and weathered loess may contain up to 60% clay. The mineralogy of Pleistocene loess deposits reflects that of source rocks: silica is usually dominant, except in loess derived from carbonate or volcanic rocks, and the main minerals present are typically quartz, feldspars, mica, calcite, dolomite and phyllosilicate clay minerals. Unweathered loess has an open microfabric in which silt-sized grains are linked by clay or carbonate bridges. This microstructure imparts strength to loess deposits so that they can form stable vertical faces (Figure 14.11), though softening or eluviation of clay bridges reduces intergranular cohesion, so that steep exposures may collapse when wetted. Pleistocene loess forms a surface mantle a few centimetres to a few hundreds of metres thick. Where loess deposits are relatively thin, their surface morphology is dictated by underlying topography, but thick loess deposits may completely bury the underlying land surface.

Though many Pleistocene loess deposits are of glacigenic origin (derived from silt produced by glacial abrasion, and subsequently deposited on proglacial outwash plains), others appear to have a non-glacigenic origin, principally as desert loess derived from unglaciated arid or semi-arid regions. The loess of the Great Plains in North America, for example, appears to be mainly of non-glacigenic origin (Aleinikoff et al., 2008). The vast loess deposits of China are often cited as an example of desert loess, as loess thickness and grain size diminish downwind from adjacent desert basins, but research by Sun (2002) suggests a more complex scenario. Sun showed that the deserts of southern Mongolia and neighbouring areas were the primary sources of Chinese loess. He also argued, however, that these 'source' areas mainly acted as reservoirs for silt produced in adjacent mountains by other processes, including weathering, glacial abrasion and fluvial comminution, so that much 'desert' loess in China ultimately originated in glaciated mountain areas. Subsequent studies have emphasized the role of the Yellow River in transporting sediment from the mountains of Tibet to the Mu Us desert, from where windblown dust has been carried to the Chinese Loess Plateau (Stevens et al., 2013). Irrespective of origin, however, there is widespread agreement that Pleistocene loess entrainment, transport and deposition in mid-latitude areas were greatest during periods of cold, relatively arid climate that correspond closely with the timing of glacial stages.

Loess and loess-derived deposits have been estimated to cover about 10% of the Earth's land surface, mainly in middle latitudes, though the area covered by primary unreworked loess is probably closer to 5%. Late Quaternary

Figure 14.11 Late Pleistocene loess deposits, Port Hills, South Island, New Zealand. The exposure is 3.5 m high.

loess in the coterminous United States is estimated to cover more than 4.5 million km^2, with the most extensive belt occupying the Mississippi Basin between Colorado and Ohio and extending southwards to Louisiana, and with separate extensive covers in eastern Washington and Oregon, and in the Snake River Plain and adjacent Idaho uplands (Figure 14.12). Thick loess deposits in the United States are largely restricted to the zone south of the limit of the last Laurentide Ice Sheet, though thin loess derived from outwash deposits occurs north of this limit (Schaetzl and Attig, 2013). Loess deposited during the last glacial stage in North America is known as the Peoria Loess, and forms the largest continuous loess sheet in the United States. It reaches depths of over 48 m in Nebraska and is typically 10–20 m thick in the Mississippi Valley (Bettis *et al.*, 2003). Though loess in the United States has both glacigenic and non-glacigenic sources, most is probably derived from the floodplains of rivers that drained the Laurentide Ice Sheet. In South America, extensive Pleistocene loess deposits mainly occur in two extensive zones, the Pampean Loess of central Argentina and the Chaco Loess of northern Argentina, though the

true extent of South American loess deposits remains to be established (Zárate, 2003).

In continental Europe, loess and loess-derived deposits extend discontinuously eastwards from northern France to the Black Sea, and across much of Ukraine and Belarus to the foothills of the Urals and Caucasus (Figure 14.13). European loess cover is almost exclusively limited to the zone south of the last Fennoscandian Ice Sheet and north of the Alpine Ice Sheet (Haase *et al.*, 2007), and generally becomes more widespread eastwards. The loess deposits of the eastern European plains between latitude 44–56° N and longitude 24–50° E are amongst the most extensive in the world (Velichko, 1990). Much of the loess in northwest Europe appears to be derived from the floors of the North Sea Basin and English Channel (which were exposed by sea-level fall during ice-sheet maxima), though farther east outwash deposits and unvegetated alluvial plains formed the main sources of airborne dust. Beyond the Urals, the Eurasian loess belt extends across Khazakstan and Russia, occupying large swathes of the Kirgiz Steppe and the West Siberian Plain before terminating in the headwater valleys of the Lena River.

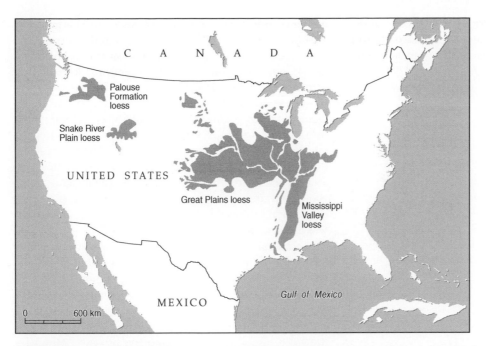

Figure 14.12 Extent of loess deposits in the coterminous United States. *Source:* Muhs *et al.* (2003). Reproduced with permission of Elsevier.

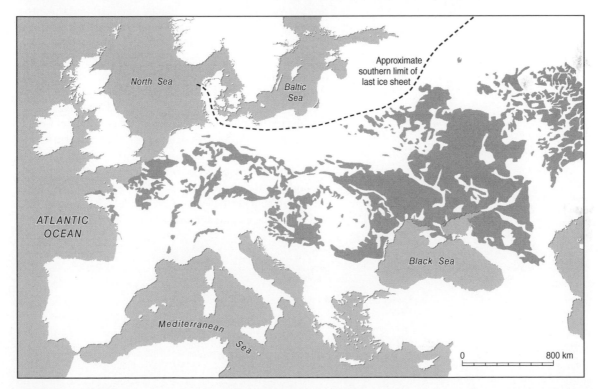

Figure 14.13 Extent of primary and reworked loess deposits in continental Europe. *Source:* Adapted from Muhs (2013). Reproduced with permission of Elsevier.

Remarkably thick loess deposits occur around the desert margins of Central Asia, and particularly in China, where loess and reworked loess deposits occupy an area of more than 1 million km². In central China, thick loess caps a broad dissected plateau, the Loess Plateau, which lies mainly within the great northwards loop of the Yellow River (Huang He), where it emerges from the Qinghai-Tibet Plateau. The Loess Plateau exceeds 273 000 km² in area, and supports Pleistocene loess deposits that are typically 80–120 m thick, but which reach thicknesses of 300–400 m in its central and western parts (Porter, 2001). Loess deposition during the last glacial stage is here represented by the Malan Loess, which exceeds 30 m in depth in the western Loess Plateau but declines in thickness eastwards, away from desert source areas (Figure 14.14).

14.5.4 The Depositional History of Pleistocene Loess Deposits

In some parts of the world, loess accumulation spans the entire Quaternary Era. In China, accumulation occurred at the very beginning of the Pleistocene (~2.58 Ma), though Porter (2013) noted that the underlying 'Red Clay' beds are also mainly loessic, and extend back to the late Miocene (6.5–7.2 Ma). There is evidence that the oldest loess in Central Asia also began to accumulate at the beginning of the Pleistocene (Dodonov, 2013),

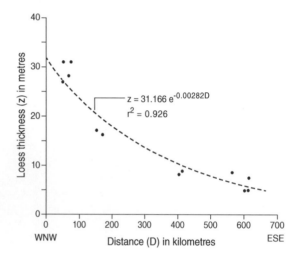

Figure 14.14 Exponential decline in the thickness of Malan (last glacial stage) Loess along a WNW–ESE transect across the southern Loess Plateau in China. *Source:* Data from Porter (2001).

with a marked increase in the rate of accumulation after 0.8–1.0 Ma. In Europe, North America and Central Asia, most loess deposits are no older than the penultimate glacial stage, though there are exceptions: loess in the Carpathian Basin in Hungary spans the last 800 000 years (Újvári *et al.*, 2014) and loess in the middle and lower Danube basin extends back roughly a million years (Markovic *et al.*, 2011; Fitzsimmons *et al.*, 2012). In general, rates of loess accumulation appear to have increased during the Pleistocene, and particularly after the 'Mid-Pleistocene Revolution', which marked the onset of more prolonged and extensive mid-latitude ice-sheet glaciation around 0.9 Ma; in many areas, loess accumulation was greatest during the last glacial stage.

Loess deposits constitute one of the most important terrestrial records of Quaternary climate change, particularly in the form of *loess–palaeosol* sequences within loess accumulations. Palaeosols are relict or buried soil horizons, in this case formed during periods when loess accumulation was reduced and pedogenesis (soil formation) was accentuated. The simplest interpretation of a single loess–palaeosol couplet is that the former represents cold, arid conditions (glacial periods) and the latter relatively warm, wetter conditions (interglacial periods), so that a stacked sequence of loess–palaeosol couplets represents alternating glacial and interglacial conditions (Figure 14.15). But 'loess stratigraphy is rarely simple' (Muhs, 2013). In part, this is because reduced dust accumulation may persist during periods of warmer, wetter climate, so that palaeosols do not represent distinct breaks in the stratigraphic record; loess accumulation and pedogenesis are thus competing rather than mutually exclusive processes (Kemp, 2001; Muhs *et al.*, 2004). An additional complication is that shorter-term warming periods during glacial stages (interstades) may permit limited pedogenesis, and climatic changes during interglacials

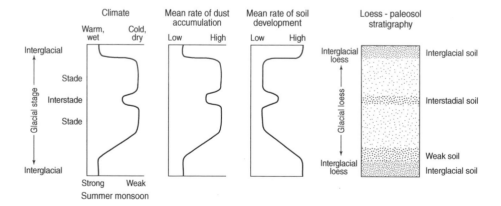

Figure 14.15 Model of the effect of glacial–interglacial climatic changes on loess accumulation rates and soil development in the loess region of central China. A strong summer monsoon generates warm, moist conditions that promote soil development, whilst a weak summer monsoon is associated with cold, dry conditions, increased dust influx and loess accumulation. *Source:* Porter (2001). Reproduced with permission of Elsevier.

may result in renewed loess accumulation (Figure 14.15). In consequence, many loess sequences exhibit a continuum of deposits, ranging from relatively unmodified loess and weakly developed leached layers to intensely weathered palaeosols and pedocomplexes. Unravelling such complex stratigraphy, however, is aided by measurements of the magnetic susceptibility of loess–palaeosol sequences, as loess has low magnetic susceptibility but that of palaeosols is relatively high, enabling correlation between sites and with other palaeoclimatic records, such as marine oxygen isotope ($\delta^{18}O$) stages. Dating of loess sequences poses an additional challenge. Though the ages of numerous mid- and late-Pleistocene sequences have been successfully determined by radiocarbon dating, and particularly luminescence dating (e.g. Roberts *et al.*, 2003; Pierce *et al.*, 2011; Muhs *et al.*, 2013; Pigati *et al.*, 2013), these techniques are limited to the dating of sediments <100 ka in age; dating of older loess is primarily achieved by reference to the timing of palaeomagnetic reversals.

Despite these challenges, remarkable progress has been made in deciphering loess archives and in relating major periods of loess accumulation to global-scale climatic changes. By far the most complete loess–palaeosol record is that for the Loess Plateau in China, where 32 loess–palaeosol couplets have been recorded above the level of the Matuyama-Gauss palaeomagnetic reversal at ~2.5 Ma and shown to relate closely to the marine $\delta^{18}O$ record of glacial–interglacial oscillations (Porter, 2001; Sun and Zhu, 2010; Zhang *et al.*, 2016; Figure 14.16). The loess–palaeosol couplets on the Loess Plateau are thought to relate to the alternating dominance of the Mongolian high-pressure atmospheric system during glacial stages and the East Asian summer monsoon during interglacials. The former drives cold dry air across China, depositing airborne dust as loess, whereas the latter is associated with landward airflow, gentle winds and warmer, humid conditions that favour pedogenesis. This scenario, though, simplifies the Chinese loess record, which includes numerous weaker soil-forming events during glacial periods and variation in loess accumulation rates during interglacials. Moreover, analysis of changes in the magnetic susceptibility and grain size of loess in a 416 m-long core retrieved from the western Loess Plateau suggests changes in airmass circulation during

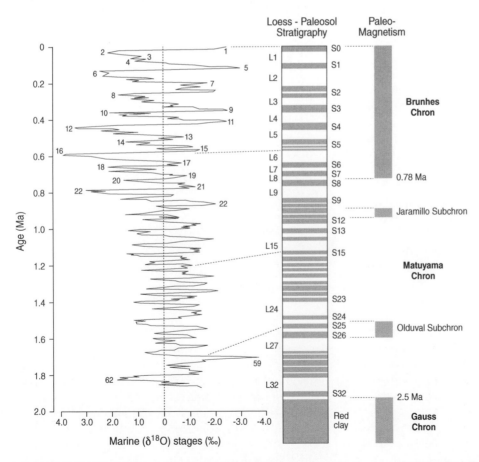

Figure 14.16 The loess–palaeosol stratigraphy of the Chinese Loess Plateau over the last 2.5 Ma, correlated with the marine oxygen isotope stages and the timing of palaeomagnetic reversals. In the marine isotope record, glacial stages have even numbers and interglacials odd ones. Loess units are prefaced by the letter L and palaeosols by the letter S. Both are numbered from the top of the sequence downward. *Source:* Adapted from Porter (2001). Reproduced with permission of Elsevier.

the Pleistocene, with evidence for stepwise enhancement of the East Asian summer monsoon at ~1.24, ~0.87 and ~0.62 Ma, trends attributed by Zhang *et al.* (2016) to uplift of the Qinghai-Tibet Plateau.

Most loess–palaeosol sequences in North America and Europe are limited to events spanning the last two glacial–interglacial cycles. In interior Alaska, the inclusion of tephra (fine volcanic ash) layers of known age within loess demonstrates accumulation from ~150 ka to the Holocene, implying that the oldest loess is of Late Illinoian age (Jensen *et al.*, 2016). Similarly, in the mid-continental United States, the Middle Pleistocene (Illinoian) Loveland Loess is the oldest (~160–140 ka) at most locations, but is rarely more than a few metres thick. It is succeeded in turn by the Sangamon Geosol (palaeosol) of last interglacial age, a thin early Wisconsinan loess (Roxana Silt or Pisgah Loess), an organic-rich palaeosol (Farmdale Geosol) and the Late Wisconsinan Peoria Loess, which accumulated between ~28 and ~13 ka (Bettis *et al.*, 2003; Muhs *et al.*, 2013). Not all North American loess stratigraphies follow this exact pattern, however; loess in the Snake River valley of western Wyoming, for example, exhibits eight discrete loess units deposited since ~150 ka (Pierce *et al.*, 2011). The non-glacigenic loess of the Great Plains is also mainly of last glacial age, and is locally overlain by Holocene loess (Roberts *et al.*, 2003; Miao *et al.*, 2005). Mid-to-late Pleistocene loess stratigraphy in Europe is broadly similar: a Middle Pleistocene (Saalian) loess is overlain by an interglacial (Eemian) palaeosol, the Rocourt palaeosol, which is covered by extensive Weichselian loess, most of which was deposited between ~70 and ~13 ka, and particularly during the period 28–15 ka (Frechen *et al.*, 2003).

Loess deposits of the last glacial stage also exhibit rhythmic variations that correspond to global millennial-scale climate fluctuations. Grain size peaks in the Malan Loess of China correspond to the timing of large-scale iceberg rafting episodes (Heinrich Events) in the North Atlantic Ocean, showing that quasi-synchronous changes in climate occurred throughout the northern hemisphere (Chen *et al.*, 1997). Similarly, grain-size variations in European loess of last glacial age show a strong correspondence to dust content in Greenland ice cores and a cyclicity that corresponds to millennial-scale Dansgaard–Oeschger events, with coarser loess layers representing colder episodes and finer layers (with associated palaeosol development) representing more temperate conditions (Vandenberghe and Nugteren, 2001; Rousseau *et al.*, 2002; Figure 14.17).

Though loess rarely preserves pollen grains, insect remains or most other proxy indicators of palaeoclimate, the shells of land snails are common in some loess deposits. Not only do these have potential for radiocarbon dating (Pigati *et al.*, 2013), but the presence of molluscan species with cold-climate affinities provides additional

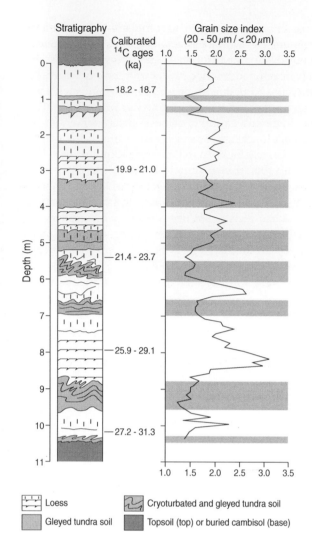

Figure 14.17 Stratigraphy of the last glacial age loess sequence at Nussloch in Germany, showing millennial-scale variations in grain size. Coarser grain sizes are inferred to represent more arid conditions and increased wind strength, whereas finer grain sizes correspond to periods of gleyed tundra soil formation, suggesting wetter conditions. *Source:* Adapted from Rousseau *et al.* (2002). Reproduced with permission of Elsevier.

evidence of air temperatures at the time of loess accumulation (Rousseau *et al.*, 2000). Moreover, because loess thickness, grain size and carbonate content all tend to decrease in a downwind direction, loess deposits also indicate dominant wind direction at the time of accumulation. The southeastwards decline in the thickness of the Malan Loess in China (Figure 14.14) indicates dominant dust-bearing winds from the northwest quadrant (Porter, 2001), and an easterly or southeasterly decline in loess thickness and grain size in the midcontinental United States likewise indicates dominant winds from the west or northwest (Bettis *et al.*, 2003). Detailed analyses of the Peoria Loess in Iowa by Muhs *et al.* (2013) have shown that the coarsest grain sizes and highest rates of loess accumulation occurred at ~23 ka, when the Laurentide Ice Sheet reached its maximum southern extent. This correspondence suggests that increased dust transport during the last glacial period was driven by increased gustiness, itself related to a steepened meridional atmospheric temperature gradient caused by the southwards advance of the ice sheet.

Finally, it is notable that because significant infall of airborne dust continues to the present in some regions, loess–palaeosol sequences constitute an important archive of Holocene climate change, for example on the northeastern Qinghai-Tibet Plateau, the Chinese Loess Plateau and northern Mongolia (Jia *et al.*, 2011; Lehmkuhl *et al.*, 2012, 2014). In particular, palaeosols within Holocene loess deposits provide evidence of periods of increased humidity that can be related to lake-level changes, and indicate periods of increased precipitation.

14.6 Synthesis

Many present-day periglacial environments experience strong winds and gusty conditions, but significant aeolian erosion, transportation and deposition of sediment are mainly limited to polar deserts, and more locally to unvegetated or partly vegetated terrain such as outwash plains and glacier forelands. Typical cold-climate erosional features include deflation surfaces, blowouts and ventifacts, and actively accumulating aeolian features include sand dunes, sand sheets, fluvio-aeolian and niveo-aeolian deposits, and loess. During Pleistocene glacial periods, however, windblown sand and dust were deposited under cold conditions across more than 10 million km^2 of the northern hemisphere mid-latitudes in the form of sand sheets (coversands), dunefields and loess. Such deposits constitute a key palaeoenvironmental archive that in some areas spans much or all of the Quaternary, providing information on periods of aridity, palaeowind directions and switching airmass dominance, all of which can be related to the timing of glacial and interglacial periods and, at least over the past million years, to the expansion and contraction of the great northern hemisphere ice sheets. Globally, Pleistocene aeolian sediments probably represent the most extensive periglacial deposits on Earth. As many constitute areas of fertile, well-drained soils and high agricultural productivity, they are also of crucial importance to global food security.

15

Periglacial Coasts

15.1 Introduction

The distinguishing characteristic of periglacial coasts is the presence of ice at the shore for much of the year. During winter and spring, ice covers the intertidal zone, preventing erosion by wave action. Farther offshore, drifting pack ice inhibits wave development and reduces fetch length, even in summer. Conversely, ice floes may erode bedrock and raft, push and scour sediment at the coastline, and storm waves acting on coasts underlain by ice-rich permafrost may cause rapid shoreline erosion during summer open-water conditions. Here we first consider the nature of periglacial coasts and the effects of ice on coastal processes, then focus on several arctic and subarctic coastal types: ice-rich permafrost coasts, thermokarst coasts, barrier coasts, tidal flats and salt-marshes, and rock coasts. The chapter concludes with a brief account of lake shorelines in periglacial environments. Forbes and Hansom (2011) have summarized the characteristics of polar coasts in both hemispheres, and the Arctic Coastal Dynamics Project (Lantuit *et al.*, 2012b; Overduin *et al.*, 2014) provides a synthesis of the geomorphology of coasts fronting the Arctic Ocean and recent changes in coastal dynamics.

15.2 The Nature of Periglacial Coasts

Periglacial coasts make up more than a third of the Earth's coastlines, and Lantuit *et al.* (2013) cite a figure of 30–34% for permafrost coasts alone. In the northern hemisphere, ice-affected coasts encircle the Arctic Ocean and in places extend south of 60°N, notably around the Bering Sea, Sea of Okhotsk, Baltic Sea, Labrador, Newfoundland, Gulf of St Lawrence and Hudson Bay (Figure 15.1). However, only the northernmost coasts of the Arctic Ocean tend to be ice-bound throughout the year. Other periglacial coasts experience summer open-water conditions for periods ranging from a few weeks in the Canadian Arctic Archipelago to 3–4 months along the circum-arctic coastline that extends from the White

Sea to the Beaufort Sea, and 6 months on the east coast of Labrador. In the southern hemisphere, the Antarctic continent is surrounded by an extensive girdle of sea ice cover in winter, but becomes almost entirely free of sea ice during the summer. Thus, whereas 75–85% of pack ice in the Arctic Ocean is multi-year ice, often exceeding 2 m in thickness, almost all of the sea ice around Antarctica comprises 1–2 m-thick single-year ice that is dispersed into the Southern Ocean during the summer months.

Continuous permafrost underlies the Arctic Ocean coasts (apart from the shores of the Barents Sea and White Sea, which are warmed by the waters of the North Atlantic Drift) and occurs under intertidal sediments at depths of 0.4–1.0 m. Permafrost is also present along unglacierized coasts in Antarctica and the Antarctic Islands (Bockheim, 1995). Subsea permafrost underlies much of the continental shelf that occurs offshore between the Kara Sea and the Beaufort Sea, extending up to 600 km offshore in the Laptev and East Siberian Seas (Overduin *et al.*, 2007). Such submarine permafrost apparently formed before the shelf was inundated by rising sea levels caused by the melting of the last Pleistocene ice sheets (Winterfield *et al.*, 2011), and has survived because of very cold sea-bottom temperatures. The residual effects of Pleistocene glaciation on periglacial coasts are also evident in other ways. In areas that were formerly occupied by thick ice sheets, such as Hudson Bay and the Gulf of Bothnia, continuing glacio-isostatic uplift of up to 10 mm a^{-1} means that the sea is gradually receding from the shore. The Antarctic coast is also predominantly emergent. Conversely, much of the arctic coastline from the Laptev Sea to the Beaufort Sea has experienced Holocene marine transgression, and continuing slow (up to 4 mm a^{-1}) sea-level rise in some areas makes coastlines particularly vulnerable to marine erosion (Manson *et al.*, 2005). The sediment budgets of many arctic and subarctic coasts, moreover, are influenced by paraglacial inputs of Pleistocene glacigenic sediments, which are either deposited in the nearshore zone by rivers or reworked by wave action at the shore-face (Forbes, 2011).

Figure 15.1 Ice-affected coasts of the northern hemisphere, showing the approximate limit of winter sea ice and the limit of summer (multi-year) sea ice in 2009. The main zone of ice-rich permafrost coastline extends from Banks Island via the Beaufort Sea, Chukchi Sea and East Siberian Sea to the Laptev Sea.

Despite the defining influence of ice on the evolution of periglacial coasts, wave action associated with storm events during the open water season is often the dominant agent of coastline modification and littoral sediment transport. Consequently, many coastal landforms in arctic and subarctic environments – beaches, spits, bars, barrier islands, headlands, salt marshes, tidal flats, cliffs and rock-cut shore platforms – differ only in detail from those in ice-free coastal environments. Only coasts developed through erosion of bluffs composed of ice-rich sediment, and particularly those where littoral erosion affects areas of thermokarst, are distinctly different from those at lower latitudes.

15.3 The Role of Ice in Shoreline Evolution

Ice both limits and enhances erosion of periglacial coasts. Ice frozen to the shoreface in winter protects the shore from wave action for much of the year, and sea ice limits the period of open-water conditions when wave action is possible. Conversely, mobile sea ice driven by wind stress can erode and transport sediment, forming a number of distinctive landforms on beaches, tidal flats and the sea floor.

At the onset of the annual freezeback, *bottomfast ice* or *anchor ice* (ice frozen to the underlying substrate) forms on the foreshore. Ice formation initially results from freezing of wave spray, swash and backwash in the upper part of the intertidal zone, but subsequent ice build-up may incorporate slush ice, ice floes and ice blocks carried ashore by storm waves and frozen on to the growing ice accumulation. The resultant zone of bottomfast ice is termed the *ice foot*. Where layers of bottomfast ice alternate with layers of sediment carried inshore from an ice-free surf zone, the ice foot deposit is referred to as *kaimoo*. In general, a low-gradient shoreface and large tidal range promote the development of a wide ice foot, and *vice versa* (Wiseman *et al.*, 1981). The ice foot may extend above high-tide level due to freezing of swash and spray during storms, and bottomfast ice may extend seawards to depths of 2–3 m. At its seaward edge, the ice foot abuts *strand ice* that floats as the tide rises and grounds as the tide falls, the boundary between the bottomfast and floating ice being marked by a hinge zone of tidal cracks. Ice connected to the shore (*shorefast ice*) may extend several kilometres offshore, eventually merging with drifting *pack ice* at a zone of pressure-ridge development.

In summer, the ice foot often persists on polar beaches long after the onset of open-water conditions, and may survive until the following winter, protecting the shore from wave action. In some circumstances, the ice foot also exerts a more subtle influence on the coastal environment. In the microtidal environment of Manitounuk Strait in subarctic Quebec, deep penetration of frost in tidal flats under the ice foot impedes winter drainage, causing emergence of springs and formation of icings, frost blisters and lens-shaped bodies of intrusive ice at the upper margin of the tidal zone. At the same location, segregated ice develops annually in the marine clays under the ice foot, causing liquifaction and flow of sediment during thaw (Allard *et al.*, 1998).

Not all coastal ice has a protective role (Are *et al.*, 2008). Pack ice driven ashore by storms may ride up over the ice foot or buckle upwards to form an ice mound several metres high in the upper part of the intertidal zone. Ice floes driven inshore during such events may bulldoze beach sediment, producing irregular *ice-pushed ridges* and *scour marks* that may survive above the upper limit of wave action. Exceptionally, ice-pushed ridges of beach sediment can reach heights of 5 m or more, or extend up to 100 m inland of the intertidal zone, but most are much smaller, more ephemeral and closer to the high-tide mark. If coarse debris is rafted shorewards in ice floes or pushed ashore by pack ice, *boulder pavements* and *boulder ridges* or *boulder ramparts* may develop in the upper intertidal zone, due to stranding of ice-pushed or ice-rafted boulders when the ice melts (Hansom, 1983a, 1986; Figure 15.2). Ice push and pile-up processes are thus locally important in shaping the shoreface profile and transporting sediment landwards, above the normal reach of wave action. Scouring of sediment by the keels of floating pressure-ridge ice occurs at water depths of up to 65 m, and seabed surveys of the floating shorefast ice zone along the Beaufort Sea coast suggest that ice scour excavates and mobilizes an estimated 3000–6000 $m^3 km^{-2} a^{-1}$ of sediment, much of which enters suspension transport in the surf zone (Rearic *et al.*, 1990).

Also formed by ice movement in the intertidal zone are *boulder barricades*, low ridges of boulders along the outer margins of tidal flats. The formation of boulder barricades is generally attributed to break-up of boulder-rich intertidal bottomfast ice occurring before that of floating shorefast ice, so that the latter prevents seawards transport of coarse debris. Accumulation of boulders in this ice-free zone may reflect bulldozing of boulders by ice floes, seawards rafting of boulders on ice or longshore rafting and deposition of boulders where ice floes ground near the low-water mark (Gilbert, 1990; Dionne, 2002). Other exotic (but ephemeral) landforms produced by ice on tidal flats include *strudel scours*, which form where snowmelt runoff over shorefast ice pours through cracks in the ice, excavating cone-shaped depressions in underlying sediment; *ice-wallow topography* of ridges and depressions formed in sediment by grounded ice floes that are raised and lowered by the tide; and elongate

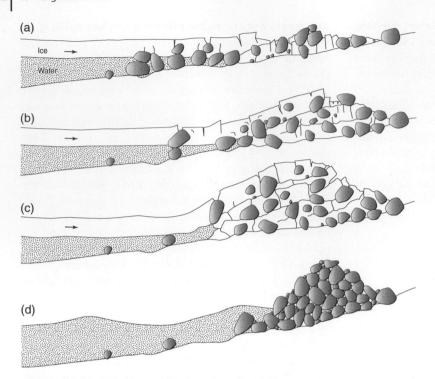

Figure 15.2 Boulder ridge formation by ice push. (a) Boulders freeze on to the base of pack ice. (b) Lateral thrusting stacks boulder-bearing ice on the shore. (c) Boulder-rich ice ridge. (d) Boulder ridge after melt of ice. Most boulder ridges develop as a result of repeated ridge-building events. *Source:* Barnes (1982), http://arctic.journalhosting.ucalgary.ca/arctic/index.php/arctic/article/view/2330. Used under CC BY 4.0 https://creativecommons.org/licenses/by/4.0/.

Table 15.1 Estimates of sediment concentration in sea ice.

Area	Fine sediment	Coarse sediment	Concentration (t km^{-2})
Arctic Canada			
Frobisher Bay, Baffin Island	x	x	63 750
Pangnirtung, Baffin Island	x	–	90 000
South Bay, Southampton Island	x	x	2500
Foxe Basin, Melville Peninsula	x	–	20 000
Beaufort Sea			
North Shore	x	–	243
Prudhoe Bay	x	–	800
Harrison Bay	x	–	2950
Makkovik Bay (Labrador)	x	x	13 000
Norton Sound (Bering Sea)	x	–	31 000
Barents Sea	x	–	27 272

Source: Adapted from Dionne (1993).

depressions caused by tidal scour around grounded ice blocks (Forbes and Taylor, 1994).

The importance of sea ice as an agent of sediment transport is spatially and temporally variable. Depictions of the ice foot or shorefast ice in arctic environments often show clean ice with little incorporated sediment, but estimated sediment concentrations in sea ice elsewhere (Table 15.1) suggest that it plays a moderate to very important role in transporting sediment offshore. Some sediment is delivered to sea ice from bluffs or cliffs by rockfall, debris falls, block collapse, mudflows and retrogressive thaw slumps (Solomon, 2005); some is of aeolian origin, and deposited by wind blowing offshore; some is entrained through freezing of turbid sea water, particularly in the surf zone or in areas of high suspended sediment concentration; and some is entrained through freezing of sediment to the base of bottomfast ice that subsequently floats offshore. Movement of large boulders by sea ice may reflect rafting (freezing of boulders to shore ice that subsequently floats, carrying the attached boulders) or pushing, where wind-driven lateral movement of pack ice drives boulders across tidal flats by sliding or rolling, leaving furrow-like depressions in their wake (Gilbert, 1990).

15.4   Ice-rich Permafrost Coasts

Much of the arctic coastline from the Kara Sea to the Beaufort Sea consists of low bluffs of ice-rich yedoma sediments, sometimes fronted by narrow beaches. Most of these *ice-rich permafrost coasts* currently experience net erosion (Lantuit *et al.*, 2012b, 2013). The volumetric ice content of coastal bluffs around the Arctic Ocean averages about 18.4%, but varies greatly; some unlithified coastal cliffs cut in yedoma sediments contain 80–90% ice by volume, mainly in the form of wedge and segregated ice (Romanovskii *et al.*, 2000; Are *et al.*, 2005; Overduin *et al.*, 2007; Streletskaya *et al.*, 2008; Figure 15.3). Some also contain high concentrations of largely undecomposed organic matter and organic carbon, which get released into the Arctic Ocean as coasts are eroded (Lynn *et al.*, 2008; Ping *et al.*, 2011; Sánchez-Garcia *et al.*, 2014). The granulometry of the sediments influences rates of erosion, as fine-grained muds are likely to be carried offshore in suspension, whereas sand and gravel are entrained by longshore drift, and may form protective beaches, spits and barrier islands (Héquette and Ruz, 1991).

Several studies have shown that erosion of ice-rich permafrost coasts is largely limited to storm events, usually in the late summer open-water interval, with intervening periods of very limited coastal change (e.g. Dallimore *et al.*, 1996b; Wolfe *et al.*, 1998; Overeem *et al.*, 2011; Barnhart *et al.*, 2014a). During the most violent storms, wind-driven waves may raise water levels up to two metres above the high tide mark, inundating low-lying tundra several kilometres inland (Harper *et al.*, 1988).

Figure 15.3 Ice-rich yedoma sediments exposed by coastal erosion, Muostakh Island, southern Laptev Sea. The large syngenetic ice wedges are thought to be of Late Pleistocene age. Coastal retreat is ~50 m a^{-1} in this area. A colour version of this figure appears in the plates section. *Source:* Courtesy of Hans Hubberten.

Erosion of ice-rich coastal bluffs during storms occurs in several ways. The most distinctive is the development of a *thermo-erosional niche* that undercuts bluffs at high-tide level, with resultant block collapse of sections of coastline, usually along ice wedges. Niche erosion, sometimes referred to as *thermoabrasion*, is caused by a combination of mechanical erosion by waves and tidal currents and thaw of ice-rich permafrost through contact with waves and spray. Block collapse may take the form of toppling or sliding failure; Hoque and Pollard (2009) have suggested that the former is more likely on low bluffs, the latter on high bluffs. Once a block of ice-rich frozen soil has collapsed into the surf zone, it is rapidly dismembered through thaw of ice and mechanical removal of loosened particles by wave action. Niche development and block collapse represent the dominant mode of coastline retreat on many ice-rich permafrost coasts (Vasiliev *et al.*, 2005), but in areas of high ice content, retrogressive thaw slumps may also make a major contribution to sediment loss (Figure 15.4). Thaw slumping forms large cuspate thermokarst scars, which

are periodically reactivated as the coastline retreats (Lantuit and Pollard, 2008; Chapter 8), with mudflows over frozen ground feeding sediment into the nearshore zone. Several other ground ice-related processes are also involved more locally in coastal erosion during storms, including thermal gullying by snowmelt runoff and active-layer detachment failure (Solomon, 2005). A more indirect contribution to shore erosion results from degradation of ice-rich permafrost in the nearshore zone. In the Canadian Beaufort Sea, for example, the permafrost table declines to 15–20 m depth within 1 km of the shoreline (Hill and Solomon, 1999), and thaw of ground ice in this zone has caused seafloor subsidence of 3 m or more (Wolfe *et al.*, 1998). Such subsidence not only creates accommodation space for sediment eroded from the shoreface or carried to the coast by rivers, but also results in changes to the shoreface profile that may enhance the erosive effectiveness of waves during storms (Dallimore *et al.*, 1996a).

Rates of retreat of ice-rich permafrost coasts have been calculated using sequential aerial photographs, satellite

Figure 15.4 Retrogressive thaw slumps on the southeast coast of Herschel Island, Beaufort Sea Coast, arctic Canada. The large slump extends ~500 m inland and is ~400 m wide; the headwall reaches a height of ~25 m. A colour version of this figure appears in the plates section. *Source:* Courtesy of Hugues Lantuit.

imagery and periodically resurveyed ground transects. Retreat rates vary greatly, both spatially and temporally. Héquette and Ruz (1990) found that retreat rate is positively correlated with ground-ice content and wave power, and Lantuit *et al.* (2008) detected a weak positive relationship between retreat rate and ground ice content. Such relationships, however, are complicated by such factors as relative sea-level change, nearshore topography, longshore sediment drift and exposure to prevailing winds. Temporal variability in coastline retreat is also strongly conditioned by the frequency, magnitude and timing of major storm events (Wolfe *et al.*, 1998; Overeem *et al.*, 2011; Barnhart *et al.*, 2014a, 2014b).

The rates of coastline retreat on ice-rich permafrost coasts are amongst the most rapid in the world, despite the limitations imposed by an abbreviated open-water season and the influence of offshore pack ice in limiting fetch length and thus wave energy. Annual retreat of the Laptev Sea coast ranges from a few centimetres to $20\,\text{m}$ per year, and in most areas averages 2–$6\,\text{m a}^{-1}$ (Rachold *et al.* 2000), while Solomon (2005) showed that coasts in the Beaufort Sea area retreated at average rates of 0.6–$22.5\,\text{m a}^{-1}$ between 1972 and 2000. Rates of coastal retreat are so rapid on the yedoma coasts of the Eastern Siberian shelf than numerous islands have disappeared within the last millennium, and some within the last 50 years (Gavrilov *et al.*, 2003). Muostakh Island in the Laptev Sea, for example, lost 24% of its area between 1951 and 2013, and is predicted to disappear within a century (Günther *et al.*, 2013). Rachold *et al.* (2000) estimated that average sediment yield due to erosion of the Laptev Sea coast ($11\,200\,\text{t km}^{-1}\text{a}^{-1}$) is roughly twice that estimated for the Beaufort Sea coast (5000–$7000\,\text{t km}^{-1}\text{a}^{-1}$), reflecting higher average coastal retreat rates due to the greater ice content of Laptev Sea coastal bluffs (Table 15.2).

A more nuanced picture of coastal erosion rates has been provided by Jorgenson and Brown (2005), who classified coastal types of the Alaskan Beaufort Sea coast and presented data on mean erosion rates associated with each.

Table 15.2 Sediment input into the Arctic Ocean through coastal erosion.

Location	Length of coast (km)	Total coastal erosion ($10^6\,\text{t a}^{-1}$)	Average coastal erosion ($10^3\,\text{t km}^{-1}\text{a}^{-2}$)
Laptev Sea, northern Siberia	5200	58.4	11.2
Beaufort Sea, northern Alaska	1957	3.3	2.7
Beaufort Sea, northwest Canada	1150	5.6	5.0

Source: Data from Rachold *et al.* (2000) and Jorgenson and Brown (2005).

They found, for example, that average recent erosion rate for the entire coastline has been roughly $1.6\,\text{m a}^{-1}$, but that for exposed bluffs has been much greater ($2.4\,\text{m a}^{-1}$) than that for sheltered lagoon sites ($0.7\,\text{m a}^{-1}$). Many permafrost coasts have also experienced significant recent changes in erosion rate. For the coasts of Herschel Island in the Beaufort Sea, Lantuit and Pollard (2008) found that average erosion rates decreased from $0.61\,\text{m a}^{-1}$ over the period 1952–70 to $0.45\,\text{m a}^{-1}$ in 1970–2000. Conversely, Jones *et al.* (2009) calculated that mean coastal erosion rates along a rapidly retreating segment of the Beaufort Sea coast increased from $6.8\,\text{m a}^{-1}$ in 1955–79 to $8.7\,\text{m a}^{-1}$ in 1979–2002 to $13.6\,\text{m a}^{-1}$ in 2002–07. Similarly, Ping *et al.* (2011) calculated that for 48 sites distributed along various coastal settings on the Alaskan Beaufort Sea coast, the average rate of coastal retreat doubled between ~1950–80 (~$0.6\,\text{m a}^{-1}$) and ~1980–2000 (~$1.2\,\text{m a}^{-1}$). Recent acceleration of coastal erosion is consistent with warming of permafrost and increasing summer sea-surface temperatures, both of which promote more rapid thaw of exposed ice-rich sediments. The most important driver of increased coastal erosion on ice-rich permafrost coasts, however, is recent (post-1970) shrinkage of the extent of summer pack ice in the Arctic Ocean (Serreze *et al.*, 2007; Stroeve *et al.*, 2007, 2012, 2014), which has increased average fetch over open water and therefore wave energy during storms (Overeem *et al.*, 2011; Barnhart *et al.*, 2014a, 2014b). Such trends are worrying, partly because they imply an increase in release of organic carbon from frozen littoral sediments (Vonk *et al.*, 2012), but also because many settlements are located on ice-rich permafrost coasts. At Tuktoyaktuk on the Beaufort Sea coast, for example, parts of the townsite experienced more than $100\,\text{m}$ of coastal retreat between 1935 and 1971, necessitating abandonment and relocation of buildings (Wolfe *et al.*, 1998).

15.5 Thermokarst Coasts

Thermokarst coasts represent a subset of ice-rich permafrost coasts, and occur along the Beaufort Sea coasts and northern coasts of Siberia and Yakutia. They are characterized by opening of thermokarst lakes and drained lake basins to the sea by coastal erosion (Ruz *et al.*, 1992; Dallimore *et al.*, 1996a; Romanovskii *et al.*, 2000; Solomon, 2005). This produces a distinctive coastal configuration of narrow headlands separated by circular or oval lagoons, rather like a cross-section through an Emmental cheese (Figure 15.5). As with other ice-rich permafrost coasts, thermokarst coasts are eroded during storms through thermo-erosional niche excavation and block collapse, thaw slumping, active-layer detachment failure, gullying and mudflows, with rapid erosion of exposed headlands and generally slower retreat within

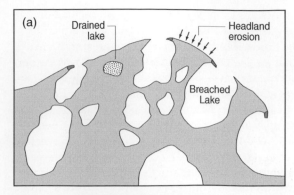

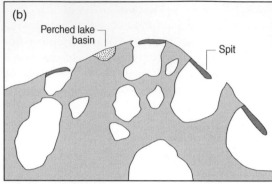

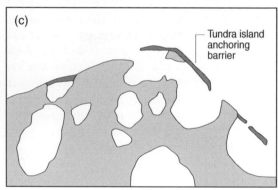

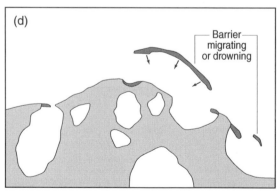

Figure 15.5 Coastline evolution in thermokarst terrain.
(a) Embayments and headlands formed from transgressive coastal erosion of thermokarst lakes. (b) Formation of spits composed of sediment derived from headlands. (c) Isolation of offshore barrier islands by coastal retreat. (d) Migration or drowning of barriers. All of these stages occur along the Beaufort Sea coast. *Source:* Adapted from Hill and Solomon (1999). Reproduced with permission of the Coastal Education and Research Foundation.

sheltered lagoons, which represent the sites of drained thermokarst lakes (Jorgenson and Brown, 2005). Breaching of thermokarst lakes by coastal retreat results in invasion by the sea if the lake floor lies below sea level and tidal flooding if it lies within the altitudinal range of the intertidal zone (Solomon *et al.*, 2000). Where the altitude of the lake floor lies above that of the tidal zone, breaching results in lake drainage and sometimes in the formation of hydrostatic pingos in the drained lake basin (Mackay, 1988a; Romanovskii *et al.*, 2000).

Along the coast of the Beaufort Sea, longshore movement of sediment released by erosion of headlands on thermokarst coasts has locally resulted in the development of a protective fringe of beaches, barrier-spit complexes and sandflats (Ruz *et al.*, 1992; Hill and Solomon, 1999). In new embayments formed by lake drainage, a spit usually develops from erosion of an updrift headland and longshore transport of the eroded sediment (Figure 15.5), and it may grow to span the entrance of the embayment. Barrier islands form either through breaching of spits or through progressive erosion of small islands isolated by coastline retreat. Spits and barrier islands fringing the thermokarst coastlines of the Beaufort Sea are migrating landwards as the coastline itself retreats, the main causes of migration being storm overwash and transfer of nearshore and beach sediments to the back barrier zone. The rate of onshore migration is partly controlled by sediment supply, being greatest when there is a dearth of sediment to feed spit or barrier growth (Héquette and Ruz, 1991). This model, however, is inapplicable to the thermokarst coasts of the Laptev Sea, where the ground ice content of permafrost is greater and there is consequently insufficient sediment eroded from headlands to permit spit and barrier-island formation (Romanovskii *et al.*, 2000).

15.6 Barrier Coasts

Barrier coasts develop where sediment is abundant in the nearshore zone, and are characterized by the presence of beaches, baymouth bars, spits and barrier islands. As outlined above, these features are commonly associated with sediment derived from eroding headlands on thermokarst coasts bordering the Beaufort Sea; nearly 30% of the coastline east of the Mackenzie Delta consists of spits and barrier islands fed by sediment derived from erosion of headlands or islands of ice-rich sediment (Figure 15.6). Spits and barrier islands also occur on outwash coasts, where glacial and periglacial rivers provide an abundant supply of sediment that is reworked by waves, tidal currents and longshore drift, for example in northwest Svalbard and the Gulf of Alaska (Héquette and Ruz, 1990; Forbes, 2011). Barrier coasts are also found

Figure 15.6 Calton Point, a gravel spit extending from the Yukon coast south of Hershel Island, arctic Canada. Similar gravel spits and barrier islands extend westwards along the Beaufort Sea coast to Point Barrow, Alaska. *Source:* Courtesy of Hugues Lantuit.

in high-arctic locations, despite the brief ice-free period, and appear to be relatively resilient; a study of an emergent gravel-dominated coastline in the Canadian Arctic Archipelago by St-Hilaire-Gravel *et al.* (2012) found evidence for recent (1958–2006) net shoreline aggradation, and suggested that the impacts of storm events are short-lived. Within the Canadian Arctic Archipelago, beach-ridge complexes occur almost exclusively on emergent coastlines, where isostatic uplift has exceeded eustatic sea-level rise during the Holocene, though exceptions occur where sediment supply has been abundant (St-Hilaire-Gravel *et al.*, 2015).

Barrier coasts in periglacial environments differ little from those at lower latitudes, but those in high latitudes tend to be dominated by gravel deposits rather than sand, and beach gravels are typically more angular and less well sorted than is the norm elsewhere. A study of gravel beach-ridge systems in the Antarctic (Lindhorst and Schutter, 2014) showed that these are composed of two units: a strand plain formed by net progradation due to swash sediment accretion under comparatively calm conditions, probably during periods of sustained sea-ice cover and limited storminess, and gravel beach ridges built by reworking of strand plain sediments during recurrent extreme storm events. Polar beaches, spits and barrier islands commonly exhibit evidence of ice action in the form of ice-push ridges and ramparts, ice scour marks and melt-out hollows, but such forms tend to be ephemeral unless preserved above the upper limit of wave action. In some arctic locations, ice-pushed sediment nevertheless represents a significant component of beach nourishment (Reimnitz *et al.*, 1990). Areas of coastal sand dunes tend to be limited because of overwashing of spits and barrier islands during storms, though low embryonic dunes sometimes occur on the proximal parts of spits, and dunes may occur on eroding headlands and inland from lagoons (Ruz, 1993).

15.7 Salt Marshes and Tidal Flats

Intertidal *salt marshes* are common in sheltered embayments on subarctic coasts such as those of Labrador and Hudson Bay, but are also found along the Beaufort Sea coastline, in the Canadian Arctic Archipelago and along the Russian Arctic coast from Chukotka to the White Sea. Most are small fringing forms at the heads of large embayments. The general characteristics of tidal marshes are similar to those of lower latitudes, though most support boulders embedded within or resting on the marsh deposits, reflecting either frost heave of boulders to the surface or ice-rafting of coarse debris during periods of high tide.

Tidal flats are low-gradient, low-relief, sparsely vegetated intertidal landforms, built largely or entirely of unconsolidated sediments, that occupy sheltered coastal areas. Study of tidal flats developed under macrotidal conditions at the head of Frobisher Bay (Baffin Island) by Dale *et al.* (2002) has shown that though they are subject to sediment movement by waves and fast-flowing tidal currents during the open-water season, prolonged ice cover and ice action are important determinants of both tidal-flat microrelief and sediment characteristics. Annual freeze-up is marked by incorporation of sediment of all sizes within ice in the intertidal zone, and this sediment is redistributed by movement of ice floes during the break-up period. The tidal-flat sediments therefore tend to be poorly sorted, and sorting trends across the flats are weak or absent. Additionally, transport of boulders across the tidal flat by floating ice has resulted in the formation of boulder ridges, boulder garlands, boulder pavements, boulder mounds and boulder barricades, particularly near high tide levels where ice floes are grounded by onshore winds. By contrast, in the sheltered microtidal environment of Manitounuk Sound on the eastern Hudson Bay coastline, sea-ice action appears to play a relatively minor role in tidal-flat evolution, as there is no extensive intertidal zone of floating ice. This is an area of rapid coastal emergence ($10\,\mathrm{mm\,a^{-1}}$), and the tidal flats represent erosional platforms cut across earlier marine deposits, with a superficial cover of sandy sediments that coarsens away from the shore. Boulders and cobbles are scattered over the surface of the flat, but appear to represent clasts derived from former till deposits rather than ice-rafted or ice-pushed debris (Ruz *et al.*, 1998).

15.8 Rock Coasts

At high latitudes in both hemispheres, extensive areas of coastline consist of rock cliffs and headlands that plunge below sea level so that no intertidal landforms are developed. Some periglacial coasts, however, are fringed by horizontal or gently sloping intertidal *shore platforms* cut across bedrock (Figure 15.7). These can

Figure 15.7 Intertidal rock platform, northwest Hornsund, Svalbard. *Source:* Courtesy of Zuzanna Swirad.

attain impressive dimensions: those on the southern shore of the Byers Peninsula in the South Shetland Islands average 200–300 m in width and extend continuously alongshore for up to 8 km (Hansom, 1983b), and platforms 700–800 m wide fringe both banks of the upper St Lawrence Estuary (Dionne and Brodeur, 1988a). Rock-cut shore platforms have a global distribution, but those in periglacial environments have characteristics that suggest that sea ice and frost weathering have played an important role in their evolution.

The concept of distinctive periglacial shore platforms formed by ice action was developed by Hansom (1983b) with reference to platforms on the South Shetland Islands. Unlike the sloping ramp-like rock platforms produced by storm wave activity, those on the sheltered southern coast of the South Shetlands are almost horizontal, with an abrupt drop in level at the low-tide mark and an equally abrupt steepening at the high-tide mark. They extend across a range of lithologies and are widest in sheltered areas of restricted fetch, where shorefast and floating ice persists longest. Hansom argued that ice in the intertidal zone has played an important role in recent platform evolution through both scouring and abrasion of the platform surface by mobile, debris-charged ice floes and freezing of rock fragments to the base of fast ice, which subsequently floats and moves away, dragging the attached debris with it. Similar processes have been invoked to explain the development or modification of near-horizontal intertidal rock platforms cut across slates and shales in the upper St Lawrence Estuary (Dionne and Brodeur, 1988a, 1988b), and evidence for the detachment of rock slabs by sea ice on the coast of Ungava Bay has been outlined by Fournier and Allard (1992). Dionne and Brodeur (1988a, p. 117) described the role of ice on the shore platforms of the St Lawrence Estuary as 'abrasion, dislodgement and removing, pushing, scraping and levelling, plucking, scouring, and transport of debris', leaving little doubt as to its efficacy as an erosional agent.

Though the erosional and transport roles of sea ice in the modification and development of rock-cut shore platforms are widely acknowledged, the contribution of frost weathering to shore platform development is more contentious (Trenhaile, 1983), particularly in the light of uncertainties regarding the efficacy of frost weathering in breaking up intact, low-porosity bedrock (Chapter 10). Inferences regarding the role of frost weathering in shore platform development have been based mainly on platform morphology and the nature of detached debris, rather than on process measurements. Trenhaile and Rudakas (1981) concluded that weathering by frost may play an important role in the breakdown of cliffs of shale and argillite at the rear of intertidal platforms on the Gaspé Peninsula, Quebec, but they found no evidence for frost weathering on the platform itself. Similarly,

Hansom (1983b) observed large volumes of 'frost-shattered scree' at the rear of shore platforms on the South Shetland Islands, suggesting recent cliff retreat due to frost weathering. On the sedimentary rocks of the St Lawrence Estuary, there is evidence both for lateral extension of platforms through frost weathering of backing cliffs and removal of weathered debris by sea ice, and for detachment by frost wedging of flakes and slabs of rock from the intertidal platform itself (Dionne and Brodeur, 1988a, 1988b; Dionne, 1993). Fournier and Allard (1992) have also described evidence for frost wedging operating along joints in gneissic rocks on the coast of Ungava Bay. Frost weathering of intertidal rock platforms may even occur in temperate latitudes: during the unusually severe winter of 2008–09, detachment of chalk fragments amounting to average rock platform lowering of 0.8 ± 0.5 mm occurred in response to 25 freeze–thaw cycles involving temperatures below $-2.5\,°C$ in northwest France (Dewez et al., 2015). Collectively, the above evidence suggests that frost weathering probably contributes to platform extension through both cliff retreat and release of debris from the platform surface, at least on fissile sedimentary rocks; at issue is the contribution of frost weathering to shore platform development relative to that of storm waves, tidal currents and erosion by mobile shore ice. The answer probably depends, at least in part, on the susceptibility of the local bedrock to frost weathering.

15.9 Raised and Inherited Shorelines

A further complication in interpreting shore platform development on ice-affected coasts is the possibility that some intertidal rock platforms are ancient, inherited landforms that are being modified, rather than created, by present-day processes. High-latitude coastlines that were covered by ice sheets during the last glacial maximum (26.5–19.0 ka) have experienced glacio-isostatic uplift due to removal of overlying glacier ice. In locations where glacio-isostatic uplift exceeded eustatic sea-level rise caused by melting of the last Pleistocene ice sheets, the overall effect has been one of marine regression. This process continues today near the centres of former ice sheets, for example in Hudson Bay and around the Gulf of Bothnia. Shorelines were formed at times when the rates of sea-level rise and isostatic uplift were locally in balance, and in areas where subsequent isostatic uplift outstripped eustatic sea-level rise they have been elevated above present sea level. Such emergent shorelines commonly take the form of raised beaches, tidal flats and deltas, but raised rock-cut platforms are sometimes also present (Figure 15.8). Some authors have attributed such raised shore platforms to postglacial formation through a combination of frost weathering of bedrock and rapid

Figure 15.8 Raised rock platform on the west coast of Scotland. Formation of this platform has been attributed to a combination of frost weathering of bedrock and removal of debris by storm waves or sea ice during the Younger Dryas Stade (~12.9–11.7 ka), when permafrost extended to sea level. *Source:* Courtesy of Murray Gray.

removal of frost-weathered debris by storm waves or sea ice (e.g. Dawson, 1988), but this interpretation requires very rapid erosion and evacuation of resistant bedrock during relatively brief periods of relative sea-level stability. An alternative interpretation is that both raised and present-day intertidal shore platforms are pre-Holocene landforms, possibly developed over several glacial–interglacial cycles, that have escaped elimination by glacial erosion during cold stages and have since been modified by postglacial intertidal processes (e.g. Guilcher *et al.*, 1986; Dawson *et al.*, 2013).

The term *strandflat* was originally applied to the uneven, partly submerged rock platform that extends seawards from the coastal mountains of Norway, but has now achieved wider currency in describing similar coastal features elsewhere. The Norwegian explorer Fridtjof Nansen (1922) proposed that the strandflat of high-latitude coasts represents prolonged frost weathering of sea cliffs under periglacial conditions and removal of resultant debris by wave action and sea ice, an interpretation that still commands some support (Holtedahl, 1998). This view, however, implies that the strandflat developed when former sea level was fairly similar to that of the present, and survived recurrent periods of Pleistocene glacial erosion. A paper addressing the 'strandflat problem' in various parts of the world (Guilcher *et al.* 1986) revealed limited agreement amongst researchers as to the age and origin of these coastal landforms, apart from consensus that strandflats have not developed *ab initio* since Late Pleistocene

deglaciation. Guilcher *et al.* (1986) proposed that strandflats and other high- or mid-latitude shore platforms represent inherited landforms of much greater antiquity that have been modified by wave action, ice action and, in some cases, glacial erosion and frost weathering. Similarly, Dawson *et al.* (2013) have speculated that the strandflat of western Scotland essentially formed during the Pliocene. The 'strandflat problem' remains unresolved, but it is unlikely that all strandflats or periglacial shore platforms represent the outcome of a common evolutionary history.

15.10 Lake Shorelines

Many of the features that characterize periglacial sea coasts are also found along the shores of lakes in cold environments, the main differences being that tides are absent and that storm wave fetch is limited not only by ice cover but also by lake extent. The absence of tides means that there is no lacustrine equivalent to tidal flats or tidal marshes, that the zone affected by wave action and bottomfast ice is much narrower and that development of spits and barrier islands is limited. There is nevertheless great variety in the nature of lake shorelines in periglacial environments. In Svalbard, Åkerman (1992) identified nine shoreline types, namely rock shores, block (boulder-dominated) shores, bog shores, patterned ground shores (characterized by the development of sorted nets, circles or polygons), ridge

Figure 15.9 Evidence for frost weathering of gneissic bedrock and formation of an incipient shore platform, Böverbrevatnet, Jotunheimen, Norway. Partial drainage of the lake has exposed platform fragments above present lake level. *Source:* Courtesy of Rick Shakesby.

shores (dominated by ice-push ridges), thermokarst shores, delta shores, talus shores and terrace shores (indicating fluctuating lake levels). To this list should be added beach or barrier shores and rock-platform shores.

During winter freezeback, an ice foot develops on lakeshores, protecting beach sediments from wave action. This often incorporates sediment deposited by wave spray or frozen on to the base of the ice. Lake ice can also move on to the shore, either as wind-driven ice floes or due to thermal expansion of an intact ice cover in response to an increase in air temperature (Åkerman, 1992; Sasaki, 1992). As on sea coasts, onshore movement of ice results in the development of small-scale ice-push landforms (boulder ramparts or ridges, boulder pavements and furrows excavated in beach sediment by ice blocks or ice-pushed boulders), as well as ice-rafted features such as perched boulders and, more rarely, lacustrine boulder barricades (Gilbert, 1990), though wave action driven by onshore winds during the open water season tends to destroy or modify all but the more robust of such features. In areas of ice-rich permafrozen sediments, lacustrine shore erosion occurs through development of thermo-erosional niches (and associated block collapse) and retrogressive thaw slumping, both of which are particularly effective in extending the dimensions of thermokarst lakes (Chapter 8), though otherwise lake shoreline retreat tends to be much slower than on the sea coast, as wave energy is often limited by restricted fetch.

There is also evidence that rock-cut platforms can develop at the margins of lakes in cold environments. Aarseth and Fossen (2004) have described a lacustrine rock platform up to 20 m wide that is cut across metamorphic rocks around the shores of a lake in western

Norway, and argued that this developed during the last ~6000 years through a combination of frost wedging acting on fractured bedrock at the lake margin and removal of debris by drifting ice. Matthews *et al.* (1986a) have documented the case of a platform up to 5.3 m wide formed over 73–125 years along the shores of Böverbrevatnet, a small high-altitude lake in Norway (Figure 15.9). This platform and the associated bedrock notch are about 2 m above present lake level, and apparently developed when the lake was dammed by glacier ice, so that lake level was higher than now. Shakesby and Matthews (1987) also found that a rock platform averaging 3.6 m in width has apparently formed at shallow depth since Böverbrevatnet drained to its present level, implying a post-lowering shore erosion rate of 9–36 mm a^{-1}. The development of rock platforms around Böverbrevatnet was attributed by Matthews *et al.* (1986a) to deep penetration of the annual freezing cycle, movement of lake water towards the freezing plane and growth of ice lenses in cracks and fissures at the lake–bedrock interface (Figure 15.10), with removal of slabs of detached rock through freezing to bottomfast ice and subsequent ice rafting. It is not clear, however, whether this processes is widespread along the shores of other lakes in areas of deep seasonal freezing or permafrost, or whether it is limited by bedrock characteristics such as joint density.

15.11 Synthesis

Although the distinguishing characteristic of periglacial coasts is persistence of ice at the shore for much of the year, the geomorphology of many periglacial coasts

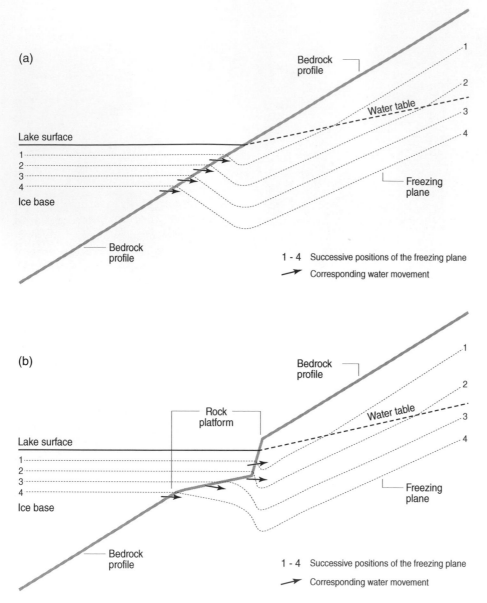

Figure 15.10 Model of lacustrine rock-platform development due to frost weathering. (a) Lake-ice thickening, frost penetration into bedrock and movement of unfrozen water towards the freezing plane. (b) The same processes superimposed on a developing rock platform. *Source:* Matthews *et al.* (1986a). Reproduced with permission of Wiley.

differs only in detail from that of their mid-latitude counterparts. Despite the limited length of the open water season, storm waves (and, on mesotidal and macrotidal sea coasts, tidal currents) are the main determinants of coastal form. The most distinctive and vulnerable periglacial coasts are those developed in ice-rich permafrozen sediments. In such areas, rates of shoreline retreat often exceed $1\,\mathrm{m\,a^{-1}}$ as a result of undercutting, block collapse and retrogressive thaw slumping. Along the circum-arctic coastline, there is evidence for recent warming of permafrost and marked diminution of summer sea-ice cover, both of which are extremely likely to result in accelerated erosion of ice-rich permafrost coasts. Moreover, future sea-level rise resulting from warming of the global oceans and melt of glacier ice will inevitably increase the rate of erosion of ice-rich permafrost coastlines of the Arctic Ocean, a scenario that is considered further in Chapter 17.

16

Past Periglacial Environments

The present is the key to the past
Archibald Geikie, 1835–1924

16.1 Introduction

A major focus of research in periglacial geomorphology is the use of relict periglacial features to reconstruct past environmental conditions, particularly those that pertained in mid-latitudes during cold stages of the Quaternary Period. As outlined in Chapter 1, the Quaternary comprises two geological epochs, the Pleistocene (~2.58 Ma to ~11.7 ka) and the Holocene (~11.7 ka to the present). Some authors have argued for the addition of a third epoch, the Anthropocene, which acknowledges the role of human activity in influencing recent environmental change and some geological processes (Steffen *et al.*, 2011; Zalasiewicz *et al.*, 2011). There is limited agreement, however, as to where the Holocene–Anthropocene boundary occurs, and many scientists question the validity of the concept in stratigraphic terms and consider that it should only be employed informally (Ruddiman, 2013; Gibbard and Walker, 2014). The Pleistocene epoch was marked by the alternation of cold *glacial stages* and warmer *interglacial stages*; shorter intervals of cold conditions are referred to as *stades* or *stadials*, and shorter intervals of warmer climate as *interstades* or *interstadials*. The Holocene effectively represents the present interglacial, and in the absence of human activity would inevitably be succeeded at some future time by another glacial stage.

Globally, the alternation of glacial and interglacial periods is recorded by the oxygen isotope ($^{18}O/^{16}O$ or $\delta^{18}O$) record in marine microfossils recovered from deep-ocean cores. The oxygen isotope signal is largely controlled by fluctuations in the global volume of land ice, and hence represents a continuous record of the alternation of cold glacial stages (large global ice volume) and relatively warm interglacial stages (restricted global ice volume, as at present). A total of 103 marine isotope stages have been identified within the 2.58 Ma of the Quaternary, and 22 within the past ~880 ka (Lisiecki and Raymo, 2005), the latter representing 11 glacial stages and a similar number of interglacials (or near-interglacials). By convention, each stage is numbered from the most recent, with glacial stages being allocated even numbers and interglacials being given odd numbers so that the present (Holocene) interglacial is marine isotope stage 1 (MIS 1), the Late Wisconsinan/Weichselian glacial is MIS 2 and so on. Since the beginning of the Middle Pleistocene at ~0.9 Ma (MIS 22), glacial stages have been much more prolonged (~10^5 years) than interglacials (~10^4 years), and maximum global ice volumes have been reached near the end of each glacial stage (Figure 16.1). With the exception of loess-palaeosol sequences (Chapter 14), the terrestrial record of Quaternary glacial–interglacial fluctuations tends to be much more fragmented than the marine record, not least because the last ice sheets were particularly extensive, covering roughly 30% of the present land area of the globe and eroding or burying sediments relating to earlier glacial or interglacial stages. The period of maximum glacier ice volume during MIS 2 is referred to as the *last glacial maximum* (LGM), and occurred within the period ~26.5–19.0 ka (Clark *et al.*, 2009), though the timing of the maximum extent of different ice sheets (or even different sectors of the same ice sheet) was asynchronous (Hughes *et al.*, 2013). Vandenberghe *et al.* (2014) have introduced the term *last permafrost maximum* (LPM) to describe the period of greatest permafrost extent during the last (Late Weichselian/Wisconsinan) glacial stage, though this is also very unlikely to have been globally synchronous and does not necessarily coincide with the LGM.

Reconstruction of past environments and environmental changes during the Quaternary Period is based on *proxy evidence* of former conditions, and scientists involved in reconstruction of Quaternary environments have developed a formidable arsenal of techniques that permit retrodiction of former climates, vegetation cover and habitats, together with a range of dating methods that allow for increasing accuracy and precision in establishing the timing of environmental changes. Lowe and Walker (2015) have identified three different classes of proxy evidence used to reconstruct former environmental conditions: (i) geomorphological evidence, such as

Periglacial Geomorphology, First Edition. Colin K. Ballantyne.
© 2018 John Wiley & Sons Ltd. Published 2018 by John Wiley & Sons Ltd.

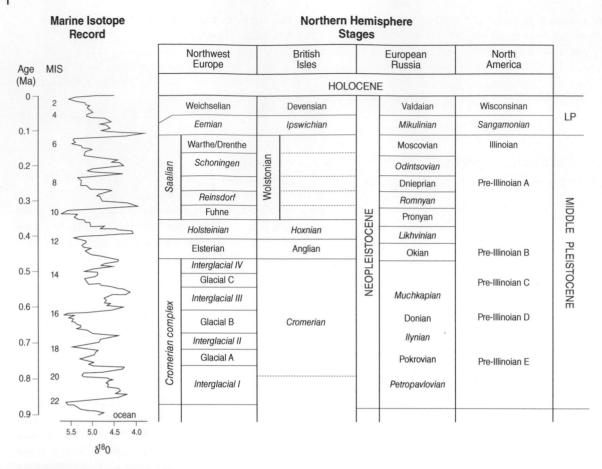

Figure 16.1 The marine oxygen isotope (marine isotope stage, MIS) record for the last 0.9 Ma, based on a composite of deep-ocean cores and the equivalent terrestrial Middle and Late Quaternary stratigraphy of the northern hemisphere. LP, Late Pleistocene. The even numbers in the MIS record represent glacial stages, and interglacial stages are shown in italics. *Source:* Adapted from Lowe and Walker (2015) with permission from Routledge.

glacial landforms that indicate the former extent of glaciers and ice sheets; (ii) lithological and stratigraphic evidence, such as that provided by glacial deposits, palaeosols, lake and bog deposits, deep-sea sediment accumulations and deep ice cores extracted from the Greenland and Antarctic Ice Sheets; and (iii) a wide range of biological evidence, based on the palaeoenvironmental implications of, for example, assemblages of subfossil pollen grains, diatoms, plant macrofossils, insect remains, molluscs, foraminifera, marine microfossils and terrestrial vertebrate remains.

Relict periglacial features contribute to both the geomorphological evidence for former cold conditions (e.g. relict frost polygons, lithalsa scars and relict rock glaciers) and the lithological or stratigraphic evidence for periglacial palaeoenvironments (e.g. gelifluctate, coversand and loess deposits, and soil structures such as ice-wedge pseudomorphs and cryoturbations). A wide range of dating techniques are employed to establish the age of Quaternary landforms, sediments and biological material within Quaternary stratigraphic successions (Walker, 2005; Lowe and Walker, 2015), but only three

have been routinely employed to establish the absolute age of Pleistocene periglacial features: radiocarbon dating of organic material, luminescence dating of sediments and cosmogenic isotope dating, primarily employed to date the exposure ages of boulders and rock surfaces. Stratigraphic methods have, however, been widely used to establish the relative age (chronological order) of periglacial sediments and structures in vertical exposures, as outlined below.

16.2 Palaeoenvironmental Reconstruction Based on Periglacial Features

16.2.1 Preservation Potential

Throughout Chapters 6 to 15, an attempt has been made to identify the key characteristics of periglacial landforms, sediments and soil structures that survive in relict form in mid-latitude regions, and to assess the potential and problems associated with the employment of such features in reconstructing past periglacial environments. The range of periglacial features that have been employed in palaeoenvironmental reconstructions is limited by their long-term *preservation potential*, which represents the likelihood of their surviving climate warming, permafrost degradation, erosion by slope, fluvial and littoral processes and burial by peat growth or sediment accumulation. Some periglacial features, such as ephemeral frost mounds and icings, usually leave little or no morphological signature. Landforms such as collapsed palsas or peat plateaus, small sorted patterns and most forms of nonsorted patterned ground have limited preservation potential over millennial timescales, and hence have rarely been employed in Pleistocene palaeoenvironmental reconstructions. The same is true of the small-scale thermokarst landforms described in Chapter 8 (thermokarst pits, troughs, gullies, mounds and sinkholes), which are likely to have disappeared or experienced radical modification as the ice-rich permafrost in which they developed thawed, and so have left little diagnostic trace of their former existence or status. Distinctive littoral or riparian landforms, such as thermo-erosional niches, thermokarst coasts and beaded streams, are also likely to have disappeared as a consequence of ground-ice thaw and erosion by waves or rivers. In general, only the most robust relict periglacial landforms, such as thermal-contraction polygons, pingo and lithalsa scars, alases, blockfields, snow avalanche landforms and rock glaciers, have survived the transition from cold to temperate climates and have resisted burial or erosion during the ~11 700 years that separate the end of the Pleistocene epoch from the present.

Conversely, many periglacial sediments, such as gelifluctates, periglacial fluvial deposits, coversands and loess, have persisted in almost pristine form (though altered by pedogenesis or decalcification) across broad swathes of mid-latitude environments. Note that the terms 'periglacial sediments' and 'periglacial deposits' are here used to describe sediments laid down under periglacial climatic conditions, even if the depositional processes are identical to those operating in non-periglacial environments (Vandenberghe, 2011). Many periglacial soil structures, such as ice-wedge pseudomorphs, sand wedges, thermokarst involutions, cryoturbations, fragipans and small-scale soil structures produced by freezing and thawing of the ground have also proved resilient over millennial timescales. Because periglacial sediments and soil structures often occur in aggradational sediment sequences, they locally provide evidence not only for environmental conditions during the last glacial stage, but also for changing environmental conditions earlier during the Pleistocene epoch. The classic examples of this are the loess-palaeosol record of the Chinese Loess Plateau, which spans the entire Pleistocene (Sun and Zhu, 2010; Porter, 2013; Zhang *et al.*, 2016), and

European loess deposits that began to accumulate at roughly 0.8–1.0 Ma (Markovic *et al.*, 2011; Fitzsimmons *et al.*, 2012; Újvári *et al.*, 2014; Chapter 14). Some aggradational fluvial and aeolian sediment sequences contain several generations of periglacial sediments and soil structures, sometimes separated by palaeosols, that provide evidence for changing environmental conditions during sediment accumulation (e.g. Rose *et al.*, 2000; Antoine *et al.*, 2003a, 2009, 2016; Murton *et al.*, 2003; Renssen and Vandenberghe, 2003). Other sedimentary sequences provide snapshots of changing conditions during particular intervals as far back as the Middle Pleistocene (e.g. Rose *et al.*, 1985; Murton *et al.*, 1995, 2015b), or even the Early Pleistocene (e.g. Kasse, 1993). Inevitably, however, the use of the relict periglacial phenomena in palaeoenvironmental reconstruction has focused on the last glacial stage (MIS 5d–2), for which evidence is most abundant, and in particular on the period of the Late Weichselian/Wisconsinan glacial substage (MIS 2), which extended from ~31 ka to the onset of rapid warming at ~11.7 ka. Greenland ice-core records, however, show that the last glacial stage, far from being a period of continuously cold climate, was characterized by stadial–interstadial oscillations separated by periods of rapid warming or cooling (Blockley *et al.*, 2012; Rasmussen *et al.*, 2014; Figure 16.2). Many relict periglacial phenomena, particularly in northwest Europe, are attributable to the Younger Dryas Stade (~12.9–11.7 ka), the most recent period of permafrost development in this region.

Collectively, relict periglacial landforms, sediments and soil structures potentially constitute a rich archive of evidence for reconstructing Pleistocene palaeoenvironments, particularly when employed in conjunction with other forms of geomorphological, lithological, biological and stratigraphic evidence. This is particularly true in areas that remained outside the maximum reach of the last (Late Wisconsinan/Weichselian; MIS 2) ice sheets, where the landform and sedimentary record has not been truncated by glacial erosion or modification. To unlock this archive successfully, however, we need to be aware of potential problems relating to the identification, dating and interpretation of relict periglacial phenomena.

16.2.2 Identification of Relict Periglacial Features

Though identification of many relict periglacial features is straightforward, ambiguity may arise in a number of key situations. Frost-wedge pseudomorphs, for example, are the most important structures for the identification of former permafrost in mid-latitude areas, but even where similar structures of non-permafrost origin can be eliminated, it may still be difficult to differentiate ice-wedge pseudomorphs from relict sand wedges,

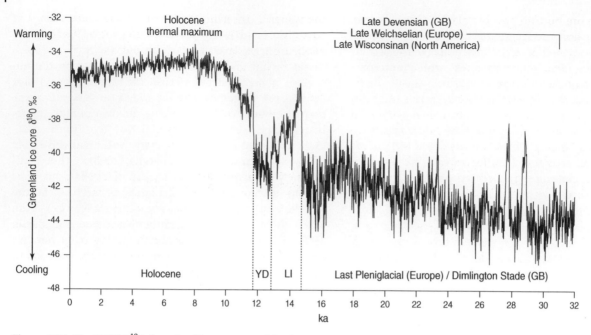

Figure 16.2 The NGRIP δ^{18}O Greenland ice core record for the past 32 ka, showing the major subdivisions of the last (Late Devensian/Weichselian/Wisconsinan) glacial substage. YD, Younger Dryas Stade; LI, Lateglacial Interstade (Bølling–Allerød Interstade in Europe); GB, Great Britain.

relict composite wedges or even seasonal soil wedges (Kolstrup, 2004; Ghysels and Heyse, 2006; Ribolini *et al.*, 2014; Chapter 6). Similarly, ramparted ground-ice depressions may represent either pingo scars or collapsed lithalsas; both indicate former permafrost, but their palaeotemperature implications differ (e.g. Pissart, 2000, 2003; Ross *et al.*, 2011; Chapter 7). Many soil structures formerly interpreted as cryoturbations formed by freezing and thawing of a former active layer should probably be reinterpreted as thermokarst involutions formed during permafrost thaw (Murton and French, 1993a; Murton *et al.*, 1995), and therefore provide no indication of former active-layer depth. Even large-scale landforms are prone to misidentification. Harrison *et al.* (2008), for example, collated evidence for putative 'relict rock glaciers' in the British Isles, most if not all of which have subsequently been reinterpreted, particularly as rockslide runout deposits (Ballantyne *et al.*, 2009; Jarman *et al.*, 2013). It is likely that some supposed 'relict rock glaciers' in other mid-latitude mountain areas have also been misinterpreted, such as those on the Faroe Islands (Humlum, 1998b) and in Ottadalen, Norway (Matthews *et al.*, 2013), which in both cases are probably rockslide runout deposits. The same is true of numerous supposed relict pronival (protalus) ramparts, including some originally identified as such by the author (Ballantyne and Kirkbride, 1986), which have subsequently been reinterpreted as arcuate rockslide deposits (e.g. Ballantyne and Stone, 2009).

As these cautionary examples illustrate, correct identification of relict periglacial phenomena requires not only forensic examination of their characteristics, setting and relationship with other landforms, sediments and structures, but also consideration of all possible alternative origins. If such alternatives cannot be eliminated, extreme caution must be applied to interpretation of the palaeoenvironmental significance of such features.

16.2.3 Relative Age Dating: Stratigraphy and Morphostratigraphy

Relict periglacial features are of limited use in palaeoenvironmental reconstruction if their age is unknown. There have been two complementary approaches to determining the ages of relict landforms, sediments and soil structures: *relative age dating*, where particular features are placed in an age sequence relative to that of underlying, overlying or host landforms or sediments; and *absolute age dating*, such as radiocarbon dating, which enables the approximate 'real' age of a feature to be determined. Many studies have employed a combination of the two.

Where periglacial sediments such as gelifluctate, coversands, loess and fluvial deposits occur in stratigraphic successions, or sediments of periglacial or non-periglacial origin contain periglacial structures such as ice-wedge pseudomorphs or cryoturbations, the principles of stratigraphic analysis can be applied to determine the relative ages of sediments or contained structures. There are four basic principles. The *principle of superposition* states that in any undisturbed depositional sequence, the sediment layers young progressively upwards, with the oldest at the bottom and the youngest at the top. This principle holds even if there is a break in deposition and/or partial erosion of strata before the overlying sediments are laid down. Such a break is termed an *unconformity* and represents a gap in the upward-younging sequence. The *principle of original horizontality* assumes that sediments were originally deposited in horizontal layers, for example on a lakefloor or floodplain. This principle holds for most alluvial and aeolian deposits and for lakefloor deposits settling from suspension, but is vitiated by prograding sediment sequences such as the slip faces on dunes, prograding lacustrine deposits and delta foreset beds, and by all slope deposits, such as talus, stratified slope deposits and gelifluctate. The *principle of included fragments* holds that particles included within undisturbed sediment must be derived from rocks, soils or sediments that pre-date the sediment. This principle is important in the case of organic inclusions such as peat blocks or plant remains, which are potentially datable using radiocarbon techniques, and thus provide an upper limiting age for the sediment body in which they occur. Finally, the *principle of cross-cutting relationships* states that strata or structures that interrupt the orderly sequence of sediments are younger than those sediments. This is relevant in the case of relict cryostructures such as involutions or epigenetic ice-wedge pseudomorphs, but does not apply in the case of syngenetic structures that are of the same age as the host sediments, such as syngenetic sand wedges.

Figure 16.3 illustrates the combination of stratigraphic and absolute age data to reconstruct a sequence of events. The lithostratigraphy comprises glacial till overlain by glaciofluvial sands and gravels, coversand,

then loess. Interdigitation of gelifluctate with the coversand and loess shows that coversand and loess deposition was accompanied by solifluction on slopes. Radiocarbon ages of 42.3 ka and 32.0 ka bracket the timing of sand and gravel deposition, and the latter provides a maximum age for the onset of gelifluctate deposition. Sand wedges in the glaciofluvial deposits indicate that permafrost later developed in these sediments, and imply thermal contraction cracking under very cold winter temperatures. The sand wedge infill, dated at one location to 23.1 ka, also suggests arid, windswept conditions. Subsequent permafrost degradation is indicated by the formation of large thermokarst involutions. Luminescence ages of 18.1 ka and 15.7 ka bracket the ensuing period of coversand deposition, but as the coversands contain only small cryoturbations that may reflect only seasonal ground freezing, there is no evidence for re-formation of permafrost during this period. Partial erosion of the coversand deposit is indicated by an unconformity, and a luminescence age implies that loess accumulation began shortly before 12.8 ka, but terminated before 11.4 ka, as determined by a radiocarbon age from the base of the overlying peat. Return of permafrost during this period is demonstrated by the presence of large ice-wedge pseudomorphs in the loess and the formation of a ramparted depression that represents the site of a former pingo or lithalsa.

Less frequently acknowledged is the role of *morphostratigraphy* in palaeoenvironmental reconstructions. Morphostratigraphy employs landforms as stratigraphic units. At its simplest, where one particular landform overlies another, as for example where a rock glacier or

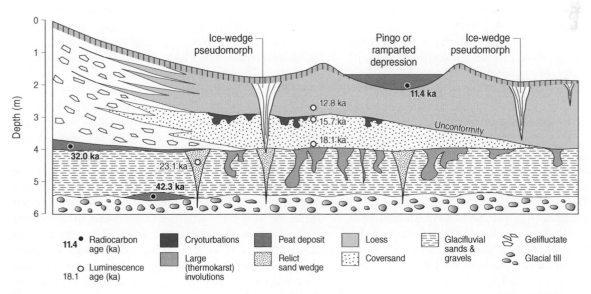

Figure 16.3 Schematic reconstruction of the ages and relative ages of periglacial landforms, sediments and structures. For details, see text.

debris cone has partly buried an underlying moraine, the upper landform must post-date the lower. The same principle applies where a landform overlies a sedimentary unit; for example, in Figure 16.3, the ramparted depression is developed in the loess deposit, and hence must post-date the deposition of the underlying loess. Morphostratigraphy is also important when applied to the distribution of landforms. For example, relict frost polygons or lithalsa scars that occur inside the limits of the last ice sheets represent landforms that must have developed since ice-margin retreat, whereas those outside ice-sheet limits may have formed earlier.

16.2.4 Absolute Age Dating

Whereas relative age dating usually provides only broad constraints on the ages of relict periglacial phenomena, absolute dating provides numerical age estimates of the time of formation, or limiting age estimates of the time of formation (after and/or before a certain date). The main dating methods hitherto employed in palaeoenvironmental reconstructions based on periglacial features are radiocarbon dating of contained, overlying or underlying organic material, luminescence dating of sediments (particularly aeolian and fluvial deposits) and cosmogenic isotope dating of the exposure age of boulder and bedrock surfaces. The principles, potential and problems associated with these techniques were introduced in Chapter 1 and are considered in detail in Lowe and Walker (2015). Additionally, *tephrostratigraphy* has been employed in a number of periglacial contexts, notably to date loess-palaeosol sequences (e.g. Jensen *et al.*, 2016). This approach involves identification of thin layers of volcanic ash shards within deposits. Tephra layers relating to particular eruptions can be identified from their distinctive chemical composition, so if the age of a particular tephra layer can be determined at one location (e.g. by radiocarbon dating of underlying or overlying peat), then the same age can be assumed for the same tephra layer at all other sites where it is present.

Other techniques employed to estimate the age of bouldery periglacial landforms include *lichenometry* and assessment of the degree of boulder surface weathering. Lichenometry involves the establishment of a local growth rate for a particular lichen species (usually the slow-growing lichen *Rhizocarpon geographicum*), then measurement of the largest lichens present on boulder surfaces to estimate the age of boulder emplacement. This technique has been employed to establish the spatial and temporal patterns of recent rockfall on to talus surfaces (McCarroll *et al.*, 1998, 2001; Sass, 2010) and the ages of debris-flow deposits (Innes, 1983), but establishment of a reliable growth rate is difficult, and in most environments die-back of lichens limits the use of this technique to the past 500 years or less. The degree of surface weathering is sometimes assessed by *Schmidt hammer exposure dating*. The Schmidt hammer measures the percentage rebound of a controlled blow of a spring-propelled mass on a steel rod held against a boulder surface. This technique assumes that rebound response declines through time as boulder surfaces become progressively weathered, and large numbers of measurements are required to characterize the mean rebound response of boulders deposited at the same time. Though it is useful for differentiating landforms of very different ages, its use as an absolute dating tool is questionable, though it has been employed to estimate the approximate exposure ages of boulders on supposed relict rock glaciers (Matthews *et al.*, 2013) and pronival ramparts (Matthews and Wilson, 2015). As boulder-surface weathering rates tend to attenuate through time, this approach is probably effective only for Lateglacial and Holocene bouldery deposits.

16.2.5 Palaeoenvironmental Interpretation

Because some active periglacial landforms occupy a particular environmental niche that is controlled by such factors as the presence of permafrost, annual temperature regime, precipitation, vegetation cover and wind velocity, it should theoretically be possible to identify the climatic boundary conditions that control their present distribution and to use the past distribution of equivalent relict features to retrodict environmental conditions at the time of their formation. Moreover, as the climatic constraints on the development of particular periglacial features differ, recognition of coeval assemblages of relict forms may provide us with insights into the nature of the climate at the time of their formation. In practice, matters are more complex, for five main reasons. First, the present-day climatic controls on some periglacial phenomena are not well established. Second, many periglacial features produced by ground freezing (or freezing and thawing) reflect the ground-surface thermal regime, which differs from the air-temperature regime by a surface temperature offset that varies seasonally depending on such factors as vegetation cover, aspect, and depth and persistence of winter snowcover. Third, the present distribution of some periglacial features is partly determined by soil texture, hydrology or topography, not climate, so some landforms (such as pingos or rock glaciers) are absent from areas that are climatically suitable for their formation. Fourth, some supposedly 'active' periglacial phenomena may have developed under a colder climate than exists now, so that use of recent climate statistics to constrain their climatic implications yields inaccurate results. Finally, present high-latitude environments differ from Pleistocene mid-latitude periglacial environments, as the latter received much higher insolation and were subject to rapid, drastic climate switches (stadial–interstadial oscillations) that have no counterpart within the past few millennia.

These caveats are usefully illustrated with reference to the interpretation of ice-wedge pseudomorphs, which conclusively demonstrate the former presence of permafrost, and in most cases continuous permafrost. Traditionally ice-wedge pseudomorphs were interpreted as indicating former mean annual air temperatures (MAATs) lower than −6°C to −8°C, but there are accounts of wedge cracking or growth under MAATs as high as −3.5°C to −4.0°C (Hamilton *et al.*, 1983; Burn, 1990). Because the critical tensile stress leading to permafrost cracking is a function of several variables, including surface-temperature offset, ground-temperature gradient, sediment texture and ice content, as well as the rate of winter cooling (Chapter 6), ice-wedge pseudomorphs offer at best a 'warm-side' (maximal) proxy estimate of 'average' climatic conditions, which in any case are likely to have changed markedly over millennial timescales. In view of these caveats, numerous authors have cautioned against

quantitative association of frost-wedge pseudomorphs with climatic averages (Harry and Gozdzik, 1988; Burn, 1990; Murton and Kolstrup, 2003; Plug and Werner, 2008; Murton, 2013a). It is likely that most ice-wedge pseudomorphs represent ice wedges that formed under MAATs lower than about −4°C in fine sediments and lower that about −6 to −8°C in coarse sediments, but these are no more than approximate upper estimates.

Tables 16.1, 16.2 and 16.3 outline the palaeoenvironmental implications of relict periglacial landforms, sediments and soil structures. These are based on studies of both active features and their relict counterparts, details of which are given in Chapters 6–15, and partly on palaeotemperature thresholds suggested by Huijzer and Vandenberghe (1998); they should be regarded as indicative rather than definitive in view of the caveats outlined above. For palaeoenvironmental reconstructions, the most useful relict periglacial features

Table 16.1 Palaeoenvironmental implications of thermal contraction crack features, relict ground ice landforms and structures, and patterned ground. Evidence for continuous permafrost (CFP) usually implies MAAT less than −5°C to −8°C. Evidence for discontinuous or sporadic permafrost (DPF) usually implies MAAT in the range −1°C to −6°C.

Relict feature	Palaeoenvironmental implications	Chapter
Frost polygons and ice-wedge pseudomorphs	Usually CPF; DPF in fine sediments	6
	MAAT ≤ −4°C in fine sediments	
	MAAT ≤ −8°C to ≤ −6°C in coarse sediments	
	Rapid winter cooling	
	Temperature of coldest month ≤ −20°C	
Large relict sand wedges and composite wedge pseudomorphs	As above Aeolian sand infill implies unvegetated source and strong winds May indicate aridity	6
Seasonal frost cracks and soil wedges	Deep seasonal ground freezing	6
	MAAT ≤ 0°C	
	Rapid winter cooling	
Pingo scars	Hydrostatic pingos: CPF and site of former lake	7
	MAAT ≤ −6°C to ≤ −4°C	
	Hydraulic pingos: CPF or DPF	
	MAAT ≤ −3°C	
Palsas	DPF or sporadic permafrost	7
	MAAT ≤ −1°C	
Lithalsas	Extensive DPF	7
	MAAT typically −4 to −6°C	
Thermokarst lake basins	Deep basins (alases): ice-rich syngenetic CPF	8
	Shallow basins: ice-rich epigenetic permafrost	
Thermokarst valleys	Ice-rich syngenetic CPF (yedoma)	8
Thermokarst lake sediments	Ice-rich permafrost, probably CPF	8
Retrogressive thaw slumps	Ice-rich CPF, possibly DPF	8

(Continued)

Table 16.1 (Continued)

Relict feature	Palaeoenvironmental implications	Chapter
Thermokarst involutions	Thaw of ice-rich permafrost	8
	MAAT $\geq -4\,°C$	
Upfrozen clasts	Seasonally frozen ground or former active layer	9
Large sorted patterned ground	Probably former active layer above permafrost	9
	May indicate former active-layer depth	
Small sorted patterned ground	Seasonally frozen ground or short-term ground freezing and thawing	9
Large cryoturbations	Former active layer above permafrost	9
	May indicate former minimum active-layer depth	
Small cryoturbations	Seasonally frozen ground or former active layer	9
Small-scale soil structures	Former permafrost or seasonally frozen ground	9
	May indicate former active-layer depth	
Fragipans	Former permafrost and active-layer depth	9
	May have non-periglacial origin	

Table 16.2 Palaeonvironmental implications of landforms, deposits and structures formed by periglacial weathering and mass movement.

Relict feature	Palaeoenvironmental implications[a]	Chapter
Tors	Residual rock towers	10
	Azonal	
	May have survived glacial erosion under cold-based ice	
Autochthonous blockfields	Predominantly mechanical (frost) weathering	10
	Possibly indicate depth of former active layer	
Blockstreams	Possibly permafrost; considerable uncertainty	10
Periglacial trimlines	Upper altitudinal limit of effective glacial erosion	10
Brecciated bedrock	Ice segregation in former permafrost or possibly deep seasonally frozen ground	10
Solifluction sheets and lobes	Seasonal freeze–thaw of soil, in both permafrost and nonpermafrost environments	11
Granular gelifluctate ('head')	As above	11
Stratified slope deposits	Polygenetic and probably azonal	11
	Often attributed to solifluction, can be produced by other processes	
Active-layer failures	Ice-rich permafrost	11
Low-angle shears in clays	Active-layer sliding over former permafrost	11
Cambering and valley bulging	Ice-rich permafrost in argillaceous rocks (?)	11
Nivation features	Former persistent snowcover	11
Cryoplanation terraces	Former solifluction; probably permafrost	11
Cryopediments	Inherited features, modified by periglaciation	11
Rockfall talus	Azonal; often paraglacial in glaciated areas	12
Stratified talus deposits	Azonal; may reflect former periglacial conditions	12
Snow avalanche landforms	Increased snow accumulation	12
Debris-flow landforms/deposits	Azonal; extreme rainstorm events	12
Talus rock glaciers	Permafrost; MAAT $< -2\,°C$; high debris input	12
Glacigenic rock glaciers	Permafrost; former debris-covered glacier	12
Pronival (protalus) ramparts	Perennial firn fields; cool summers	12

[a] The term *azonal* is applied to landforms and deposits that form in both periglacial and non-periglacial environments.

Table 16.3 Palaeonvironmental implications of periglacial alluvial, littoral and aeolian landforms and deposits.

Relict feature	Palaeoenvironmental implications[a]	Chapter
River terraces	Floodplain incision	13
	May reflect base-level lowering, temperate-to-cold climate transitions or fluvial downcutting under cold, arid permafrost conditions	
Braided gravel rivers	Represented by relict channel patterns and deposits	13
	Glacial or nival regime in rivers draining high relief	
Anastomosing rivers	Represented by relict channel patterns and deposits	13
	Nival or wetland regime in areas of low relief	
Ventifacts	Former dominant wind direction	14
Coversands and dunefields	Former dominant wind direction	14
	May represent regional aridity or proximity to a sediment source	
Loess deposits	Former dominant wind direction	14
	Strong winds	
	Relative aridity	
Loess-palaeosol stratigraphy	Climatic oscillations	14
	Changes in dominant wind direction and atmospheric pressure systems	
Raised shore platforms	Azonal. May reflect enhanced frost weathering	15

[a] The term *azonal* is applied to landforms and deposits that form in both periglacial and non-periglacial environments.

are probably (i) periglacial sediments that form stratigraphic sequences interrupted by palaeosols, thus representing a time-transgressive terrestrial record of environmental change, such as loess-palaeosol sequences, and (ii) periglacial features that indicate former permafrost and thus provide an indication of former maximum MAAT. Such information is particularly valuable when combined with mean July temperature estimates based on coeval subfossil insect assemblages, such as *coleoptera* (beetles) and *chironomids* (midges), as the combination of mean annual and mean July palaeotemperature permits an estimate of the former seasonal temperature regime and thus the degree of continentality (annual temperature range) that prevailed during Quaternary cold stages (Figure 16.4).

To illustrate the use of periglacial landforms, sediments and soil structures in reconstruction of Quaternary (mainly Late Pleistocene) palaeoenvironments, we consider below their use at two different spatial scales. The relict periglacial features in the British Isles are used to illustrate their representation and interpretation over a comparatively small area that contains lowland terrain that remained unglaciated throughout the Pleistocene, terrain that was covered by successive ice sheets, and high ground where periglacial processes continued to operate during the Holocene and are active at present. The regional periglaciation of Europe and North America during the last (MIS 2) cold stage is then

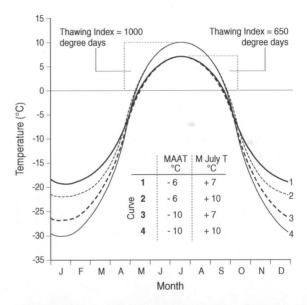

Figure 16.4 Mean monthly air temperature regime reconstructed from evidence for continuous permafrost (assumed MAATs of −6 or −10 °C) and mean July temperatures, implied by subfossil insect assemblages, of +7 and +10 °C. The equivalent approximate thawing index (degree days) is also shown. *Source:* Reproduced from Ballantyne and Harris (1994) with permission from Cambridge University Press.

outlined more broadly. Permafrost distribution and other aspects of palaeoenvironmental change on other continents during this period are summarized by Zhao *et al.* (2014) for China, Saito *et al.* (2016) for South America,

Astakhov (2014) for northern Russia and Saito *et al.* (2014) for northeast Asia. The chapter concludes by considering global permafrost distribution at this time.

16.3 Past Periglacial Environments of the British Isles

16.3.1 Chronological Framework

The last glacial stage in Britain is termed the Devensian, and the preceding interglacial is termed the Ipswichian; the chronological equivalents in northwestern Europe are the Weichselian Glacial Stage and the Eemian Interglacial (Figure 16.1). The Devensian is conventionally subdivided into the Early Devensian (MIS 5d–4; ~116–58 ka), Middle Devensian (MIS 3; ~58–31 ka) and Late Devensian (MIS 2; ~31.0–11.7 ka). The Middle Devensian was not a 'true' interglacial, but a period of stadial–interstadial oscillations when the climate remained colder than now. The Late Devensian is conventionally subdivided into the Dimlington Stade (~31.0–14.7 ka) and Loch Lomond Stade (~12.9–11.7 ka), separated by the Lateglacial (Windermere) Interstade (~14.7–12.9 ka). The Loch Lomond Stade is equivalent to the Younger

Dryas Stade of northwestern Europe, so 'Younger Dryas' is hereafter employed to describe this time interval for all sites, including the British Isles. Pre-Ipswichian stages are poorly differentiated in the stratigraphic record for the British Isles, apart from the Anglian Glacial Stage (MIS 12), the preceding Cromerian Interglacial (MIS 13) and the succeeding Hoxnian Interglacial (MIS 11). The fragmentary stratigraphic record of events between the Hoxnian and Ipswichian Interglacials (MIS 10–6) is informally termed 'Wolstonian' (Figure 16.5). In some British literature, the Holocene is referred to as the 'Flandrian', but this term is now rarely used.

16.3.2 Periglacial Regions in the British Isles

The British Isles can be subdivided into *periglacial regions* on the basis of the extent of glacier ice at various times. Region 1 (Figure 16.5) consists of areas occupied by glacier ice during the Younger Dryas Stade of ~12.9–11.7 ka, and in such areas periglacial processes have affected the landscape only since the retreat and final disappearance of Younger Dryas glaciers, which occurred within the period ~12.5–11.6 ka. Region 2 is the most extensive, comprising all areas that were covered by the last British–Irish Ice Sheet (BIIS) during the Dimlington Stade.

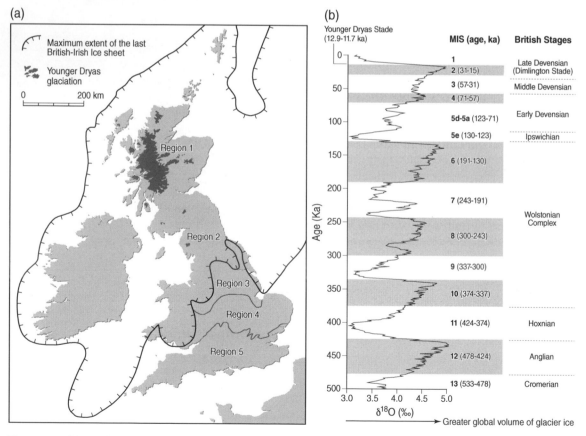

Figure 16.5 (a) Periglacial regions of the British Isles, defined by the maximum extent of former ice sheets. For details, see text. (b) Subdivision of Late and Middle Pleistocene glacial and interglacial stages in Britain, plotted against the marine δ¹⁸O record and marine isotope stages (MIS). *Source:* Murton and Ballantyne (2017). Reproduced with permission of the Geological Society.

During the LGM, this ice sheet extended to the Atlantic shelf break in the west, across the Celtic Sea in the south and was confluent with the Fennoscandian Ice Sheet in the North Sea Basin, though different sectors of the BIIS achieved their maximum extent at different times (Clark *et al.*, 2012). The last BIIS covered all of Scotland and Ireland, and most of Wales. In England, the ice sheet terminated along a line between the Bristol Channel and north Yorkshire, though the ice margin in the North Sea Basin reached as far south as The Wash (Figure 16.5). Exposure dating of perched boulders and erratics on mountain summits has shown that all high ground was covered by the last BIIS (Fabel *et al.*, 2012; Ballantyne and Stone, 2015; Hughes *et al.*, 2016). Within region 2, surface and nearsurface periglacial features must have developed after the retreat of the last ice sheet, which progressively exposed the land surface to periglacial conditions between ~23.0 and ~14.7 ka.

Region 3 was apparently last covered by glacier ice during MIS 6 at ~160 ka (Gibbard *et al.*, 2009), though the extent of this glaciation is poorly defined. Region 4 was last glaciated during MIS 12 at ~430 ka, during the Anglian Glacial Stage (the Elsterian of mainland Europe). The Anglian ice sheet was apparently the most extensive ice sheet in the British Isles, covering much of the English Midlands and East Anglia (Rose, 2009). Within zones 3 and 4, severe periglacial conditions pertained during all later cold periods, with associated development of permafrost, slope instability and, during intervening warming episodes, the development of thermokarst. Region 5 represents the 'never-glaciated' part of England, where periglacial conditions of varying severity pertained during successive Quaternary cold stages. In this zone, there is evidence for former glaciation only on Dartmoor, which may have supported a plateau ice cap during the Dimlington Stade, and possibly during earlier glacial stages.

The mid-latitude location of the British Isles (~50–60° N) at the eastern margin of the Atlantic Ocean means that these islands experienced extreme environmental changes during the Pleistocene. Rose (2010) has suggested that these fall into three states: (i) cool maritime conditions during interglacials, such as those of the present; (ii) cold maritime conditions, when the Atlantic Ocean was largely free of ice cover so that relatively high precipitation persisted across much of the British Isles, encouraging ice-sheet expansion; and (iii) cold continental conditions, when the Atlantic oceanic polar front extended far to the south, allowing extensive sea-ice cover to develop across the northeastern Atlantic Ocean and causing arid conditions and extremely cold winters in Britain and Ireland. Under the present cool maritime regime, periglacial activity is confined to high ground, generally above 500–700 m, and this is likely to have been true of earlier interglacials (Ballantyne and Harris, 1994).

Rose (2010) argued that the cold maritime regime would have resulted in rapid expansion of glacier ice, highly active periglacial slope processes and enhanced fluvial activity, whereas the importance of these processes would have diminished under cold arid conditions but wind action (deflation and re-deposition of sediment as coversand and loess) would have increased. He also emphasized landscape changes at climate transitions, notably paraglacial effects accompanying ice-sheet retreat, changes in fluvial regime and the development of thermokarst as permafrost degraded.

The evidence for past permafrost and periglacial activity in Great Britain was reviewed by Ballantyne and Harris (1994), and subsequent research is described in Murton and Ballantyne (2017). The following account focuses on the implications of two sets of features: those that indicate past permafrost, and aeolian deposits. These are considered here in terms of three time periods: pre-Late Devensian, the Dimlington Stade (~31.0–14.7 ka) and the Younger Dryas Stade (~12.9–11.7 ka), followed by some general comments about the role of periglacial features in palaeoenvironmental reconstructions for the British Isles.

16.4 Pre-Late Devensian Periglacial Features in the British Isles

Evidence for pre-Late Devensian periglaciation in the British Isles is largely restricted to regions 3–5 in Figure 16.5, because elsewhere such evidence has been largely destroyed by glacial erosion or buried under glacigenic deposits. Even in areas of England that lay beyond the reach of the last ice sheet, pre-Late Devensian landforms and deposits were subject to subsequent modification, erosion and burial during the Late Devensian and Holocene.

16.4.1 Evidence for Past Permafrost

The earliest evidence for permafrost in the British Isles comprises a sequence of ice-wedge pseudomorphs and involutions in pre-Anglian (pre-MIS 12) sediments on the East Anglian coast. These occur both above and below interglacial deposits and represent two or possibly three periods of extreme cooling and permafrost development (West, 1980; Whiteman, 2002; Lee *et al.*, 2003; Candy *et al.*, 2011). The development of continuous permafrost during the early Anglian stage is also attested by ice-wedge pseudomorphs, sand wedges and involutions in the Barham palaeosol, which underlies Anglian glacigenic deposits in Suffolk and Essex (Rose *et al.*, 1985), and by glacitectonically-deformed sediments on the Norfolk coast that indicate that the advancing Anglian ice sheet over-rode permafrost (Waller *et al.*, 2009, 2011).

The stratigraphic positions of ice-wedge pseudomorphs in alluvial sediments in southern and eastern England also imply the development of continuous permafrost at various times during the 'Wolstonian' period (MIS 10–6), but it is often difficult to assign these to particular cold stages within this long time interval (Ballantyne and Harris, 1994, pp. 54–58). A notable exception occurs at a site at Marsworth in south-central England, where the stratigraphic context of ice-wedge pseudomorphs and brecciation of the underlying chalk by ice segregation demonstrates development of permafrost at this southerly location (51.8°N) during the MIS 6 cold stage (Murton *et al.*, 2015b). The sediment sequence at this site has also produced evidence for coeval reworking of brecciated chalk, loess and pre-existing sediments by mass movement (solifluction) and slopewash. Permafrost in East Anglia during MIS 6 is also implied by ground-ice depressions (probably lithalsa scars) that pre-date the Ipswichian Interglacial (West, 2015). The presence of permafrost in southernmost England during successive Middle and Late Pleistocene cold stages is also indicated in a study by Brown *et al.* (2015) of quasi-continuous fluvial aggradation in the Axe Valley in southwest England from MIS 10 and possibly MIS 12 until the Late Devensian. These authors interpreted arcuate scars along springlines as relict retrogressive ground-ice slumps that fed sediment to the river during cold stages over the past 300–400 ka, implying that ice-rich continuous permafrost extended to the coast of southwest England.

The evidence for extensive continuous permafrost in Great Britain during the Early Devensian (MIS 4) and possibly the Middle Devensian (MIS 3) has been summarized by Ballantyne and Harris (1994, pp. 58–61). In Norfolk, for example, ice-wedge pseudomorphs, ground-ice depressions and involutions in fluvial sediments directly overlying Ipswichian interglacial deposits are probably of Early Devensian age. At Chelford in the Cheshire Plain and Four Ashes in Staffordshire, ice-wedge pseudomorphs in stratigraphic sequences underlying Late Devensian till also date from the Early Devensian. Similarly, in a sediment sequence exposed at Kirkhill in northeast Scotland, ice-wedge pseudomorphs in a gelifluctate deposit underlying a till of inferred Early Devensian age imply development of permafrost prior to advance of an Early Devensian ice sheet across the site.

Though the stratigraphic record of pre-Late Devensian permafrost as outlined above is fragmentary, it implies that widespread extension of permafrost across much or all of the British Isles outside the limits of successive ice sheets has been a recurrent feature of Pleistocene cold stages, with persuasive evidence for development of continuous permafrost at least as far south as 51–52°N during MIS 12, MIS 6 and MIS 4. It seems reasonable to infer that permafrost was also widespread outside the margins of ice sheets during all extended cold stages of the Middle and Late Pleistocene.

16.4.2 Loess and Coversands

Cold-climate aeolian deposits in the British Isles occur mainly in parts of Wales and England; few have been identified in Ireland, and none in Scotland. Almost all are of Late Devensian age, but thermoluminence dating of loess and loessic deposits at sites in southern and southeast England by Parks and Rendell (1992) has provided tantalizing evidence for the survival of pockets of much older windblown silts (Figure 16.6), suggesting that deposition of loess was widespread during earlier cold stages but that the resulting deposits have been buried under later sediments or eroded and incorporated into alluvial and slope sediments. The oldest documented loess in England, the Barham Loess, is of MIS 12 age (~430 ka). Other loess deposits in buried sediment sequences are certainly of pre-Devensian age, though Parks and Rendell (1992) cautioned that the ages they obtained for pre-Devensian loesses (116–230 ka; Figure 16.6) should be regarded as minimal, so that attribution to particular cold stages is uncertain. The sediment sequence at Marsworth in south-central England contains evidence for loess deposition during MIS 6 (Murton *et al.*, 2015b). A few loessic deposits have also been dated to the Early Devensian: loessic valley infill at Thanet, Kent has produced luminescence ages of ~88–75 ka (Murton *et al.*, 2003),

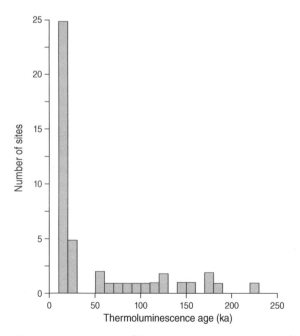

Figure 16.6 Histogram of the thermoluminescence ages of 46 samples from loess and loessic deposits at 26 sites in southern and southeast England, showing the preponderance of Late Devensian (~31.0–11.7 ka) ages. *Source:* Parks and Rendell (1992). Reproduced with permission of Wiley.

and samples obtained by Parks and Rendell (1992) from loess at three sites on the south coast of England have also yielded Early Devensian luminescence ages. An excavation near Heathrow Airport (London) has produced a remarkable record of loess deposition that spans the entire Devensian. At the base of this site are alluvial gravels of MIS 6 age, overlain by an Ipswichian Interstadial palaeosol, then by three distinct sediment units. The lowest of these is an aeolian sandy silt, attributed by Rose *et al.* (2000) to the Early Devensian; the middle unit of laminated silt and sand, they considered to be a product of aeolian rainout and slopewash deposition during the Middle Devensian; and the uppermost unit they interpreted as loess deposited in a cold arid environment during the Dimlington Stade and modified by ice-lens growth and cryoturbation during the Younger Dryas Stade. The stratigraphy at this site suggests that southeast England experienced relatively arid periglacial conditions during much of the Early and Late Devensian, interrupted by generally wetter conditions during the Middle Devensian. It also suggests that Early Devensian loessic deposits may be widespread, but buried under later sediments.

Few coversand deposits of pre-Late Devensian age have been identified in the British Isles, through aeolian sands and sand wedges occur in association with the Barham palaeosol of MIS 12 age (Rose *et al.*, 1985) and coversands of MIS 6 age are present at Marsworth (Murton *et al.*, 2015b). Aeolian sand deposits incorporated in patterned ground in East Anglia have yielded OSL ages that indicate multiple periods of deposition over the last 90 ka, with evidence of enhanced deposition at ~60–55 ka and 35–31 ka, towards the end of the Early Devensian and Middle Devensian respectively (Bateman *et al.*, 2014).

Collectively, this evidence suggests that loess and coversand deposition was a characteristic feature of Middle and Late Pleistocene cold stages, but that such deposits have been extensively eroded, reworked or buried under later sediments so that they have been identified only at a few sites in southern and eastern England. Together with the evidence for widespread permafrost outside the margins of coeval ice sheets, it implies that periglacial conditions during the Dimlington Stade (see below) provide a realistic approximation for conditions during successive cold stages stretching back to the onset of prolonged glacial stages at ~0.9 Ma (MIS 22) and possibly earlier.

16.5 The Dimlington Stade in the British Isles

The Dimlington Stade represents the last period of ice-sheet glaciation in the British Isles, when only England south of a line from the Bristol Channel to north Yorkshire and a small area of south Wales escaped encroachment by glacier ice. Analysis of subfossil coleopteran assemblages from sites in the British Isles suggests that during the final part of the stade, MAATs fell to between −5 °C and −10 °C and mean January temperatures fell to between −20 °C and −25 °C and possibly lower, implying a temperature range of 30–35 °C between the warmest and coldest months, and thus much more continental conditions than those of the present (Atkinson *et al.*, 1987). Data for earlier in the stade are more fragmentary, but seem to suggest rather warmer mean January temperatures of around −16 °C during the period of ice-sheet advance prior to ~24 ka. In general terms, this appears consistent with the two-state concept of Rose (2010), which envisages cold maritime conditions during ice-sheet expansion succeeded by cold continental (and more arid) conditions during ice-sheet shrinkage.

16.5.1 Evidence for Past Permafrost

Evidence for continuous permafrost during the Late Devensian (~31.0–11.7 ka) is widespread throughout the British Isles (Figure 16.7), except in the area occupied by glaciers during the Younger Dryas Stade (region 1 in Figure 16.5). The most widespread evidence takes the form of nearsurface relict frost wedges exposed in section and relict frost polygons visible as crop marks on aerial photographs, though other relict ground ice features, cryoturbations, thermokarst involutions and large-scale patterned ground features also indicate permafrost during this period. There are, however, two caveats involved in the interpretation of this evidence. First, few sites are securely dated. Though most permafrost indicators depicted in Figure 16.7 are probably of Late Devensian age, some outside of the limit of Late Devensian glaciation could be older, and undated landforms and structures inside the Late Devensian glacial limit could be of either Dimlington Stadial or Younger Dryas age. The second caveat is that the Dimlington Stade was a period of climatic variability (Figure 16.2), so the evidence for past permafrost during this stade may be asynchronous; the southernmost limit of permafrost and the continuous–discontinuous permafrost boundary probably oscillated across a few degrees of latitude over millennial or shorter timescales.

16.5.2 Relict Frost Wedges and Frost Polygons

Many structures originally identified as ice-wedge pseudomorphs in the British Isles may actually represent composite wedge pseudomorphs or relict sand wedges, but as all three types of structure represent thermal contraction cracking of continuous permafrost, they are here collectively referred to as 'relict frost wedges'. Most

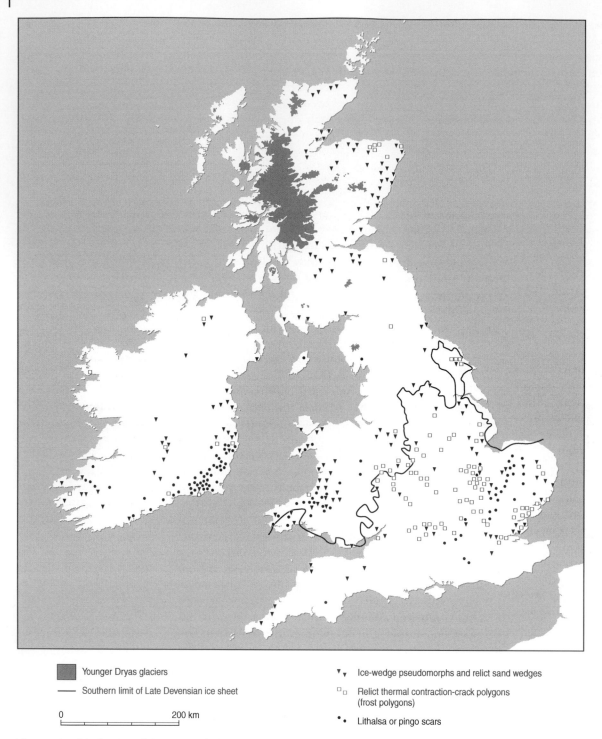

Figure 16.7 Distribution of documented relict permafrost features of inferred Late Devensian age in the British Isles. *Source:* Based on multiple sources, particularly Lewis (1985) and Ballantyne and Harris (1994).

relict frost wedges of inferred Dimlington Stadial age are typically up to 3 m long and 1 m wide (Ballantyne and Harris, 1994); longer wedges may represent syngenetic ice-wedge growth in accumulating sand and gravel deposits. Frost polygons (thermal contraction crack polygons) and relict frost wedges of inferred Late Devensian age are widespread both outside and inside the limit of

Late Devensian glaciation, implying that continuous permafrost developed in terrain exposed by retreat of the last ice sheet after ~23 ka, consistent with the air temperature regime (see above) inferred for the final part of the stade from coleopteran assemblages (Atkinson *et al.*, 1987). Radiocarbon and luminescence dating have confirmed that relict frost wedges formed in eastern and central England during the Dimlington Stade (e.g. Briggs *et al.*, 1975; West, 1993; Briant *et al.*, 2004; Briant and Bateman, 2009), suggesting that similar features recorded on deglaciated terrain in northern Britain and Ireland also formed during this period.

16.5.3 Relict Ground-Ice Features: Pingos, Lithalsas and Thermokarst Landforms

Clusters of shallow circular or oval ground depressions up to about 150 m in diameter and partly enclosed by low ramparts occur in various parts of the British Isles, notably across southern Ireland, southern Wales and East Anglia (Figure 16.7). Such features occur both inside and outside the limit of the last ice sheet, and were initially interpreted as the remnants of hydraulic pingos (e.g. Lewis, 1985; Coxon and O'Callaghan, 1987), though some authors more cautiously refer to them as 'ramparted ground ice depressions', leaving open the question of origin (Sparks *et al.*, 1972; Bryant and Carpenter, 1987). Several researchers have suggested that most of these features should be reinterpreted as collapsed lithalsas, pointing out that the clustering of depressions and evidence for overlap of different generations of ramparts is consistent with that for extant lithalsas (Chapter 7), and that these depressions tend to be developed in frost-susceptible sediments conducive to the formation of ice lenses on freezing (Gurney, 1995, 2000; Worsley *et al.*, 1995; Pissart, 2000, 2003; Clay, 2015). Though most are located in valley-floor locations, sometimes corresponding to springlines, many occur in lowland areas, where it is questionable that the local relief could have permitted generation of groundwater pressure gradients sufficient for the formation of hydraulic pingos. Some ramparted depressions may, however, be attributed to melt-out of blocks of glacier ice embedded in glacilacustrine sediments, or represent the remnants of small, hybrid permafrost mounds, small hydraulic pingos or seasonal frost mounds formed at zones of groundwater seepage (Gurney *et al.*, 2010; Ross *et al.*, 2011; Clay 2015; West, 2015).

Ramparted depressions that occur inside the limit of the last ice sheet must be of Late Devensian age, and it is likely that this is true also of those south of the ice-sheet limit in eastern and southern England. The oldest organic deposits retrieved from within depressions are of early Holocene age, suggesting that the parent landforms could have developed during the Younger Dryas Stade, as discussed below. The southerly distribution of almost all reported clusters (Figure 16.7), however, may imply that that they developed during the Dimlington Stade, either outside the limit of the last ice sheet or during the early stages of ice-sheet retreat.

Distinct from such ramparted depressions are large, flat-floored depressions in the Fenland Basin of eastern England. These are inset into Devensian alluvial terraces that support relict frost polygons, generally exceed ~1 km in diameter, are bordered by degraded scarps and have narrow outlets. Together with a single example in the Vale of York, and smaller enclosed basins in East Anglia, these landforms have been interpreted as former thermokarst lake basins that formed through degradation of ice-rich permafrost (Burton, 1987; West, 1991, 2015), probably during the Dimlington Stade. Arcuate embayments in the basin margins have been interpreted as the sites of retrogressive thaw slumps.

16.5.4 Permafrost Extent and Thickness

The widespread distribution of Late Devensian landforms and structures indicative of former continuous permafrost (Figure 16.7) suggests that permafrost developed extensively on terrain exposed by retreat of the last BIIS, apart perhaps from areas of the Scottish Highlands that remained occupied by glacier ice until the rapid warming that ushered in the Lateglacial Interstade at ~14.7 ka. This conclusion, however, requires verification by dating of relict frost wedges inside the limits of the last ice sheet. More conclusively, the distribution of dated relict frost wedges suggests that during the coldest parts of the Dimlington Stade, continuous permafrost extended southwards to at least 52° N, and probably to the south coast of England. Some authors have argued that permafrost may have been absent or discontinuous in southwest England during the stade despite reports of nearsurface relict frost wedges in this region. Vandenberghe *et al.* (2014) placed the southern boundary of continuous permafrost at latitude ~47° N in France, implying that all unglaciated parts of England almost certainly lay within the continuous permafrost zone during the coldest parts of the stade, but others have concluded that continuous permafrost at this time extended no farther than ~50° N (Andrieux *et al.*, 2016).

Busby *et al.* (2016) have attempted to reconstruct the thickness of permafrost at six sites in Great Britain over the past ~130 ka, using a one-dimensional heat-conduction model driven by step changes in assumed MAATs. The results suggest that Dartmoor and East Anglia were largely or completely permafrost-free throughout much of the Early and Middle Devensian but were underlain by shallow (<80 m-deep), warm (> −2.5 °C) permafrost during much or all of the Dimlington Stade. For more northerly sites, their modelling is complicated by assumptions regarding the timing and duration of former glacier cover and subglacial temperature regime, and yields improbable outcomes.

16.5.5 Loess and Coversands

Loess and loessic deposits in Britain have traditionally been referred to as *brickearths*, reflecting their use in brick manufacture, though some deposits designated as brickearths are of alluvial origin. The great majority of surficial loesses and loessic deposits in the British Isles are of Dimlington Stadial age (Parks and Rendell, 1992; Antoine *et al.* 2003a; Figure 16.6) and are thinner and more fragmented than coeval loesses in continental Europe. They are, moreover, largely restricted to England and Wales, where they occur in three main provinces. The most extensive group comprises the loesses that overlie Triassic and younger rocks outside the limit of the Late Devensian ice sheet in southeast and southern England (Figure 16.8). Much more restricted areas of loess are present in southwest England (Cornwall and the Scilly Isles) and around the Bristol Channel, and small patches of loess occur inside the Late Devensian ice-sheet limit in north Wales and northwest England. Patches of undated loess and loessic soils also occur in karstic uplands in west Ireland, and probably reflect winnowing of silts from outwash in the adjacent Galway Bay during a period of low sea level.

According to Antoine *et al.* (2003a), the loess deposits of eastern and southern England typically have a median grain size of 25–35 μm, lack stratification and often exhibit prismatic structure in vertical exposures. The loess deposits of Cornwall, north Wales and northwest England are generally coarser, and are mineralogically distinct from those farther south and east. Most English loess deposits are <1 m thick and completely decalcified, though thicker deposits occur along the Sussex coastal plain, in Essex and in Kent (Figure 16.9), attaining a thickness of ~4 m at Pegwell Bay in east Kent (Murton *et al.*, 2003) and up to 8 m in south Essex. In many coastal areas, loessic deposits are incorporated in gelifluctate ('head') deposits, suggesting that they pre-date or are coeval with periods of solifluction activity (Ealey and James, 2011).

Though luminescence ages obtained for Dimlington Stadial loess deposits at various sites span almost the entire stade, the great majority fall within the period ~18–15 ka, suggesting that conditions were most propitious for loess transport and accumulation long after the

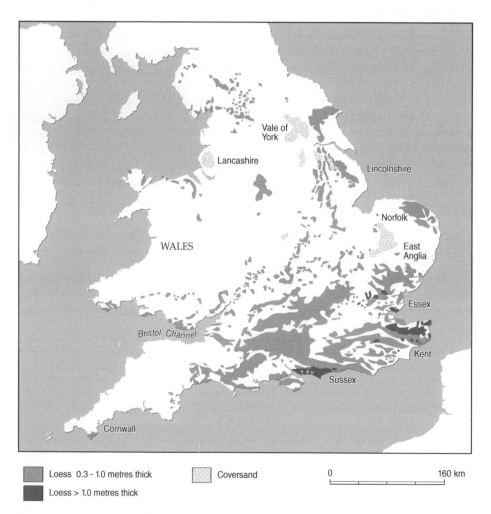

Loess 0.3 - 1.0 metres thick

Coversand

0   160 km

Loess > 1.0 metres thick

Figure 16.8 Distribution of loess and coversand deposits in England and Wales. *Source:* Reproduced from Ballantyne and Harris (1994) with permission from Cambridge University Press.

Figure 16.9 Loess exposure at Pegwell Bay, Kent, southeast England. Up to 3 m of loess overlies a brecciated and involuted chalk horizon above chalk bedrock. *Source:* Courtesy of Mark Bateman.

last BIIS reached its maximum extent. This temporal focus could reflect aridity during the later part of the stade, but it may also be a result of the increased area of potential loess sources on exposed continental shelves vacated by retreat of the ice sheet. Across eastern and southern England, Dimlington Stadial loess deposits exhibit a general westwards decrease in grain size. This trend has been interpreted as indicating derivation from glacifluvial outwash plains on the exposed floor of the North Sea Basin and transportation across southeast England by dominantly easterly to northeasterly winds (Antoine *et al.*, 2003a; Clarke *et al.*, 2007). However, the distinctive 'western' loess deposits of northwest England and north Wales appear to have been sourced from outwash on the exposed Irish Sea Basin, and those of southwest England from the Irish Sea Basin or Celtic Sea shelf, implying dominant westerly winds. The different palaeowind directions implied by loess deposits of eastern and western provenance might be explained by seasonal or longer-term changes in dominant wind direction, the timing of sea-level changes and consequent exposure of silt-rich source sediments, or even differences in the timing of the advance and retreat of glacier ice in the Irish Sea and North Sea Basins (Ballantyne and Harris, 1994, pp. 283–284). Antoine *et al.* (2009) have proposed an alternative solution, suggesting that high atmospheric pressure over the last BIIS forced low-pressure cells (cyclones) to track eastwards through the English Channel, promoting first strong northerly to northwesterly winds across southwest England, then northeasterly winds over southeast England. The thin loesses of north Wales and northwest England presumably developed later, after shrinkage of the ice sheet and resumption of dominant westerly airflow across these areas.

Evidence of coversand accumulation during the Dimlington Stade is limited to a few sites in eastern and southeastern England. Luminescence dating of aeolian sand incorporated in involutions and patterned ground suggests that coversands began to accumulate in Kent at 24–21 ka, and that in East Anglia a major period of deposition occurred at ~22–20 ka, both age ranges being consistent with that of the Older Coversand I of continental Europe (Koster, 2005; Kasse *et al.*, 2007). Intact coversand deposits have yielded ages of 18–14 ka in northern Lincolnshire and ~15.5 ka in Kent, corresponding to the Older Coversand II and Younger Coversand I depositional episodes in the European sand belt (Murton *et al.*, 2003; Bateman *et al.*, 2014). The apparently very limited spatial distribution of Dimlington Stadial coversands probably reflects subsequent erosion of coversand deposits, their incorporation into alluvial and slope deposits, and aeolian reworking of sand sheets during the Younger Dryas Stade.

16.6 The Younger Dryas (Loch Lomond) Stade in the British Isles

The Younger Dryas Stade (~12.9–11.7 ka) represents the final period of severe cold to affect northwest Europe. During this period, North Atlantic margins experienced cooler summers and much colder winters than during the preceding Lateglacial Interstade (~14.7–12.9 ka), a climatic shift widely attributed to weakening or cessation of thermohaline circulation in the North Atlantic Ocean as a result of a rapid influx of fresh water (Broeker, 2006; Carlson, 2010). The resulting southwards movement of the North Atlantic oceanic polar front resulted in the British Isles being enveloped by cold polar waters and winter sea ice. This comparatively brief return to stadial conditions led to the growth of an extensive icefield in the Western Highlands of Scotland and numerous smaller icefields and glaciers in other mountain areas (Golledge, 2010).

Analyses of the palaeotemperature implications of subfossil chironomid assemblages in Ireland, northern England and Scotland consistently suggest that mean July air temperatures during the coldest part of the stade were around 7–8 °C near sea level (Lang *et al.*, 2010; Watson *et al.*, 2010; Brooks *et al.*, 2012; Van Asch *et al.*, 2012). With less confidence, the mutual climatic range of Younger Dryas coleopteran assemblages from sites south of 55° N suggests that January temperatures during the coldest part of the stade were between −15 °C and −20 °C (Atkinson *et al.*, 1987). Together, these two ranges imply that MAATs over much of the British Isles during the thermal nadir of the Younger Dryas lay between −3.5 °C and −6.5 °C, though colder temperatures probably prevailed in central and northern Scotland, particularly in areas close to glacier margins and exposed to katabatic winds blowing off the ice. It might therefore be expected that continuous permafrost extended to low ground in northern parts of Britain and Ireland, and that at least shallow discontinuous permafrost formed during the stade in southern Britain and Ireland.

16.6.1 Evidence for Past Permafrost in Upland Areas During the Younger Dryas Stade

During the Younger Dryas Stade, a rich variety of periglacial landforms developed in the mountains of the British Isles (Figure 16.10). Some of these, particularly relict bouldery solifluction lobes and large sorted patterned ground features, are indicative of (but not conclusive evidence of) past permafrost. Accounts of Younger Dryas rock glaciers in the British Isles are now largely discredited, as most or all putative examples appear to have a different origin, mainly as rockslide runout debris (Ballantyne *et al.*, 2009; Jarman *et al.*, 2013). Harrison *et al.* (2010) have suggested that valley-fill deposits that accumulated in upland valleys during the Younger Dryas Stade may have been emplaced, at least in part, by active-layer detachment slides, implying past permafrost; though plausible, this suggestion remains conjectural. It is also notable that there is no evidence that the rapid warming that terminated the stade at ~11.7 ka resulted in a pulse of rock-slope failure attributable to thaw of permafrost ice in rock joints (Ballantyne *et al.*, 2014).

16.6.2 Evidence for Past Permafrost in Lowland Areas During the Younger Dryas Stade

Definitive evidence for the extent of permafrost in lowland areas during the Younger Dryas Stade has proved elusive.

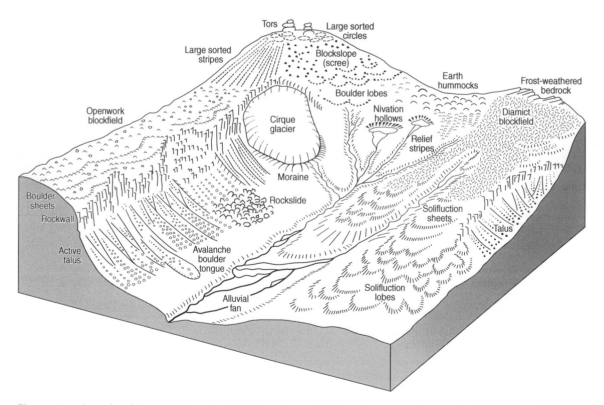

Figure 16.10 Lateglacial (Younger Dryas) periglacial and paraglacial landform assemblage typical of British and Irish mountains.

This is mainly because nearsurface relict frost wedges and crop marks representing frost polygons inside the limits of the last ice sheet (Figure 16.7) could represent thermal contraction cracking of permafrost either during the final part of the Dimlington Stade, after ice-sheet retreat, or during the Younger Dryas Stade. Most accounts of supposed relict frost wedges of Younger Dryas age are equivocal because of uncertainties regarding the age of the host sediments or interpretation of the wedge structures (e.g. Hutchinson, 2010). The strongest case is made by Worsley (2014), who described ice-wedge pseudomorphs in gravels of Younger Dryas age in Lincolnshire. This implies that permafrost extended southwards to latitude 53°N during the stade.

As noted earlier, the oldest organic deposits hitherto found within ramparted ground-ice depressions in England, Wales and Ireland are of early Holocene age, suggesting that these depressions may represent the remnants of Younger Dryas lithalsas (or, less likely, hydraulic pingos). Though these features could be older, stratigraphic evidence reported by Sparks *et al.* (1972) supports a Younger Dryas age for a 'younger generation' of such features in East Anglia. If this could be confirmed for ramparted ground-ice depressions, the distribution of these features would imply that a belt of discontinuous permafrost extended westwards from East Anglia through southern Wales and across southern Ireland at this time (latitudes 51.5–53.0°N; Figure 16.7), implying MAATs of roughly −4°C to −6°C (Pissart, 2000, 2003), and thus consistent with palaeotemperature estimates (MAAT −3.5°C to −6.5°C) based on insect assemblages. Apparent confirmation that permafrost was present in southern England during the Younger Dryas Stade comes from clay-rich 'head' deposits that overlie organic deposits of Lateglacial Interstadial age. These deposits contain or are underlain by slope-parallel or emergent shear surfaces, and have been interpreted as low-gradient active-layer slides over permafrost. They have been documented at various localities, and as far south as Kent (e.g. Skempton and Weeks, 1976; Chandler, 1976; Spink, 1991; Hutchinson, 2010), suggesting that discontinuous permafrost locally extended to the south coast of England, consistent with the pan-European reconstruction of Younger Dryas permafrost proposed by Isarin (1997b).

16.6.3 Younger Dryas Aeolian Deposits

There is limited evidence for Younger Dryas loess deposition in the British Isles. Conversely, most British coversands are of Younger Dryas age, and thus stratigraphically equivalent to the Younger Coversand II deposits in continental Europe (Koster, 2005). Moreover, soils across large areas of England contain an allochthonous sand component, suggesting more widespread aeolian sand deposition than that represented by surviving coversands. Younger Dryas coversands are widespread in the Vale of York, northern Lincolnshire, central East Anglia and southwest Lancashire (Bateman, 1995; Figure 16.8), though Antoine *et al.* (2003a) noted additional areas of (undated) coversands south of the Bristol Channel and on the Mendip Hills, and limited areas of Younger Dryas coversand occur in Kent (Murton *et al.*, 2003) and West Norfolk (Hoare *et al.*, 2002). The coversands in these areas generally comprise well-sorted fine to medium sand, with evidence of wind abrasion in the form of rounded, spherical grains and occasional ventifacts. Bedding is largely horizontal or subhorizontal, and slip-face bedding is rare. Stratigraphically, Younger Dryas coversands are locally underlain by peat deposits of Lateglacial Interstadial age and overlain by 'drift sand' deposits that represent reworking of coversands during periods of Holocene climatic deterioration or human disturbance.

The best-preserved Younger Dryas coversands are those within and east of the Trent Valley in northern Lincolnshire (Bateman, 1998; Bateman *et al.*, 2000; Baker *et al.*, 2013). These represent the remnants of a regionally extensive sand sheet, are generally 1–7 m thick, and often overlie peat deposited under wet tundra conditions. This basal peat has yielded radiocarbon ages of ~13.9–12.0 ka, and thermoluminescence dating of the coversand indicates that most was deposited in the interval 12.9–11.4 ka. Pollen grains from thin organic laminae within Younger Dryas coversands indicate a treeless landscape characterized by bare soils, open habitats and cold conditions, and associated coleoptera have arctic and tundra affinities. Dunes are generally indistinct, and locally represent Holocene reworking of earlier aeolian deposits. Bateman (1998) argued that the main sources of windblown sand were outwash deposits and local sandstone outcrops; as the coversands are thickest on west-facing scarp slopes, he concluded that the dominant palaeowinds responsible for coversand deposition were westerlies. Dominantly westerly palaeowinds during the Younger Dryas Stade were also inferred by Baker *et al.* (2013) for the coversand deposits in the Trent Valley, which are banked against west-facing slopes but exhibit an overall eastwards decline in thickness. The coversands of southwest Lancashire also appear to have been transported eastwards, in this case from outwash deposits in the Irish Sea Basin (Bateman, 1995). A dominantly westerly airflow during the Younger Dryas is supported by a marked eastwards and northeastwards rise in the equilibrium-line altitudes of Younger Dryas glaciers across the Scottish Highlands, which has been interpreted as indicating an eastwards decline in snowfall, particularly east of the icefield that occupied the West Highlands (Sissons, 1979; Ballantyne and Harris, 1994; Benn and Ballantyne, 2005).

16.7 Past Periglacial Environments of the British Isles: Commentary

Though the above accounts focus mainly on indicators of past permafrost and wind action, and exclude evidence relating to other manifestations of past periglacial activity such as frost weathering, slope processes, fluvial activity and cryogenesis of soils, they highlight both the strengths and the weaknesses of employing periglacial features in palaeoenvironmental reconstructions. The most pervasive limitation is poor dating resolution. For example, the wide extent of relict frost wedges indicative of continuous permafrost and permafrost cracking under extremely cold winter temperatures during the Dimlington Stade is consistent with palaeotemperature reconstructions based on subfossil coleopteran assemblages. However, with few exceptions, the dating resolution of relict frost wedges and associated frost polygons is inadequate to demonstrate when these features were active *within* the ~15 000-year duration of the stade. Similarly, the inference that continuous permafrost extended to ~53° N and that a belt of discontinuous permafrost existed at 51.5–53.0° N during the Younger Dryas Stade, though consistent with palaeoclimatic evidence, rests on slender chronological control from a handful of sites and probably reflects conditions during only the coldest part of the stade.

A second limitation is ambiguity of palaeoenvironmental interpretation. The concentration of luminescence ages for loess deposits within the late Dimlington Stade (~18–15 ka), for example, suggests a period of cold, arid continental conditions that may explain the retreat of the last ice sheet at that time (due to reduction in snowfall) and is consistent with the palaeotemperature regime implied by coleoptera for this interval. Alternatively, however, retreat of the ice-sheet margin may have exposed increased sources of silt for aeolian transport, and thus enhanced loess deposition. Similarly, taken in isolation, the Younger Dryas coversand deposits described above provide sound evidence for aeolian sand transport by dominant westerly winds, but limited evidence concerning the environmental conditions under which they were entrained, transported and deposited.

The British Isles case study suggests that the most informative Pleistocene palaeoenvironmental reconstructions based on relict periglacial phenomena in mid-latitude environments have three characteristics. The first is good *absolute* dating control, capable of constraining interpretations based on stratigraphic principles and allowing correlation both between sites and with more general records of environmental change, such as the Greenland ice-core $\delta^{18}O$ record (Blockley *et al.*, 2012; Rasmussen *et al.*, 2014; Figure 16.2). The second is employment of multi-proxy approaches to the investigation of periglacial

sediment sequences: research based on the characteristics and sequence of periglacial sediments and structures, in combination with analyses of subfossil insect remains, molluscs, ostracods, pollen assemblages and/or plant macrofossils, has generally produced fuller and less ambiguous interpretations of Middle and Late Pleistocene palaeoenvironmental conditions than those based on periglacial features alone (e.g. Briggs *et al.*, 1975; West *et al.*, 1999; Murton *et al.*, 2001a, 2015b; Briant *et al.*, 2004). Finally, the most useful studies from the perspective of identifying environmental *changes* within cold stages are those that correlate stratigraphic sequences across an area and relate them to the wider pattern of environmental change.

An excellent example of periglacial stratigraphic correlation is a reconstruction of the Devensian periglacial record on Thanet in Kent (51.3° N) by Murton *et al.* (2003). In this area, unweathered chalk is overlain in turn by brecciated chalk, an involuted layer, and pebbly sand that infills the troughs of large-scale patterned ground and elsewhere forms an irregular sand sheet. The timing of events represented by this stratigraphy was constrained by luminescence dating. For the Dimlington Stade, the results imply that an early phase of aeolian sand deposition and brecciation of chalk by ice segregation in permafrost (~24–21 ka) was succeeded by active-layer deepening, represented by thermokarst involutions and the formation of patterned ground (~22–21 ka), then by active-layer thinning due to climatic cooling, represented by small cryoturbation structures, by solutional truncation of the involuted layer (~21–18 ka), and finally by deposition of aeolian sand and loess (~15.5 ka). Importantly, Murton *et al.* (2003) were able to correlate the aeolian deposits at Thanet with those of the European coversand and loess stratigraphies, as well as with coversands elsewhere in England. This study illustrates the potential for identifying regionally significant changes in Pleistocene periglacial environments across the British Isles, though realization of this potential requires not only a high level of forensic analysis of sediments and soil structures, but also serendipitous discovery of sites where sequences of periglacial sediments and soil structures are exposed in section.

16.8 Late Weichselian Periglacial Environments in Continental Europe

Reconstruction of past periglacial environments in Europe involves a wider canvass, one that is here painted with a broader brush than that employed in the survey of past periglacial features of the British Isles. The Late Weichselian in Europe is conventionally divided into three periods: the Late (or Upper) Pleniglacial (~31.0–14.7 ka), the Bølling–Allerød (= Lateglacial) Interstade (~14.7–12.9 ka) and the

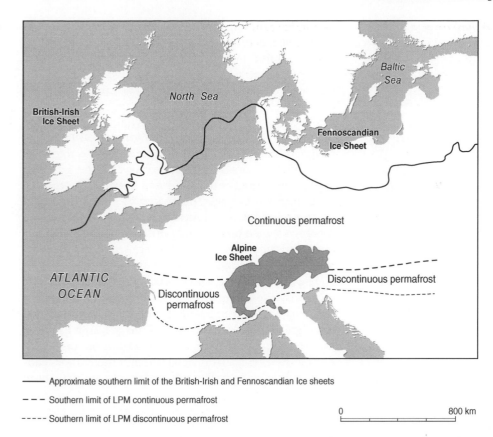

——— Approximate southern limit of the British-Irish and Fennoscandian Ice sheets

– – – Southern limit of LPM continuous permafrost

- - - - - Southern limit of LPM discontinuous permafrost

0   800 km

Figure 16.11 Maximum extent of the last (Late Weichselian) ice sheets in western and central Europe and maximum extent of continuous and discontinuous permafrost during the last permafrost maximum (LPM). Other researchers (e.g. Andrieux *et al.*, 2016) have inferred that the LPM permafrost limits in western Europe lay farther north than those depicted here. Smaller areas of last glacial maximum (LGM) glaciation are not shown. *Source:* Adapted from Vandenberghe *et al.* (2014). Reproduced with permission of Wiley.

Younger Dryas Stade (~12.9–11.7 ka). During the LGM, the Fennoscandian Ice Sheet covered all of Scandinavia, northern and eastern Denmark and a broad swath of the North European Plain, reaching its maximum southern extent at ~53° N in northern Germany (Figure 16.11). Southwards expansion of the ice sheet was driven by temperature decline. Coleopteran remains recovered from a fluvial succession in central Poland (51.3° N) indicate that during the period ~30–28 ka, the mean temperature of the warmest month was around 8 °C and that of the coldest month was probably as low as –27 °C, implying a MAAT of about –9 °C (Kasse *et al.*, 1998) and suggesting that a belt of permafrost extended southwards in advance of the expanding ice sheet.

The periglacial features of Europe have been the subject of detailed investigation since the 1940s, generating a vast literature on indicators of past permafrost (relict frost wedges and polygons, cryoturbations and involutions, fragipans, pingo scars, lithalsa scars and other ground-ice depressions) cold-climate aeolian sediments (loess, coversands, fluvio-aeolian deposits and dunefields), periglacial slope deposits, cryoplanation terraces and changes in river behaviour (cycles of aggradation, incision and planform change). Encapsulating the full

richness of this heritage would require another book. Here we focus on recent developments in the study of past permafrost and aeolian sequences, and how these relate to wider Late Weichselian climatic and environmental changes.

16.8.1 Late Weichselian Permafrost in Europe

Attempts to delineate the extent of past permafrost in Europe on the basis of periglacial landforms, sediments and, particularly, structures (relict frost wedges and involutions) have a long history, with notable contributions by, for example, Poser (1948) for western Europe, Dylik (1956) for Poland and Velichko (1972, 1982) for all of Europe. More recent research has combined periglacial evidence with that provided by biotic proxies (principally coleopteran and pollen assemblages) to reconstruct both the likely extent of permafrost and associated palaeotemperatures. A landmark paper by Huijzer and Vandenberghe (1998) employed periglacial, aeolian, fluvial, botanical and coleopteran evidence from numerous sites across northern Europe, from Ireland to Poland, to reconstruct Upper Pleniglacial climatic conditions and permafrost distribution. Their analysis suggested that during the period from ~31 to ~24 ka (recalibrated from their radiocarbon-based chronology), northern Europe experienced MAATs of –4 °C to –8 °C (based on the distribution of ice-wedge pseudomorphs) and mean January temperatures (based on coleopteran assemblages) of –20 °C to –25 °C, with the continuous–discontinuous permafrost boundary located along the south coast of England and the border between France and Belgium (49–51° N). They inferred that during the period from ~24 to ~19 ka, extremely cold conditions persisted, finding evidence for sporadic or discontinuous permafrost from northern France to the Netherlands, where ice-wedge pseudomorphs in loess indicate MAATs ≤ –4 °C, and for continuous permafrost in Denmark, northern Germany and Poland, where relict frost wedges in coarse sediments indicate MAATs ≤ –8 °C. On terrain exposed by the shrinking Fennoscandian Ice Sheet, relict frost wedges in outwash deposits demonstrate aggradation of continuous permafrost in the wake of the retreating ice margin (e.g. Böse, 1992; Kozarski, 1993; Ewertowski, 2009), suggesting that the latitudinal belts of continuous and discontinuous permafrost migrated northwards as the ice-sheet margin receded (Figure 16.12). During the final part of the Upper Pleniglacial (19–15 ka), only frost fissures and small cryoturbations appear to have developed in Belgium, the Netherlands and Germany, suggesting that by this time these areas experienced only seasonal ground freezing, though relict permafrost may have persisted at depth (Delisle, 1998). Evidence for the formation of

ice-wedge polygons in northern Poland, however, suggests that a belt of permafrozen terrain may have persisted south of the ice-sheet margin until the end of the Upper Pleniglacial.

A recent development has been the incorporation of evidence for Upper Pleniglacial permafrost in a geographic information system (GIS) database, a procedure that allows the distribution of periglacial features to be analysed both regionally and in terms of possible local controlling factors, such as substrate, altitude, gradient, aspect and presence of loess and coversand deposits. This approach has been pioneered by French researchers (Bertran *et al.*, 2014; Andrieux *et al.*, 2016), who have recorded a wide range of potentially diagnostic features (frost polygons, nets, soil stripes, ice-wedge and composite-wedge pseudomorphs, relict primary sand wedges and involutions) on terrain below ~300 m. Their database incorporates existing inventories, published sources and new data based on archaeological excavations and aerial photographs. The primary aim of this research has been to determine past permafrost distribution in France during the Upper Pleniglacial, though the authors caution that only a small number of features have been securely dated.

The distribution of documented periglacial features (Figure 16.13) suggests three distinct regions. Region 1 comprises the extreme southwest of France (southern Aquitaine) and the western Mediterranean coastal zone (Languedoc), which contain no unequivocal evidence for

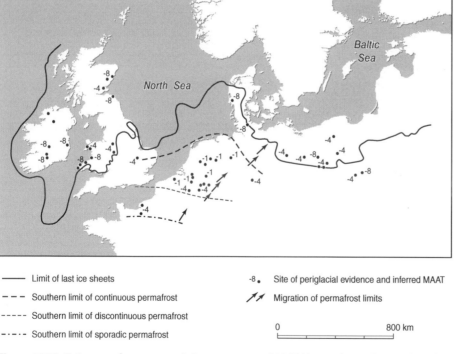

————— Limit of last ice sheets

– – – Southern limit of continuous permafrost

- - - - - Southern limit of discontinuous permafrost

–·–·– Southern limit of sporadic permafrost

-8• Site of periglacial evidence and inferred MAAT

⤴⤴ Migration of permafrost limits

0   800 km

Figure 16.12 Estimates of mean annual air temperatures (MAATs) in northwest Europe, based on periglacial evidence for the period ~24–15 ka, and northwards migration of permafrost zones. *Source:* Adapted from Huijzer and Vandenberghe (1998). Reproduced with permission of Wiley.

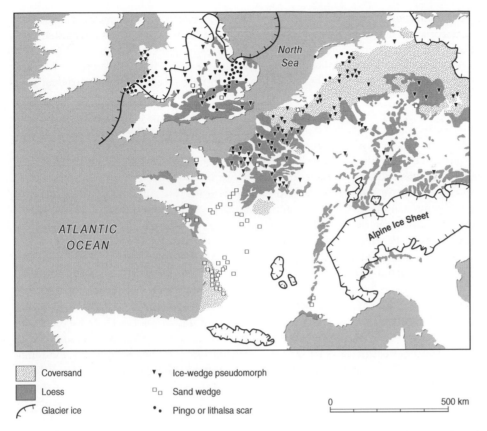

Figure 16.13 Late Weichselian periglacial features in western Europe, showing also the distribution of loess and coversand and the maximum extent of glacier ice. *Source:* Andrieux *et al.* (2016). Reproduced with permission of Wiley.

past permafrost, though soil veins indicate seasonal ground freezing. Region 2 includes the rest of southwest France (northern Aquitaine) and the lower Rhône Valley, which contain large polygons, relict sand wedges and possible composite wedges, often in proximity to coversand deposits. Sand-wedge infill in Aquitaine has been OSL dated to ~34–24 ka, and this assemblage extends southwards to 43.5° N in western France and 43° N in the Rhône Valley. Finally, region 3 lies north of 47° N and contains an assemblage of large frost polygons, numerous ice-wedge pseudomorphs and involutions up to ~2 m in amplitude, as well as large soil stripes, typically 6–12 m wide, developed mainly in coverloams on the eastern margin of the Paris Basin. The distribution of ice-wedge pseudomorphs in this zone is strongly correlated with that of loess deposits (Figure 16.13). Andrieux *et al.* (2016) have shown that the length and width of the largest relict frost wedges tend to increase with latitude, and that south of 47° N, wedges are typically <2 m long and <1 m wide (Figure 16.14).

The interpretation of these patterns is contentious. Bertran *et al.* (2014) inferred that they imply that a large part of France was underlain by permafrost that extended southwards to 43–44° N, and that during the coldest part of the Late Weichselian, ice-rich permafrost was present across all of the Paris Basin and France north of 47° N.

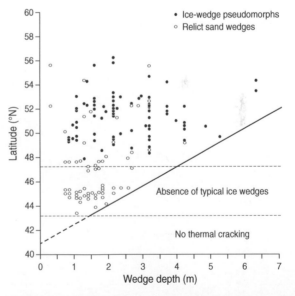

Figure 16.14 Length of ice-wedge pseudomorphs and sand wedges in France and elsewhere in western Europe, plotted against latitude. *Source:* Andrieux *et al.* (2016). Reproduced with permission of Wiley.

This interpretation was adopted by Vandenberghe *et al.* (2014) in constructing the LPM map of the northern hemisphere (Figure 16.10). Andrieux *et al.* (2016), however, subsequently concluded that none of France was underlain by continuous permafrost during the Late Weichselian, but that region 3 (47–51° N) was underlain by discontinuous permafrost, implying MAATs of –3 °C to –4 °C at its southern boundary, and that region 2 was characterized by sporadic permafrost and seasonally frozen ground, with MAATs close to 0 °C to –1 °C at latitude 43.5° N. This interpretation appears conservative, given the abundance of relict sand wedges, the large size of relict frost polygons and the occurrence of possible composite wedges in region 2: all of these features are usually interpreted as indicators of at least discontinuous permafrost, and imply thermal contraction cracking of the ground under very cold winter temperatures. This GIS approach nonetheless highlights both the value of such detailed regional inventories and the difficulties that still exist in interpreting past permafrost extent and palaeotemperature implications from the available evidence.

Fewer uncertainties afflict interpretation of Late Weichselian permafrost extent in Europe east of the Alps. For example, in Hungary, which lies at the southern margin of LPM permafrost as depicted by Vandenberghe *et al.* (2014), relict sand wedges with infill dated to ~23–15 ka, relict frost polygons and occasional ice-wedge pseudomorphs occur north of about 47° N, but the southern part of the country appears to support only soil wedges and cryoturbation or involution structures. Because the sand wedges occur in coarse sediments, Fábián *et al.* (2014) concluded that continuous permafrost formed in northern Hungary during the coldest part of the Late Weichselian, but that central and southern Hungary may have been underlain only by seasonally frozen ground.

It is important to stress that the evidence summarized here relates to permafrost extent during the coldest part or parts of the Late Weichselian. Greenland ice-core data demonstrate that the period ~31.0–14.7 ka incorporated episodes of sustained cold conditions punctuated by brief, relatively warm interstades (Figure 16.2), and there is stratigraphic evidence, particularly in loess and cover-sand sequences, for two or more phases of permafrost aggradation and degradation during this long interval (e.g. Van Huissteden *et al.*, 2000; Antoine *et al.*, 2016). It is probable that equilibrium permafrost disappeared from western and central Europe south of latitude ~60° N during the Bølling–Allerød Interstade of ~14.7–12.9 ka, except around the margins of the retreating Fennoscandian Ice Sheet, though relict permafrost may locally have persisted at depth throughout this period (Delisle, 1998). During the Younger Dryas Stade, however, equilibrium permafrost returned to extensive areas of northern Europe. Evaluation of the periglacial evidence attributable to this period allowed Isarin (1997a, 1997b) to reconstruct the extent of Younger Dryas permafrost in western and central Europe. He concluded that the southern limit of continuous permafrost (and the –8 °C MAAT isotherm) extended across Europe at latitude ~54° N, and that the southern margin of discontinuous permafrost (and the –1 °C MAAT isotherm) lay at ~50° N (Figure 16.15). For Isarin's continuous permafrost zone,

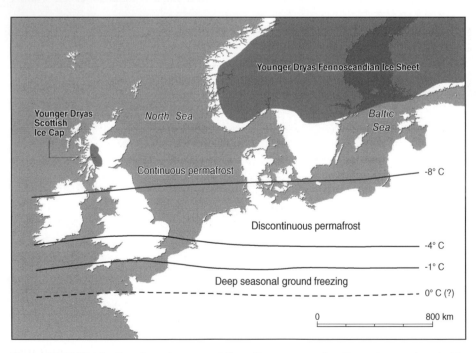

Figure 16.15 Distribution of continuous and discontinuous permafrost in western and central Europe during the Younger Dryas Stade and associated isotherms of inferred mean annual air temperature (MAAT). The discontinuous permafrost zone extends from the –8 °C isotherm to the –1 °C isotherm. *Source:* Adapted from Isarin (1997b). Reproduced with permission of Wiley.

mean July temperatures inferred from chironomid assemblages (Brooks and Langdon, 2014) were 7–9 °C, implying mean January temperatures below –23 °C; for the discontinuous permafrost zone, chironomid assemblages suggest mean July temperatures of 9–12 °C, implying that mean January temperatures ranged from ≤ –23 °C at ~54° N to about –10 °C to –13 °C at 50° N. It therefore appears that there was a much steeper meridional thermal gradient in winter (>10 °C) than in summer (~3 °C) over the 550 km separating the southern boundaries of continuous and discontinuous permafrost during the Younger Dryas Stade.

Isarin *et al.* (1998) noted that the largest differences between Younger Dryas temperatures and those of the present occurred during winter, particularly along the western seaboard of Europe, where winter temperatures are presently moderated by warm maritime airmasses. Aeolian landforms and model simulations (Isarin *et al.*, 1997) and an eastwards rise in the altitudes of Younger Dryas glaciers in Scotland (Sissons, 1979; Ballantyne and Harris, 1994) demonstrate that atmospheric circulation during the Younger Dryas was dominated by strong westerly winds blowing from the Atlantic Ocean. Isarin *et al.* (1998) reasoned from these observations that extensive winter sea-ice cover in the cold polar waters of the North Atlantic had cooled the air blowing eastwards across northern Europe, depressing winter air temperatures, and had steered Atlantic storm tracks along more southerly routes than occur now. They concluded that the permafrost boundaries in Europe during the Younger Dryas were strongly controlled by the latitude of the southern margin of winter sea ice in the North Atlantic Ocean, itself determined by reduction or cessation of North Atlantic thermohaline circulation (Broeker, 2006; dos Santos *et al.*, 2010) and by southwards movement of the oceanic polar front (Ruddiman and McIntyre, 1981).

This argument was developed by Renssen and Vandenberghe (2003), who compared atmospheric model simulations of sea-ice cover in the North Atlantic Ocean with reconstructed European MAATs derived from permafrost indicators. Their results suggested that during Weichselian cold phases, the southern limit of permafrost was strongly controlled by the latitude of the winter sea-ice margin, implying that during both the LGM and the Younger Dryas Stade, winter sea ice extended southwards to at least 50° N. Subsequent research using more sophisticated modelling (Vandenberghe *et al.*, 2012) confirmed a close relationship between North Atlantic winter sea-ice cover and the southernmost permafrost limit in Europe during the LGM. There is evidence, however, that the position of the North Atlantic oceanic polar front (and thus the extent of winter sea-ice cover) was linked to iceberg discharge events in the North Atlantic, and oscillated markedly during the Late Weichselian

(Eynaud *et al.*, 2009), so the southern limit of equilibrium permafrost in Europe seems likely to have migrated in response.

16.8.2   Late Weichselian Coversands in Europe

Late Weichselian aeolian sand deposits are widespread in northern Europe, in a sand belt that stretches across the North European Plain from England to Poland and farther east (Figure 14.9a). The stratigraphy of these deposits appears to be fairly consistent throughout much of this region, though not all units are present everywhere (e.g. Kolstrup, 2007; Zielinski *et al.*, 2014). The western European coversand has been studied in detail and dated using OSL techniques at sites in the eastern and southern Netherlands (Kasse *et al.*, 2007; Vandenberghe *et al.*, 2013). In these areas, it comprises six stratigraphic units (Figure 16.16): Older Coversand I (OCS I), which is overlain or truncated by the Beuningen Gravel Bed (BGB), which is overlain in turn by the Older Coversand II (OCS II), Younger Coversand I (YCS I), Usselo palaeosol and Younger Coversand II (YCS II). The sequence is locally overlain by recent drift sands that reflect anthropogenic disturbance of vegetation cover (Beerten *et al.*, 2014). Because of the dating imprecision of OSL techniques, the timing of the onset and termination of each unit is approximate and probably diachronous from site to site, but there is broad consistency in OSL ages and the attribution of individual units to particular time periods.

In valleys throughout northwest Europe, the OCS I unit generally overlies fluvial sediments, typically sandy bar deposits and silt drapes, which probably indicate an ephemeral nival runoff regime (Figure 16.16). These grade upwards to fluvio-aeolian deposits that appear to have begun to accumulate around 26–25 ka, then to aeolian sands that contain ice-wedge pseudomorphs and involutions and have been interpreted as indicating permafrost conditions and progressively increasing aridity during and after the LGM (Meinsen *et al.*, 2014). The coversands of OCS I are truncated by the BGB, which represents a stratigraphic marker throughout much of the European sand belt, but may be diachronous: some OSL age ranges for sand samples bracketing the BGB suggest that it formed between ~18–17 ka and ~16–15 ka (e.g. Kasse *et al.*, 2007), but others imply earlier formation (Bateman and Van Huissteden, 1999) or formation over more than one time interval (Vandenberghe *et al.*, 2013). In places, the BGB contains fluvially deposited gravelly sand overlain by a gravel lag layer; elsewhere, only the gravel lag is present. The BGB has generally been interpreted as representing a gravel lag deposit produced by deflation under cold desert conditions, and it may correspond to iceberg discharge event H1 in the North Atlantic Ocean.

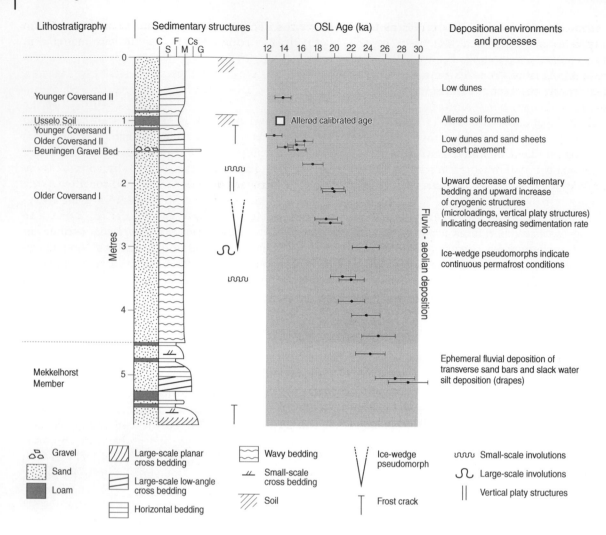

Figure 16.16 Lithostratigraphy, sedimentary environments and OSL ages of the Late Weichselian coversand succession at Grubbenvorst in the southern Netherlands. C, clay; S, silt; F, fine sand; M, medium sand; Cs, coarse sand; G, gravel. *Source:* Kasse *et al.* (2007). Reproduced with permission of Wiley.

The OCS II unit directly overlies the BGB and represents coversand deposition during the final millennia of the Late Pleniglacial, though at some sites deposition apparently continued into the early Lateglacial period after ~14.7 ka. An apparent lack of permafrost indicators in this unit suggests that it was deposited under less severe conditions, though frost cracks imply deep seasonal ground freezing. However, Buylaert *et al.* (2009) have described well-developed sand wedges and composite-wedge pseudomorphs in silty coversands above the BGB in Belgium, implying the existence of permafrost and severe winter freezing, through the resolution of their OSL dates did not allow them to specify whether these developed during the final part of the Late Pleniglacial or during the Younger Dryas Stade. At some sites, the coversands of OCS II grade imperceptibly into those of YCS I (Kasse *et al.*,

2007; Kolstrup, 2007), but at others the two are separated by a silt-rich layer. Vandenberghe *et al.* (2013) have suggested that this layer, where present, represents a northwards shift in loess deposition during the relatively warm Bølling period (~14.7–14.1 ka). As the Usselo palaeosol overlying YCS I deposits is of Allerød age (~14.0–12.9 ka), this suggests that YCS I may mainly have been deposited during a brief cooling event at ~14.0 ka (the Older Dryas oscillation), but as Kolstrup (2007) has noted, the timing of the transition from older to younger coversand differs between localities. OSL ages from various sites have confirmed that the YCS II deposits are of Younger Dryas age, though locally they grade into early Holocene aeolian sand deposits. Deposition ceased as complete vegetation cover became established (Kaiser *et al.*, 2009a; Zielinski *et al.*, 2015). Low-angle cross-bedding within YCS II deposits implies the formation of low dunes at this time, possibly reflecting increased vegetation cover during the preceding interstade, and indicating renewed sand drifting in response to marked cooling, increased aridity and opening of vegetation cover.

16.8.3 Late Weichselian Loess Deposits in Europe

Loess and loess-derived deposits in Europe extend discontinuously eastwards from northern France to the Black Sea, in a broad corridor bordered by the coversand belt to the north and the Alpine Ice Cap to the south (Haase *et al.*, 2007; Figure 14.13). Within this belt, many loess-palaeosol sequences span the entire last glacial–interglacial cycle above a palaeosol of last interglacial (Eemian) age, and consequently the loess belt provides the most extensive and continuous archive of Weichselian environmental change in Europe. From dated loess profiles across the European loess belt, Frechen *et al.* (2003) reconstructed mean accumulation rates of Late Weichselian loess deposits for the periods 28–18 ka and 18–13 ka (Figure 16.17). These show no consistent trend across the region, but highlight the importance of local loess sources, notably the Rhine River system in southern Germany and the Danube Basin in eastern Europe.

In western Europe, where loess deposition by westerly winds was strongly influenced by climatic fluctuations over the North Atlantic Ocean, Weichselian loess sequences record stratigraphic changes that can be related to millennial-scale climatic changes (Dansgaard–Oeschger cycles) evident in Greenland ice cores, and also a correlation between particle size parameters and ice-core dust content (e.g. Rousseau *et al.*, 2002, 2007; Moine *et al.*, 2008;

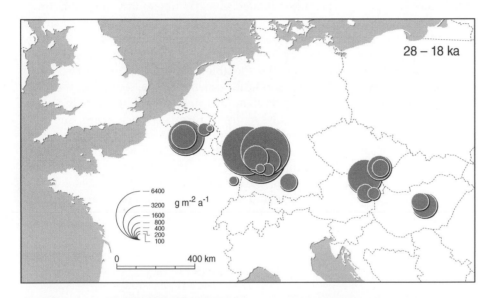

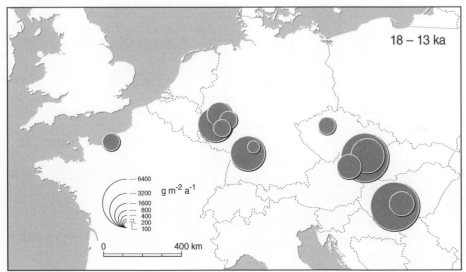

Figure 16.17 Mean accumulation rates of primary and reworked loess at European sites during the periods 28–18 ka (top) and 18–13 ka (bottom). *Source:* Data from Frechen *et al.* (2003).

Antoine *et al.*, 2009, 2013; Sima *et al.*, 2009). The thickest Weichselian loess accumulations, like that at Nussloch in southwest Germany (up to ~23 m), record not only the most rapid loess accumulation rates but also the greatest number of distinct stratigraphic units and subunits (e.g. Antoine *et al.*, 2009; Gocke *et al.*, 2014). Weichselian loess-palaeosol stratigraphy is nevertheless remarkably consistent across a wide area from northwestern France to western Germany and beyond, allowing correlation of particular units between sites. Antoine *et al.* (2016) have summarized the major elements of this common stratigraphy, covering the entire Weichselian Glacial Stage (Figure 16.18). Here we consider the Late Weichselian (Upper Pleniglacial) components of this stratigraphy and their palaeoenvironmental implications with reference to their synthesis of loess stratigraphy in northern France.

At the base of the sequence (Unit 7 in Figure 16.18) is a cryoturbated tundra gley palaeosol, OSL-dated to ~32–31 ka. This contains 2–3 m-deep loess-filled ice-wedge pseudomorphs that indicate development of ice-rich permafrost, followed by rapid permafrost degradation and infill of degrading ice wedges by loess. The lower part of the overlying unit (6) comprises loess and intercalated tundra gleys, suggesting climatic instability, overlain by calcareous loess that began to be deposited at ~30.6 ka, coeval with an increase in dust content in Greenland ice cores. Unit 5, dated to ~28 ka, comprises two tundra gley palaeosols with an intervening loess layer. The lower palaeosol contains loess-filled ice-wedges with thaw structures, indicating first the renewed formation of permafrost, then permafrost degradation in

response to warming. The upper band of tundra gley contains a later generation of superimposed wedge structures, indicating a further period of cooling and permafrost formation, but in this case these lack thaw structures, suggesting that loess was the primary infill. Unit 4 comprises finely laminated loess exhibiting synsedimentary cryodesiccation microcracks. This loess was deposited between ~27 and ~23 ka and has been interpreted as indicating open, unvegetated conditions. The overlying unit (3) comprises the most recent tundra gley palaeosol, which is cut by the youngest ice-wedge pseudomorphs, typically 0.7–0.8 m deep. This palaeosol is locally overlain by a thin layer of decalcified loess, but elsewhere is integrated with the surface soil, so the uppermost loess layer cannot be detected. A notable feature of this sequence is the apparent absence of Younger Dryas loess, as in the British Isles, suggesting that vegetation cover limited silt entrainment by wind at that time.

As this example demonstrates, the loess-palaeosol sequences of western Europe have potential not only to provide a teleconnection to other Weichselian palaeoclimatic proxies (notably the Greenland ice core record), but also to indicate millennial-scale changes in the status of permafrost in marginal areas during this period. Remarkably, many features of the western European loess stratigraphy are also recorded in central Europe, much farther from North Atlantic climatic influence. An excellent example is the loess-palaeosol record in the Czech Republic at Dolní Vestoni (~49° N), which is over 1700 km east of the Atlantic ocean and ~300 km south of the limit of the Fennoscandian Ice Sheet. The stratigraphy at this site spans the entire Weichselian, though

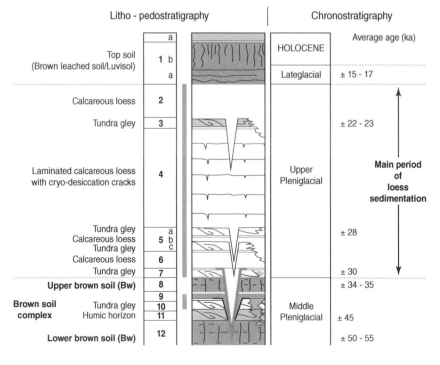

Figure 16.18 Synthesis of Late Weichselian (Middle and Upper Pleniglacial) loess-palaeosol stratigraphy for northern France. The V-shaped symbol represents ice-wedge pseudomorphs. *Source:* Adapted from Antoine *et al.* (2016). Reproduced with permission of Elsevier.

there is a hiatus (unconformity) between a Middle Weichselian interstadial soil and the overlying Late Weichselian loess sequence (Antoine *et al.*, 2013). The latter is underlain by a palaeosol dated to ~30.6 ka, and comprises a ~6 m thick layer of sandy calcareous loess that represents a drastic increase in sedimentation rate after ~27 ka; it is equivalent to unit 4 in the northern France sequence. The grain size of this loess unit exhibits a general coarsening upwards trend, superimposed on which is a pattern of coarse loess events that can be correlated with similar episodes in the loess sequences of western Europe, for example at Nussloch in the Rhine Valley (Antoine *et al.*, 2009) and even as far east as Ukraine (Rousseau *et al.*, 2011). Conversely, episodes of relatively fine-grained loess deposition appear to be linked to brief interstadial events evident in the Greenland ice-core stratigraphy. High-resolution multi-proxy deciphering of the European loess stratigraphy therefore provides a unique level of insight into environmental changes *during* the Late Weichselian on millennial or shorter timescales, and one that can potentially be correlated across much of the continent.

16.9 Late Wisconsinan Periglacial Environments in North America

Most research on past periglacial environments in North America is comparatively recent, and the published evidence relating to past periglaciation is slighter than is the case for Europe. During the LGM (~26.5–19 ka), the Late Wisconsinan Laurentide Ice Sheet covered almost all of Canada except for the interior Yukon, parts of the Beaufort Sea coast and the western islands of the Canadian Arctic Archipelago, areas that were too arid to support glacier ice cover at that time. In these areas, relict Pleistocene permafrost has survived to the present. In western North America, the ice sheet extended to just south of the Canadian border (roughly 48° N), but farther east it penetrated southwards to latitude 40° N between Illinois and northern New Jersey. Analyses of subfossil coleopteran assemblages of LGM age from various sites south of the ice-sheet limit in Iowa, Missouri and Illinois suggest that mean July temperatures were within the range 12.9 °C to 15.4 °C, and that mean January temperatures were within the range −13 °C to −30 °C (Elias, 1999); the beetle assemblage at one site in Iowa at ~40° N, dated to ~21.4 ka, indicates a mean July temperature estimate of 12.9 °C and mean January temperature estimate of −27 °C to −29 °C, implying that MAAT was roughly −6.5 °C. Pollen assemblages from sites near the ice-sheet limit in Indiana and Ohio demonstrate that terrain outside the LGM ice-sheet margin was dominated by tundra or peatland, with scattered spruce (Heusser *et al.*, 2002); areas farther from the ice margin supported boreal forest or parkland.

16.9.1 Evidence for Late Wisconsinan Permafrost in North America

The approximate extent of LGM permafrost during and after the Late Wisconsinan ice-sheet maximum in North America has been reconstructed by French and Millar (2014), principally from the recorded distribution of ice-wedge pseudomorphs and relict sand wedges (Figure 16.19), though they also included possible indicators of past permafrost such as solifluction deposits, patterned ground features, cryoplanation terraces, blockstreams and cryoturbations in their analysis. As they acknowledged, however, the dearth of recorded evidence over wide areas means that their map of LGM permafrost extent is provisional, and that the available evidence does not permit differentiation of continuous, discontinuous and sporadic past permafrost. Some of the most detailed evidence is available from southern New Jersey (39° N), where relict sand wedges dated to ~18–15 ka and thermokarst involutions were interpreted by French *et al.* (2003) as representing discontinuous permafrost that developed under MAATs of roughly −3 °C to −4 °C. Similar features, together with sediment-filled wedges and a widespread fragipan, provide evidence for possible past permafrost farther south, in Delaware, Maryland, and along the Atlantic coastal plain (Lemcke and Nelson, 2004; French *et al.*, 2009). Gao (2014) concluded that relict thermal contraction-crack polygons on the coastal plain demonstrate that a zone of continuous permafrost stretched 250 km south of the ice-sheet margin to 38.5° N, and imply a Late Wisconsinan MAAT of −6 °C or lower. West of the Appalachians, evidence for Late Wisconsinan permafrost (probably continuous permafrost) in the form of ice-wedge pseudomorphs, relict sand wedges and frost polygons occurs in a zone roughly parallel to the limit of the Laurentide Ice Sheet (e.g. Johnson, 1990; Wayne, 1991; Walters, 1994; Doolittle and Nelson, 2009), then broadens southwards across the higher ground of eastern Wyoming (Nissen and Mears, 1990). Extensive former alpine permafrost in the Rocky Mountains and western Cordillera is indicated by apparent examples of relict rock glaciers, though French and Millar (2014) cautioned that some of these features might be of non-permafrost origin.

An interesting contrast between the Late Weichselian permafrost zone south of the Fennoscandian Ice Sheet in Europe and the Late Wisconsinan permafrost belt south of the Laurentide Ice Sheet in North America is that the latter appears to have been much narrower (typically 100–250 km wide) than the former (up to 1000 km wide). This contrast may reflect under-representation of the true extent of Late Wisconsinan permafrost in North America, but is at least partly explicable in terms of the much deeper southwards penetration of the last Laurentide Ice Sheet (to ~40–48° N) compared with the last

Area covered by the last ice sheet during Late Wisconsinan glaciation

Lowland areas of past permafrost (continuous, discontinuous or sporadic)

Areas of montane or alpine past permfrost

▼ ▼ Ice-wedge or composite-wedge pseudomorphs, relict sand wedges or
relict thermal contraction-crack polygons of Late Wisconsinan age

----- Southern limit of LGM ice

——— Limit of LGM past permafrost

0 ————————— 1000 km

Figure 16.19 Extent of past (LGM) permafrost south of the limit of the Late Wisconsinan ice sheet in North America, as estimated by French and Millar (2014). Relict frost wedges and frost polygons outside the ice-sheet limit are inferred to be of LGM age, whereas those inside the limit represent the development of permafrost on terrain vacated by the retreating ice sheet. *Source:* French and Millar (2014). Reproduced with permission of Wiley.

Fennoscandian Ice Sheet (~53° N). This probably implies a steeper atmospheric thermal gradient south of the ice-sheet margin in North America, and consequently a narrower zone of permafrost formation during the LGM.

There is abundant evidence, mainly in the form of relict frost wedges and frost-wedge polygons, that permafrost formed under terrain vacated by the retreating Laurentide Ice Sheet (Figure 16.19). Moreover, where this evidence is securely dated, it indicates persistence of continuous permafrost several millennia after the LGM. Lusch *et al.* (2009), for example, have described large polygons, apparently formed by thermal contraction cracking of permafrost between ~14.8 ka and ~14.3 ka, in the lowlands southeast of Lake Huron, and Gao (2005) has outlined convincing evidence for ice-wedge pseudomorphs in southern Ontario, dated on the basis of till stratigraphy to between ~15 and ~13 ka. Such evidence

suggests that the belt of Late Wisconsinan permafrost persisted, at least locally, adjacent to the margin of the retreating ice sheet. This interpretation is supported by Fisher (1996), who documented relict sand wedges in northwest Saskatchewan (56° N), which he inferred to have formed between ~12.9 ka and ~12.3 ka, immediately after ice-sheet deglaciation. Evidence for permafrost aggradation during the Younger Dryas period (~12.9–11.7 ka) is present on the Magdalen Islands (47.5° N) in the Gulf of St Lawrence, where ice-wedge and composite-wedge pseudomorphs have been dated by Rémillard *et al.* (2015), and on Newfoundland, where ice-wedge pseudomorphs have been dated to between 14.0–12.8 ka and ~12.2 ka (Liverman *et al.*, 2000). Both studies confirm that the Atlantic Provinces of Canada experienced marked cooling at this time, similar to that in the British Isles and northwest Europe.

16.9.2 Late Wisconsinan Aeolian Deposits in North America

Outside of the mountainous areas of North America, which support a wide range of relict periglacial features (tors, blockfields, blockstreams, cryoplanation terraces, solifluction landforms and deposits, and large-scale frost-sorted patterned ground), by far the most widespread periglacial deposits are extensive areas of loess, windblown sand and dunefields, most of which were deposited beyond the limits of the last ice sheet. These were introduced in Chapter 14, and here we focus in more detail on those of Late Wisconsinan age in the mid-latitude United States and Canada. Ventifacts have also been reported from areas close to the ice-sheet margin during the LGM (Demitroff, 2016) and during its retreat (Fisher, 1996), and provide evidence of cold, dry, windy conditions and sparse vegetation cover.

By far the largest area of Late Wisconsinan loess occurs in mid-continental North America, and stretches southwards from the limit of the Laurentide Ice Sheet to the lower Mississippi Valley and eastwards from Colorado to Ohio (Figure 16.20). Loess cover north of the glacial limit is generally thin, and is thought to be derived from long-distance transport (Jacobs *et al.*, 2011), the drained floors of proglacial lakes (Schaetzl, 2012) or the braidplains of rivers draining the retreating ice margin (Schaetzl and Attig, 2013). By contrast, the Late Wisconsinan loess of the mid-continental region, the Peoria Loess, is generally 1–5 m thick, and exceeds 10 m close to its sources, in Nebraska and east of the Missouri and Mississippi Rivers (Figure 16.20). Within this region, Bettis *et al.* (2003) distinguished a Great Plains province (Nebraska, Kansas and east Colorado) and a Central Lowlands province (roughly the Missouri–Mississippi Valley and areas farther east).

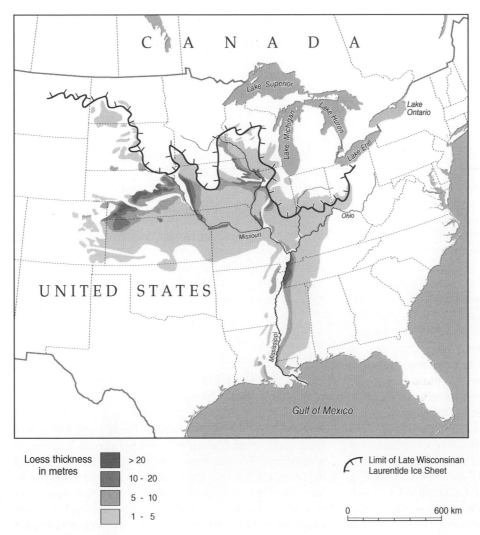

Loess thickness in metres
- \> 20
- 10 - 20
- 5 - 10
- 1 - 5

Limit of Late Wisconsinan Laurentide Ice Sheet

0   600 km

Figure 16.20 Distribution and thickness of loess, mainly Peoria Loess of Late Wisconsinan age, in central North America. *Source:* Muhs (2013). Reproduced with permission of Academic Press.

In the former, the loess deposits were apparently sourced mainly from volcanoclastic siltstones; in the latter, the outwash plains of rivers draining the Laurentide Ice Sheet furnished the main source of loess, together with a component of far-travelled aeolian dust from nonglacigenic sources on the Great Plains (Aleinikoff *et al.*, 2008; Muhs *et al.*, 2008, 2013). Abundant dating evidence suggests that the Peoria Loess began to accumulate between 31 ka and 25 ka, or locally earlier, and that accumulation largely ceased by 14–13 ka. The basal age of Peoria Loess decreases eastwards of the Missouri River, implying that there was progressive overlap of older by younger loess deposits (Bettis *et al.*, 2003).

The mid-continental loess deposits provide conclusive evidence that during the Late Wisconsinan glacial stage, the dominant dust-carrying palaeowinds blew across this area from the northwest or west. This is evident from an eastwards reduction in loess thickness (Figure 16.20), changes in geochemical characteristics and an eastwards decline in modal grain size (Bettis *et al.*, 2003). From detailed analyses of the timing and characteristics of Peoria Loess at a site in western Iowa, Muhs *et al.* (2013) demonstrated that the coarsest loess and highest mean accumulation rates (10 m in ≤2000 years) occurred between ~24 and ~21 ka, when the Laurentide Ice Sheet reached its southernmost limit. They attributed this remarkable rate of accumulation to strong westerly winds generated by the steep meridional temperature gradient in the zone south of the ice-sheet margin, supporting the concept that 'increased gustiness enhanced dustiness' (McGee *et al.*, 2010) at that time. Though this is an attractive idea, it may not be valid everywhere: Roberts *et al.* (2003) found that the maximum loess accumulation rates in western Nebraska occurred not during the LGM, but in the final part of the Late Wisconsinan stage, after the onset of ice-sheet retreat.

Farther west, thin (generally <2 m thick) loess deposits of Late Wisconsinan age are present on the Colorado Plateau and the Snake River Plain (Columbia Plateau), probably sourced mainly from the floodplains of the San Juan River and Snake River respectively (Figure 14.12). The loess deposits of the Snake River Plain began to accumulate at ~28–25 ka or earlier, and accumulation terminated at ~16–12 ka; like the mid-continental loess deposits, these were deposited by palaeowinds blowing from the west or northwest (Bettis *et al.*, 2003). The extensive (~50 000 km²) upper loess deposits in the Palouse region of eastern Washington and adjacent parts of Idaho and Oregon (Figure 14.12) are also of Late Wisconsinan age, but date to the period after the Laurentide Ice Sheet began to retreat. This is because they were sourced from sediments deposited by the cataclysmic floods released by drainage of proglacial Lake Missoula as the ice barrier damming the lake thinned following the LGM.

The most extensive areas of aeolian sand deposits (coversands) and dunefields in North America occur in the central and western United States, south of the limit of the Laurentide Ice Sheet, with smaller areas south and southwest of the Great Lakes (Figure 14.9b) and northern outliers in areas formerly occupied by the ice sheet in Saskatchewan and Alberta. The most extensive area of aeolian deposits, the Nebraska Sand Hills, covers ~50 000 km² and contains barchanoid ridges up to 20 km long with superimposed parabolic dunes and areas of longitudinal dunes, often superimposed on earlier dune forms. The aeolian sand deposits of adjacent states (Wyoming, Kansas and Colorado) contain areas of parabolic dunes separated by sand sheets. All are presently stabilized by vegetation cover.

Though there is evidence for aeolian deposition and dune formation in these areas during the Late Wisconsinan glacial stage, indicating transport by westerly or northwesterly winds at that time (Muhs *et al.*, 1996), dated palaeosols at sites throughout this region indicate extensive renewed sand deposition and sand reworking at various times during the Holocene. Sun and Muhs (2013) concluded that the extent of Holocene sand deposits in the Great Plains region is much greater than that of Late Wisconsinan sand deposits, so that the orientation of dunes cannot be used to interpret the direction of Late Wisconsinan palaeowinds.

More limited evidence of Late Wisconsinan aeolian sand deposition is also found at sites along the Atlantic Coastal Plain between Maryland and Georgia, where parabolic dunes and sand hills interpreted as dune remnants extend for short distances (often <3 km) east and northeast of major rivers. These deposits have yielded luminescence ages of ~40–15 ka, indicating entrainment of floodplain sediments and deposition by westerly winds throughout much of the Late Wisconsinan (Ivester and Leigh, 2003; Swezey *et al.*, 2013). A small number of OSL ages also suggest reactivation of dunes during the Younger Dryas Stade.

16.10 Permafrost Extent in the Northern Hemisphere During the Last Glacial Stage

An ambitious recent development has been the use of local and regional inventories of periglacial features to reconstruct the maximum extent of permafrost during the MIS 2 glacial stage for the entire northern hemisphere. Such reconstructions acknowledge that the maximum extent of permafrost was asynchronous across the entire hemisphere, and assume that undated indicators of past permafrost are likely to have been active during the period of most intense cold. Vandenberghe *et al.* (2014)

LPM permafrost on present land area

LPM permafrost on exposed shelves

Extent of continental ice sheets during LGM

- - - Approximate limit of LPM winter sea ice

0   5000 km

Figure 16.21 Extent of northern hemisphere permafrost during the last permafrost maximum (LPM). Maximum permafrost extent was probably not synchronous across all areas. *Source:* Adapted from Vandenberghe *et al.* (2014). Reproduced with permission of Wiley.

based their map of LPM permafrost (Figure 16.21) on the documented distribution of ice-wedge pseudomorphs, relict sand wedges and large cryoturbations, and, where possible, attempted to discriminate between continuous and discontinuous permafrost. A GIS-based compilation of similar archival evidence by Lindgren *et al.* (2016), using only ice-wedge pseudomorphs and relict sand wedges, resulted in a map of the maximum extent of LGM permafrost that differs from that of Vandenberghe *et al.* (2014) only in detail. Both teams noted the difficulty of distinguishing the southern continuous–discontinuous permafrost boundary on the basis of existing evidence. Vandenberghe *et al.* (2014) estimated the total area of LPM permafrost as 36.5 million km^2; Lindgren *et al.* (2016) calculated the probable area of past permafrost as 34.5 million km^2, but noted numerous areas of uncertainty, particularly for permafrost extent on continental

shelves that were exposed by the fall of sea level (~125 m) that accompanied growth of the last ice sheets. Lindgren *et al.* (2016) presented a breakdown of maximum permafrost area by continent (Table 16.4), though it should be noted that this includes non-permafrost terrain in areas of discontinuous and sporadic permafrost, so the true areas for permafrozen ground are slightly less than those shown. Interestingly, they estimated that areas of discontinuous or sporadic permafrost occupied only ~15% of the total, much less than the present 53% (Brown *et al.*, 2014). This is also evident from the map produced by Vandenberghe *et al.* (2014), who depicted only a narrow discontinuous permafrost belt in Europe (Figure 16.11). The areas calculated by Lindgren *et al.* (2016) suggest that the maximum LGM permafrost extent was ~33% greater than that of the present. This figure demonstrates that though permafrost subsequently formed in arctic

Table 16.4 Area estimates (10^3 km^2) of the northern circumpolar permafrost region during the last glacial maximum (LGM). Estimates include zones underlain by discontinuous and sporadic permafrost. *Source:* Lindgren *et al.* (2016). Reproduced with permission of Wiley.

	Likely	Minimum	Maximum	Present
Under present-day land areas:				
Asia	21 600	20 450	21 600	14 210
North America	3000	2780	3230	8590
Europe	5200	4760	5380	800
Total	29 800	27 990	30 210	23 600
Sea shelves	4700	4700	5100	2400
Total permafrost	34 500	32 690	35 310	26 000
Glacier ice	28 200	–	–	2130
Subglacial permafrost	7350	5220	7350	1700

and subarctic areas exposed by the retreat of northern hemisphere ice sheets, this gain in permafrost area failed to compensate for the loss of permafrost in mid-latitude areas as climate warming drove the southern boundary of equilibrium permafrost northwards towards its present limits.

In tandem with attempts to reconstruct the extent of past permafrost from proxy periglacial evidence, various attempts have been made to model past permafrost dynamics (rates of aggradation and degradation) or depth for particular areas (e.g. Delisle, 1998; Kitover *et al.*, 2013; Busby *et al.*, 2016). By coupling a permafrost model with an Earth-system model, Kitover *et al.* (2016) have derived the first simulations of both permafrost extent and thickness for the northern hemisphere during the LGM. Their model underestimates permafrost extent in comparison with the reconstruction of Vandenberghe *et al.* (2014), but for much of Asia it appears to approximate the boundary of continuous permafrost; the spatial resolution of the model is insufficient to identify areas of discontinuous permafrost, but shows them as having mean annual ground surface temperatures (MAGST) $\geq 0\,°C$. For various parts of Asia north of 60°N, it produces a very wide range of simulated permafrost thicknesses, up to 1030 m for the central Siberian Plateau and central Siberian Coastal Plain. As the authors admit, the results 'are not wholly congruent with observational data', but the approach should be regarded as a first step towards refining models at different scales to produce convergence between model outcomes and empirically based reconstructions.

16.11 Concluding Comments

The preceding brief reviews of last glacial stage periglacial environments in Europe and North America highlight several features. As with the review of the past periglaciation of the British Isles, the importance of absolute dating and multi-proxy approaches emerges keenly. Analyses of the distribution of periglacial features within GIS databases (Andrieux *et al.*, 2016; Lindgren *et al.* 2016) present great potential for refining and progressively updating the distribution of key permafrost features and for establishing regional associations between landforms, deposits, sediment structures and terrain characteristics. The relationship between past permafrost extent in Europe and winter sea-ice cover (Isarin *et al.*, 1998; Renssen and Vandenberghe, 2003) not only provides a causal link between the latitude of the North Atlantic oceanic polar front, winter sea-ice cover, permafrost extent and palaeotemperature regime, but also opens a window on the future consequences of sea-ice shrinkage in the Arctic Ocean on circum-arctic permafrost (Vandenberghe *et al.*, 2013). The Weichselian loess and coversand stratigraphies of Europe reveal evidence for quasi-synchronous environmental changes across the continent, and provide event stratigraphies that can be linked to those in the Greenland ice-core record. Unravelling of the sources of Wisconsinan aeolian deposits in North America throws light on palaeowind vectors, and future analyses on the European model may allow emergence of a similar event stratigraphy, though it is notable that whereas most European loess and coversand deposits occupy areas of past permafrost, those of North America south of the Laurentide Ice Sheet limits were mainly deposited outside the permafrost zone. Finally, there is still debate over the palaeotemperature significance of key indicators of past permafrost, particularly ice-wedge pseudomorphs in host sediments of contrasting texture, relict frost polygons and relict sand wedges; this is evident, for example, in contrasting interpretations of the southern limit of Late Weichselian permafrost in France (Vandenberghe *et al.*, 2014; Andrieux *et al.*, 2016). Such disagreements can probably only be resolved through further investigation of present-day thermal contraction cracking in areas of discontinuous permafrost and seasonally frozen ground.

Reconstructing the boundaries of LPM permafrost is of more than academic significance. The rapid warming that accompanied the Lateglacial–Holocene climatic transition resulted not only in shrinkage of ice sheets and sea-ice cover, but also in the final disappearance of equilibrium permafrost from vast areas of northern-hemisphere mid-latitudes, with concomitant thaw of soil and ground ice. The frozen soils of the LPM stored large amounts of organic carbon in the form of partially decomposed plant matter, and subsequent thaw of organic-rich soil released carbon to the atmosphere (Zimov *et al.*, 2006, 2009; Walter *et al.*, 2007; Ciais *et al.*, 2012; DeConto *et al.* 2012), contributing to increases in atmospheric greenhouse gases (carbon dioxide and methane) and thus probably accelerating climate warming at the last glacial–interglacial transition. We began this chapter with a phrase coined by the Scottish geologist Archibald Geikie: 'the present is the key to the past'. As we enter a new era of rapid climate change, these dramatic events suggest that the past might equally hold the key to the future. The potential consequences for the periglacial realm are explored in the final chapter.

17

Climate Change and Periglacial Environments

17.1 Introduction

The climate of the Earth is constantly changing. As we saw in the previous chapter, the Quaternary Period has been dominated by the alternation of glacial and interglacial stages, interrupted by shorter-term stadial and interstadial oscillations. During much of the present (Holocene) interglacial, climatic shifts have been less pronounced and superimposed on a gradual cooling trend (Figure 16.2). In the early Holocene, northern high latitudes experienced a period of relatively high temperatures, roughly 1–3 °C above the 20th-century average, though this *Holocene thermal maximum* (sometimes termed the *climatic optimum* or *hypsithermal*) was spatially asynchronous: in Alaska and northwest Canada, it occurred at ~11–9 ka, in the Canadian Arctic Archipelago, Greenland and Iceland around 9 ka, and in the Hudson Bay region as late as ~7 ka (Kaufman *et al.*, 2004). Data for the Arctic Ocean coasts of Siberia suggest that average temperatures were about 2–4 °C higher than the 20th-century average during the Holocene thermal maximum. Since then, middle and high latitudes in the northern hemisphere have experienced gradual but oscillatory cooling, culminating in the *Little Ice Age* of the 16th to 19th centuries AD, a period marked by expansion of glaciers and renewed permafrost aggradation.

After the end of the Little Ice Age, global surface air temperatures warmed gradually (though with marked interannual and interdecadal variability) until about 1960, after which they have undergone a much steeper increase (Figure 17.1). Globally-averaged surface temperature data show a warming of 0.85 °C between 1880 and 2012, and in the northern hemisphere the three decades from 1983 to 2012 probably represent the warmest 30 year period for at least 1400 years (IPCC, 2013). This recent warming has been attributed to an increase in the concentrations of well-mixed greenhouse gases in the atmosphere, particularly carbon dioxide (CO_2), methane (CH_4) and nitrous oxide (N_2O) as a result of anthropogenic emissions of these gases, mainly through burning of fossil fuels and land-use changes. Pre-1750 concentrations of these gases in the troposphere (lower atmosphere)

were ~280 ppm (parts per million), ~722 ppb (parts per billion) and ~270 ppb respectively. By 2016, these concentrations had risen to 400 ppm, 1834 ppb and 328 ppb, increases of 43%, 154% and 21%. Though the concentrations of these gases are small, they absorb re-radiated longwave energy at particular wavelengths, and thus alter the rate of energy exchange in the troposphere. The extent to which greenhouse gases change the tropospheric energy balance is termed *radiative forcing*, and the increased radiative forcing attributable to the increased concentrations of CO_2, CH_4 and N_2O since 1750 has been estimated (in 2016) to be 1.94, 0.50 and 0.20 W m^{-2} respectively. Though the atmospheric concentrations of CH_4 and N_2O are much smaller than those of CO_2, these gases have much greater warming potential: the radiative efficiency of CH_4 is about 26.5 times greater than that of CO_2, and that of N_2O is about 221 times greater (IPCC, 2013). The average residence time of CH_4 in the atmosphere is only about 12 years, however, whereas CO_2 may persist for ~100–300 years.

Maps of recent surface temperature changes show that warming has tended to be greatest over land, and has been strikingly greater at high latitudes, particularly in the Arctic (Figure 17.2), where the rise in nearsurface air temperatures has been roughly twice the global average. This effect is known as *Arctic amplification*, and is an inherent characteristic of the global climate system. This phenomenon is still incompletely understood, though most authors acknowledge that there are multiple interrelated causes (Serreze and Barry, 2011). These include thermal feedback effects, reduction in the extent and duration of summer and autumn sea-ice cover and snowcover, increased poleward transport of heat and moisture to the Arctic, changes in cloud cover and water vapour that alter the longwave radiation flux to the surface, and even the effect of 'black carbon' (soot) in reducing the albedo (reflectivity) of snowcover. Decreasing summer sea-ice cover certainly appears to play a major role in enhancing Arctic temperatures: as summer sea-ice cover across the Arctic Ocean diminishes, heat stored in the upper ocean during the summer is released during autumn and winter to cause nearsurface atmospheric

Periglacial Geomorphology, First Edition. Colin K. Ballantyne.
© 2018 John Wiley & Sons Ltd. Published 2018 by John Wiley & Sons Ltd.

warming. Screen and Simmonds (2010) have shown that the vertical structure of seasonal temperature amplification in the lower atmosphere is consistent with this explanation. However, climate model simulations by Pithan and Mauritsen (2014) indicate that the primary cause of Arctic amplification is a thermal feedback effect:

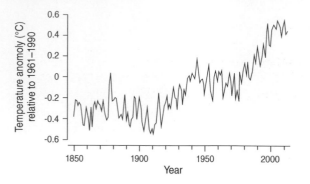

Figure 17.1 Globally averaged combined land and ocean surface temperature anomaly, 1850–2012, relative to 1961–90 mean surface temperature.

during periods of surface warming, less energy is radiated back to space at high latitudes than at low latitudes, though they concede that this effect is enhanced by surface albedo feedback effects due to reduction in summer sea-ice cover and reduction in snowcover on land. Recent warming of northern hemisphere high latitudes appears particularly marked when compared with a 2000 year record of summer air temperatures for sites polewards of 60° N, derived from proxy temperature records (Kaufman *et al.*, 2009). This shows that warming after the mid 20th century terminated a long-term decline in Arctic summer temperatures (Figure 17.3), and that these are now higher than at any time in the past two millennia. In the southern hemisphere, temperature amplification appears to be limited over much of the Antarctic Ice Sheet, but there is evidence for enhanced warming of the Antarctic Peninsula and possibly West Antarctica (Steig *et al.*, 2009); the Antarctic Peninsula is one of the most rapidly warming regions on the planet, with an increase of mean annual air temperature (MAAT) of 0.56 °C per decade between 1950 and 2000 (Turner *et al.*, 2005).

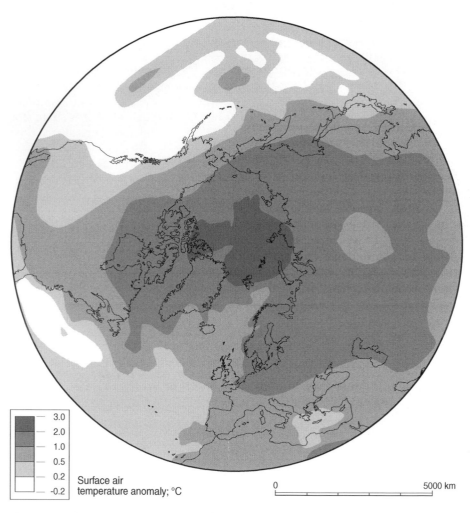

Figure 17.2 The warming Arctic: annual surface air-temperature anomalies, averaged for 2005–09, relative to the 1951–2000 mean. *Source:* Adapted from AMAP (2011) with permission from the Arctic Monitoring and Assessment Programme, Oslo.

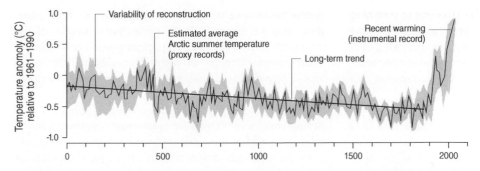

Figure 17.3 Estimated Arctic summer temperature anomalies, relative to the 1960–90 mean, for the past 2000 years, and the instrumental record of recent warming. The summer temperatures are based on proxy records obtained from lake sediments, ice cores and tree rings from 23 sites. *Source:* Reproduced from Kaufman *et al.* (2009) with permission from the American Association for the Advancement of Science.

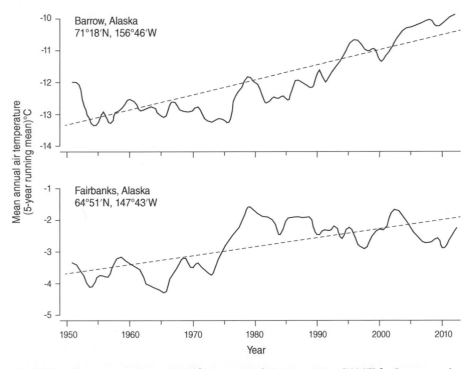

Figure 17.4 Five-year running means of mean annual air temperature (MAAT) for Barrow on the northern coast of Alaska and Fairbanks in interior Alaska, 1949–2014. The dashed line shows the overall trend. *Source:* Data from Alaska Climate Research Center.

Recent warming of northern high latitudes has been more pronounced in some locations than others, and the timing of warming has not been synchronous. Figure 17.4 compares 5-year running means of MAATs for Barrow, on the Beaufort Sea Coast of northern Alaska, and Fairbanks, 715 km farther south in interior Alaska. The Barrow record shows no evidence of warming over the period 1950–75, then rapid warming of about 3 °C from 1975 to 2010, partly due to decreasing sea-ice cover and storage of heat in the adjacent sea. The Fairbanks record shows marked warming in the 1960s and 1970s, but no warming trend thereafter.

Predictions of future climate change are based on model simulations that assume particular future greenhouse gas emission scenarios and the associated changes in radiative forcing. All plausible scenarios envisage an increase in CO_2 concentrations throughout the 21st century, as CO_2 has a long atmospheric residence time (~100–300 years), so that even under the assumption of reduced CO_2 emissions, the cumulative effect is one of increased CO_2 concentration and thus increased radiative forcing. According to the latest model predictions of the Intergovernmental Panel on Climate Change (IPCC, 2013), the increase in global mean temperatures for 2081–2100 relative to 1986–2005 ranges from a minimum of 0.3–1.7 °C to a maximum of 2.6–4.8 °C, depending largely on assumed future anthropogenic greenhouse gas emissions. All models predict greater surface warming over land than over the oceans, and more rapid warming of the Arctic than elsewhere. A report by the Arctic Monitoring and Assessment Programme (AMAP, 2011) concludes that average Arctic autumn and winter temperatures will rise by 3–6 °C even if greenhouse gas emissions stabilize.

Atmospheric warming is only one component of recent climate change, as differential warming at different latitudes and between the land and ocean surfaces also affects atmospheric circulation and causes changes in precipitation patterns and cloud cover. Prediction of changes in precipitation is complex, but almost all models suggest increased precipitation over most of the Arctic by 2050, with the greatest increase in winter snowfall over parts of Siberia. However, because of higher spring temperatures, average snowcover duration across the Arctic is projected to decline by 10–20% over the same timescale, with the greatest reduction (30–40%) over Alaska and northern Scandinavia (AMAP, 2011). Though snowcover duration varies greatly both regionally and annually, satellite data show that for the Arctic as a whole annual snowcover duration has already decreased by an average of about 12 days since the early 1970s, mainly as a result of earlier snowmelt: May and June snowcover extent decreased 14% and 46% respectively over the pan-Arctic region between 1967 and 2008 (Brown *et al.*, 2010). In combination with elevated summer temperatures, a more prolonged snow-free period is likely to increase surface evaporation, leading to drying of some wetland areas. Future changes in snow depth and snowcover duration are also likely to alter the hydrologic regime of arctic rivers, with earlier arrival of spring snowmelt floods.

As this summary indicates, many periglacial environments have already experienced significant climatic changes over the past 50 years, and such changes are likely to continue, and probably accelerate, throughout the coming decades. This chapter considers the present and likely future effects of such changes from a geomorphological perspective. Inevitably, attention focuses on the circumpolar north, where permafrost terrain is likely to be most affected, but the consequences of future climate changes in alpine and montane periglacial environments are also considered. The chapter concludes by considering how future changes in periglacial environments may feed back into the climate system, primarily through release of carbon dioxide and methane to the atmosphere as a result of microbial decomposition of thawed organic matter that is presently frozen in permafrost.

17.2   Permafrost Degradation

As we saw in Chapter 4, equilibrium permafrost on land occurs where there is a thermal balance between the existing (below 0 °C) mean annual ground surface temperature (MAGST) and geothermal heat flux. Assuming no change in the latter, perturbation of the former by an increase in air temperature is likely to cause a progressive change in permafrost thermal state. In areas of cold permafrost, the effect may simply be progressive downwards warming of the permafrost, accompanied by deepening of the active layer. In areas of warm permafrost, the effect of sustained ground surface warming may be more drastic, involving permafrost degradation through development of a suprapermafrost talik and, if warmer ground-surface temperatures persist, complete thaw and replacement of permafrost by seasonally frozen ground. Callaghan *et al.* (2011) have provided a useful overview of current and projected changes in permafrost in the Arctic, and some of the geomorphological and hydrological consequences.

17.2.1   Recent Trends in the Thermal State of Permafrost

A major inventory of the thermal state of permafrost was made during the International Polar Year (2007–09). The overall results were summarized by Romanovsky *et al.* (2010a) and regional syntheses were produced for North America (Smith *et al.*, 2010), Russia (Romanovsky *et al.*, 2010b), the Nordic area including Greenland (Christiansen *et al.*, 2010), the Antarctic (Vieira *et al.*, 2010) and Central Asia (Zhao *et al.*, 2010). This survey was based on vertical temperature profiles recorded in boreholes, and in addition to providing a global snapshot of permafrost temperatures, yielded valuable information on changes of the thermal state of permafrost in response to recent climate warming. A key finding of this programme is that mean annual ground temperatures (MAGTs) measured in the northern-hemisphere discontinuous permafrost zone tend to be warmer than –3 °C, and usually warmer than –2 °C, whereas in the continuous permafrost zone they are generally colder (–2 °C to –15 °C). Most sites for which sequential measurements of permafrost temperature are available show a gradual warming trend, consistent with response to a general warming of air temperatures since the 1970s (Figure 17.5). Recent warming is also evident in changes in vertical temperature profiles through time. Figure 17.6 shows recent temperature changes in two boreholes at Alert on Ellesmere Island (82° 30′ N). Warming has penetrated to depths at or below the depth of zero annual amplitude, causing the upper part of the annual temperature profile to bend towards higher temperatures. At other sites where only a single year's measurements are available, similar positive inflection of the ground-temperature profile provides conclusive evidence of recent top-down permafrost warming. The past thermal evolution of permafrost has also been reconstructed by using borehole temperature data to calibrate heat-conduction models

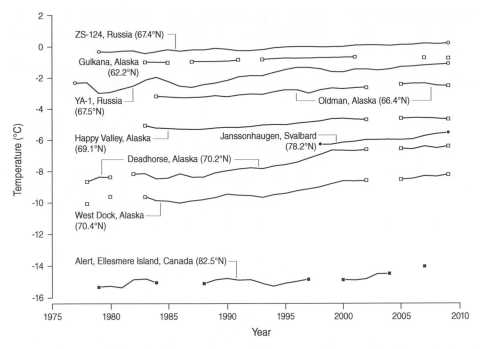

Figure 17.5 Time series of mean annual ground temperatures (MAGTs) at depths of 10–20 m for selected boreholes in the circumpolar north. *Source:* Adapted from Romanovsky *et al.* (2010a). Reproduced with permission of Wiley.

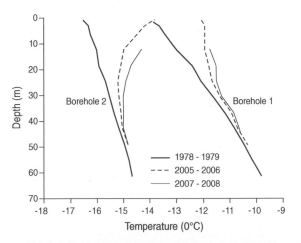

Figure 17.6 Mean annual ground temperature (MAGT) profiles for two deep boreholes at Alert on Ellesmere Island in the Canadian Arctic Archipelago (82° 30′ N), showing evidence of progressive downwards warming over 3 decades. *Source:* Smith *et al.* (2010). Reproduced with permission of Wiley.

and by establishing relationships between ground-surface temperatures and longer-term air-temperature records. Using this approach, Etzelmüller *et al.* (2011) showed that permafrost on Svalbard warmed by around 1 °C between the late 19th century and 1990, and since then has warmed by a further 0.5–1.0 °C. A similar approach was employed by Burn and Zhang (2009) to show that since the beginning of the 20th century, MAGTs on Herschel Island near the Beaufort Sea coast have increased by 2.6 °C at the top of the permafrost and by 1.9 °C at 20 m depth. Most sites for which sequential

data are available provide evidence of warming amounting to ~0.5–2.0 °C at the depth of zero annual amplitude since the 1970s.

The amount of recent permafrost warming, however, has varied temporally, regionally and locally. In Russia, for example, most warming occurred between the 1970s and 1990s, with little evidence of warming from the 1990s until 2008 (Romanovsky *et al.*, 2010b). In North America, permafrost warming has been generally greater north of the treeline (Smith *et al.*, 2010). Locally, thermal buffering factors such as depth and duration of snow-cover, the nature and completeness of vegetation cover, soil organic-layer thickness and the water or ice content of the soil have created marked variation in both the thermal state and the rate of warming of permafrost. The most important regional contrast is that cold permafrost (below −2 °C to −3 °C) in the continuous permafrost zone has generally warmed more than the warm permafrost in the discontinuous permafrost zone (Figure 17.5). This is because of latent heat effects. These are negligible in cold permafrost, but when permafrost temperatures approach 0 °C much of the energy (heat) penetrating ice-rich soil is consumed in the phase change from ice to liquid water, lowering the apparent thermal diffusivity of the soil and ultimately producing isothermal conditions in the ground. Warm ice-rich permafrost at the southern margin of the discontinuous permafrost zone can therefore persist for extended periods under a warming climate, especially if there is an overlying insulating layer of organic soil or peat, but disturbance of this layer may lead to rapid permafrost thaw (Shur and Jorgenson, 2007).

The trend of permafrost temperatures in Antarctica is poorly known. As measured permafrost temperatures in continental Antarctica range from about −8 °C to about −24 °C, much Antarctic permafrost is at no immediate risk of degradation; warm permafrost at temperatures slightly below 0 °C occurs only near sea level in the South Shetland Islands and probably on the northern part of the Antarctic Peninsula, and is more sensitive to thawing, particularly as this area is one of the most rapidly warming regions on Earth (Vieira *et al.*, 2010). By contrast, the temperature of much of the high montane permafrost of Central Asia is close to 0 °C, and hence sensitive to degradation under a warming climate. There is evidence that permafrost warming in this vast region is already underway: Zhao *et al.* (2010) noted that the temperature at the top of the permafrost on the Qinghai-Tibet Plateau warmed by an average of 0.6 °C in the first decade of the 21st century, and that increases in permafrost temperatures in the northern Tien Shan from 1974 to 2009 ranged from 0.3 °C to 0.6 °C.

17.2.2 Active-layer Thickness

In addition to causing changes in permafrost thermal state, atmospheric warming over decadal timescales and the resultant increase in ground-surface temperatures also influence the end-of-summer thickness of the active layer. The response of the active layer to warmer summers is conditioned by a range of site factors. Some of these are effectively static over short timescales (vegetation cover and thickness of organic soil or peat cover), whilst others vary annually (depth and duration of winter snowcover, soil moisture content, temperature and ice content of the active layer at the onset of spring thaw). As a result, two summers with similar air-temperature characteristics can potentially result in different depths of seasonal ground thawing. Moreover, where the active layer is underlain by ice-rich permafrost, notably an ice-rich transition zone, thaw of the upper parts of the permafrost slows because latent heat effects reduce the apparent thermal diffusivity of the ground.

In 1991, an international programme, the Circumpolar Active Layer Monitoring (CALM) network was established to monitor the long-term response of the active layer and nearsurface permafrost to changes in climate; it now incorporates more than 200 sites located in both hemispheres. Despite the complications outlined above, the CALM programme has demonstrated a significant increase in active-layer thickness at many arctic and subarctic sites, though this trend is interrupted by marked interannual variability and differs both locally and regionally (Figure 17.7). A number of studies have demonstrated

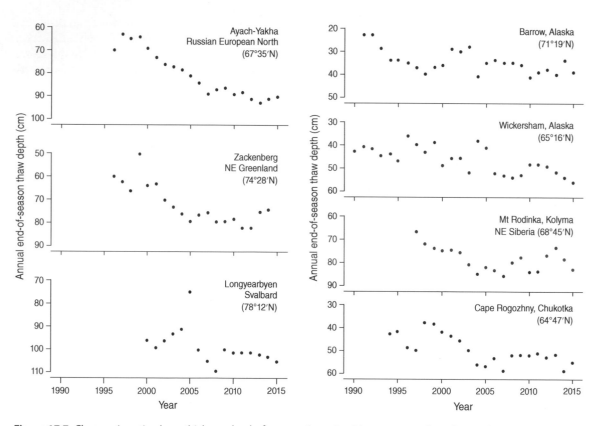

Figure 17.7 Changes in active-layer thickness (end-of-season thaw depth) at seven northern-hemisphere locations. All sites exhibit a general increase in thaw depth over the measurement period, but also considerable interannual variation. *Source:* Data from Circumpolar Active Layer Monitoring (CALM) website.

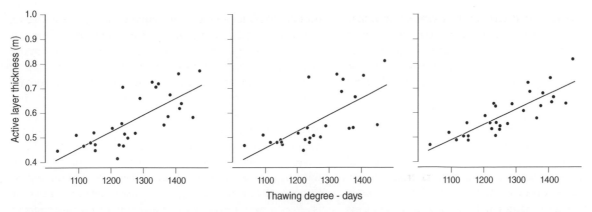

Figure 17.8 Relationship between active-layer thickness and thawing degree-days (°C) measured over 29 years at three sites in northern Sweden. Six other sites showed similar relationships. *Source:* Adapted from Åkerman and Johansson (2008). Reproduced with permission of Wiley.

that such trends are attributable to warming summer air temperatures (Smith *et al.*, 2009). Åkerman and Johansson (2008), for example, employed a 29-year record (1978–2006) of active-layer thickness measurements for nine mire sites in northernmost Sweden to demonstrate not only general thickening of the active layer at all sites at averaged rates of 7–13 mm a^{-1}, but also statistically significant correlations between active-layer thickness and mean summer temperature, and between active-layer thickness and thawing degree-days (Figure 17.8) for all nine sites. For the Bellsund region of Svalbard, Marsz *et al.* (2013) identified a strong relationship between active-layer thickness and May and June air temperatures, and argued that this reflects the timing of seasonal snowmelt, which regulates the duration of the summer thaw season. Regionally, a progressive increase in active-layer depth prior to 2010 has been detected at sites in Scandinavia, Greenland, the Russian European North, eastern Siberia and Chukotka, though absent or compromised by large interannual fluctuations in parts of Alaska and the Canadian Arctic (Callaghan *et al.*, 2011; IPCC, 2013).

17.2.3 Recent Changes in Permafrost Extent

Sporadic and discontinuous permafrost located along the southern margins of the circum-arctic permafrost zone is characteristically warm (MAGT >−2 °C) and a few metres to tens of metres thick. In some such areas (and even at the southern margin of the continuous permafrost zone), recent ground warming has resulted in thaw of the uppermost permafrost, producing a closed talik between the zone of seasonal ground freezing and the top of the permafrost, or in thickening of existing suprapermafrost taliks. In Russia, for example, these changes have been documented in the permafrost belt west of the Urals, but areas farther east appear less affected (Oberman, 2008). In the Pechora lowlands of the Russian European North, complete thawing of permafrost has resulted in a northwards migration of the

permafrost boundary by 30–40 km during the period 1970–2005, and by up to 80 km on the foothills in the eastern part of the region. Over the same time period, the southern boundary of continuous permafrost has shifted northwards by 15–20 km in the lowlands and 30–50 km in the foothills (Oberman and Liygin, 2009). From a study of palsa and thermokarst distribution in the James Bay area of northern Quebec (51° 45′–55° 00′ N), based on sequential aerial images and ground survey, Thibault and Payette (2009) inferred that the southern boundary of isolated permafrost had migrated northwards by 130 km in the previous 50 years. Subsequent research based on evidence of vegetation changes, palsa decline and thermokarst based on sequential satellite imagery of the area east of Hudson Bay (~56° 30′ N) suggests that permafrost degradation continues to affect much of subarctic Quebec (Beck *et al.*, 2015).

Overall, therefore, in the northern circum-arctic zone, there is presently a contrast between (i) generally warming but stable cold permafrost, primarily in the continuous permafrost zone, where the main effect of climate change over the past few decades has been active-layer thickening, and (ii) more vulnerable warm permafrost around the southern margins of the northern permafrost zone, where the warming trend of the past few decades has resulted not only in permafrost warming and active-layer thickening, but also in the development of closed taliks, the disappearance of isolated patches of permafrost, and at least localized northwards retreat of the permafrost boundary and the southern margin of continuous permafrost. The sleeping giant of the circum-arctic permafrost zone has not yet woken, but is certainly stirring in his sleep.

17.2.4 Projections

Models of future global climate warming are strongly conditioned by assumptions concerning the future trajectory of anthropogenic greenhouse gas emissions.

The Intergovernmental Panel for Climate Change has modelled future global warming on the basis of different emissions scenarios (IPCC, 2013). Relative to average global air temperatures for 1986–2005, the minimum rise of global temperatures (at 95% confidence) is predicted to be 0.4–1.6°C in 2046–65 and 0.3–1.7°C in 2081–2100; the maximum predicted rise is 1.4–2.6°C for 2046–65 and 2.6–4.8°C for 2081–2100. All models predict greatly enhanced warming in the Arctic (Figure 17.9). Under all scenarios, therefore, the circumpolar north will experience continued warming throughout this century.

Assuming the accuracy of such predictions, it is likely that further warming of permafrost will be ubiquitous and that degradation and retreat of permafrost will be widespread across the circumpolar north (Lawrence *et al.*, 2012). Prediction of the rate and scale of permafrost warming is difficult, not only because of the uncertainty inherent in MAAT projections, but also because permafrost degradation is likely to be accompanied by changes in ground-surface conditions, such as snowcover thickness and duration, subsidence, waterlogging, desiccation and vegetation change, which may enhance or retard further change. Ground thermal models that incorporate the physical variables characterizing the state of permafrost have been applied to map future permafrost distribution and derive ground-temperature evolution for

(a)

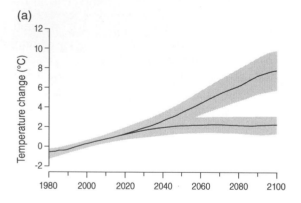

(b)

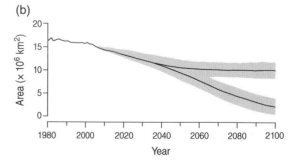

Figure 17.9 (a) Maximum and minimum scenarios of mean change in surface air temperature over present-day permafrost areas. (b) Projected change in global sustainable permafrost area under the same scenarios. Shaded zones represent 95% confidence limits. *Source:* Adapted from Slater and Lawrence (2013) with permission from the American Meteorological Society.

particular regions (e.g. Westermann *et al.*, 2013; Zhang, 2013; Zhang *et al.*, 2013), and the most recent models incorporate resultant hydrological changes (Westermann *et al.*, 2016). Global models of projected changes in permafrost extent (Figure 17.9) nevertheless remain characterized by wide uncertainty, primarily but not exclusively due to the uncertainty in future air-temperature predictions (Koven *et al.*, 2013). Depending on the assumed emissions scenario, Slater and Lawrence (2013) have projected that by 2080–99 the area underlain by nearsurface equilibrium permafrost could shrink by between 37 ± 11% and 81 ± 12%. Deep permafrost, which is effectively buffered from changes in the surface energy balance, is expected to degrade much more slowly, over a timescale of centuries to millennia.

The implications of climate warming for permafrost extent and thermal state at regional scales can be illustrated through a study by Jafarov *et al.* (2012), who developed a transient numerical model of changes in Alaskan permafrost at decadal intervals throughout the 21st century. The model inputs are spatial datasets of mean monthly temperature and precipitation, prescribed thermal properties of the multilayered soil column and specified water content for each soil class and location. Climate change is based on a composite of five IPCC global circulation models at 2 km × 2 km resolution for all of Alaska. The model is calibrated using borehole temperature data and active-layer thickness data from CALM observation stations, and tuned by adding simulated organic layers to minimize differences between observed and modelled ground-temperature profiles. The results are striking. For ground depths of 2 m, 5 m and 20 m, the model projects that areas with MAGT warmer than 0°C will increase at rates of 3.7%, 3.5% and 2.4% per decade; the model also indicates that the proportion of Alaska in which the ground at 2 m depth has a mean annual temperature at or below 0°C will decline from ~58% in 2010 to ~43% by 2050 and ~27% by the end of this century. The implication is that widespread permafrost degradation after 2040 is likely to affect the vast area south of the Brooks Range, apart from high altitude areas where ground temperatures will probably remain sufficiently low to permit survival of equilibrium permafrost.

17.3 Geomorphological Implications of Climate Change in the Circumpolar North

The most obvious landscape changes attributable to climate change in the Arctic and Subarctic concern the development and extension of thermokarst due to thaw of ice-rich permafrost, though not all thermokarst formation is due to warming climate: thermokarst may be autogenic (such as the collapse of pingos), anthropogenic (mainly through degradation or destruction of vegetation cover) or triggered by fire or non-climatogenic hydrological changes. There is nonetheless considerable evidence that climate changes over the past few decades have triggered, enhanced or extended thermokarst development over wide areas of the circumpolar north, mainly from analysis of sequential aerial photographs and satellite images. Jorgenson and Osterkamp (2005) have provided a useful typology of thermokarst development following post Little Ice Age warming and the response of boreal ecosystems to such changes.

17.3.1 Ice-wedge Degradation and Thermokarst Gullying

In the continuous permafrost zone, shallow permafrost degradation is evident from the thaw of the upper parts of large ice wedges. From comparative analysis of aerial photographs taken in 1945, 1982 and 2001, Jorgenson *et al.* (2006) showed that there has been an abrupt increase in the development of water-filled thermokarst pits along wedge troughs and intersections due to melt of the upper portions of ice wedges in two areas of the Arctic Coastal Plain of northern Alaska. Between 1945 and 1982, the spatial density of pits increased slightly (from 88 to $128\,pits\,km^{-2}$), but between 1982 and 2001 pit density increased to $1336\,km^{-2}$, an increase that they attributed to initiation of thaw during an exceptionally warm and wet summer in 1989 and sustained by relatively warm summers during the following decade. Ponding of water in ice-wedge troughs enhances thaw of the underlying ice, but they also observed that the accumulation of vegetation and organic matter in pits subsequently limits the depth of thaw. More generally, recent degradation of ice wedges is evident from transformation of low-centred ice-wedge polygons to high-centred polygons as a result of ice-wedge melt and consequent subsidence of polygon borders. From remotely-sensed images of the Kobuk Valley area of Alaska, Necsoiu *et al.* (2013) detected limited change in polygon topography from 1951 to 1978, but widespread development of high-centred polygons between 1978 and 2005 in response to recent warming. Research by Steedman *et al.* (2016) has shown that in the Tuktoyaktuk Coastlands of arctic Canada, high-centred polygons with depressed borders occupy 10% of the terrestrial landscape, though these authors detected only a small recent increase in the area occupied by water-filled polygon margins.

Further degradation of ice wedges results in the formation of polygonal mounds surrounded by water-filled thermokarst troughs that overlie thawing ice wedges and meet at thermokarst pits. Subsequent collapse of polygon margins creates a hummocky terrain of thermokarst mounds surrounded by indistinct troughs (Jorgenson

and Osterkamp, 2005; Jorgenson *et al.*, 2006). Such terrain may be further modified by flowing water and fluviothermal erosion to create beaded streams crossing low-gradient terrain and thermokarst gullies on slopes. The latter rapidly widen and propagate upslope, creating a badland landscape of deep, wide gullies (Godin and Fortier, 2012). In the discontinuous permafrost zone of interior Alaska, thermokarst gullying was apparently triggered by an exceptional rainstorm that led to the development of subsurface pipeflow, subsequent pipe collapse and headwards gully extension (Toniolo *et al.*, 2009).

Under continued climate warming, it is likely that further development of high-centred polygons, hummocky thermokarst mounds and thermokarst gullies will be extensive but not ubiquitous. The research of Jorgenson *et al.* (2006) suggests that in areas of cold permafrost, ice-wedge networks beheaded by thaw may stabilize under an insulating accumulation of organic matter that inhibits further degradation. The onset of water flow within degraded networks, however, probably constitutes a critical threshold beyond which recovery is unlikely.

17.3.2 Degradation of Ground-ice Mounds

As pingo collapse is an inevitable consequence of pingo growth, pingos seem likely to be unaffected by recent climate change, though collapse may be accelerated by ground warming and active-layer thickening. More frequent drainage of thermokarst lakes may even promote the development of new hydraulic (closed-system) pingos in areas of continuous permafrost.

Conversely, palsas and peat plateaus in the zone of sporadic permafrost or near the southern margin of the discontinuous permafrost zone, many of which probably formed under the relatively cold conditions of the Little Ice Age, appear to be degrading rapidly as a result of recent climate change. Mapping by Luoto and Seppälä (2003) of extant palsas and thermokarst ponds representing collapsed palsas in northernmost Finland, for example, suggests that the former distribution of palsas was three times greater than today, implying that recent palsa collapse has outstripped palsa formation. Similarly, Zuidhoff and Kolstrup (2000) reported a 50% decrease in the area of palsas at a bog in northern Sweden between 1960 and 1997, with no evidence for recent palsa formation, a trend they attributed to a combination of increasing air temperatures and greater snowfall after the 1930s. A study by Vallée and Payette (2007) found that along the Boniface River in subarctic Quebec, the area occupied by palsas decreased by 23% and that occupied by thermokarst ponds increased by 76% between 1957 and 2001; they noted that palsa degradation was greatest on the river floodplain, and suggested that changes in water level may have initiated palsa collapse. In a particularly interesting study, Payette *et al.* (2004) investigated the progressive degradation of a peat plateau that formed during the Little Ice Age and subsequently fragmented into individual palsas. By 1957, only 18% of the original plateau had degraded, but by 2003 only 13% survived, the remainder having been replaced by thermokarst ponds and fens. In this case, they concluded that the main cause of peat plateau degradation had been increased snowfall after 1957, as MAATs in this area were generally stable until the 1990s. Lithalsas tend to be located in the colder, more northerly part of the discontinuous permafrost zone (Wolfe *et al.*, 2014), but appear to be similarly affected by recent climate change; Beck *et al.* (2015) detected a 6% decrease in the area occupied by lithalsas in subarctic Quebec between 2004 and 2009.

As palsas and peat plateaus often represent the southernmost extent of permafrost terrain, these studies suggest not only progressive elimination of these landforms as a result of recent climate change, but also northwards migration of the southern limit of sporadic and discontinuous permafrost (Thibault and Payette, 2009). It might be argued that northwards migration of permafrost could engender new palsa, peat plateau and possibly lithalsa formation poleward of their present range. This seems unlikely, as these landforms develop as a result of cooling (as in the Little Ice Age), not under conditions of progressive warming, though local circumstances such as diminution of snowcover may favour transient regeneration of palsas in some areas. As Payette *et al.* (2004) and Sannel and Kuhry (2011) have shown, palsas and permafrost plateaus are likely to be replaced by thermokarst ponds, bogs and fens, which may be rapidly obscured by peat accumulation due to vegetation succession and terrestrialization.

17.3.3 Thermokarst Lakes, Ponds, Bogs and Fens

As outlined in Chapter 8, the response of thermokarst lakes to climate change has been complex, with a broad distinction between areas of thick, continuous permafrost, where climate warming is likely to promote lake formation and expansion through degradation of ice-rich permafrost, and areas of discontinuous permafrost, where lakes are more likely to drain through the establishment of sublacustrine taliks (Yoshikawa and Hinzman, 2003). Such contrast are evident in the study by Smith *et al.* (2005) of changes in Siberian lakes between 1973 and 1998: areas of continuous permafrost exhibited a 12% increase in lake area (mainly due to lake expansion), but lakes occupying discontinuous and sporadic permafrost exhibited a ~12% decline in area. Similarly, a study by Riordan *et al.* (2006) showed that shallow ponds (>0.2 ha) underlain by continuous permafrost in Alaska exhibited no change from the 1950s to 2002, whereas those in areas underlain by discontinuous permafrost decreased in area by 4–31% and in number by 5–54%,

changes they attributed to a combination of enhanced drainage and increased evaporation. However, there is also evidence that even in the continuous permafrost zone there may be a recent net decline in the area occupied by thermokarst lakes: Jones *et al.* (2011) showed that the total area of water bodies >1 ha in diameter on the continuous permafrost of the northern Seward Peninsula of Alaska declined by 14.9% between 1950/51 and 2006/07, primarily due to lateral breaching, and that averaged lake expansion rates changed little during this period. A study of thermokarst lakes in the continuous permafrost zone of the Kolyma Lowland in northern Siberia has also demonstrated significant reduction in lake area (by 5–6% overall) between 1973 and 2001 (Veremeeva and Gubin, 2009).

As these examples illustrate, it is difficult to encapsulate the recent and future trajectory of thermokarst lake and pond initiation, expansion, drainage and decline. Disappearance of thermokarst lakes due to sublacustrine drainage or breaching, or reduction in lake area through increased evaporation and terrestrialization (the colonization of lake margins by fens), may be at least partly offset by the formation of new water bodies through degradation of ice-rich permafrost and possibly accelerated lake expansion due to rising water and permafrost temperatures. Thermokarst lakes and ponds in areas of sporadic and discontinuous warm permafrost appear to be most vulnerable to drainage and shrinkage, whereas those in areas of cold continuous permafrost appear to be generally more stable (Smith *et al.*, 2005; Riordan *et al.*, 2006). A continent-wide satellite-based study of arctic lakes in Canada by Carroll *et al.* (2011), however, showed a net reduction in lake area of over 6700 km^2 between 2000 and 2009, focused particularly in the Canadian north. Though this study includes all lakes, and not just thermokarst water bodies, it implies much greater reduction in lake area in tundra environments than previous studies of more limited spatial scope have suggested.

On ice-rich yedoma terrain, widespread formation of deep thermokarst lakes was initiated during the Pleistocene–Holocene transition at ~13–11 ka, and locally reached a peak during the early Holocene thermal maximum (e.g. Romanovskii *et al.*, 2000, 2004; Morgenstern *et al.*, 2013; Kanevskiy *et al.*, 2014). This raises the possibility that future warming may result in similar extensive thermal degradation of yedoma terrain and deep thermokarst lake formation, which would initiate deep permafrost degradation over wide areas even if atmospheric warming were to slow or cease. Morgenstern *et al.* (2011) have pointed out, however, that such dramatic permafrost degradation and ground subsidence would be limited to residual yedoma plateaus, as drained lake basins (alases) contain much lower ground-ice content. On Kurungnakh Island in the Lena Delta area of northern Yakutia, at least 71% of all thermokarst lakes and alas basins have already subsided to the yedoma base, implying that further subsidence is likely to be minimal. In their inventory of types of thermokarst terrain in arctic Alaska, Farquharson *et al.* (2016a) have also highlighted the likelihood that residual yedoma terrain will experience marked thermokarst development under a future warming climate.

Similar complexities affect the recent and future evolution of thermokarst wetlands, particularly in the boreal forest zone. Though there is concern that atmospheric warming, active-layer thickening, earlier snowmelt and enhanced evaporation may cause reduction or drying of wetlands or drought stress in boreal forest communities (Goetz *et al.*, 2007), permafrost degradation over recent decades has locally resulted in increased waterlogging (Johansson *et al.*, 2006). In the Tanana Flats region of Alaska, where degradation of discontinuous permafrost is widespread, analysis of sequential imagery indicates a 7% increase in thermokarst bogs and fens at the expense of birch forest stands over the past 60 years (Lara *et al.*, 2016). The short-term response to atmospheric warming in areas of degrading permafrost and poor drainage therefore may be one of expansion of thermokarst-generated wetlands, though further warming may induce wetland contraction as a result of increased evapotranspiration.

17.3.4 Slope Processes

The effects of climate change on solifluction processes acting on arctic and subarctic slopes are uncertain. Active-layer thickening in warm summers is likely to increase the depth of mobile soil and cause thawing of the ice-rich layers at the base of the active layer or top of the permafrost, leading to an increase in the rate of movement and volume of soil transported annually (Matsuoka, 2001c; Harris *et al.*, 2009). This outcome is supported by long-term (1972–2002) monitoring of annual surface displacement rates on Svalbard by Åkerman (2005), who showed that these are strongly positively correlated with active-layer depth, which in turn reflects the number of thawing degree-days. Conversely, however, earlier snowmelt and increased evaporation as summer temperatures increase may reduce the moisture content of the soil at the end of the thaw season, so that seasonal ice content in the active layer (or in seasonally frozen ground where permafrost is absent) is reduced, causing a reduction in the rates of both frost creep and gelifluction.

Active-layer thickening in response to warming air temperatures is, however, likely to increase the incidence of active-layer detachment failures (Blais-Stevens *et al.*, 2014). Such failures are triggered when late-summer thaw results in melt of ground ice at the base of the active layer or in the transition zone at the top of the underlying permafrost, resulting in an increase in pore-water pressures if the rate of meltwater production at the thaw

front exceeds that of meltwater drainage through the overlying thaw-consolidated soil. Rising pore-water pressures cause a decrease in the effective shear strength of the soil, potentially initiating sliding of the full depth of the active layer over comparatively low-gradient slopes (Lewkowicz and Harris, 2005a, 2005b; Niu *et al.*, 2016). Enhanced summer warming, possibly allied to increased summer rainfall, is therefore likely to increase the incidence of active-layer detachment failure as the frequency of deep end-of-season thaw increases. Moreover, progressive thickening of the active layer in some areas (Figure 17.7) enhances the likelihood of thaw of the extremely ice-rich intermediate layer at the base of the transition zone (Shur *et al.*, 2005).

Increased frequency of active-layer detachment under a warming climate is also likely to enhance initiation of retrogressive thaw slumps, at least at sites where removal of the active layer exposes ice-rich permafrost or massive ground ice (Lacelle *et al.*, 2010). Slump initiation on hillslopes has been linked to exceptionally warm summers (Balser *et al.*, 2014), and there is convincing evidence that atmospheric warming has already increased not only the frequency of slump initiation, but also the rate of slump expansion through headwall retreat and sediment removal. Along the coast of Hershel Island in the Beaufort Sea, Lantuit and Pollard (2008) have documented a 63% increase in active slumps between 1952 and 2000, and a 160% increase in the area of thaw slumping over the same period, though some of this increase may reflect enhanced coastal erosion. On the Richardson Mountains of northwest Canada, rates of slump initiation in 1985–2004 were almost double those in 1954–71 (Lacelle *et al.*, 2010). Growth of slumps bordering thermokarst lakes and ponds in the Mackenzie Delta area in 1979–2004 was about 1.4 times faster than in 1953–70, and the mean rate of headwall retreat approximately doubled, changes that Lantz and Kokelj (2008) showed were probably due to warming air temperatures. Increases in precipitation also appear to drive increased slump activity. On the ice-cored terrain of the Peel Plateau in northwest Canada, the number and size of large active thaw slumps increased markedly from the 1980s to 2011 despite no increase in summer temperatures, but in concert with a significant increase in rainfall magnitude and intensity. This correspondence suggests that acceleration in thaw slump activity has been mainly driven by increased summer rainfall, which has accelerated sediment movement and removal across slump floors (Kokelj *et al.*, 2015).

17.3.5 Arctic Rivers

As outlined in Chapter 13, the runoff regime of arctic and subarctic rivers is usually dominated by the annual snowmelt flood. In small catchments, the snowmelt flood typically lasts a few days to a few weeks, but longer in large catchments where snowmelt inputs from various parts of the catchment arrive at different times. Though rainstorms may reinvigorate river discharge later in the thaw season, for many high-latitude rivers up to about 90% of water discharge and sediment transport occurs during the nival flood period.

Climate change is likely to affect river runoff regime in several ways. There are already indications that annual precipitation is increasing over some parts of the Arctic and Subarctic, and all modelling simulations predict increasing precipitation north of 60° N latitude during the 21st century, irrespective of the assumed emissions scenario. Under a moderate emissions scenario, for example, pan-arctic cold-season precipitation is projected to increase by ~25% and warm-season precipitation by ~15% above 1986–2005 averages (IPCC, 2013). Such projections imply increased winter snowfall in many areas, but as this will be accompanied by rising air temperatures, snowcover duration will be reduced, particularly because of earlier spring thaw. Data from satellite imagery show that during the period 1972–2009, pan-arctic snowcover diminished by about 12 days, at a rate of 3.4 days per decade; snowcover extent in May and June decreased by 18% between 1966 and 2008, though there are marked regional variations. Across the Arctic, annual snowcover duration is projected to decrease by a further 10–20% by 2050, and by up to 30–40% over Alaska and northern Scandinavia (AMAP, 2011). For river systems in northern high latitudes, these changes imply that snowmelt floods will begin earlier but probably extend over a longer period, potentially reducing the peak flood discharges of some rivers. Increases in precipitation, however, are liable to be at least partly offset by increases in evaporation due to longer snow-free seasons and warmer summer temperatures, suggesting that baseflow after the snowmelt flood may be reduced, causing small streams to dry up earlier unless sustained by summer rainfall. In the subarctic boreal forest zone, snowmelt-dominated flow patterns may be replaced by a pluvial (rainfall-dominated) or pluvio-nival regime (Woo *et al.*, 2008b; Woo, 2012).

There is persuasive evidence that circum-arctic river systems are already responding to increased precipitation linked to atmospheric warming. An analysis of river-monitoring data for the six largest Eurasian rivers that discharge into the Arctic Ocean by Peterson *et al.* (2002) demonstrated a 7% increase in aggregated discharge between 1936 and 1999, a finding corroborated by Shiklomanov *et al.* (2006). Their analysis also demonstrated a statistically significant correlation between increasing discharge to the Arctic Ocean and global MAAT, suggesting that rising temperatures have directly or indirectly augmented annual discharge. Analysis of the monthly discharges for 19 circum-arctic rivers of

varying catchment area ($16–295 \times 10^3\,km^2$) showed that 15, representing 98% of the aggregated drainage area, experienced an increase in discharge over the period 1977–2007 (Overeem and Syvitski, 2010). The annual discharge of the Yukon River increased by ~7% during this period, and those of the Mackenzie, Lena and Yenisei rivers by 13–14%. Overeem and Syvitski's analysis also indicated earlier snowmelt floods and that discharges during the 'melt month' (the earliest month of strong discharge increases, usually May) increased on average by 66%, though with a decrease averaging 6.4% during the month of peak discharge. However, the trend of increasing annual discharge is not observed everywhere; two of the rivers in their analysis exhibited a decline in discharge. Déry and Wood (2005) found that aggregated data for 64 rivers in northern Canada indicated a 10% decline in annual discharge over the period 1964–2003, attributable to an average decline in precipitation across the same domain, though it is notable that their aggregated discharge data appear to show a general increase after about 1989. McClelland *et al.* (2006) have related these apparently conflicting trends to regional differences, showing that over the period 1964–2000 the combined annual discharge of rivers flowing north from Eurasia into the Arctic Ocean exhibited a strong positive trend, whereas those of North American rivers flowing to the Arctic Ocean showed a weak negative trend and those draining into Hudson, James and Ungava Bays demonstrated a strong negative trend. An updated synthesis for the largest Eurasian and North American rivers flowing into the Arctic Ocean suggests that despite large interannual variability, there is a general positive trend in annual discharge (Holmes *et al.*, 2015; Figure 17.10).

Projections of future changes in hydrological regime in smaller catchments suggest that these will respond differently to climate change in different regions. For a $63\,km^2$ headwater basin in an upland region near Inuvik, Canada (68° 45′ N), Pohl *et al.* (2007) used outputs from eight climate models to predict changes in runoff regime for 2040–69 and 2070–99. For both periods, all simulations indicated earlier runoff initiation and earlier peak runoff, and all suggested that freeze-up and runoff cessation will eventually occur up to a month later than now. Annual discharge was projected to increase in response to increased precipitation (offset to varying extent by increased evaporation), but the projections suggested limited change in spring peak discharge. By contrast, end-of-century runoff predictions for a small ($8\,km^2$) basin on Melville Island (74° 55′ N) suggest a 50–100% increase in annual peak discharge, due mainly to rapid melt of increased winter snow accumulation, and a similar increase in total annual runoff, due to greater annual precipitation (Lewis and Lamoureux, 2010). Both of the models employed in these predictions indicated considerable

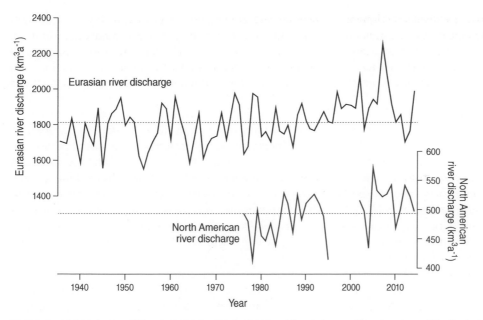

Figure 17.10 Total annual discharge to the Arctic Ocean of the six largest Eurasian rivers (Ob, Yenisei, Lena, Severnaya Dvina, Pechora and Kolyma) and two largest North American rivers (Yukon and Mackenzie). Dashed line represents the mean of all measurements. *Source:* Adapted from Holmes *et al.* (2015). Reproduced with permission of Dr. Max Holmes.

interannual variability in both parameters, as might be expected for a small catchment with limited storage capacity. Lewis and Lamoureux also suggested limited influence of increased evaporation compared with the projections of Pohl *et al.* (2007), consistent with the more northerly latitude of their catchment. Future changes in the regime of the Tana River in Subarctic Scandinavia (basin area ~16 000 km^2) have been simulated for 2070–99 by Lotsari *et al.* (2010), based on three different climate models and three emissions scenarios. As with the other two projections discussed here, Lotsari *et al.* found that the onset of the annual snowmelt floods is likely to become earlier (by 2–3 weeks), but in this case they also found that average annual snowmelt flood discharges are likely to be markedly less, and maximum flood discharges slightly less, than those in the 1971–2000 control period. Given the different simulation protocols adopted by these three studies and the differences in catchment area and relief, it is difficult to draw general conclusions. All three predict earlier snowmelt and a longer runoff season, but the outcomes in terms of peak snowmelt discharge vary.

The geomorphological and sediment-transport implications of recent and anticipated changes in runoff regime depend on a complex interplay of factors, such as catchment size and relief, sediment availability and the degree to which changes in precipitation, snow accumulation and evaporation affect a particular catchment. In general, assuming unlimited sediment availability, non-dissolved sediment flux scales nonlinearly with increases in water discharge. Syvitski (2002) has applied a stochastic

model that predicts sediment flux in ungauged catchments for 46 arctic and subarctic rivers, from which he derived projections of the change in sediment load under a changing climate: for every 2 °C increase in MAAT, the model predicts an eventual 22% increase in sediment flux, and for every 20% increase in runoff, it predicts an eventual 10% increase in sediment transport. Applying this model to the sediment load of the six largest Russian rivers flowing into the Arctic Ocean under various climate change scenarios, Gordeev (2006) suggested that by 2100 the sediment reaching their estuaries will have increased by 30–122%. The validity of these estimates is difficult to judge, particularly because large amounts of sediment may be deposited on floodplains or trapped upstream in lakes. For a small (8 km^2) catchment on Melville Island in the Canadian Arctic Archipelago, Lewis and Lamoureux (2010) projected increased sediment yields of 100–600% by the end of the 21st century by establishing a rating curve that defines the present relationship between suspended sediment concentration and discharge, then applying this to projected future runoff scenarios. They noted, however, that even this broad range of estimates is probably minimal, because the effects of ground warming in generating additional sediment supply to the river (as a result, for example, of ground-ice thaw and active-layer detachment failures) are not included. Subsequent research based on the same catchment has suggested a further complication: that sediment introduced into stream channels by permafrost disturbance events may be stored within the channel rather than immediately evacuated (Favaro and Lamoureux, 2015).

Within much larger catchments, increased sediment entrainment in headwater streams may also be mitigated by increased sediment storage in lakes, wetlands and floodplains, so that nondissolved sediment evacuated from large catchments terminating in the sea may be markedly less than that entrained in headwaters.

How changes in discharge regime and sediment flux might affect channel planform and floodplain evolution will probably depend largely on local circumstances. In their projection of the future flow characteristics of the subarctic Tana River of northern Scandinavia, Lotsari *et al.* (2010) calculated that though flow velocity, bed shear stress and stream power per unit area are likely to decrease due to reduced water discharge, the river will still be competent to transport the fine sediments that form the river bed, suggesting limited future change. The future consequences for rivers that are projected to experience enhanced flood discharges and associated increases in flow velocity, stream power and sediment load remain to be established.

17.3.6 Arctic Coasts

As outlined in Chapter 15, coastlines comprising bluffs of ice-rich sediment are particularly vulnerable to accelerated erosion as a result of climate warming and its indirect effects, notably sea-level rise, decrease in summer sea-ice cover (and consequent increase in open-water fetch), warming seawater temperatures and warming of permafrost at the shoreface. Ice-rich permafrost coasts make up most of the coastline of the Arctic Ocean from the Kara Sea to the Beaufort Sea, and thus form over 50% of all circum-arctic coastlines.

Rising sea level threatens ice-rich permafrost coastlines around the Arctic Ocean because high-tide storm waves will access more of the shoreface, increasing the zone affected by both thermal and mechanical erosion. According to the IPCC (2013), global sea levels rose about 0.19 m between 1901 and 2010, at an average rate of $1.7 \pm 0.2\,\mathrm{mm\,a^{-1}}$, but satellite altimetry data show that the rate of global sea-level rise was greater $(3.2 \pm 0.4\,\mathrm{mm\,a^{-1}})$ between 1992 and 2010. Depending on the greenhouse gas emission scenario adopted, global sea level is projected to rise above 1985–2005 levels by 0.26 m to 0.98 m before the end of this century. Most of this projected rise is due to warming and thermal expansion of the oceans, coupled with contributions from melting glaciers and ice sheets. Sea-level rise will not be uniform, however, and around the Arctic Ocean will be strongly conditioned by differences in rates of glacio-isostatic uplift. Where uplift is still occurring, such as along coasts that were occupied by the last Barents–Kara Ice Sheet, continuing crustal uplift is likely to offset sea-level rise. All sea-level projections that take uplift into account suggest that relative sea-level rise this century will be

very limited along the coasts of the Barents and Kara seas, but that they will increase eastwards, reaching a maximum of ~0.5–1.0 m along the Beaufort Sea coast by the end of this century (IPCC, 2013).

Reduction in sea-ice cover in the Arctic Ocean affects erosion of ice-rich permafrost coasts in several ways. Earlier removal of bottomfast ice and the ice foot on beaches and bluffs exposes the shoreface to a more prolonged period of wave erosion, and increased absorption of heat by open water during the summer increases seawater temperature and thus the potential for thermal erosion of ice-rich bluffs. The most important effect of shrinking summer sea-ice cover is an increase in fetch (the distance over which winds operate to generate storm waves), which means that storm waves reaching the shoreface have greater energy (Overeem *et al.*, 2011; Barnhart *et al.*, 2014a, 2014b).

Reconstructions of summer sea-ice extent across the Arctic Ocean show no clear trend until about 1970, after which the trend is downwards (Figure 17.11). According to the IPCC (2013), minimum (end-of-season) summer sea-ice cover in the Arctic declined at a rate of 9.4–13.6% per decade over the period 1979–2012. For the Arctic Ocean as a whole, the melt season increased by ~5 days per decade from 1979 to 2013, a pattern dominated by progressively later freeze-back in the Kara, Laptev, East Siberian, Chukchi and Beaufort Seas. Moreover, because of the additional heat stored in the upper part of the Arctic Ocean through enhanced absorption of energy by open water, sea-surface temperatures increased by 1.0–1.5 °C in the first decade of this century (Stroeve *et al.*, 2014). All IPCC model projections predict further decline in summer sea-ice extent before the end of this century: for the period 2081–2100, compared with 1986–2005, the Arctic Ocean sea-ice cover is projected to decline by 43–94%, and under the most pessimistic projection a nearly ice-free Arctic Ocean in September may occur before 2050. Within a decade or two, it should be possible to kayak to the North Pole.

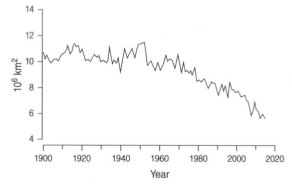

Figure 17.11 Annual variability of summer sea-ice extent across the Arctic Ocean, 1900–2013. Decline after about 1970 has been monitored from satellite observations.

The effects of recent permafrost warming, ocean warming and enhanced thermal and mechanical erosion due to increased storm-wave energy are already evident on some ice-rich permafrost coasts. For a 60 km stretch of the Beaufort Sea coast, for example, averaged coastal erosion rates increased from 6.8 m a^{-1} during the period 1955–79 to 8.7 m a^{-1} in 1979–2002 and 13.6 m a^{-1} in 2002–07 (Jones *et al.*, 2009). Similarly, for 48 coastal sites of various types (barrier islands, bluffs, deltas, bays, lagoons, etc.) along the Alaskan Beaufort Sea coast, Ping *et al.* (2011) found that averaged erosion rates had increased from 0.6 m a^{-1} during ~1950–80 to 1.2 m a^{-1} in ~1980–2000. For Simpson Point at the eastern end of Herschel Island, Radosavljevic *et al.* (2016) detected little difference in averaged coastline retreat for the period 1970–2000 (~0.5 m a^{-1}) compared with 1952–70 (~0.6 m a^{-1}), but a marked increase to ~1.3 m a^{-1} for 2000–11. Günther *et al.* (2013) found that recent short-term erosion rates at sites along the Laptev Sea coast (5.3 ± 1.3 m a^{-1}) are much higher than the averaged rate for the period ~1951–2011 (2.2 ± 0.1 m a^{-1}). Averaged rates of coastline retreat, however, conceal marked spatial variations within different geomorphological contexts, with generally greater erosion rates at exposed ice-rich coastal bluffs than in sheltered locations such as lagoons and embayments (Jorgenson and Brown, 2005). Moreover, an overview of coastal changes in the Arctic by Overduin *et al.* (2014) cautions that studies of recent changes in coastal erosion rates have tended to focus on the Beaufort Sea coast and on areas of rapid erosion, and may not be representative for the full spectrum of permafrost coasts. All recent commentators nevertheless stress the inevitability of future enhanced erosion of ice-rich permafrost coasts and release of sediment and organic matter to the shallow shelf seas that border the Arctic Ocean.

17.4 Geomorphological Implications of Climate Change in High Mountain Environments

A number of studies have provided evidence that recent temperature warming in high mountain environments exceeds global averages (e.g. Beniston, 2007), and all projections suggest atmospheric warming during the 21st century will markedly exceed that during the 20th (IPCC, 2013). For the European Alps, a long-term reconstruction of MAATs shows gradual warming from ~1880 to ~1980, then an unprecedented temperature rise of about 1.5 °C thereafter (EEA, 2009; Figure 17.12). Projections of future warming during this century suggest that it will be more pronounced in some mountain systems than in others; high-latitude mountains are projected to

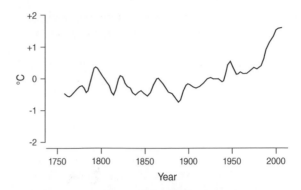

Figure 17.12 Mean annual air temperatures (MAATs; 20-year running mean) for the European Alpine region, relative to the 1851–2000 average, showing gradual warming from ~1880 to ~1980 and subsequent rapid warming of ~1.5 °C. *Source:* Adapted from EEA (2009) with permission from the European Environment Agency, Copenhagen.

experience the greatest warming, and low-latitude mountains, together with those of southern Africa, Australia and New Zealand, the least (Nogués-Bravo *et al.*, 2007). The same analysis suggests that in comparison with 1961–90 averages, mid-latitude mountains in Europe and North America will see MAAT increases of 2.3–3.3 °C by 2055 and 2.9–5.4 °C by 2085, and that those in Asia will experience increases of 2.7–3.8 °C and 3.6–6.5 °C respectively. The European Environmental Agency (EEA, 2009) has forecast that by the end of this century, MAATs in the European Alps will have increased by 3.9 °C, with a slightly greater increase (4.2 °C) for terrain above 1500 m.

17.4.1 Climate Change and High-Alpine Permafrost

High-alpine permafrost in mid-latitude mountains is characteristically warmer than –3 °C, except on the highest summits, and the first-order response to an atmospheric warming trend is likely to be active-layer thickening and a gradual rise in the lower altitudinal limit of equilibrium permafrost. The response of high-alpine permafrost to anticipated future warming is complicated by several factors, however, notably strong topographic control and complex patterns of snow-cover and vegetation cover. As outlined in Chapter 4, MAGSTs may be several degrees warmer on steep south-facing slopes than on north-facing ones, so the lower altitudinal limit of permafrost may be hundreds of metres lower on the latter (Gruber *et al.*, 2004b; Noetzli *et al.*, 2007; Hipp *et al.*, 2014). Near the lower altitudinal limit of mountain permafrost, variations in snowcover create a mosaic of permafrost and permafrost-free terrain (Isaksen *et al.*, 2011), and because most mountains support bedrock at the surface or under a thin debris cover, active layers are typically several metres thick.

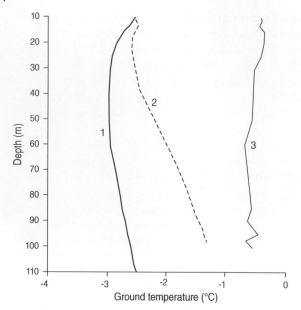

Figure 17.13 Temperature profiles in permafrost at three high mountain sites. (1) Juvasshøe, Jotunheimen, Norway (61°41′N, altitude 1894 m). (2) Stockhorn, Swiss Alps (45°59′N, altitude 3410 m). (3) Schilthorn, Swiss Alps (46°34′N, altitude 2909 m). Inflection of the top of the temperature profiles reflects recent warming. *Source:* Harris *et al.* (2009). Reproduced with permission of Elsevier.

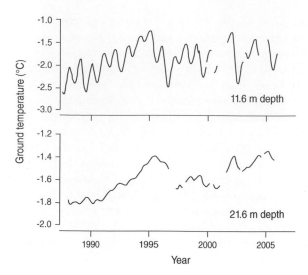

Figure 17.14 Ground-temperature record 1987–2006 for the Murtèl–Corvatsch rock glacier borehole at 2670 m in the Swiss Alps. *Source:* Harris *et al.* (2009). Reproduced with permission of Elsevier.

The effects of recent warming on mountain permafrost are evident from borehole temperature profiles, which generally show an inflection towards warmer temperatures in the upper part of the permafrost, indicative of recent warming (Figure 17.13). At relatively shallow depths, decadal trends in permafrost temperatures may be obscured by large interannual variations in snow-cover, but depths close to that of zero annual amplitude often show a general warming trend (Zenklusen Mutter *et al.*, 2010). Figure 17.14 illustrates this difference for a borehole through the Murtèl–Corvatsch rock glacier at 2670 m altitude in the Swiss Alps. At 11.6 m depth, an initial warming trend from 1987 to 1994 is succeeded by a period of increased annual temperature amplitude, whereas at 21.6 m depth a more general warming trend is evident. Much of the interannual variation at 11.6 m depth is due to variation in the thickness of winter snow-cover at this site (Harris *et al.*, 2009; Haeberli *et al.*, 2010).

For some high-altitude areas, a rise in the lower limit of permafrost due to recent warming has already been established. According to Cheng and Wu (2007), the lower altitudinal limit of permafrost on the Qinghai-Tibet Plateau rose on average 25 m in the north and 50–80 m in the south as a result of atmospheric warming since the 1970s. Li *et al.* (2008) estimated that the area underlain by permafrost on the Qinghai-Tibet Plateau shrank by ~100 000 km² between the 1970s and 1990s, and projected that one-third to one-half of the permafrost in this vast area will have degraded by the end of this century.

Other predictions are equally drastic: for the Bolivian Andes, Rangecroft *et al.* (2016) have suggested that 95% of present permafrost could disappear by the mid-21st century. Similarly, for the Jotunheimen Massif in south-central Norway (~61°30′N), where permafrost underwent significant warming between 1999 and 2009 (Isaksen *et al.*, 2011), Hipp *et al.* (2014) have suggested that a 2°C rise in MAAT would result in an increase in the lower limit of permafrost from 1200–1300 m to ~1700 m on north-facing rockwalls and from 1600–1700 m to ~2100 m on south-facing ones, eliminating equilibrium permafrost from a large part of this massif.

Recent and future warming and thaw of permafrost in mountain areas have major implications for slope stability and slope processes. Harris *et al.* (2009) and Stoffel *et al.* (2014) have reviewed geomorphological responses to climate change and warming of mountain permafrost in a European context, and Stoffel and Huggel (2012) have provided a more general overview of the effects of climate change on mass movements in mountain environments. Below we consider three geomorphological consequences of climate change in high-alpine environments: rock-slope instability; the implications for talus accumulation, protalus ramparts and rock glaciers; and the frequency and magnitude of debris-flow events. Most research on these topics has been carried out in the European Alps.

17.4.2 Rock-slope Instability: Rockfalls and Rock Avalanches

Several studies have reported an enhanced frequency during recent decades of rock-slope failures from steep mountain rockwalls underlain by permafrost. These

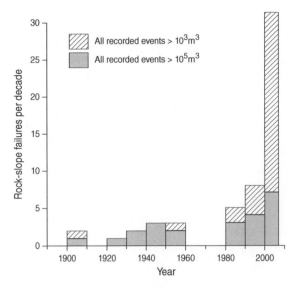

Figure 17.15 Number of recorded rock-slope failures per decade for an area of 25 000 km^2 in the central Alps, 1900–2007. Events <10^5 m^3 are probably under-represented prior to 1980, but the data show an increase in larger (and better documented) events after that year. *Source:* Fischer, http://www.nat-hazards-earth-syst-sci.net/12/241/2012/. Used under CC BY 3.0 https://creativecommons.org/licenses/by/3.0/.

range in magnitude from small-scale debris falls (<10 m^3) to catastrophic rock avalanches involving more than a million cubic metres of rock and ice (e.g. Sass, 2005b; Rabatel *et al.*, 2008; Huggel *et al.*, 2010; Krautblatter *et al.*, 2010; note that most studies refer to such events as 'rockfalls', irrespective of the magnitude or mode of rock detachment). An inventory of rockfalls with volumes of 500–65 000 m^3 from a north-facing rockwall in the Mont Blanc massif compiled by Ravanel and Deline (2010), for example, showed that of 42 events since 1947, over 70% occurred after 1990. Similarly, a study by Fischer *et al.* (2012) of 56 rock-slope failures exceeding 1000 m^3 in the central European Alps for the period 1900–2007 showed a marked increase in such events after 1980, and particularly after 2000; of these, only eight large-magnitude (>10^5 m^3) events occurred between 1900 and 1980, but 13 occurred between 1980 and 2007, an almost eightfold increase in frequency (Figure 17.15). For 24 recorded rockfalls in the Swiss Alps, Allen and Huggel (2013) found that 14 were preceded by one or more exceptionally warm days in the previous week, suggesting that even extreme short-term warming could trigger failure, though they noted that no similar relationship emerged for an inventory of recent rockfalls in the Southern Alps of New Zealand. There is widespread consensus that a recent increase in rockfall activity is related to warming and thaw of ice in rockwalls underlain by permafrost.

Establishing a definitive causal relationship between warming, permafrost degradation and increased rockfall activity is difficult, however, in part because of the relatively short timeframe of recorded rockfalls, but also because the distribution, depth and thermal properties of permafrost under rockwalls have been monitored at very few sites, though modelling of ground thermal regime in steep rockwalls offers an alternative approach (e.g. Noetzli *et al.*, 2007; Salzmann *et al.*, 2007; Noetzli and Gruber, 2009). In some cases, moreover, recent rockfalls on high-alpine terrain probably represent a paraglacial response to downwastage of adjacent valley glaciers or the disappearance of hanging glaciers from rock faces (Fischer *et al.*, 2006, 2010, 2012; Allen *et al.*, 2011; McColl, 2012), and hence are unrelated to permafrost degradation. There is nevertheless strong circumstantial evidence that implicates warming-driven permafrost degradation as a major cause of the observed recent increase in the frequency of rockfall and rock-avalanche events. Data compiled by Fischer *et al.* (2012) indicate a strong seasonal bias in the timing of such events: winter failures are very rare, and though large-scale (>10^5 m^3) events tend to span the period March–November, smaller events (10^3–10^4 m^3) are strongly focused within the summer months, when the most rapid rockwall warming occurs. Tellingly, frequent rockfall events are associated with periods of summer warming (Huggel *et al.*, 2010), the classic example being the hot summer of 2003 in the Alps, which witnessed unprecedented levels of rockfall activity. For this period, Gruber *et al.* (2004a) modelled deeper rockwall thaw depths than at any time during the previous two decades, and concluded that the unusually high incidence of rockfalls had been due to the rapid thermal response of rockwalls to warming, with consequent destabilization of ice-filled discontinuities (fractures) within the rock. Enhanced rockfall was particularly pronounced on north-facing slopes, where direct insolation is minimal, so that warming-induced thaw penetrated to greater depths than usual. It is also notable that the elevation of many rock detachment zones tends to be near the lower limit of mountain permafrost, where permafrost temperatures are close to 0 °C, and often correspond with convex topography (buttresses, ridges, spurs and summits) where ground warming is not simply one-dimensional, as on planar slopes, but bi- or multi-sided, leading to more rapid subsurface warming (Noetzli *et al.*, 2007).

In both permafrost and non-permafrost environments, failure of steep bedrock slopes is usually related to loss of strength along an existing fracture (joint) network. In rockwalls underlain by permafrost, most fractures are filled with ice, which can be seen in fissures exposed in failure scars immediately after rockfall events, in tunnels and, less commonly, amongst recent rockfall runout debris. In an important paper, Gruber and Haeberli (2007) argued that warming and thaw of ice-filled discontinuities could trigger rockwall destabilization, though several effects are probably involved. During freezing, volumetric expansion that accompanies the phase change from water to ice may widen fractures (Matsuoka, 2001a, 2008) and ice segregation operating within bedrock at temperatures below 0 °C may create new discontinuities (Murton *et al.*, 2006). Thermomechanical forcing caused by temperature-driven expansion and contraction of inter-fracture rock probably also causes progressive dilation of rock fractures, particularly if an ice infill impedes fracture closure (Hasler *et al.*, 2012). Ice-filled joints tend to be stable at temperatures below −2 °C, giving rise to the concept of 'ice-cemented' fractures. The strength of ice in joints decreases as it warms, however, becoming most rapid as temperatures approach 0 °C (Davies *et al.*, 2001), resulting in either ductile deformation of the ice, breaking of the contact between the ice and the adjacent rock, or both (Günzel, 2008). When warming initiates ice melting, there is further loss of bonding between the ice and fracture walls, and hydrostatic pressure may build up within the fracture network if meltwater escape is impeded. Shearing along fractures may thus represent strength reduction at rock–rock contacts, at rock–ice contacts or within fracture ice (Hasler *et al.*, 2012). An important insight of Gruber and Haeberli (2007) is that though conductive heat transfer through rock may initiate warming and thaw of ice in fractures, advective heat transfer by percolating meltwater can accelerate permafrost warming within a rock mass (Hasler *et al.*, 2011), leading to the development of thaw corridors within the fracture network and potentially destabilizing much greater volumes of rock than the effects of conductive heat transfer alone. Thus, whereas small rockfalls may be caused by rapid conductive warming and thaw of ice in nearsurface fractures, larger failure events may reflect the effects of deep advective heat transfer by water moving through the joint system.

Although much remains to be learnt about the response of rockwalls underlain by permafrost to warming (Krautblatter *et al.*, 2012b), the established connections between exceptionally warm summer weather, permafrost warming and enhanced rockfall activity in alpine environments indicate a potential hazard to infrastructure and settlements located at the foot of steep rockwalls underlain by permafrost. The possibility of catastrophic rock avalanches into lakes poses the additional threat of displacement waves engulfing lakeside communities. The incidence of rockfall events driven by warming episodes is likely to increase. Modelling of the frequency of future high-temperature events on the basis of future climate projections for the central Swiss Alps indicates that by 2050 extreme warming events lasting 5, 10 or 30 days are likely to increase in frequency by at least 1.5–4.0 times in comparison with a 1950–2000 reference period (Huggel *et al.*, 2010), emphasizing the need to identify rockwalls most at risk of collapse.

17.4.3 Talus, Protalus Ramparts and Rock Glaciers

One implication of the evidence for the recent climate-driven increase in the frequency of rockfall events in high-alpine areas is that recently measured rockwall rates (Table 12.1) may overestimate long-term (centennial or millennial) averages. Enhanced rockfall rates, and particularly an increased incidence of larger (>100 m³) rockfall events, also imply increased rates of talus accumulation. The future of protalus ramparts formed by the accumulation of rockfall debris at the foot of perennial firn fields is less certain. The most likely scenario is that under warming air temperatures, firn fields will shrink and disappear. Slow shrinkage of firn fields under conditions of continued or increasing rockfall activity is likely to result in debris accumulation in the zone immediately upslope of the original rampart, potentially transforming ramparts with an outer ridge into bench-like debris accumulations. Disappearance of firn fields may ultimately result in complete burial of protalus ramparts by talus accumulation. Conversely (though less likely under a warming climate), increased snowfall may promote persistence or even transient thickening of firn fields, and enhanced rockfall due to permafrost degradation under high-alpine rockwalls may accelerate rampart accumulation.

Increased rockfall activity may also increase the supply of coarse debris to rock glaciers, but the response of rock glaciers to atmospheric warming is complex. In high-alpine environments, many rock glaciers are located near the lower altitudinal boundary of permafrost, and because the coarse debris cover on rock glaciers promotes ground cooling due to enhanced heat transfer to the atmosphere, some rock glaciers extend below the local permafrost limit at altitudes where MAATs are in the region −3 °C to 0 °C. Though the coupling between air temperatures and ground temperatures in rock glaciers is complicated by thermal offset effects due to seasonal snowcover and the surface debris cover, a plot of the maximum surface velocity of a global sample of rock glaciers shows that this increases approximately exponentially with increasing MAAT (Figure 17.16); the wide scatter of points is explained by other factors that determine creep rate, such as slope, the thickness of the deforming layer(s), debris content and distribution, water content and marginal friction (Kääb *et al.*, 2007). This relationship suggests that warming air temperatures are likely to lead, at least initially, to accelerated rock-glacier movement, and that the effect of warming air temperatures will be most pronounced as MAATs approach 0 °C. There is evidence that such acceleration is already occurring in response to warming in the past few decades. Recent acceleration of rock glaciers in the European Alps has been documented by several researchers (e.g. Lambiel and Delaloye, 2004; Kääb *et al.*, 2007;

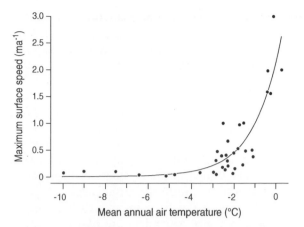

Figure 17.16 Maximum surface movement rate of a global sample of rock glaciers plotted against mean annual air temperature (MAAT). The line is an exponential fit through all points. *Source:* Adapted from Kääb *et al.* (2007). Reproduced with permission of Elsevier.

Hartl *et al.*, 2016). A study by Roer *et al.* (2005) showed that the individual velocities of 14 rock glaciers in the western Swiss Alps in the period 1993–2001 were 16–350% greater than in 1975–1993, a change that appears to have been driven by increasing air temperatures since the early 1990s. Acceleration of some rock glaciers is marked by the formation and widening of transverse crevasse-like cracks on their surfaces, particularly on frontal tongues (Roer *et al.*, 2008).

The manner in which increasing air temperatures generate accelerated rock-glacier velocities is incompletely understood. Numerical modelling of the potential response of rock-glacier creep to changes in ground temperature by Kääb *et al.* (2007) produced results broadly consistent with the above observations and suggests that accelerated movement is due mainly to enhanced temperature-dependent deformation of ice: as the ice within a rock glacier warms, its viscosity decreases and it deforms more rapidly under a given stress. They noted, however, that this explanation is insufficient to account for the observed acceleration of some rock glaciers, and suggested that the presence of water in rock glaciers close to the melting point may be critical in accounting for acceleration beyond rates attributable to enhanced deformation of warming ice (Ikeda *et al.*, 2008). Sorg *et al.* (2015) have suggested that rapid coupling between the atmosphere and thermally inert permafrost in rock glaciers could be caused by heat advection resulting from enhanced lateral or vertical water percolation. Interestingly, Kääb *et al.* (2007) also described evidence that some previously stable, non-deforming ground ice underlying debris-covered slopes had recently begun to deform, suggesting that air-temperature warming may locally result in the birth of new rock glaciers.

Kääb *et al.* (2007) also proposed that if warming continues, the period of accelerated movement will be succeeded by melting of ice in the rock-glacier core, leading to deceleration and eventual stagnation. There is evidence for this in the reconstructed movement rates of four rock glaciers in the Tien Shan of central Asia (Sorg *et al.*, 2015). A long-term reconstruction of the activity of these rock glaciers shows the expected correspondence between periods of above-average summer air temperatures and enhanced rock-glacier activity, but their response to air-temperature warming since the 1970s has been mixed: one has accelerated, but the other three exhibited reduced activity, suggesting that deactivation has commenced due to melting and shrinking of the ice core. This is most likely to occur with thickening of the active layer and concomitant reduction of the ice core near the fronts of rock glaciers (Trombotto and Borzotta, 2009), and may be associated with the development and widening of transverse cracks, frontal instability and progressive subsidence, though under a warming air temperature regime the insulating effects of winter snowcover may also be responsible for permafrost degradation and subsidence nearer source areas (Zhou *et al.*, 2015). Because of the insulating effect of the surface debris layer and latent heat effects, complete deactivation of rock glaciers due to melt of internal ice is likely to be prolonged.

17.4.4 Debris-flow Activity

As outlined in Chapter 12, debris flows are generally triggered by high pore-water pressures associated with transient high groundwater levels. Initial failure on open slopes often takes the form of translational sliding of a sediment body, followed by soil liquefaction as the soil fabric is disturbed by movement (hillslope debris flows), though debris flows originating within gullies are triggered by liquefaction of loose sediment (valley-confined debris flows). In both cases, sediment liquefaction initiates rapid downslope or downvalley flow of water-saturated debris (Iverson, 1997). It follows that two factors control the frequency and magnitude of most debris flows in high-alpine environments: sediment availability and rainstorms that generate a rapid rise in pore-water pressures. The rainstorms that trigger debris flows tend to be either high-intensity convectional storms of relatively brief duration or advective rainstorms of moderate intensity but prolonged duration (Guzzetti *et al.*, 2008). If climate change affects the frequency or timing of such rainstorms then there is likely to be change in the frequency and magnitude of debris-flow events.

Because debris flows are triggered by rainstorms that may affect limited areas, reconstructions of changes in the magnitude or frequency of past debris-flow events tend to be site-specific, so that past trends identified in one area may conflict with trends identified in others.

Site conditions in the source areas of debris flows (such as morphometry, sediment availability, vegetation cover and the presence or absence of permafrost) also condition terrain susceptibility to debris-flow initiation (Jomelli *et al.*, 2007, 2009, 2015). Additionally, there is greater uncertainty regarding future regional precipitation patterns than there is for climate warming, and different mountain areas are likely to experience different trends in seasonal rainfall patterns and in the frequency of debris-flow generating rainstorms. Case studies based on particular areas are therefore not generally applicable globally or even regionally, but nevertheless provide insights into possible future scenarios.

As with other aspects of the impact of climate change in high-alpine environments, most research has been carried out in the European Alps. A record of 565 valley-confined debris-flow events for the period 1970–2005 in the French Alps is of particular interest, as it covers a much greater area ($>16\,000\,km^2$) than other studies, and hence is less sensitive to localized effects. Within both the northern and southern parts this area, the frequency of debris-flow events exhibited a significant increase after the mid-1980s (Figure 17.17), mainly attributable to an increase in the frequency of summer convectional rainstorms (Pavlova *et al.*, 2014). Conversely, an analysis (based on tree-ring data) of the timing of 296 valley-confined debris-flow events over the past 150 years in the Zermatt valley of western Switzerland detected no significant temporal trends but showed that the decade 2000–09 witnessed the lowest frequency of debris-flow events for a century (Bollschweiler and Stoffel, 2010).

To complicate matters further, an analysis of 428 hillslope debris-flow events in the Massif des Ecrins (French Alps) by Jomelli *et al.* (2007) demonstrated an overall decrease in event frequency since the 1970s, but also showed that response differed between sandstone and granite areas, and that it further depended on the origin of the source sediment and the altitude of flow initiation. As the Massif des Ecrins lies within the area studied by Pavlova *et al.* (2014), the apparently conflicting results obtained by these two studies suggest that there may be differences in the response of hillslope and valley-confined debris flows to recent and possibly future climate change.

Secular changes in the frequency and magnitude of debris-flow activity are dependent not only on changes in rainstorm frequency, but also on antecedent soil moisture conditions, depth of unfrozen ground and seasonal temperature changes. From their 150 year record of debris-flow events in the Zermatt valley, where valley-confined debris flows are sourced in high-altitude terrain underlain by permafrost, Schneuwly-Bollschweiler and Stoffel (2012) showed that the debris-flow season has lengthened over the 20th century, from 4 months (June–September) to 6 (May–October); with continued climatic warming, further lengthening of the annual period of debris-flow activity is likely. This analysis suggests that debris-flow events occurring between May and August are initiated mainly by brief high-intensity convectional rainstorms, whereas those later in the year tend to be triggered by prolonged advective precipitation. Schneuwly-Bollschweiler and Stoffel also demonstrated that events earlier in the debris-flow season tend to be

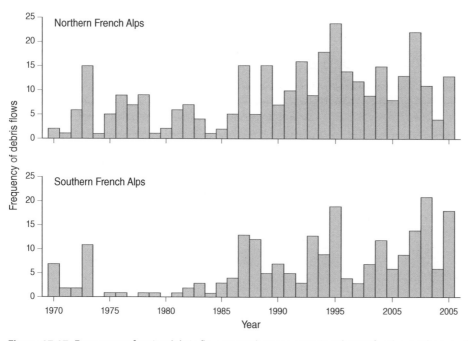

Figure 17.17 Frequency of major debris-flow events between 1970 and 2005 for the northern and southern subdivisions of an area exceeding $16\,000\,km^2$ in the French Alps. *Source:* Pavlova *et al.* (2014). Reproduced with permission of Elsevier.

triggered by lower rainstorm totals than those in August–October. In part, this may be due to the additional contribution of nival meltwater as a result of spring rain-on-snow events, but Stoffel *et al.* (2011) have suggested that it also reflects shallow thaw depth (and thus limited water storage and more rapid runoff) early in the debris-flow season.

On the basis of future temperature and precipitation projections based on regional climate models informed by local meteorological records, Markus Stoffel and his co-workers have devised a scenario of future debris-flow activity for the Zermatt valley, and in particular for the Ritigraben torrent, a steep catchment with its source above 2600 m altitude (and partly occupied by an active rock glacier) and its outlet on a debris fan at 1500–1800 m (Stoffel and Beniston, 2006; Stoffel, 2010; Stoffel *et al.*, 2011, 2014). At present, the main sources of sediment feeding debris flows in this catchment include debris at the margin of the Ritigraben rock glacier, landslides, rockfalls, partial collapse of channel walls and active-layer failures on the rock-glacier snout; the latter are often triggered by the same rainfall events that initiate debris flows (Lugon and Stoffel, 2010). Their projections suggest that the area will experience drier summers but possibly increased rainstorm events in spring, autumn and early winter, though the latter may not compensate for the former. As a result, only limited change is expected in the frequency of debris-flow events before 2050. However, a possible increase in the magnitude of rainstorm events, coupled with increased sediment delivery to source channels, could result in debris-flow events of much greater magnitude than any in the past 150 years, particularly if there is destabilization of the Ritigraben rock glacier as a result of permafrost degradation. Stoffel (2010) has suggested that under these circumstances, future debris flows incorporating ~50000 m^3 of sediment are possible. A huge (~500000 m^3) debris flow in the central Swiss Alps in August 2005, the largest in Switzerland for 20 years, may provide a foretaste of things to come. Roughly 40% of this flow represents mobilization of recently-exposed glacigenic sediments in the source area, and the remainder was entrained on the valley-floor debris fan (Stoffel and Huggel, 2012). A similar event in a settled area could have disastrous consequences.

17.5 Climate Change, Permafrost Degradation and Greenhouse Gas Emissions

This chapter has so far considered the effects of recent and future climate change on periglacial environments. We conclude by briefly considering the feedback effects that surface and nearsurface changes in periglacial environments may have on future climate. Some of these, notably shrinking summer sea-ice cover and lengthening of the arctic snow-free season, have the effect of reducing surface albedo and thus increasing the amount of heat absorbed at the sea or land surface. Other effects relate to changes in vegetation cover under a warming climate, but these are complex: trapping of snowcover and increased shading by trees and shrubs tend to lower the amount of incident radiation reaching the ground surface, but some possible future changes in ground cover may increase summer heating. Chapin *et al.* (2005), for example, have estimated that complete conversion to shrub tundra in arctic Alaska has the potential to increase summer heating by ~6–21 W m^{-2} per decade, and complete conversion of shrub tundra to tree cover by ~26 W m^{-2} per decade. By far the greatest concern, however, is that microbial breakdown of organic carbon released from thawing permafrost will release globally significant quantities of carbon dioxide (CO_2) and methane (CH_4), thereby accelerating greenhouse gas-induced climate change. An overview of this topic is included in IPCC (2013), and recent research has been summarized by Schuur *et al.* (2015).

17.5.1 Permafrost Carbon: Reserves, Emissions and Projections

Organic carbon in permafrost represents the remnants of plant matter that became frozen into aggrading permafrost, usually in depositional settings or through peat accumulation, throughout the Holocene and, in some areas, the Late Pleistocene. By far the largest reserve of permafrost carbon is located in the Arctic and Subarctic. According to Schuur *et al.* (2015), the total pool of terrestrial carbon in the northern permafrost zone is thought to amount to 1330–1580 Gt (1 Gt = 1 billion metric tons). Much of this is distributed within 3 m of the surface (1035 ± 150 Gt), the rest as deep carbon in yedoma sediments, estimated to amount to 211 ± 70 Gt (Strauss *et al.*, 2013) or 456 ± 45 Gt (Walter Anthony *et al.*, 2014), or in major Arctic river deltas (91 ± 39 Gt; Hugelius *et al.*, 2013). An additional ~350–465 Gt is estimated to exist within other deep terrestrial permafrost sediments, and a potentially huge quantity of organic carbon is stored within subsea permafrost. Collectively, these stores contain roughly two to three times as much carbon as the global atmosphere.

The rate at which CO_2 and CH_4 are released from permafrost carbon stocks depends partly on the rate of permafrost thaw (discussed earlier in this chapter) and partly by the decomposability (propensity for microbial decomposition) of organic carbon. Initial rates of permafrost carbon loss are potentially rapid, but tend to decline exponentially through time as more labile constituents become exhausted (Knoblauch *et al.*, 2013). Laboratory studies of thawed permafrost soils suggest that decade-long carbon losses fall within a very wide range, from 1% to 76%. A major cause of this wide variation has been linked to the carbon-to-nitrogen ratio of the organic matter: soils with >20% carbon have mean decade-long losses of 17–34%, whereas those with <20% carbon have mean decade-long losses of 6–13% (Schädel *et al.*, 2014). The rate of carbon loss also depends on whether decomposition takes place under anaerobic (water-saturated) or aerobic (unsaturated) conditions. Over a one-year period, cumulative carbon emissions under anaerobic conditions are on average about 80% lower than those under aerobic conditions. However, specialized microbes release methane as well as carbon dioxide in anaerobic environments, and because methane is a much more potent greenhouse gas, this partly offsets the lower CO_2 emissions from waterlogged soils in terms of radiative efficiency. Across the northern permafrost zone, there are therefore marked contrasts in the decomposability of thawing organic-rich soil, depending not only on the carbon content of the soil but also on whether it exists above or below the water table.

Modelled projections of potential carbon release from thawing permafrost by 2100 suggest that it could be as low as ~37 Gt (Zhuang *et al.*, 2006) or as high as ~174 Gt (Burke *et al.*, 2013), with an average projection of 92 ± 17 Gt across seven different scenarios. Of this, it is estimated that only ~2.3% is contributed by methane, but because of its much greater radiative efficiency this small contribution may increase the warming potential of released carbon by 35–48% over a century (Schuur *et al.*, 2013). Some models suggest that a combination of warmer, longer growing seasons, elevated atmospheric carbon dioxide levels and nutrient release from decomposing organic carbon will stimulate increased uptake of atmospheric carbon by vegetation, offsetting or exceeding carbon release due to thaw of frozen organic matter for several decades as the climate warms (Koven *et al.*, 2011; Schaefer *et al.*, 2011; MacDougall *et al.*, 2012), so that in the immediate future the tundra regions across the Arctic may continue to act as a weak sink of atmospheric CO_2 (McGuire *et al.*, 2009). Over longer timescales, however, microbial release of carbon seems certain to outstrip the capacity for plant carbon uptake (Vogel *et al.*, 2009). Modelled net carbon emissions due to permafrost degradation under various warming scenarios indicate 0.13–0.27 °C additional global warming by 2100, but are poorly constrained (Schuur *et al.*, 2015).

17.5.2 Thermokarst Development and Carbon Emissions

One source of uncertainty in predicting carbon emissions resulting from microbial decomposition of thawed soils in the northern permafrost zone is the role played by rapid permafrost degradation and resultant thermokarst development. Evidence for recent increases in thermokarst development was outlined earlier in this chapter: ice-wedge degradation and thermokarst gullying; collapse of palsas, lithalsas and peat plateaus; formation, expansion and drainage of thermokarst lakes, ponds, bogs and fens; increased incidence of active-layer failures and thaw slumping; and accelerated erosion of ice-rich permafrost coasts. All of these are manifestations of relatively rapid warming-induced changes in permafrost landscapes.

Rapid thaw associated with thermokarst development not only exposes previously frozen organic carbon to microbial decomposition, but also alters the balance of carbon dioxide and methane emissions through changes in local hydrology. Some of the highest levels of methane emission are associated with thermokarst lakes, ponds and wetlands (Walter *et al.*, 2006; Olefeldt *et al.*, 2013), but there is also evidence that accumulation of new carbon under anaerobic conditions in lake sediments and saturated peat may offset emissions in some circumstances (Walter Anthony *et al.*, 2014). Conversely, lake drainage and lowered water tables expose previously saturated organic material to decomposition under aerobic conditions, accelerating carbon dioxide release. Erosion and redeposition of organic matter into lakes and the sea is also thought to accelerate microbial breakdown, in part through exposure to sunlight (Cory *et al.*, 2013). A key issue is therefore not only how widespread thermokarst will become under a future warming climate, but also whether it will tend to result in net landscape drying (aerobic conditions) or waterlogging (anaerobic conditions). At present, modelled projections of carbon dioxide emissions in the northern permafrost zone consider only gradual permafrost degradation, not the effects of rapid thaw associated with thermokarst processes. A further source of uncertainty is the possible increased frequency of fires, particularly in the boreal forest zone. Fires not only cause combustion of organic matter, but also remove the protective organic layer above permafrost, exposing the surviving organic material to decomposition as a result of soil warming and permafrost thaw (Rocha and Shaver, 2011).

17.5.3 Greenhouse Gas Emissions from Subsea Permafrost

Frozen organic-rich sediments occur on Arctic Ocean shelves that were submerged by rising sea levels as the last Pleistocene ice sheets melted, and it is thought that these have been subject to gradual thaw throughout the Holocene, reducing the carbon pool in subsea permafrost to ~400 Gt, though this is partly replenished by organic sediments released through erosion of ice-rich permafrost coasts or deposited on arctic shelves by rivers

(Vonk *et al.*, 2012); fluvial transport of organic carbon to the Arctic Ocean basin is thought to be about 10 times greater than the global average (McGuire *et al.*, 2009). In addition to organic carbon in subsea permafrost, methane has accumulated in Arctic Ocean shelf sediments, either as free gas or as methane clathrates (methane hydrates). These are compounds ($4CH_4 \cdot 23H_2O$) in which methane is trapped within a crystal structure of water, forming solids similar to ice. Methane clathrates are stable at the pressures and temperatures on the Arctic seabed, and are thought to be prevented from destabilizing and releasing methane gas by being locked in permafrost. There is evidence from the East Siberian shelf that warming sea temperatures and increased storm activity may have accelerated methane release, releasing ~17 Mt of methane per year (Shakhova *et al.*, 2010, 2014), though this figure does not include sudden releases of methane, which are likely to accompany clathrate destabilization. It is not known whether future increases in methane release from Arctic Ocean sediments pose an immediate threat or will be significant only over centennial timescales. However, the amount of methane that could theoretically be released in the future is enormous. Callaghan *et al.* (2011) estimated that the total amount of carbon preserved within the Arctic Ocean shelves could be around 1300 Gt, of which ~800 Gt is previously formed methane that could be released suddenly as appropriate pathways develop. They noted that release of only 1% of this huge methane reservoir would more than triple the atmospheric mixing ratio of methane, probably triggering abrupt climate change.

17.5.4 Permafrost Degradation and the Global Carbon Cycle

Following a measured review of the evidence, Schuur *et al.* (2015) have estimated that 5–15% of the terrestrial permafrost carbon pool (very approximately ~70–220 Gt) is vulnerable to release as greenhouse gases this century under the current warming trajectory, with the majority of released gas being carbon dioxide. This is much less than current fossil-fuel emissions (9.7 ± 0.5 Gt per year in 2012), and they concluded that projected emissions of carbon dioxide and methane from thawing terrestrial permafrost are unlikely to cause short-term abrupt climate change, but over decadal timescales will nevertheless accelerate climate change attributable to anthropogenic greenhouse-gas emissions (IPCC, 2013). The prospect of a short-term 'climate bomb' due to warming and thaw of circum-arctic permafrost (Whiteman *et al.*, 2013) thus appears unlikely, even though the effects of rapid permafrost degradation (thermokarst development) and

potential release of methane from subsea permafrost are wild cards that could potentially accelerate arctic and subarctic carbon emissions beyond the scope of recent modelling scenarios. Moreover, even if anthropogenic emissions are curtailed, the northern permafrost zone will continue to warm and release carbon dioxide and methane through microbial decomposition of the vast stocks of organic matter presently locked in permafrost, and will thereby continue to contribute to atmospheric warming, probably for centuries. The Arctic has been described as the thermostat of mid-latitude climate, and the setting on this thermostat is likely to change in the foreseeable and unforeseeable future.

17.6 Conclusion

In recent geological time, the periglacial environments of the Earth have experienced two periods of rapid change. The first occurred during the final millennia of the Pleistocene and the early Holocene, when the last Pleistocene ice sheets shrank, the southern limit of northern-hemisphere permafrost retreated northwards and permafrost degradation triggered widespread thermokarst formation. The second is now. In both the circumpolar north and mid-latitude mountains, permafrost is warming and slowly thawing, triggering a cascade of landscape changes ranging from accelerated thermokarst development to destabilization of mountain rockwalls. Summer sea-ice cover is shrinking in the Arctic Ocean, the duration of the snow-free season on land is increasing, the runoff regime of Arctic rivers is changing, and there is evidence that ice-rich permafrost coasts are being eroded at increasing rates. The timeline for investigation of most of these recent changes is short – about 50 years – but already there has been a subtle shift in the emphasis of much periglacial research. Until the present century, this focused mainly on the interpretation of past periglacial phenomena and attempts to understand the geomorphic processes operating in present periglacial environments: processes that have formed the fascinating landforms and landscapes of the periglacial realm. As this book demonstrates, huge advances have been made, and continue to be made, in our understanding of periglacial processes and landforms. A major challenge is now to understand how these processes, landforms and landscapes are responding and will continue to respond to climate change, and the effects of their responses on future climate. The battle to limit global climate change has barely begun, but periglacial environments are in the front line of a conflict that will affect humanity and ecosystems throughout the world.

Appendix

Text Abbreviations, Units and Symbols Employed in Equations

Text Abbreviations

BTS	basal temperature of snow
DEM	digital elevation model
GIS	geographical information system
LGM	last (global) glacial maximum
LPM	last permafrost maximum
MAAT	mean annual air temperature
MAP	mean annual precipitation
MAGT	mean annual ground temperature
MAGST	mean annual ground surface temperature
PISR	potential incoming solar radiation

Units

Temperature

°C	degrees Celsius
K	Kelvin

Length

km	kilometres or kilometers (10^3 m)
m	metres or meters
cm	centimetres or centimeters (10^{-2} m)
mm	millimetres or millimeters (10^{-3} m)
μm	microns (10^{-6} m)

Mass

Mt	megatons (10^6 kg)
t	metric tons (10^3 kg)
kg	kilograms
g	grams (10^{-3} kg)

Time

Ma	millions of years (10^6 a)
ka	thousands of years (10^3 a)
a	years
d	days
s	seconds

Compound Units

Pa	pascals ($kg\,m^{-1}\,s^{-2}$)
kPa	kilopascals (10^3 Pa)
MPa	megapascals (10^6 Pa)
J	joules ($kg\,m^2\,s^{-2}$)
kJ	kilojoules (10^3 J)
MJ	megajoules (10^6 J)
W	Watts ($J\,s^{-1}$)

Symbols

A	amplitude of a temperature cycle
C	volumetric heat capacity ($J\,m^{-3}\,°C^{-1}$)
C_v	coefficient of consolidation
c	cohesion (kPa)
c'	effective cohesion (kPa)
D	displacement or downslope movement
E	internal energy of a system
ET	evapotranspiration
F	force
Fn	frost number
F_s	factor of safety
G	Gibbs free energy
G_i	Gibbs free energy of ice
G_w	Gibbs free energy of water
g	gravitational acceleration ($9.81\,m\,s^{-2}$)
H	enthalpy
h	amount of frost heave
I_{FA}	screen temperature freezing index in degree-days
I_{FT}	screen temperature thaw index in degree-days
I_{sf}	surface freezing index in degree-days
I_{st}	surface thaw index in degree-days
i_E	excess ice content
i_G	gravimetric ice content
i_V	volumetric ice content
k	thermal conductivity ($W\,m^{-1}\,°C^{-1}$)
k_f	thermal conductivity of frozen soil
k_t	thermal conductivity of thawed soil
k^*	soil permeability
L	length

Periglacial Geomorphology, First Edition. Colin K. Ballantyne.

L_f	latent heat of fusion	T_z	temperature at depth z
L_m	latent heat of melting	t	time
M	mass	t^*	phase lag of temperature cycle
n_f	freezing period n-factor	u	pore-water pressure (Pa or kPa)
n_t	thawing period n-factor	V	volume (m^3)
P	pressure (Pa or kPa)	V_i	(specific) volume of ice
P_i	ice pressure	V_w	(specific) volume of water
P_w	water pressure	v	velocity
Pn	precipitation	W	weight of soil
p	period of a temperature cycle	z	depth (m)
Q	runoff or water discharge ($m^3 s^{-1}$)	z_p	depth of permafrost
Q_g	heat flow downwards from the surface	α	slope angle
R	thaw consolidation ratio	γ	unit weight of soil
R_r	rockwall retreat rate	γ_w	unit weight of water
r	radius	ε	strain
rk	thermal conductivity ratio	κ	thermal diffusivity ($m^2 s^{-1}$)
S	internal entropy of a system	κ_a	apparent thermal diffusivity ($m^2 s^{-1}$)
T	temperature (°C or K)	ρ	density ($kg\,m^{-3}$)
T_m	melt temperature of ice	σ_t	tensile strength (Pa or kPa)
T_0	freezing point of water under atmospheric pressure (=0°C)	σ'	effective (normal) stress (Pa or kPa)
T_s	temperature at the ground surface	τ	shear stress (Pa or kPa)
T_t	temperature at time t	ϕ'	effective friction angle
T_{TOP}	mean annual temperature at the top of permafrost	ψ	coefficient of thermal expansion or contraction

References

Aarseth, I. and Fossen, H. (2004) A Holocene lacustrine rock platform around Storavatnet, Osterøy, western Norway. *The Holocene* 14, 589–596.

Abramov, A., Gruber, S. and Gilichinsky, D.A. (2008) Mountain permafrost on active volcanoes: field data and statistical mapping, Klyuchevskaya volcano group, Kamchatka, Russia. *Permafrost and Periglacial Processes* 19, 261–277.

Ackert, R.P (1998) A rock-glacier/debris-covered glacier system at Galena Creek, Absaroka Mountains, Wyoming. *Geografiska Annaler* 80A, 267–276.

Ackroyd, P. (1986) Debris transport by avalanche, Torlesse Range, New Zealand. *Zeitschrift für Geomorphologie* 30, 1–14.

Ahlbrandt, T.S. and Andrews, S. (1978) Distinctive sedimentary features of cold climate eolian deposits, North Park, Colorado. *Palaeogeography, Palaeoclimatology, Palaeoecology* 25, 327–351.

Ahnert, F. (1994) Modelling the development of non-periglacial sorted nets. *Catena* 23, 43–63.

Aide, M. and Marshaus, A. (2002) Fragipan genesis in two Alfisols in east central Missouri. *Soil Science* 167, 453–464.

Aitken, M.J. (1998) *An Introduction to Optical Dating.* Oxford University Press, Oxford.

Akagawa, S. and Fukuda, M. (1991) Frost heave mechanism in welded tuff. *Permafrost and Periglacial Processes* 2, 301–309.

Åkerman, H.J. (1982) Observations of palsas within the continuous permafrost zone in eastern Siberia and in Svalbard. *Geografisk Tidsskrift* 82, 45–51.

Åkerman, H.J. (1984) Notes on talus slope morphology and processes in Spitsbergen, *Geografiska Annaler* 66A, 267–284.

Åkerman, H.J. (1992) Hydrographic characteristics of the Strokdammane Plain, west Spitsbergen, Svalbard. *Geografiska Annaler* 74A, 169–182.

Åkerman, H.J. (1996) Slow mass movements and climatic relationships, 1972–1996. Kapp Linné, West Spitsbergen. In Anderson, M.G. and Brooks, S.M. (eds) *Advances in Hillslope Processes.* John Wiley & Sons, Chichester, 1219–1256.

Åkerman, H.J. (2005) Relationships between slow slope processes and active-layer thickness 1972–2002, Kapp Linné, Svalbard. *Norsk Geografisk Tidsskrift* 59, 116–128.

Åkerman, H.J. and Johansson, M. (2008) Thawing permafrost and thicker active layers in sub-arctic Sweden. *Permafrost and Periglacial Processes* 19, 279–292.

Åkerman, H.J. and Malmström, B. (1986) Permafrost mounds in the Abisko area, northern Sweden. *Geografiska Annaler* 68A, 155–165.

Aleinikoff, J.N., Muhs, D.R., Bettis, E.A. III *et al.* (2008) Isotopic evidence for the diversity of late Quaternary loess in Nebraska: glaciogenic and non-glaciogenic sources. *Geological Society of America Bulletin* 120, 1362–1377.

Allard, M. and Kasper, J.N. (1998) Temperature conditions for ice wedge cracking: field measurements from Salluit, northern Québec. In Lewkowicz, A.G. and Allard, M. (eds) *Permafrost: Proceeding of the 7th International Conference.* Université Laval, Québec, 5–11.

Allard, M. and Rousseau, L. (1999) The internal structure of a palsa and a peat plateau in the Rivière Boniface region, Québec: inferences on the formation of ice-segregation mounds. *Géographie Physique et Quaternaire* 53, 373–387.

Allard, M., Seguin, M.K. and Lévesque, R. (1987) Palsas and mineral permafrost mounds in northern Québec. In Gardner, V. (ed.) *International Geomorphology, 1986.* John Wiley & Sons, Chichester, 285–309.

Allard, M., Caron, S. and Begin, T.Y. (1996) Climatic and ecological controls on ice segregation and thermokarst: the case history of a permafrost plateau in northern Québec. *Permafrost and Periglacial Processes* 7, 207–227.

Allard, M., Michaud, Y., Ruz, M.-H. and Héquette, A. (1998) Ice foot, freeze-thaw of sediments, and platform erosion in a subarctic microtidal environment, Manitounuk Strait, northern Quebec, Canada. *Canadian Journal of Earth Sciences* 35, 965–979.

Allen, C.E., Darmody, R.G., Thorn, C.E. *et al.* (2001) Clay mineralogy, chemical weathering and landscape evolution in arctic-alpine Sweden. *Geoderma* 99, 277–294.

Periglacial Geomorphology, First Edition. Colin K. Ballantyne.
© 2018 John Wiley & Sons Ltd. Published 2018 by John Wiley & Sons Ltd.

Allen, C.R., O'Brien, R.M.G. and Sheppard, S.M.F. (1976) The chemical and isotopic characteristics of some northeast Greenland surface and pingo waters. *Arctic and Alpine Research* 8, 297–317.

Allen, S.K. and Huggel, C. (2013) Extremely warm temperatures as a potential cause of recent high mountain rockfall. *Global and Planetary Change* 107, 59–69.

Allen, S.K., Cox, S.C. and Owens, I.F. (2011) Rock avalanches and other landslides in the central Southern Alps of New Zealand: a regional study considering possible climate change impacts. *Landslides* 8, 33–48.

Allison, R.J. and Davies, K.C. (1996) Ploughing blocks as evidence of downslope sediment transport in the English Lake District. *Zeitschrift für Geomorphologie Supplementband* 106, 199–219.

Amadei, B. and Stephansson, O. (1997) *Rock Stress and its Measurement*. Chapman and Hall, London.

AMAP (2011) *Snow, Water, Ice and Permafrost in the Arctic (SWIPA): Climate Change and the Cryosphere*. Arctic Monitoring and Assessment Programme (AMAP), Oslo.

An, W. and Allard, M. (1995) A mathematical approach to modelling palsa formation: insights on processes and growth conditions. *Cold Regions Science and Technology* 23, 231–244.

Andersland, O.B. and Ladanyi, B. (2003) *Frozen Ground Engineering*, 2nd edn. John Wiley & Sons, New York.

Anderson, R.S. (1998) Near-surface thermal profiles in alpine bedrock: implications for the frost weathering of bedrock. *Arctic and Alpine Research* 30, 362–372.

Anderson, R.S. (2002) Modeling the tor-dotted crests, bedrock edges, and parabolic profiles of high alpine surfaces of the Wind River Range, Wyoming. *Geomorphology* 46, 35–58.

Anderson, R.S., Anderson, S.P. and Tucker, G.E. (2013) Rock damage and regolith transport by frost: an example of climate modulation of the geomorphology of the critical zone. *Earth Surface Processes and Landforms* 38, 299–316.

Anderson, S.P. (1988) The upfreezing process: experiments with a single clast. *Geological Society of America Bulletin* 100, 609–621.

Andersson, J.G. (1906) Solifluction: a component of subaerial denudation. *Journal of Geology* 14, 91–112.

André, M.-F. (1990a) Geomorphic impact of spring avalanches in Northwest Spitsbergen. *Permafrost and Periglacial Processes* 1, 97–110.

André, M.-F. (1990b) Frequency of debris flows and slush avalanches in Spitsbergen: a tentative evaluation from lichenometry. *Polish Polar Research* 11, 345–363.

André, M.-F. (1993) *Les Versants du Spitsberg*. Presses Universitaires de Nancy, Nancy.

André, M.-F. (1997) Holocene rockwall retreat in Svalbard: a triple-rate evolution. *Earth Surface Processes and Landforms* 22, 423–440.

André, M.-F. (2002) Rates of postglacial rock weathering on glacially-scoured outcrops (Abisko-Riksgränsen area, 68°N). *Geografiska Annaler* 84A, 139–150.

André, M.F. (2003) Do periglacial landscapes evolve under periglacial conditions? *Geomorphology* 52, 149–164.

André, M.-F. (2004) The geomorphic impact of glaciers as indicated by tors in North Sweden (Aurivaara, 68°N) *Geomorphology* 57, 403–421.

André M.-F. (2009) From climatic to global change geomorphology: contemporary shifts in periglacial geomorphology. *Geological Society, London, Special Publication* 320, 5–28.

André, M.-F. and Hall, K. (2005) Honeycomb development on Alexander Island, glacial history of George IV Sound and palaeoclimatic implications. *Geomorphology* 65, 117–138.

André, M.-F., Hall, K., Bertran, P. and Arocena, J. (2008) Stone runs in the Falkland Islands: periglacial or tropical? *Geomorphology* 95, 524–543.

Andreev, A.A., Schirrmeister, L., Siegert, C. *et al.* (2002) Palaeoenvironmental changes in north-eastern Siberia during the Late Quaternary – evidence from pollen records of the Bykovsky Peninsula. *Polarforschung* 70, 13–25.

Andrieux, E., Bertran, P. and Saito, K. (2016) Spatial analysis of French Pleistocene permafrost by a GIS database. *Permafrost and Periglacial Processes* 27, 17–30.

Angillieri, M.Y.E. (2010) Application of frequency ratio and logistic regression to active rock glacier occurrence in the Andes of San Juan, Argentina. *Geomorphology* 114, 396–405.

Antoine, P., Catt, J., Lautridou, J.-P. and Sommé, J. (2003a) The loess and coversands of northern France and southern England. *Journal of Quaternary Science* 18, 309–318.

Antoine, P., Munaut, A.-V., Limondin-Lozouet, N. *et al.* (2003b) Response of the Selle River to climate modifications during the Lateglacial and Early Holocene (Somme Basin – Northern France). *Quaternary Science Reviews* 22, 2061–2076.

Antoine, P., Rousseau, D.-D., Moine, O. *et al.* (2009) Rapid and cyclic aeolian deposition during the Last Glacial in European loess: a high-resolution record from Nussloch, Germany. *Quaternary Science Reviews* 28, 2955–2973.

Antoine, P., Rousseau, D.-D., Degeai, J.-P. *et al.* (2013) High resolution record of the environmental response to climatic variations during the last interglacial-glacial cycle in Central Europe: the loess-palaeosol sequence of Dolní Vestonici (Czech Republic) *Quaternary Science Reviews* 67, 17–38.

Antoine, P., Coutard, S., Guerin, G. *et al.* (2016) Upper Pleistocene loess-palaeosol records from Northern France in the European context: environmental background and dating of the Middle Palaeolithic. *Quaternary International* 411, 4–24.

Aoyama, M. (2005) Rock glaciers in the northern Japanese Alps: palaeoenvironmental implications since the Late Glacial. *Journal of Quaternary Science* 20, 471–484.

Arcone, S.A., Chacho, E.F. and Delaney, A.J. (1998) Seasonal structure of taliks beneath arctic streams determined with ground-penetrating radar. In Lewkowicz, A.G. and Allard, M. (eds) *Permafrost: Proceedings of the 7th International Conference.* Université Laval, Qeébec, 19–24.

Are, F.E., Grigoriev, M.N., Hubberten, H-W and Rachold, V. (2005) Using thermoterrace dimensions to calculate the coastal erosion rate. *Geo-Marine Letters* 25, 121–126.

Are, F.E., Reimnitz, E., Grigoriev, M. *et al.* (2008) The influence of cryogenic processes on the erosional arctic shoreface. *Journal of Coastal Research* 24, 110–121.

Arenson, L.U. and Springman, S. (2005) Mathematical descriptions for the behaviour of ice-rich frozen soils at temperatures close to 0 °C. *Canadian Geotechnical Journal* 42, 431–442.

Arenson, L.U., Hoelzle, M. and Springman, S. (2002) Borehole deformation measurements and internal structure of some rock glaciers in Switzerland. *Permafrost and Periglacial Processes* 13, 117–135.

Arenson, L.U., Azmatch, T.F. and Sego, D.C. (2008) A new hypothesis on ice lens formation in frost-susceptible soils. In Kane, D.L. and Hinkel, K.M. (eds) *Ninth International Conference on Permafrost.* University of Alaska, Fairbanks, 59–64.

Arnalds, O. (2000) The Icelandic 'Rofabard' soil erosion features. *Earth Surface Processes and Landforms* 25, 17–28.

Arnalds, O., Gisladottir, F.O. and Sigurjonsson, H. (2001) Sandy deserts of Iceland: an overview. *Arid Environments* 47, 359–371.

Arnalds, O., Gisladottir, F.O. and Orradottir, B. (2012) Determination of aeolian transport rates of volcanic soils in Iceland. *Geomorphology* 167–168, 4–12.

Arocena, J.M., Zhu, L.P. and Hall, K. (2003) Mineral accumulations induced by biological activity on granitic rocks in Qinghai Plateau, China. *Earth Surface Processes and Landforms* 28, 1429–1437.

Arp, C.D., Whitman, M.S., Jones, B.M. *et al.* (2015) Distribution and biophysical processes of beaded streams in Arctic permafrost landscapes. *Biogeosciences* 12, 29–47.

Ascaso, C., Sancho, L.G. and Rodriguez-Pascual, C. (1990) The weathering action of saxicolous lichens in the maritime Antarctic. *Polar Biology* 11, 33–39.

Astakhov, V.I. (2014) The postglacial Pleistocene of the northern Russian mainland. *Quaternary Science Reviews* 92, 388–408.

Astakhov, V.I. and Svendsen, J.I. (2002) Age and remnants of a Pleistocene glacier in the Bolshezemelskaya tundra. *Doklady Earth Sciences* 384, 468–472.

Astakhov, V.I., Kaplyanskaya, F.A. and Tarnogradskiy, V.D. (1996) Pleistocene permafrost of west Siberia as a deformable glacier bed. *Permafrost and Periglacial Processes* 7, 165–191.

Atkinson, T.C., Briffa, K.R. and Coope, G.R. (1987) Seasonal temperatures in Britain during the past 22 000 years, reconstructed using beetle remains. *Nature* 325, 587–592.

Augustinus, P.C. and Selby, M.J. (1990) Rock slope development in McMurdo Oasis, Antarctica, and implication for interpretations of glacial history. *Geografiska Annaler* 72A, 55–62.

Avirmed, D., Ishikawa, M., Iijima, Y. and Yamkin, J. (2014) Temperature regimes of the active layer and seasonally frozen ground under a forest-steppe mosaic, Mongolia. *Permafrost and Periglacial Processes* 25, 295–306.

Ayling, B.F. and McGowan, H.A. (2006) Niveo-aeolian sediment deposits in coastal south Victoria Land, Antarctica: indicators of regional variability in weather and climate. *Arctic, Antarctic and Alpine Research* 38, 313–324.

Baker C.A., Bateman, M., Bateman, P. and Jones, H. (2013) The aeolian sand record in the Trent Valley. *Mercian Geologist* 18, 108–118.

Ballantyne, C.K. (1978a) The hydrologic significance of nivation features in permafrost areas. *Geografiska Annaler* 60A, 51–54.

Ballantyne, C.K. (1978b) Variations in the size of coarse clastic particles over the surface of a small sandur, Ellesmere Island, N.W.T., Canada. *Sedimentology* 25, 141–147.

Ballantyne, C.K. (1979) Patterned ground on an active medial moraine, Jotunheimen, Norway. *Journal of Glaciology* 22, 396–401.

Ballantyne, C.K. (1985) Nivation landforms and snowpatch erosion on two massifs in the Northern Highlands of Scotland. *Scottish Geographical Magazine* 101, 40–49.

Ballantyne, C.K. (1986a) Nonsorted patterned ground on mountains in the Northern Highlands of Scotland. *Biuletyn Peryglacjalny* 30, 15–34.

Ballantyne, C.K. (1986b) Late Flandrian solifluction on the Fannich Mountains, Ross-shire. *Scottish Journal of Geology* 22, 395–406.

Ballantyne, C.K. (1987a) The present-day periglaciation of upland Britain. In Boardman, J. (ed.) *Periglacial Processes and Landforms in Britain and Ireland.* Cambridge University Press, Cambridge UK, 113–126.

Ballantyne, C.K. (1987b) Some observations on the morphology and sedimentology of two active protalus ramparts, Lyngen, northern Norway. *Arctic and Alpine Research* 19, 167–174.

Ballantyne, C.K. (1989) Avalanche impact landforms on Ben Nevis, Scotland. *Scottish Geographical Magazine* 106, 38–42.

Ballantyne, C.K. (1995) Paraglacial debris cone formation on recently-deglaciated terrain. *The Holocene* 5, 25–33.

Ballantyne, C.K. (1996) Formation of miniature sorted patterns by shallow ground freezing: a field experiment. *Permafrost and Periglacial Processes* 7, 409–424.

Ballantyne, C.K. (1998a) Age and significance of mountain-top detritus. *Permafrost and Periglacial Processes* 9, 327–345.

Ballantyne, C.K. (1998b) Aeolian deposits on a Scottish mountain summit: characteristics, provenance, history and significance. *Earth Surface Processes and Landforms* 23, 625–641.

Ballantyne, C.K. (2001a) The sorted stone stripes of Tinto Hill. *Scottish Geographical Journal* 117, 313–324.

Ballantyne, C.K. (2001b) Measurement and theory of ploughing boulder movement. *Permafrost and Periglacial Processes* 12, 267–288.

Ballantyne, C.K. (2002) Paraglacial geomorphology. *Quaternary Science Reviews* 21, 1935–2017.

Ballantyne, C.K. (2003) Paraglacial landform succession and sediment storage in deglaciated mountain valleys: theory and approaches to calibration. *Zeitschrift für Geomorphologie, Supplementband* 132, 1–18.

Ballantyne, C.K. (2010) A general model of autochthonous blockfield evolution. *Permafrost and Periglacial Processes* 21, 289–300.

Ballantyne, C.K. (2013a) Paraglacial geomorphology. In Elias, S. (ed.) *Encyclopedia of Quaternary Science*, 2nd edn. Elsevier, Amsterdam, 553–565.

Ballantyne, C.K. (2013b) Trimlines and palaeonunataks. In Elias, S. (ed.) *Encyclopedia of Quaternary Science*, 2nd edn. Elsevier, Amsterdam, 918–929.

Ballantyne, C.K. (2013c) A 35-year record of solifluction in a maritime periglacial environment. *Permafrost and Periglacial Processes* 24, 56–66.

Ballantyne, C.K. and Benn, D.I. (1994a) Paraglacial slope adjustment and resedimentation following recent glacier retreat, Fåbergstølsdalen, Norway. *Arctic and Alpine Research* 26, 255–269.

Ballantyne, C.K. and Benn, D.I. (1994b) Glaciological constraints on protalus rampart development. *Permafrost and Periglacial Processes* 5, 145–153.

Ballantyne, C.K. and Harris, C. (1994) *The Periglaciation of Great Britain*. Cambridge University Press, Cambridge UK, 330 pp.

Ballantyne, C.K. and Kirkbride, M.P. (1986) The characteristics and significance of some Lateglacial protalus ramparts in upland Britain. *Earth Surface Processes and Landforms* 11, 659–671.

Ballantyne, C.K. and Kirkbride, M.P. (1987) Rockfall activity in upland Britain during the Loch Lomond Stadial. *Geographical Journal* 153, 86–92.

Ballantyne, C.K. and Matthews, J.A. (1982) The development of sorted circles on recently deglaciated terrain, Jotunheimen, Norway. *Arctic and Alpine Research* 14, 341–354.

Ballantyne, C.K. and Matthews, J.A. (1983) Desiccation cracking and sorted polygon development, Jotunheimen, Norway. *Arctic and Alpine Research* 15, 339–349.

Ballantyne, C.K. and Morrocco, S.M. (2006) The windblown sands of An Teallach. *Scottish Geographical Journal* 122, 149–159.

Ballantyne, C.K. and Stone, J.O. (2009) Rock-slope failure at Baosbheinn, Wester Ross, NW Scotland: age and interpretation. *Scottish Journal of Geology* 45, 177–181.

Ballantyne, C.K. and Stone, J.O. (2015) Trimlines, blockfields and the vertical extent of the last ice sheet in southern Ireland. *Boreas* 44, 277–287.

Ballantyne, C.K. and Whittington, G.W. (1987) Niveo-aeolian sand deposits on An Teallach, Wester Ross, Scotland. *Transactions of the Royal Society of Edinburgh: Earth Sciences* 78, 51–63.

Ballantyne, C.K., Black, N.M. and Finlay, D.P. (1989) Enhanced boulder weathering under late-lying snowpatches. *Earth Surface Processes and Landforms* 14, 745–750.

Ballantyne, C.K., Schnabel, C. and Xu, S. (2009) Exposure dating and reinterpretation of coarse debris accumulations ('rock glaciers') in the Cairngorm Mountains, Scotland. *Journal of Quaternary Science* 24, 19–31.

Ballantyne, C.K., Sandeman, G.F., Stone, J.O. and Wilson, P. (2014) Rock-slope failure following Late Pleistocene deglaciation on tectonically stable mountainous terrain. *Quaternary Science Reviews* 86, 144–157.

Balme, M.R., Gallagher, C.J. and Hauber, E. (2013) Morphological evidence for geologically young thaw of ice on Mars: a review of recent studies using high-resolution imaging data. *Progress in Physical Geography* 37, 289–324.

Balser, A.W., Jones, J.B. and Gens, R. (2014) Timing of retrogressive thaw slump initiation in the Noatak Basin, northwest Alaska. *Journal of Geophysical Research: Earth Surface* 119, 1106–1120.

Baranov, I.Y. (1964) Geographical distribution of seasonally frozen ground and permafrost. *National Research Council of Canada, Technical Translation* 1121, 30 pp.

Barnes, P.W. (1982) Marine ice-pushed boulder ridge, Beaufort Sea, Alaska. *Arctic* 35, 312–316.

Barnhart, K.R., Anderson, R.S., Overeem, I. *et al.* (2014a) Modeling erosion of ice-rich permafrost bluffs along the Alaskan Beaufort Sea coast. *Journal of Geophysical Research: Earth Surface* 119, 1155–1179.

Barnhart, K.R., Overeem, I. and Anderson, R.S. (2014b) The effect of changing sea ice on the physical vulnerability of Arctic coasts. *The Cryosphere* 8, 1777–1799.

Barrows, T.T., Stone, J.O. and Fifield, L.K. (2004) Exposure ages for Pleistocene periglacial deposits in Australia. *Quaternary Science Reviews* 23, 697–708.

Barry, R.G. (2008) *Mountain Weather and Climate*, 3rd edn. Cambridge University Press, Cambridge.

Barsch, D. (1996) *Rockglaciers. Indicators for the Present and Former Geoecology in High Mountain Environments.* Springer-Verlag, Berlin.

Barsch, D. and Jakob, M. (1998) Mass transport by active rockglaciers in the Khumbu Himalaya. *Geomorphology* 26, 215–222.

Barsch, D. and Zick, W. (1991) Die Bewegungen des Blockgletchers Macun 1 von 1965–1988 (Unterengadin, Graübunden, Schweiz). *Zeitschrift für Geomorphologie,* 35, pp 9–14.

Barsch, D., Gude, M., Mäusbacher, R. *et al.* (1993) Slush stream phenomena – process and geomorphic impact. *Zeitschrift für Geomorphologie, Supplementband* 92, 39–53.

Bateman, M.D. (1995) Thermoluminescence dating of the British coversand deposits. *Quaternary Science Reviews* 14, 791–798.

Bateman, M.D. (1998) The origin and age of coversand in North Lincolnshire, UK. *Permafrost and Periglacial Processes* 9, 313–325.

Bateman, M.D. (2008) Luminescence dating of periglacial sediments and structures. *Boreas* 37, 574–588.

Bateman, M.D. and Murton, J.B. (2006) The chronostratigraphy of Late Pleistocene glacial and periglacial aeolian activity in the Tuktoyaktuk Coastlands, NWT, Canada. *Quaternary Science Reviews* 25, 2552–2568.

Bateman, M.D. and Van Huissteden, J. (1999) The timing of last-glacial periglacial and aeolian events, Twente, eastern Netherlands. *Journal of Quaternary Science* 14, 277–283.

Bateman, M.D., Murton, J.B. and Crowe, W. (2000) Late Devensian and Holocene depositional environments associated with the coversand around Caistor, north Lincolnshire, UK. *Boreas* 29, 1–15.

Bateman, M.D., Hitchens, S., Murton, J.B. *et al.* (2014) The evolution of periglacial patterned ground of central East Anglia, UK. *Journal of Quaternary Science* 29, 301–317.

Bates, M.R., Keen, D.H. and Lautridou, J.-P. (2003) Pleistocene marine and periglacial deposits of the English Channel. *Journal of Quaternary Science* 18, 319–337.

Beck, I., Ludwig, R., Bernier, M. *et al.* (2015) Assessing permafrost degradation and land cover changes (1986–2009) using remote sensing data over Umiujaq, sub-Arctic Québec. *Permafrost and Periglacial Processes* 26, 129–141.

Beerten, K., Vandersmissen, N., Deforce, K. and Vandenberghe, N. (2014) Late Quaternary (15 ka to present) development of a sandy landscape in the Mol area, Campine region, north-east Belgium. *Journal of Quaternary Science* 29, 433–444.

Beget, J.E., Layer, P, Stone, D. *et al.* (2008) Evidence of permafrost formation two million years ago in Alaska. In Kane, D.L. and Hinkel, K.M. (eds) *Ninth International Conference on Permafrost.* University of Alaska, Fairbanks, 95–100.

Bégin, C., Michaud, Y. and Filion, L. (1995) Dynamics of a Holocene cliff-top dune along Mountain River, Northwest Territories, Canada. *Quaternary Research* 44, 392–404.

Beilman, D.W., Vitt, D.H. and Halsey, L.A. (2001) Localized permafrost peatlands in western Canada: definitions, distribution and degradation. *Arctic, Antarctic and Alpine Research* 33, 70–77.

Bélanger, S and Filion, L. (1991) Niveo-aeolian sand deposition in subarctic dunes, eastern coast of Hudson Bay, Québec, Canada. *Journal of Quaternary Science* 6, 27–37.

Bell, I., Gardner, J. and de Scally, F. (1990) An estimate of snow avalanche debris transport, Kaghan Valley, Himalaya, Pakistan. *Arctic and Alpine Research* 22, 317–321.

Benedict, J.B. (1970) Downslope soil movement in a Colorado alpine region: rates, processes and climatic significance. *Arctic and Alpine Research* 2, 165–226.

Benedict, J.B. (1976) Frost creep and gelifluction features: a review. *Quaternary Research* 6, 55–76.

Benedict, J.B. (1993) Influence of snow upon rates of granodiorite weathering, Colorado Front Range, USA. *Boreas* 22, 87–92.

Beniston, M. (2007) Entering into the 'greenhouse century': recent record temperatures in Switzerland are comparable to the upper temperature quantiles in a greenhouse climate. *Geophysical Research Letters* 34, L16710.

Benn, D.I. and Ballantyne, C.K. (2005) Palaeoclimatic inferences from reconstructed Loch Lomond Readvance glaciers, West Drumochter Hills, Scotland. *Journal of Quaternary Science* 20, 577–592.

Benn, D.I. and Evans, D.J.A. (2010) *Glaciers and Glaciation,* 2nd edn. Hodder, London.

Bennett, L.P. and French, H.M. (1990) In situ permafrost creep, Melville Island, and implications for global change. In *Permafrost Canada: Proceedings of the 5th Canadian Permafrost Conference.* Université Laval, Québec, 119–123.

Bennett, L.P. and French, H.M. (1991) Solifluction and the role of permafrost creep, eastern Melville Island, NWT, Canada. *Permafrost and Periglacial Processes* 2, 95–102.

Bennett, M.R. (2001) The morphology, structural evolution and significance of push moraines. *Earth-Science Reviews* 53, 197–236.

Bennett, M.R., Huddart, D., Hambrey, M.J. and Ghienne, J.F. (1998) Modification of braided outwash surfaces by aufeis: an example from Pedersenbreen, Svalbard. *Zeitschrift für Geomorphologie* 42, 1–20.

Berger, J., Krainer, K. and Mostler, W. (2004) Dynamics of an active rock glacier (Ötztal Alps, Austria). *Quaternary Research* 62, 233–242.

Berrisford, M.S. (1991) Evidence for enhanced mechanical weathering associated with seasonally late-lying and perennial snow patches, Jotunheimen, Norway. *Permafrost and Periglacial Processes* 2, 331–340.

Berthling, I. (2011) Beyond confusion: rock glaciers as cryo-conditioned landforms. *Geomorphology* 131, 98–105.

Berthling, I. and Etzelmüller, B. (2011) The concept of cryo-conditioning in landscape evolution. *Quaternary Research* 75, 378–384.

Berthling, I., Etzelmüller, B, Eiken, T. and Sollid, J.L. (1998) The rock glaciers on Prins Karl Forland, Svalbard. I: Internal structure, flow velocity and morphology. *Permafrost and Periglacial Processes* 9, 135–145.

Berthling, I., Etzelmüller, B., Isaksen, K. and Sollid, J.L. (2000) Rock glaciers on Prins Karl Forland, Svalbard. II: GPR soundings and the development of internal structures. *Permafrost and Periglacial Processes* 11, 357–369.

Berthling, I., Eiken, T., Madsen, H. and Sollid, J.L. (2001a) Downslope displacement rates of ploughing boulders in a mid-alpine environment, Finse, southern Norway. *Geografiska Annaler* 83, 103–116.

Berthling, I., Eiken, T. and Sollid, J.L. (2001b) Frost heave and thaw consolidation of ploughing boulders in a mid-alpine environment, Finse, southern Norway. *Permafrost and Periglacial Processes* 12, 165–177.

Bertran, P. and Fabre, R. (2005) Pleistocene cryostructures and landslide at Petit-Bost (southwestern France, 45°N). *Geomorphology* 71, 344–356.

Bertran, P. and Jomelli, V. (2000) Discussion: post-glacial colluvium in western Norway: depositional processes, facies and palaeoclimatic record. *Sedimentology* 47, 1053–1068.

Bertran, P., Coutard, J.-P., Francou, B. *et al.* (1992) Données nouvelles sur l'origine du litage des grèzes: implications paléoclimatiques. *Géographie Physique et Quaternaire* 46, 97–112.

Bertran, P., Francou, B. and Texier, J.P. (1995) Stratified slope deposits: the stone-banked sheets and lobes model. In Slaymaker, O. (ed.) *Steepland Geomorphology*. John Wiley & Sons, Chichester, 147–169.

Bertran, P., Andrieux, E., Antoine, P. *et al.* (2014) Distribution and chronology of Pleistocene permafrost features in France: database and first results. *Boreas* 43, 699–711.

Beskow, G. (1935) Tjälbildningen och tjällyftningen. *Sveriges Geologiske Undersökning, Series C*, 375, *Årbok* 26, 242 pp. English translation by Osterberg, J.O. (1947) Soil freezing and frost heaving with special application to roads and railroads. Reprinted (1991) in *Cold Regions Research and Engineering Laboratory (CRREL) Report*, 91/23, 37–157.

Bettis, E.A. III, Muhs, D.R., Roberts, H.M. and Wintle, A.G. (2003) Last glacial loess in the coterminous USA. *Quaternary Science Reviews* 22, 1907–1946.

Beylich, A.A. (2011) Mass transfers, sediment budgets and relief development in cold environments: results of long-term geomorphologic drainage basin studies in Iceland, Swedish Lapland and Finnish Lapland. *Zeitschrift für Geomorphologie* 55, 145–174.

Beylich, A.A. and Gintz, D. (2004) Effects of high magnitude/low frequency fluvial events generated by intense snowmelt or heavy rainfall in arctic periglacial environments in northern Swedish Lapland and northern Siberia. *Geografiska Annaler* 86A, 11–29.

Beylich, A.A. and Laute, K. (2012) Spatial variations of surface water chemistry and chemical denudation in the Erdalen drainage basin, Nordfjord, western Norway. *Geomorphology* 167–168, 77–90.

Beylich, A.A., Kolstrup, E., Thysted, T. and Gintz, D. (2004) Water chemistry and its diversity in relation to local factors in the Latnjavagge drainage basin, arctic-oceanic Swedish Lapland. *Geomorphology* 58, 125–143.

Beylich, A.A., Molau, U., Luthbom, K. and Gintz, D. (2005) Rates of chemical and mechanical fluvial denudation in an arctic oceanic periglacial environment, Latnjavagge drainage basin, northernmost Swedish Lapland. *Arctic, Antarctic and Alpine Research* 37, 75–8.

Billings, W.D. and Peterson, K.M. (1980) Vegetational change and ice-wedge polygons through the thaw lake cycle in Arctic Alaska. *Arctic and Alpine Research* 12, 413–432.

Bird, J.B. (1967) *The Physiography of Arctic Canada*. John Hopkins Press, Baltimore, 336 pp.

Biskaborn, B.K., Herzschuh, U., Bolshiyanov, D.Y. *et al.* (2013) Thermokarst processes and depositional events in a tundra lake, northeastern Siberia. *Permafrost and Periglacial Processes* 24, 160–174.

Bithell, M., Richards, K.S. and Bithell, E.G. (2014) Simulation of scree-slope dynamics: investigating the distribution of debris avalanche events in an idealized two-dimensional model. *Earth Surface Processes and Landforms* 39, 1601–1610.

Bjornson, J. and Lauriol, B. (2001) Météorisation des blocs de granite à la surface des pédiments dans le nord du Yukon, Canada. *Permafrost and Periglacial Processes* 12, 289–298.

Blais-Stevens, A., Kremer, M., Bonnaventure, P.P. *et al.* (2014) Active layer detachment slides and retrogressive thaw slumps susceptibility mapping for current and future permafrost distribution, Yukon Alaska highway corridor. In *Engineering Geology for Society and Territory*, vol. 1. Springer International Publishing, Switzerland, 449–453.

Blikra, L.H. and Christiansen, H.H. (2014) A field-based model of permafrost-controlled rockslide deformation in northern Norway. *Geomorphology* 208, 34–49.

Blikra, L.H. and Nemec, W. (1998) Postglacial colluvium in western Norway: depositional processes, facies and palaeoenvironmental record. *Sedimentology* 45, 909–959.

Blikra, L.H. and Selvik, S.F. (1998) Climatic signals recorded in snow avalanche-dominated colluvium in western Norway: deposition facies successions and pollen records. *The Holocene* 8, 631–658.

Blockley, S.P.E., Lane, C.S., Hardiman, M. *et al.* (2012) Synchronisation of palaeoenvironmental records over the last 60 000 years, and an extended INTIMATE event stratigraphy to 48 000 b2k. *Quaternary Science Reviews* 36, 2–10.

Bockheim, J.G. (1995) Permafrost distribution in the southern circumpolar region and its relation to environment: a review and recommendations for further research. *Permafrost and Periglacial Processes* 6, 27–45.

Bockheim, J.G. (2007) Importance of cryoturbation in redistributing organic carbon in permafrost related soils. *Soil Science Society of America Journal* 71, 1335–1342.

Bockheim, J.G. (2015) Global distribution of cryosols with mountain permafrost: an overview. *Permafrost and Periglacial Processes* 26, 1–12.

Bockheim, J.G. and Hartemink, A.E. (2013) Soil with fragipans in the USA. *Catena* 104, 233–242.

Bockheim, J.G. and Hinkel, K.M. (2012) Accumulation of excess ground ice in an age sequence of drained thermokarst lake basins, Alaska. *Permafrost and Periglacial Processes* 23, 231–236.

Bockheim, J.G., Campbell, I.A. and McLeod, M. (2007) Permafrost distribution and active-layer depths in the McMurdo Dry Valleys, Antarctica. *Permafrost and Periglacial Processes* 18, 217–227.

Bockheim, J.G., Kurz, M.D., Soule, S.A. and Burke, A. (2009) Genesis of active sand-filled polygons in lower and central Beacon Valley, Antarctica. *Permafrost and Periglacial Processes* 20, 295–308.

Boelhouwers, J. (1999) Relict periglacial slope deposits in the Hex River Mountains, South Africa: observations and palaeoenvironmental implications. *Geomorphology* 30, 245–258.

Boelhouwers, J. and Jonsson, M. (2013) Critical assessment of the 2 °C min^{-1} threshold for thermal stress weathering. *Geografiska Annaler* 95A, 285–293.

Boelhouwers, J., Holness, S., Meiklejohn, I. and Sumner, P. (2002) Observations on a blockstream in the vicinity of Sani Pass, Lesotho Highlands, Southern Africa. *Permafrost and Periglacial Processes* 13, 251–257.

Boelhouwers, J., Holness, S. and Sumner, P. (2003) The maritime subantarctic: a distinct periglacial environment. *Geomorphology* 52, 39–55.

Boike, J., Ippisch, O., Overduin, P.P. *et al.* (2008) Water, heat and soil dynamics of a mud boil, Spitsbergen. *Geomorphology* 95, 61–73.

Bollmann, E., Girstmair, A., Mitterer, S. *et al.* (2015) A rock glacier activity index based on rock glacier thickness changes and displacement rates inferred from airborne laser scanning. *Permafrost and Periglacial Processes* 26, 347–359.

Bollschweiler, M. and Stoffel, M. (2010) Changes and trends in debris-flow frequency since AD 1850: results from the Swiss Alps. *The Holocene* 20, 907–916.

Bonnaventure, P.P. and Lewkowicz, A.G. (2008) Mountain permafrost probability mapping using the BTS method in two climatically dissimilar locations, northwest Canada. *Canadian Journal of Earth Sciences* 45, 443–455.

Bonnaventure, P.P., Lewkowicz, A.G., Kremer, M. and Sawada, M.C. (2012) A Permafrost probability model for the southern Yukon and northern British Columbia, Canada. *Permafrost and Periglacial Processes* 23, 52–68.

Böse, M. (1992) Late Pleistocene sand wedge formation in the hinterland of the Brandenburg Stade. *Sveriges Geologiska Undersökning, Series C* 81, 59–63.

Boucher, É., Bégin, Y. and Arseneault, D. (2009) Impacts of recurring ice jams on channel geometry and geomorphology in a small high-boreal watershed. *Geomorphology* 108, 273–281.

Boulton, G.S. (1999) The sedimentary and structural evolution of a recent push moraine complex: Holmstrømbreen, Spitsbergen. *Quaternary Science Reviews* 18, 339–371.

Boulton, G.S. and Dent, D.L. (1974) The nature and rate of post-depositional changes in recently-deposited till from south-east Iceland. *Geografiska Annaler* 56A, 121–134.

Bourke, M.C., Ewing, R.C., Finnegan, D. and McGowan, H.A. (2009) Sand dune movement in the Victoria Valley, Antarctica. *Geomorphology* 109, 148–160.

Bovis, M.J. and Dagg, B.R. (1992) Debris flow triggering by impulsive loading: mechanical modelling and case studies. *Canadian Geotechnical Journal* 29, 345–352.

Bovis, M.J. and Thorn, C.E. (1981) Soil loss variations within a Colorado alpine area. *Earth Surface Processes and Landforms* 6, 151–163.

Bowden, W.B., Gooseff, M.N., Balser, A. *et al.* (2008) Sediment and nutrient delivery from thermokarst features in the foothills of the North Slope, Alaska: potential impacts on headwater stream systems. *Journal of Geophysical Research* 113, G02026.

Bray, M., French, H.M. and Shur Y. (2006) Further cryostratigraphic observations in the CRREL permafrost tunnel, Fox, Alaska. *Permafrost and Periglacial Processes* 17, 233–244.

Brazier, V., Kirkbride, M.P. and Owens, I.F. (1998) The relationship between climate and rock glacier distribution in the Ben Ohau Range, New Zealand. *Geografiska Annaler* 80A, 193–207.

Brenning, A., Gruber, S. and Hoezle, M. (2005) Sampling and statistical analysis of BTS measurements. *Permafrost and Periglacial Processes* 16, 383–393.

Briant, R.M. and Bateman, M.D. (2009) Luminescence dating indicates radiocarbon age underestimation in late Pleistocene fluvial deposits from eastern England. *Journal of Quaternary Science* 24, 916–927.

Briant, R.M., Coope, G.R., Preece, R.C. *et al.* (2004) Fluvial system response to Late Devensian (Weichselian) aridity, Baston, Lincolnshire, England. *Journal of Quaternary Science* 19, 479–495.

Bridgland, D.R. and Westaway, R. (2008) Climatically controlled river terrace staircases: a worldwide Quaternary phenomenon. *Geomorphology* 98, 285–315.

Briggs, D.J., Coope, G.R. and Gilbertson, D.D. (1975) Late Pleistocene terrace deposits at Beckford, Worcestershire, England. *Geological Journal* 10, 1–16.

Briner, J.P., Miller, G.H., Davis, P.T. *et al.* (2003) Last glacial maximum ice-sheet dynamics in Arctic Canada inferred from young erratics perched on ancient tors. *Quaternary Science Reviews* 22, 437–444.

Bristow, C.S., Jol, H.M., Augustinus, P. and Wallis, I. (2010) Slipfaceless 'whaleback' dunes in a polar desert, Victoria Valley, Antarctica: insights from ground-penetrating radar. *Geomorphology* 114, 361–372.

Broeker, W.S. (2006) Was the Younger Dryas triggered by a flood? *Science* 312, 1146–1148.

Broll, G., Tarnocai, C. and Mueller, G. (1999) Interactions between vegetation, nutrients and moisture in soils in the Pangirtung Pass area, Baffin Island, Canada. *Permafrost and Periglacial Processes* 10, 265–277.

Brook, G.A. and Ford, D.C. (1978) The nature of labyrinth karst and its implications for climate-specific models of tower karst. *Nature* 280, 383–385.

Brooks, S.J. and Langdon, P.G. (2014) Summer temperature gradients in northwest Europe during the Lateglacial to early Holocene transition (15–8 ka BP) inferred from chironomid assemblages. *Quaternary International* 341, 80–90.

Brooks, S.J., Matthews, I.P., Birks, H.H. *et al.* (2012) High resolution Lateglacial and early-Holocene summer air temperature records from Scotland inferred from chironomid assemblages. *Quaternary Science Reviews* 41, 67–82.

Brookfield, M.E. (2011) Aeolian processes and features in cool climates. *Geological Society, London, Special Publication* 354, 241–258.

Brown, A.G., Keough, M.K. and Rice, J. (1994) Floodplain evolution in the East Midlands, United Kingdom: the lateglacial and Flandrian alluvial record from the Soar and Nene Valleys. *Philosophical Transactions of the Royal Society* 348A, 261–293.

Brown, A.G., Basell, L.S. and Toms, P.S. (2015) A stacked Late Quaternary fluvio-periglacial sequence from the Axe valley, southern England, with implications for landscape evolution and Palaeolithic archaeology. *Quaternary Science Reviews* 116, 106–121.

Brown, G. (1980) Palsas and other permafrost features of the Lower Rock Creek Valley, west-central Alberta. *Arctic and Alpine Research* 12, 31–40.

Brown, J., Ferrians, O., Heginbottom, J.A. and Melnikov, E. (2014) *Circum-Arctic Map of Permafrost and Ground-Ice Conditions.* National Snow and Ice Data Center, Boulder, Colorado, USA.

Brown, R.D., Derksen, C. and Wang, L. (2010) A multi-dataset analysis of variability and change in Arctic spring snow cover extent, 1967–2008. *Journal of Geophysical Research* 115, D16111.

Brown, R.J.E. (1970) *Permafrost in Canada.* University of Toronto Press, Toronto, 234 pp.

Bryant, R.H. and Carpenter, C.P. (1987) Ramparted ground ice depressions in Britain and Ireland. In Boardman, J. (ed.) *Periglacial Processes and Landforms in Britain and Ireland.* Cambridge University Press, Cambridge, 183–190.

Büdel, J. (1953) Die 'periglaziale' morphologischen Wirkungen des Eiszeitklimas auf der Ganzen Erde. *Erdekunde* 249–266.

Büdel, J. (1982) *Climatic Geomorphology.* Princeton University Press, Princeton, New Jersey.

Burger, K.C., Degenhardt, J.J. Jr and Giardino, J.R. (1999) Engineering geomorphology of rock glaciers. *Geomorphology* 31, 93–132.

Burke, E.J., Jones, C.D. and Koven, C.D. (2013) Estimating the permafrost-carbon climate response in the CMIP5 climate models using a simplified global approach. *Journal of Climatology* 26, 4897–4909.

Burn, C.R. (1988) The development of near-surface ground ice during the Holocene at sites near Mayo, Yukon Territory, Canada. *Journal of Quaternary Science* 3, 31–38.

Burn, C.R. (1990) Implications for palaeoenvironmental reconstruction of recent ice-wedge development at Mayo, Yukon Territory. *Permafrost and Periglacial Processes* 1, 3–14.

Burn, C.R. (1997) Cryostratigraphy, palaeogeography and climate change during the early Holocene warm interval, western Arctic coast, Canada. *Canadian Journal of Earth Sciences* 34, 912–935.

Burn, C.R. (1998) The active layer: two contrasting definitions. *Permafrost and Periglacial Processes* 9, 411–416.

Burn, C.R. (2000) The thermal regime of a retrogressive thaw slump near Mayo, Yukon Territory. *Canadian Journal of Earth Sciences* 37, 967–981.

Burn, C.R. (2002) Tundra lakes and permafrost, Richards Island, western Arctic coast, Canada. *Canadian Journal of Earth Sciences* 39, 1281–1298.

Burn, C.R. (2004) A field perspective on modelling 'single ridge' ice-wedge polygons. *Permafrost and Periglacial Processes* 15, 59–65.

Burn, C.R. (2013) Thermokarst topography. In Shroder, J., Giardino, J. and Harbor, J. (eds) *Treatise on Geomorphology. Volume 8: Glacial and Periglacial Geomorphology.* Academic Press, San Diego, CA, 574–581.

Burn, C.R. and Friele, P.A. (1989) Geomorphology, vegetation succession, soil characteristics and permafrost in retrogressive thaw slumps near Mayo, Yukon Territory. *Arctic* 42, 31–40.

Burn, C.R. and Michel, F.A. (1988) Evidence for recent temperature-induced water migration into permafrost from the tritium content of ground ice near Mayo, Yukon Territory, Canada. *Canadian Journal of Earth Sciences* 25, 909–915.

Burn, C.R. and Smith, M.W. (1990) Development of thermokarst lakes during the Holocene at sites near Mayo, Yukon Territory. *Permafrost and Periglacial Processes* 1, 161–176.

Burn, C.R. and Zhang, Y. (2009) Permafrost and climate change at Herschel Island (Qikiqtaruk), Yukon Territory, Canada. *Journal of Geophysical Research* 114, F02001.

Burn, C.R., Michel, F.A. and Smith, M.W. (1986) Stratigraphic, isotopic and mineralogical evidence for an early Holocene thaw unconformity at Mayo, Yukon Territory. *Canadian Journal of Earth Sciences* 23, 794–803.

Burton, R.G.O. (1987) The role of thermokarst in landscape development in eastern England. In Boardman, J. (ed.) *Periglacial Processes and Landforms in Britain and Ireland*. Cambridge University Press, Cambridge, 203–208.

Busby, J.P., Lee, J.R. Kender, S. *et al.* (2016) Regional modelling of permafrost thickness over the past 130 ka: implications for permafrost development in Great Britain. *Boreas* 45, 46–60.

Butler, D.R. (1988) Neoglacial climatic inferences from rock glaciers and protalus ramparts, southern Lemhi mountains, Idaho. *Physical Geography* 9, 71–88.

Butler, D.R. and Walsh (1990) Lithologic, structural and topographic influences on snow avalanche path location, eastern Glacier National Park, Montana. *Annals of the Association of American Geographers* 80, 362–378.

Butler, D.R., Malanson, G.P. and Walsh, S.J. (1992) Snow avalanche paths: conduits from the periglacial-alpine to the subalpine depositional zone. In: Dixon, J.C. and Abrahams, A.D. (eds) *Periglacial Geomorphology: Proceedings of the 22nd Annual Binghamton Symposium in Geomorphology*. John Wiley & Sons, Chichester, 185–202.

Buttle, J.M. and Fraser, K.E. (1992) Hydrochemical fluxes in a high arctic wetland basin during spring snowmelt. *Arctic and Alpine Research* 24, 153–164.

Buylaert, J.P., Ghysels, G., Murray, A.S. *et al.* (2009) Optical dating of relict sand wedges and composite-wedge pseudomorphs in Flanders, Belgium. *Boreas* 38, 160–175.

Cailleux, A. and Taylor, G. (1954) *Cryopédologie: étude des sols gelées*. Hermann et Cie, Paris, 218 pp.

Caine, N. (1963) Movement of low-angle scree slopes in the Lake District, northern England. *Revue de Géomorphologie Dynamique* 14, 171–177.

Caine, N. (1974) The geomorphic processes of the alpine environment. In Ives, J.D. and Barry, R.G. (eds) *Arctic and Alpine Environments*. Methuen, London, 721–748.

Caine, N. (1979) Rock weathering rates at the soil surface in an alpine environment. *Catena* 6, 131–144.

Caine, N. (1986) Sediment movement and storage on alpine slopes in the Colorado Rocky Mountains. In Abrahams, A.D. (ed.) *Hillslope Processes*. Allen and Unwin, Boston, 115–137.

Caine, N. (1992a) Sediment transfer on the floor of the Martinelli snowpatch, Colorado Front Range, USA. *Geografiska Annaler* 74A, 133–144.

Caine, N. (1992b) Spatial patterns in geochemical denudation in a Colorado Alpine environment. In Dixon, J.C. and Abrahams, A.D. (eds) *Periglacial Geomorphology: Proceedings of the 22nd Annual Binghamton Symposium in Geomorphology*. John Wiley & Sons, Chichester, 63–88.

Caine, N. and Jennings, J.N. (1968) Some blockstreams of the Toolong Range, Kosciusko State Park, New South Wales. *Journal and Proceedings, Royal Society of New South Wales* 101, 93–103.

Cairns, D.D. (1912) Differential erosion and equiplanation in portions of Yukon and Alaska. *Bulletin, Geological Society of America* 23, 333–348.

Calkin, P.E., Haworth, L.A. and Ellis, J.M. (1987) Rock glaciers of central Brooks Range, Alaska. In Giardino, J.R., Schroder, J.F. and Vitek, J.D. (eds) *Rock Glaciers*. Allen and Unwin, London, 65–82.

Callaghan, T.V., Johansson, M., Anisimov, O. *et al.* (2011) Changing permafrost and its impacts. In *Snow, Water, Ice and Permafrost in the Arctic (SWIPA): Climate Change and the Cryosphere*. Arctic Monitoring and Assessment Programme (AMAP), Oslo, 5.1–5.62.

Calmels, F. and Allard, M. (2008) Structural interpretation of the palsa/lithalsa growth mechanism through the use of CT scanning. *Earth Surface Processes and Landforms* 33, 209–225.

Calmels, F. and Coutard, J.-P. (2000) Expérience du laboratoire sur la mobilisation par cryoreptation d'un dépôt de pente schisteux, grossier et hétérométrique. *Permafrost and Periglacial Processes* 11, 207–218.

Calmels, F., Allard, M. and Delisle, G. (2008) Development and decay of a lithalsa in Northern Québec: a geomorphological history. *Geomorphology* 97, 287–299.

Calmels, F., Froese, D.G. and Clavano, W.R. (2012) Cryostratigraphic record of permafrost degradation and recovery following historic (1898–1992) surface disturbances in the Klondike region, central Yukon Territory. *Canadian Journal of Earth Sciences* 49, 938–952.

Campbell, I.B. and Claridge, G.G.C. (1987) *Antarctica: Soils, Weathering Processes and Environment*. Elsevier, Amsterdam.

Campbell, I.B. and Claridge, G.G.C. (1992) Soils of cold climate regions. In Martin, I.P. and Chesworth, W. (eds) *Weathering, Soils and Palaeosols*. Elsevier, Amsterdam, 183–201.

Campbell, I.B. and Claridge, G.G.C. (2006) Permafrost properties, patterns and processes in the Transantarctic Mountains region. *Permafrost and Periglacial Processes* 17, 215–232.

Campbell, S.W., Dixon, J.C., Darmody, R.G. and Thorn, C.E. (2001) Spatial variations of early season water surface water chemistry in Kärkevagge, Swedish Lapland. *Geografiska Annaler* 83A, 169–178.

Campbell, S.W., Dixon, J.C., Thorn, C.E. and Darmody, R.G. (2002) Chemical denudation rates in Kärkevagge, Swedish Lapland. *Geografiska Annaler* 84A, 179–185.

Candy, I., Silva, B. and Lee, J. (2011) Climates of the early Middle Pleistocene in Britain: environments of the earliest humans in Northern Europe. In Ashton, N., Lewis, S.G. and Stringer, C. (eds) *The Ancient Human Occupation of Britain*. Elsevier, Amsterdam, 11–22.

Capps, S.R. (1910) Rock glaciers in Alaska. *Journal of Geology* 18, 359–375.

Carey, S.K. and Woo, M.-K. (2000) The role of soil pipes as a slope runoff mechanism, subarctic Yukon, Canada. *Journal of Hydrology* 233, 206–222.

Carlson, A.E. (2010) What caused the Younger Dryas cold event? *Geology* 38, 383–384.

Carroll, M.L., Townshend, J.G.R., DiMiceli, C.M. *et al.* (2011) Shrinking lakes of the Arctic: spatial relationships and trajectory of change. *Geophysical Research Letters* 38, L20406.

Catto, N.R. (1993) Morphology and development of an alluvial fan in a permafrost region, Aklavik Range, Canada. *Geografiska Annaler* 75A, 83–93.

Certini, G., Ugolini, F.C., Taina, I. *et al.* (2007) Cues of the genesis of a discontinuously distributed fragipan in the northern Apennines, Italy. *Catena* 69, 161–169.

Chandler, R.J. (1976) The history and stability of two Lias clay slopes in the upper Gwash Valley, Rutland. *Philosophical Transactions of the Royal Society of London* A283, 463–491.

Chapin, F.S., Sturm, M., Serreze, M.C. *et al.* (2005) Role of land-surface changes in Arctic warming. *Science* 310, 657–660.

Chasmer, L., Quinton, W., Hopkinson, C. *et al.* (2011) Vegetation canopy and radiation controls on permafrost plateau evolution within the discontinuous permafrost zone, Northwest Territories, Canada. *Permafrost and Periglacial Processes* 22, 199–213.

Chen, F.H., Bloemendal, J., Wang, J.M. *et al.* (1997) High-resolution multi-proxy climate records from Chinese loess: evidence for rapid climatic changes over the last 75 kyr. *Palaeogeography, Palaeoclimatology, Palaeoecology* 130, 323–335.

Chen, H., Nan, Z., Zhao, L. *et al.* (2015) Noah modelling of the permafrost distribution and characteristics in the West Kunlun area, Qinghai-Tibet Plateau, China. *Permafrost and Periglacial Processes* 26, 160–174.

Chen, M., Rowland, J.C., Wilson, C.J. *et al.* (2013) The importance of natural variability in lake areas on the detection of permafrost degradation: a case study in the Yukon Flats, Alaska. *Permafrost and Periglacial Processes* 24, 224–240.

Cheng, G. (1983) The mechanism of repeated-segregation for the formation of thick-layered ground ice. *Cold Regions Science and Technology* 8, 57–66.

Cheng, G. and Wu, T. (2007) Responses of permafrost to climate change and their environmental significance, Qinghai-Tibet Plateau. *Journal of Geophysical Research* 112, F02S03.

Christiansen, H.H. (1995) Observations on open-system pingos in a marsh environment, Mellomfjord, Disko, Central West Greenland. *Danish Journal of Geography* 95, 42–48.

Christiansen, H.H. (1996) Effect of nivation on periglacial landscape evolution in western Jutland, Denmark. *Permafrost and Periglacial Processes* 7, 111–138.

Christiansen, H.H. (1998a) Nivation forms and processes in unconsolidated sediments, NE Greenland. *Earth Surface Processes and Landforms* 23, 751–760.

Christiansen, H.H. (1998b) 'Little Ice Age' nivation activity in northeast Greenland. *The Holocene* 8, 719–728.

Christiansen, H.H. (2004) Windpolished boulders and bedrock in the Scottish Highlands: evidence and implications of Late Devensian wind activity. *Boreas* 33, 82–94.

Christiansen, H.H. (2005) Thermal regime of ice-wedge cracking in Adventdalen, Svalbard. *Permafrost and Periglacial Processes* 16, 87–98.

Christiansen, H.H. and Svensson, H. (1998) Windpolished boulders as indicators of a Late Weichselian wind regime in Denmark in relation to neighbouring areas. *Permafrost and Periglacial Processes* 9, 1–21.

Christiansen, H.H., Murray, A.S., Mejdahl, V. and Humlum, O. (1999) Luminescence dating of Holocene geomorphic activity on Ammassalik Island, S.E. Greenland. *Quaternary Science Reviews* 18, 191–205.

Christiansen, H.H., French, H.M. and Humlum, O. (2005) Permafrost in the Gruve-7 mine, Adventdalen, Svalbard. *Norsk Geografisk Tiddskrift* 59, 109–115.

Christiansen, H.H., Etzelmüller, B., Isaksen, K. *et al.* (2010) The thermal state of permafrost in the Nordic area during the International Polar Year 2007–2009. *Permafrost and Periglacial Processes* 21, 156–181.

Chueca, L. (1992) A statistical analysis of the spatial distribution of rock glaciers, Spanish central Pyrenees. *Permafrost and Periglacial Processes* 3, 261–265.

Church, M. (1972) Baffin Island sandurs: a study of arctic fluvial processes. *Geological Survey of Canada Bulletin* 216, 208 pp.

Church, M. (1974) Hydrology and permafrost with reference to northern North America. In *Permafrost Hydrology: Proceedings of Workshop Seminar*. Environment Canada, Ottawa, 7–20.

Church, M. (2002) Fluvial sediment transfer in cold regions. In: Hewitt, K. *et al.* (eds) *Landscapes of Transition*. Kluwer Academic Publishers, Rotterdam, 93–117.

Church, M. and Miles, M.J. (1987) Meteorological antecedents to debris flow in southwestern British Columbia: some case studies. *Geological Society of America, Reviews in Engineering Geology* 7, 63–79.

Church, M. and Ryder, J.M. (1972) Paraglacial sedimentation: a consideration of fluvial processes conditioned by glaciation. *Geological Society of America Bulletin* 83, 3059–3071.

Church, M. and Slaymaker, O. (1989) Disequilibrium of Holocene sediment yield in glaciated British Columbia. *Nature* 337, 452–454.

Church, M., Stock, R.F. and Ryder, J.M. (1979) Contemporary sedimentary environments on Baffin Island, NWT, Canada: debris slope accumulations. *Arctic and Alpine Research* 11, 371–402.

Ciais, P., Tagliabue, A., Cuntz, M. *et al.* (2012) Large inert carbon pool in the terrestrial biosphere during the Last Glacial Maximum. *Nature Geoscience* 5, 74–79.

Clark, C.D., Hughes, A.L.C., Greenwood, S.L. *et al.* (2012) Pattern and timing of retreat of the last British-Irish Ice Sheet. *Quaternary Science Reviews* 44, 112–146.

Clark, D.H., Steig, E.J., Potter, N and Gillespie, A.R. (1998) Genetic variability of rock glaciers. *Geografiska Annaler* 80A, 175–182.

Clark, G.M. and Ciolkosz, E.J. (1988) Periglacial geomorphology of the Appalachian Highlands and Interior Highlands south of the glacial border – a review. *Geomorphology* 1, 191–220.

Clark, G.M. and Hedges, J. (1992) Origin of certain high-elevation local broad uplands in the central Appalachians south of the glacial border, USA – a palaeoperiglacial hypothesis. In Dixon, J.C. and Abrahams, A.D. (eds) *Periglacial Geomorphology: Proceedings of the 22nd Annual Binghamton Symposium in Geomorphology*. John Wiley & Sons, Chichester, 31–61.

Clark, M.J. (1988) Periglacial hydrology. In Clark, M.J. (ed.) *Advances in Periglacial Geomorphology*. John Wiley & Sons, Chichester, 415–462.

Clark, M.J. and Seppälä, M. (1988) Slushflows in a subarctic environment, Kilpisjärvi, Finnish Lappland. *Arctic and Alpine Research* 30, 97–105.

Clark, D.H., Steig, E.J., Potter, N and Gillespie, A.R. (1998) Genetic variability of rock glaciers. *Geografiska Annaler* 80A, 175–182.

Clark, P.U., Dyke, A.S., Shakun, J.D. *et al.* (2009) The Last Glacial Maximum. *Science* 325, 710–714.

Clarke, M.L., Milodowski, A.E., Bouch, J.E. *et al.* (2007) New OSL dating of UK loess: indications of two phases of Late Glacial dust accretion in SE England and climate implications. *Journal of Quaternary Science* 22, 361–371.

Clay, P. (2015) The origin of relict cryogenic mounds at East Walton and Thompson Common, Norfolk, England. *Proceedings of the Geologists' Association* 126, 522–535.

Cogley, J.G. (1972) Processes of solution in an arctic limestone terrain. *Institute of British Geographers, Special Publication* 4, 201–211.

Cogley, J.G. and McCann, S.B. (1976) An exceptional storm and its effects in the Canadian high Arctic. *Arctic and Alpine Research* 8, 105–110.

Colucci, R.R., Boccali, C., Zebre, M. and Guglielmin, M. (2016) Rock glaciers, protalus ramparts and pronival ramparts in the south-eastern Alps. *Geomorphology* 269, 112–121.

Corner, G.D. (1980) Avalanche impact landforms in Troms, north Norway. *Geografiska Annaler* 62A, 1–10.

Corte, A.E. (1966) Particle sorting by repeated freezing and thawing. *Biuletyn Peryglacjany* 15, 175–240.

Cory, R.M., Crump, B.C., Dobkowski, J.A. and Kling, G.W. (2013) Surface exposure to sunlight stimulates CO_2 release from permafrost soil carbon in the Arctic. *Proceedings of the National Academy of Sciences* 110, 3429–3434.

Costard, F., Dupreyat, L., Gautier, E. and Carey-Gailhardis, E. (2003) Fluvial thermal erosion investigations along a rapidly eroding river bank: application to the Lena River (Central Siberia). *Earth Surface Processes and Landforms* 28, 1349–1359.

Costard, F., Gautier, E., Fedorov, A. *et al.* (2014) An assessment of the erosion potential of the fluvial thermal processes during ice breakups of the Lena River (Siberia). *Permafrost and Periglacial Processes* 25, 162–171.

Costin, A.B., Thom, B.G., Wimbush, D.J. and Stuiver, M. (1967) Nonsorted steps in the Mt. Kosciusko area, Australia. *Bulletin, Geological Society of America* 78, 979–992.

Côté, M.M. and Burn, C.R. (2002) The oriented lakes of Tuktoyaktuk Peninsula, western Arctic coast, Canada: a GIS-based analysis. *Permafrost and Periglacial Processes* 13, 61–70.

Coussot, P. and Meunier, M. (1996) Recognition, classification and mechanical description of debris flows. *Earth-Science Reviews* 40, 209–227.

Coutard, J.-P. and Mücher, H.J. (1985) Deformation of laminated silt loam due to repeated freezing and thawing cycles. *Earth Surface Processes and Landforms* 10, 309–320.

Coutard, J.-P., Ozouf, J.-C., and Gabert, P. (1996) Modalités de la cryoreptation dans les Massifs du Chambeyron et de la Mortice, Haute-Ubaye, Alpes Françaises du Sud. *Permafrost and Periglacial Processes* 7, 21–51.

Coxon, P. and O'Callaghan, P. (1987) The distribution and age of pingo remnants in Ireland. In Boardman, J. (ed.) *Periglacial Processes and Landforms in Britain and Ireland*. Cambridge University Press, Cambridge, 195–202.

Cremaschi, M. and Van Vliet-Lanoë, B. (1990) Traces of frost activity and ice segregation in Pleistocene loess deposits and till of northern Italy: deep seasonal freezing or permafrost? *Quaternary International* 5, 39–48.

Cunningham, A. and Wilson, P. (2004) Relict periglacial boulder sheets and lobes on Slieve Donard, Mountains of Mourne, Ireland. *Irish Geography* 37, 187–201.

Curry, A.M. (1999) Paraglacial modification of slope form. *Earth Surface Processes and Landforms* 24, 1213–1228.

Curry, A.M. and Ballantyne, C.K. (1999) Paraglacial modification of glacigenic sediment. *Geografiska Annaler* 81A, 409–419.

Curry, A.M. and Black, R. (2003) Structure, sedimentology and evolution of rockfall talus, Mynydd Du, south Wales. *Proceedings of the Geologists' Association* 114, 49–64.

Curry, A.M. and Morris, C.J. (2004) Lateglacial and Holocene talus slope development and rockwall retreat on Mynydd Du, UK. *Geomorphology* 58, 85–106.

Czudek, T. (1995) Cryoplanation terraces – a brief review and some remarks. *Geografiska Annaler* 77A, 95–105.

Czudek, T. and Demek, J. (1970) Thermokarst in Siberia and its influence on the development of lowland relief. *Quaternary Research* 1, 103–120.

Czudek, T. and Demek, J. (1971) Pleistocene cryoplanation in the Ceská Vpocina highlands, Czechoslovakia. *Transactions of the Institute of British Geographers* 52, 95–112.

Dahms, D.E. and Rawlins, C.L. (1996) A two-year record of eolian sedimentation in the Wind River Range, Wyoming, U.S.A. *Arctic and Alpine Research* 28, 210–216.

Dale, J.E., Leech, S., McCann, S.B. and Samuelson, G. (2002) Sedimentary characteristics, biological zonation and physical processes of the tidal flats of Iqaluit, Nunavut. In: Hewitt, K., Byrne, M.-L., English, G. and Young, G. (eds) *Landscapes of Transition*. Kluwer, Amsterdam, 205–234.

Dallimore, S.R., Nixon, F.M., Egginton, P.A. and Bisson, J.G. (1996a) Deep-seated creep of massive ground ice, Tuktoyaktuk, NWT, Canada. *Permafrost and Periglacial Processes* 7, 337–347.

Dallimore, S.R., Wolfe, S.A. and Solomon, S.M. (1996b) Influence of ground ice and permafrost on coastal evolution, Richards Island, Beaufort Sea coast, NWT. *Canadian Journal of Earth Sciences* 33, 664–675.

Darmody, R.G., Thorn, C.E., Harder, R.L. *et al.* (2000) Weathering implications of water chemistry in an arctic-alpine environment, northern Sweden. *Geomorphology* 34, 89–100.

Darmody, R.G., Campbell, S.W., Dixon, J.C. and Thorn, C.E. (2002) Enigmatic efflorescence in Kärkevagge, Swedish Lapland: the key to chemical weathering? *Geografiska Annaler* 84A, 187–192.

Darmody, R.G., Thorn, C.E. and Allen, C.E. (2005) Chemical weathering and boulder mantles, Kärkevagge, Swedish Lapland. *Geomorphology* 67, 159–170.

Darmody, R.G., Thorn, C.E., Seppälä, M. *et al.* (2008) Age and weathering status of granite tors in arctic Finland (~68°N). *Geomorphology* 94, 10–23.

Dash, J.G., Rempel, A.W. and Wettlaufer, J.S. (2006) The physics of premelted ice and its environmental consequences. *Reviews of Modern Physics* 78, 695–740.

Davidson, G.P. and Nye, J.F. (1985) A photoelastic study of ice pressure in rock cracks. *Cold Regions Science and Technology* 11, 143–153.

Davies, D.A., Berrisford, M.S. and Matthews, J.A. (1990) Boulder-paved river channels: a case study of a fluvio-periglacial landform. *Zeitschrift für Geomorphologie* 84, 213–231.

Davies, M., Hamza, O. and Harris, C. (2001) The effect of rise in mean annual temperature on the stability of rock slopes containing ice-filled discontinuities. *Permafrost and Periglacial Processes* 12, 137–144.

Davis, N. (2000) *Permafrost: A Guide to Frozen Ground in Transition*. University of Alaska Press, Fairbanks, 339 pp.

Dawson, A.G. (1988) The main rock platform (Main Lateglacial Shoreline) in Ardnamurchan and Moidart, western Scotland. *Scottish Journal of Geology* 24, 163–174.

Dawson, A.G., Dawson, S., Cooper, A.G. *et al.* (2013) A Pliocene age and origin for the strandflat of the Western Isles of Scotland: a speculative hypothesis. *Geological Magazine* 150, 360–366.

DeConto, R.M., Galeoyyi, Pagani, M. *et al.* (2012) Past extreme warming events linked to massive carbon release from thawing permafrost. *Nature* 484, 87–91.

De Gans, W. (1982) Location, age and origin of pingo remnants in the Drentsche Aa valley area (the Netherlands). *Geologie en Mijnbouw* 62, 147–158.

De Gans, W. (1988) Pingo scars and their identification. In Clark, M.J. (ed.) *Advances in Periglacial Geomorphology*. John Wiley & Sons, Chichester, 299–322.

De Gans, W. and Sohl, H. (1981) Weichselian pingo remnants and permafrost on the Drente Plateau (the Netherlands). *Geologie en Mijnbouw* 60, 447–452.

Degenhardt, J.J. (2009) Development of tongue-shaped and multilobate rock glaciers in alpine environments – interpretations from ground-penetrating radar surveys. *Geomorphology* 109, 94–107.

DeGraff, J.V. (1993) The geomorphology of some debris flows in the southern Sierra Nevada, California. *Geomorphology* 10, 231–252.

De Klerk, P., Donner, N., Karpov, N.S. *et al.* (2011) Short-term dynamics of a low-centred polygon near

Chokurdakh (NE Yakutia, NE Siberia) and climate change during the last ca 1250 years. *Quaternary Science Reviews* 30, 3013–3031.

Delaloye, R. and Lambiel, C. (2005) Evidence of winter ascending air circulation throughout talus slopes and rock glaciers situated in the lower belt of alpine discontinuous permafrost. *Norwegian Journal of Geography* 59, 194–203.

Delisle, G. (1998) Numerical simulation of permafrost growth and decay. *Journal of Quaternary Science* 13, 325–333.

Demitroff, M. (2016) Pleistocene ventifacts and ice-marginal conditions, New Jersey, USA. *Permafrost and Periglacial Processes* 27, 123–137.

Derbyshire, E. and Owen, L.A. (1990) Quaternary alluvial fans in the Karakoram Mountains. In Rachocki, A.H. and Church, M (eds) *Alluvial Fans: a Field Approach.* John Wiley & Sons, Chichester, 27–53.

Déry, S.J. and Wood, E.F. (2005) Decreasing river discharge in northern Canada. *Geophysical Research Letters* 32, L10401.

De Vet, S.J. and Cammeraat, E.L.H. (2012) Aeolian contributions to the development of hillslopes and scree sediments in Graenagil, Torfajökull, Iceland. *Geomorphology* 175–176, 74–85.

Dewez, T.J.B., Regard, V., Duperret, A. and Lasseur, E. (2015) Shore platform lowering due to frost shattering during the 2009 winter at Mesnil Val, English channel coast, NW France. *Earth Surface Processes and Landforms* 40, 1688–1700.

Deynoux, M. (1982) Periglacial polygonal structures and sand wedges in the Late Precambrian glacial formation of the Taoudoni basin in the Adrar of Mauritania (West Africa). *Palaeogeography, Palaeoclimatology, Palaeoecology* 39, 55–70.

Dietrich, R.V. (1977) Impact abrasion of harder by softer materials. *Journal of Geology* 85, 242–246.

Dijkmans, J.W.A. (1990) Niveo-aeolian sedimentation and resulting sedimentary structures: Søndre Strømfjord Area, western Greenland. *Permafrost and Periglacial Processes* 1, 83–96.

Dijkmans, J.W.A. and Törnqvist, T.E. (1991) Modern periglacial eolian deposits and landforms in the Søndre Strømfjord area, West Greenland, and their palaeoenvironmental implications. *Meddelelser om Grønland, Geoscience* 25, 1–39.

Dionne, J.-C. (1983) Frost-heaved bedrock features: a valuable permafrost indicator. *Géographie Physique et Quaternaire* 37, 241–251.

Dionne, J.-C. (1993) Influence glacielle dans le façonnement d'une plate-forme rocheuse intertidale, Estuaire du Saint-Laurent, Québec. *Revue de Géomorphologie Dynamique* 42, 1–10.

Dionne, J.-C. (2002) The boulder barricade at Cap à la Baleine, North Shore of Gaspé Peninsula (Québec): nature of boulders, origin and significance. *Journal of Coastal Research* 18, 652–661.

Dionne, J.-C. and Brodeur, D. (1988a) Érosion des plates-formes rocheuses littorales par affouillement glaciel. *Zeitschrift für Geomorphologie* 32, 101–115.

Dionne, J.-C. and Brodeur, D. (1988b) Frost weathering and ice action in shore platform development with particular reference to Québec, Canada. *Zeitschrift für Geomorphologie, Supplementband* 71, 117–130.

Dionne, J.-C. and Gérardin, V. (1988) Observations sur les buttes organiques de la côte nord du Golfe du Saint-Laurent, Québec *Géographie Physique et Quaternaire* 42, 289–301.

Dixon, J.C. and Thorn, C.E. (2005) Chemical weathering and landscape development in mid-latitude alpine environments. *Geomorphology* 67, 127–145.

Dixon, J.C., Thorn, C.E. and Darmody, R.G. (1984) Chemical weathering processes on the Vantage Peak nunatak, Juneau Icefield, Southern Alaska. *Physical Geography* 5, 111–131.

Dixon, J.C., Thorn, C.E., Darmody, R.G. and Schlyter, P. (2001) Weathering rates of fine pebbles at the soil surface in Kärkevagge, Swedish Lapland. *Catena* 45, 273–286.

Dixon, J.C., Thorn, C.E., Darmody, R.G. and Campbell, S.W. (2002) Weathering rinds and rock coatings from an Arctic alpine environment, northern Scandinavia. *Geological Society of America, Bulletin* 114, 226–238.

Dodonov, A.E. (2013) Loess records: Central Asia. In Elias, S. (ed.) *Encyclopedia of Quaternary Science*, 2nd edn. Elsevier, Amsterdam, 585–594.

Doolittle, J. and Nelson, F. (2009) Characterising relict cryogenic macrostructures in mid-latitude areas of the USA with three-dimensional ground-penetrating radar. *Permafrost and Periglacial Processes* 20, 257–268.

Doran, P.T., McKay, C.P., Clow, G.D. *et al.* (2002) Valley floor climate observations from McMurdo dry valleys, Antarctica, 1986–2000. *Journal of Geophysical Research* 107, ACL 13.1–13.12.

Dos Santos, R.A.L., Prange, M., Castañeda, I.S. *et al.* (2010) Glacial–interglacial variability in Atlantic meridional overturning circulation and thermocline adjustments in the tropical North Atlantic. *Earth and Planetary Science Letters* 300, 407–414.

Douglas, G.R., Whalley, W.B. and McGreevy, J.P. (1991) Rock properties as controls on free-face debris fall activity. *Permafrost and Periglacial Processes* 2, 311–319.

Dredge, L.A. (1992) Breakup of limestone bedrock by frost shattering and chemical weathering, Eastern Canadian Arctic. *Arctic and Alpine Research* 24, 314–323.

Dredge, L.A. (2000) Age and origin of upland blockfields on the Melville Peninsula, eastern Canadian Arctic. *Geografiska Annaler* 82A, 443–454.

Dubikov, G.I. (2002) *Composition and cryogenic structure of permafrost in West Siberia* (in Russian). GEOS, Moscow, 246 pp.

Duca, S., Occhiena, C., Mattone, M. *et al.* (2014) Feasibility of ice segregation location by acoustic emission detection: a laboratory test in gneiss. *Permafrost and Periglacial Processes* 25, 208–219.

Dugan, H.A., Lamoureux, S.F., Lafrenière, M.J. and Lewis, T. (2009) Hydrological and sediment yield response to summer rainfall in a small high Arctic watershed. *Hydrological Processes* 23, 1514–1526.

Dupeyrat, L., Costard, F., Randriamazaoro, R. *et al.* (2011) Effects of ice content on the thermal erosion of permafrost: implications for coastal and fluvial erosion. *Permafrost and Periglacial Processes* 22, 179–187.

Dyke, A.S. and Evans, D.J.A. (2003) Ice-marginal terrestrial landsystems: northern Laurentide and Innuitian ice sheet margins. In Evans, D.J.A. (ed.) *Glacial Landsystems*. Arnold, London, 143–165.

Dyke, A.S. and Zoltai, S.C. (1980) Radiocarbon-dated mudboils, central Canadian Arctic. *Geological Survey of Canada Paper* 80-1B, 271–275.

Dyke, L.S. (1984) Frost heaving of bedrock in permafrost regions. *Bulletin of the Association of Engineering Geologists* 21, 389–405.

Dylik, J. (1956) Coup d'oeil sur la Pologne périglaciaire. *Biuletyn Peryglacjalny* 4, 195–238.

Eakin, W.M. (1916) The Yukon-Koyukuk region, Alaska. *United States Geological Survey Bulletin* 631, 67–88.

Ealey, P.J. and James, H.C.L. (2011) Loess of the Lizard Peninsula, Cornwall, SW Britain. *Quaternary International* 231, 55–61.

Eckerstorfer, M. and Christiansen, H.H. (2012) Meteorology, topography and snowpack conditions causing two extreme mid-winter slush and wet slab avalanche periods in high Arctic maritime Svalbard. *Permafrost and Periglacial Processes* 23, 15–25.

Eckerstorfer, M., Christiansen, H.H., Vogel, S. and Rubensdotter, L. (2013) Snow cornice dynamics as a control on plateau edge erosion in central Svalbard. *Earth Surface Processes and Landforms* 38, 466–476.

Eden, D.N. and Hammond, A.P. (2003) Dust accumulation in the New Zealand region since the last glacial maximum. *Quaternary Science Reviews* 22, 2037–2052.

Edwards, M., Grosse, G., Jones, B.M. and McDowell, P. (2016) The evolution of a thermokarst-lake landscape: Late Quaternary permafrost degradation and stabilization in interior Alaska. *Sedimentary Geology* 340, 3–14.

EEA (2009) Regional climate change and adaptation – the Alps facing the challenge of changing water resources. *European Environment Agency Report* 8/2009, 143 pp.

Egginton, P.A. (1979) Mudboil activity, central district of Keewatin. *Geological Survey of Canada Paper* 78-1B, 349–356.

Egginton, P.A. and Dyke, L.S. (1982) Density gradients and injection structures in mudboils in central district of Keewatin. *Geological Survey of Canada Paper* 82-1B, 173–176.

Egginton, P.A. and French, H.M. (1985) Solifluction and related processes, eastern Banks Island, NWT. *Canadian Journal of Earth Sciences* 22, 1671–1678.

Elias, S.A. (1999) Mid-Wisconsinan seasonal temperatures reconstructed from fossil beetle assemblages in eastern North America: comparisons with other proxy records from the northern hemisphere. *Journal of Quaternary Science* 14, 255–262.

Elconin, R.F. and LaChapelle, E.R. (1997) Flow and internal structure of a rock glacier. *Journal of Glaciology* 43, 238–244.

Elder, K. and Kattelmann, R. (1993) A low-angle slushflow in the Kirgiz Range, Kirgizstan. *Permafrost and Periglacial Processes* 4, 301–310.

Elliott, G. (1996) Microfabric evidence for podzolic soil inversion by solifluction processes. *Earth Surface Processes and Landforms* 21, 467–476.

Elliott, G. and Worsley, P. (1999) The sedimentology, stratigraphy and ^{14}C dating of a turf-banked solifluction lobe: evidence for slope instability at Okstindan, northern Norway. *Journal of Quaternary Science* 14, 175–188.

Elton, C.S. (1928) The nature and origin of soil polygons in Spitzbergen. *Quarterly Journal of the Geological Society, London* 83, 163–192.

Embleton, C. and King, C.A.M. (1968) *Glacial and Periglacial Geomorphology*. Arnold, London, 608 pp.

Emmerton, C.A., Lesack, L.F.W. and Marsh, P. (2007) Lake abundance, potential water storage, and habitat distribution in the Mackenzie River delta, western Canadian Arctic. *Water Resources Research* 43, W05419.

Erkins, G., Hoffmann, T., Gerlach, R. and Klostermann, J. (2011) Complex fluvial response to Lateglacial and Holocene allogenic forcing in the Lower Rhine Valley (Germany). *Quaternary Science Reviews* 30, 611–627.

Etienne, S. (2002) The role of biological weathering in periglacial areas: a study of weathering rinds in south Iceland. *Geomorphology* 47, 75–86.

Etienne, S. and Dupont, J. (2002) Fungal weathering of basaltic rocks in a cold oceanic environment (Iceland): comparison between experimental and field observations. *Earth Surface Processes and Landforms* 27, 737–748.

Etlicher, B. and Lautridou, J.-P. (1999) Gélifraction expérimentale d'arènes de roches cristallines: bilan d'essais de longue durée. *Permafrost and Periglacial Processes* 10, 1–16.

Etzelmüller, B. and Frauenfelder, R. (2009) Factors controlling the distribution of mountain permafrost in the northern hemisphere and their influence on sediment transport. *Arctic, Antarctic and Alpine Research* 41, 48–58.

Etzelmüller, B. Ødegård, R.S., Berthling, I. and Sollid, J.L. (2001) Terrain parameters and remote sensing data in the analysis of permafrost distribution and periglacial processes: examples from southern Norway. *Permafrost and Periglacial Processes* 12, 79–92.

Etzelmüller, B., Heggem, E.S.F., Sharkhuu, N. *et al.* (2006) Mountain permafrost distribution modelling using a multi-criteria approach in the Hövsgöl area, northern Mongolia. *Permafrost and Periglacial Processes* 17, 91–104.

Etzelmüller, B., Farbrot, H., Gudmundson, Á. *et al.* (2007) The regional distribution of mountain permafrost in Iceland. *Permafrost and Periglacial Processes* 18, 185–199.

Etzelmüller, B., Schuler, T.V., Isaksen, K. *et al.* (2011) Modeling the evolution of Svalbard permafrost during the 20th and 21st century. *The Cryosphere* 5, 67–79.

Evans, D.J.A. and Twigg, D.R. (2002) The active temperate glacial landsystem: a model based on Breidamerkurjökull and Fjallsjökull, Iceland. *Quaternary Science Reviews* 21, 2143–2177.

Evin, M. (1987) Lithology and fracturing control of rock glaciers in south-western Alps of France and Italy. In Giardino, J.R., Schroder, J.F. and Vitek, J.D. (eds) *Rock Glaciers*. Allen and Unwin, London, 83–106.

Ewertowski, M. (2009) Ice-wedge pseudomorphs and frost-cracking structures in Weichselian sediments, central-west Poland. *Permafrost and Periglacial Processes* 20, 316–330.

Eyles, N. and Kocsis, S. (1988) Sedimentology and clast fabric of subaerial debris flow facies in a glacially-influenced alluvial fan. *Sedimentary Geology* 59, 15–28.

Eynaud, F., de Abrieu, L., Voelker, A. *et al.* (2009) Position of the Polar Front along the western Iberian margin during key cold episodes of the last 45 ka. *Geochemistry, Geophysics, Geosystems* 10, 1–21.

Fabel, D., Ballantyne, C.K. and Xu, S. (2012) Trimlines, blockfields, mountain-top erratics and the vertical dimensions of the last British-Irish Ice Sheet in NW Scotland. *Quaternary Science Reviews* 55, 91–102.

Fábián, S.A., Kovács, J., Varga, G. *et al.* (2014) Distribution of relict permafrost features in the Pannonian Basin, Hungary. *Boreas* 43, 722–732.

Fahey, B.D. (1981) Origin and age of upland schist tors in central Otago, New Zealand. *New Zealand Journal of Geology and Geophysics* 24, 399–413.

Fahey, B.D. and Dagesse, D.F. (1984) An experimental study of the effect of humidity and temperature variations on the granular disintegration of argillaceous carbonate rocks in cold climates. *Arctic and Alpine Research* 16, 291–298.

Falaschi, D., Castro, M., Masiokas, M. *et al.* (2014) Rock glacier inventory of the Valles Calchaquíes Region (~25°S), Salta, Argentina. *Permafrost and Periglacial Processes* 25, 69–75.

Farbrot, H., Isaksen, K., Etzelmüller, B. and Gisnås, K. (2013) Ground thermal regime and permafrost distribution under a changing climate in northern Norway. *Permafrost and Periglacial Processes* 24, 20–38.

Farquharson, L.M., Mann, D.H., Grosse, G. *et al.* (2016a) Spatial distribution of thermokarst terrain in Arctic Alaska. *Geomorphology* 273, 116–133.

Farquharson, L.M., Walter Anthony, K.M., Bigelow, N. *et al.* (2016b) Facies analysis of yedoma thermokarst lakes on the northern Seward Peninsula, Alaska. *Sedimentary Geology* 340, 25–37.

Favaro, E.A. and Lamoureux, S.F. (2015) Downstream patterns of suspended sediment transport in a High Arctic river influenced by permafrost disturbance and recent climate change. *Geomorphology* 246, 359–369.

Feuillet, T. and Mercier, D. (2012) Post-Little Ice Age patterned ground development on two Pyrenean proglacial areas: from deglaciation to periglaciation. *Geografiska Annaler* 94A, 363–376.

Feuillet, T., Mercier, D., Decaulne, A. and Cossart, E. (2011) Classification of sorted patterned ground areas based on their environmental characteristics (Skagafjördur, Northern Iceland). *Geomorphology* 139–140, 577–587.

Filion, L., Saint-Laurent, D., Desponts, M. and Payette, S. (1991) The late Holocene record of aeolian and fire activity in northern Québec, Canada. *The Holocene* 1, 201–208.

Firpo, M., Guglielmin, M. and Queirolo, C. (2006) Relict blockfields in the Ligurian Alps (Mount Beigua, Italy). *Permafrost and Periglacial Processes* 17, 71–78.

Fischer, L., Kääb, A., Huggel, C. and Noetzli, J. (2006) Geology, glacier retreat and permafrost degradation as controlling factors of slope instabilities in a high-mountain rock wall: the Monte Rosa east face. *Natural Hazards and Earth System Sciences* 6, 761–772.

Fischer, L., Amann, F., Moore, J.R. and Huggel, C. (2010) Assessment of periglacial slope stability for the 1988 Tschierva rock avalanche. *Engineering Geology* 116, 32–43.

Fischer, L., Eisenbeiss, H., Kääb, A. *et al.* (2011) Monitoring topographic changes in a periglacial high-mountain face using high-resolution DTMs, Monte Rosa East Face, Italian Alps. *Permafrost and Periglacial Processes* 22, 140–152.

Fischer, L., Purves, R.S., Huggel, C. *et al.* (2012) On the influence of topographic, geological and cryospheric factors on rock avalanches and rockfalls in high-mountain areas. *Natural Hazards and Earth Systems Sciences* 12, 241–254.

Fisher, T.G. (1996) Sand wedge and ventifact palaeoenvironmental indicators in northwest Saskatchewan, 11 ka to 9.9 ka BP. *Permafrost and Periglacial Processes* 7, 391–408.

Fitzharris, B.B. and Owens, I.F. (1984) Avalanche tarns. *Journal of Glaciology* 30, 308–312.

Fitzsimmons, K.E., Markovic, S.B. and Hambach, U. (2012) Pleistocene environmental dynamics recorded in the loess of the middle and lower Danube Basin. *Quaternary Science Reviews* 41, 104–118.

Fitzsimons, S.J. (2006) Mechanical behaviour and structure of the basal ice layer. In Knight, P.G. (ed.) *Glacier Science and Environmental Change*. Blackwell, Oxford, 329–334.

Fjellanger, J., Sørbel, L., Linge, H. *et al.* (2006) Glacial survival of blockfields on the Varanger Peninsula, northern Norway. *Geomorphology* 82, 255–272.

Forbes, A.C. and Lamoureux, S.F. (2005) Climatic controls on streamflow and suspended sediment transport in three large middle Arctic catchments, Boothia Peninsula, Nunavut, Canada. *Arctic, Antarctic and Alpine Research* 37, 304–315.

Forbes, D.L. (2011) Glaciated Coasts. In Flemming, B.M. and Hansom, J.D. (eds) *Treatise on Estuarine and Coastal Science, Volume 3: Estuarine and Coastal Geology and Geomorphology*. Academic Press, London, 223–243.

Forbes, D.L. and Hansom, J.D. (2011) Polar Coasts. In Flemming, B.M. and Hansom, J.D. (eds) *Treatise on Estuarine and Coastal Science, Volume 3: Estuarine and Coastal Geology and Geomorphology*. Academic Press, London, 245–283.

Forbes, D.L. and Taylor, R.B. (1994) Ice in the shore zone and the geomorphology of cold coasts. *Progress in Physical Geography* 18, 59–89.

Ford, D.C. (1971) Characteristics of limestone solution in the southern Rocky Mountains and Selkirk Mountains, Alberta and British Columbia. *Canadian Journal of Earth Sciences* 8, 587–609.

Ford, D.C. (1973) Development of the canyons of the South Nahanni River, NWT. *Canadian Journal of Earth Sciences* 10, 366–378.

Ford, D.C. (1983) The physiography of the Castleguard Karst and Columbia Icefields area, Alberta, Canada. *Arctic and Alpine Research* 15, 427–436.

Ford, D.C. (1987) Effects of glaciations and permafrost upon the development of karst in Canada. *Earth Surface Processes and Landforms* 12, 507–521.

Ford, D.C. (1996) Karst in a cold climate: effects of glaciation and permafrost conditions upon karst landform systems of Canada. In McCann, S.B. and Ford, D.C. (eds) *Geomorphology Sans Frontières*. John Wiley & Sons, Chichester, 153–179.

Ford, D.C. and Williams, P.W. (2007) *Karst Hydrogeology and Geomorphology*. John Wiley & Sons, Chichester.

Fortier, D. and Allard, M. (2004) Late Holocene syngenetic ice wedge polygon development, Bylot Island, Canadian Arctic Archipelago. *Canadian Journal of Earth Sciences* 41, 997–1012.

Fortier, D. and Allard, M. (2005) Frost-cracking conditions, Bylot Island, eastern Canadian Arctic Archipelago. *Permafrost and Periglacial Processes* 16, 145–161.

Fortier, R., Levesque, R., Seguin, M.K. and Allard, M. (1991) Caractérisation du pergélisol de buttes cryogènes à l'aide de diagraphies électriques au Nunavik, Québec. *Permafrost and Periglacial Processes* 2, 79–93.

Fortier, D., Allard, M. and Pivot, F. (2006) A late-Holocene record of loess deposition in ice-wedge polygons reflecting wind activity and ground moisture conditions, Bylot Island, eastern Canadian Arctic. *The Holocene* 16, 635–646.

Fortier, D., Allard, M. and Shur, Y. (2007) Observation of rapid drainage system development by thermal erosion of ice wedges on Bylot Island, Canadian Arctic Archipelago. *Permafrost and Periglacial Processes* 18, 229–243.

Fortier, D., Kanevskiy, M. and Shur, Y. (2008a) Genesis of reticulate-chaotic cryostructure in permafrost. In Kane, D.L. and Hinkel, K.M. (eds) *Proceedings of the Ninth International Conference on Permafrost*. University of Alaska, Fairbanks, 451–456.

Fortier, R., LeBlanc, A.-M., Allard, M. *et al.* (2008b) Internal structure and conditions of permafrost mounds at Umiujak in Nunavik, Canada, inferred from field investigation and electrical resistivity tomography. *Canadian Journal of Earth Sciences* 45, 367–387.

Fournier, A. and Allard, M. (1992) Periglacial shoreline erosion of a rocky coast, George River estuary, northern Quebec. *Journal of Coastal Research* 8, 926–942.

Francou, B. (1988) *L'Éboulisation en Haute Montagne (Alpes et Andes)*. Editec, Caen.

Francou, B. (1990) Stratification mechanisms in slope deposits in high subequatorial mountains. *Permafrost and Periglacial Processes* 1, 249–263.

Francou, B. and Bertran, P. (1997) A multivariate analysis of clast displacement rates on stone-banked sheets, Cordillera Real, Bolivia. *Permafrost and Periglacial Processes* 8, 371–382.

Francou, B. and Manté, C. (1990) Analysis of the segmentation in the profile of alpine talus slopes. *Permafrost and Periglacial Processes* 1, 53–60.

Fraser, T.A. and Burn, C.R. (1997) on the nature and origin of 'muck' deposits in the Klondike area, Yukon Territory. *Canadian Journal of Earth Sciences* 34, 1333–1344.

Frauenfelder, R., Haeberli, W., Hoelzle, M. and Maisch, M. (2001) Using relict rockglaciers in GIS-based modelling to reconstruct Younger Dryas permafrost distribution patterns in the Err-Julier area, Swiss Alps. *Norsk Geografisk Tidsskrift* 55, 195–202.

Frechen, M., Oches, E.A. and Kohfeld, K.E. (2003) Loess in Europe – mass accumulation rates during the Last Glacial Period. *Quaternary Science Reviews* 22, 1835–1857.

Frehner, M., Ling, A.H.M. and Gärtner-Roer, I. (2015) Furrow-and-ridge morphology on rockglaciers explained by gravity-driven buckle folding: a case study from the Murtèl Rockglacier (Switzerland). *Permafrost and Periglacial Processes* 26, 57–66.

French, H.M. (1971) Slope asymmetry of the Beaufort Plain, northwest Banks Island, NWT, Canada. *Canadian Journal of Earth Sciences* 8, 717–731.

French, H.M. (1973) Cryopediments on the chalk of southern England. *Biuletyn Peryglacjalny* 22, 149–156.

French, H.M. (1976) *The Periglacial Environment*, 1st edn. Longman, London, 308 pp.

French, H.M. (1986) Periglacial involutions and mass displacement structures, Banks Island, Canada. *Geografiska Annaler* 68A, 167–174.

French, H.M. (1996) *The Periglacial Environment*, 2nd edn. Addison Wesley Longman, Harlow, 341 pp.

French, H.M. (2000) Does Łozinski's periglacial realm exist today? A discussion relevant to modern usage of the term 'periglacial'. *Permafrost and Periglacial Processes* 11, 35–42.

French, H.M. (2003) The development of periglacial geomorphology: 1 – up to 1965. *Permafrost and Periglacial Processes* 14, 29–60.

French, H.M. (2007) *The Periglacial Environment*, 3rd edn. John Wiley & Sons, Chichester, 458 pp.

French, H.M. (2008) Recent contributions to the study of past permafrost. *Permafrost and Periglacial Processes* 19, 179–194.

French, H.M. (2016) Do periglacial landscapes exist? A discussion of the upland landscapes of northern interior Yukon, Canada. *Permafrost and Periglacial Processes* 27, 219–228.

French, H.M. and Demitroff, M. (2001) Cold-climate origin of the enclosed depressions and wetlands ('spungs') of the Pine Barrens, Southern New Jersey, USA. *Permafrost and Periglacial Processes* 12, 337–350.

French, H.M. and Guglielmin, M. (1999) Observations on the ice-marginal periglacial geomorphology of Terra Nova Bay, Northern Victoria Land, Antarctica. *Permafrost and Periglacial Processes* 10, 331–347.

French, H.M. and Guglielmin, M. (2000) Cryogenic weathering of granite, northern Victoria Land, Antarctica. *Permafrost and Periglacial Processes* 11, 305–314.

French, H.M. and Harry, D.G. (1990) Observations on buried glacier ice and massive segregated ice, western Arctic coast, Canada. *Permafrost and Periglacial Processes* 1, 31–43.

French, H.M. and Harry, D.G. (1992) Pediments and cold-climate conditions, Barn Mountains, unglaciated northern Yukon, Canada. *Geografiska Annaler* 72A, 145–157.

French, H.M. and Millar, S.W.S. (2014) Permafrost at the time of the Last Glacial Maximum (LGM) in North America. *Boreas* 43, 667–677.

French, H.M. and Shur, Y.L. (2010) The principles of cryostratigraphy. *Earth-Science Reviews* 101, 190–206.

French, H.M., Bennett, L. and Hayley, D.W. (1986) Ground ice conditions near Rea Point and on Sabine Peninsula, eastern Melville Island. *Canadian Journal of Earth Sciences* 23, 1389–1400.

French, H.M., Demitroff, M. and Forman, S.L. (2003) Evidence for Late-Pleistocene permafrost in the New Jersey Pine Barrens (Latitude 39°N), Eastern USA. *Permafrost and Periglacial Processes* 14, 259–274.

French, H.M., Demitroff, M. and Forman, S.L. (2005) Evidence for Late-Pleistocene thermokarst in the New Jersey Pine Barrens (latitude 39°N), eastern USA. *Permafrost and Periglacial; Processes* 16, 173–186.

French, H.M., Demitroff, M. and Newell, W.L. (2009) Past permafrost on the Mid-Atlantic Coastal Plain, eastern United States. *Permafrost and Periglacial Processes* 20, 285–294.

Fried, G., Heinrich, J., Nagel, G. and Semmel, A. (1993) Periglacial denudation in formerly unglaciated areas of the Richardson Mountains (NW Canada). *Zeitschrift für Geomorphologie Supplementband* 92, 55–69.

Froese, D.G., Westgate, J.A., Reyes, A.V. *et al.* (2008) Ancient permafrost and a future, warmer Arctic. *Science* 321, 1648.

Froese, D.G., Zazula, G.D., Westgate, J.A. *et al.* (2009) The Klondike goldfields and Pleistocene environments of Beringia. *GSA Today* 19, 4–10.

Fukui, K., Fujii, Y., Mikhailov, N. *et al.* (2007) The lower limit of mountain permafrost in the Russian Altai Mountains. *Permafrost and Permafrost Processes* 18, 129–136.

Fukui, K., Sone, T., Strelin, J.A. *et al.* (2008a) Dynamics and GPR stratigraphy of a polar rock glacier on James Ross Island, Antarctic Peninsula. *Journal of Glaciology* 54, 445–451.

Fukui, K., Sone, T., Yamagata, K. *et al.* (2008b) Relationships between permafrost distribution and surface organic layers near Esso, central Kamchatka, Russian Far East. *Permafrost and Periglacial Processes* 19, 85–92.

Furdada, G., Martínez, P., Oller, P. and Vilaplana, J.M. (1999) Slushflows at El Port del Comte, northeast Spain. *Journal of Glaciology* 45, 555–558.

Galloway, J.P. and Carter, L.D. (1994) Palaeowind directions for Late Holocene dunes on the western Arctic coastal plain. *United States Geological Survey Bulletin* 2107, 27–30.

Gao, C. (2005) Ice-wedge casts in Late Wisconsinan glaciofluvial deposits, southern Ontario, Canada. *Canadian Journal of Earth Sciences* 42, 2117–2126.

Gao, C. (2014) Relict thermal-contraction-crack polygons and past permafrost south of the Late Wisconsinan glacial limit in the mid-Atlantic Coastal Plain, USA. *Permafrost and Periglacial Processes* 25, 144–149.

Garcia-Ruiz, J.M., Valero-Garcés, B., González-Sampériz, P. *et al.* (2001) Stratified scree in the central Spanish Pyrenees: palaeoenvironmental implications. *Permafrost and Periglacial Processes* 12, 233–242.

Gardner, J.S. (1979) The movement of material on debris slopes in the Canadian Rocky Mountains, *Zeitschrift für Geomorphologie* 23, 45–57.

Gardner, J.S. (1983a) Rockfall frequency and distribution in the Highwood Pass area, Canadian Rocky Mountains. *Zeitschrift für Geomorphologie* 27, 311–324.

Gardner, J.S. (1983b) Observations on erosion by wet snow avalanches, Mt. Rae area, Alberta, Canada. *Arctic and Alpine Research* 15, 271–275.

Gardner, T.W., Ritter, J.B., Shuman, C.A. *et al.* (1991) A periglacial stratified slope deposit in the Valley and Ridge Province of central Pennsylvania, USA: sedimentology, stratigraphy and geomorphic evolution. *Permafrost and Periglacial Processes* 2, 141–162.

Gärtner-Roer, I. (2012) Sediment transfer rates of two active rockglaciers in the Swiss Alps. *Geomorphology* 167–168, 45–50.

Gautier, E. and Costard, F. (2000) Les systèmes fluviaux à chenaux anastomosés en milieu périglaciaire, la Lena et ses principaux affluents (Sibérie Centrale). *Géographie Physique et Quaternaire* 54, 327–342.

Gautier, E., Brunstein, D., Costard, F. and Lodina, R. (2003) Fluvial dynamics in a deep permafrost zone: the case of the middle Lena River (central Yakutia). In Phillips, M., Springman, S.M. and Arenson, L.U. (eds) *Permafrost: Proceedings of the 8th International Conference.* Balkema, Rotterdam, 271–275.

Gavrilov, A.V., Romanovskii, N.N., Romanovsky, V.E. *et al.* (2003) Reconstruction of ice complex remnants on the eastern Siberian shelf. *Permafrost and Periglacial Processes* 14, 187–198.

Gerrard, A.J. (1988) Periglacial modification of the Cox Tor-Staples Tors area of western Dartmoor, England. *Physical Geography* 9, 280–300.

Ghysels, G. and Heyse, I. (2006) Composite-wedge pseudomorphs in Flanders, Belgium. *Permafrost and Periglacial Processes* 17, 145–161.

Giardino, J.R., Shroder, J.F. and Vitek, J.D. (eds) (1987) *Rock Glaciers.* Allen and Unwin, London.

Gibbard, P.L. and Walker, M.J.C. (2014) The term 'Anthropocene' in the context of formal geological classification. *Geological Society, London, Special Publication* 395, 29–37.

Gibbard, P.L., Pasanen, A.H., West, R.G. *et al.* (2009) Late Middle Pleistocene glaciation in East Anglia, England. *Boreas* 38, 504–528.

Gibson, J.J., Edwards, T.W.D. and Prowse, T.D. (1993) Runoff generation in a high boreal wetland in northern Canada. *Nordic Hydrology* 24, 213–224.

Gilbert, R. (1990) A distinction between ice-pushed and ice-lifted landforms on lacustrine and marine coasts. *Earth Surface Processes and Landforms* 15, 15–24.

Gilichinsky, D.A., Nolte, E., Basilyan, A.E. *et al.* (2007) Dating of syngenetic ice wedges in permafrost with ^{36}Cl. *Quaternary Science Reviews* 26, 1547–1556.

Gillies, J.A., Nickling, W.G. and Tilson, M. (2009) Ventifacts and wind-abraded rock features in the Taylor Valley, Antarctica. *Geomorphology* 107, 149–160.

Gleason, K.J., Krantz, W.B., Caine, N. *et al.* (1986) Geometrical aspects of sorted patterned ground in recurrently frozen soil. *Science* 232, 216–230.

Glenn, M.S. and Woo, M.-K. (1997) Spring and summer hydrology of a valley-bottom wetland, Ellesmere Island, Northwest Territories, Canada. *Wetlands* 17, 321–329.

Gocke, M., Hambach, U., Eckmeier, E. *et al.* (2014) Introducing an improved multi-proxy approach for palaeoenvironmental reconstruction of loess-palaeosol archives applied on the Late Pleistocene Nussloch sequence (SW Germany). *Palaeogeography, Palaeoclimatology, Palaeoecology* 410, 300–315.

Godin, E. and Fortier, D. (2012) Geomorphology of a thermo-erosion gully, Bylot Island, Nunavut, Canada. *Canadian Journal of Earth Sciences* 49, 979–986.

Goetz, S.J., Mack, M.C., Gurney, K.R *et al.* (2007) Ecosystem responses to recent climate change and fire disturbances at northern high latitudes: observations and model results contrasting northern Eurasia and North America. *Environmental Research Letters* 2, 045031.

Goldthwait, R.P. (1976) Frost-sorted patterned ground: a review. *Quaternary Research* 6, 27–35.

Golledge, N.R. (2010) Glaciation of Scotland during the Younger Dryas Stadial: a review. *Journal of Quaternary Science* 25, 550–566.

Gómez, A., Palacios, D., Ramos, M. *et al.* (2001) Location of permafrost in marginal regions: Corral del Veleta, Sierra Nevada, Spain. *Permafrost and Periglacial Processes* 12, 93–110.

Good, T.R. and Bryant, I.D. (1985) Fluvio-aeolian sedimentation – an example from Banks Island, NWT, Canada. *Geografiska Annaler* 67A, 33–46.

Goodfellow, B.W. (2007) Relict nonglacial surfaces in formerly glaciated landscapes. *Earth Science Reviews* 80, 47–73.

Goodfellow, B.W. (2012) A granulometry and secondary mineral fingerprint of chemical weathering in periglacial landscapes and its application to blockfield origins. *Quaternary Science Reviews* 57, 121–135.

Goodfellow, B.W., Fredin, I.O., Derron, M.-H. and Stroeven A.P. (2009) Weathering processes and Quaternary origin of an alpine blockfield in Arctic Sweden. *Boreas* 38, 379–398.

Goodfellow, B.W., Skelton, A., Martel, S.J. *et al.* (2014a) Controls on tor formation, Cairngorm Mountains, Scotland. *Journal of Geophysical Research: Earth Surface* 119, F002862.

Goodfellow, B.W., Stroeven, A.P., Fabel, D. *et al.* (2014b) Arctic-alpine blockfields in the northern Swedish Scandes: late Quaternary – not Neogene. *Earth Surface Dynamics* 2, 383–401.

Goodrich, L.E. (1982) An introductory review of numerical methods for ground thermal calculations. *National Research Council of Canada, Division of Building Research Paper* 1061, 32 pp.

Gorbunov, A.P. (1991) Ploughing blocks of the Tien Shan. *Permafrost and Periglacial Processes* 2, 237–243.

Gorbunov, A.P. and Seversky, E.V. (1999) Solifluction in the mountains of Central Asia: distribution, morphology, processes. *Permafrost and Periglacial Processes* 10, 81–89.

Gorbunov, A.P., Marchenko, S.S. and Seversky, E.V. (2004) The thermal environment of blocky materials in the mountains of central Asia. *Permafrost and Periglacial Processes* 15, 95–98.

Gordeev, V.V. (2006) Fluvial sediment flux to the Arctic Ocean. *Geomorphology* 80, 94–104.

Gordon, L.S. and Ballantyne, C.K. (2006) 'Protalus ramparts' on Navajo Mountain, Utah, USA: reinterpretation as blockslope-sourced rock glaciers. *Permafrost and Periglacial Processes* 17, 179–187.

Gore, D.B., Creagh, D.C., Burgess, J.S. *et al.* (1996) Composition, distribution and origin of surficial salts in the Vestfold Hills, East Antarctica. *Antarctic Science* 8, 73–84.

Goryachkin, S.A., Karavaeva, N.A., Targulian, V.O. and Glazov, M.V. (1999) Arctic soils: spatial distribution, zonality and transformation due to global change. *Permafrost and Periglacial Processes* 10, 235–250.

Gosse, J.C. and Phillips, F.M. (2001) Terrestrial *in situ* cosmogenic nuclides: theory and applications. *Quaternary Science Reviews* 20, 1475–1560.

Götz, J., Otto, J.-C. and Schrott, L. (2013) Postglacial sediment storage and rockwall retreat in a semi-closed inner-alpine sedimentary basin (Gradenmoos, Hohe Tauern, Austria). *Geografia Fisica e Dinamica Quaternaria* 36, 63–80.

Goudie, A.S. and Viles, H. (1997) *Salt Weathering Hazards*. John Wiley & Sons, Chichester.

Grab, S.W. (1994) Thufur in the Mohlesi Valley, Lesotho, southern Africa. *Permafrost and Periglacial Processes* 5, 111–118.

Grab, S.W. (2002) Characteristics and palaeoenvironmental significance of relict sorted patterned ground, Drakensberg plateau, southern Africa. *Quaternary Science Reviews* 21, 1729–1744.

Grab, S.W. (2005) Aspects of the geomorphology, genesis and environmental significance of earth hummocks (thúfur, pounus): miniature cryogenic mounds. *Progress in Physical Geography* 29, 139–155.

Grab, S.W., Dickinson, K.J.M., Mark, A.F. and Maegli, T. (2008) Ploughing boulders on the Rock and Pillar Range, south-central New Zealand: their geomorphology and alpine plant associations. *Journal of the Royal Society of New Zealand* 38, 51–70.

Gray, J.T. and Lauriol, B. (1993) Karst morphology and hydrology in a permafrost environment: the case of Akpatok Island, Ungava Bay, Eastern arctic Canada. In *Permafrost: Proceedings of the 6th International Conference.* South China University of Technology Press, Wushan Guangzhou, China, 192–197.

Gray, J.T. and Seppälä, M. (1991) Deeply dissected tundra polygons on a glacio-fluvial outwash plain, northern Ungava Peninsula, Québec. *Géographie Physique et Quaternaire* 45, 111–117.

Greenland, D. (1989) The climate of Niwot Ridge, Front Range, Colorado, U.S.A. *Arctic and Alpine Research* 21, 380–391.

Grom, J.D. and Pollard, W.H. (2008) A study of retrogressive thaw slump dynamics, Eureka Sound Lowlands, Ellesmere Island. In Kane, D.L. and Hinkel, K.M. (eds) *Proceedings of the Ninth International Conference on Permafrost.* University of Alaska, Fairbanks, 545–550.

Grosse, G. and Jones, B.M. (2011) Spatial distribution of pingos in northern Asia. *The Cryosphere* 5, 13–33.

Grosse, G., Schirrmeister, L., Kunitsky, V.V., Hubberten, H.-W. (2005) The use of CORONA images in remote sensing of periglacial geomorphology: an illustration from the NE Siberian coast. *Permafrost and Periglacial Processes* 16, 163–172.

Grosse, G., Schirrmeister, L. and Malthus, T. (2006) Application of Landsat-7 satellite data and a DEM for the quantification of thermokarst-affected terrain types in the periglacial Lena-Anabar coastal lowland. *Polar Research* 25, 51–68.

Grosse, G., Schirrmeister, L., Siegert, C. *et al.* (2007) Geological and geomorphological evolution of a sedimentary periglacial landscape in northeast Siberia during the Late Quaternary. *Geomorphology* 86, 25–51.

Grosse, G., Jones, B. and Arp, C. (2013a) Thermokarst lakes, drainage and drained basins. In Shroder, J., Giardino, J. and Harbor, J. (eds) *Treatise on Geomorphology. Volume 8: Glacial and Periglacial Geomorphology.* Academic Press, San Diego, CA, 325–353.

Grosse, G., Robinson, J.E., Bryant, R. *et al.* (2013b) Distribution of late Pleistocene syngenetic permafrost of the Yedoma Suite in east and central Siberia. *US Geological Survey Open File Report* 2013, 1078.

Grosso, S.A. and Corte, A.E. (1991) Cryoplanation surfaces in the Central Andes at latitude 35°S. *Permafrost and Periglacial Processes* 2, 49–58.

Gruber, S. (2012) Derivation and analysis of a high-resolution estimate of global permafrost zonation. *The Cryosphere* 6, 221–133.

Gruber, S. and Haeberli, W. (2007) Permafrost in steep bedrock slopes and its temperature-related destabilization following climate change. *Journal of Geophysical Research* 112, F02S18.

Gruber, S. and Haeberli, W. (2009) Mountain permafrost. In Margesin, R. (ed.) *Permafrost Soils.* Springer-Verlag, Berlin and Heidelberg, 33–44.

Gruber, S., Hoezle, M. and Haeberli, W. (2004a) Permafrost thaw and destabilization of Alpine rockwalls in the hot summer of 2003. *Geophysical Research Letters* 31, L13504.

Gruber, S., King, L., Kohl, T. *et al.* (2004b) Interpretation of geothermal profiles perturbed by topography: the alpine permafrost boreholes at Stockhorn plateau, Switzerland. *Permafrost and Periglacial Processes* 15, 349–357.

Gude, M. and Scherer, D. (1998) Snowmelt and slushflows: hydrological and hazard implications. *Annals of Glaciology* 26, 381–384.

Guglielmin, M., Cannone, N., Strini, A. and Lewkowicz, A.G. (2005) Biotic and abiotic processes on granite weathering landforms in a cryotic environment, northern Victoria land, Antarctica. *Permafrost and Periglacial Processes* 16, 69–85.

Guglielmin, M., Balks, M.R., Adlam, L.S. and Baio, F. (2011a) Permafrost thermal regime from two 30 m deep boreholes in southern Victoria Land, Antarctica. *Permafrost and Periglacial Processes* 22, 129–139.

Guglielmin, M., Favero-Longo, S.E., Cannone, N. *et al.* (2011b) Role of lichens in granite weathering in cold and arid environments of continental Antarctica. *Geological Society, London, Special Publication* 354, 195–203.

Guhl, A., Bertran, P., Zielhofer, C. and Fitzsimmons, K.E. (2012) Optically stimulated luminescence (OSL) dating of sand-filled wedge structures and their fine-grained host sediment from Jonzac, SW France. *Boreas* 42, 317–332.

Guilcher, A., Bodéré, J.-C., Coudé, A. *et al.* (1986) Le problème des strandflats en cinq pays de hautes latitudes. *Revue de Géologie Dynamique et Géographie Physique* 27, 47–79.

Gullentops, F., Janssen, J. and Paulissen, E. (1993) Saalian nivation activity in the Bosbeek Valley, NE Belgium. *Geologie en Mijnbouw* 72, 125–130.

Gunnell, Y., Jarman, D., Braucher, R. *et al.* (2013) The granite tors of Dartmoor, Southwest England: rapid and recent emergence revealed by Late Pleistocene cosmogenic apparent ages. *Quaternary Science Reviews* 61, 62–76.

Günther, F., Overduin, P.P., Sandakov, A.V. *et al.* (2013) Short- and long-term thermo-erosion of ice-rich permafrost coasts in the Laptev Sea region. *Biogeosciences* 10, 4297–4318.

Günzel, F.K. (2008) Shear strength of ice-filled rock joints. In Kane, D.L. and Hinkel, K.M. (eds) *Proceedings of the Ninth International Conference on Permafrost.* University of Alaska, Fairbanks, 581–586.

Gurney, S.D. (1995) A reassessment of the relict Pleistocene 'pingos' of west Wales: hydraulic pingos or mineral palsas? *Quaternary Newsletter* 77, 6–16.

Gurney, S.D. (1998) Aspects of the genesis and geomorphology of pingos: perennial permafrost mounds. *Progress in Physical Geography* 22, 307–324.

Gurney, S.D. (2000) Relict cryogenic mounds in the UK as evidence of climate change. In McLaren, S.J. and Kniveton, D.R. (eds) *Linking Climate Change to Land Surface Change.* Kluwer, Dordrecht, 209–229.

Gurney, S.D. (2001) Aspects of the genesis, geomorphology and terminology of palsas: perennial cryogenic mounds. *Progress in Physical Geography* 25, 249–260.

Gurney, S.D. and Worsley, P. (1996) Genetically complex and morphologically diverse pingos in the Fish Lake area of southwest Banks Island. *Geografiska Annaler* 79A, 41–56.

Gurney, S.D. and Worsley, P. (1997) A discussion on pingos in Mellemfjord, Disko, Central West Greenland. *Danish Journal of Geography* 97, 154–156.

Gurney, S.D., Astin, T.R. and Griffiths, G.H. (2010) Origin and structure of Devensian depressions at Letton, Herefordshire. *Mercian Geologist* 14, 14–21.

Guzzetti, F., Peruccacci, S., Ross, M. and Stark, C.P. (2008) The rainfall intensity-duration control of shallow landslides and debris flows: an update. *Landslides* 5, 3–17.

Haase, D., Fink, J., Haase G. *et al.* (2007) Loess in Europe – its spatial distribution based on a European Loess Map, scale 1 : 2 500 000. *Quaternary Science Reviews* 26, 1301–1312.

Haeberli, W. (1983) Permafrost-glacier relationships in the Swiss Alps – today and in the past. *Proceedings, Fourth International Conference on Permafrost.* National Academy Press, Washington DC, 695–700.

Haeberli, W. (1985) Creep of mountain permafrost: internal structure and flow of alpine rock glaciers. *Mitteilungen der Versuchsanstalt für Wasserbau, Hydrologie und Glaziologie* 77, 142 pp.

Haeberli, W. and Vonder Mühll, D. (1996) On the characteristics and possible origins of ice in rock glacier permafrost. *Zeitschrift für Geomorphologie Supplementband* 104, 43–57.

Haeberli, W., Hoezle, M., Kääb, A. *et al.* (1998) Ten years after drilling through the permafrost of the active rock glacier Murtèl, eastern Swiss Alps: unanswered questions and new perspectives. In Lewkowicz, A.G. and Allard, M. (eds) *Permafrost: Proceedings of the 7th International Conference.* Université Laval, Québec, 403–410.

Haeberli, W., Kääb, A., Wagner, S. *et al.* (1999) Pollen analysis and 14C age of moss in a permafrost core recovered from the active rock glacier Murtèl-Corvatsch, Swiss Alps: geomorphological and glaciological implications. *Journal of Glaciology* 45, 1–8.

Haeberli, W., Hallet, B., Arenson, L. *et al.* (2006) Permafrost creep and rock glacier dynamics. *Permafrost and Periglacial Processes* 17, 189–214.

Haeberli, W., Noetzli, J., Arenson, L. *et al.* (2010) Mountain permafrost: development and challenges of a young research field. *Journal of Glaciology* 56, 1043–1058.

Haldorsen, S., Heim, M., Dale, B. *et al.* (2010) Sensitivity to long-term climate change of subpermafrost groundwater systems on Svalbard. *Quaternary Research* 73, 393–402.

Hales, T.C. and Roering, J.J. (2005) Climate-controlled variations in scree production, Southern Alps, New Zealand. *Geology* 33, 701–704.

Hales, T.C. and Roering, J.J. (2007) Climatic controls on frost cracking and implications for the evolution of bedrock landscapes. *Journal of Geophysical Research* 112, F02033.

Hall, A.M. and Phillips, W.M. (2006) Glacial modification of granite tors in the Cairngorms, Scotland. *Journal of Quaternary Science* 21, 811–830.

Hall, A.M. and Sugden, D.E. (2007) The significance of tors in glaciated lands: a view from the British Isles. In André, M.-F., Etienne, S., Lageat, Y. *et al.* (eds) *Du continent au basin versant. Théories et practiques en géographie physique. (Hommage au Professeur Alain Godard)*. Presses Universitaires Blaise-Pascal, Clermont-Ferrand, 301–311.

Hall, K.J. (1985) Some observations on ground temperature and transport processes at a nivation site in northern Norway. *Norsk Geografisk Tidsskrift* 39, 27–37.

Hall, K.J. (1989) Wind-blown particles as weathering agents? An Antarctic example. *Geomorphology* 2, 405–410.

Hall, K.J. (1993) Enhanced bedrock weathering in association with late-lying snowpatches: evidence from Livingstone Island, Antarctica. *Earth Surface Processes and Landforms* 18, 121–129.

Hall, K.J. (1997a) Rock temperatures and implications for cold region weathering. 1: New data from Viking Valley, Alexander Island, Antarctica. *Permafrost and Periglacial Processes* 8, 69–90.

Hall, K.J. (1997b) Observations on 'cryoplanation' benches in Antarctica. *Antarctic Science* 9, 181–187.

Hall, K.J. (1998a) Rock temperatures and implications for cold region weathering. II: New data from Rothera, Adelaide Island, Antarctica. *Permafrost and Periglacial Processes* 9, 47–55.

Hall, K.J. (1998b) Nivation or cryoplanation: different terms, same features? *Polar Geography* 22, 1–16.

Hall, K.J. (1999) The role of thermal stress fatigue in the breakdown of rock in cold regions. *Geomorphology* 31, 47–63.

Hall, K.J. (2004) Evidence for freeze-thaw events and their implications for rock weathering in northern Canada. *Earth Surface Processes and Landforms* 29, 43–57.

Hall, K.J. (2007) Evidence for freeze-thaw events and their implications for rock weathering in northern Canada, II: the temperature at which water freezes in rock. *Earth Surface Processes and Landforms* 32, 249–259.

Hall, K.J. and André, M.-F. (2001) New insights into rock weathering from high-frequency rock temperature data:

an Antarctic study of weathering by thermal stress. *Geomorphology* 41, 23–35.

Hall, K.J. and André, M.-F. (2003) Rock thermal data at the grain scale: applicability to granular disintegration in cold environments. *Earth Surface Processes and Landforms* 28, 823–836.

Hall, K.J. and André, M.-F. (2006) Temperature observations in Antarctic tafoni: implications for weathering, biological colonization, and tafoni formation. *Antarctic Science* 18, 377–384.

Hall, K.J. and Meiklejohn, I. (1997) Some observations regarding protalus ramparts. *Permafrost and Periglacial Processes* 8, 245–250.

Hall, K.J. and Otte, W. (1990) A note on biological weathering on nunataks of the Juneau Icefield, Alaska. *Permafrost and Periglacial Processes* 1, 189–196.

Hall, K.J. and Thorn, C.E. (2011) The historical legacy of spatial scales in freeze–thaw weathering: misrepresentation and resultant misdirection. *Geomorphology* 130, 83–90.

Hall, K.J. and Thorn, C.E. (2014) Thermal fatigue and thermal shock in bedrock: an attempt to unravel the geomorphic processes and products. *Geomorphology* 206, 1–13.

Hall, K.J., Boelhouwers, J. and Driscoll, K. (2001) Some morphometric measurements on ploughing blocks in the McGregor Mountains, Canadian Rockies. *Permafrost and Periglacial Processes* 12, 219–225.

Hall, K.J., Thorn, C.E., Matsuoka, N. and Prick, A. (2002) Weathering in cold regions: some thoughts and perspectives. *Progress in Physical Geography* 26, 577–603.

Hall, K.J., Guglielmin, M. and Strini, A. (2008) Weathering of granite in Antarctica II: thermal data at the grain scale. *Earth Surface Processes and Landforms* 33, 295–307.

Hallet, B. (1990) Spatial self-organisation in geomorphology: from periodic bedforms and patterned ground to scale-invariant topography. *Earth-Science Reviews* 29, 57–75.

Hallet, B. (1998) Measurements of soil motion in sorted circles, western Spitsbergen. In Lewkowicz, A.G. and Allard, M. (eds) *Permafrost: Proceedings of the Seventh International Conference*. Université Laval, Québec, 415–420.

Hallet, B. (2008) The rich contribution of A.L. Washburn to permafrost and periglacial studies. In Kane, D.L. and Hinkel, K.M. (eds) *Ninth International Conference on Permafrost*. University of Alaska, Fairbanks, 625–630.

Hallet, B. (2013) Stone circles: form and soil kinematics. *Philosophical Transactions of the Royal Society* A371, 20120357.

Hallet, B. and Prestrud, S. (1986) Dynamics of periglacial sorted circles in western Spitsbergen. *Quaternary Research* 26, 81–99.

Hallet, B. and Waddington, E.D. (1992) Buoyancy forces induced by freeze-thaw in the active layer: implications for diapirism and soil circulation. In Dixon, J.C. and Abrahams, A.D. (eds) *Periglacial Geomorphology*. John Wiley & Sons, Chichester, 252–279.

Hallet, B., Walder, J.S. and Stubbs, C.W. (1991) Weathering by segregation ice growth in microcracks at sustained sub-zero temperatures: verification from an experimental study using acoustic emissions. *Permafrost and Periglacial Processes* 2, 283–300.

Hallet, B., Hunter, L. and Bogen, J. (1996) Rates of erosion and sediment evacuation by glaciers: a review of field data and their implications. *Global and Planetary Change* 12, 213–235.

Hambrey, M.J., Dowdeswell, J.A., Murray, T. and Porter, P.R. (1996) Thrusting and debris entrainment in a surging glacier: Bakaninbreen, Svalbard. *Annals of Glaciology* 22, 241–248.

Hamilton, J. and Ford, D.C. (2002) Karst geomorphology and hydrogeology of the Bear Rock Formation – a remarkable dolostone and gypsum megabreccia in the continuous permafrost zone of Northwest Territories, Canada. *Carbonates and Evaporites* 17, 54–56.

Hamilton, T.D., Ager, T.A. and Robinson, S.W. (1983) Late Holocene ice wedges near Fairbanks, Alaska, USA: environmental setting and history of growth. *Arctic and Alpine Research* 15, 157–168.

Hansom, J.D. (1983a) Ice-formed intertidal boulder pavements in the sub-Antarctic. *Journal of Sedimentary Petrology* 53, 135–145.

Hansom, J.D. (1983b) Shore-platform development in the South Shetland Islands, Antarctica. *Marine Geology* 53, 211–229.

Hansom, J.D. (1986) Intertidal forms produced by floating ice in Vestfirdir, Iceland. *Marine Geology* 71, 289–298.

Hansom, J.D., Evans, D.J.A., Sanderson, D.C.W. *et al.* (2008) Constraining the age and formation of stone runs on the Falkland Islands using optically-stimulated luminescence. *Geomorphology* 94, 117–130.

Hara, Y. and Thorn, C.E. (1982) Preliminary quantitative study of alpine subnival boulder pavements, Colorado Front Range, USA. *Arctic and Alpine Research* 14, 361–367.

Harada, K., Wada, K. and Fukuda, M. (2000) Permafrost mapping by transient electromagnetic method. *Permafrost and Periglacial Processes* 11, 71–84.

Harper, J.R., Henry, R.F. and Stewart, G.G. (1988) Maximum storm surge elevations in the Tuktoyaktuk region of the Canadian Beaufort Sea. *Arctic* 41, 48–52.

Harris, C. (1985) Geomorphological applications of soil micromorphology with particular reference to periglacial sediments and processes. In Richards, K.S., Arnett, R.R. and Ellis, S. (eds) *Geomorphology and Soils*. Allen and Unwin, London, 219–232.

Harris, C. (1986) Some observations concerning the morphology and sedimentology of a protalus rampart, Okstindan, Norway. *Earth Surface Processes and Landforms* 11, 673–676.

Harris, C. (1987) Solifluction and related periglacial deposits in England and Wales. In Boardman, J. (ed.) *Periglacial Processes and Landforms in Britain and Ireland*. Cambridge University Press, Cambridge, 209–223.

Harris, C. (1991) Glacial deposits at Wylfa Head, Anglesey, North Wales: evidence for Late Devensian deposition in a non-marine environment. *Journal of Quaternary Science* 6, 67–77.

Harris, C. (1996) Physical modelling of periglacial solifluction: review and future strategy. *Permafrost and Periglacial Processes* 7, 349–360.

Harris, C. (1998) The micromorphology of paraglacial and periglacial slope deposits: a case study from Morfa Bychan, west Wales, U.K. *Journal of Quaternary Science* 13, 73–84.

Harris, C. (2013) Slope deposits and forms. In Elias, S. (ed.) *Encyclopedia of Quaternary Science*, 2nd edn, vol. 3. Elsevier, Amsterdam, 481–489.

Harris, C. and Davies, M.C.R. (2000) Gelifluction: observations from large-scale laboratory simulations. *Arctic, Antarctic and Alpine Research* 32, 202–207.

Harris, C. and Ellis, S. (1980) Micromorphology of soils in soliflucted materials, Okstindan, northern Norway. *Geoderma* 23, 11–29.

Harris, C. and Lewkowicz, A.G. (1993) Form and internal structure of active-layer detachment slides, Fosheim Peninsula, Ellesmere Island, Northwest Territories, Canada. *Canadian Journal of Earth Sciences* 30, 1708–1714.

Harris, C. and Lewkowicz, A.G. (2000) An analysis of the stability of thawing slopes, Ellesmere Island, Nunavut, Canada. *Canadian Geotechnical Journal* 37, 449–462.

Harris, C. and Matthews, J.A. (1984) Some observations on boulder-cored frost boils. *Geographical Journal* 150, 63–73.

Harris, C. and Murton, J.B. (2005) Experimental simulation of ice-wedge casting: processes, products and palaeoenvironmental significance. *Geological Society, London, Special Publication* 242, 131–143.

Harris, C. and Smith, J.S. (2003) Modelling gelifluction processes: the significance of frost heave and slope gradient. In Phillips, M., Springman, S.M. and Arenson, L.U. (eds) *Proceedings of the Eighth International Conference on Permafrost*. Balkema, Lisse, 355–360.

Harris, C., Davies, M.C.R. and Coutard, J.-P. (1995) Laboratory simulation of periglacial solifluction: significance of porewater pressures, moisture contents and undrained shear strengths during soil thawing. *Permafrost and Periglacial Processes* 6, 293–311.

Harris, C., Davies, M.C.R. and Coutard, J.-P. (1997) Rates and processes of periglacial solifluction: an experimental approach. *Earth Surface Processes and Landforms* 22, 849–868.

Harris, C., Murton, J.B. and Davies, M.C.R. (2000a) Soft-sediment deformation during thawing of ice-rich frozen soils: results of scaled centrifuge modelling experiments. *Sedimentology* 47, 687–700.

Harris, C., Rea, B.R. and Davies, M.C.R. (2000b) Geotechnical centrifuge modelling of gelifluction processes: validation of a new approach to periglacial slope studies. *Annals of Glaciology* 31, 263–268.

Harris, C., Rea, B.R. and Davies, M.C.R. (2001) Scaled physical modelling of mass movement processes on thawing slopes. *Permafrost and Periglacial Processes* 12, 125–135.

Harris, C., Davies, M.C.R. and Rea, B.R. (2003) Periglacial solifluction: viscous flow or plastic creep? *Earth Surface Processes and Landforms* 28, 1289–1301.

Harris, C., Murton, J.B. and Davies, M.C.R. (2005) An analysis of mechanisms of ice-wedge casting based on geotechnical centrifuge simulations. *Geomorphology* 71, 328–343.

Harris, C., Luetschg, M., Davies, M.C.R. *et al.* (2007) Field instrumentation for real-time monitoring of periglacial solifluction. *Permafrost and Periglacial Processes* 18, 105–114.

Harris, C., Kern-Luetschg, M., Murton, J. *et al.* (2008a) Solifluction processes on permafrost and non-permafrost slopes: results of a large-scale laboratory simulation. *Permafrost and Periglacial Processes* 19, 359–378.

Harris, C., Kern-Luetschg, M., Smith, F. and Isaksen, K. (2008b) Solifluction processes in an area of seasonal ground freezing, Dovrefjell, Norway. *Permafrost and Periglacial Processes* 19, 31–47.

Harris, C., Smith, J.S., Davies, M.C.R. and Rea, B. (2008c) An investigation of periglacial slope stability in relation to soil properties based on physical modeling in the geotechnical centrifuge. *Geomorphology* 93, 437–459.

Harris, C., Arenson, L.U., Christiansen, H.H. *et al.* (2009) Permafrost and climate in Europe: monitoring and modelling thermal, geomorphological and geotechnical responses. *Earth Science Reviews* 92, 117–171.

Harris, C., Kern-Luetschg, M., Christiansen, H.H. and Smith, F. (2011) The role of interannual climate variability in controlling solifluction processes, Endalen, Svalbard. *Permafrost and Periglacial Processes* 22, 239–255.

Harris, S.A. (1993) Palsa-like mounds in a mineral substrate, Fox Lake, Yukon Territory. *Proceedings, Sixth International Conference on Permafrost*. South China University Press, Wushan Guangzhou, 238–243.

Harris, S.A. (2002) Causes and consequences of rapid thermokarst development in permafrost or glacial terrain. *Permafrost and Periglacial Processes* 13, 237–242.

Harris, S.A. and Gustafson, C.A. (1993) Debris flow characteristics in an area of continuous permafrost, St Elias Range, Yukon Teritory. *Zeitschrift für Geomorphologie* 37, 41–56.

Harris, S.A. and Pedersen, D.E. (1998) Thermal regimes beneath coarse blocky material. *Permafrost and Periglacial Geomorphology* 9, 107–120.

Harris, S.A. and Prick, A. (2000) Condition of formation of stratified screes, Slims River Valley, Yukon Territory: a possible analogue with some deposits from Belgium. *Earth Surface Processes and Landforms* 25, 463–481.

Harris, S.A., French, H.M., Heginbottom, J.A. *et al.* (1988) Glossary of permafrost and related ground-ice terms. *National Research Council of Canada, Technical Memorandum* 142, 156 pp.

Harris, S.A., Schmidt, I.H. and Krouse, H.R. (1992) Hydrogen and oxygen isotopes and the origin of ice in peat plateaux. *Permafrost and Periglacial Processes* 3, 19–27.

Harris, S.A., Cheng, G., Zhao, X. and Yongqin, D. (1998) Nature and dynamics of an active block stream, Kunlun Pass, Qinghai Province, People's Republic of China. *Geografiska Annaler* 80A, 123–133.

Harrison, S. (2002) Lithological variability of Quaternary slope deposits in the Cheviot Hills, U.K. *Proceedings of the Geologists' Association* 113, 121–138.

Harrison, S., Whalley, B. and Anderson, E. (2008) Relict rock glaciers and protalus lobes in the British Isles: implications for Late Pleistocene mountain geomorphology and palaeoclimate. *Journal of Quaternary Science* 23, 287–304.

Harrison, S., Bailey, R.M., Anderson, E. *et al.* (2010) Optical dates from British Isles 'solifluction sheets' suggests rapid landscape response to Late Pleistocene climate change. *Scottish Geographical Journal* 126, 101–111.

Harry, D.G. and Gozdzik, J.S. (1988) Ice wedges: growth, thaw transformation and palaeoenvironmental significance. *Journal of Quaternary Science* 3, 39–58.

Harry, D.G., French, H.M. and Pollard, W.H. (1988) Massive ground ice and ice-cored terrain near Sabine Point, Yukon coastal plain. *Canadian Journal of Earth Sciences* 25, 1846–1856.

Hartl, L., Fischer, A., Stocker-Waldhuber, M. and Abermann, J. (2016) Recent speed-up of an Alpine rock glacier: an updated chronology of the kinematics of outer Hochebenkar rock glacier based on geodetic measurements. *Geografiska Annaler* 98A, 129–141.

Hasler, A., Gruber, S., Font, M. and Dubois, A. (2011) Advective heat transport in frozen rock clefts: conceptual model, laboratory experiments and numerical simulation. *Permafrost and Periglacial Processes* 22, 378–389.

Hasler, A., Gruber, S. and Beutel, J. (2012) Kinematics of steep mountain permafrost. *Journal of Geophysical Research* 117, F01016.

Hauck, C. (2013) New concepts in geophysical surveying and data interpretation for permafrost terrain. *Permafrost and Periglacial Processes* 24, 131–137.

Hauck, C. and Kneisel, C. (eds) (2008) *Applied Geophysics in Periglacial Environments*. Cambridge University Press, Cambridge, 240 pp.

Hauck, C., Guglielmin, M., Isaksen, K. and Vonder Mühll, D. (2001) Applicability of frequency-domain and time-domain electromagnetic methods for mountain permafrost studies. *Permafrost and Periglacial Processes* 12, 39–52.

Hauck, C., Isaksen, D., Vonder Mühll, D. and Sollid, J.L. (2004) Geophysical surveys designed to delineate the altitudinal limit of mountain permafrost: an example from Jotunheimen, Norway. *Permafrost and Periglacial Processes* 15, 191–205.

Haugland, J.E. (2004) Formation of patterned ground and fine-scale soil development within two late Holocene glacial chronosequences: Jotunheimen, Norway. *Geomorphology* 61, 287–301.

Haugland, J.E. (2006) Short-term periglacial processes, vegetation succession and soil development within sorted patterned ground: Jotunheimen, Norway. *Arctic, Antarctic and Alpine Research* 38, 82–89.

Hausmann, H., Krainer, K., Brükl, E. and Mostler, W. (2007) Internal structure and ice content of Reichenkar rock glacier (Stubai Alps, Austria) assessed by geophysical investigations. *Permafrost and Periglacial Processes* 18, 351–367.

Hausmann, H., Krainer, K., Brückl, E. and Ullrich, C. (2012) Internal structure, ice content and dynamics of Ölgrube and Kaiserberg rock glaciers (Ötztal Alps, Austria) determined from geophysical surveys. *Austrian Journal of Earth Sciences* 102, 12–31.

Hedding, D.W. and Sumner, P.D. (2013) Diagnostic criteria for pronival ramparts: site, morphological and sedimentological characteristics. *Geografiska Annaler* 95A, 315–322.

Hedding, D.W., Meiklejohn, K.I., Le Roux, J.J. *et al.* (2010) Some observations on the formation of an active pronival rampart at Grunehogna Peaks, Western Dronning Maud Land, Antarctica. *Permafrost and Periglacial Processes* 21, 355–361.

Hegginbottom, J.A. (2002) Permafrost mapping: a review. *Progress in Physical Geography* 26, 623–642.

Henriksen, M., Mangerud, J., Matiouchkov, A. *et al.* (2003) Lake stratigraphy implies an 80 000 yr delayed melting of buried ice in northern Russia. *Journal of Quaternary Science* 18, 663–679.

Héquette, A. and Ruz, M.-H. (1990) Sédimentation littorale en bordure de plaines d'épandage fluvioglaciaire au Spitsberg nord-occidental. *Géographie Physique et Quaternaire* 44, 77–88.

Héquette, A. and Ruz, M.-H. (1991) Spit and barrier island migration in the southeastern Canadian Beaufort Sea. *Journal of Coastal Research* 7, 677–698.

Hestnes, E. (1998) Slushflow hazard – where, why and when? 25 years of experience with slushflow consulting and research. *Annals of Glaciology* 26, 370–376.

Hétu, B. (1991) Éboulis stratifiés actifs près de Manche-d'Épée, Gaspésie (Québec, Canada): rôle de la sédimentation nivéo-éolian et des transits supranivaux. *Zeitschrift für Geomorphologie* 35, 439–461.

Hétu, B. (1992) Coarse cliff-top aeolian sedimentation in northern Gaspésie, Québec (Canada). *Earth Surface Processes and Landforms* 17, 95–108.

Hétu, B. (1995) Le litage des éboulis stratifiés cryonivaux en Gaspésie (Québec, Canada): rôle de la sédimentation nivéo-éolienne et des transits supranivaux. *Permafrost and Periglacial Processes* 6, 147–171.

Hétu, B. and Gray, J.M. (2000) Effects of environmental change on scree slope development throughout the postglacial period in the Chic-Choc Mountains in the northern Gaspé Peninsula, Québec. *Geomorphology* 32, 335–355.

Hétu, B., Van Steijn, H. and Vandelac, P. (1994) Les coulées de pierres glacées: un nouveau type de coulées de pierraille sur les talus d'éboulis. *Géographie Physique et Quaternaire* 48, 3–22.

Heusser, L., Maenza-Gmelch, T., Lowell, T. and Hinnefeld, R. (2002) Late Wisconsin periglacial environments of the southern margin of the Laurentide Ice Sheet reconstructed from pollen analyses. *Journal of Quaternary Science* 17, 773–780.

Hill, P.R. and Solomon, S.M. (1999) Geomorphic and sedimentary evolution of a transgressive thermokarst coast, Mackenzie Delta region, Canadian Beaufort Sea. *Journal of Coastal Research* 15, 1011–1029.

Hinchliffe, S. and Ballantyne, C.K. (1999) Talus accumulation and rockwall retreat, Trotternish, Isle of Skye, Scotland. *Scottish Geographical Journal* 115, 53–70.

Hinchliffe, S. and Ballantyne, C.K. (2009) Talus structure and evolution on sandstone mountains in NW Scotland. *The Holocene* 19, 139–144.

Hinchliffe, S., Ballantyne, C.K. and Walden, J. (1998) The structure and sedimentology of relict talus, Trotternish, northern Skye, Scotland. *Earth Surface Processes and Landforms* 23, 545–560.

Hinkel, K.M. (1988) Frost mounds formed by degradation at Slope Mountain, Alaska, USA. *Arctic and Alpine Research* 20, 76–85.

Hinkel, K.M., Doolittle, J.A., Bockheim, J.G. *et al.* (2001a) Detection of subsurface permafrost features with ground-penetrating radar, Barrow, Alaska. *Permafrost and Periglacial Processes* 12, 179–190.

Hinkel, K.M., Paetzold, F., Nelson, F.E. and Bockheim, J.G. (2001b) Patterns of soil temperature and moisture in the active layer and upper permafrost at Barrow, Alaska: 1993–1999. *Global and Planetary Change* 29, 293–309.

Hinkel, K.M., Eisner, W.R., Bockheim, J.G. *et al.* (2003) Spatial extent, age and carbon stocks in drained thaw lake basins on the Barrow Peninsula, Alaska. *Arctic, Antarctic and Alpine Research* 35, 291–300.

Hinkel, K.M., Frohn, R.C., Nelson, F.E. *et al.* (2005) Morphometric and spatial analysis of thaw lakes and drained thaw lake basins in the western Arctic coastal plain, Alaska. *Permafrost and Periglacial Processes* 16, 327–341.

Hinkel, K.M., Jones, B.M., Eisner, W.R. *et al.* (2007) Methods to assess natural and anthropogenic thaw lake drainage on the western Arctic coastal plain of northern Alaska. *Journal of Geophysical Research* 112, F02S16.

Hinkel, K.M., Sheng, Y., Lenters, J.D. *et al.* (2012) Thermokarst lakes on the Arctic coastal plain of Alaska: geomorphic controls on bathymetry. *Permafrost and Periglacial Processes* 23, 218–230.

Hipp, T., Etzelmüller, B. and Westermann, S. (2014) Permafrost in alpine rock faces from Jotunheimen and Hurringane, southern Norway. *Permafrost and Periglacial Processes* 25, 1–13.

Hjort, J., Luoto, M. and Seppälä, M. (2007) Landscape scale determinants of periglacial features in subarctic Finland: a grid-based modelling approach. *Permafrost and Periglacial Processes* 18, 115–127.

Hoare, P.G., Stevenson, C.R. and Godby, S.P. (2002) Sand sheets and ventifacts: the legacy of aeolian action in west Norfolk. *Proceedings of the Geologists' Association* 113, 301–317.

Hoch, A.R., Reddy, M.M. and Drever, J.I. (1999) Importance of mechanical disaggregation in chemical weathering in a cold alpine environment, San Juan Mountains, Colorado. *Geological Society of America, Bulletin* 111, 304–314.

Hodgkins, R., Cooper, R., Wadham, J. and Tranter, M. (2003) Suspended sediment fluxes in a high-Arctic glacierised catchment: implications for fluvial sediment storage. *Sedimentary Geology* 162, 105–117.

Hodgson, D.A. and Nixon, F.M. (1998) Ground ice volumes determined from shallow cores from western Fosheim Peninsula, Ellesmere Island, Northwest Territories. *Geological Survey of Canada, Bulletin* 507, 178 pp.

Hoezle, M. (1992) Permafrost occurrence from BTS measurements and climatic parameters in the Eastern Swiss Alps. *Permafrost and Periglacial Processes* 3, 143–147.

Hoffmann, T. and Schrott, L. (2002) Modelling sediment thickness and rockwall retreat in an Alpine valley using 2D-seismic refraction (Reintal, Bavarian Alps). *Zeitschrift für Geomorphologie Supplementband* 127, 175–196.

Höfner, T. (1995) Fluvial dynamics in the periglacial belt of the Austrian Alps. *Zeitschrift für Geomorphologie Supplementband* 100, 159–166.

Högbom, B. (1914) Über die geologische Bedeutung des Frostes. *Uppsala University, Geological Institute Bulletin* 12, 251–389.

Holloway, J.E., Lamoureux, S.F., Montross, S.N. and Lafrenière, M.J. (2016) Climate and terrain characteristics linked to mud ejection occurrence in the Canadian High Arctic. *Permafrost and Periglacial Processes* 27, 204–218.

Holmes, R.M., Shiklomanov, A.I., Tank, S.E. *et al.* (2015) River discharge. In *Arctic Report Card: Update for 2015*. Available from: http://www.arctic.noaa.gov/report-card (last accessed 28/06/2017).

Holness, S.D. (2004) Sediment movement rates and processes on cinder cones in the maritime Subantarctic (Marion Island). *Earth Surface Processes and Landforms* 29, 91–103.

Holtedahl, H. (1998) The Norwegian strandflat – a geomorphological puzzle. *Norsk Geologisk Tidsskrift* 78, 47–66.

Hopkins, D.M. (1949) Thaw lakes and thaw sinks in the Imuruk Lake area, Seward Peninsula, Alaska. *Journal of Geology* 57, 119–130.

Hopkinson, C. and Ballantyne, C.K. (2014) Age and origin of blockfields on Scottish Mountains. *Scottish Geographical Journal* 130, 116–141.

Hoque, M.A. and Pollard, W.H. (2009) Arctic coastal retreat through block failure. *Canadian Geotechnical Journal* 46, 1103–1115.

Hu, X. and Pollard, W.H. (1997) The hydrologic significance and modelling of river icing growth, North Fork Pass, Yukon Territory, Canada. *Permafrost and Periglacial Processes* 8, 279–294.

Huang S.L., Aughenbaugh, N.B. and Wu, M.-C. (1986) Stability study of the CRREL permafrost tunnel. *Journal of Geotechnical Engineering* 112, 777–790.

Hubberten, H.W., Andreev, A., Astakhov, V.I. *et al.* (2004) The periglacial climate and environment in northeastern Eurasia during the last glaciation. *Quaternary Science Reviews* 23, 1333–1357.

Hugelius, G., Tarnocai, C., Broll, G. *et al.* (2013) The northern circumpolar soil carbon database: spatially distributed datasets of soil coverage and soil carbon storage in the northern permafrost regions. *Earth Systems Science Data* 5, 3–13.

Hugenholtz, C.H. and Lewkowicz, A.G. (2002) Morphometry and environmental characteristics of turf-banked solifluction lobes. Kluane Range, Yukon Territory, Canada. *Permafrost and Periglacial Processes* 13, 301–313.

Huggel, C., Salzmann, N., Allen, S. *et al.* (2010) Recent and future warm extreme events and high-mountain slope stability. *Philosophical Transactions of the Royal Society* A368, 2435–2459.

Huggel, C., Clague, J.J. and Korup, O. (2012) Is climate change responsible for changing landslide activity in high mountains? *Earth Surface Processes and Landforms* 37, 77–91.

Hughes, P.D., Gibbard, P.L. and Woodward, J.C. (2003) Relict rock glaciers as indicators of Mediterranean palaeoclimate during the last glacial maximum (Late Würmian) in northeast Greece. *Journal of Quaternary Science* 18, 431–440.

Hughes, P.D., Gibbard, P.L. and Ehlers, J. (2013) Timing of glaciations during the last glacial cycle: evaluating the concept of a global 'Last Glacial Maximum' (LGM). *Earth-Science Reviews* 125, 171–198.

Hughes, P.D., Glasser, N.F. and Fink, D. (2016) Rapid thinning of the Welsh Ice Cap at 20–19 ka based on ^{10}Be ages. *Quaternary Research* 85, 107–117.

Huijzer, B. and Vandenberghe, J. (1998) Climatic reconstruction of the Weichselian Pleniglacial in northwestern and central Europe. *Journal of Quaternary Science* 13, 391–417.

Huisink, M. (2000) Changing river styles in response to Weichselian climate changes in the Vecht valley, eastern Netherlands. *Sedimentary Geology* 133, 115–134.

Huisink, M., De Moor, J.J.W., Kasse, C. and Virtanen, T. (2002) Factors influencing periglacial fluvial morphology in the northern European Russian tundra and taiga. *Earth Surface Processes and Landforms* 27, 1223–1235.

Humlum, O. (1996) Origin of rock glaciers: observations from Mellemfjord, Disko Island, central west Greenland. *Permafrost and Periglacial Processes* 7, 361–380.

Humlum, O. (1997) Active layer thermal regime at three rock glaciers in Greenland. *Permafrost and Periglacial Processes* 8, 383–408.

Humlum, O. (1998a) The climatic significance of rock glaciers. *Permafrost and Periglacial Processes* 9, 375–395.

Humlum, O. (1998b) Rock glaciers on the Faroe Islands, the North Atlantic. *Journal of Quaternary Science* 13, 293–307.

Humlum, O. (2000) The geomorphic significance of rock glaciers: estimates of rock glacier debris volumes and headwall recession rates in West Greenland. *Geomorphology* 35, 41–67.

Humlum, O., Instanes, A. and Sollid, J.L. (2003) Permafrost in Svalbard: a review of research history, climatic background and engineering challenges. *Polar Research* 22, 191–215.

Humlum, O., Christiansen, H.H. and Juliussen, H. (2007) Avalanche-derived rock glaciers in Svalbard. *Permafrost and Periglacial Processes* 18, 75–88.

Hürlimann, M., McArdell, B.W. and Rickli, C. (2015) Field and laboratory analyses of the runout characteristics of hillslope debris flows in Switzerland. *Geomorphology* 232, 20–32.

Hutchinson, J.N. (1991) Periglacial and slope processes. *Geological Society, London, Engineering Geology Special Publication* 7, 283–331.

Hutchinson, J.N. (2010) Relict sand wedges in soliflucted London Clay at Wimbledon, London, UK. *Proceedings of the Geologists' Association* 121, 444–454.

Hutchinson, J.N. and Coope, G.R. (2002) Cambering and valley bulging, periglacial solifluction and Lateglacial Coleoptera at Dowdeswell, near Cheltenham. *Proceedings of the Geologists' Association* 113, 291–300.

Iannicelli, M. (2003) Reinterpretation of the original DeKalb mounds in Illinois. *Physical Geography* 24, 170–182.

IJmker, J., Stauch, G., Pötsch, S. *et al.* (2012) Dry periods on the NE Tibetan Plateau during the late Quaternary. *Palaeogeography, Palaeoclimatology, Palaeoecology* 346–347, 108–119.

Ikeda, A. (2006) Combination of conventional geophysical methods for sounding the composition of rock glaciers in the Swiss Alps. *Permafrost and Periglacial Processes* 17, 35–48.

Ikeda, A. and Matsuoka, N. (2006) Pebbly versus bouldery rock glaciers: morphology, structure and processes. *Geomorphology* 73, 279–296.

Ikeda, A., Matsuoka, N. and Kääb, A. (2008) Fast deformation of perennially-frozen debris in a warm rock-glacier in the Swiss Alps: an effect of liquid water. *Journal of Geophysical Research* 113, F01021.

Imhof, M., Pierrehumbert, G., Haeberli, W. and Kienholz, H. (2000) Permafrost investigation in the Schilthorn Massif, Bernese Alps, Switzerland. *Permafrost and Periglacial Processes* 11, 189–206.

Innes, J.L. (1983) Lichenometric dating of debris flow deposits in the Scottish Highlands. *Earth Surface Processes and Landforms* 8, 579–588.

IPCC (2013) *Climate Change 2013: The Physical Science Basis*. Contribution of Working Group 1 to the Fifth Assessment Report of the Intergovernmental Panel on Climate Change, Stocker, T.F., Qin, D., Plattner, G.-K. *et al.* (eds) Cambridge University Press, Cambridge and New York, 1535 pp.

Isaksen, K., Ødegård, R.S., Eiken, T. and Sollid, J.L. (2000) Composition, flow and development of two tongue-shaped rock glaciers in the permafrost of Svalbard. *Permafrost and Periglacial Processes* 11, 241–257.

Isaksen, K., Hauck, C., Gudevang, E. *et al.* (2002) Mountain permafrost distribution in Dovrefjell and Jotunheimen, southern Norway, based on BTS and DC resistivity tomography data. *Norwegian Journal of Geography* 56, 122–136.

Isaksen, K., Ødegård, R.S., Etzelmüller, B. *et al.* (2011) Degrading mountain permafrost in southern Norway: spatial and temporal variability of mean ground temperatures, 1999–2009. *Permafrost and Periglacial Processes* 22, 361–377.

Isarin, R.F.B. (1997a) *The climate in north-western Europe during the Younger Dryas: a comparison of multi-proxy climate reconstructions with simulation experiments.* Drukkerij Elinkwijk, Utrecht, 160 pp.

Isarin, R.F.B. (1997b) Permafrost distribution and temperature in Europe during the Younger Dryas. *Permafrost and Periglacial Processes* 8, 313–333.

Isarin, R.F.B. and Bohncke, S.P.J. (1999) Mean July temperatures during the Younger Dryas in Northwestern and Central Europe as inferred from climate indicator plant species. *Quaternary Research* 51, 58–173.

Isarin, R.F.B., Renssen, H. and Koster, E.A. (1997) Surface wind climate during the Younger Dryas in Europe as inferred from aeolian records and model simulations. *Palaeogeography, Palaeclimatology, Palaeoecology* 134, 127–148.

Isarin, R.F.B., Renssen, H. and Vandenberghe, J. (1998) The impact of the North Atlantic Ocean on the Younger Dryas climate in northwestern Europe. *Journal of Quaternary Science* 13, 447–453.

Ishikawa, M. (2003) Thermal regimes at the snow-ground interface and their implications for permafrost investigation. *Geomorphology* 52, 105–120.

Ishikawa, M. and Hirakawa, K. (2000) Mountain permafrost distribution based on BTS measurements and DC resistivity soundings in the Daisetsu Mountains, Hokkaido, Japan. *Permafrost and Periglacial Processes* 11, 109–123.

Ishikawa, M., Sharkhuu, N., Zhang, Y. *et al.* (2005) Ground thermal and moisture conditions at the southern boundary of discontinuous permafrost, Mongolia. *Permafrost and Periglacial Processes* 16, 209–216.

Iverson, R.M. (1997) The physics of debris flows. *Reviews of Geophysics* 35, 245–296.

Ivester, A.H. and Leigh, D.S. (2003) Riverine dunes on the coastal plain of Georgia, USA. *Geomorphology* 51, 289–311.

Iwahana, G., Fukui, K., Mikhailov, N. *et al.* (2012) Internal structure of a lithalsa in the Akkol Valley, Russian Altai Mountains. *Permafrost and Periglacial Processes* 23, 107–118.

Izmailow, B. (1984) Eolian processes in alpine belts of the High Tatra Mountains, Poland. *Earth Surface Processes and Landforms* 9, 143–151.

Jacobs, P.M., Mason, J.A. and Hanson, P.R. (2011) Mississippi Valley regional source of loess on the southern Green Bay Lobe land surface, Wisconsin. *Quaternary Research* 75, 574–583.

Jacobsen, N.K. (1987) Studies on soils and potential for soil erosion in the sheep farming area of south Greenland. *Arctic and Alpine Research* 19, 498–507.

Jaesche, P., Huwe, B., Stingl, H. and Veit, H. (2002) Temporal variability of alpine solifluction: a modelling approach. *Geographica Helvetica* 57, 157–169.

Jaesche, P., Veit, H. and Huwe, B. (2003) Snowcover and soil moisture controls on solifluction in an area of seasonal frost, eastern Alps. *Permafrost and Periglacial Processes* 14, 399–410.

Jafarov, E.E., Marchenko, S.S. and Romanovsky, V.E. (2012) Numerical modelling of permafrost dynamics in Alaska using a high spatial resolution dataset. *The Cryosphere* 6, 613–624.

Jahn, A. (1961) Quantitative analysis of some periglacial processes in Spitsbergen. *Nauka O Ziem II, Seria B* 5, 3–34.

Jahn, A. (1975) *Problems of the Periglacial Zone.* Polish Scientific Publishers, Warsaw, 221 pp.

Jakob, M. (1992) Active rock glaciers and the lower limit of discontinuous alpine permafrost, Khumbu Himalaya, Nepal. *Permafrost and Periglacial Processes* 3, 253–256.

Janke, J.R. (2005) The occurrence of alpine permafrost in the Front Range of Colorado. *Geomorphology* 67, 375–389.

Janke, J.R. (2007) Colorado Front Range rock glacers: distribution and topographic characteristics. *Arctic, Antarctic and Alpne Research* 39, 74–83.

Janke, J.R., Williams, M.W. and Evans, A. (2012) A comparison of permafrost prediction models along a section of Trail Ridge Road, Rocky Mountain National Park, Colorado, USA. *Geomorphology* 138, 111–120.

Jarman D., Wilson, P. and Harrison, S. (2013) Are there any rock glaciers in the British mountains? *Journal of Quaternary Science* 28, 131–143.

Jaworski, T. and Niewiarowski, W. (2012) Frost peat mounds on Hermansenøya (Oscar II Land, NW Svalbard) – their genesis, age and terminology. *Boreas* 41, 660–672.

Jean, M. and Payette, S. (2014a) Dynamics of active layer in wooded palsas of northern Quebec. *Geomorphology* 206, 87–96.

Jean, M. and Payette, S. (2014b) Effect of vegetation cover on the ground thermal regime of wooded and non-wooded palsas. *Permafrost and Periglacial Processes* 25, 281–294.

Jennings, J.N. and Costin, A.B. (1978) Stone movement through snow creep, 1963–1975, Mount Twynam, Snowy Mountains, Australia. *Earth Surface Processes* 3, 3–22.

Jensen, B.J.L., Evans, M.E., Froese, D.G. and Kravchinsky, V.A. (2016) 150 000 years of loess accumulation in central Alaska. *Quaternary Science Reviews* 135, 1–23.

Jepsen, S.M., Voss, C.I., Walvoord, M.A. *et al.* (2013) Linkages between lake shrinkage/expansion and sublacustrine permafrost distribution determined from remote sensing of interior Alaska, USA. *Geophysical Research Letters* 40, 882–887.

Jerwood, L.C., Robinson, D.A. and Williams, R.B.G. (1990) Experimental frost and salt weathering of chalk – 1. *Earth Surface Processes and Landforms* 15, 611–624.

Jetchick, E. and Allard, M. (1990) Soil wedge polygons in northern Québec: description and palaeoclimatic significance. *Boreas* 19, 353–367.

Jia, H., Xiang, W. and Krautblatter, M. (2015) Quantifying rock fatigue and decreasing compressive and tensile strength after repeated freeze-thaw cycles. *Permafrost and Periglacial Processes* 26, 368–377.

Jia, Y., Huang, C. and Mao, L. (2011) OSL dating of a Holocene loess-palaeosol sequence in the Southern Loess Plateau, China. *Environmental Earth Science* 64, 1071–1079.

Johansson, T., Malmer, N., Crill, P.M. *et al.* (2006) Decadal vegetation changes in a northern peatland, greenhouse gases and net radiative forcing. *Global Change Biology* 12, 2352–2369.

Johnson, A.L. and Smith, D.J. (2010) Geomorphology of snow avalanche impact landforms in the southern Canadian Cordillera. *Canadian Geographer* 54, 87–103.

Johnston, G.H., Ladanyi, B., Morgenstern, N.R. and Penner, E. (1981) Engineering characteristics of frozen and thawing soils. In Johnston, G.H. (ed.) *Permafrost: Design and Engineering*. John Wiley & Sons, Toronto, ON, 73–147.

Johnson, W.H. (1990) Ice-wedge casts and relict patterned ground in central Illinois and their environmental significance. *Quaternary Research* 33, 51–72.

Jomelli, V. and Bertran, P. (2001) Wet snow avalanche deposits in the French Alps: structure and sedimentology. *Geografiska Annaler* 83A, 15–28.

Jomelli, V. and Francou, B. (2000) Comparing the characteristics of rockfall talus and snow avalanche landforms in an Alpine environment using a new methodological approach: Massif des Ecrins, French Alps. *Geomorphology* 35, 181–192.

Jomelli, V., Brunstein, D., Grancher, D. and Pech, P. (2007) Is the response of hill slope debris flows to recent climate change univocal? A case study in the Massif des Ecrins (French Alps). *Climatic Change* 85, 119–137.

Jomelli, V., Brunstein, D., Déqué, M. *et al.* (2009) Impacts of future climate change (2070–2099) on the potential occurrence of debris flows: a case study in the Massif des Ecrins (French Alps). *Climatic Change* 97, 171–191.

Jomelli, V., Pavlova, I., Eckert, N. *et al.* (2015) A new hierarchical Bayesian approach to analyse environmental and climatic influences on debris flow occurrence. *Geomorphology* 250, 407–421.

Jones, B.M. and Arp, C.D. (2015) Observing a catastrophic thermokarst lake drainage in northern Alaska. *Permafrost and Periglacial Processes* 26, 119–128.

Jones, B.M., Arp, C.D., Jorgenson, M.T. *et al.* (2009) Increase in the rate and uniformity of coastline erosion in Arctic Alaska. *Geophysical Research Letters* 36, L03503.

Jones, B.M., Grosse, G., Arp, C.D. *et al.* (2011) Modern thermokarst lake dynamics in the continuous permafrost zone, northern Seward Peninsula, Alaska. *Journal of Geophysical Research – Biogeosciences* 116, G00M03.

Jones, B.M., Grosse, G., Hinkel, K.M. *et al.* (2012a) Assessment of pingo distribution and morphometry using an IfSAR derived digital surface model, western Arctic Coastal Plain, Northern Alaska. *Geomorphology* 138, 1–14.

Jones, M.C., Grosse, G., Jones, B.M., Walter Anthony, K.M. (2012b) Peat accumulation in drained thermokarst lake basins in continuous ice-rich permafrost, northern Seward Peninsula, Alaska. *Journal of Geophysical Research* 117, G00M07.

Jorgensen, M.T. (2013) Thermokarst terrains. In Shroder, J., Giardino, J. and Harbor, J. (eds) *Treatise on Geomorphology. Volume 8: Glacial and Periglacial Geomorphology.* Academic Press, San Diego, CA, 313–324.

Jorgenson, M.T. and Brown, J. (2005) Classification of the Alaska Beaufort Sea coast and estimation of carbon and sediment inputs from coastal erosion. *Geo-Marine Letters* 25, 69–80.

Jorgenson, M.T. and Osterkamp, T.E. (2005) Response of boreal ecosystems to varying modes of permafrost degradation. *Canadian Journal of Forest Research* 35, 2100–2111.

Jorgensen, M.T. and Shur, Y. (2007) Evolution of lakes and basins in northern Alaska and discussion of the thaw lake cycle. *Journal of Geophysical Research* 112, F02S17.

Jorgenson, M.T., Racine, C.H., Walters, J.C. and Osterkamp, T.E. (2001) Permafrost degradation and ecological changes associated with a warming climate in central Alaska. *Climatic Change* 48, 551–579.

Jorgenson, M.T., Shur, Y. and Pullman, E.R. (2006) Abrupt increase in permafrost degradation in Arctic Alaska. *Geophysical Research Letters* 33, L02503.

Jorgenson, M.T., Romanovsky, V., Harden, J. *et al.* (2010) Resilience and vulnerability of permafrost to climate change. *Canadian Journal of Forest Research* 40, 1219–1236.

Julián, A. and Chueca, J. (2007) Permafrost distribution from BTS measurements (Sierra de Telera, Central Pyrenees, Spain): assessing the importance of solar radiation in a mid-elevation shaded mountainous area. *Permafrost and Periglacial Processes* 18, 137–149.

Juliussen, H. and Humlum, O. (2007) Towards a TTOP ground temperature model for mountainous terrain in central-eastern Norway. *Permafrost and Periglacial Processes* 18, 161–184.

Juliussen, H. and Humlum, O. (2008) Thermal regime of openwork blockfields on the mountains Elgåhogna and Sølen, central-western Norway. *Permafrost and Periglacial Processes* 19, 1–18.

Kääb, A. (2013) Rock glaciers and protalus forms. In Elias, S. (ed.) *Encyclopedia of Quaternary Science*, 2nd edn. Elsevier, Amsterdam, 535–541.

Kääb, A. and Haeberli, W. (2001) Evolution of a high-mountain thermokarst lake in the Swiss Alps. *Arctic, Antarctic and Alpine Research* 33, 385–390.

Kääb, A. and Kneisel, C. (2006) Permafrost creep within a recently deglaciated glacier foreland: Muragl, Swiss Alps. *Permafrost and Periglacial Processes* 17, 79–85.

Kääb, A. and Reichmuth, T. (2005) Advance mechanisms of rockglaciers. *Permafrost and Periglacial Processes* 16, 187–193.

Kääb, A. and Weber, M. (2004) Development of transverse ridges on rock glaciers. *Permafrost and Periglacial Processes* 15, 379–391.

Kääb, A., Isaksen, K., Eiken, T. and Farbrot, H. (2002) Geometry and dynamics of two lobe-shaped rock glaciers in the permafrost of Svalbard. *Norwegian Journal of Geography* 56, 152–160.

Kääb, A., Frauenfelder, R. and Roer, I. (2007) On the response of rock glacier creep to surface temperature increase. *Global and Planetary Change* 56, 172–187.

Kääb, A., Girod, L. and Berthling, I. (2014) Surface kinematics of periglacial sorted circles using structure-from-motion technology. *The Cryosphere* 8, 1041–1056.

Kachurin, S.P. (1964) Cryogenic physico-geological phenomena in permafrost regions. *Canadian National Research Council, Technical Translation* 1157, 91 pp.

Kade, A. and Walker, D.A. (2008) Experimental alteration of vegetation on nonsorted circles: effects of cryogenic activity and implications for climate change in the arctic. *Arctic, Antarctic and Alpine Research* 40, 96–103.

Kade, A., Romanovsky, V.E. and Walker, D.A. (2006) The n-factor of nonsorted circles along a climate gradient in arctic Alaska. *Permafrost and Periglacial Processes* 17, 279–289.

Kaiser, C., Meyer, H., Biasi, C. *et al.* (2007) Conservation of soil organic matter through cryoturbation in arctic soils in Siberia. *Journal of Geophysical Research* 112, G02017.

Kaiser, K., Hilgers, A., Schlaak, N. *et al.* (2009a) Palaeopedological marker horizons in northern central Europe: characteristics of Lateglacial Usselo and Finow soils. *Boreas* 38, 591–609.

Kaiser, K., Lai, Z., Schneider, B. *et al.* (2009b) Stratigraphy and palaeoenvironmental implications of Pleistocene and Holocene aeolian sediments in the Lhasa area (Tibet). *Palaeogeography, Palaeoclimatology, Palaeoecology* 271, 329–342.

Kane, D.L., Hinzman, L.D., Benson, C.S. and Liston, G.E. (1991) Snow hydrology of a headwater arctic basin. 1: physical measurements and process studies. *Water Resources Research* 27, 1099–1109.

Kane, D.L., Soden, D.J., Hinzman, L.D. and Gieck, R.E. (1998) Rainfall runoff of a nested watershed in the Alaskan Arctic. In: Lewkowicz, A.G. and Allard, M. (eds) *Permafrost: Proceedings of the 7th International Conference*. Université Laval, Québec, 539–543.

Kane, D.L., Hinkel, K.M., Goering, D.J. *et al.* (2001) Non-conductive heat transfer associated with frozen soils. *Global and Planetary Change* 29, 275–292.

Kanevskiy, M., Fortier, D., Shur, Y. *et al.* (2008) Detailed cryostratigraphic studies of syngenetic permafrost in the winze of the CRREL permafrost tunnel, Fox, Alaska. In Kane, D.L. and Hinkel, K.M. (eds) *Proceedings of the Ninth International Conference on Permafrost*. University of Alaska, Fairbanks, 889–894.

Kanevskiy, M., Shur, Y., Fortier, D. *et al.* (2011) Cryostratigraphy of Late Pleistocene syngenetic permafrost (yedoma) in northern Alaska, Itkillik River exposure. *Quaternary Research* 75, 584–596.

Kanevskiy, M., Shur, Y., Krzewinski, T. and Dillon, M. (2013) Structure and properties of ice-rich permafrost near Anchorage, Alaska. *Cold Regions Science and Technology* 93, 1–11.

Kanevskiy, M., Jorgenson, M.T., Shur, Y. *et al.* (2014) Cryostratigraphy and permafrost evolution in the lacustrine lowlands of west-central Alaska. *Permafrost and Periglacial Processes* 25, 14–34.

Kanevskiy, M., Shur, Y., Strauss, J. *et al.* (2016) Patterns and rates of riverbank erosion involving ice-rich permafrost (yedoma) in northern Alaska. *Geomorphology* 253, 370–384.

Kaplar, C.W. (1965) Stone migration by freezing of soil. *Science* 149, 1520–1521.

Kariya, Y. (2002) Geomorphic processes at a snowpatch hollow on Gassan Volcano, northern Japan. *Permafrost and Periglacial Processes* 13, 107–116.

Kariya, Y. (2005) Holocene landscape evolution of a nivation hollow on Gassan volcano, northern Japan. *Catena* 62, 57–76.

Kasper, J.N. and Allard, M. (2001) Late Holocene climatic changes as detected by the growth and decay of ice wedges on the southern shore of Hudson Strait, Northern Québec, Canada. *The Holocene* 11, 563–577.

Kasse, C. (1993) Periglacial environments and climatic development during the early Pleistocene Tiglian stage (Beerse Glacial) in northern Belgium. *Geologie en Mijnbouw* 72, 107–123.

Kasse, C. (1997) Cold-climate aeolian sand-sheet formation in North-Western Europe (*c.* 14–12.4 ka); a response to permafrost degradation and increased aridity. *Permafrost and Periglacial Processes* 8, 295–311.

Kasse, C. (2002) Sandy aeolian deposits and environments and their relation to climate during the Last Glacial Maximum and Lateglacial in northwest and central Europe. *Progress in Physical Geography* 26, 507–532.

Kasse, C. and Vandenberghe, J. (1998) Topographic and drainage control on Weichselian ice-wedge and sand-wedge formation, Vennebrügge, German-Dutch border. *Permafrost and Periglacial Processes* 9, 95–106.

Kasse, C., Huijzer, A.S., Krzyszkowski, D. *et al.* (1998) Weichselian Late Pleniglacial and Lateglacial depositional environments, Coleoptera and periglacial climatic records from central Poland (Belchatów). *Journal of Quaternary Science* 13, 455–469.

Kasse, C., Vandenberghe, J., Van Huissteden, J. *et al.* (2003) Sensitivity of Weichselian fluvial systems to climate change (Nochten mine, eastern Germany). *Quaternary Science Reviews* 33, 2141–2156.

Kasse, C., Vandenberghe, D., De Corte, F. and Van Den Haute, P. (2007) Late Weichselian fluvio-aeolian sands and coversands of the type locality Grubbenvorst (southern Netherlands): sedimentary environments, climate record and age. *Journal of Quaternary Science* 22, 695–708.

Katamura, F., Fukuda, M., Bosikov, N.P *et al.* (2006) Thermokarst formation and vegetation dynamics inferred from a palynological study in central Yakutia, eastern Siberia, Russia. *Arctic, Antarctic and Alpine Research* 38, 561–570.

Katamura, F., Fukuda, M., Bosikov, N.P., Desyatkin. R.V. (2009) Charcoal records from thermokarst deposits in central Yakutia, eastern Siberia: implications for forest fire history and thermokarst development. *Quaternary Research* 71, 36–40.

Katasonov, E.M. (1973) Present-day ground ice and ice veins in the region of the Middle Lena. *Biuletyn Peryglacjalny* 23, 81–89.

Kattelmann, R. (1991) Peak flows from snowmelt runoff in the Sierra Nevada, USA. *International Association of Hydrological Sciences Publication* 205, 203–211.

Kaufman, D.S., Ager, T.A., Anderson, N.J. *et al.* (2004) Holocene thermal maximum in the western Arctic (0–180°W). *Quaternary Science Reviews* 23, 529–560.

Kaufman, D.S., Schneider, D.P., McKay, N.P. *et al.* (2009) Recent warming reverses long-term arctic cooling. *Science* 325, 1236–1239.

Kawashima, K., Yamada, T. and Wakahama, G. (1993) Investigations of internal structure and transformational processes from firn to ice in a perennial snow patch. *Annals of Glaciology* 18, 117–122.

Keller, K., Blum, J.D. and Kling, G.W. (2007) Geochemistry of soil and streams on surfaces of varying ages in arctic Alaska. *Arctic, Antarctic and Alpine Research* 39, 84–98.

Kemp, R.A. (2001) Pedogenic modification of loess: significance for palaeoclimatic reconstructions. *Earth-Science Reviews* 54, 145–156.

Kennedy, B.A. and Melton, M.A. (1972) Valley asymmetry and slope forms of a permafrost area in the Northwest Territories, Canada. *Institute of British Geographers Special Publication* 4, 107–121.

Kenner, R., Phillips, M., Danioth, C. *et al.* (2011) Investigation of rock and ice loss in a recently deglaciated mountain rock wall using terrestrial laser scanning. *Cold Regions Science and Technology* 67, 157–164.

Kerguillec, R. (2014) Recent patterned grounds development on a glacier surface (Dovrefjell, central Norway): an ephemeral periglacial activity in a paraglacial context. *Geografiska Annaler* 96A, 1–7.

Kern-Luetschg, M. and Harris, C. (2008) Centrifuge modelling of solifluction processes: displacement profiles associated with one-sided and two-sided active layer freezing. *Permafrost and Periglacial Processes* 19, 379–392.

Kern-Luetschg, M., Harris, C., Cleall, P. *et al.* (2008) Scaled centrifuge modeling in permafrost and seasonally-frozen soils. In Kane, D.L. and Hinkel, K.M. (eds) *Proceedings, Ninth International Conference on Permafrost.* University of Alaska, Fairbanks, 919–924.

Kessler, M.A. and Werner, B.T. (2003) Self organization of sorted patterned ground. *Science* 299, 380–383.

Kessler, M.A., Murray, A.B., Werner, B.T. and Hallet B. (2001) A model for sorted circles as self-organized patterns. *Journal of Geophysical Research* 106 (B7), 13 287–13 306.

Keys, J.R. and Williams, K. (1981) Origins of crystalline, cold desert salts in the McMurdo Region, Antarctica. *Geochimica et Cosmochimica Acta* 45, 2299–2309.

Killingbeck, J. and Ballantyne, C.K. (2012) Earth hummocks in west Dartmoor, SW England: characteristics, age and origin. *Permafrost and Periglacial Processes* 23, 153–161.

Kimble, J.M. (2004) (ed.) *Cryosols: Permafrost-affected Soils.* Springer-Verlag, Berlin, 726 pp.

Kinnard, C. and Lewkowicz, A.G. (2005) Movement, moisture and thermal conditions at a turf-banked solifluction lobe, Kluane Range, Yukon Territory. *Permafrost and Periglacial Processes* 16, 261–275.

Kinnard, C. and Lewkowicz, A.G. (2006) Frontal advance of turf-banked solifluction lobes, Kluane Range, Yukon Territory, Canada. *Geomorphology* 73, 261–276.

Kirkby, M.J. (1984) Modelling cliff development in South Wales: Savigear re-viewed. *Zeitschrift für Geomorphologie* 28, 405–426.

Kirkby, M.J. (1987) General models of long-term slope evolution through mass movement. In Anderson, M.G. and Richards, K.S. (eds) *Slope Stability.* John Wiley & Sons, Chichester, 359–380.

Kirkby, M.J. and Statham, I. (1975) Surface stone movement and scree formation, *Journal of Geology* 83, 349–362.

Kitover, D.C., van Balen, R.T., Roche, D.M. *et al.* (2013) New estimates of permafrost evolution during the last 21 k years in Eurasia using numerical modelling. *Permafrost and Periglacial Processes* 24, 286–303.

Kitover, D.C., van Balen, R.T., Vandenberghe, J. *et al.* (2016) LGM permafrost thickness and extent in the northern hemisphere derived from the Earth system model *iLOVECLIM*. *Permafrost and Periglacial Processes* 27, 31–42.

Klaminder, J., Yoo, K., Olid, C. *et al.* (2014) Using short-lived radionuclides to estimate rates of soil motion in frost boils. *Permafrost and Periglacial Processes* 25, 184–193.

Kleman, J. and Glasser, N. (2007) The subglacial thermal organization (STO) of ice sheets. *Quaternary Science Reviews* 25, 585–597.

Klene, A.E., Nelson, F.E., Shiklomanov, N.I. and Hinkel, K.M. (2001) The n-factor in natural landscapes: variability of air and soil-surface temperatures, Kuparak River Basin, Alaska, USA. *Arctic, Antarctic and Alpine Research* 33, 140–148.

Kling, J. (1997) Observations on sorted circle development, Abisko, northern Sweden. *Permafrost and Periglacial Processes* 8, 447–453.

Kluiving, S.J., Verbers, A.L.L.M. and Thijs, W.J.F. (2010) Lithological analysis of 45 presumed pingo remnants in the northern Netherlands (Friesland): substrate control and fill sequences. *Netherlands Journal of Geosciences* 89, 61–75.

Kneisel, C., Hauck, C., Fortier, R. and Moorman, B. (2008) Advances in geophysical methods for permafrost investigations. *Permafrost and Periglacial Processes* 19, 157–178.

Knight, J. (2008) The environmental significance of ventifacts: a critical review. *Earth-Science Reviews* 86, 89–105.

Knoblauch, C., Beer, C., Sosnin, A. *et al.* (2013) Predicting long-term carbon mineralization and trace gas production from thawing permafrost of Northeast Siberia. *Global Change Biology* 19, 1160–1172.

Kohout, T., Bucko, M.S., Rasmus, K. *et al.* (2014) Non-invasive geophysical investigation and thermodynamic analysis of a palsa in Lapland, northwest Finland. *Permafrost and Periglacial Processes* 25, 45–52.

Koiwa, N. (2003) Thee-dimensional structure of involutions formed in a Pleistocene tephra layer, northeastern Japan. *Geomorphology* 52, 131–140.

Kojima, S. (1994) Relationships of vegetation, earth hummocks and topography in the high arctic environment of Canada. *Polar Biology* 7, 256–269.

Kokelj, S.V. and Burn, C.R. (2004) Tilt of spruce trees near ice wedges, Mackenzie Delta, Northwest Territories, Canada. *Arctic, Antarctic and Alpine Research* 36, 615–623.

Kokelj, S.V. and Burn, C.R. (2005) Geochemistry of the active layer and near-surface permafrost, Mackenzie delta region, Northwest Territories, Canada. *Canadian Journal of Earth Sciences* 42, 37–48.

Kokelj, S.V. and Jorgensen, M.T. (2013) Advances in thermokarst research. *Permafrost and Periglacial Processes* 24, 108–119.

Kokelj, S.V. and Lewkowicz, A.G. (1998) Long-term influence of active-layer detachment sliding on permafrost slope hydrology, Hot Weather Creek, Ellesmere Island, Canada. In: Lewkowicz, A.G. and Allard, M. (eds) *Permafrost: Proceedings of the 7th International Conference.* Université Laval, Québec, 583–589.

Kokelj, S.V., Jenkins, R.E., Milburn, D. *et al.* (2005) The influence of thermokarst disturbance on the water quality of small upland lakes, Mackenzie Delta region, Northwest Territories, Canada. *Permafrost and Periglacial Processes* 16, 343–353.

Kokelj, S.V., Burn, C.R. and Tarnocai, C. (2007) The structure and dynamics of earth hummocks in the subarctic forest near Inuvik, Northwest Territories, Canada. *Arctic, Antarctic and Alpine Research* 39, 99–109.

Kokelj, S.V., Lantz, T.C., Kanigan, J. *et al.* (2009) Origin and polycyclic behaviour of tundra thaw slumps, Mackenzie Delta Region, Northwest Territories, Canada. *Permafrost and Periglacial Processes* 20, 173–184.

Kokelj, S.V., Lacelle, D., Lantz, T.C. *et al.* (2013) Thawing of massive ground ice in mega slumps drives increase in stream sediment and solute flux across a range of watershed scales. *Journal of Geophysical Research: Earth Surface* 118, 681–692.

Kokelj, S.V., Tunnicliffe, J., Lacelle, D. *et al.* (2015) Increased precipitation drives mega-slump development and destabilization of ice-rich permafrost terrain, northwestern Canada. *Global and Planetary Change* 129, 56–68.

Kolstrup, E. (1986) Reappraisal of the upper Weichselian periglacial environment from Danish frost wedge casts *Palaegeography, Palaeoclimatology, Palaeoecology* 56, 237–249.

Kolstrup, E. (2004) Stratigraphic and environmental implications of a large ice-wedge cast at Tjaereborg, Denmark. *Permafrost and Periglacial Processes* 15, 31–40.

Kolstrup, E. (2007) Lateglacial older and younger coversand in northwest Europe: chronology and relation to climate and vegetation. *Boreas* 36, 65–75.

Kolstrup, E. and Thyrsted, T. (2010) Stone heave field experiment in clayey silt. *Geomorphology* 117, 90–105.

Kolstrup, E. and Thyrsted, T. (2011) Stone heave field experiment in sand. *Geomorphology* 129, 361–375.

Konishchev, V.N. and Rogov, V.V. (1993) Investigation of cryogenic weathering in Europe and northern Asia. *Permafrost and Periglacial Processes* 4, 49–64.

Konrad, S.K., Humphrey, N.F., Steig, E.J. *et al.* (1999) Rock glacier dynamics and palaeoclimatic implications. *Geology* 27, 1131–1134.

Koster, E.A. (1988) Ancient and modern cold-climate aeolian sand deposition: a review. *Journal of Quaternary Science* 3, 69–83.

Koster, E.A. (2005) Recent advances in luminescence dating of late Pleistocene (cold-climate) aeolian sand and loess deposits in Western Europe. *Permafrost and Periglacial Processes* 16, 131–143.

Koster, E.A. and Dijkmans, J.W.A. (1988) Niveo-aeolian deposits and denivation forms, with special reference to the Great Kobuk sand dunes, northwestern Alaska. *Earth Surface Processes and Landforms* 13, 153–170.

Kotarba, A. (1992) High energy geomorphologic events in the Polish Tatra Mountains, *Geografiska Annaler* 74A, 123–131.

Kotarba, A. and Strömquist, L. (1984) Transport, sorting and deposition processes of alpine debris slope deposits in the Polish Tatra Mountains. *Geografiska Annaler* 66A, 285–294.

Kotarba, A., Kaszowski, L. and Krzemien, K. (1987) *High-Mountain Denudational System of the Polish Tatra Mountains*. Polish Academy of Sciences,Wroclaw.

Kotler, E. and Burn, C.R. (2000) Cryostratigraphy of the Klondike 'muck' deposits, west-central Yukon Territory. *Canadian Journal of Earth Sciences* 37, 849–861.

Koven, C.D., Ringeval, B., Friedlingstein, P. (2011) Permafrost carbon-climate feedbacks accelerate global warming. *Proceedings of the National Academy of Sciences* 108, 14769–14774.

Koven, C.D., Riley, W.J. and Stern, A. (2013) Analysis of permafrost thermal dynamics and response to climate change in the CMIP5 Earth system models. *Journal of Climatology* 26, 1877–1900.

Kozarski, S. (1993) Late Plenivistulian deglaciation and the expansion of the periglacial zone in NW Poland. *Geologie en Mijnbouw* 72, 143–157.

Krainer, K. and Mostler, W. (2000) Reichenkar rock glacier: a glacier-derived debris-ice system in the western Stubai Alps, Austria. *Permafrost and Periglacial Processes* 11, 267–275.

Krainer, K. and Ribis, M. (2012) A rock glacier inventory of the Tyrolean Alps (Austria). *Austrian Journal of Earth Sciences* 105, 32–47.

Krantz, W.B. (1990) Self-organization manifest as patterned ground in recurrently frozen soils. *Earth-Science Reviews* 29, 117–130.

Krautblatter, M. and Dikau, R. (2007) Towards a uniform concept for the comparison and extrapolation of rockwall retreat and rockfall supply. *Geografiska Annaler* 29A, 21–40.

Krautblatter, M. and Moser, M. (2009) A nonlinear model coupling rockfall and rainfall intensity based on a four year measurement in a high Alpine rock wall (Reintal, German Alps). *Natural Hazards and Earth System Sciences* 9, 1425–1432.

Krautblatter, M., Verleysdonk, S., Flores-Orozco, A. and Kemna, A. (2010) Temperature-calibrated imaging of seasonal changes in permafrost rock walls by quantitative electrical resistivity tomography (Zugspitze, German/Austrian Alps). *Journal of Geophysical Research* 115, F02003.

Krautblatter, M., Moser, M., Schrott, L. *et al.* (2012a) Significance of rockfall magnitude and carbonate dissolution for rock slope erosion and geomorphic work on Alpine limestone cliffs. *Geomorphology* 167–168, 21–34.

Krautblatter, M., Huggel, C., Deline, P. and Hasler, A. (2012b) Research perspectives on unstable high-Alpine bedrock permafrost: measurement, modelling and process understanding. *Permafrost and Periglacial Processes* 23, 80–88.

Krautblatter, M., Funk, D. and Günzel, F.K. (2013) Why permafrost rocks become unstable: a rock-ice-mechanical model in time and space. *Earth Surface Processes and Landforms* 38, 876–887.

Krawczyk, W.E. and Pettersson, L.-E. (2007) Chemical denudation rates and carbon dioxide drawdown in an ice-free polar karst catchment: Londonelva, Svalbard. *Permafrost and Periglacial Processes* 18, 337–350.

Kudryavstsev, V.A. (1965) Temperature, thickness and discontinuity of permafrost. *Canadian National Research Council, Technical Translation* 1187, 75 pp.

Kuhry, P., Grosse, G., Harden, J.W. *et al.* (2013) Characterisation of the permafrost carbon pool. *Permafrost and Periglacial Processes* 24, 146–155.

Kujala, K., Seppälä, M. and Holappa, T. (2008) Physical properties of peat and palsa formation. *Cold Regions Science and Technology* 52, 408–414.

Kurylyk, B.L. (2015) Discussion of 'a simple thaw-freeze algorithm for a multi-layered soil using the Stefan equation' by Xie and Gough (2013). *Permafrost and Periglacial Processes* 26, 200–206.

Kurylyk, B.L. and Hayashi, M. (2016) Improved Stefan equation correction factors to accommodate sensible heat storage during soil freezing or thawing. *Permafrost and Periglacial Processes* 27, 189–204.

Lacelle, D. and Vasil'chik, Y.K. (2013) Recent progress (2007–2012) in permafrost isotope geochemistry. *Permafrost and Periglacial Processes* 24, 138–145.

Lacelle, D., Bjornson, J., Lauriol, B. *et al.* (2004) Segregated-intrusion ice of subglacial meltwater origin in retrogressive thaw flow headwalls, Richardson Mountains, NWT, Canada. *Quaternary Science Reviews* 23, 681–696.

Lacelle, D., Lauriol, B., Clark, I.D. *et al.* (2007) Nature and origin of a Pleistocene-age massive ground-ice body exposed in the Chapman Lake moraine complex, central Yukon Territory, Canada. *Quaternary Research* 68, 249–260.

Lacelle, D., Juneau, V., Pellerin, A. *et al.* (2008) Weathering regime and geochemical conditions in a polar desert environment, Haughton impact structure region, Devon Island, Canada. *Canadian Journal of Earth Sciences* 45, 1139–1157.

Lacelle, D., Bjornson, J. and Lauriol, B. (2010) Climatic and geomorphic factors affecting contemporary (1950–2004) activity of retrogressive thaw slumps on the Aklavik Plateau, Richardson Mountains, NWT, Canada. *Permafrost and Periglacial Processes* 21, 1–15.

Lacelle, D., Brooker, A., Fraser, R.H. and Kokelj, S.V. (2015) Distribution and growth of thaw slumps in the Richardson Mountains-Peel Plateau region, northwestern Canada. *Geomorphology* 235, 40–51.

Lachenbruch, A.H. (1962) Mechanics of thermal contraction cracks and ice-wedge polygons in permafrost. *Geological Society of America Special Paper* 70, 69 pp.

Lachniet, M.S., Lawson, D.E. and Sloat, A. (2012) Revised ^{14}C dating of ice-wedge growth in interior Alaska to MIS2 reveals cold palaeoclimate and carbon recycling in ancient permafrost terrain. *Quaternary Research* 78, 217–225.

Ladanyi, B., Foriero, A., Dallimore, S.R. *et al.* (1995) Modelling of deep seated creep in massive ice, Tuktoyaktuk Coastlands, NWT. In *Proceedings of the 48th Canadian Geotechnical Conference, Vancouver*, 1023–1030.

Lafrenière, M.J. and Lamoureux, S.F. (2013) Thermal perturbation and rainfall runoff have greater impact on seasonal solute loads than physical disturbance of the active layer. *Permafrost and Periglacial Processes* 24, 241–251.

Lagerbäck, R. and Rodhe, L. (1985) Pingos in northernmost Sweden. *Geografiska Annaler* 67A, 239–245.

Lagerbäck, R. and Rodhe, L. (1986) Pingos and palsas in northernmost Sweden – preliminary notes on recent investigations. *Geografiska Annaler* 68A, 149–154.

Lai, Z., Kaiser, K. and Brückner, H. (2009) Luminescence-dated aeolian deposits of late Quaternary age in the southern Tibetan Plateau and their implications for landscape history. *Quaternary Research* 72, 421–430.

Laignel, B., Quesnel, F., Spencer, C. *et al.* (2003) Slope clay-with-flints (*biefs à silex*) as indicators of Quaternary periglacial dynamics in the western part of the Paris Basin, France. *Journal of Quaternary Science* 18, 295–299.

Lambiel, C. and Delaloye, R. (2004) Contribution of real-time kinematic GPS in the study of creeping mountain permafrost: examples from the western Swiss Alps. *Permafrost and Periglacial Processes* 15, 229–241.

Lambiel, C. and Pieracci, K. (2008) Permafrost distribution in talus slopes located within the alpine periglacial belt, Swiss Alps. *Permafrost and Periglacial Processes* 19, 293–304.

Lambiel, C. and Reynaud, E. (2001) Regional modelling of present, past and future potential distribution of discontinuous permafrost based on a rock glacier inventory in the Bagnes Hérémence area (Western Swiss Alps). *Norsk Geografisk Tiddskrift* 55, 219–223.

Lamirande, I., Lauriol, B., Lalonde, A.E. and Clark, I.D. (1999) La production de limon sur des terrasses de cryoplanation dans les Monts Richardson, Canada. *Canadian Journal of Earth Sciences* 36, 1645–1654.

Lamoureux, S.F. and Lafrenière, M.J. (2009) Fluvial impact of extensive active layer detachments, Cape Bounty, Melville Island, Canada. *Arctic, Antarctic and Alpine Research* 41, 59–68.

Lamoureux, S.F., Lafrenière, M.J. and Favoro, E.A. (2014) Erosion dynamics following localized permafrost slope disturbances. *Geophysical Research Letters* 41, 5499–5505.

Lang, B., Brooks, S.J., Bedford, A. *et al.* (2010) Regional consistency in chironomid-inferred temperatures from five sites in north-west England. *Quaternary Science Reviews* 29, 1528–1538.

Lantuit, H. and Pollard, W.H. (2008) Fifty years of coastal erosion and retrogressive thaw slump activity on Herschel Island, southern Beaufort Sea, Yukon Territory, Alaska. *Geomorphology* 95, 84–102.

Lantuit, H., Overduin, P.P., Couture, N. and Ødegård, R.S. (2008) Sensitivity of coastal erosion to ground ice contents: an arctic-wide study based on the ACD classification of arctic coasts. In Kane, D.L. and Hinkel, D.M. (eds) *Ninth International Conference on Permafrost*. University of Alaska, Fairbanks, 1025–1030.

Lantuit, H., Pollard, W.H., Couture, N. *et al.* (2012a) Modern and Late Holocene retrogressive thaw slump activity on the Yukon coastal plain and Herschel Island, Yukon Territory, Canada. *Permafrost and Periglacial Processes* 23, 39–51.

Lantuit, H., Overduin, P.P., Couture, N. *et al.* (2012b) The Arctic Coastal Dynamics database: a new classification scheme and statistics on Arctic permafrost coastlines. *Estuaries and Coasts* 35, 383–400.

Lantuit, H., Overduin, P.P. and Wetterich, S. (2013) Recent progress regarding permafrost coasts. *Permafrost and Periglacial Processes* 24, 120–130.

Lantz, T.C. and Kokelj, S.V. (2008) Increasing rates of retrogressive thaw slump activity in the Mackenzie Delta Region, NWT, Canada. *Geophysical Research Letters* 35, L06502.

Lantz, T.C., Kokelj, S.V., Gergel, S.E. and Henry, G.H.R. (2009) Relative impacts of disturbance and temperature: persistent changes in microenvironment and vegetation in retrogressive thaw slumps. *Global Change Biology* 15, 1664–1675.

Lara, M.J., Genet, H., McGuire, A.D. *et al.* (2016) Thermokarst rates intensify due to climate change and forest fragmentation in an Alaskan boreal forest lowland. *Global Change Biology* 22, 816–829.

Larocque, S.J., Hétu, B. and Filion, L. (2001) Geomorphic and dendroecological impacts of slushflows in central Gaspé Peninsula (Québec, Canada). *Geografiska Annaler* 83A, 191–201.

Larsson, S. (1982) Geomorphological effects on the slopes of Longyear Valley, Spitsbergen, after a heavy rainstorm in July 1972. *Geografiska Annaler* 64A, 105–125.

Laurain, M., Guérin, H., Marre, A. and Richard, J. (1995) Processus génétiques a l'origine des formations de pente à graviers de craie en Champagne. *Permafrost and Periglacial Processes* 6, 103–108.

Lauriol, B. (1990) Cryoplanation terraces, northern Yukon. *Canadian Geographer* 34, 347–351.

Lauriol, B. and Godbout, L. (1988) Les terrasses de cryoplanation dans le nord de Yukon: distribution, genèse et age. *Géographie Physique et Quaternaire* 42, 303–313.

Lauriol, B. and Gray, J.T. (1990) Drainage karstique en milieu de pergélisol: le cas de l'Île d'Akpatok. *Permafrost and Periglacial Landforms* 1, 129–144.

Lauriol, B., Ford, D.C. Cinq-Mars, J. and Morris, W.A. (1997a) The chronology of speleothem deposition in northern Yukon and its relationships to permafrost. *Canadian Journal of Earth Sciences* 34, 902–911.

Lauriol, B., Lalonde, A.E. and Dewez, V. (1997b) Weathering of quartzite on a cryoplanation terrace in northern Yukon, Canada. *Permafrost and Periglacial Processes* 8, 147–153.

Lauriol, B., Hétu, B., Cote, D. and Gwyn, H. (1985) Phenomènes karstiques et périglaciales dans un lac de niveau variable de l'Île d'Anticosti, Québec, Canada. *Zeitschrift für Geomorphologie* 29, 252–265.

Lauriol, B., Lamirande, I. and Lalonde, A.E. (2006) The giant steps of Bug Creek, Richardson Mountains, NWT, Canada. *Permafrost and Periglacial Processes* 17, 267–275.

Lauriol, B., Lacelle, D., Duguay, C.R. *et al.* (2009) Holocene evolution of lakes in the Bluefish Basin, Northern Yukon, Canada. *Arctic* 62, 212–224.

Laute, K. and Beylich, A.A. (2014) Morphometric and meteorological controls on recent snow avalanche distribution and activity at hillslopes in steep mountain valleys in western Norway. *Geomorphology* 218, 16–24.

Lautridou, J.-P. (1988) Recent advances in cryogenic weathering. In Clark, M.J. (ed.) *Advances in Periglacial Geomorphology*. John Wiley & Sons, Chichester, 33–47.

Lautridou, J.-P. and Ozouf, J.-C. (1982) Experimental frost-shattering: 15 years of research at the Centre de Géomorphologie du CNRS. *Progress in Physical Geography* 6, 215–232.

Lawler, D.M. (1988) Environmental limits of needle ice: a global survey. *Arctic and Alpine Research* 20, 137–159.

Lawrence, D., Slater, A. and Swenson, S. (2012) Simulation of present-day and future permafrost and seasonally frozen ground conditions in CCSM4. *Journal of Climatology* 25, 2207–2225.

Lawson, D.E. (1983) Ground ice in perennially frozen sediments, Northern Alaska. *Proceedings, Fourth International Conference on Permafrost*. National Academy Press, Washington DC, 695–700.

Lawson, D.E. (1986) Response of permafrost terrain to disturbance: a synthesis of observations from northern Alaska, USA. *Arctic and Alpine Research* 18, 1–17.

Lea, P.D. (1990) Pleistocene periglacial eolian deposits in southwestern Alaska: sedimentary facies and depositional processes. *Journal of Sedimentary Petrology* 60, 582–591.

Lea, P.D. and Waythomas, C.F. (1990) Late Pleistocene eolian sand sheets in Alaska. *Quaternary Research* 34, 269–281.

Leckie, D.A. and McCann, S.B. (1982) Active, small-scale periglacial features on the south coast of Newfoundland. *Géographie Physique et Quaternaire* 36, 327–329.

Lee, J., Brown, E.J., Rose, J. *et al.* (2003) A reply to 'Implications of a Middle Pleistocene ice wedge cast at Trimingham, Norfolk, Eastern England' (Whiteman, 2002). *Permafrost and Periglacial Processes* 14, 75–77.

Leffingwell, E.K. (1915) Ground-ice wedges, the dominant form of ground ice on the north coast of Alaska. *Journal of Geology* 23, 635–654.

Legros, J.P. (1992) Soils of Alpine mountains. In Martin, I.P. and Chesworth, W. (eds) *Weathering, Soils and Palaeosols*. Elsevier, Amsterdam, 155–181.

Lehmkuhl, F., Klinge, M., Rees-Jones, J. and Rhodes, E.J. (2000) Late Quaternary aeolian sedimentation in central and south-eastern Tibet. *Quaternary International* 68–71, 117–132.

Lehmkuhl, F., Hülle, D. and Knippertz, M. (2012) Holocene geomorphic processes and landscape evolution in the lower reaches of the Orkhon Valley (northern Mongolia). *Catena* 98, 17–28.

Lehmkuhl, F., Schulte, P., Zhau, H. *et al.* (2014) Timing and spatial distribution of loess and loess-like sediments in the mountain areas of the northeastern Tibetan Plateau. *Catena* 117, 23–33.

Leibman, M.O. (1995) Cryogenic landslides on the Yamal Peninsula, Russia: preliminary observations. *Permafrost and Periglacial Processes* 6, 259–264.

Leland, J., Reid, M.R., Burbank, D.W. *et al.* (1998) Incision and differential bedrock uplift along the Indus River near Nanga Parbat, Pakistan Himalaya, from Be and Al exposure dating of bedrock straths. *Earth and Planetary Science Letters* 154, 93–107.

Lemcke, M.D. and Nelson, F.E. (2004) Cryogenic sediment-filled wedges in northern Delaware, USA. *Permafrost and Periglacial Processes* 15, 319–326.

Lenz, J., Wetterich, S., Jones, B.M. *et al.* (2016) Evidence of multiple thermokarst lake generations from an 11 800-year-old permafrost core on the northern Seward Peninsula, Alaska. *Boreas* 45, 584–603.

Leopold, M., Williams, M.W., Caine, N. *et al.* (2011) Internal structure of the Green Lake 5 rock glacier, Colorado Front Ranges, USA. *Permafrost and Periglacial Processes* 22, 107–119.

Lev, A. and King, R.H. (1999) Spatial variation of soil development in a High Arctic soil landscape: Truelove Lowland, Devon Island, Nunavut, Canada. *Permafrost and Periglacial Processes* 10, 289–307.

Levy, J.S., Fountain, A.G., Welch, K.A. and Berry Lyons, W. (2012) Hypersaline 'wet patches' in Taylor Valley, Antarctica. *Geophysical Research Letters* 39, L05402.

Lewin, J. and Gibbard, P.L. (2010) Quaternary river terraces in England: forms, sediments and processes. *Geomorphology* 120, 293–311.

Lewis, C.A. (1985) Periglacial features. In Edwards, K.J. and Warren, W.P. (eds) *The Quaternary History of Ireland*. Academic Press, London, 95–113.

Lewis, G.C., Krantz, W.B. and Caine, N. (1993) A model for the initiation of patterned ground owing to differential frost heave. In: *Permafrost: Proceedings of the Sixth International Conference*. South China University of Technology Press, Wushan Guangzhou, China, 1044–1049.

Lewis, T. and Lamoureux, S.F. (2010) Twenty-first century discharge and sediment yield predictions in a small high Arctic watershed. *Global and Planetary Change* 71, 27–41.

Lewis, T., Lafrenière, M.J. and Lamoureux, S.F. (2012) Hydrochemical and sedimentary responses of paired high Arctic watersheds to unusual climate and permafrost disturbance, Cape Bounty, Melville Island, Canada. *Hydrological Processes* 26, 2003–2008.

Lewkowicz, A.G. (1983) Erosion by overland flow, central Banks Island, western Canadian arctic. *Proceedings, Fourth International Conference on Permafrost*. National Academy Press, Washington DC, 701–706.

Lewkowicz, A.G. (1992) Factors influencing the distribution and initiation of active-layer detachment slides on Ellesmere Island, Arctic Canada. In Dixon, J.C. and Abrahams, A.D. (eds) *Periglacial Geomorphology*. John Wiley & Sons, Chichester, 223–250.

Lewkowicz, A.G. (1994) Ice-wedge rejuvenation, Fosheim Peninsula, Ellesmere Island, Canada. *Permafrost and Periglacial Processes* 5, 251–268.

Lewkowicz, A.G. (2001) Temperature regime of a small sandstone tor, latitude 80°N, Ellesmere Island, Nunavut, Canada. *Permafrost and Periglacial Processes* 12, 351–366.

Lewkowicz, A.G. (2007) Dynamics of active-layer detachment failures, Fosheim Peninsula, Ellesmere Island, Nunavut, Canada. *Permafrost and Periglacial Processes* 18, 89–103.

Lewkowicz, A.G. (2011) Slope hummock development, Fosheim Peninsula, Ellesmere Island, Nunavut, Canada. *Quaternary Research* 75, 334–346.

Lewkowicz, A.G. and Bonnaventure, P.P. (2011) Equivalent elevation: a new method to incorporate variable lapse rates into mountain permafrost modelling. *Permafrost and Periglacial Processes* 22, 153–162.

Lewkowicz, A.G. and Clarke, S. (1998) Late-summer solifluction and active layer depths, Fosheim Peninsula, Ellesmere Island, Canada. In Lewkowicz, A.G. and Allard, M. (eds) *Permafrost: Proceedings of the 7th International Conference*. Université Laval, Québec, 641–646.

Lewkowicz, A.G. and Ednie, M. (2004) Probability mapping of permafrost using the BTS method, Wolf Creek, Yukon Territory, Canada. *Permafrost and Periglacial Processes* 15, 67–80.

Lewkowicz, A.G. and French, H.M. (1982) Downslope water movement and solute concentrations within the active layer, Banks Island, N.W.T. *Proceedings of the Fourth Canadian Permafrost Conference*. National Research Council of Canada, Ottawa, 163–172.

Lewkowicz, A.G. and Harris, C. (2005a) Frequency and magnitude of active-layer detachment failures in discontinuous and continuous permafrost, northern Canada. *Permafrost and Periglacial Processes* 16, 115–130.

Lewkowicz, A.G. and Harris, C. (2005b) Morphology and geotechnique of active-layer detachment failures in discontinuous and continuous permafrost, northern Canada. *Geomorphology* 69, 275–297.

Lewkowicz, A.G. and Kokelj, S.V. (2002) Slope sediment yield in arid lowland continuous permafrost environments, Canadian Arctic Archipelago. *Catena* 46, 261–283.

Lewkowicz, A.G. and Wolfe, P.M. (1994) Sediment transport in Hot Weather Creek, Ellesmere Island, N.W.T., Canada, 1990–1991. *Arctic and Alpine Research* 26, 213–226.

Lewkowicz, A.G. and Young, K.L. (1990) Hydrology of a perennial snowbank in the continuous permafrost zone, Melville Island, Canada. *Geografiska Annaler* 72A, 13–21.

Lewkowicz, A.G. and Young, K.L. (1991) Observations of aeolian transport and niveo-aeolian deposition at three lowland sites, Canadian Arctic Archipelago. *Permafrost and Periglacial Processes* 2, 197–210.

Lewkowicz, A.G., Etzelmüller, B. and Smith, S.L. (2011) Characteristics of discontinuous permafrost based on ground temperature measurements and electrical resistivity tomography, southern Yukon, Canada. *Permafrost and Periglacial Processes* 22, 320–342.

Lewkowicz, A.G., Bonnaventure, P.P., Smith, S.L. and Kuntz, Z. (2012) Spatial and thermal characteristics of mountain permafrost, northwest Canada. *Geografiska Annaler* 94A, 193–213.

Li, H., Wang, W., Wu, F. *et al.* (2014) A new sand-wedge-forming mechanism in an extra-arid area. *Geomorphology* 211, 43–51.

Li, X., Cheng, G., Jin, H. *et al.* (2008) Cryospheric change in China. *Global and Planetary Change* 62, 210–218.

Liestøl, O. (1977) Pingos, springs and permafrost in Spitsbergen. *Norsk Polarinstitutt Årbok* 1975, 7–29.

Lilleøren, K.S. and Etzelmüller, B. (2011) A regional inventory of rock glaciers and ice-cored moraines in Norway. *Geografiska Annaler* 93A, 175–191.

Lilleøren, K.S., Etzelmüller, B., Gärtner-Roer, I. *et al.* (2013) The distribution, thermal characteristics and dynamics of permafrost in Tröllaskagi, northern Iceland, as inferred from the distribution of rock glaciers and ice-cored moraines. *Permafrost and Periglacial Processes* 24, 322–335.

Lilly, E.K., Kane, D.L., Hinzman, L.D. and Gieck, R.E. (1998) Annual water balance for three nested watersheds on the North Slope of Alaska. In: Lewkowicz, A.G. and Allard, M. (eds) *Permafrost: Proceedings of the 7th International Conference.* Université Laval, Québec, 669–674.

Lin, Z., Niu, F, Xu, Z. *et al.* (2010) Thermal regime of a thermokarst lake and its influence on permafrost, Beiluhe Basin, Qinghai-Tibet Plateau. *Permafrost and Periglacial Processes* 21, 315–324.

Lin, Z., Burn, C.R., Niu, F. *et al.* (2015) The thermal regime, including a reversed thermal offset, of arid permafrost sites with variation in vegetation cover density, Wudaoliang Basin, Qinghai-Tibet Plateau. *Permafrost and Periglacial Processes* 26, 142–159.

Lindgren, A., Hugelius, G., Kuhry, P. *et al.* (2016) GIS-based maps and area estimates of northern hemisphere permafrost extent during the Last Glacial Maximum. *Permafrost and Periglacial Processes* 27, 6–16.

Lindhorst, S. and Schutter, I. (2014) Polar gravel beach-ridge systems: sedimentary architecture, genesis, and implications for climate reconstructions (South Shetland Islands/western Antarctic Peninsula). *Geomorphology* 221, 187–203.

Lisieki, L.E. and Raymo, M.E. (2005) A Pliocene-Pleistocene stack of 57 globally distributed benthic $\delta^{18}O$ records. *Paleoceanography* 20, PA1003.

Liu G., Xiong, H. and Cui, Z. (1995) Gelifluction in the alpine periglacial environment of the Tianshan Mountains, China. *Permafrost and Periglacial Processes* 6, 265–271.

Liu G., Cui, Z., Ge, D. and Wu, Y. (1999) The stratified slope deposits a Kunlunshan Pass, Tibet Plateau, China. *Permafrost and Periglacial Processes* 10, 369–375.

Liu, X.-J. and Lai, Z.-P. (2013) Optical dating of sand wedges and ice-wedge casts from Qinghai Lake area on the northeastern Qinghai-Tibetan Plateau and its palaeoenvironmental implications. *Boreas* 42, 333–341.

Liverman, D., Catto, N., Batterson, M. *et al.* (2000) Evidence of late glacial permafrost in Newfoundland. *Quaternary International* 68–71, 163–174.

Long, D. (1991) The identification of features due to former permafrost in the North Sea. *Geological Society, London, Engineering Geology Special Publication* 7, 369–372.

Lotsari, E., Veijalainen, N., Alho, P. and Käyhkö, J. (2010) Impact of climate change on future discharges and flow characteristics of the Tana River, Sub-Arctic northern Fennoscandia. *Geografiska Annaler* 92A, 263–284.

Lowe, J.J. and Walker, M.J.C. (2015) *Reconstructing Quaternary Environments*, 3rd edn. Routledge, London and New York, 538 pp.

Łozinski, M.W. (1909) Über die mechanische Verwitterung der Sandsteine im gemässigten Klima. *Académie des Sciences de Cracovie: Bulletin International, Sciences Mathematiques et Naturelles* 1, 1–25. Reprinted as 'On the mechanical weathering of sandstone in temperate climates' in Evans, D.J.A. (ed.) *Cold Climate Landforms.* John Wiley & Sons, Chichester, 119–134.

Łozinski, M.W. (1912) Die periglaziale Fazies der mechanischen Verwitterung. *Proceedings, 11th International Geological Congress (Stockholm, 1910)*, 1039–1053.

Luckman, B.H. (1976) Rockfalls and rockfall inventory data: some observations from Surprise Valley, Jasper National Park, Canada. *Earth Surface Processes* 1, 287–298.

Luckman, B.H. (1977) The geomorphic activity of snow avalanches. *Geografiska Annaler* 59A, 31–48.

Luckman, B.H. (1978) Geomorphic work of snow avalanches in the Canadian Rocky Mountains. *Arctic and Alpine Research* 10, 261–276.

Luckman, B.H. (1988) Debris accumulation patterns on talus slopes in Surprise Valley, Alberta. *Géographie Physique et Quaternaire* 42, 247–278.

Luckman, B.H. (1992) Debris flows and snow avalanche landforms in the Lairig Ghru, Cairngorm Mountains, Scotland. *Geografiska Annaler* 74A, 109–121.

Luckman, B.H. (2013) Processes, transport, deposition and landforms: rockfall. In Shroder, J., Marston, R.A. and Stoffel, M. (eds) *Treatise on Geomorphology, Volume 7: Mountain and Hillslope Geomorphology.* Academic Press, San Diego, CA, 174–182.

Luckman, B.H. and Fiske, C.J. (1995) Estimating long-term rockfall accretion rates by lichenometry. In Slaymaker, O. (ed.) *Steepland Geomorphology.* John Wiley & Sons, Chichester, 233–255.

Luethi, R., Gruber, S. and Ravanel, L. (2015) Modelling transient ground surface temperatures of past rockfall events: towards a better understanding of failure mechanisms in changing periglacial environments. *Geografiska Annaler* 97A, 753–767.

Luetscher, M., Jeannin, P.-Y. and Haeberli, W. (2005) Ice caves as an indicator of winter climate evolution: a case study from the Jura Mountains. *The Holocene* 15, 982–993.

Lugon, R. and Stoffel, M. (2010) Rock-glacier dynamics and magnitude-frequency relations of debris flows in a high-elevation watershed: Ritigraben, Swiss Alps. *Global and Planetary Change* 73, 201–210.

Lunardini, V.J. (1991) *Heat Transfer with Freezing and Thawing.* Elsevier, Amsterdam, 437 pp.

Luoto, M. and Seppälä, M. (2003) Thermokarst ponds as indicators of the former distribution of palsas in Finnish Lappland. *Permafrost and Periglacial Processes* 14, 19–27.

Lusch, D.P., Stanley, K.E., Schaetzl, R.J. *et al.* (2009) Characterization and mapping of patterned ground in the Saginaw Lowlands, Michigan: possible evidence for Late-Wisconsin permafrost. *Annals, Association of American Geographers* 99, 445–466.

Lynn, L.A., Ping, C.L., Michaelson, G.J. and Jorgenson, M.T. (2008) Soil properties of the eroding coastline at Barter Island, Alaska. In Kane, D.L. and Hinkel, D.M. (eds) *Ninth International Conference on Permafrost.* University of Alaska, Fairbanks, 1087–1092.

Ma, W., Zhang, L. and Yang, C. (2015) Discussion of the applicability of the generalized Clausius-Clapeyron equation and the frozen fringe process. *Earth-Science Reviews* 142, 47–59.

MacDougall, A.H., Avis, C.A. and Weaver, A.J. (2012) Significant contribution to climate warming from the permafrost carbon feedback. *Nature Geoscience* 5, 719–721.

Mackay, J.R. (1971) The origin of massive icy beds in permafrost, western Arctic coast, Canada. *Canadian Journal of Earth Sciences* 8, 397–422.

Mackay, J.R. (1972) The world of underground ice. *Annals, American Association of Geographers* 62, 1–22.

Mackay, J.R. (1974) Reticulate ice veins in permafrost, Northern Canada. *Canadian Geotechnical Journal* 11, 230–237.

Mackay, J.R. (1975) The closing of ice-wedge cracks in permafrost, Garry Island, Northwest Territories. *Canadian Journal of Earth Sciences* 12, 1668–1674.

Mackay, J.R. (1977) Pulsating pingos, Tuktoyaktuk Peninsula, NWT. *Canadian Journal of Earth Sciences* 14, 209–222.

Mackay, J.R. (1978) Sub-pingo water lenses, Tuktoyaktuk Peninsula, Northwest Territories. *Canadian Journal of Earth Sciences* 15, 1219–1227.

Mackay, J.R. (1979) Pingos of the Tuktoyaktuk Peninsula area, Northwest Territories. *Géographie Physique et Quaternaire* 33, 3–61.

Mackay, J.R. (1980) The origin of hummocks, western Arctic coast, Canada. *Canadian Journal of Earth Sciences* 17, 996–1006.

Mackay, J.R. (1981) Active layer slope movement in a continuous permafrost environment, Garry Island, Northwest Territories, Canada. *Canadian Journal of Earth Sciences* 18, 1666–1680.

Mackay, J.R. (1983) Downward water movement into frozen ground, western Arctic coast, Canada. *Canadian Journal of Earth Sciences* 20, 120–134.

Mackay, J.R. (1984) The direction of ice-wedge cracking in permafrost; downward or upward? *Canadian Journal of Earth Sciences* 21, 516–524.

Mackay, J.R. (1985) Pingo ice of the western Arctic coast, Canada. *Canadian Journal of Earth Sciences* 22, 1452–1464.

Mackay, J.R. (1987) Some mechanical aspects of pingo growth and failure, western Arctic coast, Canada. *Canadian Journal of Earth Sciences* 24, 1108–1119.

Mackay, J.R. (1988a) Catastrophic lake drainage, Tuktoyaktuk Peninsula area, District of Mackenzie. *Geological Survey of Canada Paper* 88-1D, 83–90.

Mackay, J.R. (1988b) Pingo collapse and palaeoclimatic reconstruction. *Canadian Journal of Earth Sciences* 25, 495–511.

Mackay, J.R. (1989) Massive ice: some field criteria for the identification of ice types. *Geological Survey of Canada Paper* 89-1G, 5–11.

Mackay, J.R. (1990a) Some observations on the growth and deformation of epigenetic, syngenetic and anti-syngenetic ice wedges. *Permafrost and Periglacial Processes* 1, 15–29.

Mackay, J.R. (1990b) Seasonal growth bands in pingo ice. *Canadian Journal of Earth Sciences* 27, 1115–1125.

Mackay, J.R. (1992) The frequency of ice-wedge cracking (1967–1987) at Garry Island, western Arctic coast, Canada. *Canadian Journal of Earth Sciences* 29, 236–248.

Mackay, J.R. (1993a) Air temperature, snow cover, creep of frozen ground, and the time of ice-wedge cracking, western Arctic coast. *Canadian Journal of Earth Sciences* 30, 1720–1729.

Mackay, J.R. (1993b) The sound and speed of ice-wedge cracking, Arctic Canada. *Canadian Journal of Earth Sciences* 30, 509–518.

Mackay, J.R. (1994) Pingos and pingo ice of the western Arctic coast, Canada. *Terra* 106, 1–11.

Mackay, J.R. (1995) Ice wedges on hillslopes and landform evolution in the late Quaternary, western Arctic coast, Canada. *Canadian Journal of Earth Sciences* 32, 1093–1105.

Mackay, J.R. (1998) Pingo growth and collapse, Tuktoyaktuk Peninsula area, western arctic coast, Canada: a long-term field study. *Géographie Physique et Quaternaire* 52, 271–323.

Mackay, J.R. (1999a) Periglacial features developed on the exposed lake bottoms of seven lakes that drained rapidly after 1950, Tuktoyaktuk Peninsula area, western arctic coast, Canada. *Permafrost and Periglacial Processes* 10, 39–63.

Mackay, J.R. (1999b) Cold-climate shattering (1974–1993) of 200 glacial erratics on the exposed bottom of a recently drained lake, western Arctic coast, Canada. *Permafrost and Periglacial Processes* 10, 125–136.

Mackay, J.R. (2000) Thermally induced movements in ice-wedge polygons, western Arctic coast: a long-term study. *Géographie Physique et Quaternaire* 54, 41–68.

Mackay, J.R. and Burn, C.R. (2002) The first 20 years (1978–1979 to 1998–1999) of ice-wedge growth at the Illisarvik experimental drained site, western Arctic coast, Canada. *Canadian Journal of Earth Sciences* 39, 95–111.

Mackay, J.R. and Burn, C.R. (2005) A long-term field study (1951–2003) of ventifacts formed by katabatic winds at Paulatuk, western Arctic coast, Canada. *Canadian Journal of Earth Sciences* 42, 1615–1635.

Mackay, J.R. and Burn, C.R. (2011) A century of change in a collapsing pingo, Parry Peninsula, western Arctic coast, Canada. *Permafrost and Periglacial Processes* 22, 266–272.

Mackay, J.R. and Dallimore, S.R. (1992) Massive ice of the Tuktoyaktuk area, western Arctic coast, Canada. *Canadian Journal of Earth Sciences* 29, 1235–1249.

Mackay, J.R. and Mackay, D.K. (1976) Cryostatic pressures in nonsorted circles (mud hummocks), Inuvik, North-West Territories. *Canadian Journal of Earth Sciences* 13, 889–897.

Mackay, J.R. and Mathews, W.H. (1974) Movement of sorted stripes, the Cinder Cone, Garibaldi National Park, BC, Canada. *Arctic and Alpine Research* 6, 347–359.

Maddy, D., Bridgland, D. and Westaway, R. (2001) Uplift-driven valley incision and climate-controlled river development in the Thames Valley, UK. *Quaternary International* 79, 23–36.

Makkaveyev, A.N., Bronguleev, V.V. and Karavaev, V.A. (2015) Pleistocene pingo in the central part of the East European Plain. *Permafrost and Periglacial Processes* 26, 360–367.

Manikowska, B. (1991) Vistulian and Holocene aeolian activity, pedostratigraphy and relief evolution in central Poland. *Zeitschrift für Geomorphologie Supplementband* 90, 131–141.

Mann, D.H., Heiser, P.A. and Finney, B.P. (2002) Holocene history of the Great Kobuk sand dunes, northwestern Alaska. *Quaternary Science Reviews* 21, 709–731.

Manson, G.K., Solomon, S.M., Forbes, D.L. *et al.* (2005) Spatial variability of factors influencing coastal change in the western Canadian Arctic. *Geo-Marine Letters* 25, 138–145.

Marchant, D.R., Lewis, A.R., Phillips, W.M. *et al.* (2002) Formation of patterned ground and sublimation till over Miocene glacier ice in Beacon Valley, southern Victoria Land, Antarctica. *Geological Society of America Bulletin* 114, 718–730.

Margesin, R. (2009) (ed.) *Permafrost Soils.* Springer-Verlag, Berlin and Heidelberg, 320 pp.

Margold, M., Treml, V., Petr, L. and Nyplová, P. (2011) Snowpatch hollows and pronival ramparts in the Krkonose Mountains, Czech Republic: distribution, morphology and chronology of formation. *Geografiska Annaler* 93A, 137–150.

Mark, A.F. (1994) Patterned ground activity in a southern New Zealand high-alpine complex. *Arctic and Alpine Research* 26, 270–280.

Markovic, S.B., Hambach, U., Stevens, T. *et al.* (2011) The last million years recorded at the Stari Slankamen (Northern Serbia) loess-palaeosol sequence: revised chronostratigraphy and long-term environmental trends. *Quaternary Science Reviews* 30, 1142–1154.

Marquette, G.C., Gray, J.T., Gosse, J.C. *et al.* (2004) Felsenmeer persistence under non-erosive ice in the Torngat and Kaumajet mountains, Quebec and Labrador, as determined by soil weathering and cosmogenic nuclide exposure dating. *Canadian Journal of Earth Sciences* 41, 19–38.

Marsh, B. (1987) Pleistocene pingo scars in Pennsylvania. *Geology* 15, 945–947.

Marsh, P. and Neumann, N. (2001) Processes controlling the rapid drainage of two ice-rich permafrost-dammed lakes in NW Canada. *Hydrological Processes* 15, 3433–3446.

Marsh, P. and Woo, M.-K. (1981) Snowmelt, glacier melt and high Arctic streamflow regimes. *Canadian Journal of Earth Sciences* 18, 1380–1384.

Marsh, P., Russell, M., Pohl, S. *et al.* (2009) Changes in thaw lake drainage in the Western Canadian Arctic from 1950 to 2000. *Hydrological Processes* 23, 145–158.

Marshall, S.J. and Clark, P.U. (2002) Basal temperature evolution of North American ice sheets and implications for the 100-kyr cycle. *Geophysical Research Letters* 29, 67-1–67-4.

Marsz, A.A., Stysznska, A., Pekala, K. and Repelewska-Pekalowa, J. (2013) Influence of meteorological elements on changes in active-layer thickness in the Bellsund region, Svalbard. *Permafrost and Periglacial Processes* 24, 304–312.

Martin, Y. (2003) Evaluation of bed load transport formulae using field evidence from the Vedder River, British Columbia. *Geomorphology* 53, 75–95.

Matsumoto, H., Yamada, S. and Hirakawa, K. (2010) Relationships between ground ice and solifluction: field measurements in the Daisetsu Mountains, northern Japan. *Permafrost and Periglacial Processes* 21, 78–89.

Matsuoka, N. (1990a) Mechanism of rock breakdown by frost action: an experimental approach. *Cold Regions Science and Technology* 17, 253–270.

Matsuoka, N. (1990b) Rate of bedrock weathering by frost action: field measurements and a predictive model. *Earth Surface Processes and Landforms* 15, 73–90.

Matsuoka, N. (1994) Diurnal freeze-thaw depth in rockwalls: field measurements and theoretical considerations. *Earth Surface Processes and Landforms* 19, 423–455.

Matsuoka, N. (1995) Rock weathering processes and landform development in the Sør Rondane Mountains, Antarctica. *Geomorphology* 12, 323–339.

Matsuoka, N. (1998a) The relationship between frost heave and downslope soil movement: field measurements in the Japanese Alps. *Permafrost and Periglacial Processes* 9, 121–133.

Matsuoka, N. (1998b) Modelling frost creep rates in an alpine environment. *Permafrost and Periglacial Processes* 9, 397–409.

Matsuoka, N. (1999) Monitoring of thermal contraction cracking at an ice wedge site, central Spitsbergen. *Polar Geoscience* 12, 258–271.

Matsuoka, N. (2001a) Direct observation of frost wedging in alpine bedrock. *Earth Surface Processes and Landforms* 26, 601–614.

Matsuoka, N. (2001b) Microgelivation versus macrogelivation: towards bridging the gap between laboratory and field frost weathering. *Permafrost and Periglacial Processes* 12, 299–313.

Matsuoka, N. (2001c) Solifluction rates, processes and landforms: a global review. *Earth-Science Reviews* 55, 107–134.

Matsuoka, N. (2005) Temporal and spatial variations in periglacial soil movements on alpine slope crests. *earth Surface Processes and Landforms* 30, 41–58.

Matsuoka, N. (2006) Monitoring periglacial processes: towards construction of a global network. *Geomorphology* 80, 20–31.

Matsuoka, N. (2008) Frost weathering and rockwall erosion in the southeastern Swiss Alps: long-term (1994–2006) observations. *Geomorphology* 99, 353–368.

Matsuoka, N. (2010) Solifluction and mudflow on a limestone periglacial slope in the Swiss Alps: 14 years of monitoring. *Permafrost and Periglacial Processes* 21, 219–240.

Matsuoka, N. (2014) Combining time-lapse photography and multisensory monitoring to understand frost creep dynamics in the Japanese Alps. *Permafrost and Periglacial Processes* 25, 94–106.

Matsuoka, N. and Christiansen, H.H. (2009) Ice-wedge polygon dynamics in Svalbard: high resolution monitoring by multiple techniques. In Kane, D.L. and Hinkel, D.M. (eds) *Proceedings of the Ninth International Conference on Permafrost.* University of Alaska, Fairbanks, 1149–1154.

Matsuoka, N. and Hirakawa, K. (2000) Solifluction resulting from one-sided and two-sided freezing: field data from Svalbard. *Polar Geoscience* 13, 187–201.

Matsuoka, N. and Moriwaki, K. (1992) Frost heave and creep in the Sør Rondane Mountains, Antarctica. *Arctic and Alpine Research* 24, 271–280.

Matsuoka, N. and Murton, J. (2008) Frost weathering: recent advances and future directions. *Permafrost and Periglacial Processes* 19, 195–210.

Matsuoka, N. and Sakai, H. (1999) Rockfall activity from an alpine cliff during thawing periods. *Geomorphology* 28, 309–329.

Matsuoka, N., Moriwaki, K. and Hirakawa, K. (1996) Field experiments on physical weathering and wind erosion in an Antarctic cold desert. *Earth Surface Processes and Landforms* 21, 687–699.

Matsuoka, N., Hirakawa, K., Watanabe, T. and Moriwaki, K. (1997) Monitoring of periglacial slope processes in the Swiss Alps: the first two years of frost shattering, heave and creep. *Permafrost and Periglacial Processes* 8, 155–177.

Matsuoka, N., Abe, M. and Ijiri, M. (2003) Differential frost heave and sorted patterned ground: field measurements and a laboratory experiment. *Geomorphology* 52, 73–85.

Matsuoka, N., Sawaguchi, S. and Yoshikawa, K. (2004) Present-day periglacial environments in central Spitsbergen. *Geographical Review of Japan, English Edition* 1, 54–78.

Matsuoka, N., Ikeda, A., and Date, T. (2005) Morphometric analysis of solifluction lobes and rock glaciers in the Swiss Alps. *Permafrost and Periglacial Processes* 16, 99–113.

Matsuoka, N., Thomachot, C.E., Oguchi, C.T. *et al.* (2006) Quaternary bedrock weathering and landscape evolution in the Sør Rondane Mountains. East Antarctica: re-evaluating rates and processes. *Geomorphology* 81, 408–420.

Matthews, J.A. and McCarroll, D. (1994) Snow avalanche impact landforms in Breheimen, southern Norway: origin, age and palaeoclimatic implications. *Arctic and Alpine Research* 26, 103–115.

Matthews, J.A. and Owen, G. (2011) Holocene chemical weathering, surface lowering and rock weakening rates on glacially eroded bedrock surfaces in an alpine periglacial environment, Jotunheimen, southern Norway. *Permafrost and Periglacial Processes* 22, 279–290.

Matthews, J.A. and Seppälä, M. (2014) Holocene environmental change in subarctic aeolian dune fields: the chronology of sand dune reactivation events in relation to forest fires, palaeosol development and climatic variations in Finnish Lapland. *The Holocene* 24, 149–164.

Matthews, J.A. and Wilson, P. (2015) Improved Schmidt-hammer exposure ages for active and relict pronival ramparts in southern Norway, and their palaeoenvironmental implications. *Geomorphology* 246, 7–21.

Matthews, J.A., Dawson, A.G. and Shakesby, R.A. (1986a) Lake shoreline development, frost weathering and rock platform erosion in an alpine periglacial environment, Jotunheimen, southern Norway. *Boreas* 15, 33–50.

Matthews, J.A., Harris, C. and Ballantyne, C.K. (1986b) Studies on a gelifluction lobe, Jotunheimen, Norway: [14]C chronology, stratigraphy, sedimentology and environment. *Geografiska Annaler* 68A, 345–360.

Matthews, J.A., Ballantyne, C.K., Harris, C. and McCarroll, D. (1993) Solifluction and climatic variation in the Holocene: discussion and synthesis. *Paläoklimaforschung* 11, 339–361.

Matthews, J.A., Dahl, S.O., Berrisford, M.S. and Nesje, A. (1997a) Cyclic development and thermokarst degradation of palsas in the mid-Alpine zone at Leirpullen, Dovrefjell, southern Norway. *Permafrost and Periglacial Processes* 8, 107–122.

Matthews, J.A., Dahl, S.O., Berrisford, M.S. *et al.* (1997b) A preliminary history of colluvial (debris-flow) activity, Leirdalen, Jotunheimen, Norway. *Journal of Quaternary Science* 12, 117–129.

Matthews, J.A., Shakesby, R.A., Berrisford, M.S. and McEwen, L.J. (1998) Periglacial patterned ground in the Styggedalsbreen glacier foreland, Jotunheimen, southern Norway: micro-topographical, paraglacial and geochronological controls. *Permafrost and Periglacial Processes* 9, 147–166.

Matthews, J.A., Shakesby, R.A., McEwen, L.J. *et al.* (1999) Alpine debris flows in Leirdalen, Norway, with particular reference to distal fans, intermediate-type deposits and flow types. *Arctic and Alpine Research* 31, 421–435.

Matthews, J.A., Seppälä, M. and Dresser, P.Q. (2005) Holocene solifluction, climate variation and fire in a subarctic landscape at Pippokangas, Finnish Lapland, based on radiocarbon-dated buried charcoal. *Journal of Quaternary Science* 20, 533–548.

Matthews, J.A., Shakesby, R.A., Owen, G. and Vater, A.E. (2011) Protalus rampart formation in relation to snow-avalanche activity and Schmidt hammer exposure-age dating: three case studies from southern Norway. *Geomorphology* 130, 280–288.

Matthews, J.A., Nesje, A. and Linge, H. (2013) Relict talus-foot rock glaciers at Øyberget, upper Ottadalen, southern Norway: Schmidt hammer exposure ages and palaeoenvironmental implications. *Permafrost and Periglacial Processes* 24, 336–346.

Matthews, J.A., McEwen, L.J. and Owen, G. (2015) Schmidt-hammer exposure-age dating (SHD) of snow-avalanche impact ramparts in southern Norway: approaches, results and implications for landform age, dynamics and development. *Earth Surface Processes and Landforms* 40, 1705–1718.

Mausbacher, R., Schneider, H. and Igl, M. (2001) Influence of late glacial climate changes on sediment transport in the River Werra (Thuringia, Germany). *Quaternary International* 79, 101–109.

McCann, S.B. and Cogley, J.G. (1974) The geomorphic significance of fluvial activity at high latitudes. In Fahey, B.D. and Thompson, R.D. (eds) *Research in Polar and Alpine Geomorphology*. Geo-Abstracts, Norwich, 118–135.

McCarroll, D. (1990) Differential weathering of feldspar and pyroxene in an arctic-alpine environment. *Earth Surface Processes and Landforms* 15, 641–651.

McCarroll, D. and Viles, H. (1995) Rock weathering by the lichen *Lecidea auriculata* in an arctic alpine environment. *Earth Surface Processes and Landforms* 20, 199–206.

McCarroll, D., Shakesby, R.A. and Matthews, J.A. (1998) Spatial and temporal variations of late Holocene rockfall activity on a Norwegian talus slope: a lichenometric and simulation-modelling approach. *Arctic and Alpine Research* 30, 51–60.

McCarroll, D., Shakesby, R.A. and Matthews, J.A. (2001) Enhanced rockfall activity during the Little Ice Age: further lichenometric evidence from a Norwegian talus. *Permafrost and Periglacial Processes* 12, 157–164.

McClelland, J.W., Déry, S.J., Peterson, B.J. *et al.* (2006) A pan-arctic evaluation of changes in river discharge during the latter half of the 20th century. *Geophysical Research Letters* 33, L06715.

McColl, S.T. (2012) Paraglacial rock-slope stability. *Geomorphology* 153–154, 1–16.

McDonald, D.M. and Lamoureux, S.F. (2009) Hydroclimatic and channel snowpack controls over suspended sediment and grain size transport in a High Arctic catchment. *Earth Surface Processes and Landforms* 34, 424–436.

McEwen, L.J. and Matthews, J.A. (1998) Channel form, bed material and sediment sources of the Sprongdøla, southern Norway: evidence for a distinct periglacio-fluvial system. *Geografiska Annaler* 80A, 17–36.

McEwen, L.J., Matthews, J.A., Shakesby, R.A. and Berrisford, M.S. (2002) Holocene gorge excavation linked to boulder fan formation and frost weathering in a Norwegian alpine perifluvioglacial system. *Arctic, Antarctic and Alpine Research* 34, 345–357.

McGee, D., Broeker, W.S. and Winckler, G. (2010) Gustiness: the driver of glacial dustiness? *Quaternary Science Reviews* 29, 2340–2350.

McGowan, H.A., Sturman, A.P. and Owens, I.F. (1996) Aeolian dust transport and deposition by foehn winds in an alpine environment, Lake Tekapo, New Zealand. *Geomorphology* 15, 135–146.

McGuire, A.D., Anderson, L.G., Christensen, T.R. *et al.* (2009) Sensitivity of the carbon cycle in the Arctic to climate change. *Ecological Monographs* 79, 523–555.

McKay, C.P., Molaro, J.L. and Marinova, M.M. (2009) High frequency rock temperature data from hyper-arid desert environments in the Atacama and the Antarctic Dry Valleys and implications for rock weathering. *Geomorphology* 110, 182–187.

McKenna Neuman, C. (1989) Kinetic energy transfer through impact and its role in entrainment by wind of particles from frozen surfaces. *Sedimentology* 36, 1007–1015.

McKenna Neuman, C. (1990a) Observations of winter aeolian transport and niveo-aeolian deposition at Crater Lake, Pangnirtung Pass, NWT, Canada. *Permafrost and Periglacial Processes* 1, 235–247.

McKenna Neuman, C. (1990b) Role of sublimation in particle supply for aeolian sediment transport in cold environments. *Geografiska Annaler* 72A, 329–335.

McKenna Neuman, C. (1993) A review of aeolian transport processes in cold environments. *Progress in Physical Geography* 17, 137–155.

McKenna Neuman, C. (2004) Effects of temperature and humidity upon the entrainment of sedimentary particles by wind. *Sedimentology* 51, 1–17.

McKenna Neuman C. and Gilbert, R. (1986) Aeolian processes and landforms in glacio-fluvial environments of southeastern Baffin Island, NWT, Canada. In: Nickling, W.G. (ed.) *Aeolian Geomorphology*. Allen and Unwin, London, 213–235.

McRoberts, E.C. and Morgenstern, N.R. (1974) The stability of thawing slopes. *Canadian Geotechnical Journal* 11, 447–469.

Mears, B. (1997) Pleistocene gelifluction and rock deformation on slopes in southern Wyoming. *Permafrost and Periglacial Processes* 8, 251–255.

Meiklejohn, I. and Hall, K.J. (1997) Aqueous geochemistry as an indicator of chemical weathering on southeastern Alexander Island, Antarctica. *Polar Geography* 21, 101–112.

Meinardus, W. (1912) Beobachtungen über Detritussortierung und Strukturboden auf Spitzbergen. *Gesellschaft für Erdkunde, Berlin, Zeitschrift* 1912, 250–259.

Meinsen, J., Winsemann, J., Roskosch, J. *et al.* (2014) Climate control on the evolution of Late Pleistocene alluvial-fan and aeolian sand-sheet systems in NW Germany. *Boreas* 43, 42–66.

Melnikov, V.P. and Spesivtsev, V.I. (2000) *Cryogenic formations in the Earth's lithosphere*. Scientific Publishing Center: Novosibirsk.

Mercier, D. (2011) *La Géomorphologie Paraglaciaire*. Éditions Universitaires Européennes, Saarbrücken, 256 pp.

Meyer, H., Dereviagin, A., Siegert, C. *et al.* (2002) Palaeoclimatic reconstruction on Big Lyakhovsky Island, North Siberia – Hydrogen and Oxygen Isotopes in ice wedges. *Permafrost and Periglacial Processes* 13, 91–105.

Meyer, H., Schirrmeister, L., Yoshikawa, K. *et al.* (2010) Permafrost evidence for severe winter cooling during the Younger Dryas in northern Alaska. *Geophysical Research Letters* 37, L03501.

Miao, X., Mason, J.A., Goble, R.J. and Hanson, P.R. (2005) Loess record of dry climate and aeolian activity in the early- to mid-Holocene, central Great Plains, North America. *The Holocene* 15, 339–346.

Michaud, Y. and Dionne, J.-C. (1987) Alteration des substrats rocheux et rôle du soulevement gélival dans la formation des champs de blocaille, en Hudsonie. *Géographie Physique et Quaternaire* 41, 7–18.

Michaud, Y. and Dyke, L.D. (1990) Mechanism of bedrock frost heave in permafrost regions. In *Permafrost Canada: Proceedings of the 5th Canadian Permafrost Conference*. Université Laval, Québec, 125–130.

Michaud, Y., Dionne, J.-C. and Dyke, L.D. (1989) Frost bursting: a violent expression of frost action in rock. *Canadian Journal of Earth Sciences* 26, 2075–2080.

Minsley, B.J., Abraham, J.D., Smith, B.D. *et al.* (2012) Airborne electromagnetic imaging of discontinuous permafrost. *Geophysical Research Letters* 39, L02503.

Moine, O., Rousseau, D.-D. and Antoine, P. (2008) Abrupt malacological and lithological changes along the upper Weichselian sequence of Nussloch (Rhine Valley, Germany). *Quaternary Research* 70, 91–104.

Mol, J., Vandenberghe, J. and Kasse, C. (2000) River response to variations of periglacial climate in mid-latitude Europe. *Geomorphology* 33, 131–148.

Mollard, J.D. (2000) Ice-shaped ring forms in Western Canada: their airphoto expressions and manifold polygenetic origins. *Quaternary International* 68–71, 187–198.

Monnier, S. and Kinnard, C. (2015) Reconsidering the glacier to rock glacier transformation problem: new insights from the central Andes of Chile. *Geomorphology* 238, 47–55.

Monnier, S., Camerlynck, C. and Rejiba, F. (2008) Ground penetrating radar survey and stratigraphic survey of the Plan du Lac rock glaciers, Vanoise Massif, northern French Alps. *Permafrost and Periglacial Processes* 19, 19–30.

Monnier, S., Camerlynck, C., Rejiba, F. *et al.* (2011) Structure and genesis of the Thabor rock glacier (northern French Alps) determined from morphological and ground-penetrating radar surveys. *Geomorphology* 134, 269–279.

Mooers, H.D. and Glaser, P.H. (1989) Active patterned ground at sea level, Fourcha, Nova Scotia, Canada. *Arctic and Alpine Research* 21, 425–432.

Moore, J.R., Sanders, J.W., Dietrich, W.E. and Glaser, S.D. (2009) Influence of rock mass strength on the erosion rate of alpine cliffs. *Earth Surface Processes and Landforms* 34, 1339–1352.

Moorman, B.J. and Michel, F.A. (2000) The burial of ice in the proglacial environment on Bylot Island, Arctic Canada. *Permafrost and Periglacial Processes* 11, 161–175.

Moorman, B.J., Michel, F.A. and Wilson, A. (1998) The development of tabular massive ground ice at Peninsula Point, N.W.T., Canada. In Lewkowicz, A.G. and Allard, M. (eds) *Proceedings of the 7th International Permafrost Conference*, Université Laval, Québec, 757–762.

Moran, A.P., Ivy Ochs, S., Vockenhuber, C. and Kerschner, H. (2016) Rock glacier development in the Northern Calcarious Alps at the Pleistocene-Holocene boundary. *Geomorphology* 273, 178–188.

Morgenstern, A., Grosse, G., Günther, F. *et al.* (2011) Spatial analysis of thermokarst lakes and basins in Yedoma landscapes of the Lena Delta. *The Cryosphere* 5, 849–867.

Morgenstern, A., Ulrich, M., Günther, F. *et al.* (2013) Evolution of thermokarst in East Siberian ice-rich permafrost: a case study. *Geomorphology* 201, 363–379.

Morgenstern, N.R. and Nixon, J.K. (1971) One dimensional consolidation of thawing soils. *Canadian Geotechnical Journal* 8, 558–565.

Morse, P.D. and Burn, C.R. (2013) Field observations of syngenetic ice-wedge polygons, outer Mackenzie Delta, western Arctic coast, Canada. *Journal of Geophysical Research: Earth Surface* 118, 1320–1332.

Morse, P.D. and Burn, C.R. (2014) Perennial frost blisters of the outer Mackenzie Delta, western Arctic coast, Canada. *Earth Surface Processes and Landforms* 39, 200–213.

Morse, P.D., Burn, C.R. and Kokelj, S.V. (2012) Influence of snow on near-surface ground temperatures in upland and alluvial environments of the outer Mackenzie Delta, Northwest Territories. *Canadian Journal of Earth Sciences* 49, 895–913.

Mountney, N.P. and Russell, A.J. (2004) Sedimentology of cold-climate aeolian sand sheets in the Askja region of northeast Iceland. *Sedimentary Geology* 166, 223–244.

Mugridge, S.-J. and Young, H.R. (1983) Disintegration of shale by cyclic wetting and drying and frost action. *Canadian Journal of Earth Sciences* 20, 568–576.

Muhs, D.R. (2004) Mineralogical maturity in dunefields of North America, Africa and Australia. *Geomorphology* 59, 247–269.

Muhs, D.R. (2013) Loess and its geomorphic, stratigraphic, and palaeoclimatic significance in the Quaternary. In: Shroder, J. (ed.) *Treatise on Geomorphology*, vol. 11. Academic Press, San Diego, CA, 149–183.

Muhs, D.R. and Benedict, J.B. (2006) Eolian additions to Late Quaternary alpine soils, Indian Peaks Wilderness area, Colorado Front Range. *Arctic, Antarctic and Alpine Research* 38, 120–130.

Muhs, D.R., Stafford, T.W., Cowherd, S.D. *et al.* (1996) Origin of the late Quaternary dunefields of northeastern Colorado. *Geomorphology* 17, 129–149.

Muhs, D.R., Ager, T.A., Bettis, E.A. III *et al.* (2003) Stratigraphy and palaeoclimatic significance of Late Quaternary loess-palaeosol sequences of the last interglacial-glacial cycle in central Alaska. *Quaternary Science Reviews* 22, 1947–1986.

Muhs, D.R., McGeehin, J.P., Beann, J. and Fisher, E. (2004) Holocene loess deposition and soil formation as competing processes, Matanuska Valley, southern Alaska. *Quaternary Research* 61, 265–276.

Muhs, D.R., Bettis, E.A. III, Aleinikoff, J.N. *et al.* (2008) Origin and paleoclimatic significance of late Quaternary loess in Nebraska: evidence from stratigraphy, chronology, sedimentology, and geochemistry. *Geological Society of America Bulletin* 120, 1378–1407.

Muhs, D.R., Bettis, E.A. III, Roberts, H.M. *et al.* (2013) Chronology and provenance of last-glacial (Peoria) loess in western Iowa and palaeoclimatic implications. *Quaternary Research* 80, 468–481.

Muller, S.W. (1947) *Permafrost or permanently frozen ground and related engineering problems.* J.W. Edwards, Ann Arbor, 231 pp.

Müller, J., Gärtner-Roer, I., Kenner, R. *et al.* (2014) Sediment storage and transfer on a periglacial mountain slope (Corvatsch, Switzerland). *Geomorphology* 218, 35–44.

Munroe, J.S., Doolittle, J.A., Kanevskiy, M.Z. *et al.* (2007) Application of ground-penetrating radar imagery for three dimensional visualisation of near-surface structures in ice-rich permafrost, Barrow, Alaska. *Permafrost and Periglacial Processes* 18, 309–321.

Murton, J.B. (1996a) Morphology and palaeoenvironmental significance of Quaternary sand veins, sand wedges and composite wedges, Tuktoyaktuk coastlands, western Arctic Canada. *Journal of Sedimentary Research* 66, 17–25.

Murton, J.B. (1996b) Near-surface brecciation of chalk, Isle of Thanet, south-east England: a comparison with ice-rich brecciated bedrock in Canada and Spitsbergen. *Permafrost and Periglacial Processes* 7, 153–164.

Murton, J.B. (1996c) Thermokarst-lake-basin sediments, Tuktoyaktuk Coastlands, western Arctic Canada. *Sedimentology* 43, 737–760.

Murton, J.B. (2001) Thermokarst sediments and sedimentary structures, Tuktoyaktuk coastlands, western Arctic Canada. *Global and Planetary Change* 28, 175–192.

Murton, J.B. (2005) Ground ice stratigraphy and formation at North Head, Tuktoyaktuk coastlands, western Arctic Canada: a product of glacier-permafrost interactions. *Permafrost and Periglacial Processes* 16, 31–50.

Murton, J.B. (2009) Global warming and thermokarst. In Margesin, R. (ed.) *Permafrost Soils.* Springer-Verlag, Berlin and Heidelberg, 185–203.

Murton, J.B. (2013a) Ice wedges and ice-wedge casts. In Elias, S. (ed.) *Encyclopedia of Quaternary Science*, 2nd edn, vol. 3. Elsevier, Amsterdam, 436–451.

Murton, J.B. (2013b) Rock Weathering. In Elias, S. (ed.) *Encyclopedia of Quaternary Science*, 2nd edn, vol. 3. Elsevier: Amsterdam, 500–506.

Murton, J.B. (2013c) Ground ice and cryostratigraphy. In Shroder, J., Giardino, R. and Harbor, J. (eds) *Treatise on Geomorphology. Volume 8: Glacial and Periglacial Geomorphology*. Academic Press, San Diego, CA, 173–201.

Murton, J.B. and Ballantyne, C.K. (2017) Periglacial and permafrost conceptual ground models for the British Isles. In Griffiths, J. and Martin, C. (eds) *Periglacial and Glacial Engineering Geology*. Geological Society, London, Engineering Group Special Publication. In press.

Murton, J.B. and Bateman, M.D. (2007) Syngenetic sand veins and anti-syngenetic sand wedges, Tuktoyaktuk Coastlands, western Arctic Canada. *Permafrost and Periglacial Processes* 18, 33–47.

Murton, J.B. and Belshaw, R.K. (2011) A conceptual model of valley incision, planation and terrace formation during cold and arid permafrost conditions of Pleistocene southern England. *Quaternary Research* 75, 385–394.

Murton, J.B. and French, H.M. (1993a) Thermokarst involutions, Summer Island, Pleistocene Mackenzie Delta, western Canadian Arctic. *Permafrost and Periglacial Processes* 4, 217–229.

Murton, J.B. and French, H.M. (1993b) Thaw modification of frost-fissure wedges, Richards Island, Pleistocene Mackenzie Delta, western Arctic Canada. *Journal of Quaternary Science* 8, 185–196.

Murton, J.B. and French, H.M. (1994) Cryostructures in permafrost, Tuktoyaktuk Coastlands, western Arctic Canada. *Canadian Journal of Earth Sciences* 31, 737–747.

Murton, J.B. and Kolstrup, E. (2003) Ice wedge casts as indicators of palaeotemperatures: precise proxy or wishful thinking? *Progress in Physical Geography* 27, 155–170.

Murton, J.B. and Lautridou, J.-P. (2003) Recent advances in the understanding of Quaternary periglacial features of the English Channel coastlands. *Journal of Quaternary Science* 18, 301–307.

Murton, J.B., Whiteman, C.A. and Allen, P. (1995) Involutions in the Middle Pleistocene (Anglian) Barham Soil, eastern England: a comparison with thermokarst involutions from arctic Canada. *Boreas* 24, 269–280.

Murton, J.B., French. H.M. and Lamothe, M. (1997) Late Wisconsinan erosion and aeolian deposition, Summer Island area, Pleistocene Mackenzie Delta, Northwest Territories: optical dating and implications for glacial chronology. *Canadian Journal of Earth Sciences* 34, 190–199.

Murton, J.B., Worsley, P and Gozdzik, J. (2000) Sand veins and wedges in cold aeolian environments. *Quaternary Science Reviews* 19, 899–922.

Murton, J.B., Baker, A., Bowen, D.Q. *et al.* (2001a) Late Middle Pleistocene temperate–periglacial–temperate sequence (Oxygen isotope stages 7-5e) near Marsworth, Buckinghamshire, UK. *Quaternary Science Reviews* 20, 1787–1825.

Murton, J.B., Coutard, J.-P., Lautridou, J.-P. *et al.* (2001b) Physical modelling of bedrock brecciation by ice segregation in permafrost. *Permafrost and Periglacial Processes* 12, 255–266.

Murton, J.B., Bateman, M.D., Baker, C.A. *et al.* (2003) The Devensian periglacial record on Thanet, Kent. *Permafrost and Periglacial Processes* 14, 217–246.

Murton, J.B., Waller, R.I., Hart, J.K. *et al.* (2004) Stratigraphy and glaciotectonic structures of a relict deformable bed of permafrost at the northwestern margin of the Laurentide ice sheet, Tuktoyaktuk Coastlands, western Canadian Arctic. *Journal of Glaciology* 50, 399–412.

Murton, J.B., Whiteman, C.A., Waller, R.I. *et al.* (2005) Basal ice facies and supraglacial melt-out till of the Laurentide Ice Sheet, Tuktoyaktuk Coastlands, western Arctic Canada. *Quaternary Science Reviews* 24, 681–708.

Murton, J.B., Peterson, R. and Ozouf, J.-C. (2006) Bedrock fracture by ice segregation in cold regions. *Science* 314, 1127–1129.

Murton, J.B., Goslar, T., Edwards, M.E. *et al.* (2015a) Palaeoenvironmental interpretation of yedoma silt (ice complex) deposition as cold-climate loess, Duvanny Yar, Northeast Siberia. *Permafrost and Periglacial Processes* 26, 208–288.

Murton, J.B., Bowen, D.Q., Candy, I. *et al.* (2015b) Middle and Late Pleistocene environmental history of the Marsworth area, south-central England. *Proceedings of the Geologists' Association* 126, 18–49.

Nachtergaele, F.O., Spaargaren, O., Deckers, J.A. and Ahrens, B. (2000) new developments in soil classification: World Reference Base for soil resources. *Geoderma* 96, 345–357.

Nansen, F. (1922) *The Strandflat and Isostasy*. Norges Videnskaps Akademie, Kristiania.

Necsoiu, M., Dinwiddie, C.L., Walter, G.R. *et al.* (2013) Multi-temporal image analysis of historical aerial photographs and recent satellite imagery reveals evolution of water body surface area and polygonal terrain morphology in Kobuk Valley National Park, Alaska. *Environmental Research Letters* 8, 025007.

Nelson, F.E. (1989) Cryoplanation terraces: periglacial cirque analogs. *Geografiska Annaler* 71, 31–41.

Nelson, F.E. and Anisimov, O.A. (1993) Permafrost zonation in Russia under anthropogenic climate change. *Permafrost and Periglacial Processes* 4, 137–148.

Nelson, F.E. and Outcalt, S.I. (1987) A computational method for prediction and regionalization of permafrost. *Arctic and Alpine Research* 19, 279–288.

Nelson, F.E., Hinkel, K.M. and Outcalt, F.M. (1992) Palsa-scale frost mounds. In Dixon, J.C. and Abrahams, A.D. (eds) *Periglacial Geomorphology: Proceedings of the 22nd Annual Binghamton Symposium in Geomorphology*. John Wiley & Sons, Chichester, 305–325.

Nelson, F.E., Shiklomanov, N.I., Hinkel, K.M. and Brown, J. (2008) Decadal results from the circumpolar active layer monitoring (CALM) program. In Kane, D.L. and Hinkel, K.M. (eds) *Ninth International Conference on Permafrost*. University of Alaska, Fairbanks, 1273–1280.

Nesje, A. (1989) The geographical and altitudinal distribution of blockfields in southern Norway and its significance to the Pleistocene ice sheets. *Zeitschrift für Geomorphologie, Supplementband* 72, 41–53.

Nguyen, T.-N., Burn, C.R., King, D.J. and Smith, S.L. (2009) Estimating the extent of near-surface permafrost using remote sensing, Mackenzie Delta, Northwest Territories. *Permafrost and Periglacial Processes* 20, 141–153.

Nicholas, J.R.J. and Hinkel K.M. (1996) Concurrent permafrost aggradation and degradation induced by forest clearing, central Alaska, U.S.A. *Arctic and Alpine Research* 28, 294–299.

Nicholson, D.T. (2009) Holocene microweathering rates and processes on ice-eroded bedrock, Røldal area, Hardangervidda, southern Norway. *Geological Society, London, Special Publication* 320, 29–49.

Nicholson, D.T. and Nicholson, F.H. (2000) Physical deterioration of sedimentary rocks subjected to experimental freeze-thaw weathering. *Earth Surface Processes and Landforms* 25, 1295–1307.

Nicholson, F.H. (1976) Patterned ground formation and description as suggested by low arctic and subarctic examples. *Arctic and Alpine Research* 8, 329–342.

Nickling, W.G. and Brazel, A.J. (1985) Surface wind characteristics along the Icefield Ranges, Yukon Territory, Canada. *Arctic and Alpine Research* 17, 125–134.

Nickling, W.G. and McKenna Neuman, C. (2009) Aeolian sediment transport. In Parsons, A. and Abrahams, A.D. (eds) *Geomorphology of Desert Environments*, 2nd edn. Springer, London, 517–555.

Nicolsky D.J., Romanovsky, V.E., Tipenko, G.S. and Walker, D.A. (2008) Modeling biophysical interactions in nonsorted circles in the low Arctic. *Journal of Geophysical Research* 113, G03S05.

Nieuwenhuijzen, M.E. and Van Steijn, H. (1990) Alpine debris flows and their sedimentary properties. A case study from the French Alps. *Permafrost and Periglacial Processes* 1, 111–128.

Nissen, T.C. and Mears, B. (1990) Late Pleistocene ice-wedge casts and sand-wedge relics in the Wyoming Basons, USA. *Permafrost and Periglacial Processes* 1, 201–219.

Niu, F., Cheng, G., Ni, W. and Jin, D. (2005) Engineering-related slope failure in permafrost regions of the Qinghai–Tibet plateau. *Cold Regions Science and Technology* 42, 215–225.

Niu, F., Lin, Z., Liu, H. and Lu, J. (2011) Characteristics of thermokarst lakes and their influence on permafrost in Qinghai-Tibet Plateau. *Geomorphology* 132, 222–233.

Niu, F., Luo, J., Lin, Z. *et al.* (2016) Thaw-induced slope failures and stability analyses in permafrost regions of the Qinghai-Tibet Plateau, China. *Landslides* 13, 55–65.

Noetzli, J. and Gruber, S. (2009) Transient thermal effects in alpine permafrost. *The Cryosphere* 3, 85–99.

Noetzli, J., Gruber, S., Kohl, T. *et al.* (2007) Three-dimensional distribution and evolution of permafrost temperatures in idealized high-mountain topography. *Journal of Geophysical Research* 112, F02S13.

Nogués-Bravo, D., Araújo, M.B., Errea, M.P. and Martínez-Rica, J.P. (2007) Exposure of global mountain systems to climate warming during the 21st Century. *Global Environmental Change* 17, 420–428.

Nyberg, R. (1985) Debris flows and slush avalanches in Northern Swedish Lappland. *Meddelanden från Universitets Geografiska Institutionen Avhandlingar* 97, 222 pp.

Nyberg, R. (1987) Slush avalanche erosion along stream courses in the northern Swedish mountains. In Godard, A. and Rapp, A. (eds) *Processus et Mesure de l'Érosion*. CNRS, Paris, 179–186.

Nyberg, R. (1989) Observations of slushflows and their geomorphic effects in the Swedish mountain area. *Geografiska Annaler* 71A, 185–198.

Nyberg, R. (1991) Geomorphic processes at snowpatch sites in the Abisko mountains, northern Sweden. *Zeitschrift für Geomorphologie* 35, 321–343.

Oberender, P. and Plan, L. (2015) Cave development by frost weathering. *Geomorphology* 229, 73–84.

Oberman, N.G. (2008) Contemporary permafrost degradation of northern European Russia. In Kane, D.L. and Hinkel, K.M. (eds) *Ninth International Conference on Permafrost*. University of Alaska, Fairbanks, 1305–1310.

Oberman, N.G. and Liygin, A.M. (2009) Predicting permafrost degradation: case study of North European Russia. *Exploration and Protection of Natural Resources* 7, 15–20.

Ogino, Y. and Matsuoka, N. (2007) Involutions resulting from annual freeze-thaw cycles: a laboratory simulation based on observations in northeastern Japan. *Permafrost and Periglacial Processes* 18, 323–335.

Olefeldt, D., Turetsky, M.R., Crill, P.M. and McGuire, A.D. (2013) Environmental and physical controls on northern hemisphere methane emissions across permafrost zones. *Global Change Biology* 19, 589–603.

Oliva, M. and Ruiz-Fernández, J. (2015) Coupling patterns between para-glacial and permafrost degradation responses in Antarctica. *Earth Surface Processes and Landforms* 40, 1227–1238.

Oliva, M., Schulte, L. and Ortiz, A.G. (2009) Morphometry and Late Holocene activity of solifluction landforms in the Sierra Nevada, Southern Spain. *Permafrost and Periglacial Processes* 20, 369–382.

Oliva, M., Schulte, L. and Ortiz, A.G. (2011) The role of aridification in constraining the elevation range of Holocene solifluction processes and associated landforms in the periglacial belt of the Sierra Nevada (southern Spain). *Earth Surface Processes and Landforms* 36, 1279–1291.

Olyphant, G.A. (1983) Analysis of the factors controlling cliff burial by talus within Blanca Massif, southern Colorado, USA. *Arctic and Alpine Research* 15, 65–75.

O'Neill, H.B., Burn, C.R., Kokelj, S.V. and Lantz, T.C. (2015) 'Warm' tundra: atmospheric and near-surface ground temperature inversions across an alpine treeline in continuous permafrost, western Arctic, Canada. *Permafrost and Periglacial Processes* 26, 103–118.

Ono, Y. and Watanabe, T. (1986) A protalus rampart related to alpine debris flows in the Kuranosuke Cirque, northern Japanese Alps. *Geografiska Annaler* 68A, 213–223.

Opel, T., Dereviagin, A.Y., Meyer, H. *et al.* (2011) Palaeoclimatic information from stable water isotopes from Holocene ice wedges on the Dmitrii Laptev Strait, northeast Siberia, Russia. *Permafrost and Periglacial Processes* 22, 84–100.

Orwin, J.F., Guggenmos, M.R. and Holland, P.G. (2010a) Changes in suspended sediment to solute yield ratios from an alpine basin during the transition to winter, Southern Alps, New Zealand. *Geografiska Annaler* 92A, 247–261.

Orwin, J.F., Lamoureux, S.F., Warburton, J. and Beylich, A.A. (2010b) A framework for characterizing fluvial sediment fluxes from source to sink in cold environments. *Geografiska Annaler* 92A, 155–176.

Osterkamp, T.E. (2007) Characteristics of the recent warming of permafrost in Alaska. *Journal of Geophysical Research* 112, F02S02.

Osterkamp, T.E., Viereck, L., Shur, Y. *et al.* (2000) Observations of thermokarst and its impact on boreal forests in Alaska, U.S.A. *Arctic, Antarctic and Alpine Research* 32, 303–315.

Osterkamp, T.E., Jorgenson, M.E., Schuur, E.A.G. *et al.* (2009) Physical and ecological changes associated with warming permafrost and thermokarst in interior Alaska. *Permafrost and Periglacial Processes* 20, 235–256.

Otto, J.C. and Sass, O. (2006) Comparing geophysical methods for talus slope investigations in the Turtmann Valley (Swiss Alps). *Geomorphology* 76, 257–272.

Outcalt, S.I. and Nelson, F.E. (1984) Growth mechanisms in aggradation palsas. *Zeitschrift für Geomorphologie* 20, 65–78.

Outcalt, S.I., Nelson, F.E. and Hinkel, K.M. (1990) The zero-curtain effect: heat and mass transfer across an isothermal region in freezing soil. *Water Resources Research* 26, 1509–1516.

Overduin, P.P. and Kane, D.L. (2006) Frost boils and soil ice content: field observations. *Permafrost and Periglacial Processes* 17, 291–307.

Overduin, P.P., Hubbertin, H.-W., Rachold, V. *et al.* (2007) The evolution and degradation of coastal and offshore permafrost in the Laptev and East Siberian seas during the last climate cycle. *Geological Society of America, Special Paper* 426, 97–111.

Overduin, P.P., Strzelecki, M.C., Grigoriev, M.N. *et al.* (2014) Coastal changes in the Arctic. *Geological Society, London, Special Publication* 388, 103–129.

Overeem, I. and Syvitsky, J.P.M. (2010) Shifting discharge peaks in Arctic rivers. *Geografiska Annaler* 92A, 285–296.

Overeem, I., Anderson, R.S., Wobus, C.W. *et al.* (2011) Sea ice loss enhances wave action at the Arctic coast. *Geophysical Research Letters* 38, L17503.

Owen, G., Matthews, J.A., Shakesby, R.A. and He, X. (2006) Snow-avalanche impact landforms, deposits and effects at Urdvatnet, southern Norway: implications for avalanche style and process. *Geografiska Annaler* 88A, 295–307.

Owen, L.A. (1989) Terraces, uplift and climate in the Karakoram Mountains, northern Pakistan: Karakoram intermontane basin evolution. *Zeitschrift für Geomorphologie Supplementband* 76, 117–146.

Owen, L.A. and Sharma, M.C. (1998) Rates and magnitudes of paraglacial fan formation in the Garwhal Himalaya: implications for landscape evolution. *Geomorphology* 26, 171–184.

Owen, L.A., Richards, B., Rhodes, E.J. *et al.* (1998) Relict permafrost structures in the Gobi of Mongolia: age and significance. *Journal of Quaternary Science* 13, 539–547.

Owens, P.N. and Slaymaker, O. (1997) Contemporary and post-glacial rates of aeolian deposition in the Coast Mountains of British Columbia, Canada. *Geografiska Annaler* 79A, 267–276.

Ozouf, J.-C., Texier, J.-P., Bertran, P. and Coutard, J.-P. (1995) Quelques coupes caractéristiques dans les dépôts de versant d'Aquitaine Septentrionale: facies et interprétation dynamique. *Permafrost and Periglacial Processes* 6, 89–101.

Paasche, Ø., Strømsøe, J.R., Dahl, S.O. and Linge, H. (2006) Weathering characteristics of arctic islands in northern Norway. *Geomorphology* 82, 430–452.

Paasche, Ø., Dahl, S.O., Løvlie, R. *et al.* (2007) Rockglacier activity during the Last Glacial-Interglacial transition and Holocene spring snowmelting. *Quaternary Science Reviews* 26, 793–807.

Palacios, D., de Andrés, N. and Luengo, E. (2003) Distribution and effectiveness of nivation in

Mediterranean mountains: Peñalara (Spain). *Geomorphology* 54, 157–178.

Palacios, D., Zamorano, J.J. and de Andrés, N. (2007) Permafrost distribution in tropical stratovolcanoes: Popocatépetl and Iztaccíhuatl volcanoes (Mexico). *Geophysical Research Abstracts* 9, 05615.

Pancza, A. (1998) Les bourrelets-protalus: liens entre éboulis et les glaciers rocheux. *Permafrost and Periglacial Processes* 9, 167–175.

Panda, S.K., Prakash, A., Solie, D.N. *et al.* (2010) Remote sensing and field-based mapping of permafrost distribution along the Alaska highway corridor, interior Alaska. *Permafrost and Periglacial Processes* 21, 271–281.

Park Nelson, K.J., Nelson, F.E. and Walegur, M.T. (2007) Periglacial Appalachia: palaeoclimatic significance of blockfield elevation gradients, eastern USA. *Permafrost and Periglacial Processes* 18, 61–73.

Parks, C.D. (1991) A review of the possible mechanisms of cambering and valley bulging. *Geological Society, London, Engineering Geology Special Publication* 7, 373–380.

Parks, D.A. and Rendell, H.M. (1992) Thermoluminescence dating and geochemistry of loessic deposits in south-east England. *Journal of Quaternary Science* 7, 99–107.

Parsekian, A.D., Jones, B.M., Jones, M. *et al.* (2011) Expansion rate and geometry of floating vegetation mats on the margins of thermokarst lakes, northern Seward Peninsula, Alaska, USA. *Earth Surface Processes and Landforms* 36, 1889–1897.

Pastick, N.J., Jorgenson, M.T., Wylie, B.K. *et al.* (2013) Extending airborne electromagnetic surveys for regional active layer and permafrost mapping with remote sensing or ancillary data, Yukon Flats ecoregion, central Alaska. *Permafrost and Periglacial Processes* 24, 184–199.

Paterson, T.T. (1940) The effects of frost action around Baffin Bay and in the Cambridge district. *Quarterly Journal of the Geological Society, London* 96, 99–130.

Paull, C.K., Ussler, W., Dallimore, S.R. *et al.* (2007) Origin of pingo-like features on the Beaufort Sea shelf and their possible relationship to decomposing methane gas hydrates. *Geophysical Research Letters* 34, L01603.

Pavlova, I., Jomelli, V., Brunstein, D. *et al.* (2014) Debris flow activity related to recent climate conditions in the French Alps: a regional investigation. *Geomorphology* 219, 248–259.

Pawelec, H. (2011) Periglacial evolutions of slopes – rock controls versus climate factors (Cracow Uplands, S. Poland). *Geomorphology* 132, 139–152.

Pawelec, H. and Ludwikowska-Kedzia, M. (2016) Macro- and micromorphologic interpretation of relict periglacial slope deposits from the Holy Cross Mountains, Poland. *Permafrost and Periglacial Processes* 27, 229–247.

Pawluk, S. (1988) Freeze-thaw effects on granular structure reorganization for soil materials of varying texture and moisture content. *Canadian Journal of Earth Sciences* 68, 485–494.

Payette S., Delwaide, A., Caccianga, M. and Beauchemin, M. (2004) Accelerated thawing of subarctic peatland permafrost over the last 50 years. *Geophysical Research Letters* 31, L18208.

Payton, I.R.W. (1992) Fragipan formation in argillic brown earths (Fragiudalfs) of the Milfield Plain, north-east England. 1: Evidence for a periglacial stage of development. *Journal of Soil Science* 43, 621–644.

Peppin, S.S.L. and Style, R.W. (2013) The physics of frost heave and ice-lens growth. *Vadose Zone Journal* 12, 35–46.

Pérez, F.L. (1988) Debris transport over a snow surface: a field experiment. *Revue de Géomorphologie Dynamique* 37, 81–101.

Pérez, F.L. (1992) Miniature sorted stripes in the Páramo de Piedras Blancas (Venezuelan Andes). In Dixon, J.C. and Abrahams, A.D. (eds) *Periglacial Geomorphology*. John Wiley & Sons, Chichester, 125–157.

Pérez, F.L. (1993) Talus movement in the high equatorial Andes: a synthesis of ten years of data. *Permafrost and Periglacial Processes* 4, 199–215.

Pestryakova, L.A., Herzschuh, U., Wetterich, S, and Ulrich, M. (2012) Present-day variability and Holocene dynamics of permafrost-affected lakes in central Yakutia (Eastern Siberia) inferred from diatom records. *Quaternary Science Reviews* 51, 56–70.

Peterson, B.J., Holmes, R.M., McClelland, J.W. *et al.* (2002) Increasing river discharge to the Arctic Ocean. *Science* 298, 2171–2173.

Peterson, R.A. (2011) Assessing the role of differential frost heave in the origin of nonsorted circles. *Quaternary Research* 75, 325–333.

Peterson, R.A. and Krantz, W.B. (2003) A mechanism for differential frost heave and its implications for patterned ground formation. *Journal of Glaciology* 49, 69–80.

Peterson, R.A. and Krantz, W.B. (2008) Differential frost heave model for patterned ground formation: corroboration with observations along a North American arctic transect. *Journal of Geophysical Research* 113, G03S04.

Péwé, T.L. (1966) Palaeoclimatic significance of fossil ice wedges. *Biuletyn Peryglacjalny* 15, 65–73.

Phillips, M., Mutter, E.K., Kern-Luetschg, M. and Lehning, M. (2009) Rapid degradation of ground ice in a ventilated talus slope, Flüela Pass, Swiss Alps. *Permafrost and Periglacial Processes* 20, 1–14.

Phillips, W.M., Hall, A.M., Mottram, R. *et al.* (2006) Cosmogenic [10]Be and [26]Al exposure ages of tors and erratics, Cairngorm Mountains, Scotland: timescales for the development of a classic landscape of selective linear erosion. *Geomorphology* 73, 222–245.

Pierce, K.L., Muhs, D.R., Fosberg, M.A. *et al.* (2011) A loess-palaeosol record of climate and glacial history over the past two glacial-interglacial cycles (~150 ka), southern Jackson Hole, Wyoming. *Quaternary Research* 76, 119–141.

Pigati, J.S., McGeehin, J.P., Muhs, D.R. and Bettis, E.A. III (2013) Radiocarbon dating late Quaternary loess deposits using small terrestrial gastropod shells. *Quaternary Science Reviews* 76, 114–128.

Ping, C.-L., Michaelson, G.J., Overduin, P.P. and Stiles, C.A. (2003) Morphogenesis of frost boils in the Galbraith Lake area, Arctic Alaska. In Phillips, M., Springman, S.M. and Arenson, L.U. (eds) *Proceedings of the Eighth International Conference on Permafrost.* Balkema, Lisse, 897–900.

Ping, C.-L., Michaelson, G.J., Guo, L. *et al.* (2011) Soil carbon and material fluxes across the eroding Alaska Beaufort Sea coastline. *Journal of Geophysical Research* 116, G02004.

Pissart, A. (1967) Les pingos de l'Île Prince Patrick (76°N, 120°W). *Geographical Bulletin* 9, 189–217.

Pissart, A. (1982) Déformation de cylindres de limon entourés de graviers sous l'alternance gel-dégel. *Biuletyn Periglacjalny* 29, 275–285.

Pissart, A. (2000) Remnants of lithalsas of the Hautes Fagnes, Belgium: a summary of present-day knowledge. *Permafrost and Periglacial Processes* 11, 327–355.

Pissart, A. (2002) Palsas, lithalsas and remnants of these periglacial mounds. A progress report. *Progress in Physical Geography* 26, 605–621.

Pissart, A. (2003) The remnants of Younger Dryas lithalsas on the Hautes Fagnes Plateau in Belgium and elsewhere in the world. *Geomorphology* 52, 5–38.

Pissart, A. and French, H.M. (1976) Pingo investigations, north-central Banks Island, Canadian Arctic. *Canadian Journal of Earth Sciences* 13, 937–946.

Pissart, A. and French, H.M. (1977) The origin of pingos in regions of thick permafrost, western Canadian Arctic. *Questiones Geographicae* 4, 149–160.

Pissart, A., Vincent, J.S. and Edlund, S.A. (1977) Dépôts et phénomènes éoliens sur l'île de Banks, Territoires du Nord-Ouest, Canada. *Canadian Journal of Earth Sciences* 14, 2452–2480.

Pissart, A., Harris, S.A., Prick, A. and Van Vliet-Lanoë, B. (1998) La signification paléoclimatique des lithalsas (palses minérales). *Biuletyn Peryglacjany* 37, 141–154.

Pissart, A., Calmels, F. and Wastiaux, C. (2011) The potential lateral growth of lithalsas. *Quaternary Research* 75, 371–377.

Pithan, F. and Mauritsen, T. (2014) Arctic amplification dominated by temperature feedbacks in contemporary climate models. *Nature Geoscience* 7, 181–184.

Plug, L.J. and Werner, B.T. (2001) Fracture networks in frozen ground. *Journal of Geophysical Research* 106, 8599–8613.

Plug, L.J. and Werner, B.T. (2002) Nonlinear dynamics of ice-wedge networks and resulting sensitivity to severe cooling events. *Nature* 417, 929–933.

Plug, L.J. and Werner, B.T. (2008) Modelling of ice-wedge networks. *Permafrost and Periglacial Processes* 19, 63–69.

Plug, L.J. and West, J.J. (2009) Thaw lake expansion in a two-dimensional coupled model of heat transfer, thaw subsidence, and mass movement. *Journal of Geophysical Research* 114, F01002.

Plug, L.J., Walls, C. and Scott, B.M. (2008) Tundra lake changes from 1978 to 2001 on the Tuktoyaktuk Peninsula, western Canadian Arctic. *Geophysical Research Letters* 35, L03502.

Pohl, S., Marsh, P. and Bonsal, B.R. (2007) Modeling the impact of climate change on runoff and annual water balance of an Arctic headwater basin. *Arctic* 60, 173–186.

Pohl, S., Marsh, P., Onclin, C. and Russell, M. (2009) The summer hydrology of a small upland tundra thaw lake: implications to lake drainage. *Hydrological Processes* 23, 2536–2546.

Pollard, W.H. (1988) Seasonal frost mounds. In Clark, M.J. (ed.) *Advances in Periglacial Geomorphology*. John Wiley & Sons, Chichester, 201–229.

Pollard, W.H. and French, H.M. (1980) A first approximation of the volume of ground ice, Richards Island, Pleistocene Mackenzie Delta, Northwest Territories, Canada. *Canadian Geotechnical Journal* 17, 509–516.

Pollard, W.H. and French, H.M. (1984) The groundwater hydraulics of seasonal frost mounds, North Fork Pass, Yukon Territory. *Canadian Journal of Earth Sciences* 21, 1073–1081.

Pollard, W.H. and French, H.M. (1985) The internal structure and ice crystallography of seasonal frost mounds. *Journal of Glaciology* 31, 157–162.

Pollard, W.H. and Van Everdingen, R.O. (1992) Formation of seasonal ice bodies. In Dixon, J.C. and Abrahams, A.D. (eds) *Periglacial Geomorphology*. John Wiley & Sons, Chichester, 281–304.

Porter, S.C. (2001) Chinese loess record of monsoon climate during the last glacial-interglacial cycle. *Earth-Science Reviews* 54, 115–128.

Porter, S.C. (2013) Loess records: China. In Elias, S. (ed.) *Encyclopedia of Quaternary Science*, 2nd edn, vol. 2. Elsevier, Amsterdam, 595–605.

Porter, S.C., Singhvi, A., Zhisheng, A. and Zhongping, L. (2001) Luminescence age and palaeoenvironmental implications of a Late Pleistocene ground wedge on the northeastern Tibetan Plateau. *Permafrost and Periglacial Processes* 12, 203–210.

Poser, H. (1933) Das Problem des Strukturbödens. *Geologische Rundschau* 24, 105–121.

Poser, H. (1948) Böden und Klimaverhältnisse in Mittel und Westeuropa während der Wurmeiszeit. *Erdkunde* 2, 53–68.

Poser, H. (1953–54) (ed.) Studien über die Periglazialerscheinungen in Mitteleuropa. *Göttinger*

Geographische Abhandlungen 14, 128 pp., 15, 180 pp., 16, 96 pp., 17, 147 pp.

Potter, N. and Moss, J.H. (1968) Origin of the Blue Rocks blockfield and adjacent deposits, Berks County, Pennsylvania. *Geological Society of America Bulletin* 79, 255–262.

Potter, N. Jr, Steig, E.J., Clark, D.H. *et al.* (1998) Galena Creek rock glacier revisited – new observations on an old controversy. *Geografiska Annaler* 80A, 251–266.

Price, L.W. (1974) The developmental cycle of solifluction lobes. *Annals of the Association of American Geographers* 64, 430–438.

Price, L.W. (1990) Subsurface movement on solifluction slopes in the Ruby Range, Yukon Territory, Canada – a 20 year study. *Arctic and Alpine Research* 23, 200–205.

Prick, A. (1995) Dilatometric behaviour of porous calcareous rock samples subjected to freeze–thaw cycles. *Catena* 25, 7–20.

Prick, A. (1997) Critical degree of saturation as a threshold moisture level in frost weathering of limestones. *Permafrost and Periglacial Processes* 8, 91–99.

Prick, A., Pissart, A. and Ozouf, J.-C. (1993) Variations dilatometriques de cylindres de roches calcaires subissent des cycles de gel-degel. *Permafrost and Periglacial Processes* 4, 1–15.

Priesnitz, K. (1988) Cryoplanation. In Clark, M.J. (ed.) *Advances in Periglacial Geomorphology*. John Wiley & Sons, Chichester, 49–67.

Priesnitz, K. (1990) Geomorphic activity of rivers during snowmelt and break-up, Richardson Mountains, Yukon and Northwest Territories, Canada. *Permafrost and Periglacial Processes* 1, 295–299.

Priesnitz, K. and Schunke, E. (2002) The fluvial morphodynamics of two small permafrost drainage basins, Richardson Mountains, Northwestern Canada. *Permafrost and Periglacial Processes* 13, 207–217.

Prowse, T.D. and Lalonde, V. (1996) Open-water and ice-jam flooding of a northern delta. *Nordic Hydrology* 27, 85–100.

Pufahl, D.E. and Morgenstern, N.R. (1980) The energetics of an ablating headscarp in permafrost. *Canadian Geotechnical Journal* 17, 487–497.

Putkonen, J. (2008) What dictates the occurrence of zero-curtain effect? In Kane, D.L. and Hinkel, K.M. (eds) *Ninth International Conference on Permafrost*. University of Alaska, Fairbanks, 1451–1456.

Putkonen, J., Morgan, D.J. and Balco, G. (2012) Regolith transport quantified by braking block, McMurdo Dry Valleys, Antarctica. *Geomorphology* 155–156, 80–87.

Putkonen, J., Morgan, D.J. and Balco, G. (2014) Boulder weathering in McMurdo Dry Valleys, Antarctica. *Geomorphology* 219, 192–199.

Pye, K. (1995) The nature, origin and accumulation of loess. *Quaternary Science Reviews* 14, 653–667.

Pye, K. and Tsoar, H. (2009) *Aeolian Sand and Sand Dunes*. Springer Verlag, Berlin, 457 pp.

Quinton, W.L. and Marsh, P. (1998) Meltwater fluxes, hillslope runoff and stream flow in an arctic permafrost basin. In Lewkowicz, A.G. and Allard, M. (eds) *Permafrost: Proceedings of the 7th International Conference*. Université Laval, Québec, 921–926.

Rabatel, A.P., Deline, P., Jaillet, S. and Ravanel, L. (2008) Rock falls in high-alpine rock walls quantified by terrestrial lidar measurements: a case study in the Mont Blanc area. *Geophysical Research Letters* 35, L10502.

Rachold, V., Grigoriev, M.N., Are, F.E. *et al.* (2000) Coastal erosion versus riverine sediment discharge in the Arctic shelf seas. *International Journal of Earth Sciences* 89, 450–460.

Racine, C.J., Jorgenson, M.T. and Walters, J.C. (1998) Thermokarst vegetation in lowland birch forests on the Tanana Flats, interior Alaska, U.S.A. In Lewkowicz, A.G. and Allard, M. (eds) *Proceedings, Seventh International Conference on Permafrost, Université Laval, Québec*, 927–933.

Radosavljevic, B., Lantuit, H., Pollard, W. *et al.* (2016) Erosion and flooding – threats to coastal infrastructure in the Arctic: a case study from Herschel Island, Yukon Territory, Canada. *Estuaries and Coasts* 39, 900–915.

Raffi, R. and Steni, B. (2011) Isotopic composition and thermal regime of ice wedges in northern Victoria Land, east Antarctica. *Permafrost and Periglacial Processes* 22, 65–83.

Raffi, R., Stenni, B., Flora, O. *et al.* (2004) Growth processes of an inland Antarctic ice wedge, Mesa Range, northern Victoria Land. *Annals of Glaciology* 39, 379–385.

Rampton, V.N. (1974) The influence of ground ice and thermokarst upon the geomorphology of the Mackenzie-Beaufort region. In Fahey, B.D. and Thompson, R.D. (eds) *Research in Polar and Alpine Geomorphology*. Geo-Abstracts, Norwich, 43–59.

Rampton, V.N. (1988) Origin of massive ice on Tuktoyaktuk Peninsula, Northwest Territories, Canada. In *Proceedings, Fifth International Conference on Permafrost*. Tapir, Trondheim, 850–855.

Rampton, V.N. (1991) Observations on buried glacier ice and massive segregated ice, western Arctic coast, Canada: discussion. *Permafrost and Periglacial Processes* 2, 163–165.

Ran, Y.H., Li, X., Cheng, G.D. *et al.* (2012) Distribution of permafrost in China: an overview of existing permafrost maps. *Permafrost and Periglacial Processes* 23, 322–333.

Randriamazaoro, R., Dupreyrat, L., Costard, F. and Gailhardis, E.C. (2007) Fluvial thermal erosion: heat balance integral method. *Earth Surface Processes and Landforms* 32, 1828–1840.

Rangecroft, S., Suggit, A., Anderson, K. and Harrison, S. (2016) Future climate warming and changes to mountain permafrost in the Bolivian Andes. *Climate Change* 137, 231–243.

Rapp, A. (1959) Avalanche boulder tongues in Lappland, Sweden. *Geografiska Annaler* 41, 34–48.

Rapp, A. (1960) Recent development of mountain slopes in Karkevagge and surroundings, northern Scandinavia. *Geografiska Annaler* 42, 65–200.

Rapp, A. (1985) Extreme rainfall and rapid snowmelt as causes of mass movements in high latitude mountains. In Church, M. and Slaymaker, H.O. (eds) *Field and Theory: Lectures in Geocryology*. University of British Columbia Press, Vancouver, 36–56.

Rapp, A. and Nyberg, R. (1981) Alpine debris flows in Scandinavia – morphology and dating by lichenometry. *Geografiska Annaler* 63A, 183–196.

Rasmussen, S.O., Bigler, M., Blockley, S.P. *et al.* (2014) A stratigraphic framework for abrupt climatic changes during the Last Glacial period based on three synchronized Greenland ice-core records: refining and extending the INTIMATE event stratigraphy. *Quaternary Science Reviews* 106, 14–28.

Raup, H.M. (1965) The structure and development of turf hummocks in the Mester Vig district, Northeast Greenland. *Meddelelser om Grønland* 166, 1–113.

Ravanel, L. and Deline, P. (2010) Climate influence on rockfalls in high-Alpine steep rockwalls: the north side of the Aiguilles de Chamonix (Mont Blanc Massif) since the end of the 'Little Ice Age'. *The Holocene* 21, 357–365.

Ravanel, L., Allignol, F., Deline, P. *et al.* 2010. Rock falls in the Mont Blanc Massif in 2007 and 2008. *Landslides* 7, 493–501.

Rea, B.R. (2013) Blockfields (Felsenmeer). In Elias, S. (ed.) *Encyclopedia of Quaternary Science*, 2nd edn. Elsevier, Amsterdam, 523–534.

Rea, B.R., Whalley, W.B., Rainey, M.M. and Gordon, J.E. (1996) Blockfields old or new? Evidence and implications from some plateaus in northern Norway. *Geomorphology* 15, 109–121.

Rearic, D.M., Barnes, P.W. and Reimnitz, E. (1990) Bulldozing and resuspension of shallow-shelf sediment by ice keels: implications for Arctic sediment transport trajectories. *Marine Geology* 91, 133–147.

Refsnider, K.A. and Brugger, K.A. (2007) Rock glaciers in central Colorado, USA, as indicators of Holocene climate change. *Arctic, Antarctic and Alpine Research* 39, 127–136.

Reger, R.D. and Péwé, T.L. (1976) Cryoplanation terraces: indicators of a permafrost environment. *Quaternary Research* 6, 99–109.

Regmi, D. and Watanabe, T. (2009) Rockfall activity in the Kangchenjunga area, Nepal Himalaya. *Permafrost and Periglacial Processes* 20, 390–398.

Reid, J.R. and Nesje, A. (1988) A giant ploughing boulder, Finse, southern Norway. *Geografiska Annaler* 70A, 27–33.

Reimnitz, E., Barnes, P.W. and Harper, J.R. (1990) A review of beach nourishment from ice transport of shoreface materials. *Journal of Coastal Research* 6, 439–470.

Rémillard, A.M., Hétu, B., Bernatchez, P. *et al.* (2015) Chronology and palaeoenvironmental implications of the ice-wedge pseudomorphs and composite wedge casts on the Magdalen Islands (eastern Canada). *Boreas* 44, 658–675.

Rempel, A.W. (2007) Formation of ice lenses and frost heave. *Journal of Geophysical Research* 112, F02S21.

Rempel, A.W. (2010) Frost heave. *Journal of Glaciology* 56, 1122–1128.

Rempel, A.W. (2011) Microscopic and environmental controls on the spacing and thickness of segregated ice lenses. *Quaternary Research* 75, 316–324.

Rempel, A.W., Wettlaufer, J.S. and Worster, M.G. (2001) Interfacial premelting and the thermomolecular force: thermodynamic buoyancy. *Physical Review Letters* 87, 088501.

Rempel, A.W., Wettlaufer, J.S. and Worster, M.G. (2004) Premelting dynamics in a continuum model of frost heave. *Journal of Fluid Mechanics* 498, 227–244.

Renssen, H. and Vandenberghe, J. (2003) Investigation of the relationship between permafrost distribution in NW Europe and extensive winter sea-ice cover in the North Atlantic Ocean during the cold phases of the Last Glaciation. *Quaternary Science Reviews* 22, 209–223.

Renssen, H., Seppä, H., Crosta, X. *et al.* (2012) Global characterization of the Holocene thermal maximum. *Quaternary Science Reviews* 48, 7–19.

Reyes, A.V., Froese, D.G. and Jensen, B.J.L. (2010) Permafrost response to last interglacial warming: field evidence from non-glaciated Yukon and Alaska. *Quaternary Science Reviews* 29, 3256–3274.

Ribolini, A., Guglielmin, M., Fabre, D. *et al.* (2010) The internal structure of rock glaciers and recently deglaciated slopes as revealed by geoelectrical tomography: insights on permafrost and recent glacial evolution in the central and western Alps (Italy-France). *Quaternary Science Reviews* 29, 507–521.

Ribolini, A., Bini, M., Consoloni, I. *et al.* (2014) Late Pleistocene wedge structures along the Patagonian coast (Argentina): chronological constraints and palaeo-environmental implications. *Geografiska Annaler* 96A, 161–176.

Rickenmann, D. and Zimmermann, M. (1993) The 1987 debris flows in Switzerland: documentation and analysis. *Geomorphology* 8, 175–189.

Ridefelt, H. and Boelhouwers, J. (2006) Observations on regional variation in solifluction landform morphology and environment in the Abisko Region, Northern Sweden. *Permafrost and Periglacial Processes* 17, 253–266.

Ridefelt, H., Åkerman, H.J., Beylich, A.A. *et al.* (2009) 56 years of solifluction measurements in the Abisko Mountains, northern Sweden – analysis of temporal and spatial variations of slow soil surface movement. *Geografiska Annaler* 91A, 215–232.

Ridefelt, H., Etzelmüller, B. and Boelhouwers, J. (2010) Spatial analysis of solifluction landforms and process rates in the Abisko Mountains, Northern Sweden. *Permafrost and Periglacial Processes* 21, 241–255.

Ridefelt, H., Boelhouwers, J. and Etzelmüller, B. (2011) Local variations in solifluction activity and environment in the Abisko Mountains, Northern Sweden. *Earth Surface Processes and Landforms* 36, 2042–2053.

Riordan, B., Verbyla, D. and McGuire, A. (2006) Shrinking ponds in subarctic Alaska based on 1950–2002 remotely sensed images. *Journal of Geophysical Research* 111, G04002.

Riseborough, D.W. (2002) The mean annual temperature at the top of permafrost, the TTOP model, and the effect of unfrozen water. *Permafrost and Periglacial Processes* 13, 137–142.

Riseborough, D.W., Shiklomanov, N., Etzelmüller, B. *et al.* (2008) Recent advances in permafrost modelling. *Permafrost and Periglacial Processes* 19, 137–156.

Rivkin, F.M. (1998) Regional characteristics of subfluvial talik formation and structure, Yamal Peninsula, Russia. In Lewkowicz, A.G. and Allard, M. (eds) *Permafrost: Proceedings of the 7th International Conference.* Université Laval, Québec, 943–947.

Roach, J., Griffith, B., Verbyla, D. and Jones J. (2011) Mechanisms influencing changes in lake area in Alaskan boreal forest. *Global Change Biology* 17, 2567–2583.

Roberts, H.M., Muhs, D.R., Wintle, A.G. *et al.* (2003) Unprecedented last glacial mass accumulation rates determined by luminescence dating of loess from western Nebraska. *Quaternary Research* 59, 411–419.

Robinson, D.A. and Williams, R.B.G. (2000) Experimental weathering of sandstone by combination of salts. *Earth Surface Processes and Landforms* 25, 1309–1315.

Rocha, A.V. and Shaver, G.R. (2011) Burn severity influences post-fire CO_2 exchange in arctic tundra. *Ecological Applications* 21, 477–489.

Rödder, T. and Kneisel, C. (2012) Influence of snow cover and grain size on the ground thermal regime in the discontinuous permafrost regime. *Geomorphology* 175–176, 176–189.

Rodriguez-Navarro, C. and Doehne, E. (1999) Salt weathering: influence of evaporation rate, supersaturation and crystallization pattern. *Earth Surface Processes and Landforms* 24, 191–209.

Roer, I., Kääb, A. and Dikau, R. (2005) Rockglacier acceleration in the Turtmann Valley (Swiss Alps): probable controls. *Norwegian Journal of Geography* 59, 157–163.

Roer, I., Haeberli, W., Avian, M. *et al.* (2008) Observations and considerations on destabilizing active rock glaciers in the European Alps. In Kane, D.L. and Hinkel, K.M. (eds) *Proceedings of the Ninth International Conference on Permafrost*. University of Alaska, Fairbanks, 1505–1510.

Romanovskii, N.N. (1985) Distribution of recently active ice and soil wedges in the USSR. In Church, M. and Slaymaker, O. (eds) *Field and Theory: Lectures in Geocryology.* University of British Columbia Press, Vancouver, 154–165.

Romanovskii, N.N. and Tyurin A.I. (1986) Kurums. *Biuletyn Peryglacjalny* 31, 249–259.

Romanovskii, N.N., Hubberten, H.-W., Gavrilov, A.V. *et al.* (2000) Thermokarst and land-ocean interactions, Laptev Sea region, Russia. *Permafrost and Periglacial Processes* 11, 137–152.

Romanovskii, N.N., Hubberten, H.-W., Gavrilov, A.V. *et al.* (2004) Permafrost of the east Siberian Arctic shelf and coastal lowlands. *Quaternary Science Reviews* 23, 1359–1369.

Romanovsky, V.E., Smith, S.L. and Christiansen, H.H. (2010a) Permafrost thermal state in the polar northern hemisphere during the International Polar Year 2007–2009: a synthesis. *Permafrost and Periglacial Processes* 21, 106–116.

Romanovsky, V.E., Drozdov, D.S., Oberman, N.G. *et al.* (2010b) Thermal state of permafrost in Russia. *Permafrost and Periglacial Processes* 21, 136–155.

Rose, J. (2009) Early and Middle Pleistocene landscapes of eastern England. *Proceedings of the Geologists' Association* 120, 3–33.

Rose, J. (2010) The Quaternary of the British Isles: factors forcing environmental change. *Journal of Quaternary Science* 25, 399–418.

Rose, J., Allen, P., Kemp, R.A. *et al.* (1985) The Early Anglian Barham Soil of Eastern England. In Boardman, J. (ed.) *Soils and Quaternary Landscape Evolution.* John Wiley & Sons, Chichester, 197–228.

Rose, J., Lee, J.A., Kemp, R.A. and Harding, P.A. (2000) Palaeoclimate, sedimentation and soil development during the last Glacial Stage (Devensian), Heathrow Airport, London, UK. *Quaternary Science Reviews* 19, 827–847.

Ross, N. (2013) Frost mounds: active and relict forms. In Elias, S. (ed.) *Encyclopedia of Quaternary Science*, 2nd edn, vol. 3. Elsevier, Amsterdam, 472–480.

Ross, N., Harris, C., Christiansen, H.H. and Brabham, P.J. (2005) Ground-penetrating radar investigations of open system pingos, Adventdalen, Svalbard. *Norwegian Journal of Geography* 59, 129–138.

Ross, N., Brabham, P.J., Harris, C. and Christiansen, H.H. (2007) Internal structure of open-system pingos, Adventdalen, Svalbard: the use of resistivity tomography to assess ground-ice conditions. *Journal of Environmental and Engineering Geophysics* 12, 113–126.

Ross, N., Harris, C., Brabham, P.J. and Sheppard, T.H. (2011) Internal structure and geological context of ramparted depressions, Llanpumsaint, Wales. *Permafrost and Periglacial Processes* 22, 291–305.

Rousseau, D.-D., Wu, N. and Guo, Z. (2000) The terrestrial mollusks as new indices of the Asian palaeomonsoons in the Chinese loess plateau. *Global and Planetary Change* 26, 199–206.

Rousseau, D.-D., Antoine, P., Hatté, C. *et al.* (2002) Abrupt millennial climatic changes from Nussloch (Germany) Upper Weichselian eolian records during the last glaciation. *Quaternary Science Reviews* 21, 1577–1582.

Rousseau, D.-D., Sima, A., Antoine, P. *et al.* (2007) Link between European and North-Atlantic abrupt climate changes over the last glaciation. *Geophysical Research Records* 34, L22713.

Rousseau, D.-D., Antoine, P., Gerasimenko, N. *et al.* (2011) North Atlantic abrupt climatic events of the last glacial period recorded in Ukrainian loess deposits. *Climate of the Past* 7, 221–234.

Rovey, C.W. and Balco, G. (2010) Periglacial climate at the 2.5 Ma onset of northern Hemisphere glaciation inferred from the Whippoorwill Formation, northern Missouri, USA. *Quaternary Research* 73, 151–161.

Roy-Léveillée, P. and Burn, C.R. (2016) A modified landform development model for the topography of drained thermokarst lake basins in fine-grained sediments. *Earth Surface Processes and Landforms* 41, 1504–1520.

Roy-Léveillée, P., Burn, C.R. and McDonald, I.D. (2014) Vegetation-permafrost relations within the forest-tundra ecotone near Old Crow, northern Yukon, Canada. *Permafrost and Periglacial Processes* 25, 127–135.

Ruddiman, W.F. (2013) The Anthropocene. *Annual Review of Earth and Planetary Sciences* 41, 1–24.

Ruddiman, W.F. and McIntyre, A. (1981) The North Atlantic during the last deglaciation. *Palaeogeography, Palaeoclimatology, Palaeoecology* 35, 145–214.

Ruedrich, J. and Siegesmund, S. (2007) Salt and ice crystallization in porous sandstones. *Environmental Geology* 52, 225–249.

Ruegg, G.H.J. (1983) Periglacial eolian evenly laminated sandy deposits in the Late Pleistocene of NW Europe, a facies unrecorded in modern sedimentological handbooks. In Brookfield, M.E. and Ahlbrandt, T.S. (eds) *Eolian Sediments and Processes*. Elsevier, Amsterdam, 455–482.

Ruz, M.-H. (1993) Coastal dune development in a thermokarst environment: some implications for environmental reconstruction, Tuktoyaktuk Peninsula, NWT. *Permafrost and Periglacial Processes* 4, 255–264.

Ruz, M.-H. and Allard, M. (1995) Sedimentary structures of cold-climate coastal dunes, Eastern Hudson Bay, Canada. *Sedimentology* 42, 725–734.

Ruz, M.-H., Allard, M., Michaud, Y. and Héquette, A. (1998) Sedimentology and evolution of a subarctic tidal flat along a rapidly emerging coast, eastern Hudson Bay, Canada. *Journal of Coastal Research* 14, 1242–1254.

Ruz, M.-H., Héquette, A. and Hill, P.R. (1992) A model of coastal evolution in a transgressed thermokarst topography, Canadian Beaufort Sea. *Marine Geology* 106, 251–278.

Saemundsson, T., Arnalds, O., Kneisel, C. *et al.* (2012) The Orravatnsrustir palsa site in central Iceland – palsas in an aeolian sedimentation environment. *Geomorphology* 167–168, 13–20.

Saito, K., Marchenko, S., Romanovsky, V. *et al.* (2014) Evaluation of LPM permafrost distribution in NE Asia reconstructed and downscaled from GCM simulations. *Boreas* 43, 733–749.

Saito, K., Trombotto Liaudat, D., Yoshikawa, K. *et al.* (2016) Late Quaternary permafrost distributions downscaled for South America: examinations of GCM-based maps with observations. *Permafrost and Periglacial Processes* 27, 43–55.

Salvigsen, O. and Elgersma, A. (1985) Large-scale karst features and open taliks at Vardeborgsletta, outer Isfjorden, Svalbard. *Polar Research* 3, 145–153.

Salzmann, N., Frei, C., Vidale, P.-L. and Hoezle, M. (2007) The application of Regional Climate Model output for the simulation of high-mountain permafrost scenarios. *Global and Planetary Change* 56, 188–202.

Sanborn, P.T., Smith, C.A.S., Froese, D.G. *et al.* (2006) Full-glacial paleosols in perennially frozen loess sequences, Klondike goldfields, Yukon Territory, Canada. *Quaternary Resesarch* 66, 147–157.

Sánchez-Garcia, L., Vonk, J.E., Charkin, A.N. *et al.* (2014) Characterisation of three regimes of collapsing arctic ice complex deposits on the SE Laptev Sea coast using biomarkers and dual carbon isotopes. *Permafrost and Periglacial Processes* 25, 172–183.

Sanders, D. (2010) Sedimentary facies and progradational style of a Pleistocene talus slope succession, northern calcareous Alps, Austria. *Sedimentary Geology* 228, 271–283.

Sanders, D., Ostermann, M. and Kramers, J. (2009) Quaternary carbonate-rocky talus slope successions (Eastern Alps, Austria): sedimentary facies and facies architecture. *Facies* 55, 345–373.

Sannel, A.B.K. and Brown, I.A. (2010) High resolution remote sensing identification of thermokarst lake dynamics in a subarctic peat plateau complex. *Canadian Journal of Remote Sensing* 36, S26–S40.

Sannel, A.B.K. and Kuhry, P. (2011) Warming-induced destabilization of peat plateau/thermokarst lake complexes. *Journal of Geophysical Research* 116, G03035.

Sasaki, T. (1992) The development of ice-made ramparts on Lake Kussharo, Hokkaido, Japan. *Arctic and Alpine Research* 24, 165–172.

Sass, O. (2004) Rock moisture fluctuations during freeze-thaw cycles: preliminary results from electrical resistivity measurements. *Polar Geography* 28, 13–31.

Sass, O. (2005a) Rock moisture measurements: techniques, results and implications for weathering. *Earth Surface Processes and Landforms* 30, 359–374.

Sass, O. (2005b) Spatial patterns of rockfall intensity in the northern Alps. *Zeitschrift für Geomorphologie Supplementband* 138, 51–65.

Sass, O. (2006) Determination of the internal structure of alpine talus deposits using different geophysical methods (Lechtaler Alps, Austria). *Geomorphology* 80, 45–58.

Sass, O. (2007) Geophysical bedrock detection and talus thickness in the eastern European Alps. *Journal of Applied Geophysics* 6, 254–269.

Sass, O. (2010) Spatial and temporal pattern of rockfall activity – a lichenometric approach in the Stubaier Alps, Austria. *Geografiska Annaler* 92A, 375–391.

Sass, O. and Krautblatter, M. (2007) Debris flow-dominated and rockfall-dominated talus slopes: genetic models derived from GPR measurements. *Geomorphology* 86, 176–192.

Sass, O. and Wollny, K. (2001) Investigations regarding alpine talus slopes using ground-penetrating radar (GPR) in the Bavarian Alps, Germany. *Earth Surface Processes and Landforms* 26, 1071–1086.

Sato, T., Kurashige, Y. and Hirakawa, K. (1997) Slow mass movement in the Taisetsu Mountains, Hokkaido, Japan. *Permafrost and Periglacial Processes* 8, 347–357.

Savigny, K.W. and Morgenstern, N.R. (1986) In-situ creep properties of ice-rich permafrost soil. *Canadian Geotechnical Journal* 23, 504–514.

Sawada, Y., Ishikawa, M. and Ono, Y. (2003) Thermal regime of sporadic permafrost on a block slope on Mt Nishi-Nupukasushinupuri, Hokkaido Island, northern Japan. *Geomorphology* 52, 121–130.

Sazonova, T.S. and Romanovsky, V.E. (2003) A model for regional-scale estimation of temporal and spatial variability of active layer thickness and mean annual ground temperatures. *Permafrost and Periglacial Processes* 14, 125–139.

Scapozza, C., Lambiel, C., Baron, L. *et al.* (2011) Internal structure and permafrost distribution in two alpine periglacial talus slopes, Valais, Swiss Alps. *Geomorphology* 132, 208–221.

Scapozza, C., Baron, L. and Lambiel, C. (2015) Borehole logging in Alpine periglacial talus slopes. *Permafrost and Periglacial Processes* 26, 67–83.

Schädel, C., Schuur, E.A.G., Bracho, R. *et al.* (2014) Circumpolar assessment of permafrost C quality and its vulnerability over time using long-term incubation data. *Global Change Biology* 20, 641–652.

Schaefer, K., Zhang, T., Bruhwiler, L. and Barrett, A.P. (2011) Amount and timing of permafrost carbon release in response to climate warming. *Tellus B* 63, 165–180.

Schaetzl, R.J. (2012) Mississippi Valley regional source of loess in the Southern Green Bay Lobe land surface, Wisconsin – comment on the paper published by Jacobs *et al.*, *Quaternary Research* 75 (3), 574–583, 2011. *Quaternary Research* 78, 149–151.

Schaetzl, R.J. and Attig, J.W. (2013) The loess cover of northeastern Wisconsin. *Quaternary Research* 79, 199–214.

Scheidegger, J.M., Bense, V.F. and Grasby, S.E. (2012) Transient nature of Arctic spring systems driven by subglacial meltwater. *Geophysical Research Letters* 39, L12405.

Schirrmeister, L., Kunitsky, V., Grosse, G. *et al.* (2011) Sedimentary characteristics and origin of the Late Pleistocene ice complex on north-east Siberian Arctic coastal lowlands and islands – a review. *Quaternary International* 241, 3–25.

Schirrmeister, L., Froese, D., Tumskoy, V. *et al.* (2013) Yedoma: Late Pleistocene ice-rich syngenetic permafrost of Beringia. In Elias, S. (ed.) *Encyclopedia of Quaternary Science*, 2nd edn, vol. 3. Elsevier, Amsterdam, 542–552.

Schirrmeister, L., Meyer, H., Andreev, A. *et al.* (2016) Late Quaternary palaeoenvironmental records from the Chatanika River valley near Fairbanks. *Quaternary Science Reviews* 147, 259–278.

Schleusner, P., Biskaborn, B.K., Kienst, F. *et al.* (2015) Basin evolution and palaeoenvironmental variability of the thermokarst lake El'gene-Kyuele, Arctic Siberia. *Boreas* 44, 216–229.

Schlyter, P. (1994) Paleo-periglacial ventifact formation by suspended silt or snow – site studies in south Sweden. *Geografiska Annaler* 76A, 197–201.

Schlyter, P. (1995) Ventifacts as palaeo-wind indicators in southern Scandinavia. *Permafrost and Periglacial Processes* 6, 207–219.

Schneuwly-Bollschweiler, M. and Stoffel, M. (2012) Hydrometeorological triggers of periglacial debris flows in the Zermatt valley (Switzerland) since 1864. *Journal of Geophysical Research* 117, F02033.

Schunke, E. and Zoltai, S.C. (1988) Earth hummocks (thufur). In Clark, M.J. (ed.) *Advances in Periglacial Geomorphology*. John Wiley & Sons, Chichester, 231–245.

Schuur, E.A.G., Abbott, B.W., Bowden, W.B. *et al.* (2013) Expert assessment of vulnerability of permafrost carbon to climate change. *Climate Change* 119, 359–374.

Schuur, E.A.G., McGuire, A.D., Schädel, C. *et al.* (2015) Climate change and the permafrost carbon feedback. *Nature* 520, 171–179.

Schwan, J. (1986) The origin of horizontal alternating bedding in Weichselian aeolian sands in western Europe. *Sedimentary Geology* 49, 73–108.

Scott, M.B., Dickinson, K.J.M., Barratt, B.I.P., and Sinclair, B.J. (2008) Temperature and moisture trends in non-sorted earth hummocks and stripes on the Old Man Range, New Zealand: implications for mechanisms of maintenance. *Permafrost and Periglacial Processes* 19, 305–314.

Scotti, R., Brardinoni, F., Alberti, S. *et al.* (2013) A regional inventory of rock glaciers and protalus ramparts in the central Italian Alps. *Geomorphology* 186, 136–149.

Screen, J.A. and Simmonds, I. (2010) The central role of diminishing sea ice in recent Arctic temperature amplification. *Nature* 464, 1134–1137.

Sebe, K., Csillag, G., Ruszkiczay-Rüdiger, Z. *et al.* (2011) Wind erosion under cold climate: a Pleistocene periglacial mega-yardang system in central Europe (western Pannonian Basin, Hungary). *Geomorphology* 134, 470–482.

Séjourné, A., Costard, F., Fedorov, A. *et al.* (2015) Evolution of the banks of thermokarst lakes in Central Yakutia (Central Siberia) due to retrogressive thaw slump activity controlled by insolation. *Geomorphology* 241, 31–40.

Selby, M.J. (1971) Slopes and their development in an ice-free arid area of Antarctica. *Geografiska Annaler* 53, 235–245.

Selby, M.J. (1972) Antarctic tors. *Zeitschrift für Geomorphologie, Supplementband* 13, 73–86.

Seppälä, M. (1972) Pingo-like remnants in the Peltojärvi area of Finnish Lapland. *Geografiska Annaler* 54A, 38–45.

Seppälä, M. (1987) Periglacial phenomena of northern Fennoscandia. In Boardman, J. (ed.) *Periglacial processes and landforms in Britain and Ireland*. Cambridge University Press, Cambridge, 45–55.

Seppälä, M. (1988a) Palsas and related forms. In Clark, M.J. (ed.) *Advances in Periglacial Geomorphology*. John Wiley & Sons, Chichester, 247–278.

Seppälä, M. (1988b) Rock pingos in northern Ungava Peninsula, Quebec, Canada. *Canadian Journal of Earth Sciences* 25, 629–634.

Seppälä, M. (1993) Climbing and falling dunes in Finnish Lapland. *Geological Society, London, Special Publication* 72, 269–274.

Seppälä, M. (1994) Deep snow controls palsa growth. *Permafrost and Periglacial Processes* 5, 283–288.

Seppälä, M. (1995a) How to make a palsa: field experiment on permafrost formation. *Zeitschrift für Geomorphologie Supplementband* 99, 97–106.

Seppälä, M. (1995b) Deflation and redeposition of sand dunes in Finnish Lapland. *Quaternary Science Reviews* 14, 799–809.

Seppälä, M. (1997) Piping causing thermokarst in permafrost, Ungava Peninsula, Quebec, Canada. *Geomorphology* 20, 313–319.

Seppälä, M. (1998) New permafrost formed in peat hummocks (pounus), Finnish Lapland. *Permafrost and Periglacial Processes* 9, 367–373.

Seppälä, M. (2003) Surface abrasion of palsas by wind action in Finnish Lapland. *Geomorphology* 52, 141–148.

Seppälä, M. (2004) *Wind as a geomorphic agent in cold climates*. Cambridge University Press, Cambridge, 358 pp.

Seppälä, M. (2011) Synthesis of studies of palsa formation underlining the importance of local environmental and physical characteristics. *Quaternary Research* 75, 366–370.

Seppälä, M. and Kujala, K. (2009) The role of buoyancy in palsa formation. *Geological Society, London, Special Publication* 320, 51–56.

Serrano, E. and López-Martínez, J. (2000) Rock glaciers in the South Shetland Islands, Western Antarctica. *Geomorphology* 35, 145–162.

Serrano, E., San José, J.J. and Agudo, C. (2006) Rock glacier dynamics in a marginal periglacial high mountain environment: flow, movement (1991–2000) and structure of the Argualas rock glacier, the Pyrenees. *Geomorphology* 74, 285–296.

Serrano, E., de Sanjosé, J.J., González-Trueba, J.J. (2010) Rock glacier dynamics in marginal periglacial environments. *Earth Surface Processes and Landforms* 35, 1302–1314.

Serreze, M.C. and Barry, R.G. (2011) Processes and impacts of Arctic amplification: a research synthesis. *Global and Planetary Change* 77, 85–76.

Serreze, M.C. and Barry, R.G. (2014) *The Arctic Climate System*, 2nd edn. Cambridge University Press, New York, 415 pp.

Serreze, M.C., Holland, M.M. and Stroeve, J. (2007) Perspectives on the Arctic's shrinking ice cover. *Science* 31, 1533–1536.

Shakesby, R.A. (1997) Pronival (protalus) ramparts: a review of forms, processes, diagnostic criteria and palaeoenvironmental implications. *Progress in Physical Geography* 21, 394–418.

Shakesby, R.A. and Matthews, J.A. (1987) Frost weathering and rock platform erosion on periglacial lake shorelines: a test of a hypothesis. *Norsk Geografisk Tidsskrift* 67, 197–203.

Shakesby, R.A. and Matthews, J.A. (2002) Sieve deposition by debris flow on a permeable substrate, Leirdalen, Norway. *Earth Surface Processes and Landforms* 27, 1031–1041.

Shakesby, R.A., Dawson, A.G. and Matthews, J.A. (1987) Rock glaciers, protalus ramparts and related pheomena,

Rondane, Norway: a continuum of large-scale, talus-derived landforms. *Boreas* 16, 305–317.

Shakesby, R.A., Matthews, J.A. and McCarroll, D. (1995) Pronival ('protalus') ramparts in the Romsdalsalpane, southern Norway: forms, terms, subnival processes and alternative mechanisms of formation. *Arctic and Alpine Research* 27, 271–282.

Shakesby, R.A., Matthews, J.A., McEwen, L.J. and Berrisford, M.S. (1999) Snow-push processes in pronival (protalus) rampart formation: geomorphological evidence from Smørbotn, Romsdalsalpane, Southern Norway. *Geografiska Annaler* 81A, 31–45.

Shakhova, N., Semiletov, I., Leifer, I. *et al.* (2010) Geochemical and geophysical evidence of methane release over the East Siberian Arctic Shelf. *Journal of Geophysical Research* 115, C08007.

Shakhova, N., Semiletov, I., Leifer, I. *et al.* (2014) Ebullition and storm-induced methane release from the East Siberian Arctic Shelf. *Nature Geoscience* 7, 64–70.

Shapiro, B. and Cooper, A. (2003) Beringia as an Ice Age genetic museum. *Quaternary Research* 60, 94–100.

Shearer, J.M., Macnab, R.F., Pelletier, B.R. and Smith, T.B. (1971) Submarine pingos in the Beaufort Sea. *Science* 174, 814–816.

Sher, A.V., Kuzmina, S.A., Kuznetsova, T.V. and Sulerzhitky, L.D. (2005) New insights into the Weichselian environment and climate of the East Siberian Arctic, derived from fossil insects, plants and mammals. *Quaternary Science Reviews* 24, 533–569.

Shiklomanov, N.I. (2005) From exploration to systematic investigation: development of geocryology in 19th- and early 20th-century Russia. *Physical Geography* 26, 249–263.

Shiklomanov, A.I., Yakovleva, T.I., Lammers, R.B. *et al.* (2006) Cold region river discharge uncertainty – estimates from large Russian rivers. *Journal of Hydrology* 326, 231–256.

Shilts, W.W. (1978) Nature and genesis of mudboils, central Keewatin, Canada. *Canadian Journal of Earth Sciences* 15, 1053–1063.

Shostakovitch, B. (1927) Der ewig gefrorene Böden Siberiens. *Gesellschaft für Erdkunde, Berlin, Zeitschrift* 1927, 394–427.

Shumskii, P.A. (1964) Ground (subsurface) ice. *Canadian National Research Council, Technical Translation* 1130, 118 pp.

Shur, Y.L. and Jorgenson, M.T. (1998) Cryostructure development on the floodplain of the Colville River Delta, northern Alaska. In Lewkowicz, A.G. and Allard, M. (eds) *Proceedings, 7th International Conference on Permafrost*, Université Laval, Québec, 993–999.

Shur, Y.L. and Jorgensen, M.T. (2007) Patterns of permafrost formation and degradation in relation to climate and ecosystems. *Permafrost and Periglacial Processes* 18, 7–19.

Shur, Y.L., French, H.M., Bray, M.T. and Anderson, D.A. (2004) Syngenetic permafrost growth: cryostratigraphic observations from the CRREL permafrost tunnel near Fairbanks, Alaska. *Permafrost and Periglacial Processes* 15, 339–347.

Shur, Y.L., Hinkel, K.M. and Nelson, F.E. (2005) The transient layer: implications for geocryology and climate-change science. *Permafrost and Periglacial Processes* 16, 5–18.

Siegert, C., Schirrmeister, L. and Babiy, O. (2002) The sedimentological, mineralogical and geochemical composition of Late Pleistocene deposits from the ice complex on Bykovsky Peninsula, northern Siberia. *Polarforschung* 70, 3–11.

Siewert, M.B., Krautblatter, M., Christiansen, H.H. and Eckerstorfer, M. (2012) Arctic rockwall retreat rates estimated using laboratory-calibrated ERT measurements of talus cones in Longyeardalen, Svalbard. *Earth Surface Processes and Landforms* 37, 1542–1555.

Sima, A., Rousseau, D.-D., Kageyama, M. *et al.* (2009) Imprint of North-Atlantic abrupt climate changes on western European loess deposits as viewed in a dust emission model. *Quaternary Science Reviews* 28, 2851–2866.

Sissons, J.B. (1979) Palaeoclimatic inferences from former glaciers in Scotland and the Lake District. *Nature* 278, 518–521.

Sitzia, L., Bertran, P., Bahain, J.-J. *et al.* (2015) The Quaternary coversands of southwest France. *Quaternary Science Reviews* 124, 84–105.

Sjöberg, Y., Hugelius, G. and Kuhry, P. (2013) Thermokarst lake morphometry and erosion features in two peat plateau areas of northeast European Russia. *Permafrost and Periglacial Processes* 24, 75–81.

Skempton, A.W. and Weeks, A.G. (1976) The Quaternary history of the Lower Greensand escarpment and Weald Clay vale near Sevenoaks, Kent. *Philosophical Transactions of the Royal Society of London* A283, 493–526.

Slater, A.G. and Lawrence, D.M. (2013) Diagnosing present and future permafrost from climate models. *Journal of Climate* 26, 5608–5623.

Sletten, K., Blikra, L.H., Ballantyne, C.K. *et al.* (2003a) Holocene debris flows recognized in a lacustrine sedimentary succession: sedimentology, chronostratigraphy and cause of triggering. *The Holocene* 13, 907–920.

Sletten, R.S., Hallet, B. and Fletcher, R.C. (2003b) Resurfacing time of terrestrial surfaces by the formation and maturation of polygonal patterned ground. *Journal of Geophysical Research* 108 (E4), 8044.

Sloan, V.F. and Dyke, L.D. (1998) Decadal and millenial velocities of rock glaciers, Selwyn Mountains, Canada. *Geografiska Annaler* 80A, 193–207.

Small, E.E., Anderson, R.S., Repka, J.L. and Finkel, R. (1997) Erosion rate of alpine bedrock summit surfaces deduced from *in situ* ^{10}Be and ^{26}Al. *Earth and Planetary Science Letters* 150, 413–425.

Small, E.E., Anderson, R.S. and Hancock, G.S. (1999) Estimates of the rate of regolith production using ^{10}Be and ^{26}Al from an alpine hillslope. *Geomorphology* 27, 131–150.

Smalley, I.J. and Markovic, S.B. (2014) Loessification and hydroconsolidation: there is a connection. *Catena* 117, 94–99.

Smalley, I.J., Bentley, S.P. and Markovic, S.B. (2016) Loess and fragipans: development of polygonal-crack-network structures in fragipan horizons in loess ground. *Quaternary International* 399, 228–233.

Smith, D.G. and Pearce, C.M. (2002) Ice jam-caused fluvial gullies and scour holes on northern flood plains. *Geomorphology* 42, 85–95.

Smith, D.J. (1987) Late Holocene solifluction lobe activity in the Mt Rae area, southern Canadian Rocky Mountains. *Canadian Journal of Earth Sciences* 24, 1634–1642.

Smith, D.J. (1992) Long-term rates of contemporary solifluction in the Canadian Rocky Mountains. In Dixon, J.C. and Abrahams, A.D. (eds), *Periglacial Geomorphology*. John Wiley & Sons, Chichester, 203–221.

Smith, D.J., McCarthy, D.P. and Luckman, B.H. (1994) Snow-avalanche impact pools in the Canadian Rocky Mountains. *Arctic and Alpine Research* 26, 116–127.

Smith, L.C., Sheng, Y., MacDonald, G.M. and Hinzman, L.D. (2005) Disappearing arctic lakes. *Science* 308, 1429.

Smith, M.W. and Riseborough, D.W. (1996) Ground temperature monitoring and detection of climate change. *Permafrost and Periglacial Processes* 7, 301–310.

Smith, M.W. and Riseborough, D.W. (2002) Climate and the limits of permafrost: a zonal analysis. *Permafrost and Periglacial Processes* 13, 1–15.

Smith, S.L., Wolfe, S.A., Riseborough, D.W. and Nixon, F.M. (2009) Active-layer characteristics and summer climatic indices, Mackenzie Valley, Northwest Territories, Canada. *Permafrost and Periglacial Processes* 20, 201–220.

Smith, S.L., Romanovsky, V.E., Lewkowicz, A.G. *et al.* (2010) Thermal state of permafrost in North America: a contribution to the International Polar Year. *Permafrost and Periglacial Processes* 21, 117–135.

Solomon, S.M. (2005) Spatial and temporal variability of shoreline change in the Beaufort-Mackenzie region, Northwest Territories. *Geo-Marine Letters* 25, 127–137.

Solomon, S.M., Mudie, P.J., Cranston, R. *et al.* (2000) Characterisation of marine and lacustrine sediments in a drowned thermokarst embayment, Richards Island, Beaufort Sea, Canada. *International Journal of Earth Sciences* 8, 503–521.

Sone, T. and Takahashi, N. (1993) Palsa formation in the Daisetsu Mountains, Japan. *Proceedings, 6th International Conference on Permafrost*. South China University Press, Wushan Guangzhou, 1231–1234.

Sørensen, T. (1935) Bödenformen und Pflanzendecke in Nordostgrönland. *Meddelelser om Grønland*, 93, 69 pp. English translation by Halstead, C.: Ground form and plant cover in North East Greenland. In Evans, D.J.A. (ed.) (1994) *Cold Climate Landforms*. John Wiley & Sons, Chichester, 135–175.

Sorg, A., Kääb, A., Roesch, A. *et al.* (2015) Contrasting responses of Central Asian rock glaciers to global warming. *Nature – Scientific Reports* 5, 8228.

Sparks, B.W., Williams, R.B.G. and Bell, F.G. (1972) Presumed ground-ice depressions in East Anglia. *Proceedings of the Royal Society* A327, 329–343.

Spate, A.P., Burgess, J.S. and Shevlin, J. (1995) Rates of rock surface lowering, Princess Elizabeth Land, Eastern Antarctica. *Earth Surface Processes and Landforms* 20, 567–573.

Spektor, V.B. and Spektor, V.V. (2009) Karst processes and phenomena in the perennially frozen carbonate rocks of the middle Lena River basin. *Permafrost and Periglacial Processes* 20, 71–78.

Spink, T.W. (1991) Periglacial discontinuities in Eocene clays near Denham, Buckinghamshire. *Geological Society, London, Engineering Geology Special Publication* 7, 389–396.

Springman, S.M., Arenson, L.U., Yamamoto, Y. *et al.* (2012) Multidisciplinary investigations on three rock glaciers in the Swiss Alps: legacies and future perspectives. *Geografiska Annaler* 94A, 215–243.

Statham, I. (1976) A scree slope rockfall model. *Earth Surface Processes* 1, 43–62.

Stauch, G., IJmker, J., Pötsch, S. *et al.* (2012) Aeolian sediments on the north-eastern Tibetan Plateau. *Quaternary Science Reviews* 57, 71–84.

Steedman, A.E., Lantz, T.C. and Kokelj, S.V. (2016) Spatio-temportal variation in high-centre polygons and ice-wedge melt ponds, Tuktoyaktuk Coadtlands, Northwest Territories. *Permafrost and Periglacial Processes* 10.1002/ppp.1880.

Steffen, W., Grinevald, J., Crutzen, P.J. *et al.* (2011) The Anthropocene: conceptual and historical perspectives. *Philosophical Transactions of the Royal Society* A369, 842–867.

Steig, E.J., Fitzpatrick, J.J., Potter, N. and Clark, D.H. (1998) The geochemical record in rock glaciers. *Geografiska Annaler* 80A, 277–286.

Steig, E.J., Schneider, D.P., Rutherford, S.D. *et al.* (2009) Warming of the Antarctic ice-sheet surface since the 1957 International Geophysical Year. *Nature* 457, 459–462.

Steiger, M. (2005) Crystal growth in porous materials, 1: The crystallization pressure of large crystals. *Journal of Crystal Growth* 282, 455–469.

Stemerdink, C., Maddy, D., Bridgland, D.R. and Veldkamp, A. (2010) The construction of a palaeodischarge time series for use in a study of fluvial system development of the Middle to Late Pleistocene Upper Thames. *Journal of Quaternary Science* 25, 447–460.

Stevens, T., Carter, A., Watson, T.P. *et al.* (2013) Genetic linkage between the Yellow River, the Mu Us desert and the Chinese Loess Plateau. *Quaternary Science Reviews* 78, 355–368.

St-Hilaire-Gravel, D., Forbes, D.L. and Bell, T. (2012) Multitemporal analysis of a gravel-dominated coastline in the central Canadian Arctic Archipelago. *Journal of Coastal Research* 28, 421–441.

St-Hilaire-Gravel, D., Forbes, D.L. and Bell, T. (2015) Evolution and morphodynamics of a prograded beach-ridge foreland, northern Baffin Island, Canadian Arctic Archipelago. *Geografiska Annaler* 97, 615–631.

Stiegler, C., Rode, M., Sass, O. and Otto, J.-C. (2014) An undercooled scree slope detected by geophysical investigations in sporadic permafrost below 1000 m ASL, central Austria. *Permafrost and Periglacial Processes* 25, 194–207.

Stoffel, M. (2010) Magnitude-frequency relationships of debris flows – a case study based on field surveys and tree-ring records. *Geomorphology* 116, 67–76.

Stoffel, M. and Beniston, M. (2006) On the incidence of debris flows from the early Little Ice Age to a future greenhouse climate: a case study from the Swiss Alps. *Geophysical Research Letters* 33, L16404.

Stoffel, M. and Huggel, C. (2012) Effect of climate change on mass movements in mountain environments. *Progress in Physical Geography* 36, 421–439.

Stoffel, M., Bollschweiler, M. and Beniston, M. (2011) Rainfall characteristics for periglacial debris flows in the Swiss Alps: past incidences – potential future evolution. *Climatic Change* 105, 263–280.

Stoffel, M., Tirani, D. and Huggel, C. (2014) Climate change impacts on mass movements – case studies from the European Alps. *Science of the Total Environment* 493, 1255–1266.

St-Onge, D.A. and Pissart, A. (1990) Un pingo en système fermée dans des dolomites Paléozoïques de l'Arctique Canadien. *Permafrost and Periglacial Processes* 1, 275–282.

Strauss, J., Schirrmeister, L., Grosse, G. *et al.* (2013) The deep permafrost carbon pool of the yedoma region in Siberia and Alaska. *Geophysical Research Letters* 40, 6165–6170.

Streletskaya, I.D., Vasiliev, A.A. and Kanevskiy, M.Z. (2008) Freezing of marine sediments and formation of continental permafrost at the coasts of Yenisey Gulf. In Kane, D.L. and Hinkel, D.M. (eds) *Ninth International Conference on Permafrost*. University of Alaska, Fairbanks, 1721–1726.

Streletskaya, I.D., Vasiliev, A.A. and Meyer, H. (2011) Isotopic composition of syngenetic ice wedges and palaeoclimatic reconstruction, western Taymyr, Russian Arctic. *Permafrost and Periglacial Processes* 22, 101–106.

Strini, A., Guglielmin, M. and Hall, K. (2008) Tafoni development in a cryotic environment: an example from Northern Victoria Land. *Earth Surface Processes and Landforms* 33, 1502–1519.

Stroeve, J.C., Holland, M.M., Mier, W. *et al.* (2007) Arctic sea ice decline: faster than forecast. *Geophysical Research Letters* 34, L09501.

Stroeve, J.C., Kattsov, V., Barrett, A. *et al.* (2012) Trends in Arctic sea ice extent from CMIP5, CMIP3 and observations. *Geophysical Research Letters* 39, L16502.

Stroeve, J.C., Marcus, T., Boisvert, L. *et al.* (2014) Changes in Arctic melt season and implications for sea ice loss. *Geophysical Research Letters* 41, 1216–1225.

Stroeven, A.P., Fabel, D., Hättestrand, C. and Harbor, J. (2002) A relict landscape in the centre of Fennoscandian glaciation: cosmogenic radionuclide evidence of tors preserved through multiple glacial cycles. *Geomorphology* 44, 145–154.

Strömquist, L. (1985) Geomorphic impact of snowmelt on slope erosion and sediment production. *Zeitschrift für Geomorphologie* 29, 129–138.

Strunk, H. (1992) Reconstructing debris flow frequency in the southern Alps back to AD 1500 using dendrogeo-morphological analyses. *International Association of Hydrological Sciences Publication* 209, 299–306.

Style, R.W., Peppin, S.L.S., Cocks, A.C.F. and Wettlaufer, J.S. (2011) Ice-lens formation and geometrical supercooling in soils and other colloidal materials. *Physical Review* E84, 041402.

Sun, J. (2002) Provenance of loess material and formation of loess deposits on the Chinese loess plateau. *Earth and Planetary Science Letters* 203, 845–859.

Sun, J. and Muhs, D.R. (2013) Dune fields: mid-latitudes. In Elias S. (ed.) *Encyclopedia of Quaternary Science*, 2nd edn, vol. 1, 606–622.

Sun, J. and Zhu, X. (2010) Temporal variations in Pb isotopes and trace element concentrations within Chinese eolian deposits during the past 8 Ma: implications for provenance change. *Earth and Planetary Science Letters* 290, 438–447.

Sun, J., Li, S.-H., Muhs, D.R. and Li, B. (2007) Loess sedimentation in Tibet: provenance, processes and link with Quaternary glaciations. *Quaternary Science Reviews* 26, 2265–2280.

Superson, J., Gebica, P. and Brzezinska-Wójcik, T. (2010) The origin of deformation structures in periglacial fluvial sediments of the Wislok Valley, Southeast Poland. *Permafrost and Periglacial Processes* 21, 301–314.

Svendsen, J.I., Alexanderson, H., Astakhov, V.I. *et al.* (2004) Late Quaternary ice sheet history of northern Eurasia. *Quaternary Science Reviews* 23, 1229–1272.

Svensson, H. (1974) Distribution and chronology of relict polygon patterns on the Laholm Plain, the Swedish west coast *Geografiska Annaler* 54A, 159–175.

Svensson, H. (1988) Ice wedge casts and relict polygonal patterns in Scandinavia. *Journal of Quaternary Science* 3, 57–68.

Swanson, D.K., Ping, C.-L. and Michaelson, G.J. (1999) Diapirism of soils due to thaw of ice-rich material near the permafrost table. *Permafrost and Periglacial Processes* 10, 349–367.

Swezey, C.S., Schultz, A.P., González, W.A. *et al.* (2013) Quaternary eolian dunes in the Savannah River valley, Jasper County, South Carolina, USA. *Quaternary Research* 80, 250–264.

Syvitski, J.P.M. (2002) Sediment discharge variability in Arctic rivers: implications for a warmer climate. *Polar Research* 21, 323–330.

Szewczyk, J. and Nawrocki, J. (2011) Deep-seated relict permafrost in northeastern Poland. *Boreas* 40, 385–388.

Szymanski, W., Skiba, M., Wojtun, B. and Drewnik, M. (2015) Soil properties, micromorphology, and mineralogy of cryosols from sorted and unsorted patterned grounds in the Hornsund area, SW Spitsbergen. *Geoderma* 253–254, 1–11.

Taber, S. (1929) Frost heaving. *Journal of Geology* 37, 428–461.

Taber, S. (1930) The mechanics of frost heaving. *Journal of Geology* 38, 303–317.

Takahashi, T. (1991) *Debris Flow*. Balkema, Rotterdam.

Tarbeeva, A.M. and Surkov, V.V. (2013) Beaded channels of small rivers in permafrost zones. *Geography and Natural Resources* 34, 27–32.

Tarnocai, C. (2004) Classification of cryosols in Canada. In Kimble, J.M. (ed.) *Cryosols: Permafrost-affected Soils.* Springer-Verlag, Berlin, 599–610.

Tarnocai, C. and Zoltai, S.C. (1978) Earth hummocks of the Canadian Arctic. *Arctic and Alpine Research* 10, 581–594.

Tedrow, J.C.F. and Krug, E.C. (1982) Weathered limestone accumulations in the high arctic. *Biuletyn Peryglacjalny* 29, 143–146.

Telfer, M.W., Mills, S.C. and Matheer, A.E. (2014) Extensive Quaternary aeolian deposits in the Drakensberg foothills, Rooiberge, South Africa. *Geomorphology* 219, 161–175.

Thibault, S. and Payette, S. (2009) Recent permafrost degradation in bogs of the James Bay area, Northern Quebec, Canada. *Permafrost and Periglacial Processes* 20, 383–389.

Thorn, C.E. (1975) Influence of late-lying snow on rock weathering rinds. *Arctic and Alpine Research* 7, 373–378.

Thorn, C.E. (1976) Quantitative evaluation of nivation in the Colorado Front Range. *Bulletin, Geological Society of America* 87, 1169–78.

Thorn, C.E. (1979) Ground temperatures and surficial transport in colluvium during snowpatch meltout, Colorado Front Range. *Arctic and Alpine Research* 11, 41–52.

Thorn, C.E. (1988) Nivation: a geomorphic chimera. In Clark, M.J. (ed.) *Advances in Periglacial Geomorphology.* John Wiley & Sons, Chichester, 3–31.

Thorn, C.E. (1992) Periglacial geomorphology: what, where, when? In Dixon, J.C. and Abrahams, A.D. (eds) *Periglacial Geomorphology: Proceedings of the 22nd Annual Binghamton Symposium in Geomorphology.* John Wiley & Sons, Chichester, 1–30.

Thorn, C.E. and Darmody, R.G. (1985a) Grain size samplng and characterisation of eolian lag surfaces with alpine tundra, Niwot Ridge, Front Range, Colorado, USA. *Arctic and Alpine Research* 17, 443–450.

Thorn, C.E. and Darmody, R.G. (1985b) Grain-size distribution of the insoluble component of contemporary eolian deposits in the alpine zone, Front Range, Colorado, USA. *Arctic and Alpine Research* 17, 433–442.

Thorn, C.E. and Hall, K. (2002) Nivation and cryoplanation: the case for scrutiny and investigation. *Progress in Physical Geography* 26, 533–550.

Thorn, C.E., Dixon, J.C., Darmody, R.G. and Rissing, J.M. (1989) Weathering trends in fine debris beneath a snow patch, Niwot Ridge, Front Range, Colorado. *Physical Geography* 10, 307–321.

Thorn, C.E., Darmody, R.G., Dixon, J.C. and Schlyter, P. (2001) The chemical weathering regime of Kärkevagge, arctic-alpine Sweden. *Geomorphology* 41, 37–52.

Thorn, C.E., Darmody, R.G., Dixon, J.C. and Schlyter, P. (2002) Weathering rates of buried machine-polished rock disks, Kärkevagge, Swedish Lapland. *Earth Surface Processes and Landforms* 27, 831–845.

Thorn, C.E., Darmody, R.G. and Dixon, J.C. (2011) Rethinking weathering and pedogenesis in alpine periglacial regions: some Scandinavian evidence. *Geological Society, London, Special Publication* 354, 183–193.

Throop, J., Lewkowicz, A.G. and Smith, S.L. (2012) Climate and ground temperature relations at sites across the continuous and discontinuous permafrost zones, Canada. *Canadian Journal of Earth Sciences* 49, 865–876.

Toniolo, H., Kodial, P., Hinzman, L.D. and Yoshikawa, K. (2009) Spatio-temporal evolution of thermokarst in interior Alaska. *Cold Regions Science and Technology* 56, 39–49.

Trenhaile, A.S. (1983) The development of shore platforms in high latitudes. In Smith, D.E. and Dawson, A.G. (eds) *Shorelines and Isostasy.* Academic Press, London, 77–93.

Trenhaile, A.S. and Rudakas, P.A. (1981) Freeze-thaw and shore platform development in Gaspé, Québec. *Géographie Physique et Quaternaire* 35, 171–181.

Tricart, J. (1963) *Géomorphologie des Regions Froids*. Masson, Paris, 359 pp.

Troll, C. (1944) Strukturböden, Solifluktion und Frostklimate der Erde. *Geologische Rundschau* 34, 545–694. English translation by Wright, H.E. (1958) Structure soils, solifluction and frost climates of the Earth. *United States Army Snow, Ice and Permafrost Research Establishment Translation* 43, 121 pp.

Trombotto, D. and Borzotta, E. (2009) Indicators of present global warming through changes in active-layer thickness, estimation of thermal diffusivity and geomorphic observations in the Morenas Coloradas rockglacier, Central Andes of Mendoza, Argentina. *Cold Regions Science and Technology* 55, 321–330.

Tsytovitch, N.A. (1964) Physical phenomena and processes in freezing, frozen and thawing soils. *Canadian National Research Council, Technical Translation* 1163, 34 pp.

Turner, J., Colwell, S.R., Marshall, G.J. *et al.* (2005) Antarctic climate change during the last 50 years. *Journal of Climatology* 25, 279–294.

Újvári, G., Varga, A., Raucsik, B. and Kovács, J. (2014) The Paks loess-palaeosol sequence: a record of chemical weathering and provenance for the last 800 ka in the mid-Carpathian Basin. *Quaternary International* 319, 22–37.

Ulrich, M., Grosse, G., Strauss, J. and Schirrmeister, L. (2014) Quantifying wedge-ice volumes in yedoma and thermokarst basin deposits. *Permafrost and Periglacial Processes* 25, 151–161.

Urdea, P. (1995) Quelques considerations concernant des formations de pente dans les Carpates méridionales. *Permafrost and Periglacial Processes* 6, 195–206.

Utting, N., Clark, I., Lauriol, B. *et al.* (2012) Origin and flow dynamics of perennial groundwater in continuous permafrost terrain using isotopes and noble gases: case study of the Fish Branch River, northern Yukon, Canada. *Permafrost and Periglacial Processes* 23, 91–106.

Utting, N., Lauriol, B., Lacelle, D. and Clark, I. (2016) Using noble gas ratios to determine the origin of ground ice. *Quaternary Research* 85, 177–184.

Vallée, S. and Payette, S. (2007) Collapse of permafrost mounds along a subarctic river over the last 100 years (northern Québec). *Geomorphology* 90, 162–170.

Van Asch, N., Lutz, A.F., Duijkers, M.C.H. *et al.* (2012) Rapid climate change during the Weichselian Lateglacial in Ireland: a multi-proxy record from Fiddaun, Co. Galway. *Palaeogeography, Palaeoclimatology, Palaeoecology* 315–316, 1–11.

Van der Meulen, S. (1988) Spatial facies of a group of pingo remnants on the southeast Frisian till plateau (the Netherlands). *Geologie en Mijnbouw* 67, 61–74.

Van Dijk, D. and Law, J. (2003) The rate of grain release by pore-ice sublimation in cold-aeolian environments. *Geografiska Annaler* 85A, 99–113.

Van Everdingen, R.O. (1981) Morphology, hydrology and hydrochemistry of karst in permafrost terrain near Great Bear Lake, Northwest Territories. *National Hydrology Research Institute Paper No 11*. Inland Waters Research Institute, Calgary, AB, 53 pp.

Van Everdingen, R.O. (1998) *Multi-language Glossary of Permafrost and Related Ground Ice Terms*. Revised 2005. National Snow and Ice Data Center and World Center for Glaciology, Boulder, Colorado.

Van Huissteden, J. (1990) Tundra rivers of the last glacial: sedimentation and geomorphological processes during the Middle Pleniglacial in Twente, eastern Netherlands. *Mededelingen Rijks Geologische Dienst* 44-3, 138 pp.

Van Huissteden, J., Vandenberghe, J., Van der Hammen, T. and Laan, W. (2000) Fluvial and eolian interaction under permafrost conditions: Weichselian Late Pleniglacial, Twente, eastern Netherlands. *Catena* 40, 307–321.

Van Huissteden, J., Gibbard, P.L. and Briant, R.M. (2001) Periglacial fluvial systems in northwest Europe during marine isotope stages 4 and 3. *Quaternary International* 79, 75–88.

Van Steijn, H. (1996) Debris flow magnitude-frequency relationships for mountainous regions of central and northwest Europe. *Geomorphology* 15, 259–273.

Van Steijn, H. (2002) Long-term landform evolution: evidence from talus studies. *Earth Surface Processes and Landforms* 27, 1189–1199.

Van Steijn, H. (2011) Stratified slope deposits: periglacial and other processes involved. *Geological Society, London, Special Publications* 354, 213–226.

Van Steijn, H. and Hétu, B. (1997) Rain-generated overland flow as a factor in the development of some stratified slope deposits: a case study from the Pays du Buëch (Préalpes, France). *Géographie Physique et Quaternaire* 51, 3–15.

Van Steijn, H., de Ruig, J. and Hoozemans, F. (1988) Morphological and mechanical aspects of debris flows in parts of the French Alps. *Zeitschrift für Geomorphologie* 32, 143–161.

Van Steijn, H., Bertran, P., Francou, B. *et al.* (1995) Models for the genetic and environmental interpretation of stratified slope deposits. *Permafrost and Periglacial Processes* 6, 125–146.

Van Steijn, H., Boelhouwers, J., Harris, S. and Hétu, B. (2002) Recent research on the nature, origin and climatic relations of blocky and stratified slope deposits. *Progress in Physical Geography* 26, 551–575.

Van Vliet, B. and Langohr, R. (1981) Correlations between fragipans and permafrost with special reference to silty

Weichselian deposits in Belgium and northern France. *Catena* 8, 137–154.

Van Vliet-Lanoë, B. (1988) The significance of cryoturbation phenomena in environmental reconstruction. *Journal of Quaternary Science* 3, 85–96.

Van Vliet-Lanoë, B. (1991) Differential frost heave, load casting and convection: converging mechanisms; a discussion of the origin of cryoturbations. *Permafrost and Periglacial Processes* 2, 123–139.

Van Vliet-Lanoë, B. (1998) Frost and soils: implications for palaeosols, palaeoclimates and stratigraphy. *Catena* 34, 157–183.

Van Vliet-Lanoë, B. (2010) Frost action. In Stoops, G., Marcelino, V. and Mees, F. (eds) *Interpretation of Micromorphological Features of Soils and Regoliths*. Elsevier, Amsterdam, 81–108.

Van Vliet-Lanoë, B. and Seppälä, M. (2002) Stratigraphy, age and formation of peaty earth hummocks (pounus), Finnish Lapland. *The Holocene* 12, 187–199.

Van Vliet-Lanoë, B., Coutard, J.-P. and Pissart, A. (1984) Structures caused by repeated freezing and thawing in various loamy sediments: a comparison of active, fossil and experimental data. *Earth Surface Processes and Landforms* 9, 553–565.

Van Vliet-Lanoë, B., Bourgeois, O. and Dauteuil, O. (1998) Thufur formation in northern Iceland and its relation to Holocene climate change. *Permafrost and Periglacial Processes* 9, 347–365.

Van Vliet-Lanoë, B., Fox, C.A. and Gubin, S.V. (2004a) Micromorphology of cryosols. In Kimble, J.M. (ed.) *Cryosols: Permafrost-affected Soils*. Springer, Berlin, 365–390.

Van Vliet-Lanoë, B., Magyari, A. and Meilliez, F. (2004b) Distinguishing between tectonic and periglacial deformations of Quaternary continental deposits in Europe. *Global and Planetary Change* 43, 103–127.

Vandenberghe, D.A.G., Derese, C., Kasse, C. and Van Den Haute, P. (2013) Late Weichselian (fluvio-) aeolian sediments and Holocene drift-sands of the classic type locality in Twente (E Netherlands): a high-resolution dating study using optically-stimulated luminescence. *Quaternary Science Reviews* 68, 96–113.

Vandenberghe, J. (1988) Cryoturbations. In Clark, M.J. (ed.) *Advances in Periglacial Geomorphology*. John Wiley & Sons, Chichester, 179–198.

Vandenberghe, J. (1992a) Cryoturbations: a sediment structural analysis. *Permafrost and Periglacial Processes* 3, 343–351.

Vandenberghe, J. (1992b) Geomorphology and climate of the cool oxygen isotope stage 3 in comparison with the cold stages 2 and 4 in The Netherlands. *Zeitschrift für Geomorphologie, Supplementband* 86, 65–75.

Vandenberghe, J. (2001) A typology of Pleistocene cold-based rivers. *Quaternary International* 79, 111–121.

Vandenberghe, J. (2003) Climate forcing of fluvial system development: an evolution of ideas. *Quaternary Science Reviews* 22, 2053–2060.

Vandenberghe, J. (2008) The fluvial cycle at warm-cold-warm transitions in lowland regions: a refinement of theory. *Geomorphology* 98, 275–284.

Vandenberghe, J. (2011) Periglacial sediments: do they exist? *Geological Society, London, Special Publications* 354, 205–212.

Vandenberghe, J. (2013) Cryoturbation structures. In Elias, S. (ed.) *Encyclopedia of Quaternary Science*, 2nd edn, vol. 3. Elsevier, Amsterdam, 430–435.

Vandenberghe, J. and Czudek, T. (2008) Pleistocene cryopediments on variable terrain. *Permafrost and Periglacial Processes* 19, 71–83.

Vandenberghe, J. and Nugteren, G. (2001) Rapid climatic changes recorded in loess successions. *Global and Planetary Change* 28, 1–9.

Vandenberghe, J. and Van den Broek, P. (1982) Weichselian convolution phenomena and processes in fine sediments. *Boreas* 11, 299–315.

Vandenberghe, J. and Woo, M.-K. (2002) Modern and ancient periglacial river types. *Progress in Physical Geography* 26, 479–506.

Vandenberghe, J., Isarin, R.F.B. and Renssen, H. (1999) Comments on 'Windpolished boulders as indicators of a Late Weichselian wind regime in Denmark in relation to neighbouring areas' by Christiansen and Svensson [9(1): 1–21, 1998]. *Permafrost and Periglacial Processes* 10, 199–201.

Vandenberghe, J., Zhijiu, C., Liang, Z., Wei, Z. (2004) Thermal contraction crack networks as evidence of Late-Pleistocene permafrost in Inner Mongolia, China. *Permafrost and Periglacial Processes* 15, 21–29.

Vandenberghe, J., Renssen, H., Roche, D.M. *et al.* (2012) Eurasian permafrost instability constrained by reduced sea-ice cover. *Quaternary Science Reviews* 34, 16–23.

Vandenberghe, J., French, H.M., Gorbunov, A. *et al.* (2014) The Last Permafrost Maximum (LPM) map of the Northern Hemisphere: permafrost extent and mean annual air temperatures, 25–17 ka BP. *Boreas* 43, 652–666.

Vandenberghe, J., Wand, X. and Vandenberghe, D. (2016) Very large cryoturbation structures of Last Permafrost Maximum age at the foot of the Qilian Mountains (NE Tibet Plateau, China). *Permafrost and Periglacial Processes* 27, 138–143.

Vasil'chuk, Y.K. (2013) Monograph synopsis: *Syngenetic ice wedges: cyclical formation, radiocarbon age and stable isotope records* by Yurij K. Vasil'chuk, Moscow University Press, Moscow, 2006. *Permafrost and Periglacial Processes* 24, 82–93.

Vasil'chuk, Y.K. and Vasil'chuk, A.C. (1997) Radiocarbon dating and oxygen isotope variations in Late Pleistocene syngenetic ice wedges, northern Siberia. *Permafrost and Periglacial Processes* 8, 335–345.

Vasil'chuk, Y.K. and Vasil'chuk, A.C. (2009) Dansgaard-Oeschger events on isotope plots of Siberian ice wedges. In Kane, D.L. and Hinkel, D.M. (eds) *Proceedings of the Ninth International Conference on Permafrost.* University of Alaska, Fairbanks, 1809–1814.

Vasil'chuk, Y.K., Alexeev, S.V., Arzhannikov, S.G. *et al.* (2016a) Lithalsas in the Sentsa River Valley, Eastern Sayan Mountains, Southern Russia. *Permafrost and Periglacial Processes* 27, 285–296.

Vasil'chuk, Y.K., Lawson, D.E., Yoshikawa, K. *et al.* (2016b) Stable isotopes in the closed-system Weather Pingo, Alaska, and Pestovoye Pingo, northwestern Siberia. *Cold Regions Science and Technology* 128, 13–21.

Vasiliev, A., Kanevskiy, M., Cherkashov, G. and Vanshtein, B. (2005) Coastal dynamics at the Barents and Kara Sea key sites. *Geo-Marine Letters* 25, 110–120.

Veit, H. and Höfner, T. (1993) Permafrost, gelifluction and fluvial sediment transfer in the Alpine/subnival ecotone, central Alps, Austria: present, past and future. *Zeitschrift für Geomorphologie, Supplementband* 92, 71–84.

Velichko, A.A. (1972) La morphologie cryogène relicte: caractères fondamentaux et cartographie. *Zeitschrift für Geomorphologie Supplementband* 13, 59–72.

Velichko, A.A. (1982) *Palaeogeography of Europe during the last one hundred thousand years (in Russian, with English abstract and legends).* Nauka, Moscow, 156 pp.

Velichko, A.A. (1990) Loess-palaeosol formation on the Russian plain. *Quaternary International* 7–8, 103–114.

Veremeeva, A. and Gubin, S. (2009) Modern tundra landscapes of the Kolyma Lowland and their evolution in the Holocene. *Permafrost and Periglacial Processes* 20, 399–406.

Vieira, G.T., Mora, C. and Ramos, M. (2003) Ground temperature regimes and geomorphological applications in a Mediterranean mountain (Serra da Estrela, Portugal) *Geomorphology* 52, 57–72.

Vieira, G.T., Bockheim, J.G., Guglielmin, M. *et al.* (2010) Thermal state of permafrost and active-layer monitoring in the Antarctic: advances during the International Polar Year 2007–2009. *Permafrost and Periglacial Processes* 21, 182–197.

Viklander, P. (1998) Laboratory study of stone heave in till exposed to freezing and thawing. *Cold Regions Science and Technology* 27, 141–152.

Vitt, D.H., Halsey, L.A. and Zoltai, S.C. (1994) The bog landforms of continental western Canada in relation to climate and permafrost patterns. *Arctic and Alpine Research* 26, 1–13.

Vogel, J., Schuur, E.A.G., Trucco, C. and Lee, H. (2009) Response of CO_2 exchange in a tussock tundra ecosystem to permafrost thaw and thermokarst development. *Journal of Geophysical Research* 114, G04018.

Vogt, T. and Larqué, P. (2002) Clays and secondary minerals as permafrost indicators: examples from the circum-Baikal region. *Quaternary International* 95–96, 175–187.

Völkel, J., Leopold, M. and Roberts, M.C. (2001) The radar signatures and age of periglacial slope deposits, Central Highlands of Germany. *Permafrost and Periglacial Processes* 12, 379–387.

Vonder Mühll, D.S. and Klingelé, E.E. (1994) Gravimetrical investigation of ice-rich permafrost within the rock glacier Murtèl-Corvatsch (Upper Engadin, Swiss Alps). *Permafrost and Periglacial Processes* 5, 13–24.

Vonk, J.E., Sánchez-Garcia, L., van Dongen, B.E. *et al.* (2012) Activation of old carbon by coastal and subsea permafrost in Arctic Siberia. *Nature* 489, 137–140.

Wagner, S. (1992) Creep of alpine permafrost, investigated on Murtèl rock glacier. *Permafrost and Periglacial Processes* 3, 157–162.

Walder, J.S. and Hallet, B. (1985) A theoretical model of the fracture of rock during freezing. *Geological Society of America, Bulletin* 96, 336–346.

Walder, J.S. and Hallet, B. (1986) The physical basis of frost weathering: toward a more fundamental and unified perspective. *Arctic and Alpine Research* 18, 27–32.

Walker, D.A., Walker, M.D., Everett, K.R. and Webber, P.J. (1985) Pingos of the Prudhoe Bay Region, Alaska, USA. *Arctic and Alpine Research* 17, 321–336.

Walker, D.A., Gould, W.A., Maier, H.A. and Raynolds, M.K. (2002) The Circumpolar Arctic Vegetation Map. AVHRR-derived base map, environmental conditions, and integrated mapping procedures. *International Journal of Remote Sensing* 23, 4551–4570.

Walker, D.A., Epstein, H.E., Gould, W.A. *et al.* (2004) Frost-boil ecosystems: complex interactions between landforms, soils, vegetation and climate. *Permafrost and Periglacial Processes* 15, 171–188.

Walker, H.J. (1998) Arctic deltas. *Journal of Coastal Research* 14, 718–738.

Walker, H.J. and Hudson, P.F. (2003) Hydrologic and geomorphic processes in the Colville River delta, Alaska. *Geomorphology* 56, 291–303.

Walker, M.J.C. (2005) *Quaternary Dating Methods.* John Wiley & Sons, Chichester, 286 pp.

Waller, R.I., Murton, J.B. and Whiteman, C.A. (2009) Geological evidence for subglacial deformation of Pleistocene permafrost. *Proceedings of the Geologists' Association* 120, 155–162.

Waller, R.I., Phillips, E., Murton, J.B. *et al.* (2011) Sand intraclasts as geological evidence of subglacial deformation of Middle Pleistocene permafrost, north Norfolk, U.K. *Quaternary Science Reviews* 30, 3481–3500.

Waller, R.I., Murton, J.B. and Kristensen, L. (2012) Glacier-permafrost interactions: processes, products and glaciological implications. *Sedimentary Geology* 255–265, 1–28.

Walsh, J.E., Overland, J.E., Groisman, P.Y. and Rudolf, B. (2011) Arctic climate: recent variations. In *Snow, Water, Ice and Permafrost in the Arctic (SWIPA): Climate Change and the Cryosphere*. Arctic Monitoring and Assessment Programme (AMAP), Oslo, Norway, 5.1–5.62.

Walter, K.M., Zimov, S.A., Chanton, J.P. *et al.* (2006) Methane bubbling from Siberian thaw lakes as a positive feedback to climate warming. *Nature* 443, 71–75.

Walter, K.M., Edwards, M.E., Grosse, G. *et al.* (2007) Thermokarst lakes as a source of atmospheric CH$_4$ during the last deglaciation. *Science* 318, 633–636.

Walter Anthony K.M., Zimov, S.A., Grosse, G. *et al.* (2014) A shift of thermokarst lakes from carbon sources to sinks during the Holocene epoch. *Nature* 511, 452–456.

Walters, J.C. (1994) Ice-wedge casts and related polygonal patterned ground in north-east Iowa, USA. *Permafrost and Periglacial Processes* 5, 269–282.

Wan, X., Lai, Y. and Wang, C. (2015) Experimental study on the freezing temperatures of saline silty soils. *Permafrost and Periglacial Processes* 26, 175–187.

Wang, B. (1990) Permafrost and groundwater conditions, Huola River basin, northeast China. *Permafrost and Periglacial Processes* 1, 45–52.

Wang, B. and French, H.M. (1995a) Permafrost on the Tibet Plateau, China. *Quaternary Science Reviews* 14, 255–274.

Wang, B. and French, H.M. (1995b) In-situ creep of frozen soil, Fenghuo Shan, Tibet Plateau, China. *Canadian Geotechnical Journal* 32, 545–552.

Wang, B., Paudel, B. and Li, H. (2016) Behaviour of retrogressive thaw slumps in northern Canada – three year monitoring results from 18 sites. *Landslides* 13, 1–8.

Wang, L., Shi, Z.H., Wu, G.L. and Fang, N.F. (2014) Freeze/thaw and soil moisture effects on wind erosion. *Geomorphology* 207, 141–148.

Wang, N., Zhao, Q., Li, J., Hu, G. and Cheng, H. (2003) The sand wedges of the last ice age in the Hexi Corridor, China: palaeoclimatic interpretation. *Geomorphology* 51, 313–320.

Warburton, J. (1990) Secondary sorting of sorted patterned ground. *Permafrost and Periglacial Processes* 1, 313–318.

Washburn, A.L. (1956) Classification of patterned ground and review of suggested origins. *Geological Society of America, Bulletin* 67, 823–865.

Washburn, A.L. (1967) Instrumental observations of mass-wasting in Mesters Vig district, northeast Greenland. *Meddelelser øm Grønland* 166, 318 pp.

Washburn, A.L. (1969) Weathering, frost action and patterned ground in the Mesters Vig District, northeast Greenland. *Meddelelser øm Grønland* 176, 303 pp.

Washburn, A.L. (1973) *Periglacial Processes and Environments*. Arnold, London, 320 pp.

Washburn, A.L. (1979) *Geocryology: a Survey of Periglacial Processes and Environments*. Arnold, London, 406 pp.

Washburn, A.L. (1980) Permafrost features as evidence of climatic change. *Earth-Science Reviews* 15, 327–402.

Washburn, A.L. (1989) Near-surface soil displacement in sorted circles, Resolute area, Cornwallis Island, Canadian High Arctic. *Canadian Journal of Earth Sciences* 26, 941–955.

Washburn, A.L., Smith, D.D. and Godard, R.H. (1963) Frost cracking in a middle-latitude climate. *Biuletyn Peryglacjalny* 12, 175–189.

Watanabe, T. (1985) Alpine debris flows in the Kuranosuke cirque, northern Japanese Alps, central Japan. *Transactions of the Japanese Geomorphological Union* 6, 303–316.

Watanabe, T., Matsuoka, N. and Christiansen, H.H. (2012) Mudboil and ice-wedge dynamics investigated by electrical resistivity tomography, ground remperatures and surface movements in Svalbard. *Geografiska Annaler* 94A, 445–457.

Watanabe, T., Matsuoka, N. and Christiansen, H.H. (2013) Ice- and soil-wedge dynamics in the Kapp Linné area, Svalbard, investigated by two- and three-dimensional GPR and ground thermal and acceleration regimes. *Permafrost and Periglacial Processes* 24, 39–55.

Watson, J., Brooks, S.J., Whitehouse, N.J. *et al.* (2010) Chironomid-inferred Late-glacial summer air temperatures from Loch Nadourcan, Co. Donegal, Ireland. *Journal of Quaternary Science* 25, 1200–1210.

Watts, S.H. (1983) Weathering processes and products under arid arctic conditions: a study from Ellesmere Island, Canada. *Geografiska Annaler* 65A, 85–98.

Wayne, W.J. (1991) Ice-wedge casts of Wisconsinan age in eastern Nebraska. *Permafrost and Periglacial Processes* 2, 211–223.

Wei, M., Fujun, N., Satoshi, A. and Dewu, J. (2006) Slope instability phenomena in permafrost regions of Qinghai-Tibet Plateau, China. *Landslides* 3, 260–264.

Werner, B.T. and Hallet, B. (1993) Numerical simulation of self-organized stone stripes. *Nature* 361, 142–145.

West, R.G. (1980) *The Pre-glacial Pleistocene of the Norfolk and Suffolk Coasts*. Cambridge University Press, Cambridge.

West, R.G. (1991) On the origin of Grunty Fen and other landforms in southern Fenland, Cambridgeshire. *Geological Magazine* 128, 257–262.

West, R.G. (1993) Devensian thermal contraction networks and cracks at Somersham, Cambridgeshire, UK. *Permafrost and Periglacial Processes* 4, 277–300.

West, R.G. (2015) *Evolution of a Breckland landscape: chalkland under a cold climate in the area of Beachamwell, Norfolk.* The Suffolk Naturalists' Society, Ipswich, 110 pp.

West, R.G., Andrew, R., Catt, J.A. *et al.* (1999) Late and Middle Pleistocene deposits at Somersham, Cambridgeshire, U.K.: a model for reconstructing fluvial/estuarine depositional environments. *Quaternary Science Reviews* 18, 1247–1314.

West, J.J. and Plug, L.J. (2008) Time-dependent morphology of thaw lakes and taliks in deep and shallow ground ice. *Journal of Geophysical Research* 113, 1–14.

Westermann, S., Schuler, T., Gisnås, K. and Etzelmüller, B. (2013) Transient thermal modelling of permafrost conditions in Southern Norway. *The Cryosphere* 7, 719–739.

Westermann, S., Langer, M., Boike, J. *et al.* (2016) Simulating the thermal regime and thaw processes of ice-rich permafrost ground with the land-surface model CryoGrid 3. *Geoscientific Model Development* 9, 523–546.

Westin, B. and Zuidhoff, F.S. (2001) Ground thermal conditions in a frost-crack polygon, a palsa and a mineral palsa (lithalsa) in the discontinuous permafrost zone, northern Sweden. *Permafrost and Periglacial Processes* 12, 325–335.

Wetterich, S., Kuzmina, S., Andreev, A.A. *et al.* (2008) Palaeoenvironmental dynamics inferred from late Quaternary permafrost deposits on Kurungnakh Island, Lena Delta, Northeast Siberia, Russia. *Quaternary Science Reviews* 27, 1523–1540.

Wetterich, S. Schirrmeister, L., Andreev, A.A. *et al.* (2009) Eemian and Lateglacial/Holocene palaeoenvironmental records from permafrost sequences at the Dmitry Laptev Strait (NE Siberia, Russia). *Paleogeography, Paleoclimatology, Paleoecology* 279, 73–95.

Wetterich, S., Rudaya, N., Andreev, A.A. *et al.* (2011) Last Glacial Maximum records in permafrost of the East Siberian Arctic. *Quaternary Science Reviews* 30, 3139–3151.

Wetterich, S., Tumskoy, V., Rudaya, N. *et al.* (2014) Ice complex formation in arctic East Siberia during the MIS3 Interstadial. *Quaternary Science Reviews* 84, 39–55.

Wetterich, S., Tumskoy, V., Rudaya, N. *et al.* (2016) Ice complex permafrost of MIS5 age in the Dmitry Laptev Strait coastal region (East Siberian Arctic). *Quaternary Science Reviews* 147, 298–311.

Whalley, W.B. and Martin, H.E. (1992) Rock glaciers: II. Models and mechanisms. *Progress in Physical Geography* 16, 127–186.

Whalley, W.B., Rea, B.R. and Rainey, M.M. (2004) Weathering, blockfields, and fracture systems and the implications for long-term landscape formation: some

evidence from Lyngen and Öksfjordjokelen areas in north Norway. *Polar Geography* 28, 93–119.

Whipple, K.X., Snyder, M.P. and Dollenmayer, K. (2000) Rates and processes of bedrock incision by the upper Ukak River since the 1912 Novarupta ash flow in the Valley of Ten Thousand Smokes, Alaska. *Geology* 28, 835–838.

White, S.E. (1976) Is frost action really only hydration shattering? *Arctic and Alpine Research* 8, 1–6.

Whiteman, C. (2002) Implication of a Middle Pleistocene ice-wedge cast at Trimingham, Norfolk, eastern England. *Permafrost and Periglacial Processes* 13, 163–170.

Whiteman, G., Hope, C. and Wadhams, P. (2013) Climate science: vast costs of Arctic change. *Nature* 499, 401–403.

Whitney, M.I. and Splettstoesser, J.F. (1982) Ventifacts and their formation: Darwin Mountains, Antarctica. *Catena Supplement* 1, 175–194.

Whittecar, G.R. and Ryter, D.W. (1992) Boulder steams, debris fans and Pleistocene climate change in the Blue Ridge Mountains of central Virginia. *Journal of Geology* 100, 487–494.

Wilkinson, T.J. and Bunting, B.T. (1975) Overland transport of sediment by rill water in a periglacial environment in the Canadian Arctic. *Geografiska Annaler* 57A, 105–16.

Willemse, N.W., Koster, E.A., Hoogakker, B. and van Tatenhove, F.G.M. (2003) A continuous record of Holocene eolian activity in West Greenland. *Quaternary Research* 59, 322–334.

Willerslev, E., Davison, J., Moora, M. *et al.* (2014) Fifty thousand years of arctic vegetation and megafauna diet. *Nature* 506, 47–51.

Williams, D.J. and Burn, C.R. (1996) Surficial characteristics associated with the occurrence of permafrost near Mayo, central Yukon Territory, Canada. *Permafrost and Periglacial Processes* 7, 193–206.

Williams, G.E. (1986) Precambrian permafrost horizons as indicators of palaeoclimate. *Precambrian Research* 32, 233–242.

Williams, P.J. and Smith, M.W. (1989) *The Frozen Earth: Fundamentals of Geocryology.* Cambridge University Press, Cambridge, UK, 306 pp.

Williams, R.B.G. (1965) Permafrost in England during the last glacial period. *Nature* 205, 1304–1305.

Williams, R.B.G. (1969) Permafrost and temperature conditions in England during the last glacial period. In Péwé, T.L. (ed.) *The Periglacial Environment.* McGill-Queen's University Press, Montreal, 399–410.

Williams, R.B.G. and Robinson, D.A. (1991) Frost weathering of rocks in the presence of salts – a review. *Permafrost and Periglacial Processes* 2, 347–353.

Williams, R.B.G. and Robinson, D.A. (2001) Experimental frost weathering of sandstone by various combinations of salts. *Earth Surface Processes and Landforms* 26, 811–818.

Wilson, P. (1995) Forms of unusual patterned ground: examples from the Falkland Islands, South Atlantic. *Geografiska Annaler* 77A, 159–165.

Wilson, P. (2013) Block/rock streams. In Elias, S. (ed.) *Encyclopedia of Quaternary Science*, 2nd edn, vol. 3. Elsevier: Amsterdam, 514–522.

Wilson, P. and Clark, R. (1991) Development of miniature sorted patterned ground following soil erosion in East Falkland, South Atlantic. *Earth Surface Processes and Landforms* 16, 369–376.

Wilson, P. and Edwards, E.J. (2004) Further examples of ventifacts and unusual patterned ground from the Falkland Islands, South Atlantic. *Geografiska Annaler* 86A, 107–115.

Wilson, P., Bentley, M.J., Schnabel, C. *et al.* (2008) Stone run (block stream) formation in the Falkland Islands over several cold stages, deduced from cosmogenic isotope (^{10}Be and ^{26}Al) surface exposure dating. *Journal of Quaternary Science* 23, 461–473.

Winterfield, M., Schirrmeister, L., Grigoriev, M.N. *et al.* (2011) Coastal permafrost landscape development since the Late Pleistocene in the western Laptev Sea, Siberia. *Boreas* 40, 697–713.

Wiseman, W.J., Owens, E.H. and Kahn, J. (1981) Temporal and spatial variability of ice-foot morphology. *Geografiska Annaler* 63A, 69–80.

Wolfe, S.A. (2013) Cold-climate aeolian environments. In Shroder, J., Giardino, R. and Harbor, J. (eds) *Treatise on Geomorphology*, vol. 8. Academic Press, San Diego, CA, 375–394.

Wolfe, S.A., Dallimore, S.R. and Solomon, S.M. (1998) Coastal permafrost investigations along a rapidly eroding shoreline, Tuktoyaktuk, NWT. In Lewkowicz, A.G. and Allard, M. (eds) *Permafrost: Proceedings of the 7th International Conference*. Université Laval, Québec, 1125–1131.

Wolfe, S.A., Stevens, C.W., Gaanderse, A.J. and Oldenborger, G.A. (2014) Lithalsa distribution, morphology and landscape associations in the Great Slave Lowland, Northwest Territories, Canada. *Geomorphology* 204, 302–313.

Woo, M.-K. (1983) Hydrology of a drainage basin in the Canadian high arctic. *Annals of the Association of American Geographers* 73, 577–596.

Woo, M.-K. (1986) Permafrost hydrology in North America. *Atmosphere-Ocean* 24, 201–234.

Woo, M.-K. (2012) *Permafrost Hydrology*. Springer-Verlag, New York. 563 pp.

Woo, M.-K. and Marsh, P. (1977) Effect of vegetation on limestone solution in a small arctic basin. *Canadian Journal of Earth Sciences* 14, 571–581.

Woo, M.-K. and Sauriol, J. (1980) Channel development in snow-filled valleys, Resolute, N.W.T., Canada. *Geografiska Annaler* 62A, 37–56.

Woo, M.-K. and Steer, P. (1983) Slope hydrology as influenced by thawing of the active layer, Resolute, NWT. *Canadian Journal of Earth Sciences* 20, 978–986.

Woo, M.-K. and Xia, Z. (1996) Effects of hydrology on the thermal conditions of the active layer. *Nordic Hydrology* 27, 129–142.

Woo, M.-K. and Young, K.L. (1997) Hydrology of a small drainage basin with polar oasis environment, Fosheim Peninsula, Ellesmere Island, Canada. *Permafrost and Periglacial Processes* 8, 257–277.

Woo, M.-K. and Young, K.L. (2006) High arctic wetlands: their occurrence, hydrological characteristics and sustainability. *Journal of Hydrology* 320: 432–450.

Woo, M.-K., Yang, Z., Xia, Z. and Yang, D. (1994) Streamflow processes in an alpine permafrost catchment, Tianshan, China. *Permafrost and Periglacial Processes* 5, 71–85.

Woo, M.-K., Kane, D.L., Carey, S.K. and Yang, D. (2008a) Progress in permafrost hydrology in the new millennium. *Permafrost and Periglacial Processes* 19, 237–254.

Woo, M.-K., Thorne, R., Szeto, K. and Yang, D.Q. (2008b) Streamflow hydrology in the boreal region under the influence of climate and human interference. *Philosophical Transactions of the Royal Society* B363, 2251–2260.

Woronko, B. and Hoch, M. (2011) The development of frost-weathering microstructures on sand-sized quartz grains: examples from Poland and Mongolia. *Permafrost and Periglacial Processes* 22, 214–227.

Woronko, B. and Pisarska-Jamrozy, M. (2016) Micro-scale frost weathering of sand-sized quartz grains. *Permafrost and Periglacial Processes* 27, 109–122.

Worsley, P. (1999) Context of relict Wisconsinan glacial ice at Angus Lake, SW Banks Island, western Canadian Arctic and stratigraphic implications. *Boreas* 28, 543–550.

Worsley, P. (2014) Ice-wedge growth and casting in a Late Pleistocene periglacial, fluvial succession at Baston, Lincolnshire. *Mercian Geologist* 18, 159–170.

Worsley, P. and Gurney, S.D. (1996) Geomorphology and hydrological significance of the Holocene pingos in the Karup Valley area, Traill Island, northern east Greenland. *Journal of Quaternary Science* 11, 249–262.

Worsley, P., Gurney, S.D. and Collins, P.E. (1995) Late Holocene 'mineral palsas' and associated vegetation patterns: a case study from Lac Hendry, Northern Québec, Canada, and significance for European Pleistocene thermokarst. *Quaternary Science Reviews* 14, 179–192.

Wright, J.F., Duchesne, C. and Côté, M.M. (2003) Regional-scale permafrost mapping using the TTOP ground temperature model. In Phillips, M., Springman, S.M. and Arenson, L.U. (eds) *Proceedings of the Eight*

International Conference on Permafrost. Balkema, Lisse, 1241–1246.

Wu, Z. Barosh, P.J., Hu, D. *et al.* (2005) Migrating pingos in the permafrost region of the Tibetan Plateau, China and their hazard along the Golmud-Lhasa railway. *Engineering Geology* 79, 267–287.

Wünnemann, B., Reinhardt, C., Kotlia, B.S. and Riedel, F. (2008) Observations on the relationship between lake formation, permafrost activity and lithalsa development during the last 20 000 years in the Tso Kar Basin, Ladakh, India. *Permafrost and Periglacial Processes* 19, 341–358.

Xie, C., Gough, W.A., Tam, A. *et al.* (2013) Characteristics and persistence of relict high-altitude permafrost on Mahan Mountain, Loess Plateau, China. *Permafrost and Periglacial Processes* 24, 200–209.

Yamada, S., Matsumoto, H. and Hirakawa, K. (2000) Seasonal variation in creep and temperature in a solifluction lobe: continuous monitoring in the Daisetsu Mountains, northern Japan. *Permafrost and Periglacial Processes* 11, 125–135.

Yamagishi, C. and Matsuoka, N. (2015) Laboratory frost sorting by needle ice: a pilot experiment on the effects of stone size and extent of surface stone cover. *Earth Surface Processes and Landforms* 40, 502–511.

Yang, Z., Yang Z. and Wang Q. (1991) Characteristics of hydrological processes in a small high mountain basin. *International Association of Hydrological Sciences Publication* 205, 229–236.

Yatsu, E. (1988) *The Nature of Weathering*. Sozosha, Tokyo.

Yershov, E.D. (1998) *General Geocryology*. Cambridge University Press, Cambridge UK, 580 pp.

Yoshikawa, K. (1993) Notes on open-system pingo ice, Adventdalen, Spitsbergen. *Permafrost and Periglacial Processes* 4, 327–334.

Yoshikawa, K. (1998) The groundwater hydraulics of open-system pingos. In Lewkowicz, A.G. and Allard, M. (eds) *Permafrost: Proceedings of the 7th International Conference*. Université Laval, Québec, 1177–1183.

Yoshikawa, K. and Harada, K. (1995) Observations on nearshore pingo growth, Adventdalen, Spitsbergen. *Permafrost and Periglacial Processes* 6, 361–372.

Yoshikawa, K. and Hinzman, L.D. (2003) Shrinking thermokarst ponds and groundwater dynamics in discontinuous permafrost near Council, Alaska. *Permafrost and Periglacial Processes* 14, 151–160.

Yoshikawa, K., Nakamura, T. and Igarashi, Y. (1996) Growth and collapse history of pingos, Kuganquaq, Disko Island, Greenland. *Polarforschung* 64, 109–113.

Yoshikawa, K., White, D., Hinzman, L. *et al.* (2003) Water in permafrost: case study of aufeis and pingo hydrology in discontinuous permafrost. In Phillips, M., Springman, S., Arenson, L. (eds) *Proceedings of the 8th*

International Conference on Permafrost. Balkema, Lisse, 1259–1264.

Yoshikawa, K., Leuschen, C., Ikeda, A. *et al.* (2006) Comparison of geophysical investigations for detection of massive ground ice (pingo ice). *Journal of Geophysical Research* 111, E06S19.

Yoshikawa, K., Sharkhuu, N. and Sharkuu, A. (2013) Groundwater hydrology and stable isotope analysis of an open-system pingo in northwestern Mongolia. *Permafrost and Periglacial Processes* 24, 173–183.

Young, K.L. and Woo, M.-K. (2000) Hydrological response of a patchy high arctic wetland. *Nordic Hydrology* 31, 317–338.

Zalasiewicz, J., Williams, M., Haywood, A. *et al.* (2011) The Anthropocene: a new epoch of geological time. *Philosophical Transactions of the Royal Society of London* A369, 1036–1055.

Zanina, O.G., Gubin, S.V., Kuzmina, S.A. *et al.* (2011) Late-Pleistocene (MIS 3–2) palaeoenvironments as recorded by sediments, palaeosols and ground-squirrel nests at Duvanny Yar, Kolyma lowland, northeast Siberia. *Quaternary Science Reviews* 30, 2107–2123.

Zárate, M.A. (2003) Loess of southern South America. *Quaternary Science Reviews* 22, 1987–2006.

Zenklusen Mutter, E., Blanchet, J. and Phillips, M. (2010) Analysis of ground temperature trends in Alpine permafrost using generalized least squares. *Journal of Geophysical Research* 115, F04009.

Zhang, J., Li, J., Guo, B. *et al.* (2016) Magnetostratigraphic age and monsoonal evolution recorded by the thickest Quaternary loess deposit of the Lanzhou region, western Chinese Loess Plateau. *Quaternary Science Reviews* 139, 17–29.

Zhang, T., Barry, R.G., Knowles, K. *et al.* (1999) Statistics and characteristics of permafrost and ground ice distribution in the Northern Hemisphere. *Polar Geography* 23, 132–154.

Zhang, T., Heginbottom, J.A., Barry, R.G. and Brown, J. (2000) Further statistics on the distribution of permafrost and ground ice in the northern hemisphere. *Polar Geography* 24, 126–131.

Zhang, Y. (2013) Spatio-temporal features of permafrost thaw projected from long-term high-resolution modeling for a region in the Hudson Bay Lowlands in Canada. *Journal of Geophysical Research – Earth Surface* 118, 542–552.

Zhang, Y. Wang, X., Fraser, R. *et al.* (2013) Modelling and mapping climate change impacts on permafrost at high spatial resolution for an Arctic region with complex terrain. *The Cryosphere* 7, 1121–1137.

Zhao, L., Wu, Q., Marchenko, S.S. and Sharkhuu, N. (2010) Thermal state of permafrost and active layer in Central Asia during the International Polar Year. *Permafrost and Periglacial Processes* 21, 198–206.

Zhao, L., Jin, H., Li, C. *et al.* (2014) The extent of permafrost in China during the local Last Glacial Maximum (LLGM). *Boreas* 43, 688–698.

Zhestikova, T.N. (1982) *Formation of cryogenic structures in the ground.* (in Russian). Nauka, Moscow, 209 pp.

Zhou, X., Buckli, T., Kinzelbach, W. *et al.* (2015) Analysis of thermal behaviour in the active layer of degrading mountain permafrost. *Permafrost and Periglacial Processes* 26, 39–56.

Zhuang, Q., Melillo, J.M., Sarofim, M.C. *et al.* (2006) CO_2 and CH_4 exchanges between land ecosystems and the atmosphere in northern high latitudes over the 21st century. *Geophysical Research Letters* 33, L17403.

Zielinski, P., Sokolowski, R., Fedorowicz, S. and Zaleski, I. (2014) Periglacial structures within fluvio-aeolian successions of the end of the Last Glaciation – examples from SE Poland and NW Ukraine. *Boreas* 43, 712–721.

Zielinski, P., Sokolowski, R.J., Woronko, B. *et al.* (2015) The depositional conditions of the fluvio-aeolian succession during the last climate minimum based on examples from Poland and NW Ukraine. *Quaternary International* 386, 30–41.

Zimmermann, M. and Haeberli, W. (1992) Climatic change and debris flow activity in high-mountain areas – a case study from the Swiss Alps. *Catena Supplement* 22, 59–72.

Zimov, N.S., Zimov, S.A., Zimova, A.E. *et al.* (2009) Carbon storage in permafrost and soils of the mammoth tundra steppe biome: role in the global carbon budget. *Geophysical Research Letters* 36, L02502.

Zimov, S.A., Voropaev, I., Semiletov, S. *et al.* (1997) North Siberian lakes: a methane source fueled by Pleistocene carbon. *Science* 277, 800–802.

Zimov, S.A., Schuur, E.A.G. and Chapin, F.S. III (2006) Permafrost and the global carbon budget. *Science* 312, 1612–1613.

Zoltai, S.C. and Tarnocai, C. (1981) Some nonsorted patterned ground types in northern Canada. *Arctic and Alpine Research* 13, 139–151.

Zuidhoff, F.S. and Kolstrup, E. (2000) Changes in palsa distribution in relation to climate change in Laivadalen, northern Sweden, especially 1960–1997. *Permafrost and Periglacial Processes* 11, 55–69.

Zurawek, R. (2002) Internal structure of a relict rock glacier, Sleza Massif, southwest Poland. *Permafrost and Periglacial Processes* 13, 29–42.

Index